The Quadratic Formula

MW00814658

If $a \neq 0$, then the solutions of $ax^2 + bx + c = 0$ are ...

$$\frac{}{2a}$$

If $b^2 - 4ac > 0$, then there are two real solutions.

If $b^2 - 4ac = 0$, then there is one real solution.

If $b^2 - 4ac < 0$, then there are two complex solutions.

Equations and Graphs

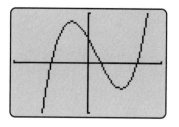

The solutions of the equation $f(x) = 0$ are the x-intercepts of the graph of $y = f(x)$.

Distance Formula

Length of segment $PQ =$

$$\sqrt{(x_1 - x_2)^2 + (y_1 - y_2)^2}$$

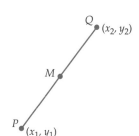

Midpoint Formula

Midpoint M of segment $PQ =$

$$\left(\frac{x_1 + x_2}{2}, \frac{y_1 + y_2}{2} \right)$$

Slope

If $x_1 \neq x_2$, slope of line $PQ =$

$$\frac{y_2 - y_1}{x_2 - x_1}$$

The equation of the straight line through (x_1, y_1) with slope m is $y - y_1 = m(x - x_1)$.
The equation of line with slope m and y-intercept b is $y = mx + b$.

Exponential Growth Functions

$$f(x) = Pa^x \quad (a > 1)$$
$$f(x) = Pe^{kx} \quad (k > 0)$$
$$f(x) = P(1 + r)^x \quad (0 < r < 1)$$

Exponential Decay Functions

$$f(x) = Pa^x \quad (0 < a < 1)$$
$$f(x) = Pe^{kx} \quad (k < 0)$$
$$f(x) = P(1 - r)^x \quad (0 < r < 1)$$

Compound Interest Formula

$$A = P(1 + r)^t$$

Contemporary
College Algebra

A Graphing Approach

Contemporary College Algebra

A Graphing Approach

Thomas W. Hungerford
Cleveland State University

THOMSON

™

BROOKS/COLE

Australia • Canada • Mexico • Singapore • Spain • United Kingdom • United States

Publisher: Emily Barrosse
Executive Editor: Angus McDonald
Acquisitions Editor: Liz Covello
Senior Marketing Strategist: Julia Conover
Developmental Editor: James D. LaPointe
Project Editor: Ellen Sklar, Ted Lewis
Production Manager: Alicia Jackson
Art Director: Paul Fry, Cara Castiglio
Text Designer: Cara Castiglio

Cover Credit: SuperStock, Inc.

Contemporary College Algebra: A Graphing Approach, 1/e
ISBN: 0-03-025621-6
Library of Congress Catalog Card Number: 00-103717

Copyright © 2001 by Thomson Learning, Inc.
Thomson Learning™ is a trademark used herein under license.

ALL RIGHTS RESERVED. No part of this work covered by the
copyright hereon may be reproduced or used in any form or by any
means — graphic, electronic, or mechanical, including photocopying,
recording, taping, Web distribution, or information storage or retrieval
systems — without the written permission of the publisher.

For information about our products, contact us:
Thomson Learning Academic Resource Center
1-800-423-0563
http://www.wadsworth.com

For permission to use material from this text, contact us by
Web: http://www.thomsonrights.com
Fax: 1-800-730-2215
Phone: 1-800-730-2214

Printed in the United States of America
10 9 8 7 6 5 4 3 2

ISBN 0-03-025621-6

To my wife

Mary Alice Ryan Hungerford

and my children

Anne Elizabeth Hungerford

Thomas Joseph Hungerford

who make it all worthwhile

Preface

This book is intended to provide a flexible approach to the college algebra curriculum that emphasizes real-world applications, integrates technology into the presentation, and is suitable for a variety of audiences. Mathematical concepts are presented in an informal manner that stresses meaningful motivation, careful explanations, and numerous examples, with an ongoing focus on real-world problem solving.

Contemporary College Algebra has relatively modest mathematical prerequisites and includes a full review of basic algebra in Chapter 0. It can be used for a variety of terminal courses for students in the liberal arts, social sciences, and business, and includes a number of interesting applications such as probability and linear programming. Furthermore, all of the (non-trigonometry) topics needed in calculus are covered here in sufficient detail to prepare a student for a business/social science calculus course.*

For readers who are familiar with my earlier book, it should be noted that this book differs from *Contemporary Precalculus* in a number of important respects (in addition to the omission of trigonometry):

The pace here is slower and gentler, with some topics that are covered in a single section of *Contemporary Precalculus* spread over several sections, and others covered in less depth.

Topics are covered in a different order, with considerably more review material, a greater emphasis on real-world applications, and the inclusion of several topics not in *Contemporary Precalculus*.

The exercise sets in this text generally have more drill exercises (when appropriate), more exercises based on real data, and fewer very hard problems.

---

*The companion volume *Contemporary College Algebra and Trigonometry*, which includes everything in this book and full coverage of trigonometry (beginning with triangle trigonometry) is suitable for preparing students for the standard science/engineering calculus sequence.

What has not changed in this book is the intelligent use of technology. The graphing calculator (or computer) is not just an optional add-on, but an essential tool that is integrated into the presentation. However, the emphasis is on using this tool effectively to develop a fuller understanding of the underlying mathematics, which is the heart of the book.

Pedagogical Features

The book contains a number of features designed to make life easier (or at least more interesting) for both students and instructors.

Graphing Explorations   Important concepts are explored from algebraic, graphical, and numerical perspectives. Students are expected to participate actively in the development of these concepts by using graphing calculators or computers, as directed in the Graphing Explorations, either to complete a particular discussion or to explore appropriate examples.

Chapter Openers   Each chapter begins with a brief example of an application of the mathematics treated in that chapter, together with a reference to an appropriate exercise. The opener also lists the titles of the sections in the chapter and includes a diagram showing their interdependence (thus making it easy for instructors to rearrange the order to suit the needs of a particular class).

Discovery Projects   Each chapter ends with an investigative problem (suitable for small group work) that enables students to apply some of the mathematics in the chapter to solve a real-world problem.

Chapter Reviews   Each chapter concludes with a list of important concepts (referenced by section and page number), a summary of important facts and formulas, and a set of review questions.

Geometry Review   Frequently used facts from plane geometry are summarized, with examples, in an appendix.

Cautions   Students are alerted to common errors and misconceptions (both mathematical and technological) by clearly marked Caution boxes.

Thinkers   Some exercise sets include problems labeled Thinkers, many of which are not difficult but simply different from what students may have seen before. A few of the Thinkers are quite challenging.

Although many (if not most) students now arrive at college with a graphing calculator, a surprisingly large number of them have almost no idea how powerful a tool it can be. Consequently, this book has several technology-assistance features to help students get the most out of their calculators.

Calculator Investigations   In the early sections of the book, there are often Calculator Investigations preceding the exercise sets that encourage students to become familiar with the capabilities and limitations of calculators.

Technology Tips   Although the discussion of technology in the text is as generic as possible, Technology Tips in the margin provide information and assistance with carrying out various procedures on specific calculator models.

Program Appendix   The appendix provides a small number of programs that are useful either for updating older calculators (such as a table maker program for the TI-85 and a RREF program for the TI-82 and Casio 9850) or for carrying out certain procedures discussed in the text (such as synthetic division).

Acknowledgments

I am particularly grateful to

Ann Steen, Santa Fe Community College

who supplied more than one hundred exercises for this book; to

Edward Miller, Lewis-Clark State College

who designed the Discovery Projects at the end of each chapter; and to our accuracy reviewer

Carolyn Robertson, University of Mississippi

who examined (and corrected where necessary) the examples and exercises. Their work has greatly improved the final project.

My sincere thanks go to the following reviewers who provided many helpful suggestions for improving the text:

Deborah Adams, Jacksonville University
Kelly Bach, University of Kansas
David Blankenbaker, University of New Mexico
Bettyann Daley, University of Delaware
Margaret Donlan, University of Delaware
Patrick Dueck, Arizona State University
Betsy Farber, Bucks County Community College
Alex Feldman, Boise State University
Betty Givan, Eastern Kentucky University
William Grimes, Central Missouri State University
Frances Gulick, University of Maryland
John Hamm, University of New Mexico
Ann Lawrance, Wake Technical Community College
Charles Laws, Cleveland State Community College
Martha Lisle, Prince George's Community College
Lonnie Hass, North Dakota State University
Mathew Liu, University of Wisconsin-Stevens Point
Sergey Lvin, University of Maine
George Matthews, Onondaga Community College
Nancy Matthews, University of Oklahoma

Ruth Meyering, Grand Valley State University
William Miller, Central Michigan University
Philip Montgomery, University of Kansas
Katherine Muhs, St. Norbert College
Roger Nelson, Lewis and Clark College
Jack Porter, University of Kansas
Carolyn Robertson, University of Mississippi
Joe Rody, Arizona State University
Robert Rogers, University of New Mexico
Barbara Sausen, Fresno City College
Marvin Stick, University of Massachusetts-Lowell
Hugo Sun, California State University at Fresno
Stuart Thomas, University of Oregon
Bettie Truitt, Black Hawk College
Jan Vandever, South Dakota University
Judith Wolbert, Kettering University

I also want to thank the people who have prepared the various supplements that are available to instructors and students who use this book:

Instructor's Resource Manual: Matt Foss, North Hennepin
Community College
Student Resource Manual: Matt Foss, North Hennepin
Community College
Test Bank: Nancy Matthews, University of Oklahoma
Graphing Calculator Manual: Joan H. McCarter, Arizona
State University
Math in Practice: An Applied Video Companion CD-ROM:
Lori Palmer, Utah Valley State College

As always, the Saunders staff has done a superb production job. My special thanks to

James D. LaPointe, Developmental Editor
Amanda Loch, Editorial Assistant
Ellen Sklar, Project Editor
Ted Lewis, Project Editor
Alicia Jackson, Production Manager

Finally, I want to thank my wife Mary Alice, once again, for her support, understanding, and love, without which I would not long survive.

Thomas W. Hungerford
Cleveland, Ohio
July 2000

Table of Contents

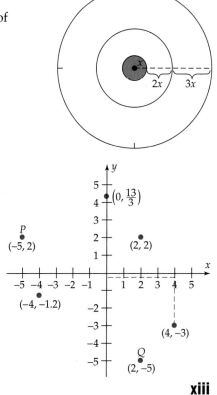

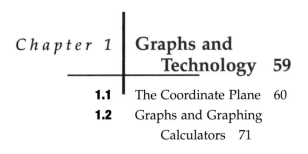

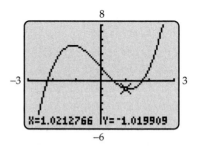

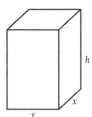

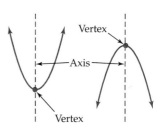

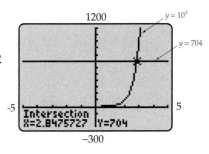

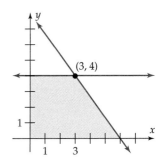

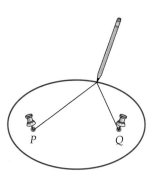

To the Instructor

The following information should assist you in planning a workable syllabus for your course.

Prerequisites Chapter 0 (Algebra Review) is a prerequisite for the entire book; it may be omitted by well-prepared classes.

Interdependence of Chapters and Sections The chart on the facing page shows the interdependence of chapters. A similar chart appears at the beginning of each chapter, showing the interdependence of sections within the chapter.

Excursions Sections labeled "Excursion" are usually related to the preceding section and are never prerequisites for other sections of the text. The "Excursion" label is designed solely to make syllabus planning easier and is not intended as any kind of value judgment on the topic in question.

In this text "calculator" means "graphing calculator." You and your students should be aware of the following facts about calculators.

Minimal Technology Requirements It is assumed that every student has either a computer with appropriate software or a calculator at least at the level of a TI-82. Among current calculator models that meet or exceed this minimal requirement are TI-82 through TI-92, Sharp 9600, HP-38 through HP-49, Casio 9850, Casio 9970, and Casio FX2.

Technology Tips To avoid much clutter, only a limited number of calculators are specifically mentioned in the Technology Tips. However, unless noted otherwise,

Technology Tips for the TI-83 also apply to the TI-82;

Technology Tips for the TI-86 also apply to the TI-85;

Technology Tips for the TI-89 also apply to the TI-92;

Technology Tips for the Casio 9850 also apply to the Casio 9970.

(text continues on p. xx)

(text continues on p

Interdependence of Chapters

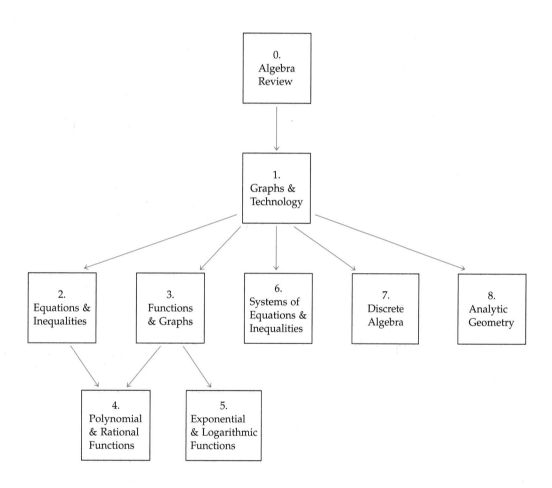

There are no Tips specifically for HP-48 and HP-49 calculators since they use operating systems entirely different from those of other calculators.

Supplements

Instructors who adopt this text may receive, free of charge, the following items:

Electronic Companion to College Algebra This dynamic and interactive CD-ROM, which is packaged with the text, covers the key concepts using multiple representations.

Instructor's Resource Manual This manual comprises two unique parts. The first part contains detailed solutions to all the exercises and end-of-chapter Review Questions to assist the instructor in the classroom and in grading assignments. Solutions to the Discovery Projects at the end of each chapter are also included. The second part of the manual ties in with the *Math in Practice: An Applied Video Companion CD-ROM* that is packaged with the manual. This innovative CD-ROM is designed to show students how and where college algebra topics arise in real life. Lori Palmer of Utah Valley State College has conducted more than 20 engaging interviews with individuals in such fields as aviation, food services, banking, and environmental science to motivate the key concepts from the text. Each vignette is accompanied by two problems written by Carolyn Hamilton, also of Utah Valley State College, to test students' understanding of the underlying mathematical ideas and skills. Printed versions of these problems are included in the manual. Answers to the problems are provided on the disk with detailed solutions within the manual.

Test Bank This manual provides sample problems for each section of the text. On average, there are at least 15 sample problems per section. These problems are both free-response and multiple choice and are broken into three levels of difficulty: easy, medium, and hard. From the sample problems, five sample tests per chapter will be created. Tests A, B, and C will be free-response. Tests D and E will be multiple choice. A complete answer section is included.

ESATEST Computerized Test Bank The computerized test bank contains all the problems from the printed test bank and allows instructors to prepare quizzes and examinations quickly and easily. Instructors may also add questions or modify existing ones. ESATEST has gradebook capabilities for recording and tracking students' grades. Instructors have the opportunity to post and administer a test over a network or on the Web. ESATEST's user-friendly printing capability accommodates all printing platforms.

Graphing Calculator Manual This manual comprises three parts. The initial part of the manual is entitled Basic Calculator Topics. This section provides a general understanding of all the capabilities available on a graphing calculator. Calculator Notes and Problems follow, in which narrative, examples, and exercises to develop the concepts in the text in association with your graphing calculator are used. The final section of the manual provides detailed instructions on 10 specific calculators.

Video Series Free to adopters, the videotape package consists of nine VHS videotapes, one for each chapter in the text. Each tape is one hour long and further develops the concepts of the chapters. On-location footage is utilized to introduce an extended application at the beginning of each tape. This application is explained fully at the end of each tape.

Web Site The Web site (www.harcourtcollege.com/mathexpress) offers additional resources to both instructors and students in conjunction with the adoption of the text.

Saunders College Publishing, a division of Harcourt College Publishers, may provide complimentary instructional aids and supplements or supplement packages to those adopters qualified under our adoption policy. Please contact your sales representative for more information. If as an adopter or potential user you receive supplements you do not need, please return them to your sales representative or send them to:

Attn: Returns Department
Troy Warehouse
465 South Lincoln Drive
Troy, MO 63379

To the Student

This text assumes the use of technology, so you should be aware of the following facts:

Terminology In this text "calculator" means "graphing calculator." All discussions of calculators, with obvious modifications, apply to graphing software for computers.

Minimal Technology Requirements In order to use this text effectively, you must have either a computer with appropriate software or a calculator at least at the level of a TI-82. Among current models that meet or exceed this minimal requirement are the TI-82 through TI-92, the Sharp 9600, HP-38, HP-39, HP-48, HP-49, and Casio 9850, 9770, and FX2.

The following features of the text will enable you to get the most out of your calculator.

Technology Tips Some of the Technology Tips in the margin tell you the proper menus or keys to be used on specific calculators to carry out procedures mentioned in the text. Other Tips offer general information or helpful advice for performing a particular task on a calculator.

As a general rule, the only calculators mentioned in the Technology Tips are the TI-83, TI-86, TI-89, Sharp 9600, HP-38, and Casio 9850*. However, unless noted otherwise,

Technology Tips for the TI-83 also apply to the TI-82;

Technology Tips for the TI-86 also apply to the TI-85;

Technology Tips for the TI-89 also apply to the TI-92;

Technology Tips for the Casio 9850 also apply to the Casio 9970.

---

*There are no Tips in the text specifically for HP-48 and HP-49 calculators, which use an entirely different operating system than other calculators.

Calculator Investigations You may not be aware of the full capabilities of your calculator (or some of its limitations). The Calculator Investigations (which appear just before the exercise sets in some of the earlier sections of the book) will help you to become familiar with your calculator and to maximize the mathematical power it provides. Even if your instructor does not assign these investigations, you may want to look through them to be sure you are getting the most you can from your calculator.

With all this talk about calculators, don't lose sight of this crucial fact:

Technology is only a *tool* for doing mathematics.

You can't build a house if you only use a hammer. A hammer is great for pounding nails, but useless for sawing boards. Similarly, a calculator is great for computations and graphing, but it is not the right tool for every mathematical task. To succeed in this course, you must develop and use your algebraic and geometric skills, your reasoning power and common sense, and you must be willing to work.

The key to success is to use all of the resources at your disposal: your instructor, your fellow students, your calculator (and its instruction manual), and this book. Here are some tips for making the most of these resources.

Ask Questions. Remember the words of Hillel:

The bashful do not learn.

There is no such thing as a "dumb question" (assuming, of course, that you have attended class, taken notes, and read the text). Your instructor will welcome questions that arise from a serious effort on your part.

Read the Book. Not just the homework exercises, but the rest of the text as well. There is no way your instructor can possibly cover the essential topics, clarify ambiguities, explain the fine points, and answer all your questions during class time. You simply will not develop the level of understanding you need to succeed in this course unless you read the text fully and carefully.

Be an Interactive Reader. You can't read a math book the way you read a novel or history book. You need pencil, paper, and your calculator at hand to work out the statements you don't understand and to make notes of things to ask your fellow students and/or your instructor.

Do the Graphing Explorations. When you come to a box labeled "Graphing Exploration," use your calculator as directed to complete the discussion. Typically, this will involve graphing one or more equations and answering some questions about the graphs. Doing these explorations as they arise will improve your understanding and clarify issues that might otherwise cause difficulties.

Do Your Homework. Remember that

Mathematics is not a spectator sport.

You can't expect to learn mathematics without doing mathematics, any more than you could learn to swim without getting wet. Like swimming or dancing or reading or any other skill, mathematics takes practice. Homework assignments are where you get the practice that is essential for passing this course and succeeding in calculus.

Supplements

The following items are available at no cost to students.

Electronic Companion to College Algebra This dynamic and interactive CD-ROM, which is packaged with the text, covers the key concepts using multiple representations.

Web Site The Web site (www.harcourtcollege.com/mathexpress) offers additional resources to both instructors and students in conjunction with the adoption of the text.

Students using *Contemporary College Algebra* may purchase the following supplements:

Student Resource Manual This manual comprises two unique parts. The first part contains detailed solutions to all the odd-numbered exercises and end-of-chapter Review Questions. Solutions to the Discovery Projects at the end of each chapter are also included. The second part of the manual ties in with the *Math in Practice: An Applied Video Companion CD-ROM* that is packaged with the manual. This innovative CD-ROM is designed to show students how and where college algebra topics arise in real life. Lori Palmer of Utah Valley State College has conducted more than 20 engaging interviews with individuals in such fields as aviation, food services, banking, and environmental science to motivate the key concepts from the text. Each vignette is accompanied by two problems written by Carolyn Hamilton, also of Utah Valley State College, to test students' understanding of the underlying mathematical ideas and skills. Printed versions of these problems are included in the manual. Answers to the problems are provided on the disk and in the manual.

Graphing Calculator Manual This manual comprises three parts. The initial part of the manual is entitled Basic Calculator Topics. This section provides a general understanding of all the capabilities available on a graphing calculator. Calculator Notes and Problems follow, in which narrative, examples, and exercises to develop the concepts in the text in association with your graphing calculator are used. The final section of the manual provides detailed instructions on 10 specific calculators.

ESATUTOR This computer software package contains hundreds of problems and answers that correspond with every section of the text. Students can complete a pretest to evaluate their level of understanding of the concepts in each chapter. Additionally, students can complete post-tests to ensure they have grasped the primary learning objectives. The software comes with a built-in graphing calculator. Students interested in purchasing this software package should refer to the marketing material inside the back cover of this text.

Core Concept Video This single videotape contains the most important topics covered in the full video series that is given to each school that uses *Contemporary College Algebra*. This take-home tutorial can be used as a preview of what is to be covered in class, as an aid to completing homework assignments, or as a tool to review for a test.

Chapter 0

Algebra Review

On a clear day, can you see forever?

If you are at the top of the Sears Tower in Chicago, how far can you see? In earlier centuries, the lookout on a sailing ship was posted on the highest mast because he could see farther from there than from the deck. How much farther? These questions, and similar ones, can be answered (at least approximately) by using basic algebra and geometry. See Example 9 on page 22 and Exercise 74 on page 24.

Chapter Outline

This chapter reviews the essential facts about real numbers, exponents, and the basic rules of algebra that are needed in this course and later ones. Well-prepared students may be able to skim over much of this material, but if you haven't used your algebraic skills for a while, you should review this chapter thoroughly. Your success in the rest of the course depends on your ability to use the fundamental algebraic tools presented here.

0.1 The Real Number System

You have been using **real numbers** most of your life. They include the **natural numbers** (or **positive integers**): 1, 2, 3, 4, … and the **integers:**

$$\ldots, -5, -4, -3, -2, -1, 0, 1, 2, 3, 4, 5, \ldots .$$

A real number is said to be a **rational number** if it can be expressed as a fraction $\frac{r}{s}$, with r and s integers and $s \neq 0$; for instance

$$\frac{1}{2}, \qquad -.983 = -\frac{983}{1000}, \qquad 47 = \frac{47}{1}, \qquad 8\tfrac{3}{5} = \frac{43}{5}.$$

Alternatively, rational numbers may be described as numbers that can be expressed as terminating decimals, such as $.25 = \frac{1}{4}$, or as nonterminating repeating decimals in which a single digit or a block of digits repeats forever, such as

$$\frac{5}{3} = 1.66666\cdots \qquad \text{or} \qquad \frac{58}{333} = .174174174\cdots .$$

A real number that cannot be expressed as a fraction with integer numerator and denominator is called an **irrational number.** Alternatively, an irrational number is one that can be expressed as a nonterminating, nonrepeating decimal (no block of digits repeats forever). For example, the number π, which is used to calculate the area of a circle, is irrational.* More information about decimal expansions of real numbers is given in Excursion 0.1.A.

The real numbers are often represented geometrically as points on a **number line,** as in Figure 0–1. We assume that there is exactly one point on the line for every real number (and vice versa) and use phrases such as "the point 3.6" or "a number on the line." This mental identification of real numbers and points on the line is often extremely helpful.

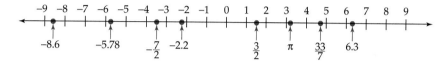

Figure 0–1

Order

The statement $c < d$, which is read **c is less than d,** and the statement $d > c$ (that is, **d is greater than c**) mean exactly the same thing:

c lies to the *left* of d on the number line.

For example, Figure 0–1 shows that $-5.78 < -2.2$ and $4 > \pi$.

The statement $c \le d$, which is read **c is less than or equal to c,** means that

Either c is less than d or c is equal to d.

Only one part of an "either … or" statement needs to be true for the entire statement to be considered true. So the statement $5 \le 10$ is true because $5 < 10$, and the statement $5 \le 5$ is true because $5 = 5$. The statement $d \ge c$ (read **d is greater than or equal to c**) means exactly the same thing as $c \le d$.

The statement $b < c < d$ means

$b < c$ and simultaneously $c < d$.

For example, $3 < x < 7$ means that x is a number that is strictly between 3 and 7 on the number line (greater than 3 and less than 7). Similarly, $b \le c < d$ means

$b \le c$ and simultaneously $c < d$,

and so on.

Order of Operations, Parentheses, and Arithmetic

To avoid ambiguity when dealing with expressions such as $6 + 3 \times 5$, mathematicians have made the following agreement (which is also followed by your calculator).

---

*This fact is difficult to prove. In the past you may have used 22/7 as π; a calculator might display π as 3.141592654. However, these numbers are just *approximations* of π (close, but not quite *equal* to π).

Order of Operations

In an expression without parentheses, multiplication and division are performed first (from left to right). Addition and subtraction are performed last (from left to right).

In light of this convention, there is only one correct way to interpret $6 + 3 \times 5$:

$$6 + 3 \times 5 = 6 + 15 = 21 \qquad \textit{[Multiplication first, addition last]}$$

On the other hand, if you want to "add $6 + 3$ and then multiply by 5," you must use parentheses:

$$(6 + 3) \cdot 5 = 9 \cdot 5 = 45.$$

This is an illustration of the first of two basic rules for dealing with parentheses.

Rules for Parentheses

1. Do all computations inside the parentheses before doing any computations outside the parentheses.
2. When dealing with parentheses within parentheses, begin with the innermost pair and work outward.

For example,

$$8 + [11 - (6 \times 3)] = 8 + (11 - 18) = 8 + (-7) = 1.$$

Inside parentheses first

We assume that you are familiar with the basic properties of real number arithmetic, particularly this fact:

Distributive Law

For all real numbers a, b, c

$$a(b + c) = ab + ac \qquad \text{and} \qquad (b + c)a = ba + ca.$$

The distributive law doesn't usually play a direct role in easy computations, such as $4(3 + 5)$. Most people don't say $4 \cdot 3 + 4 \cdot 5 = 12 + 20 = 32$. Instead, they mentally add the numbers in parentheses and say 4 times 8 is 32. But when symbols are involved, you can't do that, and the distributive law is essential. For example,

$$4(3 + x) = 4 \cdot 3 + 4x = 12 + 4x.$$

Technology Tip

To enter a negative number, such as −5, on most calculators, you must use the negation key: (−) 5. If you use the subtraction key on such calculators and enter − 5, the display reads

ANS − 5

which tells the calculator to subtract 5 from the previous answer.

Negative Numbers and Negatives of Numbers

The **positive numbers** are those to the right of 0 on the number line, that is,

$$\text{all numbers } c \text{ with } c > 0.$$

The **negative numbers** are those to the left of 0, that is,

$$\text{all numbers } c \text{ with } c < 0.$$

The **nonnegative** numbers are the numbers c with $c \geq 0$.

The word "negative" has a second meaning in mathematics. The **negative *of a* number** c is the number $-c$. For example, the negative of 5 is −5, and the negative of −3 is $-(-3) = 3$. Thus the negative of a negative number is a positive number. Zero is its own negative since $-0 = 0$. In summary,

Negatives

The negative of the number c is $-c$.

If c is a positive number, then $-c$ is a negative number.

If c is a negative number, then $-c$ is a positive number.

Square Roots

The **square root** of a nonnegative real number d is defined to be the nonnegative number whose square is d and is denoted $\sqrt{d}$. For instance,

$$\sqrt{25} = 5 \qquad \text{because} \qquad 5^2 = 25.$$

In the past you may have said that $\sqrt{25} = \pm 5$ since $(-5)^2$ is also 25. It is preferable, however, to have a single unambiguous meaning for the symbol $\sqrt{25}$. So in the real number system, the term "square root" and the radical symbol $\sqrt{}$ always denote a *nonnegative* number. To express −5 in terms of radicals, we write $-5 = -\sqrt{25}$.

Although $-\sqrt{25}$ is a real number, the expression $\sqrt{-25}$ is *not defined* in the real numbers because there is no real number whose square is −25. In fact, since the square of every real number is nonnegative,

No negative number has a square root in the real numbers.

Some square roots can be found (or verified) by hand, such as

$$\sqrt{225} = 15 \qquad \text{and} \qquad \sqrt{1.21} = 1.1.$$

Usually, however, a calculator is needed to obtain rational *approximations* of roots. For instance, we know that $\sqrt{87}$ is between 9 and 10 because $9^2 = 81$ and $10^2 = 100$. A calculator shows that $\sqrt{87} \approx 9.327379$.*

Technology Tip

To compute an expression such as $\sqrt{7^2 + 51}$ on a calculator, you must use parentheses:

$$\sqrt{\ } \ (7^2 + 51).$$

Without the parentheses, the calculator will compute

$$\sqrt{7^2} + 51 = 7 + 51 = 58$$

instead of the correct answer

$$\sqrt{7^2 + 51} = \sqrt{49 + 51}$$
$$= \sqrt{100}$$
$$= 10.$$

*$\approx$ means "approximately equal."

Absolute Value

On an informal level most students think of absolute value like this:

The absolute value of a nonnegative number is the number itself.

The absolute value of a negative number is found by "erasing the minus sign."

If $|c|$ denotes the absolute value of c, then, for example, $|5| = 5$ and $|-4| = 4$.

This informal approach is inadequate, however, for finding the absolute value of a number such as $\pi - 6$. It doesn't make sense to "erase the minus sign" here. So we must develop a more precise definition. The statement $|5| = 5$ suggests that the absolute value of a positive number ought to be the number itself. For negative numbers, such as -4, note that $|-4| = 4 = -(-4)$, that is, the absolute value of the negative number -4 is the *negative* of -4. These facts are the basis of the formal definition:

Absolute Value

The *absolute value* of a real number c is denoted $|c|$ and is defined as follows:

If $c \geq 0$, then $|c| = c$.

If $c < 0$, then $|c| = -c$.

Example 1

(a) $|3.5| = 3.5$ and $|-7/2| = -(-7/2) = 7/2$.

(b) To find $|\pi - 6|$ note that $\pi \approx 3.14$, so that $\pi - 6 < 0$. Hence, $|\pi - 6|$ is defined to be the *negative* of $\pi - 6$, that is,

$$|\pi - 6| = -(\pi - 6) = -\pi + 6.$$

(c) $|5 - \sqrt{2}| = 5 - \sqrt{2}$ because $5 - \sqrt{2} \geq 0$. ■

Here are the important facts about absolute value:

Properties of Absolute Value

1. $|c| \geq 0$ and $|c| > 0$ when $c \neq 0$.
2. $|c| = |-c|$
3. $|cd| = |c| \cdot |d|$
4. $\left|\dfrac{c}{d}\right| = \dfrac{|c|}{|d|}$ $(d \neq 0)$

Example 2   Here are examples of some of the properties listed in the box.

2. If $c = 3$, then

$$|c| = |3| = 3 \quad \text{and} \quad |-c| = |-3| = 3$$

so that $|c| = |-c|$.

3. If $c = 6$ and $d = -2$, then

$$|cd| = |6(-2)| = |-12| = 12$$

and

$$|c| \cdot |d| = |6| \cdot |-2| = 6 \cdot 2 = 12$$

so that $|cd| = |c| \cdot |d|$.

4. If $c = -5$ and $d = 4$, then

$$\left|\frac{c}{d}\right| = \left|\frac{-5}{4}\right| = \left|-\frac{5}{4}\right| = \frac{5}{4} \quad \text{and} \quad \frac{|c|}{|d|} = \frac{|-5|}{|4|} = \frac{5}{4}$$

so that

$$\left|\frac{c}{d}\right| = \frac{|c|}{|d|}. \quad \blacksquare$$

When c is a positive number, then $\sqrt{c^2} = c$, but when c is negative, this is *false*. For example, if $c = -3$, then

$$\sqrt{c^2} = \sqrt{(-3)^2} = \sqrt{9} = 3 \quad (not\ -3),$$

so that $\sqrt{c^2} \neq c$. In this case, however, $|c| = |-3| = 3$, so that $\sqrt{c^2} = |c|$. The same thing is true for any negative number c. It is also true for positive numbers (since $|c| = c$ when c is positive). In other words,

Square Roots of Squares

For every real number c,

$$\sqrt{c^2} = |c|.$$

When dealing with long expressions inside absolute value bars, do the computations inside first, and then take the absolute value.

Example 3

(a) $|5(2 - 4) + 7| = |5(-2) + 7| = |-10 + 7| = |-3| = 3.$

(b) $4 - |3 - 9| = 4 - |-6| = 4 - 6 = -2. \quad \blacksquare$

Technology Tip

To compute absolute values on a calculator, use the ABS key and parentheses. To find $|9 - 3\pi|$, for example, key in

ABS$(9 - 3\pi)$

The ABS key is on the keyboard of HP-38 and in the NUM submenu of the MATH menu of TI and Sharp 9600. It is in the NUM submenu of the Casio 9850 OPTN menu.

> **CAUTION**
> When c and d have opposite signs, $|c + d|$ is *not equal* to $|c| + |d|$. For example, when $c = -3$ and $d = 5$, then
> $$|c + d| = |-3 + 5| = 2,$$
> but
> $$|c| + |d| = |-3| + |5| = 3 + 5 = 8.$$

The Caution shows that $|c + d| < |c| + |d|$ when $c = -3$ and $d = 5$. In the general case, we have the following fact.

The Triangle Inequality

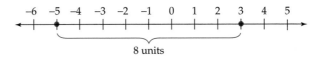

For any real numbers c and d,

$$|c + d| \le |c| + |d|.$$

Distance on the Number Line

Observe that the distance from -5 to 3 on the number line is 8 units:

$$
\begin{array}{ccccccccccccc}
-6 & -5 & -4 & -3 & -2 & -1 & 0 & 1 & 2 & 3 & 4 & 5
\end{array}
$$

8 units

Figure 0–2

This distance can be expressed in terms of absolute value by noting that $|(-5) - 3| = 8$. That is, the distance is the *absolute value of the difference* of the two numbers. Furthermore, the order in which you take the difference doesn't matter; $|3 - (-5)|$ is also 8. This reflects the geometric fact that the distance from -5 to 3 is the same as the distance from 3 to -5. The same thing is true in the general case:

Distance on the Number Line

The distance between c and d on the number line is the number

$$|c - d| = |d - c|.$$

Example 4   The distance from 4.2 to 9 is $|4.2 - 9| = |-4.8| = 4.8$ and the distance from 6 to $\sqrt{2}$ is $|6 - \sqrt{2}|$. ∎

In the special case when $d = 0$, the distance formula shows that $|c - 0| = |c|$. Hence,

Distance to Zero

$|c|$ **is the distance between c and 0 on the number line.**

Algebraic problems can sometimes be solved by translating them into equivalent geometric problems. The key is to interpret statements involving absolute value as statements about distance on the number line.

Example 5 Solve the equation $|x - 3| = 4$ geometrically.

Solution Translate the given equation into an equivalent geometric statement.

$$|x - 3| = 4$$

means

The distance between x and 3 on the number line is 4.

Figure 0–3 shows that -1 and 7 are the only numbers whose distance to 3 is 4 units.

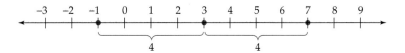

Figure 0–3

Therefore, the solutions of $|x - 3| = 4$ are $x = -1$ and $x = 7$. ∎

Example 6 Solve the equation $|x + 5| = 3$.

Solution We rewrite the equation as $|x - (-5)| = 3$. In this form it states that

*The distance between x and −5 is 3 units.**

---

*It's necessary to rewrite the equation first because the distance formula involves the *difference* of two numbers, not their sum.

Figure 0–4 shows that −8 and −2 are the only two numbers whose distance to −5 is 3 units:

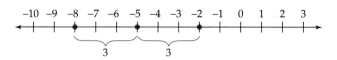

Figure 0–4

Thus $x = -8$ and $x = -2$ are the solutions of $|x + 5| = 3$. ■

Calculator Investigations 0.1

1. Edit and Replay Consider the equation $y = x^3 + 6x^2 - 5$.

  (a) Find the value of y when $x = 7$ by keying in

    (∗) $\qquad\qquad\qquad\qquad (-7)^3 + 6(-7)^2 - 5$

  and pressing ENTER.†

  (b) Find the value of y when $x = 9$ by editing your previous calculation as follows. On TI, press SECOND ENTER; on Sharp 9600, press SECOND ENTRY; on Casio 9850 press AC and the up arrow key; on HP-38, use the up arrow key until the previous computation is shaded, then press COPY. This activates the **replay** feature: Your screen now returns to the previous calculation (∗). Use the left/right arrow keys and the DEL(ete) key to move through the equation and replace −7 by 9. Then press ENTER and the result of the new computation is displayed.

  (c) Use the replay feature again to find the value of y when $x = 108$. Use the INS(ert) key to avoid unnecessary retyping.

  (d) Press the replay keys repeatedly. [On Casio, press AC once and the up arrow repeatedly.] Each time you do, one of the preceding calculations will appear. Go back to the first one (∗) and compute it again.

2. Mathematical Operations

  (a) Key in each of the following and explain why your answers are different. [See the Technology Tip on page 7.]

  ABS (−) 9 + 2 ENTER and ABS ((−) 9 + 2) ENTER.

  (b) Find INT in the NUM submenu of the MATH or OPTN menu [on HP-38, use FLOOR in the REAL submenu of the MATH menu]. Find out what this command does when you follow it by a number and ENTER. Casio users will get different answers than other brands.

†Here and throughout the book, Casio users should read "EXE" in place of "ENTER."

3. Symbolic Calculations

(a) To store the number 2 in memory A of a calculator, type

2 STO ▶ A   ENTER   [TI-86]; or
2 STO ▶ ALPHA A ENTER   [TI-83/89; Sharp 9600; HP-38]; or
2 →   ALPHA A EXE   [Casio 9850].

If you now key in ALPHA A ENTER, what does the calculator display?

(b) In a similar fashion store the number 5 in memory B and −10 in memory C. Then using the ALPHA keys display this expression on the screen: B + C/A. If you press ENTER, what happens? Explain what the calculator is doing.

(c) Experiment with other expressions, such as $B^2 - 4AC$.

4. Inequalities   Find out what happens when you key in each of these statements and press ENTER:

$$8 < 9, \qquad 8 < 5, \qquad 9 > 2, \qquad 9 > 10.$$

[Inequality symbols are in the TEST menu on the TI keyboard or the TESTS submenu of the HP-38 MATH menu or the INEQ submenu of the Sharp 9600 MATH menu.]

Exercises 0.1

1. Draw a number line and mark the location of each of these numbers: $0, -7, 8/3, 10, -1, -4.75, 1/2, -5,$ and 2.25.

In Exercises 2–14, express the given statement in symbols.

2. 5 is less than 7.

3. −4 is greater than −8.

4. −17 is less than 14.

5. π is less than 100.

6. x is nonnegative.

7. y is less than or equal to 7.5.

8. z is greater than or equal to −4.

9. t is positive.

10. d is not greater than 2.

11. c is at most 3.

12. z is at least −17.

In Exercises 13–18, fill the blank with $<$, $=$, or $>$ so that the resulting statement is true.

13. −6 _____ −2   14. 5 _____ −3

15. 3/4 _____ .75   16. 3.1 _____ π

17. 1/3 _____ .33

18. Galileo discovered that the period of a pendulum depends only on the length of the pendulum and the acceleration of gravity. The period T of a pendulum (in seconds) is

$$T = 2\pi\sqrt{\frac{l}{g}}$$

where l is the length of the pendulum in feet and $g \approx 32.2$ ft/sec^2 is the acceleration due to gravity. Find the period of a pendulum whose length is 4 ft.

In Exercises 19–20, use a calculator and list the given numbers in order from smallest to largest.

19. $\dfrac{189}{37}, \dfrac{4587}{691}, \sqrt{47}, 6.735, \sqrt{27}, \dfrac{2040}{523}$

20. $\dfrac{385}{117}, \sqrt{10}, \dfrac{187}{63}, \pi, \sqrt{\sqrt{85}}, 2.9884$

In Exercises 21–26, fill the blank so as to produce two equivalent statements. For example, the arithmetic statement "a is negative" is equivalent to the geometric statement "the point a lies to the left of the point 0."

Arithmetic Statement	**Geometric Statement**
21. _____	a lies c units to the right of b.
22. _____	a lies between b and c.
23. $a - b > 0$	_____
24. a is positive.	_____
25. _____	a lies to the left of b.
26. $a \geq b$	_____

In Exercises 27–34, simplify and write the given number without using absolute values.

27. $3 - |2 - 5|$ **28.** $-2 - |-2|$

29. $|6 - 4| + |-3 - 5|$ **30.** $|-6| - |6|$

31. $|(-13)^2|$ **32.** $-|-5|^2$

33. $|\pi - \sqrt{2}|$ **34.** $|\sqrt{2} - 2|$

In Exercises 35–40, fill the blank with $<$, $=$, or $>$ so that the resulting statement is true.

35. $|-2|$ ____ $|-5|$ **36.** 5 ____ $|-2|$

37. $|3|$ ____ $-|4|$ **38.** $|-3|$ ____ 0

39. -7 ____ $|-1|$ **40.** $-|-4|$ ____ 0

In Exercises 41–46, find the distance between the given numbers.

41. -3 and 4 **42.** 7 and 107

43. -7 and $15/2$ **44.** $-3/4$ and -10

45. π and 3 **46.** π and -3

47. A broker predicts that over the next six months, the price p of a particular stock will not vary from its current price of \$25.75 by more than \$4. Express this prediction as an inequality.

48. The following chart, from the *Wall Street Journal* (January 7, 1998), shows the anticipated sales of digital TV sets through 2002. Use it to answer these questions.
   (a) Which consecutive years project the largest percentage increase in sales?

(b) Do any pairs of consecutive years have the same percentage increase?

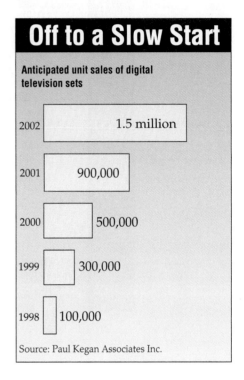

Off to a Slow Start

Anticipated unit sales of digital television sets

2002   1.5 million
2001   900,000
2000   500,000
1999   300,000
1998   100,000

Source: Paul Kegan Associates Inc.

49. Ann rode her bike down the Blue Ridge Parkway, beginning at Linville Falls (mile post 316.3) and ending at Mount Pisgah (mile post 408.6). Use absolute value notation to describe how far Ann rode.

50. At Statewide Insurance, each department's expenses are reviewed monthly. A department can fail to pass the "budget variance test" in a category if either (a) the absolute value of the difference between actual expenses and the budget is more than \$500; or (b) the absolute value of the difference between the actual expenses and the budget is more than 5% of the budgeted amount. Which of the following items fail the budget variance test? Explain your answers.

Item	Budgeted Expense	Actual Expense
Wages	\$220,750	\$221,239
Overtime	\$ 10,500	\$ 11,018
Shipping and postage	\$ 530	\$ 589

In Exercises 51–56, write the given expression without using absolute values.

51. $|t^2|$

52. $|u^2 + 2|$

53. $|(-3 - y)^2|$

54. $|-2 - y^2|$

55. $|b - 3|$ if $b \geq 3$

56. $|a - 5|$ if $a < 5$

In Exercises 57–62, express the given geometric statement about numbers on the number line algebraically, using absolute values.

57. The distance from x to 5 is less than 4.

58. x is more than 6 units from c.

59. x is at most 17 units from -4.

60. x is within 3 units of 7.

61. c is closer to 0 than b is.

62. x is closer to 1 than to 4.

In Exercises 63–66, translate the given algebraic statement into a geometric statement about numbers on the number line.

63. $|x - 3| < 2$

64. $|x - c| > 6$

65. $|x + 7| \leq 3$

66. $|u + v| \geq 2$

In Exercises 67–72, use the geometric approach explained in the text to solve the given equation.

67. $|x| = 1$

68. $|x| = 3/2$

69. $|x - 2| = 1$

70. $|x + 3| = 2$

71. $|x + \pi| = 4$

72. $\left| x - \dfrac{3}{2} \right| = 5$

Thinkers

73. Explain why the statement $|a| + |b| + |c| > 0$ is algebraic shorthand for "at least one of the numbers a, b, c is different from zero."

74. Find an algebraic shorthand version of the statement "the numbers a, b, c are all different from zero."

0.1.A **EXCURSION** # Decimal Representation of Real Numbers

Every rational number can be expressed as a terminating or repeating decimal. For instance, $3/4 = .75$. To express $15/11$ as a decimal, divide the numerator by the denominator:

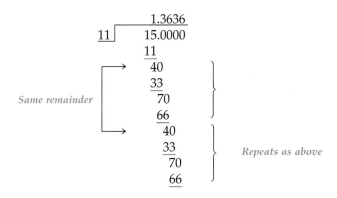

Since the remainder at the first step (namely 4) occurs again at the third step, it is clear that the division process goes on forever with the

two-digit block "36" repeating over and over in the quotient $15/11 = 1.3636363636 \cdots$.

The method used in the preceding example can be used to express any rational number as a decimal. During the division process some remainder *necessarily* repeats, as in the preceding example. If the remainder at which this repetition starts is 0, the result is a repeating decimal ending in zeros—that is, a terminating decimal (for instance, $.75000 \cdots = .75$). If the remainder at which the repetition starts is nonzero, then the result is a nonterminating repeating decimal, as in the example.

However, a typical calculator displays only the first ten digits of a number in decimal form (although it uses several additional digits in its internal computations). For example, a calculator might display the decimal expansion of $1/17$ as $.0588235294$, although the actual expansion has a repeating block of 16 digits:

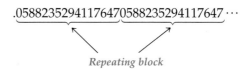

Repeating block

Thus, a calculator can contain the *exact* value of a rational number only if its decimal expansion terminates after approximately ten decimal places. Techniques for obtaining the full decimal expansion of a rational number from a calculator are discussed in Exercise 25.

Conversely, there is a simple method for converting any repeating decimal into a rational number.

Example 1   Write $d = .272727 \cdots$ as a rational number.

Solution   Assuming that the usual rules of arithmetic hold, we see that

$$100d = 27.272727 \cdots \qquad \text{and} \qquad d = .272727 \cdots$$

Now subtract d from $100d$:

$$
\begin{aligned}
100d &= 27.272727 \cdots \\
-d &= -.272727 \cdots \\
\hline
99d &= 27
\end{aligned}
$$

Dividing both sides of this last equation by 99 shows that $d = 27/99 = 3/11$.   ∎

Irrational Numbers

Many nonterminating decimals are *nonrepeating* (that is, no block of digits repeats forever), such as $.202002000200002 \cdots$ (where after each 2 there is one more 0 than before). Although the proof is too long to give here, it is in fact true that every nonterminating and nonrepeating decimal represents an *irrational* real number.

Conversely every irrational number can be expressed as a nonterminating and nonrepeating decimal (no proof to be given here). For instance, the decimal expansion of the irrational number π begins

Technology Tip

The FRAC key on most TI calculators automatically converts decimals to fractions (subject to some limitations). The same thing can be done on HP-38 by choosing FRACTION number format in the MODE menu and then entering the decimal. Conversion programs for Sharp and Casio are in the Program Appendix.

3.1415926535897 ⋯⋯. This computation has actually been carried out to more than 200 *billion* decimal places by computer.

Since a calculator can deal with only the first 12–15 digits of a number in decimal form, *a calculator cannot contain the exact value of any irrational number.* Furthermore, there are no easy calculator techniques for carrying out the decimal expansion of an arbitrary irrational number to a specified number of decimal places, as there are for rational numbers.

Since every real number is either a rational number or an irrational one, the preceding discussion can be summarized as follows.

Decimal Representation

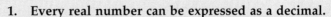

1. Every real number can be expressed as a decimal.
2. Every decimal represents a real number.
3. The terminating decimals and the nonterminating repeating decimals are precisely the rational numbers.
4. The nonterminating, nonrepeating decimals are precisely the irrational numbers.

Calculator Investigation 0.1.A

1. FRAC Key   If your calculator has a FRAC key or program (see the Program Appendix), test its limitations by entering each of the following numbers and then pressing the FRAC key.

(a) .058823529411   (b) .0588235294117

(c) .058823529411724   (d) .0588235294117985

Which of your answers are correct? [*Hint:* Look at the decimal expansion of 1/17 on page 14.]

Exercises 0.1.A

In Exercises 1–6, express the given rational number as a repeating decimal.

1. 7/9   **2.** 2/13   **3.** 23/14

4. 19/88   **5.** 1/19 (long)   **6.** 9/11

In Exercises 7–14, state whether a calculator can express the given number exactly.

7. 2/3   **8.** 7/16   **9.** 1/64   **10.** 1/22

11. $3\pi/2$ **12.** $\pi - 3$ **13.** 1/.625   **14.** 1/.16

In Exercises 15–21, express the given repeating decimal as a fraction.

15. .373737 ⋯   **16.** .929292 ⋯

17. 76.63424242 ⋯ [*Hint:* Consider $10{,}000d - 100d$, where $d = 76.63424242 ⋯$.]

18. 13.513513 ⋯ [*Hint:* Consider $1000d - d$, where $d = 13.513513 ⋯$]

19. .135135135 ⋯ [*Hint:* See Exercise 18.]

20. .33030303 ⋯   **21.** 52.31272727 ⋯

22. If two real numbers have the same decimal expansion through three decimal places, how far apart can they be on the number line?

Thinkers

23. Use the methods in Exercises 15–21 to show that both $.74999\cdots$ and $.75000\cdots$ are decimal expansions of 3/4. Every terminating decimal can also be expressed as a decimal ending in repeated 9's. It can be proved that these are the only real numbers with more than one decimal expansion.

24. *Finding remainders with a calculator* If you use long division to divide 369 by 7, you obtain.

$$
\begin{array}{r}
52 \quad \leftarrow Quotient \\
Divisor \rightarrow \quad 7\overline{)369} \quad \leftarrow Dividend \\
\underline{35} \\
19 \\
\underline{14} \\
5 \quad \leftarrow Remainder
\end{array}
$$

If you use a calculator to find $369 \div 7$, the answer is displayed as 52.71428571. Observe that the integer part of this calculator answer, 52, is the quotient when you do the problem by long division. The usual "checking procedure" for long division shows that

$$7 \cdot 52 + 5 = 369 \quad \text{or equivalently} \quad 369 - 7 \cdot 52 = 5.$$

Thus the remainder is

$$\text{Dividend} - (\text{divisor})\binom{\text{integer part of}}{\text{calculator answer}}.$$

Use this method to find the quotient and remainder in these problems:

(a) $5683 \div 9$ **(b)** $1,000,000 \div 19$
(c) $53,000,000 \div 37$

In Exercises 25–30, find the decimal expansion of the given rational number. All these expansions are too long to fit in a calculator, but can be readily found by using the hint in Exercise 25.

25. 6/17 [*Hint:* The first part of dividing 6 by 17 involves working this division problem: $6,000,000 \div 17$. The method of Exercise 24 shows that the quotient is 352,941 and the remainder is 3. Thus the decimal expansion of 6/17 begins .352941, and the next block of digits in the expansion is the quotient in the problem $3,000,000 \div 17$. The remainder when 3,000,000 is divided by 17 is 10, so the next block of digits in the expansion of 6/17 is the quotient in the problem $10,000,000 \div 17$. Continue in this way until the decimal expansion repeats.]

26. 3/19   **27.** 1/29   **28.** 3/43   **29.** 283/47

30. 768/59

31. (a) Show that there are at least as many irrational numbers (nonrepeating decimals) as there are terminating decimals. [*Hint:* With each terminating decimal associate a nonrepeating decimal.]

     (b) Show that there are at least as many irrational numbers as there are repeating decimals. [*Hint:* With each repeating decimal associate a nonrepeating decimal by inserting longer and longer strings of zeros: for instance, with $.11111111\cdots$ associate the number $.101001000100001\cdots.$]

0.2 Integral Exponents

Exponents provide a convenient shorthand for certain products. If c is a real number, then c^2 denotes cc and c^3 denotes ccc. More generally, for any positive integer n

$$c^n \text{ denotes the product } ccc\cdots c \quad (n \text{ factors}).$$

In this notation c^1 is just c, so we usually omit the exponent 1.

Example 1 $3^4 = 3 \cdot 3 \cdot 3 \cdot 3 = 81$ and

$$(-2)^5 = (-2)(-2)(-2)(-2)(-2) = -32.$$

For every positive integer n, $0^n = 0\cdots 0 = 0$. ∎

Example 2 To find $(2.4)^9$ use the $\wedge$ (or a^b or x^y) key on your calculator:

$$2.4 \wedge 9 \text{ ENTER}$$

which produces the (approximate) answer 2641.80754. ■

CAUTION

Be careful with negative bases. For instance, if you want to compute $(-12)^4$, which is a positive number, but you key in $(-)\ 12 \wedge 4$ ENTER, the calculator will interpret this as $-(12^4)$ and produce a negative answer. To get the correct answer, you must key in the parentheses:

$$(\,(-)\ 12\,) \wedge 4 \text{ ENTER}.$$

Because exponents are just shorthand for multiplication, it is easy to determine the rules they obey. For instance,

$$c^3 c^5 = (ccc)(ccccc) = c^8, \qquad \text{that is,} \qquad c^3 c^5 = c^{3+5}.$$

The same thing works in general:

Multiplication with Exponents

> **To multiply c^m by c^n, add the exponents:**
>
> $$c^m \cdot c^n = c^{m+n}.$$

For division we have, for example,

$$\frac{c^7}{c^4} = \frac{\cancel{cccc}ccc}{\cancel{cccc}} = ccc = c^3, \qquad \text{that is,} \qquad \frac{c^7}{c^4} = c^{7-4}.$$

In general,

Division with Exponents

> **To divide c^m by c^n, subtract the exponents:**
>
> $$\frac{c^m}{c^n} = c^{m-n}.$$

Finally, note that

$$(c^2)^3 = (cc)^3 = (cc)(cc)(cc) = c^6, \qquad \text{that is,} \qquad (c^2)^3 = c^{2\cdot3},$$

which suggests this rule:

Power of a Power

> To find a power of a power $(c^m)^n$, multiply the exponents:
> $$(c^m)^n = c^{mn}.$$

Example 3

(a) To find $4^2 \cdot 4^7$, we note that this is multiplication and *add* the exponents: $4^2 \cdot 4^7 = 4^{2+7} = 4^9$.

(b) To divide $2^8/2^3$, we *subtract* the exponents: $2^8/2^3 = 2^{8-3} = 2^5$.

(c) To compute $(3^5)^4$, which is a power of a power, we *multiply* the exponents: $(3^5)^4 = 3^{5 \cdot 4} = 3^{20}$. ■

When an exponent applies to the product or quotient of two numbers, use the definitions carefully. For instance,

$$(7 \cdot 19)^3 = (7 \cdot 19)(7 \cdot 19)(7 \cdot 19) = 7 \cdot 7 \cdot 7 \cdot 19 \cdot 19 \cdot 19 = 7^3 \cdot 19^3.$$

In the general case,

Product to a Power

> To find $(cd)^n$, apply the exponent to each term in parentheses:
> $$(cd)^n = c^n d^n.$$

> ## CAUTION
> A common mistake is to write an expression such as $(2x)^5$ incorrectly as $2x^5$, that is, to apply the exponent only to the second term in parentheses. Using the property in the box, however, we see that
> $$(2x)^5 = 2^5 x^5 = 32x^5.$$

A similar conclusion holds for division.

Quotient to a Power

> To find $(c/d)^n$ apply the exponent to both numerator and denominator:
> $$\left(\frac{c}{d}\right)^n = \frac{c^n}{d^n}.$$

For example,

$$\left(\frac{x}{7}\right)^2 = \frac{x}{7} \cdot \frac{x}{7} = \frac{x^2}{7^2} = \frac{x^2}{49}.$$

> ## CAUTION
> The preceding rules do *not* apply to situations such as $2^4 \cdot 3^5$ or $4^7/5^2$, where powers of two different numbers are being multiplied or divided.

Next we consider exponents other than positive integers. The symbol c^0 has no obvious meaning (what would it mean to multiply c by itself 0 times?). On the other hand, consider the rule for multiplying with exponents. For this rule to hold with a zero exponent, we would have to have

$$c^5 c^0 = c^{5+0} = c^5.$$

On the other hand, we know that

$$c^5 \cdot 1 = c^5.$$

so that it is reasonable to *define* c^0 to be 1:

Zero Exponent

If $c \neq 0$, then c^0 is defined to be the number 1.

Similarly, for the multiplication rule to hold for negative exponents we would have to have, for example,

$$c^5 c^{-5} = c^{5-5} = c^0 = 1.$$

But we know that

$$c^5 \cdot \frac{1}{c^5} = 1,$$

so that it is also reasonable to define c^{-5} to be $1/c^5$. More generally, if $c \neq 0$ and n is a positive integer,

Negative Exponents

c^{-n} is defined to be the number $\dfrac{1}{c^n}$.

Note that 0^0 and negative powers of 0 are not defined (negative powers of 0 would involve division by 0).

Example 4 $6^{-3} = 1/6^3 = 1/216$ and $(-2)^{-5} = 1/(-2)^5 = -1/32$.
A calculator shows that $(.287)^{-12} \approx 3{,}201{,}969.857$.   ■

CAUTION

When c and d are nonzero and $n \neq 1$, then

$$(c + d)^n \neq c^n + d^n.$$

For example, $(2 + 5)^3 = 7^3 = 343$, but $2^3 + 5^3 = 8 + 125 = 133$ and similarly,
$(2 + 4)^{-1} = 6^{-1} = 1/6$, but $2^{-1} + 4^{-1} = 1/2 + 1/4 = 3/4$.

Because of the way zero and negative exponents are defined,

The multiplication, division, and power rules for exponents in the boxes above hold for all integer exponents (positive, negative, or zero),

as illustrated in the following examples.

Example 5

(a)   By the multiplication and division rules,

$$\pi^{-5}\pi^2 = \pi^{-5+2} = \pi^{-3} = 1/\pi^3 \qquad \text{and} \qquad x^9/x^4 = x^{9-4} = x^5.$$

(b)   The power of a power rule shows that $(5^{-3})^2 = 5^{(-3)\cdot 2} = 5^{-6}$.

(c)   The power rule for quotients and the power of a power rule show that

$$\left(\frac{x^2}{3}\right)^4 = \frac{(x^2)^4}{3^4} = \frac{x^8}{81}.$$

(d)   By the negative exponent rule,

$$\frac{1}{x^{-5}} = \frac{1}{1/x^5} = x^5. \qquad ■$$

The exponent rules can often be used to simplify complicated expressions.

Example 6 Simplify:

(a) $(2x^2y^3z)^4$ (b) $(r^{-3}s^2)^{-2}$ (c) $\dfrac{x^5(y^2)^3}{(x^2y)^2}$

Solution

(a)   *Product to a power:* $(2x^2y^3z)^4 = 2^4(x^2)^4(y^3)^4z^4$

              Power of a power: $= 16x^8y^{12}z^4$

(b) *Product to a power:* $(r^{-3}s^2)^{-2} = (r^{-3})^{-2}(s^2)^{-2}$

Power of a power: $= r^6 s^{-4} = \dfrac{r^6}{s^4}$

(c) *Power of a power:* $\dfrac{x^5(y^2)^3}{(x^2y)^2} = \dfrac{x^5 y^6}{(x^2 y)^2}$

Product to a power: $= \dfrac{x^5 y^6}{(x^2)^2 y^2}$

Power of a power: $= \dfrac{x^5 y^6}{x^4 y^2}$

Division with exponents: $= x^{5-4} y^{6-2} = xy^4$ ∎

Using the exponent laws with negative exponents is usually more efficient than first converting to positive exponents. That conversion, if necessary, can always be done at the end.

Example 7 Simplify and express without negative exponents:

$$\frac{a^{-2}(b^2 c^3)^{-2}}{(a^{-3} b^{-5})^2 c}.$$

Solution *Product to a power:* $\dfrac{a^{-2}(b^2 c^3)^{-2}}{(a^{-3} b^{-5})^2 c} = \dfrac{a^{-2}(b^2)^{-2}(c^3)^{-2}}{(a^{-3})^2 (b^{-5})^2 c}$

Power of a power: $= \dfrac{a^{-2} b^{-4} c^{-6}}{a^{-6} b^{-10} c}$

Division with exponents: $= a^{-2-(-6)} b^{-4-(-10)} c^{-6-1}$

Negative exponent: $= a^4 b^6 c^{-7} = \dfrac{a^4 b^6}{c^7}$ ∎

Scientific Notation

Any real number may be written as the product of a power of 10 and a number between 1 and 10. For example,

$$356 = 3.56 \times 100 = 3.56 \times 10^2$$

$$1,563,427 = 1.563427 \times 1,000,000 = 1.563427 \times 10^6$$

$$.072 = 7.2 \div 100 = 7.2 \times 1/100 = 7.2 \times 10^{-2}$$

$$.000862 = 8.62 \div 10,000 = 8.62 \times 10^{-4}$$

A number written in this form is said to be in **scientific notation.** Scientific notation is very useful for computations with very large or very small numbers.

Example 8

$$(.00000002)(4{,}300{,}000{,}000) = (2 \times 10^{-8})(4.3 \times 10^9)$$
$$= 2(4.3)10^{-8+9} = (8.6)10^1 = 86. \quad \blacksquare$$

When calculators display numbers in scientific notation, they omit the 10 and use an E to indicate the exponent. For instance,

$$7.235 \times 10^{-12} \qquad \text{is displayed as} \qquad 7.235 \text{ E} -12$$

To enter a number in scientific notation, for example 5.6×10^{73}, type in 5.6, then press the EE key (labeled EEX or EXP or 10^x on some calculators) and type in 73. Calculators automatically switch to scientific notation whenever a number is too large or too small to be displayed in the standard way. If you try to enter a number with more digits than the calculator can handle, such as 45,000,000,333,222,111, a typical calculator will approximate it using scientific notation as 4.500000033 E 16, that is, as 45,000,000,330,000,000.

Example 9
Suppose you are located h feet above the ground. Because of the curvature of the earth, the maximum distance you can see is approximately d miles, where

$$d = \sqrt{1.5h + (3.587 \times 10^{-8})h^2}.$$

How far can you see from the 500-ft-high Smith Tower in Seattle and from the 1454-ft-high Sears Tower in Chicago?

Solution   For the Smith Tower, substitute 500 for h in the formula, and use your calculator:

$$d = \sqrt{1.5(500) + (3.587 \times 10^{-8})500^2} \approx 27.4 \text{ miles.}$$

For the Sears Tower, you can see almost 47 miles because

$$d = \sqrt{1.5(1454) + (3.587 \times 10^{-8})1454^2} \approx 46.7 \text{ miles.} \quad \blacksquare$$

CAUTION
Your calculator may not always obey the laws of arithmetic when dealing with very large or very small numbers. For instance, we know that

$$(1 + 10^{19}) - 10^{19} = 1 + (10^{19} - 10^{19}) = 1 + 0 = 1$$

but a calculator may round off $1 + 10^{19}$ as 10^{19} (instead of the correct number 10,000,000,000,000,000,001). So the calculator incorrectly computes $(1 + 10^{19}) - 10^{19}$ as $10^{19} - 10^{19} = 0$. Check to see how your calculator handles this problem.

Exercises 0.2

In Exercises 1–12, evaluate the expression.

1. $(-6)^2$ **2.** -6^2

3. $5 + 4(3^2 + 2^3)$ **4.** $(-3)2^2 + 4^2 - 1$

5. $\dfrac{(-3)^2 + (-2)^4}{-2^2 - 1}$ **6.** $\dfrac{(-4)^2 + 2}{(-4)^2 - 7} + 1$

7. $2^4 - 2^7$ **8.** $3^3 - 3^{-7}$

9. $(2^{-2} + 2)^2$ **10.** $(3^{-1} - 3^3)^2$

11. $\dfrac{1}{2^3} + \dfrac{1}{2^{-4}}$ **12.** $3^2\left(\dfrac{1}{3} + \dfrac{1}{3^{-2}}\right)$

In Exercises 13–28, simplify the expression. Each letter represents a nonzero real number and should appear at most once in your answer.

13. $x^2 \cdot x^3 \cdot x^5$ **14.** $y \cdot y^4 \cdot y^6$

15. $(.03)y^2 \cdot y^7$ **16.** $(1.3)z^3 \cdot z^5$

17. $(2x^2)^3 3x$ **18.** $(3y^3)^4 5y^2$

19. $(3x^2y)^2$ **20.** $(2xy^3)^3$

21. $(2w)^3(3w)(4w)^2$ **22.** $(3d)^4(2d)^2(5d)$

23. $a^{-2}b^3a^3$ **24.** $c^4d^5c^{-3}$

25. $(2x)^{-2}(2y)^3(4x)$ **26.** $(3x)^{-3}(2y)^{-2}(2x)$

27. $(2x^2y)^0(3xy)$ **28.** $(3x^2y^4)^0$

In Exercises 29–32, express the given number as a power of 2.

29. $(64)^2$ **30.** $\left(\dfrac{1}{8}\right)^3$

31. $(2^4 \cdot 16^{-2})^3$ **32.** $\left(\dfrac{1}{2}\right)^{-8}\left(\dfrac{1}{4}\right)^4\left(\dfrac{1}{16}\right)^{-3}$

In Exercises 33–44, simplify and write the given expression without negative exponents. All letters represent nonzero real numbers.

33. $\dfrac{x^4(x^2)^3}{x^3}$ **34.** $\left(\dfrac{z^2}{t^3}\right)^4 \cdot \left(\dfrac{z^3}{t}\right)^5$

35. $\left(\dfrac{e^6}{c^4}\right)^2 \cdot \left(\dfrac{c^3}{e}\right)^3$ **36.** $\left(\dfrac{x^7}{y^6}\right)^2 \cdot \left(\dfrac{y^2}{x}\right)^4$

37. $\left(\dfrac{a^6}{b^{-4}}\right)^2$ **38.** $\left(\dfrac{x^{-2}}{y^{-2}}\right)^2$

39. $\left(\dfrac{c^5}{d^{-3}}\right)^{-2}$ **40.** $\left(\dfrac{x^{-1}}{2y^{-1}}\right)\left(\dfrac{2y}{x}\right)^{-2}$

41. $\dfrac{(a^{-3}b^2c)^{-2}}{(ab^{-2}c^3)^{-1}}$ **42.** $\dfrac{(-2cd^2e^{-1})^3}{(5c^{-3}de)^{-2}}$

43. $(c^{-1}d^{-2})^{-3}$ **44.** $[(x^2y^{-1})^2]^{-3}$

In Exercises 45–50, determine the sign of the given number without calculating the product.

45. $(-2.6)^3(-4.3)^{-2}$ **46.** $(4.1)^{-2}(2.5)^{-3}$

47. $(-1)^9(6.7)^5$ **48.** $(-4)^{12}6^9$

49. $(-3.1)^{-3}(4.6)^{-6}(7.2)^7$

50. $(45.8)^{-7}(-7.9)^{-9}(-8.5)^{-4}$

In Exercises 51–54, r, s, and t are positive integers and a, b, and c are nonzero real numbers. Simplify and write the given expression without negative exponents.

51. $\dfrac{3^{-r}}{3^{-s-r}}$ **52.** $\dfrac{4^{-(t+1)}}{4^{2-t}}$ **53.** $\left(\dfrac{a^6}{b^{-4}}\right)^t$

54. $\dfrac{c^{-t}}{(6b)^{-s}}$

In Exercises 55–61, express the given numbers (based on 1998 estimates) in scientific notation.

55. Population of the United States: 270,312,000

56. Population of the world: 5,927,000,000

57. Total number of housing units in the United States: 109,000,000

58. U.S. civilian labor force: 137,523,000

59. Number of unemployed U.S. workers: 6,529,000

60. Radius of a hydrogen atom: .00000000005 meters

61. Width of a DNA double helix: .000000002 meters

In Exercises 62–65, express the given number in normal decimal notation.

62. Speed of light in a vacuum: 2.9979×10^8 meters/sec

63. Average distance from earth to sun: 1.50×10^{11} meters

64. Proton mass: 1.6726×10^{-27} kg

65. Electron charge: 1.602×10^{-19} C

In Exercises 66–72, express your answer in scientific notation.

66. One light-year is the distance light travels in a 365-day year. The speed of light is about 186,282.4 miles per second.
   (a) How long is 1 light-year (in miles)?
   (b) Light from the North Star takes 680 years to reach earth. How many miles is the North Star from earth?

67. If a single gene has an average weight of 10^{-17} gram, how many genes would you expect in an organism having cells with a DNA weight of 10^{-12} gram?

68. Johannes Kepler discovered that the ratio R^3/T^2 is the same for the moon and other objects that orbit the earth, where R is the object's average distance from the earth and T is the object's orbital period. The moon is 240,000 miles from the earth and makes one complete orbit every 27 days. Use Kepler's discovery to find the orbital period of a satellite orbiting the earth at a distance of 900,000 miles.

69. Assume that the earth is a sphere with a radius of 6400 kilometers. What is the surface area of the earth in square kilometers? [*Hint:* The surface area S of a sphere of radius r is given by $S = 4\pi r^2$.]

70. If the circumference of the earth at the equator is 25,000 miles, how many inches is it around the equator?

71. Assume that light travels at 1.86×10^5 miles per sec. Sound travels at 1.1×10^3 feet per sec. Elizabeth is at a rock concert sitting 100 ft from the band. Her friend back home—1000 miles away— is listening on the radio (radio waves travel at the speed of light). Who hears the music first?

72. Australia has 7,617,930 square kilometers of land area, and in 1996 it was estimated to have 18,438,824 people. Singapore has 622.6 square kilometers of land area, and in 1996 it was estimated to have 3,396,924 people.
   (a) Express Australia's and Singapore's populations and land areas in scientific notation.
   (b) What is the population density (people per square kilometer) for Australia? for Singapore?

73. After five years an investment of $1000 at interest rate r, compounded annually, will be worth $1000(1 + r)^5$, where r is written as a decimal. Use a calculator to fill in this table.

r	$1000(1 + r)^5$
$3\frac{1}{2}\%$	
5%	
$7\frac{1}{2}\%$	
10%	

74. Suppose you are k miles (not feet) above the ground. The radius of the earth is approximately 3960 miles, and the point where your line of sight meets the earth is perpendicular to the radius of the earth at that point, as shown in the figure.

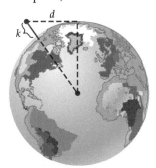

   (a) Use the Pythagorean Theorem (see the Geometry Review Appendix) to show that
$$d = \sqrt{(3960 + k)^2 - 3960^2}$$
   (b) Show that the equation in part (a) simplifies to $d = \sqrt{7920k + k^2}$.
   (c) If you are h feet above the ground, then you are $h/5280$ miles high. Why? Use this fact and the equation in part (b) to obtain the formula used in Example 9.

Errors to Avoid

In Exercises 75–78, give a numerical example to show that the statement is false *for some numbers.*

75. $a^r a^s = a^{rs}$

76. $a^r b^s = (ab)^{r+s}$

77. $c^{-r} = -c^r$

78. $\dfrac{c^r}{c^s} = c^{r/s}$

0.3 Roots, Radicals, and Rational Exponents

The square root of 5 is the positive number $\sqrt{5}$ whose square is 5. Similarly, the fourth root of 5, denoted $\sqrt[4]{5}$, is the positive number whose fourth power is 5, and in the general case:

Even Roots

If n is an even positive integer and c is any nonnegative real number, then

$\sqrt[n]{c}$ denotes the nonnegative number whose nth power is c.*

$\sqrt[n]{c}$ is called the **nth root** of c.

For example, $\sqrt[4]{81} = 3$ because $3^4 = 81$ and $\sqrt[6]{64} = 2$ because $2^6 = 64$. When $n = 2$, we shall continue to use the ordinary radical $\sqrt{}$, rather than $\sqrt[2]{}$. Since even powers of nonzero numbers are positive, no negative number can have a fourth root, sixth root, etc.

The odd roots (3ʳᵈ, 5ᵗʰ, etc.) of a number c are defined similarly, except that c need not be nonnegative.

Odd Roots

If n is an odd positive integer and c is any real number, then

$\sqrt[n]{c}$ denotes the number whose nth power is c.

$\sqrt[n]{c}$ is called the **nth root** of c.

Example 1 Find $\sqrt[3]{-27}$, $\sqrt[3]{64}$, and $\sqrt[5]{32}$.

Solution $\sqrt[3]{-27} = -3$ because $(-3)^3 = -27$ and $\sqrt[3]{64} = 4$ because $4^3 = 64$, as you can easily verify. Similarly, $\sqrt[5]{32} = 2$ because $2^5 = 32$. ∎

Example 2 Show that $\sqrt[3]{64} = \sqrt[3]{16} \cdot \sqrt[3]{4}$ without using a calculator.

Solution By the exponent rules and the definition of cube root

$$\left(\sqrt[3]{16} \cdot \sqrt[3]{4}\right)^3 = \left(\sqrt[3]{16}\right)^3 \left(\sqrt[3]{4}\right)^3 = 16 \cdot 4 = 64.$$

Thus, $\sqrt[3]{16} \cdot \sqrt[3]{4}$ is the number whose cube is 64. But $\sqrt[3]{64}$ is the number whose cube is 64. Therefore, $\sqrt[3]{64} = \sqrt[3]{16} \cdot \sqrt[3]{4}$. ∎

---

*We are assuming for now that there is exactly one such number, a fact that will be verified when we study equations and polynomial graphs later in the book.

Since $64 = 16 \cdot 4$, Example 2 illustrates the first of the following facts, which apply to all real numbers c and d that have nth roots.

Products and Quotients of Roots

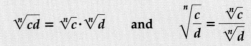

$$\sqrt[n]{cd} = \sqrt[n]{c} \cdot \sqrt[n]{d} \qquad \text{and} \qquad \sqrt[n]{\frac{c}{d}} = \frac{\sqrt[n]{c}}{\sqrt[n]{d}}$$

Example 3 Simplify:

(a) $\sqrt{63}$ (b) $\sqrt{12} - \sqrt{75}$ (c) $\dfrac{2 + 3\sqrt{20}}{2}$

Solution

(a) Factor 63 as the product of 9 (a perfect square) and 7 and then use the product of roots property.
$$\sqrt{63} = \sqrt{9 \cdot 7} = \sqrt{9}\sqrt{7} = 3\sqrt{7}$$

(b)

Factor perfect squares out of 12 and 75: $\sqrt{12} - \sqrt{75} = \sqrt{4 \cdot 3} - \sqrt{25 \cdot 3}$

Product of roots: $= \sqrt{4}\sqrt{3} - \sqrt{25}\sqrt{3}$

$= 2\sqrt{3} - 5\sqrt{3} = -3\sqrt{3}$

(c)

Factor perfect square out of 20: $\dfrac{2 + 3\sqrt{20}}{2} = \dfrac{2 + 3\sqrt{4 \cdot 5}}{2}$

Product of roots: $= \dfrac{2 + 3\sqrt{4}\sqrt{5}}{2}$

$= \dfrac{2 + 3 \cdot 2\sqrt{5}}{2} = \dfrac{2 + 6\sqrt{5}}{2}$

Factor out 2 and simplify: $= \dfrac{2(1 + 3\sqrt{5})}{2} = 1 + 3\sqrt{5}$ ■

Example 4 Simplify:

(a) $\sqrt[3]{40k^4}$ (b) $\sqrt[3]{\dfrac{y^4}{8x^6}}$

Solution

(a) Note that $8k^3$ is a factor of $40k^4$ and that $8k^3 = (2k)^3$ is a perfect cube.

Factor out $8k^3$: $\qquad \sqrt[3]{40k^4} = \sqrt[3]{8k^3 \cdot 5k}$

Product of roots: $\qquad\qquad = \sqrt[3]{8k^3}\sqrt[3]{5k}$

$\qquad\qquad\qquad\qquad\qquad = \sqrt[3]{(2k)^3}\sqrt[3]{5k} = 2k\sqrt[3]{5k}$

(b)

Quotient of roots: $\qquad \sqrt[3]{\dfrac{y^4}{8x^6}} = \dfrac{\sqrt[3]{y^4}}{\sqrt[3]{8x^6}}$

Factor out perfect cubes: $\qquad = \dfrac{\sqrt[3]{y^3 y}}{\sqrt[3]{2^3 (x^2)^3}}$

Product of roots: $\qquad = \dfrac{\sqrt[3]{y^3}\sqrt[3]{y}}{\sqrt[3]{2^3}\sqrt[3]{(x^2)^3}} = \dfrac{y\sqrt[3]{y}}{2x^2}$ ∎

CAUTION

When c and d are nonzero, then

$$\sqrt[n]{c + d} \neq \sqrt[n]{c} + \sqrt[n]{d}.$$

For example,

$$\sqrt[3]{8 + 27} = \sqrt[3]{35} \approx 3.271, \qquad \text{but} \qquad \sqrt[3]{8} + \sqrt[3]{27} = 2 + 3 = 5.$$

Similarly, the expression $\sqrt{4 + x^2}$ is *NOT* equal to $2 + x$.

Finally, we note that nth roots of nth powers behave in the same way as square roots when n is even, but not when n is odd.

nth Roots of nth Powers

$$\sqrt[n]{c^n} = |c| \quad \textbf{for } n \textbf{ even;}$$

$$\sqrt[n]{c^n} = c \quad \textbf{for } n \textbf{ odd.}$$

For example, $\sqrt[4]{(-2)^4} = \sqrt[4]{16} = 2 = |-2|$ and $\sqrt[5]{(-2)^5} = \sqrt[5]{-32} = -2$.

Rational Exponents

nth roots provide a natural way to define fractional exponents. Suppose, for example, that we want to define $5^{1/2}$ in such a way that the exponent rules continue to hold. Then we would certainly want

$$(5^{1/2})^2 = 5^{\frac{1}{2} \cdot 2} = 5^1 = 5.$$

But we know that

$$\left(\sqrt{5}\right)^2 = 5.$$

So it is reasonable to define $5^{1/2}$ to be the number $\sqrt{5}$. Similarly, for any real number c and positive integer n,

Reciprocal Exponents

$c^{1/n}$ **is defined to be the number** $\sqrt[n]{c}$ **(provided it exists).**

Example 5   Use a calculator to approximate

(a) $\sqrt[5]{40}$ (b) $\sqrt[11]{225}$.

Solution   Although most calculators have an nth root function (usually on a submenu of the MATH menu), it's easier to use fractional exponents.

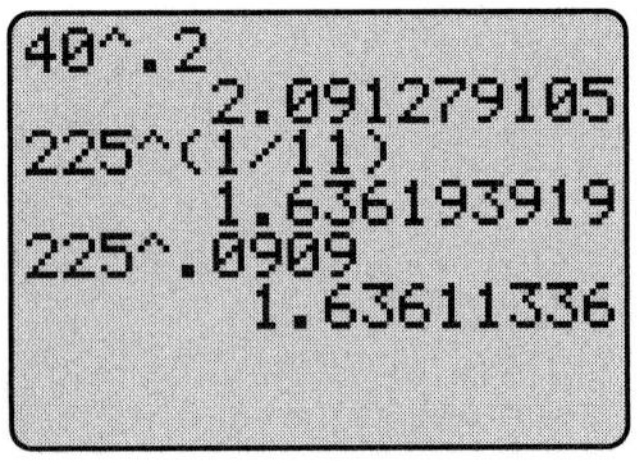

(a)   Since $\sqrt[5]{40} = 40^{1/5}$ and $1/5 = .2$, we compute $40^{.2}$, as in Figure 0–5.

(b)   To approximate $\sqrt[11]{225} = 225^{1/11}$, note that $1/11$ has an infinite decimal expansion, $1/11 = .09090909\cdots$. To ensure the highest degree of accuracy, key in 225 $\wedge$ (1 $\div$ 11). If you round off the decimal expansion of $1/11$, say to .0909, you won't get the same answer, as Figure 0–5 shows.   ∎

```
40^.2
       2.091279105
225^(1/11)
       1.636193919
225^.0909
       1.63611336
```

Figure 0–5

The next step is to give a meaning to fractional exponents for any fraction, not just those of the form $1/n$. If possible, they should be defined in such a way that the various exponent rules continue to hold. Consider, for example, how $4^{3/2}$ might be defined. The exponent $3/2$ can be written as either

$$3\cdot\left(\frac{1}{2}\right) \quad\text{or}\quad \left(\frac{1}{2}\right)\cdot 3$$

If the power of a power property $(c^m)^n = c^{mn}$ is to hold, we might define $4^{3/2}$ as either $(4^3)^{1/2}$ or $(4^{1/2})^3$. The result is the same in both cases:

$$(4^3)^{1/2} = 64^{1/2} = \sqrt{64} = 8 \quad\text{and}\quad (4^{1/2})^3 = \left(\sqrt{4}\right)^3 = 2^3 = 8$$

It can be proved that the same thing is true in the general case, which leads to this definition.

Rational Exponents

If c is a real number and t/k is a rational number in lowest terms with positive denominator, then

$c^{t/k}$ **is defined to be the number** $\sqrt[k]{c^t}$ **(provided it exists).**

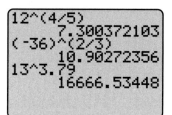

Figure 0–6

Fractional powers can sometimes be computed by hand, for example,

$$8^{2/3} = \sqrt[3]{8^2} = \sqrt[3]{64} = 4.$$

Most of the time, however, we approximate them with a calculator, as shown in Figure 0–6. Note that decimal exponents now have a meaning, since they are simply a special case of fractional exponents; for instance, $13^{3.79} = 3^{379/100}$.

CAUTION

We know that

$$(-8)^{2/3} = \sqrt[3]{(-8)^2} = \sqrt[3]{64} = 4$$

but entering $(-8)^{2/3}$ on TI-82/85, Casio 9850, HP-38, or Sharp 9600 produces either an error message or a complex number (which the calculator indicates by an ordered pair or an expression involving i). However, you can get the correct answer by keying in one of

$$[(-8)^2]^{1/3} \quad \text{or} \quad [(-8)^{1/3}]^2$$

each of which is equal to $(-8)^{2/3}$. TI-89 will produce the correct answer provided that the COMPLEX FORMAT in the MODE menu is set to "Real."

Rational exponents were defined in a way that guaranteed that one of the familiar properties of exponents remains valid. In fact, all of the exponent properties developed for integer exponents are valid for rational exponents, as summarized here and illustrated in Examples 6–8.

Exponent Laws

Let c and d be any real numbers and let r and s be any rational numbers in lowest terms for which each of the following exists. Then

1. $c^r c^s = c^{r+s}$

2. $\dfrac{c^r}{c^s} = c^{r-s} \quad (c \neq 0)$

3. $(c^r)^s = c^{rs}$

4. $(cd)^r = c^r d^r$

5. $\left(\dfrac{c}{d}\right)^r = \dfrac{c^r}{d^r} \quad (d \neq 0)$

6. $\dfrac{1}{c^{-r}} = c^r \quad (c \neq 0)$

Example 6 Simplify: $(x^{5/2}y^4)(xy^{7/4})$.

Solution

Group like terms together: $\qquad (x^{5/2}y^4)(xy^{7/4}) = (x^{5/2}x)(y^4 y^{7/4})$

Multiplication with exponents (law 1): $\qquad = x^{5/2+1}y^{4+7/4}$

$\dfrac{5}{2} + 1 = \dfrac{7}{2} \quad and \quad 4 + \dfrac{7}{4} = \dfrac{23}{4}: \qquad = x^{7/2}y^{23/4}$ ∎

CAUTION

The exponent laws deal only with products and quotients. There are no analogous properties for sums. In particular, if both c and d are nonzero, then

$$(c + d)^r \text{ is } \textbf{not} \text{ equal to}$$
$$c^r + d^r.$$

Example 7 Compute the product $x^{1/2}(x^{3/4} - x^{3/2})$.

Solution

Distributive law: $x^{1/2}(x^{3/4} - x^{3/2}) = x^{1/2}x^{3/4} - x^{1/2}x^{3/2}$

Multiplication with exponents (law 1): $= x^{1/2+3/4} - x^{1/2+3/2}$

$\dfrac{1}{2} + \dfrac{3}{4} = \dfrac{5}{4}$ *and* $\dfrac{1}{2} + \dfrac{3}{2} = 2$: $= x^{5/4} - x^2$ ∎

Example 8 Simplify $(8r^{3/4}s^{-3})^{2/3}$, and express it without negative exponents.

Solution

Product to a power (law 4): $(8r^{3/4}s^{-3})^{2/3} = 8^{2/3}(r^{3/4})^{2/3}(s^{-3})^{2/3}$

Power of a power (law 3): $= 8^{2/3}r^{(3/4)(2/3)}s^{(-3)(2/3)}$

$\dfrac{3}{4} \cdot \dfrac{2}{3} = \dfrac{1}{2}$ *and* $(-3)\dfrac{2}{3} = -2$: $= 8^{2/3}r^{1/2}s^{-2}$

Definition of negative exponents: $= \dfrac{8^{2/3}r^{1/2}}{s^2}$

$8^{2/3} = 4$ *(as shown on page 29)*: $= \dfrac{4r^{1/2}}{s^2}$ or $\dfrac{4\sqrt{r}}{s^2}$ ∎

Exercises 0.3

In Exercises 1–18, simplify the expression without using a calculator.

1. $\sqrt{80}$ **2.** $\sqrt{96}$

3. $\sqrt{147}$ **4.** $\sqrt{405}$

5. $\sqrt{6}\sqrt{12}$ **6.** $\sqrt{8}\sqrt{96}$

7. $\dfrac{-6 + \sqrt{99}}{15}$ **8.** $\dfrac{5 - \sqrt{175}}{10}$

9. $\sqrt{50} - \sqrt{72}$ **10.** $\sqrt{75} + \sqrt{192}$

11. $5\sqrt{20} - \sqrt{45} + 2\sqrt{80}$

12. $\sqrt[3]{40} + 2 \cdot \sqrt[3]{135} - 5 \cdot \sqrt[3]{320}$

13. $\sqrt{16a^8b^{-2}}$ **14.** $\sqrt{24x^6y^{-4}}$

15. $\dfrac{\sqrt{c^2d^6}}{\sqrt{4c^3d^{-4}}}$ **16.** $\dfrac{\sqrt{a^{-10}b^{-12}}}{\sqrt{a^{14}d^{-4}}}$

17. $\sqrt[3]{27a^9b^{-3}}$ **18.** $\sqrt[4]{40x^6y^{-6}}$

In Exercises 19–22, fill the blank with one of $<$, $=$, $>$ so that the resulting statement is true.

19. $\sqrt[3]{4} + \sqrt[3]{2}$ ____ $\sqrt[3]{6}$ **20.** $\sqrt{11} - \sqrt{2}$ ____ 3

21. $\sqrt[5]{7}\sqrt[5]{11}$ ____ $\sqrt[5]{77}$ **22.** $\sqrt[3]{35}\,\sqrt[4]{35}$ ____ $\sqrt{35}$

In Exercises 23–28, write the given expression without using radicals.

23. $\sqrt[3]{a^2 + b^2}$ **24.** $\sqrt[4]{a^3 - b^3}$ **25.** $\sqrt[4]{\sqrt[4]{a^3}}$

26. $\sqrt{\sqrt[3]{a^3b^4}}$ **27.** $\sqrt[5]{t}\sqrt{16t^5}$ **28.** $\sqrt{x}\left(\sqrt[3]{x^2}\right)\left(\sqrt[4]{x^3}\right)$

In Exercises 29–36, simplify the expression.

29. $(25k^2)^{3/2}(16k^{1/3})^{3/4}$ **30.** $(3x^{2/3})(4y^{3/4})(2x^{-1/3})(3y^{13/4})$

31. $\sqrt{x^7} \cdot x^{5/2} \cdot x^{-3/2}$ **32.** $(x^{1/2}y^3)(x^0y^7)^{-2}$

33. $\dfrac{(x^2)^{1/3}(y^2)^{2/3}}{3x^{2/3}y^2}$ **34.** $\dfrac{(c^{1/2})^3(d^3)^{1/2}}{(c^3)^{1/4}(d^{1/4})^3}$

35. $\dfrac{(7a)^2(5b)^{3/2}}{(5a)^{3/2}(7b)^4}$ **36.** $\dfrac{(6a)^{1/2}\sqrt{ab}}{a^2b^{3/2}}$

In Exercises 37–42, compute and simplify.

37. $x^{1/2}(x^{2/3} - x^{4/3})$ **38.** $x^{1/2}(3x^{3/2} + 2x^{-1/2})$

39. $(x^{1/2} + y^{1/2})(x^{1/2} - y^{1/2})$

40. $(x^{1/3} + y^{1/2})(2x^{1/3} - y^{3/2})$

41. $(x + y)^{1/2}[(x + y)^{1/2} - (x + y)]$

42. $(x^{1/3} + y^{1/3})(x^{2/3} - x^{1/3}y^{1/3} + y^{2/3})$

In Exercises 43–46, assume that you have invested $2000 at 5.5% interest compounded annually. The value of your investment after x years is $2000(1.055^x)$. Find the value of your investment (to the nearest penny) at the end of the given period.

43. Five years and six months

44. 21 months

45. Eight years and seven months

46. 12 years and one month

In Exercises 47–50, use the equation $y = 92.8935 \cdot x^{.6669}$, which gives the approximate distance y (in millions of miles) from the sun to a planet that takes x earth years to complete one orbit of the sun. Find the distance from the sun to the planet whose orbit time is given.

47. Mercury (.24 years)   **48.** Mars (1.88 years)

49. Saturn (29.46 years)   **50.** Pluto (247.69 years)

Between 1790 and 1860 the population y of the United States (in millions) in year x was given by $y = 3.9572(1.0299^x)$, where $x = 0$ corresponds to 1790. In Exercises 51–54, find the U.S. population in the given year.

51. 1800   **52.** 1817   **53.** 1845   **54.** 1859

55. The output Q of an industry depends on labor L and capital C according to the equation

$$Q = L^{1/4}C^{3/4}.$$

(a) Use a calculator to determine the output for the following resource combinations:

L	C	$Q = L^{1/4}C^{3/4}$
10	7	
20	14	
30	21	
40	28	
60	42	

(b) When you double both labor and capital what happens to the output? When you triple both labor and capital what happens to the output?

56. Do Exercise 55 when the equation relating output to resources is $Q = L^{1/4}C^{1/2}$.

57. Do Exercise 55 when the equation relating output to resources is $Q = L^{1/2}C^{3/4}$.

58. In Exercises 55–57, how does the sum of the exponents on L and C affect the increase in output?

59. Here are some of the reasons why restrictions are necessary when defining fractional powers of a negative number.

(a) Explain why the equations $x^2 = -4$, $x^4 = -4$, $x^6 = -4$, etc., have no real solutions. Hence we cannot define $c^{1/2}$, $c^{1/4}$, $c^{1/6}$ when $c = -4$.

(b) Since $1/3$ is the same as $2/6$, it should be true that $c^{1/3} = c^{2/6}$, that is, that $\sqrt[3]{c} = \sqrt[6]{c^2}$. Show that this is false when $c = -8$.

60. Use a calculator to find a *six*-place decimal approximation of $(311)^{-4.2}$. Explain why your answer cannot possibly be the number $(311)^{-4.2}$.

61. Each of the following formulas can be used to approximate square roots:

(i) $\sqrt{a^2 + b} \approx a + \dfrac{b}{2a + 1}$

(ii) $\sqrt{a^2 + b} \approx a + \dfrac{b}{2a}$

(iii) $\sqrt{a^2 + b} \approx a + \dfrac{b}{2a} - \dfrac{b^2}{8a^3}$

(a) Use each of the formulas to find an approximation of $\sqrt{10}$. [*Hint:* $10 = 9 + 1$.] Use a calculator to see which approximation is best.

(b) Do the same for $\sqrt{19}$.

Errors to Avoid

In Exercises 62–65, give a numerical example to show that the given statement is false for some numbers.

62. $\sqrt[3]{a + b} = \sqrt[3]{a} + \sqrt[3]{b}$

63. $\sqrt{c^2 + d^2} = c + d$

64. $\sqrt{8a} = 4\sqrt{a}$

65. $\sqrt{a}(\sqrt[3]{a}) = \sqrt[4]{a}$

0.4 Polynomials

Expressions such as

$$b + 3c^2, \qquad 3x^2 - 5x + 4, \qquad \sqrt{x^3 + z}, \qquad \frac{x^3 + 4xy - \pi}{x^2 + xy}$$

are called **algebraic expressions.** The letters in algebraic expressions represent numbers. A letter that represents a number whose value remains unchanged throughout the discussion is called a **constant.** Constants are usually denoted by letters near the beginning of the alphabet. A letter that may represent *any* number is called a **variable.** Letters near the end of the alphabet usually denote variables. Thus, in the expression $3x + 5 + c$, it is understood that x is a variable and c is a constant. The value of this expression depends on the value of the variable x. If $x = 4$, then

$$3x + 5 + c = 3(4) + 5 + c = 17 + c$$

and if $x = \frac{1}{3}$, then

$$3x + 5 + c = 3(\tfrac{1}{3}) + 5 + c = 6 + c$$

and so on.

The most common algebraic expressions in this book are **polynomials,** such as

$$4x^3 - 6x^2 + \tfrac{1}{2}x \qquad \text{or} \qquad y^{15} + y^{10} + 7 \qquad \text{or} \qquad 3z - \pi \qquad \text{or} \qquad 12.$$

Thus, a polynomial is a sum of **terms** of the form

$$(\text{constant}) \times (\text{power of the variable}),$$

such as $4x^3$ or $1y^{15} = y^{15}$. We assume that $x^0 = 1$, $y^0 = 1$, etc., so that terms such as 7 and 12 may be written as $7y^0$ and $12x^0$. In each term of a polynomial, the constant can be any number, but the exponent is required to be a *nonnegative* integer. Thus, $x^{1/2} + 5$ is *not* a polynomial, but $\tfrac{1}{4}z + 7$ is a polynomial.

Any letter may be used for the variable in a polynomial, but we shall usually use x. The constants that appear in each term of a polynomial are called the **coefficients** of the polynomial. For example, the coefficients of $3x^2 - x + 4$ are 3, -1, and 4. The coefficients of

$$5x^3 + x - 2$$

are 5, 0, 1, and -2 because this polynomial can be written as

$$5x^3 + 0x^2 + x - 2.$$

The coefficient of x^0 in a polynomial is called the **constant term.** For instance, the constant term of $3x^2 + 5x - 6$ is -6. A polynomial that consists of only a constant term, such as 12, is called a **constant polynomial.** The **zero polynomial** is the constant polynomial 0.

The *exponent* of the highest power of x that appears with *nonzero* coefficient is the **degree** of the polynomial, and the nonzero coefficient of this highest power of x is the **leading coefficient.** For example,

Polynomial	Degree	Leading Coefficient	Constant Term
$6x^7 + 4x^3 + 5x^2 - 7x + 10$	7	6	10
$-x^4 + 2x^3 + \frac{1}{2}$	4	-1	$\frac{1}{2}$
x^3	3	1	0
12	0	12	12

The degree of the zero polynomial is *not defined* since no exponent of x occurs with nonzero coefficient. First-degree polynomials are often called **linear polynomials.** Second- and third-degree polynomials are called **quadratics** and **cubics,** respectively.

Polynomial Arithmetic

The usual rules of arithmetic are valid for polynomials and other algebraic expressions.

Example 1 Use the distributive law to *combine like terms;* for instance,

$$3x + 5x - 2x = (3 + 5 - 2)x = 6x.$$

In practice, you do the middle part in your head and simply write $3x + 5x - 2x = 6x.$ ∎

Example 2 Compute: $(2x^3 - 6x^2 + x) + (4x^5 + 5x^3 - 3x - 7)$.

Solution

Eliminate parentheses:	$= 2x^3 - 6x^2 + x + 4x^5 + 5x^3 - 3x - 7$
Group like terms:	$= 4x^5 + (2x^3 + 5x^3) - 6x^2 + (x - 3x) - 7$
Combine like terms:	$= 4x^5 + 7x^3 - 6x^2 - 2x - 7$ ∎

Be careful when parentheses are preceded by a minus sign. For instance, the expression $-(b - 3)$ means $(-1)(b - 3)$ and, hence, by the distributive law

$$-(b - 3) = (-1)(b - 3) = (-1)b + (-1)(-3) = -b + 3.$$

A common mistake is to write $-(b - 3) = -b - 3$, rather than the correct answer $-b + 3$. Consequently, we have the following rules.

Rules for Eliminating Parentheses

Parentheses preceded by a plus sign (or no sign) may be deleted.

Parentheses preceded by a minus sign may be deleted *if* the sign of every term within the parentheses is changed.

Example 3 Compute: $(4x^3 - 5x^2 + 7x - 2) - (2x^3 + x^2 - 6x - 8)$.

Solution

Eliminate parentheses: $= 4x^3 - 5x^2 + 7x - 2 - 2x^3 - x^2 + 6x + 8$

Group like terms: $= (4x^3 - 2x^3) + (-5x^2 - x^2) + (7x + 6x) + (-2 + 8)$

Combine like terms: $= 2x^3 - 6x^2 + 13x + 6$ ∎

To multiply polynomials, use the distributive law repeatedly, as shown in the following examples. The net result is to *multiply every term in the first polynomial by every term in the second.*

Example 4 Compute: $(y - 2)(3y^2 - 7y + 4)$.

Solution We first apply the distributive law, treating $(3y^2 - 7y + 4)$ as a single number:

$$(y - 2)(3y^2 - 7y + 4) = y(3y^2 - 7y + 4) - 2(3y^2 - 7y + 4)$$

Distributive law: $= 3y^3 - 7y^2 + 4y - 6y^2 + 14y - 8$

Regroup: $= 3y^3 - \underbrace{7y^2 - 6y^2} + \underbrace{4y + 14y} - 8$

Combine like terms: $= 3y^3 - \quad 13y^2 \quad + \quad 18y \quad - 8.$ ∎

Example 5 Compute: $(2x - 5y)(3x + 4y)$.

Solution Treat $3x + 4y$ as a single number and use the distributive law:

$$(2x - 5y)(3x + 4y) = 2x(3x + 4y) - 5y(3x + 4y)$$

Distributive law: $= 2x \cdot 3x + 2x \cdot 4y + (-5y) \cdot 3x + (-5y) \cdot 4y$

Regroup: $= 6x^2 + \underbrace{8xy - 15xy} - 20y^2$

Combine like terms: $= 6x^2 \quad - \quad 7xy \quad - 20y^2.$ ∎

Observe the pattern in the second line of Example 5 and its relationship to the terms being multiplied:

$$(2x - 5y)(3x + 4y) = 2x \cdot 3x + 2x \cdot 4y + (-5y) \cdot 3x + (-5y) \cdot 4y$$

$(2x - 5y)(3x + 4y)$ *First terms*

$(2x - 5y)(3x + 4y)$ *Outside terms*

$(2x - 5y)(3x + 4y)$ *Inside terms*

$(2x - 5y)(3x + 4y)$ *Last terms*

This pattern is easy to remember by using the acronym FOIL (**F**irst, **O**utside, **I**nside, **L**ast). The FOIL method makes it easy to find products such as this one mentally, without the necessity of writing out the intermediate steps.

CAUTION

The FOIL method can be used only when multiplying two expressions that each have two terms.

Example 6

$$(3x + 2)(x + 5) = 3x^2 + 15x + 2x + 10 = 3x^2 + 17x + 10. \quad \blacksquare$$

First   Outside   Inside   Last

Factoring

Factoring is the reverse of multiplication: We begin with a product and find the factors that multiply together to produce this product. Factoring skills are necessary to simplify expressions, to do arithmetic with fractional expressions, and to solve equations and inequalities.

The first general rule for factoring is

Common Factors

If there is a common factor in every term of the expression, factor out the common factor of highest degree.

Example 7 In $4x^6 - 8x$, for example, each term contains a factor of $4x$, so that $4x^6 - 8x = 4x(x^5 - 2)$. Similarly, the common factor of highest degree in $x^3y^2 + 2xy^3 - 3x^2y^4$ is xy^2 and

$$x^3y^2 + 2xy^3 - 3x^2y^4 = xy^2(x^2 + 2y - 3xy^2). \quad \blacksquare$$

You can greatly increase your factoring proficiency by learning to recognize multiplication patterns that appear frequently. Here are the most common ones.

Quadratic Factoring Patterns

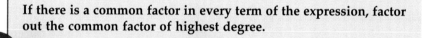

Difference of Squares	$u^2 - v^2 = (u + v)(u - v)$
Perfect Squares	$u^2 + 2uv + v^2 = (u + v)^2$
	$u^2 - 2uv + v^2 = (u - v)^2$

Example 8

(a) $x^2 - 9y^2$ can be written $x^2 - (3y)^2$, a difference of squares. Therefore, $x^2 - 9y^2 = (x + 3y)(x - 3y)$.

(b) $y^2 - 7 = y^2 - (\sqrt{7})^2 = (y + \sqrt{7})(y - \sqrt{7})$.*

(c) $36r^2 - 64s^2 = (6r)^2 - (8s)^2 = (6r + 8s)(6r - 8s)$

$$= 2(3r + 4s)2(3r - 4s) = 4(3r + 4s)(3r - 4s). \quad\blacksquare$$

Example 9 Factor: $4x^2 - 36x + 81$.

Solution Since the first and last terms of $4x^2 - 36x + 81$ are perfect squares, we try to use the perfect square pattern with $u = 2x$ and $v = 9$:

$$4x^2 - 36x + 81 = (2x)^2 - 36x + 9^2$$

$$= (2x)^2 - 2 \cdot 2x \cdot 9 + 9^2 = (2x - 9)^2. \quad\blacksquare$$

Cubic Factoring Patterns

Difference of Cubes	$u^3 - v^3 = (u - v)(u^2 + uv + v^2)$
Sum of Cubes	$u^3 + v^3 = (u + v)(u^2 - uv + v^2)$
Perfect Cubes	$u^3 + 3u^2v + 3uv^2 + v^3 = (u + v)^3$
	$u^3 - 3u^2v + 3uv^2 - v^3 = (u - v)^3$

Example 10 Factor:

(a) $x^3 - 125$ (b) $x^3 + 8y^3$ (c) $x^3 - 12x^2 + 48x - 64$

Solution

(a) $x^3 - 125 = x^3 - 5^3 = (x - 5)(x^2 + 5x + 5^2)$

$$= (x - 5)(x^2 + 5x + 25).$$

(b) $x^3 + 8y^3 = x^3 + (2y)^3 = (x + 2y)[x^2 - x \cdot 2y + (2y)^2]$

$$= (x + 2y)(x^2 - 2xy + 4y^2).$$

(c) $x^3 - 12x^2 + 48x - 64 = x^3 - 12x^2 + 48x - 4^3$

$$= x^3 - 3x^2 \cdot 4 + 3x \cdot 4^2 - 4^3$$

$$= (x - 4)^3. \quad\blacksquare$$

---

*When a polynomial has integer coefficients, we normally look only for factors with integer coefficients. But when it is easy to find other factors, as here, we shall do so.

When none of the multiplication patterns applies, use trial and error to factor quadratic polynomials. If a quadratic has two first-degree factors, then the factors must be of the form $ax + b$ and $cx + d$ for some constants a, b, c, d. The product of such factors is

$$(ax + b)(cx + d) = acx^2 + adx + bcx + bd$$
$$= acx^2 + (ad + bc)x + bd.$$

Note that ac is the *coefficient of* x^2 and bd is the *constant term* of the product polynomial. This pattern can be used to factor quadratics by reversing the FOIL process.

Example 11 Factor: $x^2 + 9x + 18$.

Solution If $x^2 + 9x + 18$ factors as $(ax + b)(cx + d)$, then we must have $ac = 1$ (coefficient of x^2) and $bd = 18$ (constant term). Thus, $a = \pm 1$ and $c = \pm 1$ (the only integer factors of 1). The only possibilities for b and d are

$$\pm 1, \pm 18 \quad \text{or} \quad \pm 2, \pm 9 \quad \text{or} \quad \pm 3, \pm 6.$$

We mentally try the various possibilities, using FOIL as our guide. For example, we try $b = 2$, $d = 9$ and check this factorization: $(x + 2)(x + 9)$. The sum of the outside and inside terms is $9x + 2x = 11x$, so this product can't be $x^2 + 9x + 18$. By trying other possibilities we find that $b = 3$, $d = 6$ leads to the correct factorization:

$$x^2 + 9x + 18 = (x + 3)(x + 6). \quad \blacksquare$$

Example 12 Factor: $6x^2 + 11x + 4$ as $(ax + b)(cx + d)$.

Solution We must find numbers a and c whose product is 6, the coefficient of x^2, and numbers b and d whose product is the constant term 4. Some possibilities are

$ac = 6$	a	± 1	± 2	± 3	± 6
	c	± 6	± 3	± 2	± 1

$bd = 4$	b	± 1	± 2	± 4
	d	± 4	± 2	± 1

Trial and error shows that $(2x + 1)(3x + 4) = 6x^2 + 11x + 4$. $\quad \blacksquare$

Occasionally these patterns can be used to factor expressions involving exponents larger than 2.

Example 13 Factor:

(a) $x^6 - y^6$ (b) $x^8 - 1$

Solution

(a) $x^6 - y^6 = (x^3)^2 - (y^3)^2 = (x^3 + y^3)(x^3 - y^3)$

$= (x + y)(x^2 - xy + y^2)(x - y)(x^2 + xy + y^2).$

(b) $x^8 - 1 = (x^4)^2 - 1 = (x^4 + 1)(x^4 - 1)$

$= (x^4 + 1)(x^2 + 1)(x^2 - 1)$

$= (x^4 + 1)(x^2 + 1)(x + 1)(x - 1).$ ■

Example 14   Factor: $x^4 - 2x^2 - 3$.

Solution   Let $u = x^2$. Then,

$x^4 - 2x^2 - 3 = (x^2)^2 - 2x^2 - 3$

$= u^2 - 2u - 3 = (u + 1)(u - 3)$

$= (x^2 + 1)(x^2 - 3)$

$= (x^2 + 1)(x + \sqrt{3})(x - \sqrt{3}).$ ■

Example 15   Factor: $3x^3 + 3x^2 + 2x + 2$.

Solution $3x^3 + 3x^2 + 2x + 2$ can be factored by regrouping and using the distributive law to factor out a common factor:

$(3x^3 + 3x^2) + (2x + 2) = 3x^2(x + 1) + 2(x + 1)$

$= (3x^2 + 2)(x + 1).$ ■

Exercises 0.4

In Exercises 1–50, perform the indicated operations and simplify your answer.

1. $x + 7x$

2. $5w + 7w - 3w$

3. $6a^2b + (-8b)a^2$

4. $-6x^3\sqrt{t} + 7x^3\sqrt{t} - 15x^3\sqrt{t}$

5. $(x^2 + 2x + 1) - (x^3 - 3x^2 + 4)$

6. $\left(u^4 - (-3)u^3 + \dfrac{u}{2} + 1\right) + \left(u^4 - 2u^3 + 5 - \dfrac{u}{2}\right)$

7. $\left(u^4 - (-3)u^3 + \dfrac{u}{2} + 1\right) - \left(u^4 - 2u^3 + 5 - \dfrac{u}{2}\right)$

8. $(6a^2b + 3a\sqrt{c} - 5ab\sqrt{c}) + (-6ab^2 - 3ab + 6ab\sqrt{c})$

9. $(4z - 6z^2w - (-2)z^3w^2) + (8 - 6z^2w - zw^3 + 4z^3w^2)$

10. $(x^5y - 2x + 3xy^3) - (-2x - x^5y + 2xy^3)$

11. $(9x - x^3 + 1) - [2x^3 + (-6)x + (-7)]$

12. $(x - \sqrt{y} - z) - (x + \sqrt{y} - z) - (\sqrt{y} + z - x)$

13. $(x^2 - 3xy) - (x + xy) - (x^2 + xy)$

14. $2x(x^2 + 2)$

15. $(-5y)(-3y^2 + 1)$

16. $x^2y(xy - 6xy^2)$

17. $3ax(4ax - 2a^2y + 2ay)$

18. $2x(x^2 - 3xy + 2y^2)$

19. $6z^3(2z + 5)$

20. $-3x^2(12x^6 - 7x^5)$

21. $3ab(4a - 6b + 2a^2b)$

22. $(-3ay)(4ay - 5y)$

23. $(x + 1)(x - 2)$

24. $(x + 2)(2x - 5)$

25. $(-2x + 4)(-x - 3)$

26. $(y - 6)(2y + 2)$

27. $(y + 3)(y + 4)$

28. $(w - 2)(3w + 1)$

29. $(3x + 7)(-2x + 5)$

30. $(ab + 1)(a - 2)$

31. $(y - 3)(3y^2 + 4)$

32. $(y + 8)(y - 8)$

33. $(x + 4)(x - 4)$

34. $(3x - y)(3x + y)$

35. $(4a + 5b)(4a - 5b)$

36. $(x + 6)^2$

37. $(y - 11)^2$

38. $(2x + 3y)^2$

39. $(5x - b)^2$

40. $(2s^2 - 9y)(2s^2 + 9y)$

41. $(4x^3 - y^4)^2$

42. $(4x^3 - 5y^2)(4x^3 + 5y^2)$

43. $(-3x^2 + 2y^4)^2$

44. $(c - 2)(2c^2 - 3c + 1)$

45. $(2y + 3)(y^2 + 3y - 1)$

46. $(x + 2y)(2x^2 - xy + y^2)$

47. $(5w + 6)(-3w^2 + 4w - 3)$

48. $(5x - 2y)(x^2 - 2xy + 3y^2)$

49. $2x(3x + 1)(4x - 2)$

50. $3y(-y + 2)(3y + 1)$

In Exercises 51–56, perform the indicated multiplication and simplify your answer if possible.

51. $\left(\sqrt{x} + 5\right)\left(\sqrt{x} - 5\right)$

52. $\left(2\sqrt{x} + \sqrt{2y}\right)\left(2\sqrt{x} - \sqrt{2y}\right)$

53. $\left(3 + \sqrt{y}\right)^2$

54. $\left(7w - \sqrt{2x}\right)^2$

55. $\left(1 + \sqrt{3x}\right)\left(x + \sqrt{3}\right)$

56. $\left(2y + \sqrt{3}\right)\left(\sqrt{5y} - 1\right)$

In Exercises 57–62, compute the product and arrange the terms of your answer according to decreasing powers of x, with each power of x appearing at most once.
Example: $(ax + b)(4x - c) = 4ax^2 + (4b - ac)x - bc.$

57. $(ax + b)(3x + 2)$

58. $(4x - c)(dx + c)$

59. $(ax + b)(bx + a)$

60. $rx(3rx + 1)(4x - r)$

61. $(x - a)(x - b)(x - c)$

62. $(2dx - c)(3cx + d)$

In Exercises 63–66, fill the blanks, using the Box Method of multiplying polynomials, which is illustrated here.

To multiply $(2x + 3)(5x - 7)$, form a 2-by-2 box, with row and column labels as shown. Then enter the products of each row label by each column label; for instance, the entry $15x$ in the lower left corner is the product of $5x$ and 3.

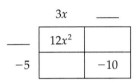

The sum of the entries in the box is the product:

$$10x^2 + 15x - 14x - 21 = 10x^2 + x - 21.$$

63. $(3x + 2)(4x - 5) = $ _____

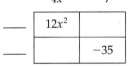

64. $(4x - 7)($_____$) = 12x^2 - x - 35$

65. $(2x + 5)(4x^2 - 10x + 25) = $ _____

66. $(4x - 3)($_____$) = 64x^3 - 27$

67. Four squares, each measuring x by x inches, are cut from the corners of a rectangular sheet of cardboard that measures 30 by 22 inches and the flaps are folded up to form an open-top box, as shown below.

(a) Express the length and width of the resulting box as polynomials in x.

(b) Express the volume of the resulting box as a polynomial.

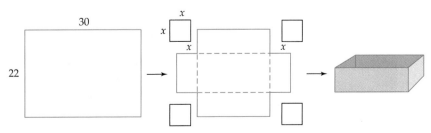

68. Do Exercise 67 when the sheet of cardboard measures 40 by 40 inches.

69. A gutter is to be constructed by folding up the edge of a 16-inch-wide piece of metal, as shown in the figure.

    (a) Express the area of the cross section of the gutter as a polynomial in x.

    (b) Fill in the table.

x	1	2	3	4	5	6	7	8
Area of cross section								

    (c) Which value of x appears to produce the largest cross-sectional area (and hence the largest volume of water in the gutter)?

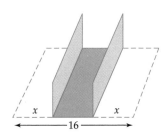

In Exercises 70–72, express the area of the shaded region as a polynomial in x.

70.

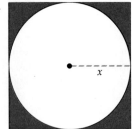

71.

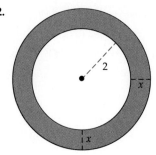

72.

In Exercises 73–122, factor the expression.

73. $x^2 - 4$

74. $x^2 + 6x + 9$

75. $9y^2 - 25$

76. $y^2 - 4y + 4$

77. $81x^2 + 36x + 4$

78. $4x^2 - 12x + 9$

79. $5 - x^2$

80. $1 - 36u^2$

81. $49 + 28z + 4z^2$

82. $25u^2 - 20uv + 4v^2$

83. $x^4 - y^4$

84. $x^2 - 1/9$

85. $x^2 + x - 6$

86. $y^2 + 11y + 30$

87. $z^2 + 4z + 3$

88. $x^2 - 8x + 15$

89. $y^2 + 5y - 36$

90. $z^2 - 9z + 14$

91. $x^2 - 6x + 9$

92. $4y^2 - 81$

93. $x^2 + 7x + 10$

94. $w^2 - 6w - 16$

95. $x^2 + 11x + 18$

96. $x^2 + 3xy - 28y^2$

97. $3x^2 + 4x + 1$

98. $4y^2 + 4y + 1$

99. $2z^2 + 11z + 12$

100. $10x^2 - 17x + 3$

101. $9x^2 - 72x$

102. $4x^2 - 4x - 3$

103. $10x^2 - 8x - 2$

104. $7z^2 + 23z + 6$

105. $8u^2 + 6u - 9$

106. $2y^2 - 4y + 2$

107. $4x^2 + 20xy + 25y^2$

108. $63u^2 - 46uv + 8v^2$

109. $x^3 - 125$

110. $y^3 + 64$

111. $x^3 + 6x^2 + 12x + 8$

112. $y^3 - 3y^2 + 3y - 1$

113. $8 + x^3$

114. $27 - t^3$

115. $x^3 + 1$

116. $x^3 - 1$

117. $8x^3 - y^3$

118. $(x - 1)^3 + 1$

119. $x^6 - 64$

120. $x^5 - 8x^2$

121. $y^4 + 7y^2 + 10$

122. $81 - y^4$

In Exercises 123–128, factor by regrouping and using the distributive law (as in Example 15).

123. $x^2 - yz + xz - xy$

124. $x^6 - 2x^4 - 8x^2 + 16$

125. $a^3 - 2b^2 + 2a^2b - ab$

126. $u^2v - 2w^2 - 2uvw + uw$

127. $x^3 + 4x^2 - 8x - 32$

128. $z^8 - 5z^7 + 2z - 10$

Errors to Avoid

In Exercises 129–136, find a numerical example to show that the given statement is false. Then find the mistake in the statement and correct it.

Example: The statement $-(b + 2) = -b + 2$ is false when $b = 5$, since $-(5 + 2) = -7$ but $-5 + 2 = -3$. The mistake is the sign on the 2. The correct statement is $-(b + 2) = -b - 2$.

129. $3(y + 2) = 3y + 2$

130. $x - (3y + 4) = x - 3y + 4$

131. $(x + y)^2 = x + y^2$ **132.** $(2x)^3 = 2x^3$

133. $y + y + y = y^3$ **134.** $(a - b)^2 = a^2 - b^2$

135. $(x - 3)(x - 2) = x^2 - 5x - 6$

136. $(a + b)(a^2 + b^2) = a^3 + b^3$

Thinkers

In Exercises 137 and 138, explain algebraically why each of these parlor tricks always works.

137. Write down a nonzero number. Add 1 to it and square the result. Subtract 1 from the original number and square the result. Subtract this second square from the first one. Divide by the number with which you started. The answer is 4.

138. Write down a positive number. Add 4 to it. Multiply the result by the original number. Add 4 to this result and then take the square root. Subtract the number with which you started. The answer is 2.

139. Invent a similar parlor trick in which the answer is always the number with which you started.

140. Show that there do *not* exist real numbers c and d such that

$$x^2 + 1 = (x + c)(x + d).$$

0.5 Fractional Expressions

Quotients of algebraic expressions are called **fractional expressions.** A quotient of two polynomials, such as

$$\frac{x^2 - 2}{x - 1} \qquad \frac{1}{x^3 + 4x^2 - x} \qquad \text{or} \qquad \frac{3x - 7}{x^3(x + 1)}$$

is called a **rational expression.** Throughout this section, we assume that all denominators are nonzero.

The basic rules for dealing with fractional expressions are essentially the same as those for ordinary numerical fractions. In particular, the usual cancellation property holds.

Cancellation

If $k \neq 0$, then $\dfrac{ka}{kb} = \dfrac{a}{b}.$

Example 1 Simplify:

(a) $\dfrac{36}{56}$ (b) $\dfrac{x^4 - 1}{x^2 + 1}$.

Solution

(a) *Factor numerator and denominator:* $\dfrac{36}{56} = \dfrac{9 \cdot 4}{7 \cdot 8} = \dfrac{9 \cdot 4}{7 \cdot 2 \cdot 4}$

Cancel the common factor 4: $= \dfrac{9}{7 \cdot 2} = \dfrac{9}{14}$.

(b) The same procedure works for the rational expression.

Factor the numerator: $\dfrac{x^4 - 1}{x^2 + 1} = \dfrac{(x^2 + 1)(x^2 - 1)}{x^2 + 1}$

Cancel the common factor $x^2 + 1$: $= \dfrac{x^2 - 1}{1} = x^2 - 1$ ∎

A fraction is in **lowest terms** if its numerator and denominator have no common factors except ±1.

Example 2 Express $\dfrac{x^2 + x - 6}{x^2 - 3x + 2}$ in lowest terms.

Solution   Proceed as we did in Example 1: factor numerator and denominator and cancel common factors:

$$\frac{x^2 + x - 6}{x^2 - 3x + 2} = \frac{(x - 2)(x + 3)}{(x - 2)(x - 1)} = \frac{x + 3}{x - 1}. \quad ∎$$

To add or subtract two fractions with the same denominator, simply add or subtract the numerators:

$$\frac{a}{b} + \frac{c}{b} = \frac{a + c}{b} \qquad \text{and} \qquad \frac{a}{b} - \frac{c}{b} = \frac{a - c}{b}.$$

Example 3

$$\frac{7x^2 + 2}{x^2 + 3} - \frac{4x^2 + 2x - 5}{x^2 + 3} = \frac{(7x^2 + 2) - (4x^2 + 2x - 5)}{x^2 + 3}$$

$$= \frac{7x^2 + 2 - 4x^2 - 2x + 5}{x^2 + 3}$$

$$= \frac{3x^2 - 2x + 7}{x^2 + 3}. \quad ∎$$

To add or subtract fractions with different denominators, you must use a **common denominator,** that is, replace the given fractions by equivalent ones that have the same denominator. For instance, to add 1/3 and 1/4, we use the common denominator 12 (the product of 3 and 4):

$$\frac{1}{3} = \frac{4}{12} \quad \text{and} \quad \frac{1}{4} = \frac{3}{12},$$

so that

$$\frac{1}{3} + \frac{1}{4} = \frac{4}{12} + \frac{3}{12} = \frac{4 + 3}{12} = \frac{7}{12}.$$

Example 4 Compute: $\dfrac{2x + 1}{3x} - \dfrac{x^2 - 2}{x - 1}$.

Solution We use the product of the denominators $3x(x - 1)$ as the common denominator. To rewrite the first fraction in terms of this denominator, we must find the right-side numerator here:

$$\frac{2x + 1}{3x} = \frac{?}{3x(x - 1)}$$

Note that to get from the left-side denominator to the right-side one, you multiply by $x - 1$. To keep the fractions equal you must multiply the left-hand numerator by $x - 1$ also:

$$\frac{2x + 1}{3x} = \frac{(2x + 1)(x - 1)}{3x(x - 1)}$$

A similar procedure works with the other fraction:

$$\frac{x^2 - 2}{x - 1} = \frac{?}{3x(x - 1)} \qquad \frac{x^2 - 2}{x - 1} = \frac{3x(x^2 - 2)}{3x(x - 1)}$$

Multiply by 3x

Now that the fractions have the same denominator, it's easy to subtract them:

$$\frac{2x + 1}{3x} - \frac{x^2 - 2}{x - 1} = \frac{(2x + 1)(x - 1)}{3x(x - 1)} - \frac{3x(x^2 - 2)}{3x(x - 1)}$$

$$= \frac{(2x + 1)(x - 1) - 3x(x^2 - 2)}{3x(x - 1)}$$

$$= \frac{2x^2 - x - 1 - 3x^3 + 6x}{3x^2 - 3x}$$

$$= \frac{-3x^3 + 2x^2 + 5x - 1}{3x^2 - 3x}. \quad \blacksquare$$

Although the product of the denominators can always be used as a common denominator, it's usually more efficient to use the **least common denominator,** that is, the smallest quantity that can be the denominator of both fractions. The next example illustrates an easy way to find the least common denominator when dealing with numbers.

Example 5 Find the least common denominator for $\dfrac{1}{100}$ and $\dfrac{1}{120}$.

Solution The two denominators can be factored as
$$100 = 2^2 \cdot 5^2 \qquad \text{and} \qquad 120 = 2^3 \cdot 3 \cdot 5.$$
The distinct factors that appear in both numbers are 2, 3, and 5. For each factor, take the highest power to which it appears in either number (namely, 2^3, 3^1, and 5^2). The product of these highest powers is the least common denominator:
$$2^3 \cdot 3 \cdot 5^2 = 600.$$
Note that $600 = 100 \cdot 6$ and $600 = 120 \cdot 5$, so that
$$\frac{1}{100} = \frac{1 \cdot 6}{100 \cdot 6} = \frac{6}{600} \qquad \text{and} \qquad \frac{1}{120} = \frac{1 \cdot 5}{120 \cdot 5} = \frac{5}{600}. \qquad \blacksquare$$

The technique of Example 5 can also be used to find the least common denominator of rational expressions.

Example 6 Find the least common denominator for
$$\frac{1}{x^2 + 2x + 1}, \qquad \frac{5x}{x^2 - x}, \qquad \frac{3x - 7}{x^4 + x^3}$$
and express each fraction in terms of this common denominator.

Solution Begin by factoring each denominator completely:
$$x^2 + 2x + 1 = (x + 1)^2 \qquad x^2 - x = x(x - 1) \qquad x^4 + x^3 = x^3(x + 1).$$
The distinct factors are x, $x + 1$, and $x - 1$. The least common denominator is
$$x^3(x + 1)^2(x - 1).$$
Therefore,
$$\frac{1}{(x + 1)^2} = \frac{1}{(x + 1)^2} \cdot \frac{x^3(x - 1)}{x^3(x - 1)} = \frac{x^3(x - 1)}{x^3(x + 1)^2(x - 1)}$$
$$\frac{5x}{x(x - 1)} = \frac{5x}{x(x - 1)} \cdot \frac{x^2(x + 1)^2}{x^2(x + 1)^2} = \frac{5x^3(x + 1)^2}{x^3(x + 1)^2(x - 1)}$$
$$\frac{3x - 7}{x^3(x + 1)} = \frac{3x - 7}{x^3(x + 1)} \cdot \frac{(x + 1)(x - 1)}{(x + 1)(x - 1)} = \frac{(3x - 7)(x + 1)(x - 1)}{x^3(x + 1)^2(x - 1)}. \qquad \blacksquare$$

Example 7 Compute: $\dfrac{1}{z} + \dfrac{3z}{z+1} - \dfrac{z^2}{(z+1)^2}$.

Solution We use the least common denominator $z(z+1)^2$:

$$\frac{1}{z} + \frac{3z}{z+1} - \frac{z^2}{(z+1)^2} = \frac{(z+1)^2}{z(z+1)^2} + \frac{3z^2(z+1)}{z(z+1)^2} - \frac{z^3}{z(z+1)^2}$$

$$= \frac{(z+1)^2 + 3z^2(z+1) - z^3}{z(z+1)^2}$$

$$= \frac{z^2 + 2z + 1 + 3z^3 + 3z^2 - z^3}{z(z+1)^2}$$

$$= \frac{2z^3 + 4z^2 + 2z + 1}{z(z+1)^2}. \quad \blacksquare$$

Multiplication of fractions is easy: Multiply corresponding numerators and denominators, then simplify your answer:

Example 8

$$\frac{x^2-1}{x^2+2} \cdot \frac{3x-4}{x+1} = \frac{(x^2-1)(3x-4)}{(x^2+2)(x+1)}$$

$$= \frac{(x-1)(x+1)(3x-4)}{(x^2+2)(x+1)} = \frac{(x-1)(3x-4)}{x^2+2}. \quad \blacksquare$$

Division of fractions is given by the rule:

Invert the divisor and multiply: $\quad \dfrac{a}{b} \div \dfrac{c}{d} = \dfrac{a}{b} \cdot \dfrac{d}{c} = \dfrac{ad}{bc}$.

Example 9

$$\frac{x^2+x-2}{x^2-6x+9} \div \frac{x^2-1}{x-3} = \frac{x^2+x-2}{x^2-6x+9} \cdot \frac{x-3}{x^2-1}$$

$$= \frac{(x+2)(x-1)}{(x-3)^2} \cdot \frac{x-3}{(x-1)(x+1)}$$

$$= \frac{x+2}{(x-3)(x+1)}. \quad \blacksquare$$

Division problems can also be written in fractional form. For instance, 8/2 means $8 \div 2$. Similarly, a compound fraction such as

$$\frac{\dfrac{x+1}{x^2-4}}{\dfrac{3x+2}{x+2}}$$

means

$$\frac{x + 1}{x^2 - 4} \div \frac{3x + 2}{x + 2}, \quad \text{which is equivalent to} \quad \frac{x + 1}{x^2 - 4} \cdot \frac{x + 2}{3x + 2}.$$

Consequently, the basic rule for simplifying a compound fraction is:

Invert the denominator and multiply it by the numerator.

Example 10 Compute:

(a) $\dfrac{\dfrac{16y^2z}{8yz^2}}{\dfrac{yz}{6y^3z^3}}$ (b) $\dfrac{\dfrac{y^2}{y + 2}}{y^3 + y}$

Solution

(a) $\dfrac{16y^2z/8yz^2}{yz/6y^3z^3} = \dfrac{16y^2z}{8yz^2} \cdot \dfrac{6y^3z^3}{yz} = \dfrac{16 \cdot 6 \cdot y^5z^4}{8y^2z^3}$

$$= 2 \cdot 6 \cdot y^{5-2}z^{4-3} = 12y^3z$$

(b) $\dfrac{\dfrac{y^2}{y + 2}}{y^3 + y} = \dfrac{y^2}{y + 2} \cdot \dfrac{1}{y^3 + y} = \dfrac{y^2}{(y + 2)(y^3 + y)}$

$$= \dfrac{y^2}{(y + 2)y(y^2 + 1)}$$

$$= \dfrac{y}{(y + 2)(y^2 + 1)} \quad \blacksquare$$

Example 11 Simplify the compound fraction

$$\frac{\dfrac{1}{x + h} - \dfrac{1}{x}}{h}.$$

Solution First write the numerator as a single fraction, then invert and multiply.

$$\frac{\dfrac{1}{x + h} - \dfrac{1}{x}}{h} = \frac{\dfrac{x}{x(x + h)} - \dfrac{x + h}{x(x + h)}}{h}$$

$$= \frac{\dfrac{x - (x + h)}{x(x + h)}}{h} = \frac{\dfrac{-h}{x(x + h)}}{h}$$

$$= \frac{-h}{x(x + h)} \cdot \frac{1}{h} = \frac{-1}{x(x + h)} \quad \blacksquare$$

Rationalizing Numerators and Denominators

When dealing with fractions in the days before calculators, it was customary to *rationalize the denominators,* that is, write equivalent fractions with no radicals in the denominator, because this made many computations easier. With calculators, of course, there is no computational advantage to rationalizing denominators. Nevertheless, rationalizing denominators or numerators is sometimes needed to simplify expressions and to derive useful formulas.

Example 12 Rationalize the denominators of

(a) $\dfrac{7}{\sqrt{5}}$ (b) $\dfrac{2}{3 + \sqrt{6}}$

Solution

(a) The key is to multiply the fraction by 1, with 1 written as a radical fraction:

$$\frac{7}{\sqrt{5}} = \frac{7}{\sqrt{5}} \cdot 1 = \frac{7}{\sqrt{5}} \cdot \frac{\sqrt{5}}{\sqrt{5}} = \frac{7\sqrt{5}}{5}.$$

(b) The same technique works here.

Multiply by 1:
$$\frac{2}{3 + \sqrt{6}} = \frac{2}{3 + \sqrt{6}} \cdot 1$$

Rewrite 1 as a radical fraction:
$$= \frac{2}{3 + \sqrt{6}} \cdot \frac{3 - \sqrt{6}}{3 - \sqrt{6}}$$

Multiply:
$$= \frac{2(3 - \sqrt{6})}{(3 + \sqrt{6})(3 - \sqrt{6})}$$

Use the multiplication pattern
$(a + b)(a - b) = a^2 - b^2$ in the denominator:
$$= \frac{6 - 2\sqrt{6}}{3^2 - (\sqrt{6})^2}$$

Simplify:
$$= \frac{6 - 2\sqrt{6}}{9 - 6} = \frac{6 - 2\sqrt{6}}{3}. \quad\blacksquare$$

Example 13 Assume $h \neq 0$ and rationalize the numerator of

$$\frac{\sqrt{x + h} - \sqrt{x}}{h};$$

that is, write an equivalent fraction with no radicals in the numerator.

Solution Begin by multiplying the fraction by 1, with 1 written as a suitable radical fraction,

$$\frac{\sqrt{x+h}-\sqrt{x}}{h} = \frac{\sqrt{x+h}-\sqrt{x}}{h}\cdot 1$$

$$= \frac{\sqrt{x+h}-\sqrt{x}}{h}\cdot\frac{\sqrt{x+h}+\sqrt{x}}{\sqrt{x+h}+\sqrt{x}}$$

Use the multiplication pattern

$(a+b)(a-b)=a^2-b^2$:

$$= \frac{(\sqrt{x+h})^2-(\sqrt{x})^2}{h(\sqrt{x+h}+\sqrt{x})} = \frac{x+h-x}{h(\sqrt{x+h}+\sqrt{x})}$$

$$= \frac{h}{h(\sqrt{x+h}+\sqrt{x})}$$

Cancel h:

$$= \frac{1}{\sqrt{x+h}+\sqrt{x}} \quad\blacksquare$$

Exercises 0.5

In Exercises 1–10, express the fraction in lowest terms.

1. $\dfrac{63}{49}$

2. $\dfrac{121}{33}$

3. $\dfrac{13\cdot 27\cdot 22\cdot 10}{6\cdot 4\cdot 11\cdot 12}$

4. $\dfrac{x^2-4}{x+2}$

5. $\dfrac{x^2-x-2}{x^2+2x+1}$

6. $\dfrac{z+1}{z^3+1}$

7. $\dfrac{a^2-b^2}{a^3-b^3}$

8. $\dfrac{x^4-3x^2}{x^3}$

9. $\dfrac{(x+c)(x^2-cx+c^2)}{x^4+c^3x}$

10. $\dfrac{x^4-y^4}{(x^2+y^2)(x^2-xy)}$

In Exercises 11–28, perform the indicated operations.

11. $\dfrac{3}{7}+\dfrac{2}{5}$

12. $\dfrac{7}{8}-\dfrac{5}{6}$

13. $\left(\dfrac{19}{7}+\dfrac{1}{2}\right)-\dfrac{1}{3}$

14. $\dfrac{1}{a}-\dfrac{2a}{b}$

15. $\dfrac{c}{d}+\dfrac{3c}{e}$

16. $\dfrac{r}{s}+\dfrac{s}{t}+\dfrac{t}{r}$

17. $\dfrac{b}{c}-\dfrac{c}{b}$

18. $\dfrac{a}{b}+\dfrac{2a}{b^2}+\dfrac{3a}{b^3}$

19. $\dfrac{1}{x+1}-\dfrac{1}{x}$

20. $\dfrac{1}{2x+1}+\dfrac{1}{2x-1}$

21. $\dfrac{1}{x+4}+\dfrac{2}{(x+4)^2}-\dfrac{3}{x^2+8x+16}$

22. $\dfrac{1}{x}+\dfrac{1}{xy}+\dfrac{1}{xy^2}$

23. $\dfrac{1}{x}-\dfrac{1}{3x-4}$

24. $\dfrac{3}{x-1}+\dfrac{4}{x+1}$

25. $\dfrac{1}{x+y}+\dfrac{x+y}{x^3+y^3}$

26. $\dfrac{6}{5(x-1)(x-2)^2}+\dfrac{x}{3(x-1)^2(x-2)}$

27. $\dfrac{1}{4x(x+1)(x+2)^3}-\dfrac{6x+2}{4(x+1)^3}$

28. $\dfrac{x+y}{(x^2-xy)(x-y)^2}-\dfrac{2}{(x^2-y^2)^2}$

In Exercises 29–40, multiply and express in lowest terms.

29. $\dfrac{3}{4}\cdot\dfrac{12}{5}\cdot\dfrac{10}{9}$

30. $\dfrac{10}{45}\cdot\dfrac{6}{14}\cdot\dfrac{1}{2}$

31. $\dfrac{3a^2c}{4ac}\cdot\dfrac{8ac^3}{9a^2c^4}$

32. $\dfrac{6x^2y}{2x}\cdot\dfrac{y}{21xy}$

33. $\dfrac{7x}{11y}\cdot\dfrac{66y^2}{14x^3}$

34. $\dfrac{ab}{c^2}\cdot\dfrac{cd}{a^2b}\cdot\dfrac{ad}{bc^2}$

35. $\dfrac{3x+9}{2x}\cdot\dfrac{8x^2}{x^2-9}$

36. $\dfrac{4x+16}{3x+15}\cdot\dfrac{2x+10}{x+4}$

37. $\dfrac{5y-25}{3}\cdot\dfrac{y^2}{y^2-25}$

38. $\dfrac{6x-12}{6x}\cdot\dfrac{8x^2}{x-2}$

39. $\dfrac{u}{u-1}\cdot\dfrac{u^2-1}{u^2}$

40. $\dfrac{t^2-t-6}{t^2-6t+9}\cdot\dfrac{t^2+4t-5}{t^2-25}$

In Exercises 41–52, compute the quotient and express in lowest terms.

41. $\dfrac{5}{12}\div\dfrac{4}{14}$

42. $\dfrac{\dfrac{100}{52}}{\dfrac{27}{26}}$

43. $\dfrac{uv}{v^2w}\div\dfrac{uv}{u^2v}$

44. $\dfrac{3x^2y}{(xy)^2} \div \dfrac{3xyz}{x^2y}$

45. $\dfrac{\dfrac{x+3}{x+4}}{\dfrac{2x}{x+4}}$

46. $\dfrac{\dfrac{(x+2)^2}{(x-2)^2}}{\dfrac{x^2+2x}{x^2-4}}$

47. $\dfrac{\dfrac{(c+d)^2}{c^2-d^2}}{cd}$

48. $\dfrac{\dfrac{1}{x}-\dfrac{3}{2}}{\dfrac{2}{x-2}+\dfrac{5}{x}}$

49. $\dfrac{\dfrac{6}{y}-3}{1-\dfrac{1}{y-1}}$

50. $\dfrac{\dfrac{1}{3x}-\dfrac{1}{4y}}{\dfrac{5}{6x^2}+\dfrac{1}{y}}$

51. $\dfrac{\dfrac{1}{(x+h)^2}-\dfrac{1}{x^2}}{h}$

52. $(x^{-1}+y^{-1})^{-1}$

In Exercises 53–58, rationalize the denominator and simplify your answer.

53. $\dfrac{3}{\sqrt{8}}$

54. $\dfrac{2}{\sqrt{6}}$

55. $\dfrac{3}{2+\sqrt{12}}$

56. $\dfrac{1+\sqrt{3}}{5+\sqrt{10}}$

57. $\dfrac{2}{\sqrt{x}+2}$

58. $\dfrac{\sqrt{x}}{\sqrt{x}-\sqrt{c}}$

In Exercises 59–62, rationalize the numerator and simplify your answer.

59. $\dfrac{\sqrt{x+h+1}-\sqrt{x+1}}{h}$

60. $\dfrac{2\sqrt{x+h+3}-2\sqrt{x+3}}{h}$

61. $\dfrac{\sqrt{(x+h)^2+1}-\sqrt{x^2+1}}{h}$

62. $\dfrac{\sqrt{(x+h)^2-(x+h)}-\sqrt{x^2-x}}{h}$

In Exercises 63 – 66, use the following formula, which gives the monthly payment required to pay off in n months a loan of P dollars at an annual interest rate i compounded monthly. The interest rate i is expressed as a decimal (for instance, 9% = .09).

$$R = \dfrac{Pi}{12-12\left[1+\dfrac{i}{12}\right]^{-n}}$$

63. (a) If you get a car loan of $12,000 for four years at 9%, what is your monthly payment? [*Hint:* 4 years = 48 months.]
  (b) How much interest will you pay on this loan? [*Hint:* The total you pay is 48 × (monthly payment); the difference between this and $12,000 is the interest.]

64. What would your monthly payment be in Exercise 63 if you wanted to pay off the loan in three years?

65. You buy a house for $80,000. Your parents give you $8000 for a down payment and you get a 30-year mortgage for $72,000 at an interest rate of 7.5%.
  (a) Find your monthly payment.
  (b) Find the total amount of interest you will pay over 30 years.

66. Do Exercise 65 for a 15-year mortgage (same amount and interest rate).

In Exercises 67–70, a dartboard is shown. Assuming that every dart actually hits the board, find the probability that the dart will land in the shaded area. This probability is the quotient

$$\dfrac{\text{Area of shaded region}}{\text{Area of dartboard}}$$

67.

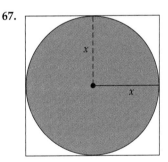

68.

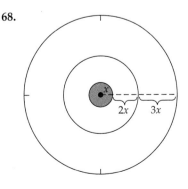

69.

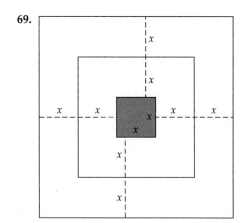

70.

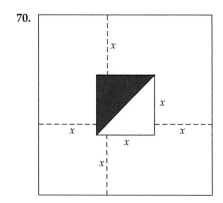

71. The concentration of a drug in a person's bloodstream t hours after its injection is modeled by the rational expression.

$$\frac{3t}{4t^2 + 5}$$

(a) What is the concentration after $2\frac{1}{2}$ hours?

(b) Complete the following table:

t (hours)	$\frac{1}{2}$	1	$1\frac{1}{2}$	2	$2\frac{1}{2}$	3
concentration						

(c) About what time is the concentration the highest?

(d) What happens to the concentration of the drug as time goes on?

72. The rational expression

$$\frac{.62x^3 - 5.95x^2 + 91.7x + 124}{x}$$

gives the average cost per case of producing x cases of pencils.

(a) What is the average cost of producing 3 cases?

(b) Complete the following table:

x	1	2	3	4	5	6	7	8	9	10
Average cost										

(c) How many cases should be produced to minimize average cost?

Errors to Avoid

In Exercises 73–79, find a numerical example to show that the given statement is false. Then find the mistake in the statement and correct it.

73. $\dfrac{1}{a} + \dfrac{1}{b} = \dfrac{1}{a + b}$

74. $\dfrac{x^2}{x^2 + x^6} = 1 + x^3$

75. $\left(\dfrac{1}{\sqrt{a} + \sqrt{b}}\right)^2 = \dfrac{1}{a + b}$

76. $\dfrac{r + s}{r + t} = 1 + \dfrac{s}{t}$

77. $\dfrac{x}{2 + x} = \dfrac{1}{2}$

78. $\dfrac{\frac{1}{x}}{\frac{1}{y}} = \dfrac{1}{xy}$

79. $\left(\sqrt{x} + \sqrt{y}\right)\dfrac{1}{\sqrt{x} + \sqrt{y}} = x + y$

Chapter 0   *Review*

Important Concepts

Important Facts and Formulas

- $|c - d|$ = distance from c to d on the number line.
- Laws of Exponents: $c^r c^s = c^{r+s}$ $(cd)^r = c^r d^r$

$$\frac{c^r}{c^s} = c^{r-s} \qquad \left(\frac{c}{d}\right)^r = \frac{c^r}{d^r}$$

$$(c^r)^s = c^{rs} \qquad c^{-r} = \frac{1}{c^r}$$

Review Questions

1. Fill the blanks with one of the symbols $<$, $=$, or $>$ so that the resulting statement is true.

   (a) 142 _____ $|-51|$

   (b) $\sqrt{2}$ _____ $|-2|$

   (c) -1000 _____ $\dfrac{1}{10}$

   (d) $|-2|$ _____ $-|6|$

   (e) $|u - v|$ _____ $|v - u|$ where u and v are fixed real numbers.

2. List two real numbers that are *not* rational numbers.

3. Express in symbols:
   (a) y is negative, but greater than -10;
   (b) x is nonnegative and not greater than 10.

4. Express in symbols:
   (a) $c - 7$ is nonnegative;   (b) .6 is greater than $|5x - 2|$.

5. Express in symbols:
   (a) x is less than 3 units from -7 on the number line;
   (b) y is farther from 0 than x is from 3 on the number line.

6. Simplify: $|b^2 - 2b + 1|$ 7. Solve: $|x - 5| = 3$

8. Solve: $|x + 2| = 4$ 9. Solve: $|x + 3| = \dfrac{5}{2}$

10. Solve: $|x - 5| = 2$

11. (a) $|\pi - 7| = $ _____   (b) $|\sqrt{23} - \sqrt{3}| = $ _____

12. If c and d are real numbers with $c \neq d$ what are the possible values of $\dfrac{c - d}{|c - d|}$?

13. A number x is one of the five numbers A, B, C, D, E on the number line shown in the figure.

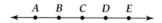

If x satisfies *both* $A \leq x < D$ *and* $B < x \leq E$, then
   (a) $x = A$ (b) $x = B$
   (c) $x = C$ (d) $x = D$
   (e) $x = E$

14. Which one of these statements is *always* true for any real numbers x, y?
   (a) $|2x| = 2x$ (b) $\sqrt{x^2} = x$
   (c) $|x - y| = x - y$ (d) $|x - y| = |y - x|$
   (e) $|x - y| = |x| + |y|$

15. Express $0.282828\cdots$ as a fraction.

16. Express $.362362362\cdots$ as a fraction.

In Questions 17–20, simplify the expression and use only positive exponents in your answer.

17. $(c^3 d^{-3} e^{-1})^5$ 18. $(c^3 d^2)^3 (c^{-1} d)^{-2}$

19. $\dfrac{(8u^5)^2 2^{-4} u^{-3}}{2u^8}$ 20. $\dfrac{(2x^2)^{-1} y z^3}{x^2 (yz)^3}$

21. Express in scientific notation:
   (a) $12{,}320{,}000{,}000{,}000{,}000$ (b) .0000000000789

22. Express in decimal notation:
   (a) 4.78×10^8 (b) 6.53×10^{-9}

In Questions 23–28, simplify the expression.

23. $\sqrt{\sqrt[3]{c^{12}}}$ 24. $\left(\sqrt[3]{4} c^3 d^2\right)^3 \left(c\sqrt{d}\right)^2$

25. $(a^{-2/3} b^{2/5})(a^3 b^6)^{4/3}$ 26. $\dfrac{(3c)^{3/5} (2d)^{-2} (4c)^{1/2}}{(4c)^{1/5} (2d)^4 (2c)^{-3/2}}$

27. $(u^{1/4} - v^{1/4})(u^{1/4} + v^{1/4})$ 28. $c^{3/2}(2c^{1/2} + 3c^{-3/2})$

In Questions 29 and 30, simplify and write the expression without radicals or negative exponents:

29. $\dfrac{\sqrt[3]{6c^4d^{14}}}{\sqrt[3]{48c^{-2}d^2}}$

30. $\dfrac{(8u^5)^{1/4}2^{-1}u^{-3}}{2u^8}$

31. Let a and b be positive real numbers. Which of these statements is *false*?

    (a) $a^r a^s = a^{r+s}$

    (b) $\sqrt{a}\sqrt{b} = \sqrt{ab}$

    (c) $\sqrt[3]{a^3} = a$

    (d) $\sqrt{a+b} = \sqrt{a} + \sqrt{b}$

    (e) $\sqrt{\sqrt{a}} = \sqrt[4]{a}$

32. The length L of an animal is related to its surface area S by the equation

$$L = \left(\frac{S}{a}\right)^{1/2},$$ where a is a constant that depends on the type of animal.

Suppose one animal has twice the surface area of another animal of the same type. How much longer is the first animal than the second?

In Questions 33–40, perform the indicated operations and simplify your answer.

33. $(-2p^3 - 5p + 7) + (-4p^2 + 8p + 2)$

34. $(-4y^2 - 3y + 8) - (2y^2 - 6y - 2)$

35. $(3z + 5)(4z^2 - 2z + 1)$

36. $(2k + 3)(4k^3 - 3k^2 + k)$

37. $(6k - 1)(2k - 3)$

38. $(8r + 3)(r - 1)$

39. $3x(2x - 2)^2$

40. $\left(\sqrt{x} + 2y\right)\left(\sqrt{x} - y\right)$

41. Suppose the cost of making x cartons of videotapes is
$$.02x^3 + x^2 - 2x + 520 \qquad (0 \le x \le 65)$$
and the revenue from selling them is
$$-.0027x^4 + .27x^3 - 5.2x^2 + 68x.$$
Find a polynomial that expresses the profit on the sale of x cartons.

42. The volume V of the frustum of a square pyramid, as shown in the figure, is given by
$$V = \frac{h(a^2 + ab + b^2)}{3}$$

Suppose the side of the base b is twice as long as the side of the top a.
(a) Find a formula for the volume in terms of a and h.
(b) Find a formula for the volume in terms of b and h.

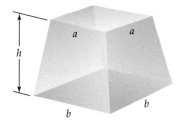

In Questions 43–54, factor the given expression.

43. $x^2 + 5x + 4$

44. $x^2 - 4x + 3$

45. $x^2 - 2x - 8$

46. $x^2 - 10x + 25$

47. $4x^2 - 9$

48. $49y^2 - \frac{1}{4}$

49. $4x^2 - 4x - 15$

50. $6x^2 - x - 2$

51. $3x^3 + 5x^2 - 2x$

52. $2x^3 + 4x^2 - 6x$

53. $x^4 - 1$

54. $x^3 - 8$

55. Which of these statements is *true*?
  (a) $(4x)^3 = 4x^3$
  (b) $(x + y)^2 = x^2 + y^2$
  (c) $(a + b)(a^2 + b^2) = a^3 + b^3$
  (d) $x - (2y + 6) = x - 2y + 6$
  (e) none of the above

56. Which of these statements is *false*?
  (a) $3x - (2x - 7) = x + 7$
  (b) $x^3 - 27 = (x - 3)(x^2 + 3x + 9)$
  (c) $\left(2\sqrt{u}\right)^4 = 16u^2$
  (d) $\left(3\sqrt[4]{v}\right)^2 = 9v^2$
  (e) None of the above

In Questions 57–70, compute and simplify.

57. $\dfrac{2a + 4a^2}{2a}$

58. $\dfrac{b^2 - b - 6}{b + 2}$

59. $\dfrac{4x - 7}{8x} - \dfrac{3x - 5}{12x}$

60. $\dfrac{3y + 6}{y} + \dfrac{y - 2}{3}$

61. $1 - \dfrac{x}{x - y}$

62. $\dfrac{2x - 3}{x - 2} + \dfrac{x}{x + 2}$

63. $\dfrac{x^2 - 3x + 2}{x^2 - 4} \cdot \dfrac{x + 2}{x - 3}$

64. $\dfrac{9x - 18}{6x + 12} \cdot \dfrac{3x + 6}{15x - 30}$

65. $\dfrac{12r + 24}{36r - 36} \div \dfrac{6r + 12}{8r - 8}$

66. $\dfrac{4a + 12}{2a - 10} \div \dfrac{a^2 - 9}{a^2 - a - 20}$

67. $\dfrac{6r - 18}{9r^2 + 6r - 24} \cdot \dfrac{12r - 16}{4r - 12}$

68. $\dfrac{k^2 - k - 6}{k^2 + k - 12} \cdot \dfrac{k^2 + 3k - 4}{k^2 + 2k - 3}$

69. $\dfrac{1 + \dfrac{1}{x}}{1 - \dfrac{1}{x}}$

70. $\dfrac{2 - \dfrac{2}{y}}{2 + \dfrac{2}{y}}$

71. Rationalize the numerator and simplify:

$$\frac{\sqrt{2x + 2h + 1} - \sqrt{2x + 1}}{h}.$$

72. Rationalize the denominator:

$$\frac{5}{\sqrt{x} - 3}.$$

Discovery Project 0

Real Numbers in the Real World

Mathematical calculations arise throughout our technological society. They may range from a simple calculation of the sales tax to the extremely complex logic and calculation used in modern aircraft navigation systems. We mostly ignore the presence of these calculations and simply do what computer software tells us to do. When the software is written, however, decisions must be made to account for the fact that measurement and commerce do not use nonrepeating nonterminating decimals; in other words, people use only rational numbers, not the whole family of real numbers. In fact, in most cases we are restricted to a very small selection of noninteger rational numbers.

1. Suppose you are constructing a triangular display case with a right angle, and that you want the legs of the triangle to be 35 cm each. How long should the long side be? Answer this question using the Pythagorean Theorem and assume that your measuring device is accurate to the nearest millimeter.

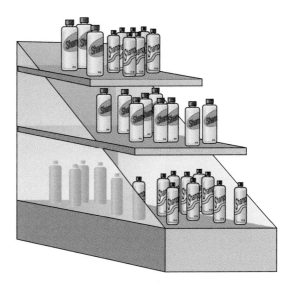

2. Suppose you are constructing a triangular display case with a right angle, and that you want the legs of the triangle to be 14 inches each. If your measuring device is accurate to the nearest sixteenth of an inch, how long should the third side be?

   In your answers to Questions 1 and 2, you had to approximate an irrational number, such as $\sqrt{70}$, and round your answer to meet the required tolerance (nearest millimeter or nearest sixteenth of an inch). In this case, rounding does not cause any difficulty, but when a rounded value is used in subsequent calculations, problems can arise. One situation in which such difficulties occur is in compound interest calculations, when amounts are rounded to the nearest penny. For example, suppose you borrow $50 at an interest rate of 1.5%, compounded monthly. At the end of the first month, the amount of interest is

$$.015 \times 50 = .75$$

so that you now owe $50.75. At the end of the second month, you owe interest on this entire amount, namely,

$$.015 \times 50.75 = .76125.$$

This can be rounded down to .76 or up to .77. Depending on which choice is made, you owe

$$\$50.75 + .76 = \$51.51 \qquad \text{or} \qquad \$50.75 + .77 = \$51.52.$$

Either way, there is a problem: the balance due is either less than or greater than the amount you actually owe. A difference of a penny may not matter now, but after a year or so, the difference may be several cents. Even that won't seriously affect you, but for a bank that has thousands of loans, the difference can be a substantial amount of money.

3. Complete the following three tables, which track the outstanding balance on a $50 loan (with deferred payment) at an interest rate of 1.5%, compounded monthly (a common rate for cash advances on a credit card).
   (a) Round interest amounts down to the next penny.

Month	Old Balance	Interest	New Balance
0	$50.00	$0.75	$50.75
1	$50.75	$0.76	$51.51
2	$51.51		
3			
4			
5			
6			
7			
8			
9			
10			
11			
12			

(b) Round interest amounts up to the next penny.

Month	Old Balance	Interest	New Balance
0	$50.00	$0.75	$50.75
1	$50.75	$0.77	$51.52
2	$51.52		
3			
4			
5			
6			
7			
8			
9			
10			
11			
12			

(c) Round to the nearest penny [.014 rounds to .01; .015 rounds to .02].

Month	Old Balance	Interest	New Balance
0	$50.00	$0.75	$50.75
1	$50.75	$0.76	$51.51
2	$51.51	$0.77	
3			
4			
5			
6			
7			
8			
9			
10			
11			
12			

(d) How do the results in parts (a)–(c) compare? Which method is best for you, as a consumer? Which one is best for the bank, and how much do they stand to make if they have 250,000 such accounts?

4. As we shall see in Section 5.2, the formula for determining the total amount owed at interest rate r per period, after n periods, is $A = P(1 + r)^n$. In our case, $r = .015$ and $n = 12$. How does the answer from the formula compare with your results in Question 3?

Chapter

1

Graphs and Technology

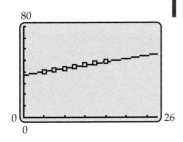

Hamburger and fries—will there be anything else?

The average number of take-out meals purchased (per person) each year increased by almost 50% from 1984 to 1996. Suppose this trend continues and you have the data for selected years during 1984–1996. Then you can use your calculator to construct a linear model to predict the number of take-out meals (per person) in future years. See Exercise 21 on page 116.

59

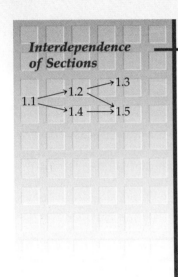

Interdependence of Sections

Chapter Outline

This chapter begins with the essential facts about the coordinate plane and graphs that will be used throughout the text. The rest of the chapter explores graphing technology and introduces techniques that will enable you to solve complicated equations easily and to deal effectively with real-world problems.

1.1 The Coordinate Plane

Just as real numbers are identified with points on the number line, ordered *pairs* of real numbers can be identified with points in the plane. To do this, draw two number lines in the plane, one vertical and one horizontal, as in Figure 1–1. The horizontal line is usually called the **x-axis** and the vertical line the **y-axis,** but other letters may be used if desired. The point where the axes intersect is the **origin.** The axes divide the plane into four regions, called **quadrants,** that are numbered as in Figure 1–1.

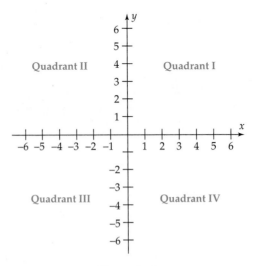

Figure 1–1

If *P* is a point in the plane, draw vertical and horizontal lines through *P* to the coordinate axes, as shown in Figure 1–2. These lines intersect the *x*-axis at some number *c* and the *y*-axis at *d*. We say that *P* has **coordinates** (*c*, *d*). The number *c* is the **x-coordinate** of *P* and *d* is the **y-coordinate** of *P*. The plane is said to have a **rectangular** (or **Cartesian**) **coordinate system.**

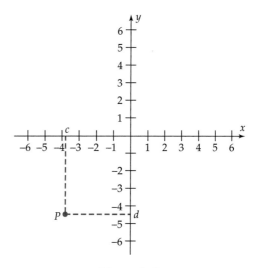

Figure 1–2

Example 1 Plot the points (4, −3), (−5, 2), (2, −5), (0, 13/3), (2, 2), and (−4, −1.2).

Solution   You can think of the coordinates of a point as directions for locating it. To find (4, −3), for instance, start at the origin and move 4 units to the right along the *x*-axis (positive direction), then move 3 units downward (negative direction), as shown in Figure 1–3. To find (−5, 2), start at the origin and move 5 units to the left along the *x*-axis (negative direction), then move 2 units upward (positive direction). The other points are plotted similarly, as shown in Figure 1–3. ■

CAUTION

The coordinates of a point are an *ordered* pair. Figure 1–3 shows that the point *P* with coordinates (−5, 2) is quite different from the point *Q* with coordinates (2, −5). The same numbers (2 and −5) occur in both cases, but in *different order.*

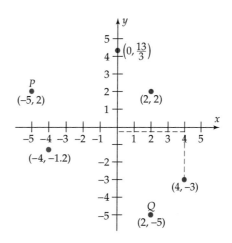

Figure 1–3

Example 2   The following table, from the Federal Election Commission, shows the total amount of money (in millions of dollars) contributed to all congressional candidates in selected years.

Year	1988	1990	1992	1994	1996
Amount	276	284	392	418	500

One way to represent this data graphically is to represent each year's total by a point; for instance, (1988, 276) and (1990, 284). Alternatively, to avoid using very large numbers, we can let x be the number of years since 1988, so that $x = 0$ is 1988, $x = 2$ is 1990, and so on. In this case, we plot the points (0, 276), (2, 284), etc., to obtain the **scatter plot** in Figure 1–4. Connecting these data points with line segments produces the **line graph** in Figure 1–5. ■

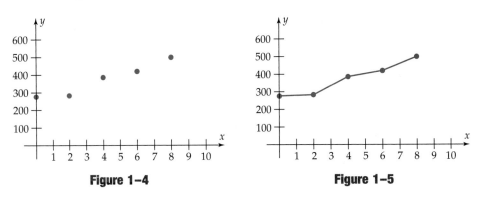

Figure 1–4   Figure 1–5

We shall often identify a point with its coordinates and refer, for example, to the point (2, 3). When dealing with several points simultaneously, it is customary to label the coordinates of the first point (x_1, y_1), the second point (x_2, y_2), the third point (x_3, y_3), and so on.* It's now easy to compute the distance between any two points:

The Distance Formula

The distance between points (x_1, y_1) and (x_2, y_2) is
$$\sqrt{(x_1 - x_2)^2 + (y_1 - y_2)^2}.$$

Before proving the distance formula, we shall see how it is used.

---

*"x_1" is read "x-one" or "x-sub-one"; it is a *single symbol* denoting the first coordinate of the first point, just as c denotes the first coordinate of (c, d). Analogous remarks apply to y_1, x_2, etc.

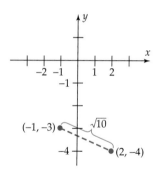

Figure 1–6

Example 3 To find the distance between the points $(-1, -3)$ and $(2, -4)$ in Figure 1–6, substitute $(-1, -3)$ for (x_1, y_1) and $(2, -4)$ for (x_2, y_2) in the distance formula:

Distance formula: $\text{distance} = \sqrt{(x_1 - x_2)^2 + (y_1 - y_2)^2}$

Substitute: $= \sqrt{(-1 - 2)^2 + (-3 - (-4))^2}$

Simplify: $= \sqrt{(-3)^2 + (-3 + 4)^2}$

$= \sqrt{9 + 1} = \sqrt{10}.$

The order in which the points are used in the distance formula doesn't make a difference. If we substitute $(2, -4)$ for (x_1, y_1) and $(-1, -3)$ for (x_2, y_2), we get the same answer:

$$\sqrt{[2 - (-1)]^2 + [-4 - (-3)]^2} = \sqrt{3^2 + (-1)^2} = \sqrt{10}. \quad \blacksquare$$

Example 4 In a Cubs game at Wrigley Field, a fielder catches the ball near the right-field corner and throws it to second base. The right-field corner is 353 feet from home plate along the right-field foul line. If the fielder is 5 feet from the outfield wall and 5 feet from the foul line, how far must he throw the ball?

Solution Imagine that the playing field is placed on the coordinate plane, with home plate at the origin and the right-field foul line along the positive x-axis, as shown in Figure 1–7 (not to scale).

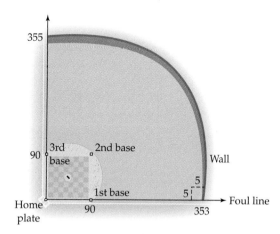

Figure 1–7

Since the four bases form a square whose sides measure 90 ft each, second base has coordinates $(90, 90)$. The fielder is located 5 ft from the wall, so his x-coordinate is $353 - 5 = 348$. His y-coordinate is 5, since he is 5 ft from the foul line. Therefore, the distance he throws is the distance from (348, 5) to (90, 90), which can be found as follows.

Distance formula: $\text{distance} = \sqrt{(x_1 - x_2)^2 + (y_1 - y_2)^2}$

Substitute: $= \sqrt{(348 - 90)^2 + (5 - 90)^2}$

Simplify: $= \sqrt{258^2 + (-85)^2}$

              $= \sqrt{73{,}789} \approx 271.6 \text{ ft.}$

Therefore, he must throw about 272 ft.   ■

Proof of the Distance Formula   Figure 1–8 shows typical points P and Q in the plane. We must find length d of line segment PQ.

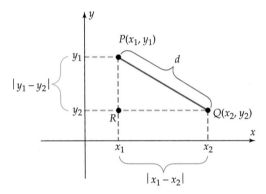

Figure 1–8

As shown in Figure 1–8, the length of RQ is the same as the distance from x_1 to x_2 on the x-axis (number line), namely, $|x_1 - x_2|$. Similarly, the length of PR is the same as the distance from y_1 to y_2 on the y-axis, namely, $|y_1 - y_2|$. According to the Pythagorean Theorem* the length d of PQ is given by:

$$(\text{length } PQ)^2 = (\text{length } RQ)^2 + (\text{length } PR)^2$$
$$d^2 = |x_1 - x_2|^2 + |y_1 - y_2|^2.$$

Since $|c|^2 = |c| \cdot |c| = |c^2| = c^2$ (because $c^2 \geq 0$), this equation becomes:

$$d^2 = (x_1 - x_2)^2 + (y_1 - y_2)^2.$$

Since the length d is nonnegative we must have

$$d = \sqrt{(x_1 - x_2)^2 + (y_1 - y_2)^2}. \quad ■$$

The distance formula can be used to prove the following useful fact (see Exercise 68).

*See the Geometry Review Appendix.

The Midpoint Formula

The midpoint of the line segment from (x_1, y_1) to (x_2, y_2) is

$$\left(\frac{x_1 + x_2}{2}, \frac{y_1 + y_2}{2} \right).$$

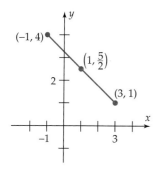

Figure 1–9

Example 5 To find the midpoint of the segment joining $(-1, 4)$ and $(3, 1)$, use the formula in the box with $x_1 = -1$, $y_1 = 4$, $x_2 = 3$, and $y_2 = 1$. The midpoint is

$$\left(\frac{x_1 + x_2}{2}, \frac{y_1 + y_2}{2} \right) = \left(\frac{-1 + 3}{2}, \frac{4 + 1}{2} \right) = \left(1, \frac{5}{2} \right)$$

as shown in Figure 1–9. ∎

Graphs

A **graph** is a set of points in the plane. Some graphs are based on data points, such as those in Example 2 above. Other graphs arise from equations, as follows. A **solution** of an equation in variables x and y is a pair of numbers such that the substitution of the first number for x and the second for y produces a true statement. For instance, $(3, -2)$ is a solution of

$$5x + 7y = 1$$

because

$$5(3) + 7(-2) = 1$$

and $(1, 4)$ is *not* a solution because $5(1) + 7(4) \neq 1$. The **graph of an equation** in two variables is the set of points in the plane whose coordinates are solutions of the equation. Thus, the graph is a *geometric picture of the solutions*.

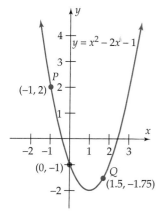

Figure 1–10

Example 6 The graph of $y = x^2 - 2x - 1$ is shown in Figure 1–10. Verify that the coordinates of P and Q actually are solutions of the equation.

Solution For the point $(-1, 2)$, substitute -1 for x in the equation:

$$y = x^2 - 2x - 1$$
$$y = (-1)^2 - 2(-1) - 1 = 1 + 2 - 1 = 2.$$

Therefore, $(-1, 2)$ is a solution. Similarly, substituting $x = 1.5$ shows that

$$y = x^2 - 2x - 1 = 1.5^2 - 2(1.5) - 1 = 2.25 - 3 - 1 = -1.75.$$

Hence, $(1.5, -1.75)$ is also a solution. ∎

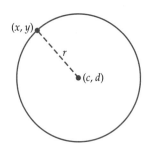

Figure 1–11

Circles

If (c, d) is a point in the plane and r a positive number, then the **circle with center (c, d) and radius r** consists of all points (x, y) that lie r units from (c, d), as shown in Figure 1–11. According to the distance formula, the statement "the distance from (x, y) to (c, d) is r units" is equivalent to:

$$\sqrt{(x - c)^2 + (y - d)^2} = r.$$

Squaring both sides shows that (x, y) satisfies this equation:

$$(x - c)^2 + (y - d)^2 = r^2.$$

Reversing the procedure shows that any solution (x, y) of this equation is a point on the circle. Therefore,

---

Equation of the Circle ▶

The circle with center (c, d) and radius r is the graph of

$$(x - c)^2 + (y - d)^2 = r^2.$$

We say that $(x - c)^2 + (y - d)^2 = r^2$ is the **equation of the circle** with center (c, d) and radius r.

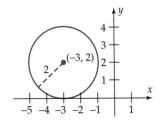

Figure 1–12

Example 7   Find the equation of the circle with center $(-3, 2)$ and radius 2 and sketch its graph.

Solution   Here the center is $(c, d) = (-3, 2)$ and the radius is $r = 2$, so the equation of the circle is:

$$(x - c)^2 + (y - d)^2 = r^2$$
$$[(x - (-3)]^2 + (y - 2)^2 = 2^2$$
$$(x + 3)^2 + (y - 2)^2 = 4.$$

Its graph is in Figure 1–12.   ■

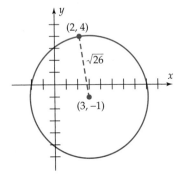

Figure 1–13

Example 8   Find the equation of the circle with center $(3, -1)$ that passes through $(2, 4)$ and sketch its graph.

Solution   We must first find the radius. Since $(2, 4)$ is on the circle, the radius is the distance from $(2, 4)$ to $(3, -1)$ as shown in Figure 1–13, namely,

$$\sqrt{(2 - 3)^2 + (4 - (-1))^2} = \sqrt{1 + 25} = \sqrt{26}.$$

The equation of the circle with center at $(3, -1)$ and radius $\sqrt{26}$ is

$$(x - 3)^2 + (y - (-1))^2 = (\sqrt{26})^2$$
$$(x - 3)^2 + (y + 1)^2 = 26$$
$$x^2 - 6x + 9 + y^2 + 2y + 1 = 26$$
$$x^2 + y^2 - 6x + 2y - 16 = 0. \quad\blacksquare$$

When the center of a circle of radius r is at the origin $(0, 0)$, its equation takes a simpler form:

$$(x - c)^2 + (y - d)^2 = r^2$$
$$(x - 0)^2 + (y - 0)^2 = r^2$$
$$x^2 + y^2 = r^2.$$

When $r = 1$, for example, the graph of the equation $x^2 + y^2 = 1$ is the circle of radius 1 with center at the origin, as shown in Figure 1–14. This circle is called the **unit circle.**

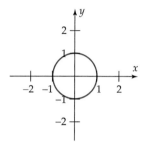

Figure 1–14

Exercises 1.1

1. Find the coordinates of points A–I.

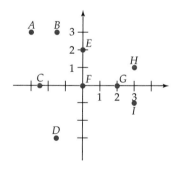

In Exercises 2–5, find the coordinates of the point P.

2. P lies 4 units to the left of the y-axis and 5 units below the x-axis.

3. P lies 3 units above the x-axis and on the same vertical line as $(-6, 7)$.

4. P lies 2 units below the x-axis and its x-coordinate is three times its y-coordinate.

5. P lies 4 units to the right of the y-axis and its y-coordinate is half its x-coordinate.

In Exercises 6–8, sketch a scatter plot and a line graph of the given data. In each case, let the x-axis run from 0 to 10, with $x = 0$ corresponding to 1990, $x = 1$ to 1991, etc.

6. The table shows the number of climbers in recent years at Mt. McKinley (*Source:* National Park Service). Mark the y-axis in units of 100.

Year	1992	1993	1994	1995	1996	1997
Climbers	935	1108	1277	1220	1148	1109

7. The table shows the number of missions by National Park rangers to rescue climbers on Mt. McKinley each year (*Source:* National Park Service).

Year	1992	1993	1994	1995	1996	1997
Missions	22	14	20	12	13	10

8. The tuition and fees at public four-year colleges in the fall of each year are shown in the table (*Source: National Center for Education Statistics (1990–1994) and the College Board (1995–1997)*).

Year	Tuition and Fees	Year	Tuition and Fees
1990	$2159	1994	$2681
1991	$2410	1995	$2811
1992	$2349	1996	$2975
1993	$2537	1997	$3111

9. The graph, which is based on data from the U.S. Department of Energy, shows approximate average gasoline prices (in cents per gallon) between 1985 and 1996, with $x = 0$ corresponding to 1985.
   (a) Estimate the average price in 1987 and in 1995.
   (b) What was the approximate percentage increase in the average price from 1987 to 1995?
   (c) In what years was the average price at least $1.10 per gallon?

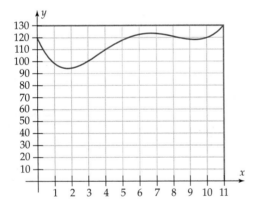

10. The graph, which is based on data from the U.S. Department of Commerce, shows the approximate amount of personal savings as a percent of disposable income between 1960 and 1995, with $x = 0$ corresponding to 1960.
    (a) In what years during this period were personal savings largest and smallest (as a percent of disposable income)?
    (b) In what years were personal savings at least 7% of disposable income?

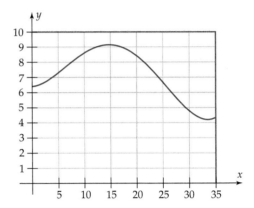

11. (a) If the first coordinate of a point is greater than 3 and its second coordinate is negative, in what quadrant does it lie?
    (b) What is the answer in part (a) if the first coordinate is less than 3?

12. In what quadrant(s) does a point lie if the product of its coordinates is
    (a) positive?   (b) negative?

13. (a) Plot the points $(3, 2)$, $(4, -1)$, $(-2, 3)$, and $(-5, -4)$.
    (b) Change the sign of the y-coordinate in each of the points in part (a) and plot these new points.
    (c) Explain how the points (a, b) and $(a, -b)$ are related graphically. [*Hint:* What are their relative positions with respect to the x-axis?]

14. (a) Plot the points $(5, 3)$, $(4, -2)$, $(-1, 4)$, and $(-3, -5)$.
    (b) Change the sign of the x-coordinate in each of the points in part (a) and plot these new points.
    (c) Explain how the points (a, b) and $(-a, b)$ are related graphically. [*Hint:* What are their relative positions with respect to the y-axis?]

In Exercises 15–22, find the distance between the two points and the midpoint of the segment joining them.

15. $(-3, 5), (2, -7)$ 16. $(2, 4), (1, 5)$

17. $(1, -5), (2, -1)$ 18. $(-2, 3), (-3, 2)$

19. $(\sqrt{2}, 1), (\sqrt{3}, 2)$ 20. $(-1, \sqrt{5}), (\sqrt{2}, -\sqrt{3})$

21. $(a, b), (b, a)$ 22. $(s, t), (0, 0)$

23. According to the Information Technology Industry Council, about 12 million personal computers

were sold in the U.S. in 1992 and about 36 million in 1998.
- **(a)** Represent the data graphically by two points.
- **(b)** Find the midpoint of the line segment joining these points.
- **(c)** How might this midpoint be interpreted? What assumptions, if any, are needed to make this interpretation?

24. Suppose a baseball playing field is placed on the coordinate plane, as in Example 4.
- **(a)** Find the coordinates of first and third base.
- **(b)** If the left fielder is at the point (50, 325), how far is he from first base?
- **(c)** How far is the left fielder in part (b) from the right fielder, who is at the point (280, 20)?

25. A standard football field is 100 yards long and $53\frac{1}{3}$ yards wide. The quarterback, who is standing on the 10-yard line, 20 yards from the left sideline, throws the ball to a receiver who is on the 45-yard line, 5 yards from the right sideline, as shown in the figure.
- **(a)** How long was the pass? [*Hint:* Place the field in the first quadrant of the coordinate plane, with the left sideline on the y-axis and the goal line on the x-axis. What are the coordinates of the quarterback and the receiver?]
- **(b)** A player is standing halfway between the quarterback and the receiver. What are his coordinates?

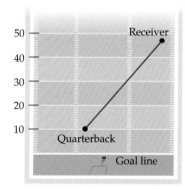

26. How far is the quarterback in Exercise 25 from a player who is on the 50-yard line, halfway between the sidelines?

In Exercises 27–32, determine whether the point is on the graph of the given equation.

27. $(1, -2)$; $3x - y - 5 = 0$
28. $(2, -1)$; $x^2 + y^2 - 6x + 8y = -15$
29. $(6, 2)$; $3y + x = 12$

30. $(1, -2)$; $3x + y = 12$
31. $(3, 4)$; $(x - 2)^2 + (y + 5)^2 = 4$
32. $(1, -1)$; $\dfrac{x^2}{2} + \dfrac{y^2}{3} = 1$

In Exercises 33–36, find the equation of the circle with given center and radius r.

33. $(-3, 4)$; $r = 2$ 34. $(-2, -1)$; $r = 3$
35. $(0, 0)$; $r = \sqrt{2}$ 36. $(5, -2)$; $r = 1$

In Exercises 37–40, sketch the graph of the equation.

37. $(x - 2)^2 + (y - 4)^2 = 1$
38. $(x + 1)^2 + (y - 3)^2 = 9$
39. $(x - 5)^2 + (y + 2)^2 = 5$
40. $(x + 6)^2 + y^2 = 4$

In Exercises 41–43, show that the three points are the vertices of a right triangle and state the length of the hypotenuse. [You may assume that a triangle with sides of lengths a, b, c is a right triangle with hypotenuse c provided that $a^2 + b^2 = c^2$.]

41. $(0, 0), (1, 1), (2, -2)$
42. $\left(\dfrac{\sqrt{2}}{2}, 0\right), \left(\dfrac{\sqrt{2}}{2}, \dfrac{\sqrt{2}}{2}\right), (0, 0)$
43. $(3, -2), (0, 4), (-2, 3)$
44. What is the perimeter of the triangle with vertices $(1, 1), (5, 4),$ and $(-2, 5)$?

In Exercises 45–52, find the equation of the circle.

45. Center (2, 2); passes through the origin.
46. Center $(-1, -3)$; passes through $(-4, -2)$.
47. Center (1, 2); intersects x-axis at -1 and 3.
48. Center (3, 1); diameter 2.
49. Center $(-5, 4)$; tangent (touching at one point) to the x-axis.
50. Center $(2, -6)$; tangent to the y-axis.
51. Endpoints of diameter are (3, 3) and $(1, -1)$.
52. Endpoints of diameter are $(-3, 5)$ and $(7, -5)$.

53. One diagonal of a square has endpoints $(-3, 1)$ and $(2, -4)$. Find the endpoints of the other diagonal.
54. Find the vertices of all possible squares with this property: Two of the vertices are (2, 1) and (2, 5). [*Hint:* There are three such squares.]
55. Find a number x such that $(0, 0), (3, 2),$ and $(x, 0)$ are the vertices of an isosceles triangle, neither of whose two equal sides lies on the x-axis.

56. Do Exercise 55 assuming that one of the two equal sides lies on the positive *x*-axis.

In Exercises 57–60, determine which of graphs A, B, C best describes the given situation.

57. You have a job that pays a fixed salary for the week. The graph shows your salary.

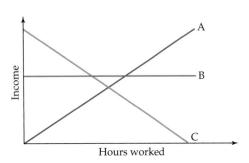

58. You have a job that pays an hourly wage. The graph shows your salary.

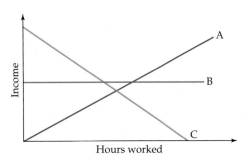

59. You inflate a birthday balloon. The graph shows the amount of air in the balloon.

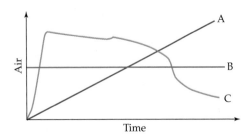

60. Alison's wading pool is filled with a hose by her big sister Emily and she plays in the pool.

When she is finished, Emily empties the pool. The graph shows the water level of the pool.

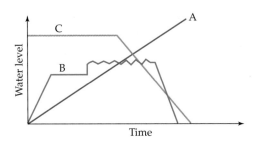

61. The graph shows the average life expectancy of an American born in selected years (*Source:* National Center for Health Statistics). Estimate the life expectancy of someone born in
  (a) 1950.   **(b)** 1990.
  (c) During what 20-year period did life expectancy increase the most?

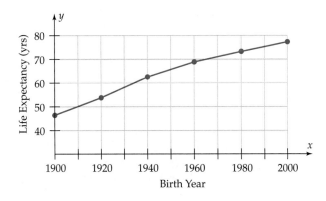

62. The graph (from the *Economist*, February 7, 1998) shows life expectancy at birth during the last half-century in several countries.
  (a) Which country had the highest life expectancy at birth in 1990? What was it?
  (b) Which country has the lowest projected life expectancy for the year 2000?
  (c) Which country seems to have suffered the greatest relative decline in life expectancy over the last 30 years?

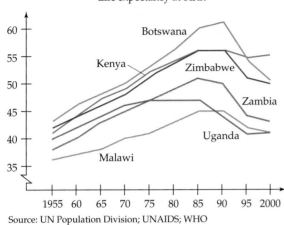

Dying Ever Younger
Life expectancy at birth

Source: UN Population Division; UNAIDS; WHO

63. Show that the midpoint M of the hypotenuse of a right triangle is equidistant from the vertices of the triangle. [*Hint:* Place the triangle in the first quadrant of the plane, with right angle at the origin so that the situation looks like the figure.]

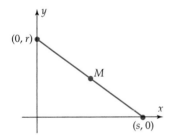

64. Show that the diagonals of a parallelogram bisect each other. [*Hint:* Place the parallelogram in the first quadrant with a vertex at the origin and one side along the x-axis, so that the situation looks like the figure.]

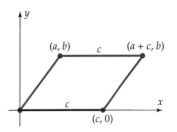

Thinkers

65. For each nonzero real number k, the graph of $(x - k)^2 + y^2 = k^2$ is a circle. Describe all possible such circles.

66. Suppose every point in the coordinate plane is moved 5 units straight up.
    (a) To what point does each of these points go: $(0, -5)$, $(2, 2)$, $(5, 0)$, $(5, 5)$, $(4, 1)$?
    (b) Which points go to each of the points in part (a)?
    (c) To what point does (a, b) go?
    (d) To what point does $(a, b - 5)$ go?
    (e) What point goes to $(-4a, b)$?
    (f) What points go to themselves?

67. Let (c, d) be any point in the plane with $c \neq 0$. Prove that (c, d) and $(-c, -d)$ lie on the same straight line through the origin, on opposite sides of the origin, the same distance from the origin. [*Hint:* Find the midpoint of the line segment joining (c, d) and $(-c, -d)$.]

68. *Proof of the Midpoint Formula.* Let P and Q be the points (x_1, y_1) and (x_2, y_2) respectively and let M be the point with coordinates $\left(\dfrac{x_1 + x_2}{2}, \dfrac{y_1 + y_2}{2} \right)$. Use the distance formula to compute the following:
    (a) The distance d from P to Q;
    (b) The distance d_1 from M to P;
    (c) The distance d_2 from M to Q.
    (d) Verify that $d_1 = d_2$.
    (e) Show that $d_1 + d_2 = d$. [*Hint:* Verify that $d_1 = \frac{1}{2}d$ and $d_2 = \frac{1}{2}d$.]
    (f) Explain why parts (d) and (e) show that M is the midpoint of PQ.

1.2 Graphs and Graphing Calculators

The traditional method of graphing an equation "by hand" is as follows: Construct a table of values with a reasonable number of entries, plot the corresponding points, and use whatever algebraic or other information is available to make an "educated guess" about the rest.

Example 1   The graph of $y = x^2$ consists of all points (x, x^2), where x is a real number. You can easily construct a table of values and plot the corresponding points, as in Figure 1–15. These points suggest that the graph looks like the one in Figure 1–16, which is obtained by connecting the plotted points and extending the graph upward.   ■

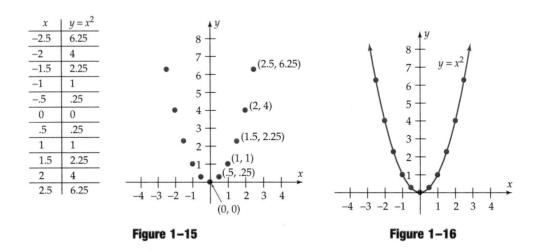

x	$y = x^2$
−2.5	6.25
−2	4
−1.5	2.25
−1	1
−.5	.25
0	0
.5	.25
1	1
1.5	2.25
2	4
2.5	6.25

Figure 1–15

Figure 1–16

Graphing technology generally improves speed, convenience, and accuracy in graphing, provided you have sufficient mathematical and technical knowledge to get the most out of your calculator or computer. The rest of this section deals with the latter issue.

Viewing Windows

The first step in graphing an equation with a calculator is to choose a **viewing window** (or **viewing rectangle**), that is, the portion of the coordinate plane that appears on the screen.

Example 2   The viewing window for the graph in Figure 1–17 is the rectangular region indicated by the dashed blue lines. It includes all points (x, y) whose coordinates satisfy $-4 \leq x \leq 5$ and $-3 \leq y \leq 6$. To display this viewing window on a calculator, press the WINDOW (or RANGE or V-WINDOW or PLOT SETUP) key* and enter the appropriate numbers, as shown in Figure 1–18 for a TI-83 (other calculators are similar). The settings Xscl = 1 and Yscl = 1 put the tick marks 1 unit apart on each axis.† This is usually the best setting for small viewing windows, but not for large ones.   ■

*On TI-85/86, press GRAPH first, then WINDOW or RANGE.
†On HP-38 "Xscl" and "Yscl" are called "Xtick" and "Ytick". Some calculators do not have an Xres setting. On those that do, it should normally be set at 1 (or at "detail" on HP-38).

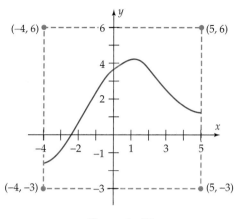

Figure 1–17   **Figure 1–18**

NOTE The heading
GRAPHING EXPLORATION
indicates that you are to use
your graphing calculator or
computer software as directed
to complete the discussion.

GRAPHING EXPLORATION

Set up a viewing window with $-200 \le x \le 200$ and $-30 \le y \le 30$ and Xscl = 1, Yscl = 1. Then press GRAPH (or PLOT) to display this window on the screen. (If a graph also appears, ignore it; just look at the axes.) Can you distinguish the tick marks on each axis? Now press WINDOW (or RANGE or PLOT SETUP) and change the settings to Xscl = 20 and Yscl = 5, so that adjacent tick marks are 20 units apart on the x-axis and 5 units apart on the y-axis, and press GRAPH (or PLOT) again. Can you distinguish the tick marks now?

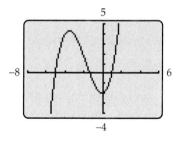

Figure 1–19

In the calculator screens in this book, the viewing window is indicated by the numbers at the ends of the axes. For instance, the window in Figure 1–19 has $-8 \le x \le 6$ and $-4 \le y \le 5$. Since a 14-unit-long segment of the x-axis is shown, the tick marks on the axis are two units apart (Xscl = 2). Similarly, the tick marks on the 9-unit-long segment of the y-axis are 1 unit apart (Yscl = 1).

Graphing with a Calculator

A graphing calculator or computer graphing program graphs in the same way you would graph by hand, but uses many more points. It plots 95 or more points and simultaneously connects them with line segments. A step-by-step procedure for calculator graphing is presented in the next example.

Example 3 Use a calculator to graph the equation

$$2x^3 - 8x - 2y + 4 = 0.$$

Solution

Step 1 *Choose a (preliminary) viewing window.* Since we don't know yet where the graph lies in the plane, we'll try the viewing window with $-10 \le x \le 10$ and $-10 \le y \le 10$. If that doesn't work, we'll try another window.

Step 2 *Enter the equation in the calculator.* Since calculators can only graph equations in the form $y =$ expression in x, we first solve the given equation for y:

$$2x^3 - 8x - 2y + 4 = 0$$

Rearrange terms: $\quad -2y = -2x^3 + 8x - 4$

Divide by -2: $\quad y = x^3 - 4x + 2.$

Now call up the **equation memory** (press Y = on TI and Sharp 9600, SYMB on HP-38, or GRAPH (Main Menu) on Casio 9850). Next, use the Technology Tip in the margin to enter the equation, as shown in Figure 1–20 for TI-83 (other calculators are similar).

Step 3 *Graph the equation.* Press GRAPH (or PLOT or DRAW) to obtain Figure 1–21. Because of the limited resolution of a calculator screen, the graph appears to consist of short adjacent line segments rather than a smooth unbroken curve.

Technology Tip

When entering an equation for graphing, use the "variable" key rather than the ALPHA X key. The variable key has a label like X,T,θ or X,T,θ,n or X,θ,T or x-VAR or X/θ/T/n. On TI-89, however, use the X key on the keyboard.

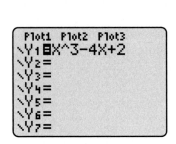

Figure 1–20

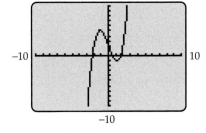

Figure 1–21

Step 4 *If necessary, adjust the viewing window for a better view.* The point where the graph crosses the y-axis isn't clear in Figure 1–21, so we change the viewing window (Figure 1–22) and press GRAPH (or PLOT or DRAW) to obtain Figure 1–23, which shows clearly that the graph crosses the y-axis at 2. ∎

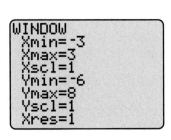

Figure 1–22

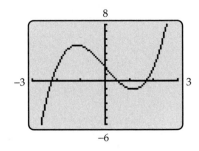

Figure 1–23

To see the points that the calculator actually plotted in Figure 1–23, press TRACE and a flashing cursor appears on the graph.* Use the *left* and *right* arrow keys to move the cursor along the graph. As you do, the coordinates of the point the cursor is on appear at the bottom of the screen (Figure 1–24).

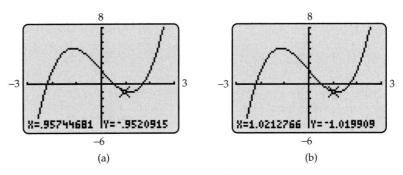

(a)   (b)

Figure 1–24

The trace cursor only displays points that the calculator plotted. In Figure 1–24, it does not land on $(1, -1)$, which is on the graph of $y = x^3 - 4x + 2$ (why?), but was not plotted. Instead it lands on two nearby points that were plotted.

An equation stays in the equation memory until you delete it. If several equations are in the memory and you want to graph only one of them, you must turn that equation "on" and the others "off," as explained in the Tip in the margin.

Now that basic graphing is clear, we'll examine some useful calculator graphing tools.

Technology Tip

On most calculators an equation is "on" if its equal sign is shaded and "off" if its equal sign is clear. On TI-89 and HP-38 an equation that is "on" has a check mark ✓ next to it.

Only the equations that are "on" will be graphed when you press GRAPH (or PLOT). To turn an equation "on" or "off" place the cursor on its equal sign and press ENTER (TI-83, Sharp 9600) or move the cursor to the equation and press SELECT (TI-86, Casio 9860) or CHECK (HP-38) or F4 (TI-89).

Example 4 The Cortopassi Computer Company can produce a maximum of 100,000 computers a year. Their annual profit is given by

$$y = -.003x^4 + .3x^3 - x^2 + 5x - 4000,$$

where y is the profit (in thousands of dollars) from selling x thousand computers. Use graphical methods to estimate how many computers should be sold in order to make the largest possible profit.

Solution We first choose a viewing window. The number x of computers is nonnegative and no more than 100,000 can be produced, so that $0 \le x \le 100$ (because x is measured in thousands). The profit y may be positive or negative (the company could lose money). So we try a window with $-25,000 \le y \le 25,000$ and obtain Figure 1–25 on the next page. For each point on the graph,

The x-coordinate is the number of thousands of computers produced;

The y-coordinate is the profit (in thousands) on that number of computers.

*The trace feature is automatically on after graphing on HP-38.

The largest possible profit occurs at the point with the largest y-coordinate, that is, the highest point in the window.

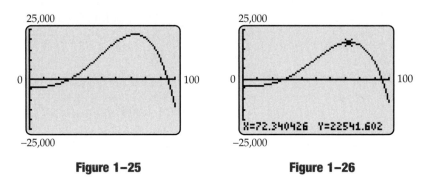

Figure 1–25   **Figure 1–26**

There are three ways to approximate the highest point.

Use the TRACE feature: Move the cursor along the graph, watching until the y-coordinate takes its largest value (Figure 1–26). Because different calculators plot different points, the trace feature on your calculator may produce a slightly different result. Because x and y are measured in thousands, the point (72.340426, 22541.602) indicates that producing about 72,340 computers gives a profit of approximately $22,541,602. Since the trace cursor displays only those points that the calculator actually plotted when it drew the graph, this point is not necessarily the highest point in the window.

Use the ZOOM feature: Select ZOOM-IN on the ZOOM menu: move the cursor to approximately the highest point, and hit ENTER. Figure 1–27 shows the result of zooming in by a factor of 4 and using TRACE again: the highest point appears to be (72.87234, 22547.542), meaning that producing about 72,872 computers gives a profit of about $22,547,542. Using larger zoom factors or zooming-in more than once produces a more accurate answer.

Technology Tip

The ZOOM menu is on the keyboard of TI-83, Sharp 9600, and Casio 9850. It is a submenu of the TI-86/89 GRAPH menu and of the HP-38 PLOT menu.

To set the factors, look for FACT or ZFACT or (SET) FACTORS in the ZOOM menu (or in its MEMORY submenu on TI-83).

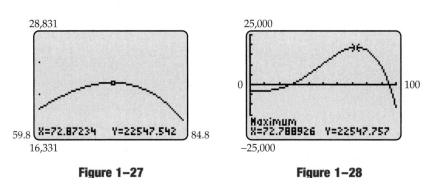

Figure 1–27   **Figure 1–28**

Use the MAXIMUM FINDER: This feature automatically finds the highest point with the greatest degree of accuracy that the calculator is capable of. The maximum finder on a TI-83 produced Figure 1–28, which shows that making about 72,789 computers results in a profit of $22,547,757. Calculus can be used to confirm that this is accurate. ∎

Technology Tip

The graphical maximum finder is in the CALC menu of TI-83 and Sharp 9600, and in the G-SOLVE menu of Casio 9850. It is in the MATH submenu of the TI-89 and TI-86 GRAPH menu (labeled FMAX in the latter). It is labeled EXTREMUM in the HP-38 PLOT FCN menu.

On some TI calculators, you must first select a left (or lower) bound, meaning an x-value to the left of the highest point, and a right (or upper) bound, meaning an x-value to its right, and make an initial guess. On other calculators, you may have to move the cursor near the point you are seeking.

GRAPHING EXPLORATION

Graph $y = .3x^3 + .8x^2 - 2x - 1$ in the window with $-5 \le x \le 5$ and $-5 \le y \le 5$. Use your maximum finder and the Tip in the margin to approximate the coordinates of the highest point to the left of the y-axis. Then use your minimum finder (in the same menu) to approximate the coordinates of the *lowest* point to the right of the y-axis. How do these answers compare with the ones you get by using the trace feature (but not zoom-in)?

Special Viewing Windows

Throughout this book and in the ZOOM menu of most calculators, the viewing window with $-10 \le x \le 10$ and $-10 \le y \le 10$ is called the **standard viewing window** (or the **default window** on Sharp 9600). Although it's often a good place to start when you don't have any other information, it may not always be the best choice.

Example 5 Graph the circle $x^2 + y^2 = 4$ on a calculator.

Solution Solving the equation for y shows that

$$y = \sqrt{4 - x^2} \quad \text{or} \quad y = -\sqrt{4 - x^2}.$$

Graphing both of these equations on the same screen produces the graph of the circle. Choose the standard viewing window and enter both equations in the equation memory (Figure 1–29); note that the equal sign is shaded in both equations, indicating that both will be graphed. Press GRAPH (or PLOT or DRAW) to obtain Figure 1–30. The circle doesn't look round, so we change the viewing window and obtain Figure 1–31, which does look like a circle. ∎

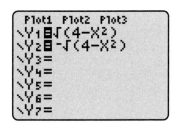

Figure 1–29

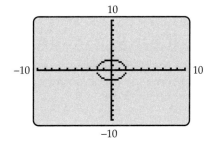

Figure 1–30

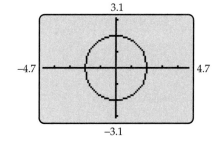

Figure 1–31

The viewing window in Figure 1–31, in which the circle looks round, is an example of a **square viewing window,** one in which a 1-unit segment on the x-axis is the same length on the screen as a 1-unit segment on the y-axis. Because calculator screens are wider than they are high, the y-axis in a square window must be shorter than the x-axis (about two-thirds as long on standard width screens, such as the TI-83 screen in Figure 1–31).* In particular, the standard viewing window ($-10 \le x \le 10$ and $-10 \le y \le 10$) is *not* square because the 20 one-unit segments on the y-axis occupy less space on the screen than the 20 one-unit segments on the x-axis. You should use square windows when you want circles to look round and perpendicular lines to look perpendicular, but use any convenient window for other graphs.

Technology Tip

To change the current viewing window to a square one, use SQUARE (or ZSQUARE or ZOOM-SQR or SQR) in the ZOOM menu.

GRAPHING EXPLORATION

The lines $y = .5x$ and $y = -2x + 2$ are perpendicular, as we shall prove in Section 1.4. Graph them in the standard viewing window. Do they look perpendicular? Now graph them in a square window (see the Tip in the margin). Do they look perpendicular?

The method used to graph the circle $x^2 + y^2 = 4$ in Example 5 can be used to graph any equation that can be solved for y.

Example 6   To graph $12x^2 - 4y^2 + 16x + 12 = 0$, solve the equation for y:

$$4y^2 = 12x^2 + 16x + 12$$

$$y^2 = 3x^2 + 4x + 3$$

$$y = \pm\sqrt{3x^2 + 4x + 3}.$$

Therefore, every point on the graph of the equation is on the graph of either

$$y = \sqrt{3x^2 + 4x + 3} \qquad \text{or} \qquad y = -\sqrt{3x^2 + 4x + 3}.$$

Technology Tip

The screen widths (in pixels) of commonly used calculators are:

TI-83 and Casio 9800: 95

TI-86 and Sharp 9600: 127

Casio 9850: 127

HP-38: 131

TI-89: 159

TI-92: 239

GRAPHING EXPLORATION

Graph the previous two equations on the same screen. The result will be the graph of the original equation.   ∎

When using the trace feature, you probably noticed that your calculator typically graphs points with first coordinates like 2.340425, rather than 2.3 or 2.34. You can sometimes rectify this situation by using the following facts. A calculator screen consists of tiny rectangles, called pixels, which are darkened to indicate a graphed point. Suppose the screen is

---

*It should be about three-fifths as long on TI-86 and about one-half as long on other wide-screen models, such as Casio 9850, HP-38, Sharp 9600, and TI-89.

127 pixels wide (see the last Tip in the margin on page 78). In a viewing window with $0 \le x \le 126$, the 127 pixels on the x-axis represent the 127 numbers $0, 1, 2, \ldots, 126$, which divide the axis into 126 pieces of length 1. Since the trace cursor moves one horizontal pixel at a time, its x-coordinate will change by 1 unit each time you press the arrow key.

Similarly, if the screen is 95 pixels wide, then the 95 pixels on the x-axis divide it into 94 equal pieces. Thus, if the x-axis has a length of 9.4 (for instance, $-4.7 \le x \le 4.7$), then the distance between adjacent pixels (the distance the trace cursor moves each time) is $9.4/94 = .1$. A screen in which the horizontal distance between adjacent pixels is .1 is sometimes called a **decimal window.**

Choosing a viewing window carefully can make the trace feature much more convenient, as the following Exploration demonstrates.

Technology Tip

For a decimal window that is also square, use ZDECIMAL or ZOOMDEC (in the TI-83/89 ZOOM menu), DECIMAL (in the Sharp 9600 ZOOM menu and the HP-38 VIEWS menu), and INIT (in the Casio 9850 V-WINDOW menu). In the TI-86 ZOOM menu, ZDECIMAL gives a decimal window that is not square.

GRAPHING EXPLORATION

Graph $y = (5/9)(x - 32)$, which relates the temperature x in degrees Fahrenheit and the temperature y in degrees Celsius, in a viewing window with $-40 \le y \le 40$ and $0 \le x \le k$, where k is chosen so that adjacent pixels are 1 unit apart (see the last Technology Tip on page 78). Use the trace feature to determine the Celsius temperatures corresponding to 20°F and 77°F.

Now use a window with $32 - k/20 \le x \le 32 + k/20$ (where k is as above). This is a decimal window with $(32, 0)$ at its center. Graph the equation again. Use the trace to determine the Celsius equivalent of 33.8°F.

Complete Graphs

A viewing window is said to display a **complete graph** if it shows all the important features of the graph (peaks, valleys, points where it touches an axis, etc.) and suggests the general shape of the portions of the graph that aren't in the window. Many different windows may show a complete graph. It's usually best to use a window that is small enough to show as much detail as possible.

In later chapters we shall develop algebraic facts that will enable us to know when certain graphs are complete. For the present, however, the best you may be able to do is to try several different windows to see which, if any, appear to display a complete graph.

Example 7 Sketch a complete graph of

$$y = .007x^5 - .2x^4 + 1.332x^3 - .004x^2 + 10.$$

Solution Four different viewing windows for this graph are shown in Figure 1–32 on the next page. Graph (a) (the standard window) is cer-

Technology Tip

Most calculators have an "auto scaling" feature. Once the range of x values has been set, the calculator selects a viewing window that includes all the points on the graph whose x-coordinates are in the chosen range. It's labeled ZOOMFIT or ZFIT in the TI ZOOM menu, AUTO in the Casio 9850 and Sharp 9600 ZOOM menus, and AUTO-SCALE in the HP-38 VIEWS menu. This feature can eliminate some guesswork, but may produce a window so large that it hides some features of the graph.

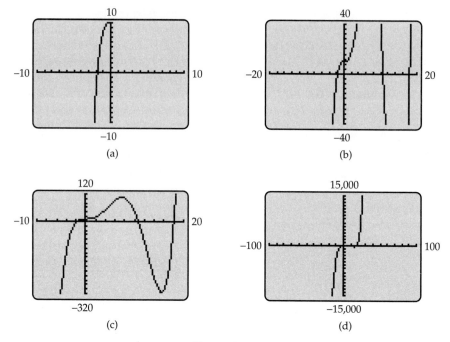

Figure 1–32

tainly not complete since it shows no points to the right of the y-axis. Graph (b) is not complete since it indicates that parts of the graph lie outside the window. Graph (d) tends to confirm what graph (c) suggests, that the graph keeps climbing sharply forever as you move to the right and that it keeps falling sharply forever as you move to the left. However, because of its large scale, graph (d) doesn't show the features of the graph near the origin. So we conclude that graph (c) is probably a complete graph since it shows important features (twists and turns) near the origin, as well as suggesting the shape of the graph farther out.   ■

Technology Tip

When you get a blank screen, press TRACE and use the left/right arrow keys. The coordinates of points on the graph will be displayed at the bottom of the screen, even though they aren't in the current viewing window. Use these coordinates as a guide for selecting a viewing window in which the graph does appear.

Example 8   If you graph $y = -2x^3 + 26x^2 + 18x + 50$ in the standard viewing window, you get a blank screen (try it!). In such cases you can usually find at least one point on the graph by setting $x = 0$ and determining the corresponding value of y. If $x = 0$ here, then $y = 50$, so the point $(0, 50)$ is on the graph. Consequently, the y-axis of our viewing window should extend well beyond 50. An alternative method of finding some points on the graph is in the Tip in the margin.

GRAPHING EXPLORATION

Find a complete graph of this equation. [*Hint:* The graph crosses the x-axis once and has one "peak" and one "valley".]   ■

As a general rule, you should follow the directions in the following box when graphing equations:

Graphing
Convention

1. Unless directed otherwise, use a calculator for graphing.
2. Complete graphs are required unless a viewing window is specified or the context of a problem indicates that a partial graph is acceptable.
3. If the directions say "obtain the graph," "find the graph," or "graph the equation," you need not actually draw the graph on paper. For review purposes, however, it may be helpful to record the viewing window used.
4. The directions "sketch the graph" mean "draw the graph on paper, indicating the scale on each axis." This may involve simply copying the display on the calculator screen, or it may require more work if the calculator display is misleading.

Calculator Investigations 1.2

1. Tick Marks

   (a) Set Xscl = 1 so that adjacent tick marks on the x-axis are 1 unit apart. Find the largest range of x values such that the tick marks on the x-axis are clearly distinguishable and appear to be equally spaced.

   (b) Do part (a) with y in place of x.

2. Viewing Windows   Look in the ZOOM menu (or the VIEWS menu on HP-38) to find out how many built-in viewing windows your calculator has. Take a look at each one.

3. Maximum/Minimum Finders   Use your minimum finder to approximate the x-coordinates of the lowest point on the graph of $y = x^3 - 2x + 5$ in the window with $0 \leq x \leq 5$ and $-3 \leq y \leq 8$. The correct answer is $x = \sqrt{\frac{2}{3}} \approx .816496580928$. How good is your approximation?

4. Square Windows   Find a square viewing window on your calculator that has $-10 \leq x \leq 10$.

5. Dot Graphing Mode   A calculator graphs by plotting points and connecting them. To see which points it plots, without the connecting segments, change your graphing mode by selecting Dot or DrawDot in the MODE menu of TI-83 or the FORMAT submenu of the TI-86 GRAPH menu or the STYLE submenu of the TI-89 Y= menu or the STYLE 1 submenu of the Sharp 9600 FORMAT menu. In the Casio 9850 SETUP menu, set the DRAWTYPE to PLOT and in the second page of the HP-38 PLOT SETUP menu, uncheck CONNECT. After changing the graphing mode, graph $y = .5x^3 - 2x^2 + 1$ in the standard window. Try some other equations as well.

Exercises 1.2

Exercises 1–4 are representations of calculator screens. State the approximate coordinates of the points P and Q.

1.

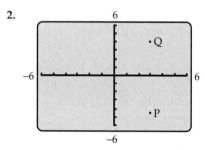

2.

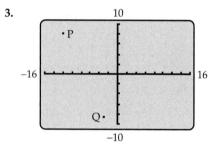

3.

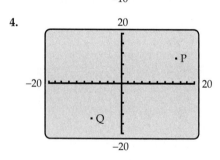

4.

In Exercises 5–10, graph the equation by hand by plotting no more than six points and filling in the rest of the graph as best you can. Then use the calculator to graph the equation and compare the results.

5. $y = |x - 2|$ **6.** $y = \sqrt{x + 5}$

7. $y = x^2 - x$ **8.** $y = x^2 + x + 1$

9. $y = x^3 + 1$ **10.** $y = \dfrac{1}{x}$

In Exercises 11–16, find the graph of the equation in the standard window.

11. $3 + y = .5x$ **12.** $y - 2x = 4$

13. $y = x^2 - 5x + 2$ **14.** $y = .3x^2 + x - 4$

15. $y = .2x^3 + .1x^2 - 4x + 1$

16. $y = .2x^4 - .2x^3 - 2x^2 - 2x + 5$

In Exercises 17–22, determine which of the following viewing windows gives the best view of the graph of the given equation.

    a. $-10 \le x \le 10;$ $-10 \le y \le 10$
    b. $-5 \le x \le 25;$ $0 \le y \le 20$
    c. $-10 \le x \le 10;$ $-100 \le y \le 100$
    d. $-20 \le x \le 15;$ $-60 \le y \le 250$
    e. None of a, b, c, d gives a complete graph.

17. $y = 18x - 3x^2$ **18.** $y = 4x^2 + 80x + 350$

19. $y = \dfrac{1}{3}x^3 - 25x + 100$ **20.** $y = x^4 + x - 5$

21. $y = x^2 + 50x + 625$ **22.** $y = .01(x - 15)^4$

23. A toy rocket is shot straight up from ground level and then falls back to earth; wind resistance is negligible. Use your calculator to determine which of the following equations has a graph whose portion above the x-axis provides the most plausible model of the path of the rocket.
    (a) $y = .1(x - 3)^3 - .1x^2 + 5$
    (b) $y = -x^4 + 16x^3 - 88x^2 + 192x$
    (c) $y = -16x^2 + 117x$
    (d) $y = .16x^2 - 3.2x + 16$
    (e) $y = -(.1x - 3)^6 + 600$

24. Monthly profits at DayGlo Tee Shirt Company appear to be given by the equation

$$y = -.00027(x - 15{,}000)^2 + 60{,}000,$$

where x is the number of shirts sold that month and y is the profit. DayGlo's maximum production capacity is 15,000 shirts per month.
    (a) If you plan to graph the profit equation, what range of x values should you use? [*Hint:* You can't make a negative number of shirts.]
    (b) The president of DayGlo wants to motivate the sales force (who are all in the profit-sharing plan) so he asks you to prepare a graph that shows DayGlo's profits increasing *dramatically* as sales increase. Using the profit equation and the x range from part (a), what viewing window is suitable?
    (c) The City Council is talking about imposing more taxes. The president asks you to prepare

a graph showing that DayGlo's profits are essentially flat. Using the profit equation and the x range from part (a), what viewing window is suitable?

In each of the applied situations in Exercises 25–28, find an appropriate viewing window for the equation (that is, a window that includes all the points relevant to the problem, but does not include large regions that are not relevant to the problem, and has easily readable tick marks on the axes). Explain why you chose this window. See the Hint in Exercise 24(a).

25. A cardiac test measures the concentration y of a dye x seconds after a known amount is injected into a vein near the heart. In a normal heart $y = -.006x^4 + .14x^3 - .053x^2 + 179x$.

26. Beginning in 1905 the deer population in a region of Arizona rapidly increased because of a lack of natural predators. Eventually food resources were depleted to such a degree that the deer population completely died out. In the equation $y = -.125x^5 + 3.125x^4 + 4000$, y is the number of deer in year x, where $x = 0$ corresponds to 1905.

27. A winery can produce x barrels of red wine and y barrels of white wine, where
$$y = \frac{200{,}000 - 50x}{2000 + x}.$$

28. The concentration of a certain medication in the bloodstream at time x hours is approximated by the equation $y = \dfrac{375x}{.1x^3 + 50}$, where y is measured in milligrams per liter. After two days the medication has no effect.

In Exercises 29–30, use zoom-in or a maximum/minimum finder to determine the highest and lowest point on the graph in the given window.

29. $y = .4x^3 - 3x^2 + 4x + 3$
($0 \le x \le 5$ and $-5 \le y \le 5$)

30. $y = .07x^5 - .3x^3 + 1.5x^2 - 2$
($-3 \le x \le 2$ and $-6 \le y \le 6$)

In Exercises 31–34, find an appropriate viewing window for the graph of the equation (which may take some experimentation) and use a maximum/minimum finder to answer the question.

31. The number y of country music radio stations in the United States is approximated by
$$y = -.44x^4 + 9.68x^3 - 91.3x^2 + 370x + 2105$$
$$(2 \le x \le 8),$$
where $x = 2$ corresponds to 1992. In what year was the number of such stations the largest?

32. The U.S. commercial catch of fish (in thousands of metric tons) is given by
$$y = 18.5x^4 - 265x^3 + 1219x^2 - 1954x + 6467$$
$$(1 \le x \le 7),$$
where $x = 1$ corresponds to 1991.
  (a) When was the catch the largest?
  (b) When was the catch the smallest?

33. The population y of New Orleans (in thousands) in year x of the 20th century is approximated by
$$y = .000046685x^4 - .0108x^3 + .7194x^2 - 9.2426x + 305$$
$$(0 \le x \le 100),$$
where $x = 0$ corresponds to 1900. According to this model, in what year was the population largest?

34. The number y of LP records sold (in millions) between 1990 and 1996 is approximated by
$$y = \frac{x^4}{120} - \frac{121x^3}{540} + \frac{149x^2}{72} - \frac{799x}{108} + \frac{311}{30}$$
$$(0 \le x \le 6),$$
where $x = 0$ corresponds to 1990. In what year were the fewest records sold?

In Exercises 35–40, use your algebraic knowledge to state whether or not the two equations have the same graph. Confirm your answer by graphing the equations in the standard window.

35. $y = |x + 3|$ and $y = |x| + 3$

36. $y = |x| - 4$ and $y = |x - 4|$

37. $y = \sqrt{x^2}$ and $y = |x|$

38. $y = \sqrt{x^2 + 6x + 9}$ and $y = |x + 3|$

39. $y = \sqrt{x^2 + 9}$ and $y = x + 3$

40. $y = \dfrac{1}{x^2 + 2}$ and $y = \dfrac{1}{x^2} + \dfrac{1}{2}$

41. (a) Confirm the accuracy of the factorization $x^2 - 5x + 6 = (x - 2)(x - 3)$ graphically. [Hint: Graph $y = x^2 - 5x + 6$ and $y = (x - 2)(x - 3)$ on the same screen. If the factorization is correct the graphs will be identical (which means that you will see only a single graph on the screen.]
  (b) Show graphically that $(x + 5)^2 \ne x^2 + 5^2$. [Hint: Graph $y = (x + 5)^2$ and $y = x^2 + 5^2$ on the same screen. If the graphs are different, then the two expressions cannot be equal.]

True or False In Exercises 42–44, use the technique of Exercise 41 to determine graphically whether the given statement is possibly true or definitely false. [We say "possibly

*true" because two graphs that appear identical on a calcula-
tor screen may actually differ by small amounts or at places
not shown in the window.]*

42. $x^3 - 7x - 6 = (x + 1)(x + 2)(x - 3)$

43. $(1 - x)^6 = 1 - 6x + 15x^2 - 20x^3 + 15x^4 - 6x^5 + x^6$

44. $x^5 - 8x^4 + 16x^3 - 5x^2 + 4x - 20 =$
$$(x - 2)^2(x - 5)(x^2 + x + 1)$$

*In Exercises 45–54, use the techniques of Examples 5 and 6
to graph the equation in a suitable square viewing window.*

45. $x^2 + y^2 = 9$ **46.** $y^2 = x + 2$

47. $3x^2 + 2y^2 = 48$

48. $25(x - 5)^2 + 36(y + 4)^2 = 900$

49. $(x - 4)^2 + (y + 2)^2 = 25$

50. $9x^2 + 4y^2 = 36$ **51.** $4x^2 - 9y^2 = 36$

52. $9y^2 - x^2 = 9$ **53.** $9x^2 + 5y^2 = 45$

54. $x = y^2 - 2$

*In Exercises 55–60, obtain a complete graph of the equation
by trying various viewing windows. List a viewing window
that produces this complete graph. [Many correct answers are
possible; consider your answer correct if your window shows
all the features in the window given in the answer section.]*

55. $y = 7x^3 + 35x + 10$ **56.** $y = x^3 - 5x^2 + 5x - 6$

57. $y = \sqrt{x^2 - x}$ **58.** $y = 1/x^2$

59. $y = -.1x^4 + x^3 + x^2 + x + 50$

60. $y = .002x^5 + .06x^4 - .001x^3 + .04x^2 - .2x + 15$

61. (a) Graph $y = x^3 - 2x^2 + x - 2$ in the standard
window.

(b) Use the trace feature to show that the portion
of the graph with $0 \le x \le 1.5$ is not actually
horizontal. [*Hint:* All the points on a
horizontal segment must have the same
y-coordinate. (Why?)]

(c) Find a viewing window that clearly shows
that the graph is not horizontal when
$0 \le x \le 1.5$.

62. (a) Graph $y = \dfrac{1}{x^2 + 1}$ in the standard window.

(b) Does the graph appear to stop abruptly part-
way along the x-axis? Use the trace feature to
explain why this happens. [*Hint:* In this
viewing window, each pixel represents a
rectangle that is approximately .32 units high.]

(c) Find a viewing window with $-10 \le x \le 10$,
which shows a complete graph that does not
fade into the x-axis.

63. Four squares, each measuring x by x inches, are
cut from the corners of a rectangular sheet of
cardboard that measures 30 by 22 inches and the
flaps are folded up to form an open-top box, as
shown in Exercise 67 of Section 0.4. In that
exercise you found a polynomial that expresses
the volume of the resulting box.

(a) Graph $y =$ volume polynomial in the window
with $0 \le x \le 11$ and $0 \le y \le 1300$.

(b) Explain what the coordinates of each point on
the graph represent.

(c) Find the value of x for which the volume of
the box is as large as possible.

64. Do Exercise 63 when the sheet of cardboard
measures 40 by 40 inches. [*Hint:* You will need a
different viewing window.]

1.3 Solving Equations Graphically and Numerically

$y = x^2 + x - 2$

Figure 1–33

Many equations can be solved algebraically, as we shall see in Chapter 2.
When algebraic techniques are inadequate, the graphical and numerical
approximation methods presented in this section will be needed. These
methods depend on the connection between equations and graphs, so we
begin with that.

In the coordinate plane all points on the x-axis have second coordinate 0. Consequently, when a graph intersects the x-axis, the intersection
point has coordinates of the form $(a, 0)$ for some real number a. The number a is called an **x-intercept** of the graph. For example, the graph of
$y = x^2 + x - 2$ in Figure 1–33 has x-intercepts at -2 and 1 because it intersects the x-axis at $(-2, 0)$ and $(1, 0)$. To say that $(1, 0)$ is on the graph
of $y = x^2 + x - 2$ means that $x = 1$ and $y = 0$ satisfy the equation:

$$y = x^2 + x - 2$$
$$0 = 1^2 + 1 - 2.$$

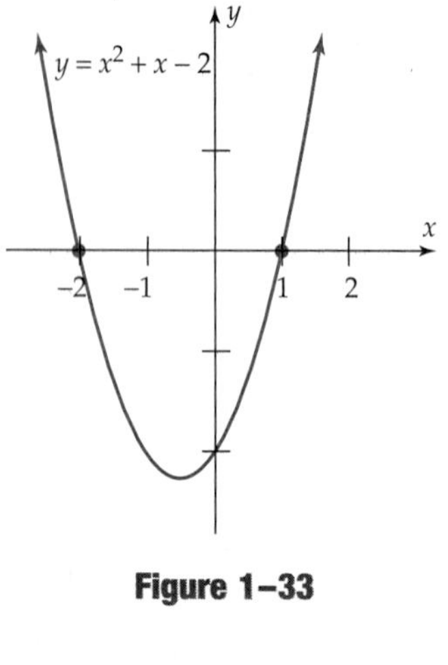

In other words, the x-intercept 1 is a *solution* of the equation

$$x^2 + x - 2 = 0.$$

Similarly, the other x-intercept, -2, is also a solution of $x^2 + x - 2 = 0$ because $(-2)^2 + (-2) - 2 = 0$. The same argument works in the general case.

Solutions and Intercepts

The real solutions of a one-variable equation of the form

expression in $x = 0$

are the x-intercepts of the graph of the two-variable equation

$$y = \text{expression in } x.$$

Therefore, to solve an equation, you need only find the x-intercepts of its graph, as illustrated in the next example.

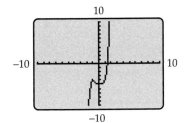

Figure 1–34

Example 1 Solve the equation $x^5 + x^2 = x^3 + 5$ graphically.

Solution We first rewrite the equation so that one side is zero:

$$x^5 - x^3 + x^2 - 5 = 0.$$

Now we graph $y = x^5 - x^3 + x^2 - 5$ in the standard viewing window (Figure 1–34). There is one x-intercept (solution of the equation) between 1 and 2. Assuming that this is the only x-intercept of the graph, this means that the original equation has exactly one real-number solution.* It may be approximated graphically in three ways.

Manual Zoom-in Repeatedly change the viewing window to show the part of the graph near the x-intercept in greater and greater detail, as shown in Figure 1–35. At each step the displayed portion of the x-axis is decreased by a factor of $1/10$ and the Xscl adjusted accordingly.

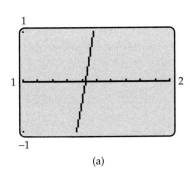

(a)

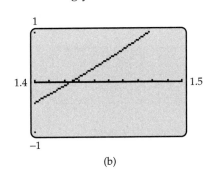

(b)

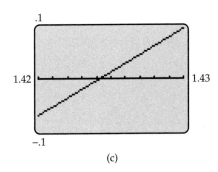

(c)

Figure 1–35

*Unless stated otherwise in the examples and exercises of this section, you may assume that the standard viewing window includes all x-intercepts of a graph.

In graph (a), the intercept (solution) is between 1.4 and 1.5. Graph (b) shows that it is between 1.42 and 1.43. Finally, graph (c) shows that the solution is between 1.424 and 1.425, at approximately 1.4242. Since adjacent tick marks in Figure 1–35(c) are .001 units apart, the maximum possible error in this approximation is .001.

Automatic Zoom-in Finding a very small window, such as Figure 1–35(c), can be done in just a few keystrokes by using either ZOOM-IN or BOX on the ZOOM menu. The procedures vary on different calculators, so check your instruction manual. After zooming in, the tick marks probably won't show, so you will have to reset the Xscl setting in order to read the approximate solution and determine its accuracy. When using ZOOM-IN, it's convenient to set the zoom factors at 10 (see the Technology Tip on page 76).

Graphical Root Finder The easiest way to find the solution (*x*-intercept) with a high degree of accuracy is to use the graphical root finder on your calculator. On some TI calculators, you must select a left (or lower) bound, meaning an *x*-value to the left of the intercept, and a right (or upper) bound, meaning an *x*-value to the right of the intercept, and to make an initial guess. On other calculators, you may have to move the cursor near the *x*-intercept you are seeking. Check your instruction manual for details. A typical root finder produced Figure 1–36, which shows that the solution (*x*-intercept) is $x \approx 1.4242577$. ∎

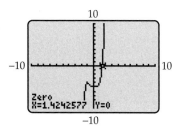

Figure 1–36

Technology Tip

The graphical root finder is labeled ROOT (or ZERO or X-INCPT) in the TI-83 CALC menu, the Sharp 9600 CALC menu, the Casio 9850 G-SOLVE menu, the MATH submenu of the TI-86/89 GRAPH menu, and the FCN submenu of the HP-38 PLOT menu.

Example 2 To approximate the solution of

$$x^4 - 4x^3 + 3x^2 - x - 2 = 0,$$

we graph $y = x^4 - 4x^3 + 3x^2 - x - 2$ (Figure 1–37). The graph shows two solutions (*x*-intercepts), one between −1 and 0, and the other between 3 and 4. A graphical root finder (Figure 1–38) shows that the negative solution is $x \approx -.52417$. ∎

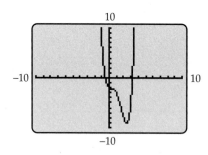

Figure 1–37

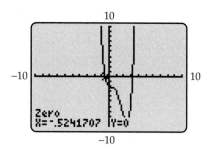

Figure 1–38

GRAPHING EXPLORATION

Use zoom-in or a graphical root finder to approximate the positive solution of the equation in Example 2.

The Intersection Method

The next two examples illustrate an alternative graphical method of approximating solutions of equations.

Example 3 Solve: $|x^2 - 4x - 3| = x^3 + x - 6$.

Solution Let $y_1 = |x^2 - 4x - 3|$ and $y_2 = x^3 + x - 6$ and graph both equations on the same screen (Figure 1–39). Consider the point where the two graphs intersect. Since it is on the graph of y_1, its second coordinate is $|x^2 - 4x - 3|$ and since it is also on the graph of y_2, its second coordinate is $x^3 + x - 6$. So for this number x, we must have $|x^2 - 4x - 3| = x^3 + x - 6$. In other words, the *x-coordinate of the intersection point is the solution of the equation.*

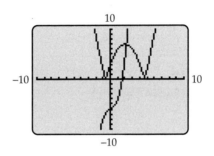

Figure 1–39   Figure 1–40

Technology Tip

The graphical intersection finder is labeled INTERSECT or ISECT or ISCT or INTSCT in the TI-83 CALC menu, the Sharp 9600 CALC menu, the Casio 9850 G-SOLVE menu, the MATH submenu of the TI-86/89 GRAPH menu, and the FCN submenu of the HP-38 PLOT menu.

This coordinate can be approximated by zooming in or by using a graphical intersection finder (see the Tip in the margin), as shown in Figure 1–40. Therefore, the solution of the original equation is $x \approx 2.207$. ■

Example 4 To solve

$$x^2 - 2x - 6 = \sqrt{2x + 7},$$

we graph $y_1 = x^2 - 2x - 6$ and $y_2 = \sqrt{2x + 7}$ on the same screen. Figure 1–41 shows that there are two solutions (intersection points). According to an intersection finder (Figure 1–42), the positive solution is $x \approx 4.3094$. ■

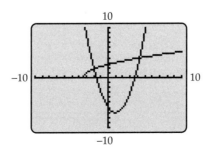
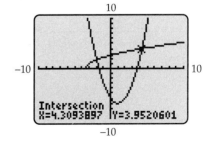

Figure 1–41   Figure 1–42

GRAPHING EXPLORATION

Use a graphical intersection finder to approximate the negative solution of the equation in Example 4.

Technology Tip

The equation solver is labeled SOLVE or SOLVER. It is on the TI-86 and Sharp 9600 keyboards, in the TI-83 MATH menu, in the TI-89 ALGEBRA menu, in the HP-38 LIB menu, and in the EQUA submenu of the Casio 9850 Main Menu. The syntax varies, depending on the calculator, so check your instruction manual.

Numerical Methods

In addition to graphical root finders, most calculators have an **equation solver** that can approximate solutions of any equation (one at a time, except on TI-89 and Casio FX2), typically by using Newton's method (which involves calculus). Usually you must enter the equation and an initial estimate of the solution, and possibly an interval in which to search.

Several calculators also have **polynomial solvers** that find all the solutions of a polynomial equation at once. Some of them, however, are limited to equations of degree 2 or 3. See the second Tip in the margin.

Example 5   When asked to search the interval $-10 \leq x \leq 10$ with an initial guess of 3, the equation solver on a TI-83 found one solution of

$$4x^5 - 12x^3 + 8x - 1 = 0,$$

namely, $x \approx .128138691376$ (Figure 1–43). The POLY solver on a TI-86 produced that solution and four others (Figure 1–44). ∎

Technology Tip

The polynomial solver is labeled POLY on the TI-86 keyboard and the Sharp 9600 TOOL menu, and EQUATION in the Casio 9850 main menu (select POLYNO-MIAL). On HP-38 use POLYROOT in the POLYNOM submenu of the MATH menu and check your instruction manual for proper syntax.

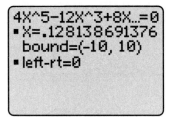

Figure 1–43

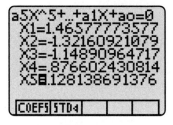

Figure 1–44

Dealing with Technological Quirks

Graphical root finders and equation solvers may fail to find some solutions of an equation, particularly when the graph of the equation touches, but does not cross, the x-axis. So if your calculator doesn't show any x-intercepts on a graph or if its root finder gives an error message, you should try an alternative approach, if possible.

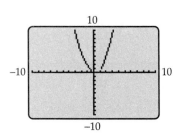

Figure 1–45

Example 6   If you attempt to solve $\sqrt{x^4 + x^2 - 2x - 1} = 0$ by graphing the equation $y = \sqrt{x^4 + x^2 - 2x - 1}$, as in Figure 1-45, and using a root finder, you may get an error message, as we did on four out of six models. If you use an equation solver instead, the results may also be disappointing (three solvers produced an error message). The difficulty can be eliminated by using the fact that the *only number whose square root is zero is zero itself.* Thus, the only time that $\sqrt{x^4 + x^2 - 2x - 1} = 0$ is when $x^4 + x^2 - 2x - 1 = 0$. So we need only solve $x^4 + x^2 - 2x - 1 = 0$, which is easily done on a calculator.

GRAPHING EXPLORATION

One solution of $x^4 + x^2 - 2x - 1 = 0$, and hence of the original equation, is $x \approx -.4046978$. Use a graphical root finder, an equation solver, or a polynomial solver to find the other solution. ∎

Even though your calculator might work directly in Example 6, there may be similar cases where it doesn't. So the technique of Example 6 should normally be used with all equations of the form "square root of a quantity $= 0$" (but not on other equations involving square roots): Set the expression under the radical equal to 0 and solve.

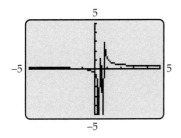

Figure 1–46

Example 7 If you try to solve $\dfrac{7x - 4}{9x^2 - 9x + 2} = 0$ by graphing

$y = \dfrac{7x - 4}{9x^2 - 9x + 2}$, you may get "garbage," as in Figure 1–46. You could zoom in for a clearer picture, but it's easier to use the fact that a *fraction is zero only when its numerator is zero and its denominator is nonzero.* Finding the numbers that make the numerator 0 means solving $7x - 4 = 0$, which is easily done.

Add 4 to both sides:	$7x = 4$
Divide both sides by 7:	$x = 4/7$

You can readily verify that $x = 4/7$ does not make the denominator 0. Hence the solution of the original equation is $4/7$. ∎

The technique of Example 7 is recommended for equations of the form "fraction $= 0$": set the numerator equal to 0 and solve. Those solutions that do not make the denominator 0 are the solutions of the original equation. A number that makes *both* numerator and denominator 0 is *not* a solution because $0/0$ is not defined.

Applications

Graphical and numerical solution methods can be very helpful in dealing with applied problems, since approximate solutions are perfectly adequate for many real-life situations.

Example 8 According to data from the U.S. Bureau of the Census, the approximate populations y (in millions) of Chicago and Los Angeles between 1950 and 2000 are given by:

Chicago $\qquad y = .00003046x^3 - .0023x^2 + .02024x + 3.62,$

Los Angeles $\qquad y = .0000113x^3 - .000922x^2 + .0538x + 1.97,$

where $x = 0$ corresponds to 1950. In what year did the two cities have the same population?

Solution We are asked to find the value of x for which

Population of Chicago = Population of Los Angeles

$.00003046x^3 - .0023x^2 + .02024x + 3.62 = .0000113x^3 - .000922x^2 + .0538x + 1.97$

We can solve this equation by graphing the left and right sides and finding their intersection point, as in Figure 1–47. Therefore, the populations are the same when $x \approx 28.77$, that is, late in the year 1978. ∎

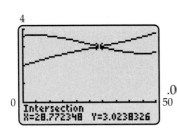

Figure 1–47

Exercises 1.3

In Exercises 1–6, determine graphically the number of solutions of the equation, but don't solve the equation. You may need a viewing window other than the standard one to find all the x-intercepts.

1. $x^5 + 5 = 3x^4 + x$ 2. $x^3 + 5 = 3x^2 + 24x$

3. $x^7 - 10x^5 + 15x + 10 = 0$

4. $x^5 + 36x + 25 = 13x^3$

5. $x^4 + 500x^2 - 8000x = 16x^3 - 32,000$

6. $6x^5 + 80x^3 + 45x^2 + 30 = 45x^4 + 86x$

In Exercises 7–20, use graphical approximation (zoom-in or a root finder or an intersection finder) to find a solution of the equation in the given interval.

7. $x^3 + 4x^2 + 10x + 15 = 0$; $(-3 < x < -2)$

8. $x^3 + 9 = 3x^2 + 6x$; $(1 < x < 2)$

9. $x^4 + x - 3 = 0$; $(x > 0)$

10. $x^5 + 5 = 3x^4 + x$; $(x < 1)$

11. $\sqrt{x^4 + x^3 - x - 3} = 0$; $(x > 0)$

12. $\sqrt{8x^4 - 14x^3 - 9x^2 + 11x - 1} = 0$; $(x < 0)$

13. $\sqrt{\frac{2}{5}x^5 + x^2 - 2x} = 0$; $(x > 0)$

14. $\sqrt{x^4 + x^2 - 3x + 1} = 0$; $(0 < x < 1)$

15. $x^2 = \sqrt{x + 5}$; $(-2 < x < -1)$

16. $\sqrt{x^2 - 1} - \sqrt{x + 9} = 0$; $(3 < x < 4)$

17. $\dfrac{2x^5 - 10x + 5}{x^3 + x^2 - 12x} = 0$; $(x < 0)$

18. $\dfrac{3x^5 - 15x + 5}{x^7 - 8x^5 + 2x^2 - 5} = 0$; $(x > 1)$

19. $\dfrac{x^3 - 4x + 1}{x^2 + x - 6} = 0$; $(x > 1)$

20. $\dfrac{4}{x + 2} - \dfrac{3}{x + 1} = 0$; $(x > 0)$ [*Hint:* Write the left side as a single fraction.]

In Exercises 21–34, use algebraic, graphical, or numerical methods to find all real solutions of the equation, approximating when necessary.

21. $2x^3 - 4x^2 + x - 3 = 0$

22. $6x^3 - 5x^2 + 3x - 2 = 0$

23. $x^5 - 6x + 6 = 0$ 24. $x^3 - 3x^2 + x - 1 = 0$

25. $10x^5 - 3x^2 + x - 6 = 0$

26. $\frac{1}{4}x^4 - x - 4 = 0$

27. $2x - \frac{1}{2}x^2 - \frac{1}{12}x^4 = 0$ 28. $\frac{1}{4}x^4 + \frac{1}{3}x^2 + 3x - 1 = 0$

29. $\dfrac{5x}{x^2 + 1} - 2x + 3 = 0$ 30. $\dfrac{2x}{x + 5} = 1$

31. $|x^2 - 4| = 3x^2 - 2x + 1$

32. $|x^3 + 2| = 5 + x - x^2$

33. $\sqrt{x^2 + 3} = \sqrt{x - 2} + 5$

34. $\sqrt{x^3 + 2} = \sqrt{x + 5} + 4$

In Exercises 35–40, find an exact solution of the equation in the given interval. [For example, if the graphical approxima-tion of a solution begins .3333, check to see if 1/3 is the ex-act solution. Similarly, $\sqrt{2} \approx 1.414$; so if your approxima-tion begins 1.414 check to see if $\sqrt{2}$ is a solution.]

35. $3x^3 - 2x^2 + 3x - 2 = 0; \quad 0 < x < 1$

36. $4x^3 - 3x^2 - 3x - 7 = 0; \quad 1 < x < 2$

37. $12x^4 - x^3 - 12x^2 + 25x - 2 = 0; \quad 0 < x < 1$

38. $8x^5 + 7x^4 - x^3 + 16x - 2 = 0; \quad 0 < x < 1$

39. $4x^4 - 13x^2 + 3 = 0; \quad 1 < x < 2$

40. $x^3 + x^2 - 2x - 2 = 0; \quad 1 < x < 2$

Exercises 41–44 deal with exponential and logarithmic equa-tions, all of which will be dealt with in detail in later chap-ters. If you are familiar with these concepts, solve each equa-tion graphically.

41. $10^x - \dfrac{1}{4}x = 28$ 42. $3^x - 2^x = 2$

43. $\ln x - x^2 + 3 = 0$ 44. $e^x - 6x = 5$

45. According to data from the National Center for Education Statistics and the College Board, the average cost y of tuition and fees (in thousands of dollars) at public four-year institutions in year x is approximated by the equation

$$y = .00044x^3 - .00039x^2 + .114x + 2.195,$$

where $x = 0$ corresponds to 1990. If this model continues to be accurate, in what year will tuition and fees reach $4,000?

46. Use the information in Example 8 to determine the year in which the population of Los Angeles reached 2.6 million.

47. According to data from the U.S. Department of Health and Human Services, the cumulative number y of AIDS cases (in thousands) diagnosed in the United States during 1982–1993 is approximated by

$$y = 3.223x^2 - 14.81x + 17.75; \quad (2 \le x \le 13),$$

where $x = 0$ corresponds to 1980. In what year did the cumulative number of cases reach 250,000?

48. **(a)** How many real solutions does the equation
$$0.2x^5 - 2x^3 + 1.8x + k = 0$$
have when $k = 0$?

   (b) How many real solutions does it have when $k = 1$?

   (c) Is there a value of k for which the equation has just one real solution?

   (d) Is there a value of k for which the equation has no real solutions?

1.4 Lines

When you move from a point P to a point Q on a line,* two numbers are involved, as illustrated in Figure 1–48:

1. The vertical distance you move (the **change in y**);

2. The horizontal distance you move (the **change in x**).

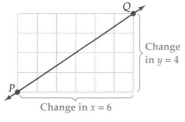

Change in $y = 4$

Change in $x = 6$

$$\frac{\text{Change in } y}{\text{Change in } x} = \frac{4}{6} = \frac{2}{3}$$

(a)

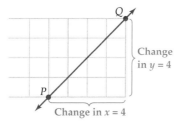

Change in $y = 4$

Change in $x = 4$

$$\frac{\text{Change in } y}{\text{Change in } x} = \frac{4}{4} = 1$$

(b)

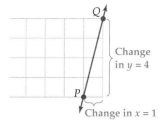

Change in $y = 4$

Change in $x = 1$

$$\frac{\text{Change in } y}{\text{Change in } x} = \frac{4}{1} = 4$$

(c)

Figure 1–48

*In this section, "line" means "straight line," and movement is from left to right.

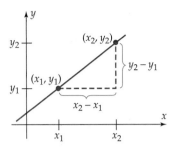

Figure 1–49

The number $\dfrac{\text{change in } y}{\text{change in } x}$ measures the steepness of the line: the steeper the line, the larger the number. In Figure 1–48, the grid allows us to measure the change in y and the change in x. When the coordinates of P and Q are given, then

The change in y is the difference of the y-coordinates of P and Q;

The change in x is the difference of the x-coordinates of P and Q;

as shown in Figure 1–49. Consequently, we have the following definition.

Slope of a Line

If (x_1, y_1) and (x_2, y_2) are points with $x_1 \neq x_2$, then the *slope* of the line through these points is the number

$$\frac{\text{change in } y}{\text{change in } x} = \frac{y_2 - y_1}{x_2 - x_1}.$$

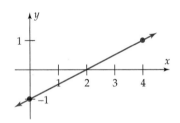

Figure 1–50

Example 1 To find the slope of the line through $(0, -1)$ and $(4, 1)$ (see Figure 1–50), we apply the formula in the preceding box with $x_1 = 0$, $y_1 = -1$ and $x_2 = 4$, $y_2 = 1$:

$$\text{slope} = \frac{y_2 - y_1}{x_2 - x_1} = \frac{1 - (-1)}{4 - 0} = \frac{2}{4} = \frac{1}{2}.$$

The order of the points makes no difference; if you use $(4, 1)$ for (x_1, y_1) and $(0, -1)$ for (x_2, y_2) we obtain the same number:

$$\text{slope} = \frac{y_2 - y_1}{x_2 - x_1} = \frac{-1 - 1}{0 - 4} = \frac{-2}{-4} = \frac{1}{2}. \quad \blacksquare$$

CAUTION

When finding slopes, you must subtract the y-coordinates and the x-coordinates in the same order. With the points (3, 4) and (1, 8), for instance, if you use $8 - 4$ in the numerator, you must use $1 - 3$ in the denominator (*not* $3 - 1$).

Slope-Intercept Form

A nonvertical line intersects the y-axis at a point with coordinates $(0, b)$ for some number b (because every point on the y-axis has first coordinate 0). The number b is called the **y-intercept** of the line. For example, the line in Figure 1–50 has y-intercept -1 because it crosses the y-axis at $(0, -1)$.

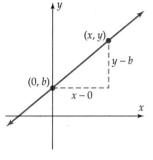

Figure 1–51

Let L be a nonvertical line with slope m and y-intercept b. Then $(0, b)$ is a point on L. Let (x, y) be any other point on L. Using points $(0, b)$ and (x, y) to compute the slope of L (see Figure 1–51), we have:

$$\text{slope of } L = \frac{y - b}{x - 0}.$$

Since the slope of L is m, this equation becomes

$$m = \frac{y - b}{x}$$

Multiply both sides by x: $mx = y - b$

Rearrange terms: $y = mx + b$

Thus, the coordinates of any point on L satisfy the equation $y = mx + b$. So we have this fact:

Slope-Intercept Form

The line with slope m and y-intercept b is the graph of the equation

$$y = mx + b.$$

A calculator can be used to graph lines, but it's usually just as easy to graph them by hand.

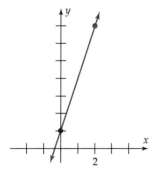

Figure 1–52

Example 2   Graph the equation $2y - 5x = 2$.

Solution   We begin by solving the equation for y:

$$2y = 5x + 2$$

Divide both sides by 2: $y = 2.5x + 1.$

Therefore, its graph is the line with slope 2.5 (coefficient of x) and y-intercept 1 (constant term), which means that $(0, 1)$ is on the line. When $x = 2$, then $y = 2.5(2) + 1 = 6$, so $(2, 6)$ is also on the line. Plotting the line through these points produces Figure 1–52.   ■

Although we don't need a calculator to graph lines, the calculator makes it easy to see some essential properties of "slope."

GRAPHING EXPLORATION

Using the standard viewing window, graph the following equations on the same screen and answer the questions below.

$$y_1 = .5x, \quad y_2 = x, \quad y_3 = 3x, \quad y_4 = 7x.$$

What are the slopes of these lines?

Which line rises least steeply (from left to right)? Which one rises most steeply?

Which line has the smallest slope? Which has the largest slope?

How is the slope of the line related to how steeply it rises?

Now graph the following equations on the same screen:

$$y_5 = 2, \quad y_6 = -x + 2, \quad y_7 = -2.5x + 2, \quad y_8 = -5x + 2.$$

What are the slopes of these lines?

Which line falls least steeply (from left to right)? Which one falls most steeply?

How is the slope of the line related to how steeply it falls?

Your answers to the questions in the preceding Exploration should indicate that slope measures the steepness of the line, as summarized here:

Properties of Slope

The graph of $y = mx + b$ is a nonvertical straight line with slope m. The slope measures how steeply the line rises or falls:

If $m > 0$, the line rises from left to right; the larger m is, the more steeply the line rises.

If $m = 0$, the line is horizontal.

If $m < 0$, the line falls from left to right; the larger $|m|$ is, the more steeply the line falls.

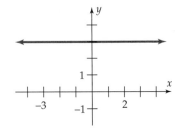

Figure 1–53

Example 3 Describe and sketch the graph of the equation $y = 3$.

Solution We can write $y = 3$ as $y = 0x + 3$. So its graph is a line with slope 0, which means the line is horizontal, and y-intercept 3, which means the line crosses the y-axis at 3. This is sufficient information to obtain the graph in Figure 1–53. ∎

Example 4 An office buys a new computer for $7000. Five years later its value is $800. Assuming linear depreciation, what was its value two years after it was purchased?

Solution Linear depreciation means that the equation that gives the value y of the computer in year x is linear, and hence of the form $y = mx + b$ for some constants m and b. Since the computer is worth $7000 new ($x = 0$), we have

$$y = mx + b$$

$$7000 = m \cdot 0 + b$$

$$b = 7000$$

so that the equation is $y = mx + 7000$. Because the computer is worth $800 after 5 years (that is, $y = 800$ when $x = 5$), we have

$$y = mx + 7000$$

$$800 = m \cdot 5 + 7000$$

Subtract 7000 from both sides: $\qquad -6200 = 5m$

Divide both sides by 5: $\qquad m = \dfrac{-6200}{5} = -1240.$

Therefore, the depreciation equation is $y = -1240x + 7000$. The value of the computer after two years ($x = 2$) is

$$y = -1240(2) + 7000 = \$4520. \quad \blacksquare$$

Example 5 A factory that makes can openers has fixed costs (for building, fixtures, machinery, etc.) of $26,000. The variable cost (materials and labor) for making one can opener is $2.75.

(a) What is the total cost of making 1000 can openers? 20,000? 40,000?

(b) What is the average cost per can opener in each case?

Solution

(a) Since each can opener costs $2.75, the variable costs for making x can openers is $2.75x$. The total cost y of making x can openers is

$$y = \text{variable costs} + \text{fixed costs} = 2.75x + 26,000.$$

The cost of making 1000 can openers is

$$y = 2.75x + 26,000 = 2.75(1000) + 26,000 = \$28,750.$$

Similarly, the cost of making 20,000 can openers is

$$y = 2.75(20,000) + 26,000 = \$81,000$$

and the cost of 40,000 is

$$y = 2.75(40,000) + 26,000 = \$136,000.$$

(b) The average cost per can opener in each case is the total cost divided by the number of can openers. So the average cost per can opener is:

For 1000: $\$28,\!750/1000 = \$28.75$ per can opener;

For 20,000: $\$81,\!000/20,\!000 = \$4.05$ per can opener;

For 40,000: $\$136,\!000/40,\!000 = \$3.40$ per can opener. ◼

We have seen that the geometrical interpretation of slope is that it measures the "steepness" of the line. Examples 4 and 5 show that slope can also be interpreted as a *rate of change*. In Example 4, the computer depreciates $1240 per year, meaning that its value changes at a rate of -1245 per year and -1245 is the slope of the depreciation equation $y = -1245x + 7000$. In Example 5, the total cost of making can openers increases at a rate of $2.75 per can opener and 2.75 is the slope of the cost equation $y = 2.75x + 26,\!000$. Rates of change will be considered further in Section 3.6.

Point-Slope Form

Suppose the line L passes through the point (x_1, y_1) and has slope m. Let (x, y) be any other point on L. Using the points (x_1, y_1) and (x, y) to compute the slope m of L (see Figure 1–54), we have

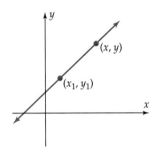

Figure 1–54

$$\frac{y - y_1}{x - x_1} = \text{slope of } L$$

$$\frac{y - y_1}{x - x_1} = m$$

Multiply both sides by $x - x_1$: $y - y_1 = m(x - x_1)$

Thus, the coordinates of every point on L satisfy the equation $y - y_1 = m(x - x_1)$ and we have this fact:

Point-Slope Form

▶ **The line with slope m through the point (x_1, y_1) is the graph of the equation**

$$y - y_1 = m(x - x_1).$$

Example 6   Find the equation of the line with slope 2 through the point $(1, -6)$.

Solution   Substitute 2 for m and $(1, -6)$ for (x_1, y_1) in the point-slope equation:

$$y - y_1 = m(x - x_1)$$

$$y - (-6) = 2(x - 1) \qquad \textit{[Point-slope form]}$$

$$y + 6 = 2x - 2$$

$$y = 2x - 8. \qquad \textit{[Slope-intercept form]} \quad ◼$$

Example 7 According to the National Center for Health Statistics, the U.S. infant mortality rate declined at an approximately linear rate from 10 per 1000 live births in 1988 to 8 per 1000 in 1994.

(a) Find an equation that gives the mortality rate y in year x.

(b) Use this equation to estimate the mortality rate in 1991 and 1997.

Solution

(a) Let $x = 0$ correspond to 1988, so that $x = 6$ corresponds to 1994. Then the given information can be represented by the points (0, 10) and (6, 8). We must find the equation of the line through these points. Its slope is

$$\frac{8 - 10}{6 - 0} = \frac{-2}{6} = -\frac{1}{3}.$$

Now we use the slope $-1/3$ and one of the points (0, 10) or (6, 8) to find the equation of the line. It doesn't matter which point, since both lead to the same equation.

$$y - y_1 = m(x - x_1) \qquad y - y_1 = m(x - x_1)$$

$$y - 10 = -\frac{1}{3}(x - 0) \qquad y - 8 = -\frac{1}{3}(x - 6)$$

$$y = -\frac{1}{3}x + 10 \qquad y - 8 = -\frac{1}{3}x + 2$$

$$y = -\frac{1}{3}x + 10$$

(b) Since 1991 corresponds to $x = 3$, the approximate mortality rate in 1993 was

$$y = -\frac{1}{3}x + 10 = -\frac{1}{3} \cdot 3 + 10 = -1 + 10 = 9 \text{ per 1000 births.}$$

and the approximate mortality rate in 1997 ($x = 10$) was

$$y = -\frac{1}{3}x + 10 = -\frac{1}{3} \cdot 10 + 10 = -\frac{10}{3} + 10 = \frac{20}{3} \approx 6.67 \text{ per 1000 births.} \quad \blacksquare$$

Parallel and Perpendicular Lines

The slope of a line measures how steeply it rises or falls. Since parallel lines rise or fall equally steeply, the following fact should be plausible (see Exercises 62–63 for a proof).

Slopes of Parallel Lines

Two nonvertical lines are parallel exactly when they have the same slope.

Example 8   Find the equation of the line L through $(2, -1)$ that is parallel to the line M whose equation is $3x - 2y + 6 = 0$.

Solution   First find the slope of M by rewriting its equation in slope-intercept form:

$$3x - 2y + 6 = 0$$

Subtract $3x + 6$ from both sides: $-2y = -3x - 6$

Divide both sides by -2: $y = \dfrac{3}{2}x + 3.$

Therefore, M has slope $3/2$. The parallel line L must have the same slope, $3/2$. Since $(2, -1)$ is on L, we can use the point-slope form to find its equation:

$$y - y_1 = m(x - x_1)$$

$$y - (-1) = \frac{3}{2}(x - 2) \qquad \textit{[Point-slope form]}$$

Multiply out right side: $y + 1 = \dfrac{3}{2}x - 3$

Subtract 1 from both sides: $y = \dfrac{3}{2}x - 4 \qquad \textit{[Slope-intercept form]}$ ■

Two lines that meet in a right angle (90° angle) are said to be **perpendicular.** As you might suspect, there is a close relationship between the slopes of two perpendicular lines.

Perpendicular Lines

> **Two nonvertical lines are perpendicular exactly when the product of their slopes is −1.**

A proof of this fact is outlined in Exercise 64.

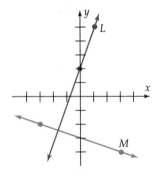

Figure 1–55

Example 9   In Figure 1–55 the line L through $(0, 2)$ and $(1, 5)$ appears to be perpendicular to the line M through $(-3, -2)$ and $(3, -4)$. We can confirm this fact by computing the slopes of these lines:

$$\text{slope } L = \frac{5 - 2}{1 - 0} = 3 \quad \text{and} \quad \text{slope } M = \frac{-4 - (-2)}{3 - (-3)} = \frac{-2}{6} = -\frac{1}{3}.$$

Since $3(-1/3) = -1$, the lines L and M are perpendicular.   ■

CAUTION

Perpendicular lines may not appear to be perpendicular on a calculator screen unless you use a square window (see Exercise 30).

Vertical Lines

The preceding discussion does not apply to vertical lines, whose equations have a different form from those above.

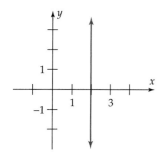

Figure 1–56

Example 10   Every point on the vertical line in Figure 1–56 has first coordinate 2. Thus, every point on the line satisfies $x + 0y = 2$. Thus, the line is the graph of the equation $x = 2$. If you try to compute the slope of this line, say using (2, 1) and (2, 4), you obtain $\dfrac{4 - 1}{2 - 2} = \dfrac{4}{0}$, which is not defined.   ■

Examples 2–10 illustrate the following facts (where A, B, C, b, and c are constants, with at least one of A or B nonzero).

Equations and Lines

> The graph of the equation $Ax + By = C$ is a straight line.
>
> A horizontal line has slope 0 and an equation of the form $y = b$.
>
> A vertical line has undefined slope and an equation of the form $x = c$.

Exercises 1.4

1. For which of the line segments in the figure is the slope
   (a) largest?   **(b)** smallest?
   (c) largest in absolute value?   **(d)** closest to zero?

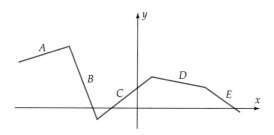

2. The doorsill of a campus building is 5 ft above ground level. To allow wheelchair access, the steps in front of the door are to be replaced by a straight ramp with constant slope 1/12, as shown in the figure. How long must the ramp be? [The answer is *not* 60 ft.]

In Exercises 3–6, find the slope and y-intercept of the line whose equation is given.

3. $2x - y + 5 = 0$ **4.** $3x + 4y = 7$

5. $3(x - 2) + y = 7 - 6(y + 4)$

6. $2(y - 3) + (x - 6) = 4(x + 1) - 2$

In Exercises 7–10, find the slope of the line through the given points.

7. $(1, 2)$; $(3, 7)$ **8.** $(-1, -2)$; $(2, -1)$

9. $(1/4, 0)$; $(3/4, 2)$ **10.** $(\sqrt{2}, -1)$; $(2, -9)$

11. Let L be a nonvertical straight line through the origin. L intersects the vertical line through $(1, 0)$ at a point P. Show that the second coordinate of P is the slope of L.

12. On one graph, sketch five line segments, not all meeting at a single point, whose slopes are five different positive numbers. Do this in such a way that the left-hand line has the largest slope, the second line from the left the next largest slope, and so on.

In Exercises 13–16, find the equation of the line with slope m that passes through the given point.

13. $m = 1$; $(3, 5)$ **14.** $m = 2$; $(-2, 1)$

15. $m = -1$; $(6, 2)$ **16.** $m = 0$; $(-4, -5)$

In Exercises 17–20, find the equation of the line through the given points.

17. $(0, -5)$ and $(-3, -2)$ **18.** $(4, 3)$ and $(2, -1)$

19. $(4/3, 2/3)$ and $(1/3, 3)$ **20.** $(6, 7)$ and $(6, 15)$

In Exercises 21–24, determine whether the line through P and Q is parallel or perpendicular to the line through R and S, or neither.

21. $P = (2, 5)$, $Q = (-1, -1)$ and $R = (4, 2)$, $S = (6, 1)$

22. $P = (0, 3/2)$, $Q = (1, 1)$ and $R = (2, 7)$, $S = (3, 9)$

23. $P = (-3, 1/3)$, $Q = (1, -1)$ and $R = (2, 0)$, $S = (4, -2/3)$

24. $P = (3, 3)$, $Q = (-3, -1)$ and $R = (2, -2)$, $S = (4, -5)$

In Exercises 25–26, determine whether the lines whose equations are given are parallel, perpendicular, or neither.

25. $2x + y - 2 = 0$ and $4x + 2y + 18 = 0$

26. $3x + y - 3 = 0$ and $6x + 2y + 17 = 0$

27. Use slopes to show that the points $(-5, -2)$, $(-3, 1)$, $(3, 0)$, and $(5, 3)$ are the vertices of a parallelogram.

28. Use slopes to show that the points $(-4, 6)$, $(-1, 12)$, and $(-7, 0)$ all lie on the same straight line.

29. Use slopes to determine if $(9, 6)$, $(-1, 2)$, and $(1, -3)$ are the vertices of a right triangle.

30. **(a)** Show that the lines $y = 2x + 4$ and $y = -.5x - 3$ are perpendicular.

(b) Graph the lines in part (a), using the standard viewing window. Do the lines look perpendicular?

(c) Find a viewing window in which the lines in part (a) appear to be perpendicular.

In Exercises 31–38, find an equation for the line satisfying the given conditions.

31. Through $(-2, 1)$ with slope 3.

32. y-intercept -7 and slope 1.

33. Through $(2, 3)$ and parallel to $3x - 2y = 5$.

34. Through $(1, -2)$ and perpendicular to $y = 2x - 3$.

35. x-intercept 5 and y-intercept -5.

36. Through $(-5, 2)$ and parallel to the line through $(1, 2)$ and $(4, 3)$.

37. Through $(-1, 3)$ and perpendicular to the line through $(0, 1)$ and $(2, 3)$.

38. y-intercept 3 and perpendicular to $2x - y + 6 = 0$.

If P is a point on a circle with center C, then the tangent line to the circle at P is the straight line through P that is perpendicular to the radius CP. In Exercises 39–42, find the equation of the tangent line to the circle at the given point.

39. $x^2 + y^2 = 25$ at $(3, 4)$ [*Hint:* Here C is $(0, 0)$ and P is $(3, 4)$; what is the slope of radius CP?]

40. $x^2 + y^2 = 169$ at $(-5, 12)$

41. $(x - 1)^2 + (y - 3)^2 = 5$ at $(2, 5)$

42. $(x + 3)^2 + (y - 4)^2 = 10$ at $(-2, 1)$

43. Let A, B, C, D be nonzero real numbers. Show that the lines $Ax + By + C = 0$ and $Ax + By + D = 0$ are parallel.

44. Let L be a line that is neither vertical nor horizontal and which does not pass through the origin. Show that L is the graph of $\dfrac{x}{a} + \dfrac{y}{b} = 1$, where a is the x-intercept and b is the y-intercept of L.

45. Sales of a software company increased linearly from \$120,000 in 1996 to \$180,000 in 1999.

(a) Find an equation that expresses the sales y in year x (where $x = 0$ corresponds to 1996).

(b) Estimate the sales in 2001.

46. Carbon dioxide (CO_2) concentration is measured regularly at the Mauna Loa observatory in Hawaii. The mean annual concentration in parts per million in various years is given in the table.

Year	Concentration (ppm)
1965	319.9
1970	325.3
1980	338.5
1988	351.3

(a) Do the data points lie on a single line? How do you know?

(b) Write a linear equation that uses the data from 1965 and 1988 to model CO_2 concentration over time. What does this model say the concentration should have been in 1970? Does the model overestimate or underestimate the concentration?

(c) Write a linear equation that uses the data from 1980 and 1988 to model CO_2 concentration over time. What does this model say the concentration should have been in 1965? Does the model overestimate or underestimate the concentration?

(d) What do the two models say about the concentration in the year 1995? Which model do you think is the most accurate? Why?

47. According to the Bureau of Debt of the U.S. Department of the Treasury, the national debt was about 2125 billion dollars in 1986 and about 5225 billion dollars in 1996.

(a) Find a linear equation that approximates the national debt y (in billions of dollars) in year x (with $x = 0$ corresponding to 1986).

(b) Use the equation of part (a) to estimate the national debt in 1991 and 2000. [For comparison purposes, the actual national debt in 1991 was 3665.3 billion dollars.]

48. At sea level, water boils at 212°F. At a height of 1100 ft, water boils at 210°F. The relationship between boiling point and height is linear.

(a) Find an equation that gives the boiling point y of water at a height of x feet.

Find the boiling point of water in each of the following cities (whose altitudes are given).

(b) Cincinnati, OH (550 ft)

(c) Springfield, MO (1300 ft)

(d) Billings, MT (3120 ft)

(e) Flagstaff, AZ (6900 ft)

49. A small plane costs $600,000 new. Ten years later, it is valued at $150,000. Assuming linear depreciation, find the value of the plane when it is 5 years old and when it is 12 years old.

50. In 1950 the death rate from heart disease was about 511 per 100,000 people. In 1996, the rate had decreased to 359 per 100,000.

(a) Assuming the rate decreased linearly, find an equation that gives the number y of deaths per 100,000 from heart disease in year x, with $x = 0$ corresponding to 1950. Round the slope of the line to one decimal place.

(b) Use the equation in part (a) to estimate the death rate in 1980 and 2000.

51. According to the National Center for Health Statistics of the U.S. Department of Health and Human Services, total health care expenditures were approximately $698 billion in 1990 and $989 billion in 1995. The growth in health care costs was approximately linear.

(a) Find an equation that gives the approximate health care costs y in year x (with $x = 0$ corresponding to 1990).

(b) Use the equation in part (a) to estimate the health care costs in 1993 and 1999.

52. The profit p (in thousands of dollars) on x thousand units of a specialty item is $p = .6x - 14.5$. The cost c of manufacturing x items is given by $c = .8x + 14.5$.

(a) Find an equation that gives the revenue r from selling x items.

(b) How many items must be sold for the company to break even (that is, for revenue to equal cost)?

53. A publisher has fixed costs of $180,000 for a mathematics text. The variable costs are $25 per book. The book sells for $40. Find equations that give

(a) the cost c of making x books.

(b) the revenue r from selling x books.

(c) the profit p from selling x books.

(d) What is the publisher's break-even point [see Exercise 52(b)]?

54. At the factory in Example 5, the cost of producing x can openers was given by $y = 2.75x + 26,000$.

(a) Write an equation that gives the average cost per can opener when x can openers are produced.

(b) How many can openers should be made to have an average cost of $3 per can opener?

55. The Whismo Hat Company has fixed costs of $50,000 and variable costs of $8.50 per hat.

(a) Find an equation that gives the total cost y of producing x hats.

(b) What is the average cost per hat when 20,000 are made? 50,000? 100,000?

56. (a) Write an equation that gives the average cost per hat when x hats are produced at the Whismo Hat Company of Exercise 55.
   (b) How many hats should be made to have an average cost of $15 per hat?

57. Suppose the cost of making x TV sets is given by $y = 145x + 120{,}000$.
   (a) Write an equation that gives the average cost per set when x sets are made.
   (b) How many sets should be made in order to have an average cost per set of $175?

Use the graph and the following information for Exercises 58–60. Rocky is an "independent" ticket dealer who markets choice tickets for Los Angeles Lakers' home games (California currently has no laws against scalping). Each graph shows how many tickets will be demanded by buyers at a particular price. For instance, when the Lakers play the Chicago Bulls, the graph shows that at a price of $160, no tickets are demanded. As the price (y-coordinate) gets lower, the number of tickets demanded (x-coordinate) increases.

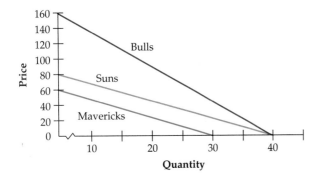

58. Write a linear equation that relates the quantity x of tickets demanded at price y when the Lakers play the
   (a) Dallas Mavericks.
   (b) Phoenix Suns.
   (c) Chicago Bulls.
   [*Hint:* In each case, use the points where the graph crosses the two axes to determine its slope.]

59. Use the equations from Exercise 58 to find the number of tickets Rocky would sell at a price of $40 for a game against the
   (a) Mavericks.   **(b)** Bulls.

60. Suppose Rocky has 20 tickets to sell. At what price could he sell them all when the Lakers play the
   (a) Mavericks?   **(b)** Suns?

61. A dietician has a "rule of thumb" for determining the target weight for her clients, based on their heights. For women, the target weight is 100 lb plus 5 lb for each inch over 5 ft in height. For men it is 106 lb plus 6 lb for each inch over 5 ft in height.
   (a) Fill in these tables.

Woman's Target Weight	
x inches over 5 ft	**Weight (lb)**
0	100
1	
2	
3	
6	
12	
15	

Man's Target Weight	
x inches over 5 ft	**Weight (lb)**
0	106
1	
2	
3	
6	
12	
15	

   (b) Write an equation that gives a woman's target weight y in terms of the number x of inches that her height exceeds 5 feet.
   (c) What is the target weight for a woman who is 5 ft 9 in. tall?
   (d) Do parts (b) and (c) for men.

Thinkers

62. Prove that nonvertical parallel lines L and M have the same slope, as follows. Suppose M lies above L and choose two points (x_1, y_1) and (x_2, y_2) on L.

(a) Let P be the point on M with first coordinate x_1. Let b denote the vertical distance from P to (x_1, y_1). Show that the second coordinate of P is $y_1 + b$.

(b) Let Q be the point on M with first coordinate x_2. Use the fact that L and M are parallel to show that the second coordinate of Q is $y_2 + b$.

(c) Compute the slope of L using (x_1, y_1) and (x_2, y_2). Compute the slope of M using the points P and Q. Verify that the two slopes are the same.

63. Show that two nonvertical lines with the same slope are parallel. [*Hint:* The equations of distinct lines with the same slope must be of the form $y = mx + b$ and $y = mx + c$ with $b \neq c$ (why?). If (x_1, y_1) were a point on both lines, its coordinates would satisfy both equations. Show that this leads to a contradiction and conclude that the lines have no point in common.]

64. This exercise provides a proof of the statement about slopes of perpendicular lines in the box on page 98. First, assume that L and M are nonvertical perpendicular lines that both pass through the origin. L and M intersect the vertical line $x = 1$ at the points $(1, k)$ and $(1, m)$ respectively, as shown in the figure at the right.

(a) Use $(0, 0)$ and $(1, k)$ to show that L has slope k. Use $(0, 0)$ and $(1, m)$ to show that M has slope m.

(b) Use the distance formula to compute the length of each side of the right triangle with vertices $(0, 0)$, $(1, k)$, and $(1, m)$.

(c) Use part (b) and the Pythagorean Theorem to find an equation involving k, m, and various constants. Show that this equation simplifies to $km = -1$. This proves one half of the statement.

(d) To prove the other half, assume that $km = -1$ and show that L and M are perpendicular as follows. You may assume that a triangle whose sides a, b, c satisfy $a^2 + b^2 = c^2$ is a right triangle with hypotenuse c. Use this fact and do the computation in part (b) in reverse (starting with $km = -1$) to show that the triangle with vertices $(0, 0)$, $(1, k)$ and $(1, m)$ is a right triangle, so that L and M are perpendicular.

(e) Finally, to prove the general case when L and M do not intersect at the origin, let L_1 be a line through the origin that is parallel to L and M_1 a line through the origin that is parallel to M. Then L and L_1 have the same slope and M and M_1 have the same slope (why?). Use this fact and parts (a)–(d) to prove that L is perpendicular to M exactly when $km = -1$.

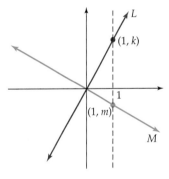

65. Show that the diagonals of a square are perpendicular. [*Hint:* Place the square in the first quadrant of the plane, with one vertex at the origin and sides on the positive axes. Label the coordinates of the vertices appropriately.]

1.5 Linear Models*

People working in business, medicine, agriculture, and other fields frequently want to know the relationship between two quantities. For instance,

How does money spent on advertising affect sales?

What effect does a fertilizer have on crop yield?

Do large doses of certain vitamins lengthen life expectancy?

*This section is optional. It will regularly be used in clearly identifiable exercises, but not elsewhere in the text.

In many such situations there is sufficient data available to construct a **mathematical model,** such as an equation or graph, which provides the desired answers or predicts the likely outcome in cases not included in the data. In this section we consider applications in which the data can be modeled by a linear equation. More complicated models will be considered in later sections.

Example 1 The following table, from the U.S. Department of Commerce, shows the poverty level for a family of four in selected years (families whose income is below this level are considered to be in poverty).

Year	1990	1991	1992	1993	1994	1995	1996
Income	$13,359	13,942	14,335	14,763	15,141	15,569	16,036

We let $x = 0$ correspond to 1990 and write the incomes in thousands (for instance, 13.359 in place of 13,359). Plotting the data points (0, 13.359), (1, 13.942), etc., we obtain the scatter plot in Figure 1–57. The fact that these points are almost in a straight line suggests that a linear equation should provide a suitable model. One such equation is

$$y = .43x + 13.43,$$

whose graph is shown in Figure 1–58.

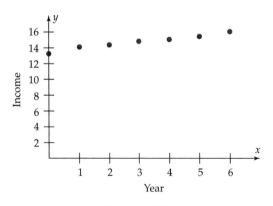

Figure 1–57

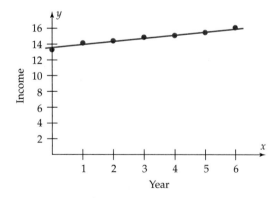

Figure 1–58

The graph appears to fit the data quite well. In fact, the poverty level given by the equation differs from the actual level each year by no more than $82, as the following table demonstrates.

Year x	0	1	2	3	4	5	6
Actual	13,359	13,942	14,335	14,763	15,141	15,569	16,036
Model	13,430	13,860	14,290	14,720	15,150	15,580	16,010

Assuming that this model remains valid after 1996, the poverty level in 2002 ($x = 12$) is approximately

$$y = .43x + 13.43 = .43(12) + 13.43 = \$18{,}590. \quad \blacksquare$$

Example 1 shows how a typical model approximates the actual data and makes it possible to predict the outcome in other cases. However, it does not deal with questions such as

How do you find an equation that models the data?

How do you determine which equation is the best model for the data?

These questions are best answered with a simplified example.

Example 2 The weekly amount spent on advertising and the weekly sales revenue of a small store over a five-week period are shown in the table.

Advertising Expenditure x (in hundreds of dollars)	1	2	3	4	5
Sales Revenue y (in thousands of dollars)	2	2	3	3	5

The data points are (1, 2), (2, 2), (3, 3), (4, 3), and (5, 5). One way to find a linear equation that models these points is to choose two of them and find the equation of the line they determine (as we did in Section 1.4). Using the points (1, 2) and (3, 3), we have:

Slope of line: $\dfrac{3 - 2}{3 - 1} = \dfrac{1}{2} = .5$

Equation:
$$y - 2 = .5(x - 1)$$
$$y = .5x - .5 + 2$$
$$y = .5x + 1.5$$

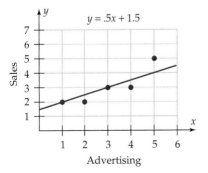

Figure 1–59

With the points (2, 2) and (5, 5) we have:

Slope of line: $\dfrac{5-2}{5-2} = 1$

Equation: $y - 2 = 1(x - 2)$

$$y = x$$

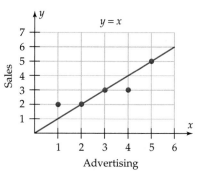

Figure 1–60

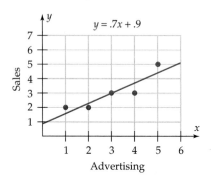

Figure 1–61

Another model, $y = .7x + 9$, which can be obtained by a process described below, is shown in Figure 1–61.

To determine the best model, we compute for each one the difference between the actual revenue r and the revenue y given by the model. If the data point is (x, r) and the corresponding point on the model is (x, y), then the difference $r - y$ measures the error in the model for that particular value of x. Graphically, this is the (positive or negative) vertical distance between the two points. Here are the errors for all data points in each model.

$y = .5x + 1.5$

Data Point (x, r)	Point on Model (x, y)	Error $r - y$
(1, 2)	(1, 2)	0
(2, 2)	(2, 2.5)	−.5
(3, 3)	(3, 3)	0
(4, 3)	(4, 3.5)	−.5
(5, 5)	(5, 4)	1
		sum 0

$y = x$

Data Point (x, r)	Point on Model (x, y)	Error $r - y$
(1, 2)	(1, 1)	1
(2, 2)	(2, 2)	0
(3, 3)	(3, 3)	0
(4, 3)	(4, 4)	−1
(5, 5)	(5, 5)	0
		sum 0

$y = .7x + .9$

Data Point (x, r)	Point on Model (x, y)	Error $r - y$
(1, 2)	(1, 1.6)	.4
(2, 2)	(2, 2.3)	−.3
(3, 3)	(3, 3)	0
(4, 3)	(4, 3.7)	−.7
(5, 5)	(5, 4.4)	.6
		sum 0

The sum of the errors in each case is 0. Although this indicates that all the models are reasonable, it doesn't help us to decide if one model is better than another. So we try another way to measure the accuracy of our models: compute the sum of the *squares* of the errors. When we do this, there is no canceling and we obtain a different sum for each model:

$y = .5x + 1.5$

Data Point (x, r)	Model Point (x, y)	Error $r - y$	Squared Error $(r - y)^2$
(1, 2)	(1, 2)	0	0
(2, 2)	(2, 2.5)	−.5	.25
(3, 3)	(3, 3)	0	0
(4, 3)	(4, 3.5)	−.5	.25
(5, 5)	(5, 4)	1	1
			sum 1.5

(a)

$y = x$

Data Point (x, r)	Model Point (x, y)	Error $r - y$	Squared Error $(r - y)^2$
(1, 2)	(1, 1)	1	1
(2, 2)	(2, 2)	0	0
(3, 3)	(3, 3)	0	0
(4, 3)	(4, 4)	−1	1
(5, 5)	(5, 5)	0	0
			sum 2

(b)

$y = .7x + .9$

Data Point (x, r)	Model Point (x, y)	Error $r - y$	Squared Error $(r - y)^2$
(1, 2)	(1, 1.6)	.4	.16
(2, 2)	(2, 2.3)	−.3	.09
(3, 3)	(3, 3)	0	0
(4, 3)	(4, 3.7)	−.7	.49
(5, 5)	(5, 4.4)	.6	.36
			sum 1.1

(c)

Using the sum of the squares of the errors as a measure of accuracy has the effect of emphasizing large errors (those with absolute value greater than 1) because the square is greater than the error and minimizing small errors (those with absolute value less than 1) because the square is less than the error. By this measure, the best of the three models is $y = .7x + .9$ because the sum of the squares of its error is smallest. ∎

It can be proved that for any set of data points there is one and only one line for which the sum of the squares of the errors is as small as possible. This line is called the **least squares regression line** and the computational process for finding its equation (which is built into most calculators) is called **linear regression.** Linear regression was used to obtain the model $y = .7x + .9$ in Example 2, as well as the linear equation in Example 1.

Example 3 The total number of farmworkers (in millions) in selected years is shown below. (*Source:* Economic Research Service of the U.S. Department of Agriculture)

Year	Workers	Year	Workers	Year	Workers
1900	29.030	1950	59.230	1985	106.210
1920	42.206	1960	67.990	1990	117.490
1930	48.686	1970	79.802	1994	120.380
1940	51.742	1980	105.060		

Use linear regression to find an equation that models this data. Use the equation to estimate the number of farmworkers in 1975 and 2000.

Solution Let $x = 0$ correspond to 1900, so that the data points are (0, 29.030), (20, 42.206), ... , (94, 120.380). We display the calculator's statistics editor and enter the data points: x-coordinates are the first list and y-coordinates (in the same order) are the second list (Figure 1–62).* Using the statistical plotting feature of the calculator to plot the data points, we obtain Figure 1–63, which shows that the data is approximately linear.

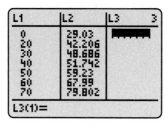

Figure 1–62

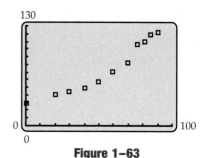

Figure 1–63

---

*Consult the Technology Tip after the example for details on how to carry out this and subsequent steps in the process.

Now we use the linear regression feature to obtain Figure 1–64, which shows that the equation of the least squares regression line is

$$y = 1.011599433x + 18.33145006.$$

When you do this on most calculators, you can simultaneously store the regression equation in the equation memory, so that it can be graphed along with the data points (Figure 1–65).

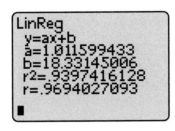

Figure 1–64

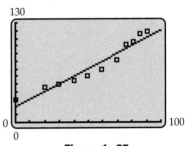

Figure 1–65

Figure 1–65 suggests that the regression line provides a reasonable model for approximating the number of farmworkers in a given year. If $x = 75$, then

$$y = 1.011599433(75) + 18.33145006 \approx 94.201$$

and if $x = 100$,

$$y = 1.011599433(100) + 18.33145006 \approx 119.491.$$

Therefore, there were approximately 94,201,000 farmworkers in 1975 and 119,491,000 in 2000. ∎

Technology Tip

Most calculators allow you to store three or more statistics graphs (identified by number); the following directions assume that the first one is used.

To call up statistics editor, use these commands:

TI-83 and Sharp 9600: STAT EDIT (lists are L_1, L_2, ...);

(Tip continues)

TI-85/86: STAT EDIT (built-in lists are x-stat, y-stat; it is usually better to create your own lists; we use L1 and L2 as list names here);*

TI-89: APPS DATA/MATRIX EDITOR NEW; then choose DATA as the TYPE, enter a VARIABLE name (we use L here), and key in ENTER (lists are C_1, C_2, ...);

Casio 9850: STAT (lists are List 1, List 2, ...);

HP-38: LIB STATISTICS (lists are C_1, C_2, ...).

To graph the data points that have been entered as lists (x-coordinates in the first list, corresponding y-coordinates in the second), choose an appropriate viewing window (TI and Sharp only). On TI-85, use STAT DRAW SCAT to plot the points determined by the lists currently chosen in the STAT EDIT screen. On other calculators, use these commands to enter the setup screen:

TI-83 and Sharp 9600: STAT-PLOT (on keyboard); choose PLOT 1;

TI-86: STAT; choose PLOT and PLOT 1;

TI-89: from the Data Editor, choose PLOT-SETUP (F_2) and DEFINE (F_1);

Casio: STAT GRPH SET;

HP-38: LIB STATISTICS PLOT-SETUP and SYMB.

Key in the appropriate information [on/off (TI-83/86 and Sharp), graph number (Casio and HP), lists to be used, graph type (scatter plot), the mark to be used for data points, and on HP-38, the viewing window]. Then press GRAPH (on TI, Sharp) or GPH 1 (on Casio) or PLOT (on HP-38).

To produce the least squares regression line, store its equation as y_1 in the equation memory, and graph it, use these commands:

TI-83: STAT CALC LinReg $(ax + b)$ L_1, L_2, Y_1;

TI-85: STAT CALC L1, L2 LINR; then press STREG and enter y_1 as the name;

TI-86: STAT CALC Lin R L1, L2, y_1;

TI-89: From the Data Editor, choose CALC (F_5); enter LIN REG as CALCULATION TYPE, C_1 as x, C_2 as y; and choose $y_1(x)$ in STORE REGEQ;

Sharp 9600: STAT REG Rg_ax + b (L_1, L_2, Y_1) [after exiting the statistics editor and returning to the home screen].

Then press GRAPH or DRAW to obtain the graph of the equation and the data points (assuming PLOT 1 has not been turned off). On HP-38, after plotting the data points, use MENU FIT to graph the regression line and SYMB to see its equation. On Casio 9850, after plotting the data points, Press X for the equation of the regression line and DRAW for its graph.

*When statistical computations are run on TI-85/86, the lists used are automatically copied into the x-stat and y-stat lists, replacing whatever was there before. So use the x-stat and y-stat lists only if you don't want to save them. See your instruction manual to find out how to create new lists in the statistics editor.

You may have noticed that Figure 1–64 on page 109 contains a number *r* (and its square), in addition to the coefficients of the regression line equation. The number *r*, which is called the **correlation coefficient,** is a statistical measure of how well the least squares regression line fits the data points. It is always between −1 and 1. The closer the absolute value of *r* is to 1, the better the fit. For instance, the regression line in Example 3 is a good fit since $r \approx .97$. When $|r| = 1$, the fit is perfect: all the data points are on the regression line. Conversely, a regression coefficient near 0 indicates a poor fit.

Example 4 The numbers of unemployed people in the labor force (in millions) for 1984–1995 are as follows. (*Source:* U.S. Department of Labor, Bureau of Labor Statistics)

Year	Unemployed	Year	Unemployed	Year	Unemployed
1984	8.539	1988	6.701	1992	9.613
1985	8.312	1989	6.528	1993	8.940
1986	8.237	1990	7.047	1994	7.996
1987	7.425	1991	8.628	1995	7.404

Is a linear equation a good model for this data?

Solution After entering the data as two lists in the statistics editor (with $x = 0$ corresponding to 1980), you can test it graphically or analytically.

Graphical: Plotting the data points (Figure 1–66) shows that they do not form a linear pattern (unemployment tends to rise and fall).

Analytical: Linear regression (Figure 1–67) produces an equation whose correlation coefficient is $r \approx .092$, a number very close to 0, which indicates that the regression line is a very poor fit for the data.

Therefore, a linear equation is not a good model for this data. ■

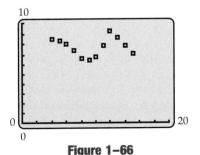

Figure 1–66

Figure 1–67

GRAPHING EXPLORATION

Enter the data from Example 4 in the statistics editor of your calculator. Graph the data points. Graph the least squares regression line on the same screen to see how poorly it fits the data.

Example 5 Forty people were randomly selected for a survey that asked their annual income and how many hours they watched TV each day. The results were as follows.

Income	TV	Income	TV	Income	TV	Income	TV
$12,000	8	$56,000	3	$48,000	3	$51,000	5
$18,000	6	$12,000	7	$20,000	4	$22,000	5
$26,000	5	$24,000	4	$14,000	4	$22,000	6
$21,000	6	$28,000	7	$96,000	0	$15,000	5
$16,000	7	$31,000	5	$33,000	4	$92,000	2
$35,000	5	$53,000	4	$29,000	6	$75,000	3
$85,000	3	$39,000	3	$16,000	7	$42,000	4
$68,000	2	$80,000	2	$64,000	0	$17,000	6
$17,000	7	$88,000	1	$77,000	5	$53,000	4
$17,000	7	$31,000	3	$45,000	3	$73,000	4

If possible, find a linear model for this data.

Solution We express incomes in thousands so that the data points are (12, 8), (18, 6), etc. We enter the data points in the statistics editor of a calculator and find that the equation of the least squares regression line (rounded) is $y = -.06x + 6.85$, with correlation coefficient $r \approx -.78$. Since $|r|$ is relatively close to 1, the regression line is a fairly good model for the data, as shown in Figure 1–68. ■

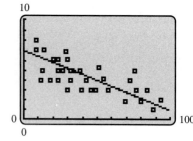

Figure 1–68

The correlation coefficient r always has the same sign as the slope of the least squares regression line. So when r is negative, as in Example 5, the regression line moves downward from left to right (Figure 1–68). In other words, as x increases, y decreases. In such cases, we say that the data has a **negative correlation.** When r is positive, as in Example 3, the regression line slopes upward from left to right (Figure 1–65) and we say that the data has a **positive correlation:** as x increases, y also increases. When r is close to 0 (regardless of sign), we say that there is **no correlation,** as in Example 4.

Exercises 1.5

1. (a) In Example 2, find the equation of the line through the data points (1, 2) and (5, 5).
 (b) Compute the sum of the squares of the errors for this line. Is it a better model than any of the models in the Example? Why?

2. The linear model in Example 1 is the least squares regression line with coefficients rounded. Find the correlation coefficient for this model.

3. (a) In Example 3, find the slope of the line through the data points for 1920 and 1994.
 (b) Find the equation of the line through these two data points.
 (c) Which model predicts the higher number of farmworkers in 2010: the line in part (b) or the regression line found in Example 3?

4. If you consider only the data for 1992–1994 in Example 4, is there a positive or negative or no correlation?

In Exercises 5–8, determine whether the given scatter plot of the data indicates that there is a positive correlation, negative correlation, or very little correlation.

5.

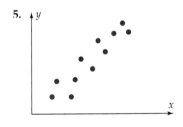

6.

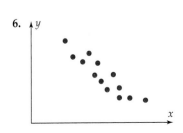

7.

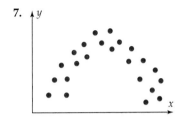

8.

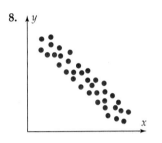

In Exercises 9–14, construct a scatter diagram for the data and answer these questions: (a) Does the data appear to be linear? (b) If so, is there a positive or negative correlation?

9. The U.S. gross domestic product (GDP) is the total value of all goods and services produced in the United States. The table shows the per capita GDP in constant 1992 dollars (adjusted for inflation). Let $x = 0$ correspond to 1990.

Year	Per Capita Gross Domestic Product
1990	$24,743
1991	24,058
1992	24,311
1993	24,638
1994	25,150
1995	25,599
1996	25,829
1997	26,622

10. The table shows the monthly premium for a term life insurance policy for female nonsmokers. Let x represent age and y premiums.

Age	Premium
25	$11.57
30	11.66
35	11.83
40	13.05
45	16.18
50	21.32
55	29.58

11. The table shows the percent of persons below the U.S. poverty level in selected years. Let $x = 0$ correspond to 1960.

Year	Percent of persons below poverty level in United States
1960	22.2
1965	17.3
1970	12.6
1975	12.3
1980	13.0
1985	14.0
1990	13.5
1991	14.2
1992	14.8
1993	15.1
1994	14.5
1995	13.8
1996	13.7

12. The vapor pressure y of water depends on the temperature x, as given in the table.

Temperature (°C)	Pressure (mm Hg)
0	4.6
10	9.2
20	17.5
30	31.8
40	55.3
50	92.5
60	149.4
70	233.7
80	355.1
90	525.8
100	760

13. The table shows the U.S. Bureau of Census population data for St. Louis, Missouri. Let $x = 0$ correspond to 1950.

Year	Population		Year	Population
1950	856,796		1990	396,685
1970	622,236		1994	368,215
1980	452,801			

14. The table shows the U.S. disposable income (personal income less personal taxes) in current dollars. (*Source:* Bureau of Economics Analysis, U.S. Dept. of Commerce) Let $x = 0$ correspond to 1990.

Year	Disposable Personal Income
1990	4179.4
1991	4356.8
1992	4626.7
1993	4829.2
1994	5052.7
1995	5355.7
1996	5608.3

15. The table gives the annual U.S. consumption (in million of pounds) of beef and poultry. (*Source:* U.S. Dept. of Agriculture)
    (a) Make scatter plots for both beef and poultry consumption, using the actual years (1990, 1991, etc.) as x in each case.
    (b) Without graphing, use your knowledge of slopes to determine which of the following equations models beef consumption and which one models poultry consumption. Confirm your answer by graphing.

$$y_1 = 717.44x - 1,405,160 \qquad y_2 = 329.86x - 632,699$$

Year	Beef	Poultry
1990	24,031	22,151
1991	24,113	23,270
1992	24,261	24,394
1993	24,006	25,099
1994	25,125	25,754
1995	25,533	25,940
1996	25,875	26,614

16. The table below gives the median weekly earnings of full-time workers 25 years and older by the amount of education. (*Source:* U.S. Bureau of Labor Statistics).
    (a) Make four scatter plots, one for each educational group, using $x = 0$ to correspond to 1980.
    (b) Four linear models are given here. Match each model with the appropriate data set.

$$y_1 = 16.06x + 312 \qquad y_2 = 7.79x + 228$$
$$y_3 = 25.64x + 379 \qquad y_4 = 11.7x + 273$$

Year	Less than 4 years of high school	High School 4 years	College 1–3 years	College 4 years or more
1980	222	266	304	376
1982	248	302	351	438
1984	263	323	382	486
1986	278	344	409	525
1988	288	368	430	585
1990	304	386	476	639

In Exercises 17–23, use the linear regression feature of your calculator to find the required model.

17. The table shows the number of deaths per 100,000 people from heart disease.

Year	1950	1960	1970	1980	1990	1996
Deaths	510.8	521.8	496.0	436.4	368.3	358.6

   (a) Find a linear model for this data, in which $x = 0$ corresponds to 1950.
   (b) In the unlikely event that the linear model in part (a) remains valid far into the future, will there be a time when death from heart disease has been completely eliminated? If so, when would this occur?

18. The table shows the share of total U.S. household income received by the poorest 20% of households and the share received by the wealthiest 5% of households. (*Source:* U.S. Census Bureau)

Shares of Household Income		
Year	Lowest 20%	Top 5%
1985	4	17.0
1990	3.9	18.6
1995	3.7	21.0
1996	3.7	21.4

(a) Find a linear model for the income share of the poorest 20% of households. Let $x = 0$ correspond to 1985.

(b) Find a linear model for the income share of the wealthiest 5% of households.

(c) What do the slopes of the two models suggest for each model?

(d) Assuming these models remain accurate, will the income gap between the wealthy and the poor grow, stay about the same, or decline in the year 2000?

19. The table shows the percent of federal aid given to college students in the form of loans in selected years at a particular college.

Year (in which school year begins)	Loans (%)
1975	18
1978	30
1984	54
1987	66
1990	78

(a) Find a linear model for this data, with $x = 0$ corresponding to 1975.

(b) Interpret the meaning of the slope and the y-intercept.

(c) If the model remains accurate in the future, what percentage of federal student aid are loans in 2000? Does this make sense?

20. The table shows the percent of federal aid given to college students in the form of grants or work-study in selected years at the college of Exercise 19.

Year (in which school year begins)	Grants and Work-Study (%)
1975	82
1978	70
1984	46
1987	34
1990	22

(a) Find a linear model for this data, with $x = 0$ corresponding to 1975.

(b) Graph the model from part (a) and the model from Exercise 19 on the same axes. What appears to be the trend in the federal share of financial aid to college students?

(c) In what year is the percent of federal aid the same for loans as for grants and work-study?

21. The table gives the average number of take-out meals purchased at restaurants per person in selected years. (*Source:* NPD Group's Crest Service)

Year	Average Number of Take-Out Meals per Person, Annually
1984	43
1986	48
1988	53
1990	55
1992	57
1994	61
1996	65

(a) Make a scatter plot of the data, with $x = 0$ corresponding to 1980.

(b) Find a linear model for the data.

(c) According to the model, what was the average number of take-out meals purchased in 1993? In 2000?

22. The table shows the median time (in months) after the application has been made for the Food and Drug Administration to approve a new drug. (*Source:* U.S. Food and Drug Administration)

Year	Median Time to Approval
1986	32.9
1987	29.9
1988	27.2
1989	29.3
1990	24.3
1991	22.1
1992	22.6
1993	23.0
1994	17.5
1995	15.9
1996	14.3

(a) Make a scatter plot of the data, with $x = 0$ corresponding to 1980.

(b) Find a linear model for the data.

(c) What are the limitations of this model? [*Hint:* What does it say about approval in the year 2005?]

23. The following data give production (x) and consumption (y) of primary energy* in quadrillion British thermal units (Btu) for a sample of countries in 1995.

Australia (7.29, 4.43)   Japan (3.98, 21.42)
Brazil (4.55, 6.76)   Mexico (8.15, 5.59)
Canada (16.81, 11.72)   Poland (3.74, 3.75)
China (35.49, 35.67)   Russia (39.1, 26.75)
France (4.92, 9.43)   Saudi Arabia (20.34, 3.72)
Germany (5.42, 13.71)   South Africa (6.08, 5.51)
India (8.33, 10.50)   United States (69.1, 88.28)
Indonesia (6.65, 3.06)   United Kingdom
                              (10.57, 9.85)
Iran (9.35, 3.90)   Venezuela (8.22, 2.53)

(a) Make a scatter plot of the data.

(b) Find a linear model for the data. Graph the model with the scatter diagram.

(c) In 1995 what three countries were the world's leading producers and consumers of energy?

(d) As a general trend what does it mean if a country is "above" the linear model?

(e) As a general trend what does it mean if a country is "below" the linear model?

(f) Identify any countries that appear to differ dramatically from most of the others.

*Production and consumption include petroleum, natural gas, coal, net hydroelectric, nuclear, geothermal, solar, wind electric power, and biofuels.

Chapter 1   *Review*

Important Concepts

Important Facts and Formulas

· *Distance Formula:* The distance from (x_1, y_1) to (x_2, y_2) is
$$\sqrt{(x_1 - x_2)^2 + (y_1 - y_2)^2}.$$

· *Midpoint Formula:* The midpoint of the line segment from (x_1, y_1) to (x_2, y_2) is
$$\left(\frac{x_1 + x_2}{2}, \frac{y_1 + y_2}{2}\right).$$

· Equation of the circle with center (c, d) and radius r:
$$(x - c)^2 + (y - d)^2 = r^2.$$

· The slope of the line through (x_1, y_1) and (x_2, y_2) (where $x_1 \neq x_2$) is
$$\frac{y_2 - y_1}{x_2 - x_1}.$$
· The equation of the line with slope m and y-intercept b is
$$y = mx + b$$
· The equation of the line through (x_1, y_1) with slope m is
$$y - y_1 = m(x - x_1).$$
· Nonvertical parallel lines have the same slope.
· Two nonvertical lines are perpendicular exactly when the product of their slopes is -1.

Review Questions

1. Find the distance from $(1, -2)$ to $(4, 5)$.

2. Find the distance from $(3/2, 4)$ to $(3, 5/2)$.

3. Find the distance from (c, d) to $(c - d, c + d)$.

4. Find the midpoint of the line segment from $(-4, 7)$ to $(9, 5)$.

5. Find the midpoint of the line segment from (c, d) to $(2d - c, c + d)$.

6. Find the equation of the circle with center $(-3, 4)$ that passes through the origin.

7. **(a)** If $(1, 1)$ is on a circle with center $(2, -3)$, what is the radius of the circle?
   (b) Find the equation of the circle in part (a).

8. Sketch the graph of $3x^2 + 3y^2 = 12$.

9. Sketch the graph of $(x - 5)^2 + y^2 - 9 = 0$.

10. Find the equation of the circle of radius 4 whose center is the midpoint of the line segment joining $(3, 5)$ and $(-5, -1)$.

11. Which of statements (a)–(d) are descriptions of the circle with center $(0, -2)$ and radius 5?
    (a) The set of points (x, y) that satisfy $|x| + |y + 2| = 5$.
    (b) The set of all points whose distance from $(0, -2)$ is 5.
    (c) The set of all points (x, y) such that $x^2 + (y + 2)^2 = 5$.
    (d) The set of all points (x, y) such that $\sqrt{x^2 + (y + 2)^5} = 5$.

12. If the equation of a circle is $3x^2 + 3(y - 2)^2 = 12$, which of the following statements is true?
    (a) The circle has diameter 3.   **(b)** The center of the circle is $(2, 0)$.
    (c) The point $(0, 0)$ is on the circle.   **(d)** The circle has radius $\sqrt{12}$.
    (e) The point $(1, 1)$ is on the circle.

13. The national unemployment rates for 1990–1996 were as follows. (*Source:* U.S. Department of Labor, Bureau of Labor Statistics)

Year	1990	1991	1992	1993	1994	1995	1996
Rate	5.6%	6.8%	7.5%	6.9%	6.1%	5.6%	5.4%

Sketch a scatter plot and a line graph for this data, letting $x = 0$ correspond to 1990.

14. The table shows the average speed (mph) of the winning car in the Indianapolis 500 race in selected years.

Year	1980	1982	1984	1986	1988	1990	1992	1994	1996
Speed	143	162	164	171	145	186	134	161	148

Sketch a scatter plot and a line graph for this data, letting $x = 0$ correspond to 1980.

In Questions 15–20,
(a) Determine which of the viewing windows a–e shows a complete graph of the equation.
(b) For each viewing window that does not show a complete graph, explain why.
(c) Find a viewing window that gives a "better" complete graph than windows a–e (meaning that the window is small enough to show as much detail as possible, yet large enough to show a complete graph).
    a. Standard viewing window
    b. $-10 \leq x \leq 10, -200 \leq y \leq 200$
    c. $-20 \leq x \leq 20, -500 \leq y \leq 500$
    d. $-50 \leq x \leq 50, -50 \leq y \leq 50$
    e. $-1000 \leq x \leq 1000, -1000 \leq y \leq 1000$

15. $y = .2x^3 - .8x^2 - 2.2x + 6$ **16.** $y = x^3 - 11x^2 - 25x + 275$

17. $y = x^4 - 7x^3 - 48x^2 + 180x + 200$

18. $y = x^3 - 6x^2 - 4x + 24$ **19.** $y = .03x^5 - 3x^3 + 69.12x$

20. $y = .00000002x^6 - .0000014x^5 - .00017x^4 + .0107x^3 + .2568x^2 - 12.096x$

21. According to data from the American Hospital Association, the number of hospital admissions y (in millions) between 1970 and 1995 can be approximated by the equation

$$y = .000185x^4 - .0064x^3 + .003x^2 + 1.08x + 31.75,$$

where $x = 0$ corresponds to 1970. In what year during this period were hospital admissions highest?

22. According to data from the U.S. Department of Education, the total enrollment y in U.S. elementary and secondary schools (in millions) between 1970 and 1995 can be approximated by

$$y = -.000139x^4 + .0087x^3 - .136x^2 + .133x + 51.25,$$

where $x = 0$ corresponds to 1970. In what year during this period was enrollment the lowest?

In Questions 23–24, determine graphically whether the statement could possibly be true or is definitely false.

23. $6x^4 - 7x^3 + 8x^2 - 7x + 2 = (2x - 1)(3x + 2)(x^2 - 1)$
24. $x^7 - 2x^6 - 6x^5 + 8x^4 + 17x^3 - 6x^2 - 20x - 8 = (x - 2)^3(x + 1)^4$

In Questions 25–28, sketch a complete graph of the equation.

25. $y = |x - 10| - 3$ **26.** $y = -|x + 8| + 12$
27. $y = x^2 + 10$ **28.** $y = x^3 - 7x + 24$

In Questions 29–34, find a solution of the equation that lies in the given interval.

29. $x^4 + x^3 - 10x^2 = 8x + 16; \quad (x \geq 0)$

30. $2x^4 + x^3 - 2x^2 + 6x + 2 = 0; \quad (x < -1)$

31. $\dfrac{x^3 + 2x^2 - 3x + 4}{x^2 + 2x - 15} = 0; \quad (x > -10)$

32. $\dfrac{3x^4 + x^3 - 6x^2 - 2x}{x^5 + x^3 + 2} = 0; \quad (x \geq 0)$

33. $\sqrt{x^3 + 2x^2 - 3x - 5} = 0; \quad (x \geq 0)$

34. $\sqrt{1 + 2x - 3x^2 + 4x^3 - x^4} = 0; \quad (-5 < x < 5)$

35. According to data from the U.S. Department of Justice, the number of prisoners y in state and federal prisons (in thousands) over the last two decades can be approximated by the equation

$$y = -.0034x^3 + 1.91x^2 + 21.7x + 315.9,$$

where $x = 0$ corresponds to 1980. In what year did the prison population first reach one million?

36. Use the information in Question 22 to estimate the number of children enrolled in U.S. elementary and secondary schools in the year 2000.

37. (a) What is the y-intercept of the graph of

$$y = x - \frac{x - 2}{5} + \frac{3}{5}?$$

(b) What is the slope of this line?

38. Find the equation of the line passing through $(1, 3)$ and $(2, 5)$.

39. Find the equation of the line passing through $(2, -1)$ with slope 3.

40. (a) Find the y-intercept of the line $2x + 3y - 4 = 0$.
(b) Find the equation of the line through $(1, 3)$ that has the same y-intercept as the line in part (a).

41. Find the equation of the line through $(-4, 5)$ that is parallel to the line through $(1, 3)$ and $(-4, 2)$.

42. Sketch the graph of the line $3x + y - 1 = 0$.

43. As a balloon is launched from the ground, the wind is blowing it due east. The conditions are such that the balloon is ascending along a straight line with slope 1/5. After 1 hour the balloon is 5000 ft vertically above the ground. How far east has the balloon blown?

44. The point (u, v) lies on the line $y = 5x - 10$. What is the slope of the line passing through (u, v) and the point $(0, -10)$?

In Questions 45–51, determine whether the statement is true or false.

45. The graph of $x = 5y + 6$ has y-intercept 6.

46. The graph of $2y - 8 = 3x$ has y-intercept 4.

47. The lines $3x + 4y = 12$ and $4x + 3y = 12$ are perpendicular.

48. Slope is not defined for horizontal lines.

49. The line in the figure has positive slope.

50. The line in the figure does not pass through the third quadrant.

51. The y-intercept of the line in the figure is negative.

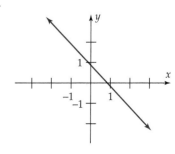

52. Consider the *slopes* of the lines shown in the figure. Which slope has the largest *absolute value*?

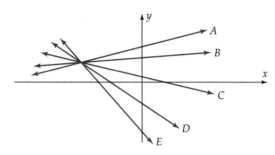

53. Which of the following lines rises most steeply from left to right?
    (a) $y = -4x - 10$ (b) $y = 3x + 4$
    (c) $20x + 2y - 20 = 0$ (d) $4x = y - 1$
    (e) $4x = 1 - y$

54. Which of the following lines is *not* perpendicular to the line $y = x + 5$?
    (a) $y = 4 - x$ (b) $y + x = -5$
    (c) $4 - 2x - 2y = 0$ (d) $x = 1 - y$
    (e) $y - x = \dfrac{1}{5}$

55. Which of the following lines does *not* pass through the third quadrant?
    (a) $y = x$ (b) $y = 4x - 7$
    (c) $y = -2x - 5$ (d) $y = 4x + 7$
    (e) $y = -2x + 5$

56. What is the y-intercept of the line $2x - 3y + 5 = 0$?

57. The average life expectancy increased linearly from 62.9 years for a person born in 1940 to 75.4 years for a person born in 1990.
    (a) Find an equation that gives the average life expectancy y of a person born in year x, with $x = 0$ corresponding to 1940.
    (b) Use the equation in part (a) to estimate the average life expectancy of a person born in 1980.

58. The population of San Diego grew in an approximately linear fashion from 334,413 in 1950 to 1,151,977 in 1994.
    (a) Find an equation that gives the population y of San Diego in year x (with $x = 0$ corresponding to 1950).
    (b) Use the equation in part (a) to estimate the population of San Diego in 1975 and 2000.

In Exercises 59–62, match the given information with the correct graph and determine the slope of the graph.

(a)
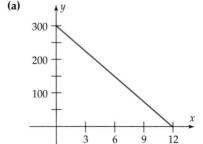
(a)

(b)
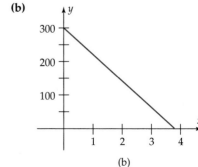
(b)

(c)

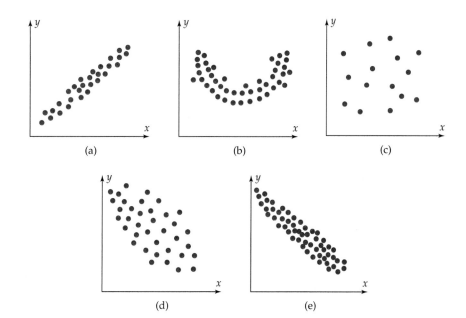

(d)

(c)   (d)

59. A salesman is paid $300 per week plus $75 for each unit sold.
60. A person is paying $25 per week to repay a $300 loan.
61. A gold coin that was purchased for $300 appreciates $20 per year.
62. A CD player that was purchased for $300 depreciates $80 per year.

63. The table shows the monthly premium for a term life insurance policy for women who smoke.

Age	25	30	35	40	45	50	55	60
Premium	19.58	20.10	20.79	25.23	34.89	48.55	69.17	98.92

(a) Make a scatter plot of the data, using x for age and y for premiums.
(b) Does the data appear to be linear?

64. For which of the following scatter plots would a linear model be reasonable? Which sets of data show positive correlation and which show negative correlation?

(a)   (b)   (c)

(d)   (e)

Exercises 65–66 refer to the following table, which shows the percentage of jobs that are classified as managerial, and the percentage of male and female employees who are managers.

Year (since 1990)	Managerial Jobs (%)	Female Managers (%)	Male Managers (%)
−8	12.32	6.28	16.81
−5	12.31	6.85	16.67
−2	12.00	7.21	16.09
0	11.83	7.45	15.64
1	11.79	7.53	15.52
3	11.43	7.65	14.79
5	11.09	7.73	14.10

65. **(a)** Make scatter plots of each data set (managerial jobs, female managers, male managers).
   (b) Match the following linear models with the correct data set. Explain your choices.

   $$y_1 = .11x + 7.34 \qquad y_2 = -.09x + 11.74 \qquad y_3 = -.21x + 15.48$$

66. **(a)** According to the models in Exercise 65, is the percentage of female or male managers increasing at the greater rate?
   (b) Use the models to predict the percentage of female managers and the percentage of male managers in the year 2000.
   (c) In what year do the models predict that the percentage of female managers will surpass the percentage of male managers?

67. The table shows the average hourly earnings of production workers. (*Source:* U.S. Bureau of Labor Statistics)

Year	1980	1982	1984	1986	1988	1990	1992	1994	1996
Earnings	6.66	7.68	8.32	8.76	9.28	10.01	10.57	11.12	11.81

   (a) Find a linear model for this data, with $x = 0$ corresponding to 1980.
   (b) Use the model to estimate the average hourly wage in 1991 and in 2001. The actual average in 1991 was $10.32. How far off is the model?

68. The table shows the winning times (in minutes) for men's 1500-meter freestyle swimming at the Olympics in selected years.

Year	1912	1924	1936	1948	1960	1972	1984	1996
Time	22.00	20.11	19.23	19.31	17.33	15.88	15.09	14.94

(a) Find a linear model for this data, with $x = 0$ corresponding to 1900.
(b) The Olympic record of 14.72 minutes was set in 1992 by Kieren Perkins of Australia. How accurately did your model estimate his time?
(c) How long is this model likely to remain accurate? Why?

69. The table shows the total amount of charitable giving (in billions of dollars) in the U.S. during recent years.

Year	Total Charitable Giving
1986	83.79
1987	89.99
1988	98.13
1989	108.73
1990	111.48
1991	117.22
1992	121.09
1993	126.46
1994	129.84
1995	143.84
1996	150.70

(a) Find a linear model for this data, with $x = 0$ corresponding to 1980.
(b) Use your model to estimate the approximate total giving in 1995 and 2002.

70. The table shows, for selected states, the percent of high school students in the class of 1997 who took the SAT and the average SAT math score.

State	Students Who Took SAT (%)	Average Math Score
Connecticut	79	507
Delaware	65	498
Georgia	60	481
Idaho	15	539
Indiana	87	497
Iowa	6	601
Montana	22	548
Nevada	32	509
New Jersey	69	508
New Mexico	12	545
North Dakota	5	595
Ohio	25	536
Pennsylvania	72	495
South Carolina	56	474
Washington	46	523

(a) Make a scatter plot of average SAT math score y and percent x of students who took the SAT.
(b) Find a linear model for the data.
(c) What is the slope of your linear model? What does this mean in the context of the problem?
(d) Here are the data on four additional states. How well does the model match the actual figures for these states?

State	Maryland	Arizona	Alaska	Hawaii
Students Taking SAT (%)	9	29	48	50
Average Math Score	566	522	517	512

Breaking Even at the Espresso Cart

A local resident owns an espresso cart and has asked you to provide an analysis based on last summer's data. To simplify things, only data for Mondays is provided. The data includes the amount the workers were paid each day, the number of cups sold, the cost of materials (frothy milk and such), and the total revenue for the day. The owner also must spend $40 each operating day for rent for her location and payment on a business loan. Sales taxes have been cleaned out of the data, so you need not consider them. Amounts have been rounded to the nearest dollar.

Date	Pay ($)	Cups Sold	Materials Cost ($)	Total Revenue ($)
June 02	68	112	55	202
June 09	60	88	42	119
June 16	66	81	33	125
June 23	63	112	49	188
June 30	63	87	38	147
July 07	59	105	45	159
July 14	57	116	49	165
July 21	61	122	52	178
July 28	64	100	48	193
August 04	58	80	36	112
August 11	65	96	42	158
August 18	57	108	52	162
August 25	64	93	47	166

The owner is, of course, interested in making a profit. She wishes to know (on the average) how many cups of espresso must be sold each day to break even without raising prices. She would also like to know how much she needs to raise prices in order to break even most days. After you tell her this information, she will decide what the best course of action is.

1. Find a linear regression model for the daily cost as a function of the number of cups sold. Be sure to include the pay for the workers, the fixed daily cost, and the cost of materials.

2. Find a linear regression model for the daily revenue as a function of the number of cups sold.

3. Use the two models you have created to locate the break-even point. That is, find the minimum number of cups for the value of the revenue model to equal or exceed the value of the cost model. Remember that the number of cups must be an integer.

4. The slope portion of the revenue model represents the average selling price of a cup of espresso. What slope in the revenue model would cause the break-even point to be less than 80 cups?

5. How much would the owner of the espresso cart need to raise her prices to break even every day, assuming that at least 80 cups will be sold each day?

2

Equations and Inequalities

How deep is that well?

Measuring the height of a cliff or the depth of a well directly can be very difficult. It can sometimes be done indirectly by using sound waves and solving an appropriate equation. The solution of many other practical problems in construction, quality control, economics, and a variety of other fields also depends on solving equations and inequalities. See Exercise 83 on page 154, Exercise 42 on page 161, and Exercise 47 on page 171.

20

−9   5

Zero
X=.41165847 Y=0

−35

Chapter Outline

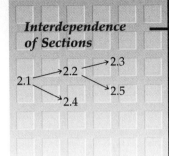

*Interdependence
of Sections*

In Chapter 1, we used graphical and numerical methods to solve equations in one variable. That approach is essential in some situations, but don't be misled into thinking that technology is always the best tool. Algebraic techniques are often easier and more accurate, and you will be expected to use them whenever appropriate. So the emphasis in the first part of this chapter is on algebraic methods of solving equations. In the last part of the chapter, we examine both algebraic and graphical methods of solving inequalities.

2.1 First-Degree Equations and Applications

Two equations are said to be **equivalent** if they have the same solutions. For example, $3x + 2 = 17$ and $x - 2 = 3$ are equivalent because 5 is the only solution of each one. The usual strategy for solving equations algebraically is to use the following principles to transform a given equation into an equivalent one whose solutions are known.

**Basic
Principles
for Solving
Equations**

> Performing any of the following operations on an equation produces an equivalent equation:
>
> 1. **Add or subtract the same quantity from both sides of the equation.**
> 2. **Multiply or divide both sides of the equation by the same *nonzero* quantity.**

The Basic Principles apply to all equations. In this section, however, we shall deal only with **first-degree equations,** which are those that involve only constants and the first power of the variable. Solving most first-degree equations is quite straightforward.

Example 1 Solve: $3x - 6 = 7x + 4$.

Solution We use the Basic Principles to transform this equation into an equivalent one whose solution is obvious:

$$3x - 6 = 7x + 4$$

Add 6 to both sides: $\qquad\qquad 3x = 7x + 10$

Subtract 7x from both sides: $\qquad -4x = 10$

Divide both sides by −4: $\qquad\quad x = \dfrac{10}{-4} = -\dfrac{5}{2}.$

Since $-5/2$ is the only solution of this last equation, $-5/2$ is the only solution of the original equation, $3x - 6 = 7x + 4$. ■

Example 2 Solve: $\dfrac{13t - 7}{2} = \dfrac{2t - 4}{3}$.

Solution The first step is to eliminate the fractions by multiplying both sides by their common denominator 6:

$$6\left(\frac{13t - 7}{2}\right) = 6\left(\frac{2t - 4}{3}\right)$$

Simplify: $\qquad\qquad\qquad 3(13t - 7) = 2(2t - 4)$

Multiply out both sides: $\qquad 39t - 21 = 4t - 8$

Subtract 4t from both sides: $\quad 35t - 21 = -8$

Add 21 to both sides: $\qquad\qquad 35t = 13$

Divide both sides by 35: $\qquad\qquad t = 13/35.$ ■

The second Basic Principle applies only to *nonzero* quantities. Multiplying both sides of an equation by a quantity involving the variable (which might be zero for some values) may lead to an **extraneous solution,** a number that does not satisfy the original equation. To avoid errors in such situations, always *check your solutions in the original equation.*

Example 3 Solve: $\dfrac{x - 3}{x - 7} = \dfrac{3x - 17}{x - 7}$.

Solution Begin by multiplying both sides by $x - 7$ to eliminate the fractions:

$$\frac{x-3}{x-7}(x-7) = \frac{3x-17}{x-7}(x-7)$$

Simplify both sides: $\quad x - 3 = 3x - 17$

Add 3 to both sides: $\quad x = 3x - 14$

Subtract 3x from both sides: $\quad -2x = -14$

Divide both sides by -2: $\quad x = 7$

Substituting 7 for x in the original equation yields

$$\frac{7-3}{7-7} = \frac{3\cdot 7 - 17}{7-7} \quad \text{or equivalently,} \quad \frac{4}{0} = \frac{4}{0}.$$

Since division by 0 is not defined, 7 is *not* a solution of the original equation. So this equation has no solutions. ■

Example 4 Solve: $\dfrac{5}{x+6} = \dfrac{3}{x-3} + \dfrac{1}{2x-6}$.

Solution We begin by multiplying both sides by a common denominator to eliminate the fractions. Note that the third denominator here is $2x - 6 = 2(x - 3)$. Hence $2(x - 3)(x + 6)$ is a common denominator for all three fractions. Multiplying both sides by this common denominator, we have

$$\frac{5}{x+6}\cdot 2(x-3)(x+6) = \frac{3}{x-3}\cdot 2(x-3)(x+6) + \frac{1}{2x-6}\cdot 2(x-3)(x+6)$$

Cancel like factors: $\quad 5\cdot 2(x-3) = 3\cdot 2(x+6) + (x+6)$

Multiply out both sides: $\quad 10(x-3) = 6(x+6) + x + 6$

$$10x - 30 = 6x + 36 + x + 6$$

Combine like terms: $\quad 10x - 30 = 7x + 42$

Subtract 7x from both sides: $\quad 3x - 30 = 42$

Add 30 to both sides: $\quad 3x = 72$

Divide both sides by 3: $\quad x = 24$

By substituting 24 for x in the original equation, it can be seen that 24 *is* a solution:

$$\frac{5}{24+6} = \frac{5}{30} = \frac{1}{6} \quad \text{and} \quad \frac{3}{24-3} + \frac{1}{2\cdot 24 - 6} = \frac{3}{21} + \frac{1}{42} = \frac{1}{6}. \quad ■$$

Algebraic methods are normally used with first-degree equations, but there are cases when the use of a calculator is advisable.

Example 5 Solve: $3.69x + \dfrac{21}{14.77} = 4.53$.

Solution We proceed as before, but to avoid round-off errors in the intermediate steps, we do all the algebra first.

Subtract 21/14.77 from both sides: $\quad 3.69x = 4.53 - \dfrac{21}{14.77}$

Divide both sides by 3.69: $\quad x = \dfrac{4.53 - \dfrac{21}{14.77}}{3.69}$

(4.53-(21/14.77))/3.6
9
                .842330366432

Figure 2–1

Now use the calculator, as in Figure 2–1, to find the solution $x \approx .8423$. Alternatively, this equation could be solved as in the Graphing Exploration below. ∎

GRAPHING EXPLORATION

Solve the equation in Example 5 by finding the point where the graphs of $y = 3.69x + 21/14.77$ and $y = 4.53$ intersect. How does your answer compare with the one in Figure 2–1?

Applications

Real-life problem situations are usually described verbally. To solve such problems, you must interpret this verbal information and express it in mathematical language (typically as an equation or inequality). The following guidelines may be helpful.

Setting Up Applied Problems

1. *Read* the problem carefully and determine what is asked for.
2. *Label* the unknown quantities by letters (variables) and, if appropriate, draw a picture of the situation.
3. *Translate* the verbal statements in the problem and the relationships between the known and unknown quantities into mathematical language.
4. *Consolidate* the mathematical information into an equation in one variable that can be solved or an equation in two variables that can be graphed in order to determine at least one of the unknown quantities.

Example 6 **(Measurement)** What is the height of a rectangular box with a square base measuring 8 inches on a side and a volume of 928 cubic inches?

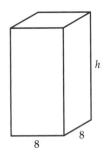

Figure 2–2

Solution   *Read:* We know the volume and the dimensions of the base and must find the height.
Label: Draw a diagram with h denoting the height (Figure 2–2).
Translate:

English Language	Mathematical Language
Height	h
Volume (Length × Width × Height)	$8 \times 8 \times h$
Volume is 928 cu in.	$8 \times 8 \times h = 928$
	$64h = 928$

Consolidate: We need only solve the equation $64h = 928$:

Divide both sides by 64: $h = \dfrac{928}{64} = 14.5 \text{ in.}$

Therefore, the box measures 8 by 8 by 14.5 inches high.   ■

Example 7   **(Counting)** A box of nickels and quarters contains 52 coins that are worth a total of $10. How many nickels and how many quarters are in the box?

Solution   *Read:* We are asked to find the number of nickels and the number of quarters.
Label: Let x be the number of nickels and y the number of quarters.
Translate:

English Language	Mathematical Language
Number of nickels	x
Number of quarters	y
Total number of coins is 52	$x + y = 52$
Value of the nickels	$.05x$
Value of the quarters	$.25y$
Value of collection is $10	$.05x + .25y = 10$

Consolidate: When you have two variables one technique is to express one variable in terms of the other and use this to obtain an equation in a single variable. In this case, we can solve $x + y = 52$ for y

(∗) $y = 52 - x$

and substitute the result in the other equation:

$$.05x + .25y = 10$$

$$.05x + .25(52 - x) = 10.$$

Now solve this last equation.

Multiply out left side:	$.05x + 13 - .25x = 10$
Combine like terms:	$-.20x + 13 = 10$
Subtract 13 from both sides:	$-.20x = -3$
Divide both sides by $-.20$:	$x = \dfrac{-3}{-.20} = 15.$

Therefore, there are 15 nickels. From equation (∗), the number of quarters is

$$y = 52 - x$$
$$y = 52 - 15 = 37. \quad\blacksquare$$

Recall that 4% is written as the decimal .04 and that "4% of 227" means ".04 *times* 227." For example, if $227 is deposited in a savings account that pays 4% annual interest, then the interest after one year is 4% of $227, that is, $.04(227) = \$9.08$. This is an example of the basic rule of annual simple interest:

Interest = Rate × Amount.

Example 8 **(Interest)** How much money must be invested at 9% to produce $128.25 interest annually?

Solution *Read and Label:* We are asked to find amount x that should be invested.
Translate and Consolidate: The basic rule of compound interest tells us that

$$\text{Rate} \times \text{Amount} = \text{Interest}$$
$$9\% \text{ of } x = 128.25$$
$$.09x = 128.25$$

Divide both sides by .09:	$x = \dfrac{128.25}{.09} = \$1425. \quad\blacksquare$

Example 9 **(Interest)** A high-risk stock pays dividends at a rate of 12% per year, and a savings account pays 6% interest per year. How much of a $9000 investment should be put in the stock and how much in the savings account in order to obtain a return of 8% per year on the total investment?

Solution *Label:* Let x be the amount invested in stock. Then the rest of the $9000, namely, $(9000 - x)$ dollars, goes in the savings account.
Translate and Consolidate: We want the total return on $9000 to be 8%, so we have

$$\left(\begin{array}{l}\text{Return on } x \text{ dollars}\\ \text{of stock at } 12\%\end{array}\right) + \left(\begin{array}{l}\text{Return on } (9000-x)\\ \text{dollars of savings at } 6\%\end{array}\right) = 8\% \text{ of } \$9000$$

$$(12\% \text{ of } x \text{ dollars}) + (6\% \text{ of } (9000 - x) \text{ dollars}) = 8\% \text{ of } \$9000$$

$$.12x + .06(9000 - x) = .08(9000)$$

$$.12x + .06(9000) - .06x = .08(9000)$$

$$.12x + 540 - .06x = 720$$

$$.12x - .06x = 720 - 540$$

$$.06x = 180$$

$$x = \frac{180}{.06} = 3000.$$

Therefore, $3000 should be invested in stock and $(9000 - 3000) = \$6000$ in the savings account. If this is done, the total return will be 12% of $3000 ($360) plus 6% of $6000 ($360), a total of $720, which is precisely 8% of $9000. ■

Example 10 **(Mixture)** A car radiator contains 12 quarts of fluid, 20% of which is antifreeze. How much fluid should be drained and replaced with pure antifreeze so that the resulting mixture is 50% antifreeze?

Solution Let x be the number of quarts of fluid to be replaced by pure antifreeze.* When x quarts are drained, there are $12 - x$ quarts of fluid left in the radiator, 20% of which is antifreeze. So we have

$$\left(\begin{array}{l}\text{Amount of antifreeze}\\ \text{in radiator after}\\ \text{draining } x \text{ quarts}\\ \text{of fluid}\end{array}\right) + \left(\begin{array}{l}x \text{ quarts of}\\ \text{antifreeze}\end{array}\right) = \left(\begin{array}{l}\text{Amount of}\\ \text{antifreeze in}\\ \text{final mixture}\end{array}\right)$$

$$20\% \text{ of } (12 - x) \qquad + \qquad x \qquad = \qquad 50\% \text{ of } 12$$

$$.20(12 - x) + x = .50(12)$$

$$2.4 - .2x + x = 6$$

$$-.2x + x = 6 - 2.4$$

$$.8x = 3.6$$

$$x = \frac{3.6}{.8} = 4.5.$$

---

*Hereafter we omit the headings Label, Translate, etc.

Therefore, 4.5 quarts should be drained and replaced with pure anti-freeze. ■

The basic formula for problems involving distance and a uniform rate of speed is

$$\textbf{Distance} = \textbf{Rate} \times \textbf{Time}.$$

For instance, if you drive at a rate of 55 mph for 2 hours, you travel a distance of $55 \cdot 2 = 110$ miles.

Example 11 (**Distance/Rate**) A car leaves Chicago traveling at an average speed of 64 kilometers per hour toward Decatur, which is 239 kilometers away. An hour later, a second car leaves Decatur for Chicago on the same road, traveling at an average speed of 86 km/h. When will the cars meet?

Solution Let t be the number of hours the first car has traveled. Since the second car left an hour later, it will have traveled $t - 1$ hours. When the cars meet, the distances they have traveled will total 239 km, as shown in Figure 2–3.

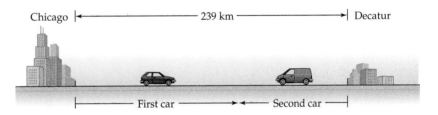

Figure 2–3

We now translate this information into an equation:

$$\begin{pmatrix} \text{Distance traveled} \\ \text{by first car} \end{pmatrix} + \begin{pmatrix} \text{Distance traveled} \\ \text{by second car} \end{pmatrix} = 239$$

$$\text{Rate} \cdot \text{Time} \quad + \quad \text{Rate} \cdot \text{Time} \quad = 239$$

$$64t \quad + \quad 86(t - 1) \quad = 239$$

$$64t + 86t - 86 = 239$$

$$64t + 86t = 239 + 86$$

$$150t = 325$$

$$t = \frac{325}{150} = 2\tfrac{1}{6} \text{ hours} \quad ■$$

Example 12 **(Work)** Tom can paint the fence in 6 hours and Huck can paint it in 4 hours. If they work together, how long will it take to paint the fence?

Solution Since Tom can paint the entire fence in 6 hours, he can paint $\frac{1}{6}$ of it in 1 hour. Similarly, Huck can paint $\frac{1}{4}$ of the fence in 1 hour. If t is the number of hours it takes to paint the fence when they work together, then together they can paint $1/t$ of the fence in 1 hour. Therefore:

$$\begin{pmatrix} \text{Part of fence} \\ \text{painted by Tom} \\ \text{in 1 hour} \end{pmatrix} + \begin{pmatrix} \text{Part of fence} \\ \text{painted by Huck} \\ \text{in 1 hour} \end{pmatrix} = \begin{pmatrix} \text{Part of fence} \\ \text{painted by both} \\ \text{in 1 hour} \end{pmatrix}$$

$$\frac{1}{6} + \frac{1}{4} = \frac{1}{t}$$

$$12t\left(\frac{1}{6} + \frac{1}{4}\right) = 12t\left(\frac{1}{t}\right)$$

$$2t + 3t = 12$$

$$5t = 12$$

$$t = \frac{12}{5} = 2.4 \text{ hours} \quad \blacksquare$$

Exercises 2.1

In Exercises 1–30, solve the equation.

1. $2x = 3$

2. $4x = 0$

3. $-x = 5$

4. $2 - x = 7$

5. $3x + 2 = 26$

6. $\frac{y}{5} - 3 = 14$

7. $3x + 2 = 9x + 7$

8. $-7(t + 2) = 3(4t + 1)$

9. $\frac{3y}{4} - 6 = y + 2$

10. $2(1 + x) = 3x + 5$

11. $\frac{1}{2}(2 - z) = \frac{3}{2} + 6z$

12. $\frac{4y}{5} + \frac{5y}{4} = 1$

13. $2x - \frac{x - 5}{4} = \frac{3x - 1}{2} + 1$

14. $2\left(\frac{x}{3} + 1\right) + 5\left(\frac{4x}{3} - 2\right) = 2$

15. $\frac{2x - 1}{2x + 1} = \frac{1}{4}$

16. $\frac{2x}{x - 3} = 1 + \frac{6}{x - 3}$

17. $\frac{1}{2t} - \frac{2}{5t} = \frac{1}{10t} - 1$

18. $\frac{1}{2} + \frac{2}{y} = \frac{1}{3} + \frac{3}{y}$

19. $\frac{2x - 7}{x + 4} = \frac{5}{x + 4} - 2$

20. $\frac{z + 4}{z + 5} = \frac{-1}{z + 5}$

21. $\frac{6x}{3x - 1} - 5 = \frac{2}{3x - 1}$

22. $1 + \frac{1}{x} = \frac{2 + x}{x} + 1$

23. $(x + 2)^2 = (x + 3)^2$

24. $x^2 + 3x - 7 = (x + 4)(x - 4)$

25. $21.31 + 41.29x = 17.51x - 8.17$

26. $2.37x + 3.1288 = 6.93x - 2.48$

27. $18.923y - 15.4228 = 10.003y + 18.161$

28. $6.31(x - 3.53) = 5.42(x + 1.07) - 21.1584$

29. $18.34x - \frac{14.21}{3} = \frac{33.41}{3} - 46.82x$

30. $\frac{423.1}{51.71} + \frac{53.19x}{21.72} = x$

In a simple model of the economy (by J. M. Keynes), equilibrium between national output and national expenditures holds when $Y = C + I$, where Y is the national income, C is consumption (which depends on the national income), and I

is the amount of investment. In Exercises 31–34, solve the equilibrium equation for Y under the given conditions.

31. $C = 70 + .8Y$ and $I = 90$

32. $C = 300 + .75Y$ and $I = 450$

33. $C = 220 + .6Y$ and $I = 300$

34. $C = 640 + .7Y$ and $I = 350$

In a more complete model of the economy (also by Keynes), the equilibrium equation is $Y = C + I + G + (X - M)$, where Y, C, and I are as in Exercises 31–34, G is government spending, X is exports, and M is imports. In Exercises 35–38, solve the equilibrium equation for Y under the given conditions.

35. $C = 120 + .9Y$, $M = 20 + .2Y$, $I = 140$, $G = 150$, and $X = 60$

36. $C = 60 + .85Y$, $M = 35 + .2Y$, $I = 95$, $G = 145$, and $X = 50$

37. $C = 300 + .875Y$, $M = 80 + .25Y$, $I = 390$, $G = 420$, and $X = 110$

38. $C = 499 + .8Y$, $M = 120 + .25Y$, $I = 695$, $G = 895$, and $X = 110$

In Exercises 39–46, solve the given equation for the indicated variable.

39. $x = 3y - 5$ for y **40.** $5x - 2y = 1$ for x

41. $\dfrac{1}{4}x - \dfrac{1}{3}y - 6 = 0$ for y

42. $3x + \dfrac{1}{2}y - z = 4$ for z

43. $A = \dfrac{h}{2}(b + c)$ for b

44. $V = \pi b^2 c$ for c

45. $V = \dfrac{\pi d^2 h}{4}$ for h **46.** $\dfrac{1}{r} = \dfrac{1}{s} + \dfrac{1}{t}$ for r

47. Find the number c such that $3x - 6 = c$ and $2 - x = 1$ have the same solutions.

48. If $1/3$ is a solution of $cx + d = 0$, then how are c and d related?

In Exercises 49–52, a problem situation is given.
(a) Decide what is being asked for and label the unknown quantities.
(b) Translate the verbal statements in the problem and the relationships between the known and unknown quantities into mathematical language, using a table as in Examples 6 and 7. You need not find an equation to be solved.

49. A student has exam scores of 88, 62, and 79. What score does he need on the fourth exam to have an average of 80?

English Language	Mathematical Language
Score on fourth exam	
Sum of scores on four exams	
Average of scores on four exams	

50. What number when added to 30 is 11 times the number?

English Language	Mathematical Language
The number	
The number added to 30	
11 times the number	

51. A rectangular garden has a perimeter of 270 ft and its length is twice its width. What are the dimensions of the garden?

English Language	Mathematical Language
Length of garden	
Width of garden	
Perimeter of garden	
The length is twice the width	

52. How many gallons of a 12% salt solution should be combined with 10 gallons of an 18% salt solution to obtain a 16% solution?

English Language	Mathematical Language
Gallons of 12% solution	
Total gallons of mixture	
Amount of salt in 10 gallons of the 18% solution	
Amount of salt in the 12% solution	
Amount of salt in the mixture	

In Exercises 53–56, set up the problem by labeling the unknowns, translating the given information into mathematical language, and finding an equation that will lead to a solution of the problem. You need not solve this equation.

53. A worker gets an 8% pay raise and now makes $1593 per month. What was the worker's salary before the raise?

54. A student has exam scores of 64, 82, and 91. What is her average? What score must she get on the fourth exam to raise her average by 3 points?

55. If 9 is added to a certain number, the result is 1 less than 3 times the number. What is the number?

56. A merchant has 5 pounds of mixed nuts that cost $30. She wants to add peanuts that cost $1.50 per pound and cashews that cost $4.50 per pound to obtain 50 pounds of a mixture that costs $2.90 per pound. How many pounds of peanuts are needed?

In the remaining exercises, solve the applied problem.

57. An item on sale costs $13.41. All merchandise was marked down 25% for the sale. What was the original price of the item?

58. A discount store sells a jacket for $28, which is 20% below the list price. What is the list price?

59. Inflation caused the price of a workbook to increase by 5%. The new price is $15.12. What was the old price?

60. Lawnmowers are priced at $300 during April. In May the price is raised by 20%. In September the price is reduced by 20%. What is the final price? Explain why it is *not* $300.

61. You have already invested $550 in a stock with an annual return of 11%. How much of an additional $1100 should be invested at 12% and how much at 6% so that the total annual return on the entire $1650 is 9%?

62. $25,000 is placed in the bank, part in an account earning 10% interest and part in an account earning 15% interest. The annual interest paid on the $25,000 is $3350. How much money is in the 10% account?

63. If you borrow $500 from a credit union at 12% annual interest and $250 from a bank at 18% annual interest, what is the *effective annual interest rate* (that is, what single rate of interest on $750 results in the same total amount of interest)?

64. Matt's credit card bill this month is $2534. He has $4000 in his savings account which earns 0.5% interest monthly. The credit card company has a service charge of 1.75% monthly on the unpaid balance.

   (a) Assuming there is no other activity on either account for the next 30-day period, how much of the credit card bill should Matt pay off with his savings so that the interest he earns on the balance in his savings will **equal** the amount of the service charge on the remainder of the credit card bill?

   (b) What will be the new balance on the credit card bill?

   (c) What will be the service charge on this bill?

   (d) What will be the new balance in Matt's savings account?

   (e) What will be the interest accrued on the new savings balance?

65. A student has an average of 72 on four exams. If the final exam is to count double, what score does she need to have an average of 80?

66. A student scored 63 and 81 on two exams, each of which accounts for 30% of the course grade. The final exam accounts for 40% of the course grade. What score must she get on the final exam to have an average grade of 76?

67. One alloy is one part copper to three parts tin. A second alloy is one part copper to four parts tin. How much of the second alloy should be combined with 24 pounds of the first alloy to obtain a new alloy that is two parts copper to seven parts tin?

68. A chemist has 10 mL of a 30% acid solution. How many milliliters of pure acid must be added in order to obtain a 50% acid solution?

69. A radiator contains 8 quarts of fluid, 40% of which is antifreeze. How much fluid should be drained and replaced with pure antifreeze so that the new mixture is 60% antifreeze?

70. Emily has 10 pounds of premium coffee that costs $7.50 per pound. She wants to mix this with ordinary coffee that costs $4 per pound to make a mixture that costs $5.50 per pound. How much ordinary coffee must be used? What will the weight of the final mixture be?

71. Two cars leave a gas station at the same time, one traveling north and the other south. The northbound car travels at 50 mph. After 3 hours the cars are 345 miles apart. How fast is the southbound car traveling?

72. A train leaves New York for Boston, 200 miles away, at 3 P.M. and averages 75 mph. Another train leaves Boston for New York on an adjacent set of tracks at 5 P.M. and averages 45 mph. At what time will the trains meet?

73. A car passes a certain point at 1 P.M. and continues along at a constant speed of 64 kilometers per hour. A second car passes the point at 2 P.M. and goes on at a constant speed of 88 kilometers per hour, following the same route as the first car. At what time will the second car catch up with the first one?

74. In a motorcycle race the winner averaged 100 mph and finished 15 minutes ahead of the loser, who averaged 95 mph. How long was the race (in miles) and what was the winner's time?

75. An airplane flew with the wind for 2.5 hours and returned the same distance against the wind in 3.5 hours. If the cruising speed of the plane was a constant 360 mph, how fast was the wind blowing? [*Hint:* If the wind speed is *r* miles per hour, then the plane travels at $(360 + r)$ mph with the wind and at $(360 - r)$ mph against the wind.]

76. A motorboat goes 15 miles downstream at its top speed and then turns around and returns 15 miles upstream at top speed. The trip upstream took twice as long as the trip downstream. If the boat's top speed is 10 mph in still water, how fast is the current? [See Exercise 75.]

77. If Charlie and Nick work together they can paint the house in 20 hours. Charlie can do the job alone in 36 hours. How long would it take Nick to do the job alone?

78. Using two lawnmowers and working together, Tom and Anne can mow the lawn in 36 minutes. It takes Anne 90 minutes if she mows the entire lawn herself. How long would it take Tom to do the job if he worked alone?

79. One pipe can fill a pool in 50 minutes. Another pipe takes 75 minutes. How long does it take for both pipes together to fill the pool? [*Hint:* This is a "work problem" in disguise: How much of the pool does the first pipe fill in 1 minute?]

80. One pipe can fill a tank in 4 hours, a second pipe can fill it in 10 hours, and a third pipe in 12 hours. The pipes are connected to the same tank. How long does it take to fill the tank if all three pipes are used together? [See Exercise 79]

81. (With apologies to Mrs. Morgan of *How Green Was My Valley*) The water faucet can fill a tub in 15 minutes. But there is a hole in the tub through which a full tub of water can drain out in 18 minutes. If the tub is empty and the faucet is turned on, how long will it take to fill the tub?

82. The length of a rectangle is 15 inches more than the width. If the perimeter is 398 inches, what are the dimensions of the rectangle?

83. The perimeter of a rectangle is 160 meters. If a new rectangle is formed by halving the length of each of one pair of opposite sides and increasing the remaining pair of sides by 30 meters each, then the new rectangle also has a perimeter of 160 meters. What were the dimensions of the original rectangle?

84. The perimeter of a rectangular garden is 72 feet. If two such gardens were placed side by side they would form a square. What are the dimensions of the garden?

Thinkers

85. Insurance on overseas shipment of goods is charged at the rate of 4% of the total value of the shipment (which is the sum of the value of the goods, the freight charges, *and* the cost of the insurance). The freight charges for $30,000 worth of goods are $4500. How much does the insurance cost?

86. A squirrel and a half eats a nut and a half in a day and a half. How many nuts do six squirrels eat in six days?

87. Diophantus of Alexandria, a third-century mathematician, lived one-sixth of his life in childhood, one-twelfth in his youth, and one-seventh as a bachelor. Then he married and five years later had a son. The son died four years before Diophantus at half the age Diophantus was when he himself died. How long did Diophantus live?

2.2   # Quadratic Equations and Applications

A **second-degree,** or **quadratic, equation** is an equation that can be written in the form

$$ax^2 + bx + c = 0$$

for some constants *a*, *b*, *c* with $a \neq 0$. There are several algebraic techniques for solving such equations. We begin with the **factoring method,** which makes use of this property of real numbers:

Zero Products

If a product of real numbers is zero, then at least one of the factors is zero; in other words,

If $cd = 0$, then $c = 0$ or $d = 0$.

Example 1 Solve: $7x^2 - 5x - 2 = 0$.

Solution   We first factor the left side:

$$7x^2 - 5x - 2 = 0$$

$$(7x + 2)(x - 1) = 0.$$

When a product of real numbers is 0, than at least one of the factors must be 0. So this equation is equivalent to

$$7x + 2 = 0 \qquad \text{or} \qquad x - 1 = 0$$
$$7x = -2 \qquad\qquad\qquad x = 1.$$
$$x = -2/7$$

Therefore, the exact solutions are $-2/7$ and 1. ∎

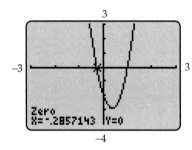

Figure 2–4

The equation in Example 1 may also be solved graphically: graph $y = 7x^2 - 5x - 2$ and use a root finder, as in Figure 2–4, to find the negative solution. In this case, the graphical method produces an *approximate* solution, rather than the exact solution found in Example 1. Thus, algebraic methods are usually preferable for quadratic equations.

Example 2 Solve: $3x^2 + x = 10$.

Solution   We first rearrange the terms to make one side 0 and then factor:

$$3x^2 + x = 10$$

Subtract 10 from each side: $3x^2 + x - 10 = 0$

Factor left side: $(3x - 5)(x + 2) = 0.$

Since the product on the left side is 0, one of its factors must be 0. Hence,

$$3x - 5 = 0 \qquad \text{or} \qquad x + 2 = 0$$
$$3x = 5 \qquad\qquad\qquad x = -2.$$
$$x = 5/3$$

The solutions are -2 and $5/3$. ∎

> **CAUTION**
> To guard against mistakes, check your solutions by substituting each one in the *original* equation to make sure it really *is* a solution.

The Quadratic Formula

The solutions of $x^2 = 7$ are the numbers whose square is 7. Although 7 has just one square root, there are *two* numbers whose square is 7, namely,

$\sqrt{7}$ and $-\sqrt{7}$. So the solutions of $x^2 = 7$ are $\sqrt{7}$ and $-\sqrt{7}$, or in abbreviated form $\pm\sqrt{7}$. A similar argument works with any positive number d in place of 7:

$$\text{The solutions of } x^2 = d \text{ are } \sqrt{d} \text{ and } -\sqrt{d}.$$

The same reasoning enables us to solve other equations.

Example 3   Solve: $(z - 2)^2 = 5$.

Solution   The equation says that $z - 2$ is a number whose square is 5. Since there are only two numbers whose square is 5, namely, $\sqrt{5}$ and $-\sqrt{5}$, we must have

$$z - 2 = \sqrt{5} \qquad \text{or} \qquad z - 2 = -\sqrt{5}$$
$$z = \sqrt{5} + 2 \qquad\qquad z = -\sqrt{5} + 2.$$

In compact notation, the solutions of the equation are $\pm\sqrt{5} + 2$. ∎

Let a, b, c be any real numbers with $a \neq 0$. We shall solve the equation

$$ax^2 + bx + c = 0.$$

We begin by noting a fact that will be needed in the solution process. By multiplying out the right side, you should verify that

$$(*) \qquad x^2 + \frac{b}{a}x + \left(\frac{b}{2a}\right)^2 = \left(x + \frac{b}{2a}\right)^2.$$

Now solve the equation as follows.*

$$ax^2 + bx + c = 0$$

Divide both sides by a:
$$x^2 + \frac{b}{a}x + \frac{c}{a} = 0$$

Subtract $\dfrac{c}{a}$ from both sides:
$$x^2 + \frac{b}{a}x = -\frac{c}{a}$$

Add $\left(\dfrac{b}{2a}\right)^2$ to both sides:
$$x^2 + \frac{b}{a}x + \left(\frac{b}{2a}\right)^2 = \left(\frac{b}{2a}\right)^2 - \frac{c}{a}$$

Factor the left side of this last equation by using equation $(*)$ above:

$$\left(x + \frac{b}{2a}\right)^2 = \left(\frac{b}{2a}\right)^2 - \frac{c}{a}$$

Find common denominator for right side:
$$\left(x + \frac{b}{2a}\right)^2 = \frac{b^2}{4a^2} - \frac{c}{a} = \frac{b^2 - 4ac}{4a^2}$$

---

*Don't be put off by all the letters here—the algebra/arithmetic is all easy. Also, don't worry about how anyone thought to do these steps; just verify that each one follows from the preceding one.

This last equation says that $x + \dfrac{b}{2a}$ is a number whose square is $\dfrac{b^2 - 4ac}{4a^2}$.

So we must have

$$x + \frac{b}{2a} = \sqrt{\frac{b^2 - 4ac}{4a^2}} \qquad \text{or} \qquad x + \frac{b}{2a} = -\sqrt{\frac{b^2 - 4ac}{4a^2}},$$

or in more compact notation,

$$x + \frac{b}{2a} = \pm\sqrt{\frac{b^2 - 4ac}{4a^2}} = \pm\frac{\sqrt{b^2 - 4ac}}{2a}.$$

Adding $\dfrac{-b}{2a}$ to both ends shows that

$$x = \frac{-b}{2a} \pm \frac{\sqrt{b^2 - 4ac}}{2a} = \frac{-b \pm \sqrt{b^2 - 4ac}}{2a}.$$

We have proved a major result:

The Quadratic Formula

> The solutions of the quadratic equation $ax^2 + bx + c = 0$ are
> $$x = \frac{-b \pm \sqrt{b^2 - 4ac}}{2a}.$$

You should memorize the quadratic formula.

Example 4   Solve: $x^2 + 3 = -8x$.

Solution   Rewrite the equation as $x^2 + 8x + 3 = 0$ and apply the quadratic formula with $a = 1$, $b = 8$, and $c = 3$:

$$\begin{aligned}
x &= \frac{-b \pm \sqrt{b^2 - 4ac}}{2a} = \frac{-8 \pm \sqrt{8^2 - 4\cdot1\cdot3}}{2\cdot1} \\[2mm]
&= \frac{-8 \pm \sqrt{52}}{2} = \frac{-8 \pm \sqrt{4\cdot13}}{2} \\[2mm]
&= \frac{-8 \pm 2\sqrt{13}}{2} = -4 \pm \sqrt{13}
\end{aligned}$$

Therefore the equation has two distinct real solutions, $-4 + \sqrt{13}$ and $-4 - \sqrt{13}$. ∎

Example 5 Solve: $x^2 - 194x + 9409 = 0$.

Solution Use a calculator and the quadratic formula with $a = 1$, $b = -194$, and $c = 9409$:

$$x = \frac{-b \pm \sqrt{b^2 - 4ac}}{2a} = \frac{-(-194) \pm \sqrt{(-194)^2 - 4 \cdot 1 \cdot 9409}}{2 \cdot 1}$$

$$= \frac{194 \pm \sqrt{37{,}636 - 37{,}636}}{2} = \frac{194 \pm 0}{2} = 97$$

Thus, 97 is the only solution of the equation. ∎

Example 6 Solve: $2x^2 + x + 3 = 0$.

Solution By the quadratic formula with $a = 2$, $b = 1$, and $c = 3$:

$$x = \frac{-b \pm \sqrt{b^2 - 4ac}}{2a} = \frac{-1 \pm \sqrt{1^2 - 4 \cdot 2 \cdot 3}}{2 \cdot 2} = \frac{-1 \pm \sqrt{1 - 24}}{4}$$

$$= \frac{-1 \pm \sqrt{-23}}{4}.$$

Since $\sqrt{-23}$ is not a real number, this equation has *no real solutions* (that is, no solutions in the real number system). ∎

The quadratic formula can be used to find exact solutions of any quadratic equation. When exact solutions are not required, approximate solutions may be found in several ways.

Technology Tip

Many calculators have built-in polynomial equation solvers that will approximate the solutions of quadratic and other polynomial equations. See Calculator Investigation 1 at the end of this section. A quadratic formula program for other calculators is in the Program Appendix.

Example 7 Approximate the solutions of $3.2x^2 + 15.93x - 7.1 = 0$.

Solution The solutions can be approximated algebraically or graphically.
Algebraic Method: First, compute $\sqrt{b^2 - 4ac} = \sqrt{15.93^2 - 4(3.2)(-7.1)}$ and store the result in memory D. By the quadratic formula, the solutions are

$$x = \frac{-b \pm \sqrt{b^2 - 4ac}}{2a} = \frac{-15.93 \pm D}{2(3.2)}.$$

They are easily approximated on a calculator:

$$x = \frac{-15.93 + D}{2(3.2)} \approx .411658467 \quad \text{and} \quad x = \frac{-15.93 - D}{2(3.2)} \approx -5.389783347.$$

Since these answers are approximations, they may not check exactly when substituted in the original equation.

Graphical Method: Graph $y = 3.2x^2 + 15.93x - 7.1$ and use a root finder to approximate the x-intercepts (solutions), as in Figure 2–5. ∎

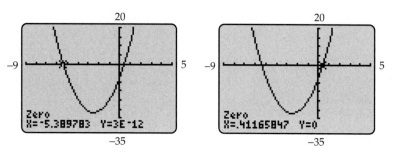

Figure 2–5

The Discriminant

The expression $b^2 - 4ac$ in the quadratic formula is called the **discriminant.** As Examples 4–7 demonstrate, the discriminant determines the *number* of real solutions of the equation $ax^2 + bx + c = 0$.

Real Solutions of a Quadratic Equation

Discriminant $b^2 - 4ac$	Number of Real Solutions of $ax^2 + bx + c = 0$	Example
> 0	Two distinct real solutions	$x^2 + 8x + 3 = 0$
$= 0$	One real solution	$x^2 - 194x + 9409 = 0$
< 0	No real solutions	$2x^2 + x + 3 = 0$

GRAPHING EXPLORATION

Determine the number of solutions of $2x^2 + 5x - 4 = 0$ by graphing $y = 2x^2 + 5x - 4$ and counting the number of x-intercepts of the graph. Confirm your answer by computing the discriminant. Which method do you prefer?

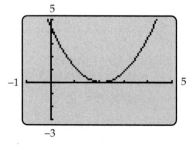

Figure 2–6

Example 8   How many solutions does this equation have?
$$x^2 - 4.2x + 4.4 = 0$$

Solution   The graph of $y = x^2 - 4.2x + 4.4$ in Figure 2–6 suggests that the equation has just one solution (x-intercept). However, the discriminant of the equation is

$$b^2 - 4ac = (-4.2)^2 - 4(1)(4.4) = .04.$$

Since the discriminant is positive, the equation has *two* solutions. The moral of this story is: Be careful when using a graph to determine the number of solutions of a quadratic equation. ■

GRAPHING EXPLORATION

Find a viewing window that clearly shows that the graph of $y = x^2 - 4.2x + 4.4$ has two x-intercepts, that is, that the equation $x^2 - 4.2x + 4.4 = 0$ has two real solutions. Then find the solutions.

Applications

The guidelines for setting up applied problems in Section 2.1 are equally useful in situations involving quadratic equations. However, setting up a problem is only half the job. You must then solve the equation you have obtained, getting exact answers whenever possible. Finally, you must

Interpret your answers in terms of the original problem. Do they make sense? Do they satisfy the required conditions?

In particular, an equation may have several solutions, some of which may not make sense in the context of the problem. For instance, distance can't be negative, the number of people in a room cannot be a proper fraction, etc.

Figure 2–7

Example 9 A rectangle of area 24.5 square inches is twice as long as it is wide. What are its dimensions?

Solution Figure 2–7 shows the situation. The given information can be translated into mathematical terms as follows.

English Language	Mathematical Language
Length and width	x and y
Length is twice the width	$x = 2y$
Area (length × width)	xy
Area is 24.5 sq. in.	$xy = 24.5$

Now substitute $x = 2y$ into the area equation and solve it.

$$xy = 24.5$$
$$(2y)y = 24.5$$
$$2y^2 = 24.5$$
$$y^2 = 12.25$$
$$y = \pm\sqrt{12.25} = \pm 3.5$$

Since the width is a positive number, $y = 3.5$ is the only relevant solution. Therefore, the rectangle has width 3.5 and length $x = 2y = 2(3.5) = 7$.   ∎

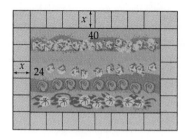

Figure 2–8

Example 10   A landscaper wants to put a cement walk of uniform width around a rectangular garden that measures 24 by 40 feet. She has enough cement to cover 660 square feet. How wide should the walk be to use up all the cement?

Solution   Let x denote the width of the walk (in feet) and draw a picture of the situation (Figure 2–8).

The length of the outer rectangle is $40 + 2x$ (the garden length plus walks on each end) and its width is $24 + 2x$.

$$\left(\begin{array}{c}\text{Area of outer}\\ \text{rectangle}\end{array}\right) - \left(\begin{array}{c}\text{Area of}\\ \text{garden}\end{array}\right) = \text{Area of walk}$$

$$\text{Length}\cdot\text{Width} - \text{Length}\cdot\text{Width} = 660$$

$$(40 + 2x)(24 + 2x) - 40\cdot24 = 660$$
$$960 + 128x + 4x^2 - 960 = 660$$
$$4x^2 + 128x - 660 = 0$$

Dividing both sides by 4 and applying the quadratic formula yields

$$x^2 + 32x - 165 = 0$$

$$x = \frac{-32 \pm \sqrt{(32)^2 - 4\cdot1\cdot(-165)}}{2\cdot1}$$

$$x = \frac{-32 \pm \sqrt{1684}}{2} \approx \left\{\begin{array}{c}4.5183\\ \text{or}\\ -36.5183\end{array}\right.$$

Only the positive solution makes sense in the context of this problem. The walk should be approximately 4.5 feet wide.   ∎

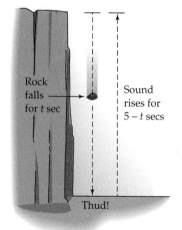

Rock falls for t sec

Sound rises for $5 - t$ secs

Thud!

Figure 2–9

Example 11   A stone is dropped from the top of a cliff. Five seconds later the sound of the stone hitting the ground is heard. Use the fact that the speed of sound is 1100 feet per second and that the distance traveled by a falling object in t seconds is $16t^2$ feet to determine the height of the cliff.

Solution   If t is the time it takes the rock to fall to the ground, then the height of the cliff is $16t^2$ feet. The total time for the rock to fall and the sound of its hitting ground to return to the top of the cliff is 5 seconds. So the sound must take $5 - t$ seconds to return to the top, as indicated in Figure 2–9. Therefore,

$$\begin{pmatrix} \text{Distance rock falls} \\ \text{in } t \text{ seconds} \end{pmatrix} = \text{Height of cliff} = \begin{pmatrix} \text{Distance sound} \\ \text{travels upward} \\ \text{in } 5 - t \text{ seconds} \end{pmatrix}$$

$$16t^2 = \text{Rate} \cdot \text{Time}$$
$$16t^2 = 1100(5 - t)$$
$$16t^2 = 5500 - 1100t$$
$$16t^2 + 1100t - 5500 = 0$$

The quadratic formula and a calculator show that:

$$t = \frac{-1100 \pm \sqrt{(1100)^2 - 4 \cdot 16 \cdot (-5500)}}{2 \cdot 16}$$

$$= \frac{-1100 \pm \sqrt{1,562,000}}{32} \approx \begin{cases} 4.6812 \\ \text{or} \\ -73.4312 \end{cases}.$$

Since the time must be a number between 0 and 5, we see that $t \approx 4.6812$ seconds. Therefore the height of the cliff is

$$16(4.6812)^2 \approx 350.6 \text{ feet.} \quad \blacksquare$$

Example 12 A pilot wants to make the 840-mile round trip from Cleveland to Peoria and back in 5 hours flying time. Going to Peoria, there will be a headwind of 30 mph, that is, a wind opposite to the direction the plane is flying. It is estimated that on the return trip to Cleveland, there will be a 40-mph tailwind (in the direction the plane is flying). At what constant speed should the plane be flown?

Solution   Let x be the engine speed of the plane. On the trip to Peoria, the actual speed will be $x - 30$ mph because of the headwind:

The distance from Cleveland to Peoria is 420 miles (half the round trip). So we have

$$\text{Rate} \cdot \text{Time} = \text{Distance}$$

$$\text{Time} = \frac{\text{Distance}}{\text{Rate}} = \frac{420}{x - 30}.$$

On the return trip, the actual speed will be $x + 40$ because of the tailwind:

so that

$$\text{Rate} \cdot \text{Time} = \text{Distance}$$

$$\text{Time} = \frac{\text{Distance}}{\text{Rate}} = \frac{420}{x + 40}.$$

Therefore,

$$5 = \left(\begin{array}{c} \text{Time from Cleveland} \\ \text{to Peoria} \end{array} \right) + \left(\begin{array}{c} \text{Time from Peoria} \\ \text{to Cleveland} \end{array} \right)$$

$$5 = \frac{420}{x - 30} + \frac{420}{x + 40}.$$

Multiplying both sides by the common denominator $(x - 30)(x + 40)$ and simplifying, we have

$$5(x - 30)(x + 40) = \frac{420}{x - 30} \cdot (x - 30)(x + 40) + \frac{420}{x + 40} \cdot (x - 30)(x + 40)$$

$$5(x - 30)(x + 40) = 420(x + 40) + 420(x - 30)$$

$$(x - 30)(x + 40) = 84(x + 40) + 84(x - 30)$$

$$x^2 + 10x - 1200 = 84x + 3360 + 84x - 2520$$

$$x^2 - 158x - 2040 = 0$$

$$(x - 170)(x + 12) = 0$$

$$x - 170 = 0 \qquad \text{or} \qquad x + 12 = 0$$

$$x = 170 \qquad\qquad\qquad x = -12.$$

Obviously, the negative solution doesn't apply. Since we multiplied both sides by a quantity involving the variable, we must check that 170 actually is a solution of the original equation. It is; so the plane should be flown at a speed of 170 mph. ■

Calculator Investigation 2.2

1. **Polynomial Equation Solvers** If you have one of the calculators mentioned below, use its polynomial equation solver to solve the equation $3.2x^2 + 15.93x - 7.1 = 0$, as follows. Bring up the solver by choosing POLY on the TI-86 keyboard, or POLY in the Sharp 9600 TOOL menu, or EQUATION in the Casio 9850 main menu (and selecting POLYNOMIAL). Enter 2 as the degree of the polynomial (called its "order" on TI), enter the coefficients of the polynomial, and press SOLVE (or EXE on Sharp 9600). On HP-38, use the POLYNOM submenu of the MATH menu, and key in POLYROOT([3.2, 15.93, −7.1]), including both parentheses and brackets; then press ENTER. On TI-89, use the ALGEBRA menu, and key in: SOLVE($3.2x^2 + 15.93x - 7.1 = 0, x$); then press ENTER. How do the answers you obtained compare with those in Example 7? Use the same technique to solve the following equations.

   (a) $x^2 + 2x - 4 = 0$ **(b)** $x^2 + 5x + 2 = 0.$

Exercises 2.2

In Exercises 1–12, solve the equation by factoring.

1. $x^2 - 8x + 15 = 0$ **2.** $x^2 + 5x + 6 = 0$

3. $x^2 - 5x = 14$ **4.** $x^2 + x = 20$

5. $2y^2 + 5y - 3 = 0$ **6.** $3t^2 - t - 2 = 0$

7. $4t^2 + 9t + 2 = 0$ **8.** $9t^2 + 2 = 11t$

9. $3u^2 + u = 4$ **10.** $5x^2 + 26x = -5$

11. $12x^2 + 13x = 4$ **12.** $18x^2 = 23x + 6$

In Exercises 13–24, use the quadratic formula to solve the equation.

13. $x^2 - 4x + 1 = 0$ **14.** $x^2 - 2x - 1 = 0$

15. $x^2 + 6x + 7 = 0$ **16.** $x^2 + 4x - 3 = 0$

17. $x^2 + 6 = 2x$ **18.** $x^2 + 11 = 6x$

19. $4x^2 - 4x = 7$ **20.** $4x^2 - 4x = 11$

21. $4x^2 - 8x + 1 = 0$ **22.** $2t^2 + 4t + 1 = 0$

23. $5u^2 + 8u = -2$ **24.** $4x^2 = 3x + 5$

In Exercises 25–30, find the number of real solutions of the equation by computing the discriminant.

25. $x^2 + 4x + 1 = 0$ **26.** $4x^2 - 4x - 3 = 0$

27. $9x^2 = 12x + 1$ **28.** $9t^2 + 15 = 30t$

29. $25t^2 + 49 = 70t$ **30.** $49t^2 + 5 = 42t$

In Exercises 31–34, solve the equation and check your answers. [Hint: First, eliminate fractions by multiplying both sides by a common denominator.]

31. $1 - \dfrac{3}{x} = \dfrac{40}{x^2}$

32. $\dfrac{4x^2 + 5}{3x^2 + 5x - 2} = \dfrac{4}{3x - 1} - \dfrac{3}{x + 2}$

33. $\dfrac{2}{x^2} - \dfrac{5}{x} = 4$ **34.** $\dfrac{x}{x - 1} + \dfrac{x + 2}{x} = 3$

In Exercises 35–44, solve the equation by any method.

35. $x^2 + 9x + 18 = 0$ **36.** $3t^2 - 11t - 20 = 0$

37. $4x(x + 1) = 1$ **38.** $25y^2 = 20y + 1$

39. $2x^2 = 7x + 15$ **40.** $2x^2 = 6x + 3$

41. $t^2 + 4t + 13 = 0$ **42.** $5x^2 + 2x = -2$

43. $\dfrac{7x^2}{3} = \dfrac{2x}{3} - 1$ **44.** $25x + \dfrac{4}{x} = 20$

In Exercises 45–48, use a calculator and the quadratic formula to find approximate solutions of the equation.

45. $4.42x^2 - 10.14x + 3.79 = 0$

46. $8.06x^2 + 25.8726x - 25.047256 = 0$

47. $3x^2 - 82.74x + 570.4923 = 0$

48. $7.63x^2 + 2.79x = 5.32$

In Exercises 49–51, find a number k such that the given equation has exactly one real solution.

49. $x^2 + kx + 25 = 0$ **50.** $x^2 - kx + 49 = 0$

51. $kx^2 + 8x + 1 = 0$

In Exercises 52–54, the discriminant of the equation $ax^2 + bx + c = 0$ (with a, b, c integers) is given. Use it to determine whether or not the solutions of the equation are rational numbers.

52. $b^2 - 4ac = 25$

53. $b^2 - 4ac = 0$

54. $b^2 - 4ac = 72$

In Exercises 55–58, solve the equation for the indicated letter. State any conditions that are necessary for your answer to be valid.

55. $E = mc^2$ for c **56.** $V = 4\pi r^2$ for r

57. $A = \pi rh + \pi r^2$ for r

58. $d = -16t^2 + vt$ for t

59. Find a number k such that 4 and 1 are the solutions of $x^2 - 5x + k = 0$.

60. Find the error in the following "proof" that 6 = 3.

	$x = 3$
Multiply both sides by x:	$x^2 = 3x$
Subtract 9 from both sides:	$x^2 - 9 = 3x - 9$
Factor each side:	$(x - 3)(x + 3) = 3(x - 3)$
Divide both sides by x − 3:	$x + 3 = 3$
Because x = 3:	$3 + 3 = 3$
Oops!	$6 = 3$

In Exercises 61–64, label the unknown quantities and fill in the rest of the right-hand column of the table. Then use the information in the table to obtain an equation in one variable that will provide the solution to the problem. Solve the equation and interpret the answers.

61. A rectangle has perimeter 45 cm and an area of 112.5 sq cm. What are its dimensions?

English Language	Mathematical Language
Length of rectangle	
Width of rectangle	
Perimeter of rectangle	
Perimeter is 45	
Area of rectangle	
Area is 112.5.	

62. A triangle has an area of 96 sq in. and its height is two-thirds of its base. What are the base and height of the triangle?

English Language	Mathematical Language
Base of triangle	
Height of triangle	
Height is two-thirds of base.	
Area of triangle	
Area is 96.	

63. A rectangular window has 5.7 square feet of glass. Its length is 2.3 ft longer than its width. What are the dimensions of the window?

English Language	Mathematical Language
Length of window	
Width of window	
Area of window	
Area of window is 5.7.	
Length is 2.3 ft longer than width.	

64. The area of a circular window (in square feet) is two and half times the length of the diameter. What is the radius of the window?

English Language	Mathematical Language
Radius of circle	
Area of circle	
Diameter of circle	
Two and a half times the diameter	
Relationship of diameter and radius	

65. The two legs of a right triangle differ in length by 1 cm and the hypotenuse is 1 cm longer than the longer leg. How long is each side?

66. The area of a triangle is 20 square feet. If the base is 2 feet longer than the height, find the height.

67. The radius of a circle is 8 cm. By what amount must the radius be decreased in order to decrease the area of the circle by 48π square centimeters?

68. A right triangle has a hypotenuse of length 12.054 meters. The sum of the lengths of the other two sides is 16.96 meters. How long is each side?

69. A 13-foot-long ladder leans on a wall, as shown in the figure. The bottom of the ladder is 5 feet from the wall. If the bottom is pulled out 3 feet farther from the wall, how far does the top of the ladder move down the wall? [*Hint:* Draw pictures of the right triangle formed by the ladder, the ground, and the wall before and after the ladder is moved. In each case, use the Pythagorean Theorem to find the distance from the top of the ladder to the ground.]

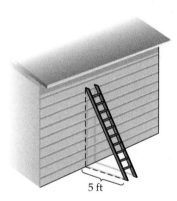

5 ft

70. A 15-foot-long pole leans against a wall. The bottom is 9 feet from the wall. How much farther should the bottom be pulled away from the wall so that the top moves the same amount down the wall?

71. A concrete walk of uniform width is to be built around a circular pool, as shown in the figure. The radius of the pool is 12 meters and enough concrete is available to cover 52π square meters. If all the concrete is to be used, how wide should the walk be?

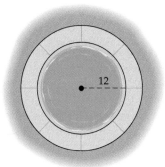

72. A decorator has 92 one-foot-square tiles that will be laid around the edges of a 12-by-15-foot room. A rectangular rug that is 3 feet longer than it is wide is to be placed over the center area where there are no tiles. To the nearest quarter foot, find the dimensions of the smallest rug that will cover the untiled part of the floor. [Assume that all the tiles are used and that none of them are split.]

73. A corner lot has dimensions of 25 by 40 yards. The city plans to take a strip of uniform width along the two sides bordering the streets in order to widen these roads. To the nearest tenth of a yard, how wide should the strip be if the remainder of the lot is to have an area of 844 square yards?

74. A box with a square base and no top is to be made from a square sheet of aluminum by cutting out 3-inch squares from each corner and folding up the sides as shown in the figure below. If the box is to have a volume of 48 cubic inches, what size should the piece of aluminum be?

75. A box with no top is to be made by taking a rectangular aluminum sheet 8 by 10 inches and cutting a square of the same size out of each corner and folding up the sides. If the area of the base is to be 24 square inches, what size squares should be cut out?

76. A box with no top is to be made by taking a rectangular piece of aluminum and cutting a square of the same size out of each corner and folding up the sides. The box is to be 2 inches deep. The length of its base is to be 5 inches more than the width. If the volume is to be 352 cubic inches, what size should the piece of aluminum be?

77. It requires 53.4 square inches of metal to make a cylindrical can whose height is 4.25 inches. To the nearest hundredth, what is the radius of the can? [*Hint:* The formula for the surface area S of a right circular cylinder of radius r and height h is $S = 2\pi r^2 + 2\pi rh$.]

78. Two trains leave the same city at the same time, one going north and the other east. The northbound train travels 20 mph faster than the eastbound one. If they are 300 miles apart after 5 hours, what is the speed of each train?

79. A student leaves the university at noon, bicycling south at a constant rate. At 12:30 a second student leaves the same point and heads west, bicycling 7 mph faster than the first student. At 2 P.M. they are 30 miles apart. How fast is each one going?

80. To get to work Sam jogs 3 kilometers to the train, then rides the remaining 5 kilometers. If the train goes 40 kilometers per hour faster than Sam's constant rate of jogging and the entire trip takes 30 minutes, how fast does Sam jog?

81. Red Riding Hood drives the 432 miles to Grandmother's house in 1 hour less than it takes the Wolf to drive the same route. Her average speed is 6 mph faster than the Wolf's average speed. How fast does each drive?

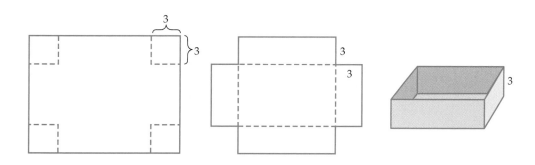

82. A canoeist paddles at a constant rate and takes 2 hours longer to go 12 kilometers upstream than to go the same distance downstream. If the current runs at 3 kilometers per hour, how long would a 12-kilometer trip take in still water?

83. Tom drops a rock into a well and 3 seconds later the sound of its splash is heard. How deep is the well? [Make the same assumptions about falling rocks and sound as in Example 11.]

Rock falls ——— Sound rises

84. A group of homeowners are to share equally in the $210 cost of repairing a bus-stop shelter near their homes. At the last moment two members of the group decide not to participate and this raises the share of each remaining person by $28. How many people were in the group at the beginning?

85. A woman and her son working together can paint a garage in 4 hours and 48 minutes. The woman working alone can paint it in 4 hours less than the son would take to do it alone. How long would it take the son to paint the garage alone?

86. When each works alone Barbara can do a job in 3 hours less time than John. When they work together it takes them 2 hours to complete the job. How long does it take each one working alone to do the job?

Background for Exercises 87–90. If an object is thrown upward, dropped, or thrown downward and travels in a vertical line subject only to gravity (with wind resistance ignored), then the height h of the object above the ground (in feet) after t seconds is given by:

$$h = -16t^2 - v_0 t + h_0,$$

where h_0 is the initial height of the object at starting time $t = 0$ and v_0 is the initial velocity (speed) of the object at time $t = 0$. The value of v_0 is taken as positive if the object starts moving upward at time $t = 0$ and negative if the object starts moving downward at $t = 0$. An object that is dropped (rather than thrown downward) has initial velocity $v_0 = 0$.

87. How long does it take an object to reach the ground if
    (a) it is dropped from the top of a 640-foot-high building?
    (b) it is thrown downward from the top of the same building, with an initial velocity of 52 feet per second?

88. You are standing on a cliff 200 feet high. How long will it take a rock to reach the ground if
    (a) you drop it?
    (b) you throw it downward at an initial velocity of 40 feet per second?
    (c) How far does the rock fall in 2 seconds if you throw it downward with an initial velocity of 40 feet per second?

89. A rocket is fired straight up from ground level with an initial velocity of 800 feet per second.
    (a) How long does it take the rocket to rise 3200 feet?
    (b) When will the rocket hit the ground?

90. A rocket loaded with fireworks is to be shot vertically upward from ground level with an initial velocity of 200 feet per second. When the rocket reaches a height of 400 feet on its upward trip the fireworks will be detonated. How many seconds after lift-off will this take place?

Thinkers

91. Two ferryboats leave opposite sides of a lake at the same time. They pass each other when they are 800 meters from the nearest shore. When it reaches the opposite side each boat spends 30 minutes at the dock and then starts back. This time the boats pass each other when they are 400 meters from the nearest shore. Assuming that each boat travels at the same speed in both directions, how wide is the lake between the two ferry docks?

92. Charlie was crossing a narrow bridge. When he was halfway across he saw a truck on the opposite side of the bridge, 200 yards away and heading toward him. He turned and ran back. The truck continued at the same speed and missed him by a hair. If Charlie had tried to cross the bridge the truck would have hit him 3 yards before he reached the other end. How long is the bridge?

2.3 Other Equations and Applications

In this section we explore both algebraic and graphical techniques for solving various kinds of equations, beginning with polynomial equations of degree greater than 2. Although graphical and numerical approximation methods usually work best for such equations, some of them can be solved algebraically.

Example 1   Solve: $4x^4 - 13x^2 + 3 = 0$.

Solution   Substitute u for x^2 and solve the resulting quadratic equation:

$$4x^4 - 13x^2 + 3 = 0$$
$$4(x^2)^2 - 13x^2 + 3 = 0$$
$$4u^2 - 13u + 3 = 0$$
$$(u - 3)(4u - 1) = 0$$

$$u - 3 = 0 \quad \text{or} \quad 4u - 1 = 0$$
$$u = 3 \quad\quad\quad 4u = 1$$
$$u = \frac{1}{4}.$$

Since $u = x^2$ we see that

$$x^2 = 3 \quad \text{or} \quad x^2 = \frac{1}{4}$$

$$x = \pm\sqrt{3} \quad\quad x = \pm\frac{1}{2}.$$

Hence the original equation has four solutions: $-\sqrt{3}, \sqrt{3}, -1/2, 1/2$.
■

Example 2   Solve: $x^4 - 4x^2 + 1 = 0$.

Solution   Let $u = x^2$; then

$$x^4 - 4x^2 + 1 = 0$$
$$u^2 - 4u + 1 = 0.$$

The quadratic formula shows that

$$u = \frac{-(-4) \pm \sqrt{(-4)^2 - 4 \cdot 1 \cdot 1}}{2 \cdot 1} = \frac{4 \pm \sqrt{12}}{2}$$

$$= \frac{4 \pm \sqrt{4 \cdot 3}}{2} = \frac{4 \pm 2\sqrt{3}}{2} = 2 \pm \sqrt{3}.$$

Since $u = x^2$, we have the equivalent statements:

$$x^2 = 2 + \sqrt{3} \qquad \text{or} \qquad x^2 = 2 - \sqrt{3}$$
$$x = \pm\sqrt{2 + \sqrt{3}} \qquad\qquad x = \pm\sqrt{2 - \sqrt{3}}.$$

Therefore, the original equation has four solutions.   ∎

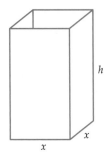

Figure 2–10

Example 3   A rectangular box with a square base and no top is to have a volume of 30,000 cubic cm. If each side of the base can be no longer than 50 cm and the surface area of the box is 6000 square cm, what are its dimensions?

Solution   Let x denote the length of each side of the base and h the height, as in Figure 2–10. We have these translations.

English Language	Mathematical Language
Length, width, and height	x, x, and h
Volume (length × width × height)	x^2h
Volume is 30,000 cu cm	$x^2h = 30{,}000$
Area of base (length × width)	x^2
Area of each side (length × height)	xh
Surface area of box (base and 4 sides)	$x^2 + 4xh$
Surface area is 6000 sq cm	$x^2 + 4xh = 6000$

Since there are two variables, we solve the volume equation for h.

$$x^2h = 30{,}000$$

$$h = \frac{30{,}000}{x^2}$$

and substitute this result into the surface area equation.

$$x^2 + 4xh = 6000$$

$$x^2 + 4x\left(\frac{30{,}000}{x^2}\right) = 6000$$

$$x^2 + \frac{120{,}000}{x} = 6000$$

$$x^3 + 120{,}000 = 6000x$$

$$x^3 - 6000x + 120{,}000 = 0.$$

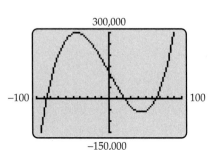

Figure 2–11

Substitution and factoring seem unlikely to work here, so we use graphical methods. The graph of $y = x^3 - 6000x + 120{,}000$ in Figure 2–11 shows that the equation has three solutions (x-intercepts). The negative one and the one between 60 and 70 do not apply here because the side of the base must be positive and less than 50 cm long.

GRAPHING EXPLORATION

Approximate the solution of the equation (x-intercept) that appears to be near 20. Then find the corresponding value of h. ■

Radical Equations

The algebraic solution of equations involving radicals depends on this fact: If two quantities are equal, say

$$x - 2 = 3,$$

then their squares are also equal:

$$(x - 2)^2 = 9.$$

Thus,

every solution of $x - 2 = 3$ is also a solution of $(x - 2)^2 = 9$.

But *be careful:* This only works in one direction. For instance, -1 is a solution of $(x - 2)^2 = 9$, but not of $x - 2 = 3$. This is an example of the Power Principle.

Power Principle

> **If both sides of an equation are raised to the same positive integer power, then every solution of the original equation is also a solution of the new equation. But the new equation may have solutions that are *not* solutions of the original one.**

Consequently, if you raise both sides of an equation to a power, you must *check your solutions* in the *original* equation. Graphing provides a quick way to eliminate most extraneous solutions. But only an algebraic computation can confirm an exact solution.

Example 4 Solve: $5 + \sqrt{3x - 11} = x$.

Solution We first rearrange terms to get the radical expression alone on one side:

$$\sqrt{3x - 11} = x - 5.$$

Then, square both sides and solve the resulting equation:

$$\left(\sqrt{3x - 11}\right)^2 = (x - 5)^2$$
$$3x - 11 = x^2 - 10x + 25$$
$$0 = x^2 - 13x + 36$$
$$0 = (x - 4)(x - 9)$$
$$x - 4 = 0 \quad \text{or} \quad x - 9 = 0$$
$$x = 4 \qquad\qquad x = 9.$$

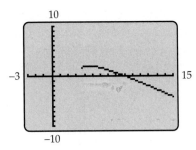

Figure 2–12

If these numbers are solutions of the original equation, they should be x-intercepts of the graph of $y = \sqrt{3x - 11} - x + 5$. (Why?) But the graph of y in Figure 2–12 doesn't have an x-intercept at $x = 4$, so 4 is not a solution of the original equation. The graph appears to show that $x = 9$ is a solution of the original equation, and direct calculation confirms this:

$$\begin{array}{ll} \textit{Left side:} & \textit{Right side:}\quad x \\[4pt] 5 + \sqrt{3x - 11} & \qquad\qquad 9 \\[4pt] 5 + \sqrt{3 \cdot 9 - 11} & \\[4pt] 5 + \sqrt{16} & \\[4pt] 9 & \end{array}$$

Hence, 9 is the only solution of the original equation. ■

Example 5 Solve: $\sqrt{2x - 3} - \sqrt{x + 7} = 2$.

Solution   We first rearrange terms so that one side contains only a single radical term:

$$\sqrt{2x - 3} = \sqrt{x + 7} + 2.$$

Then, square both sides and simplify:

$$\begin{aligned} \left(\sqrt{2x - 3}\right)^2 &= \left(\sqrt{x + 7} + 2\right)^2 \\ 2x - 3 &= \left(\sqrt{x + 7}\right)^2 + 2 \cdot 2 \cdot \sqrt{x + 7} + 2^2 \\ 2x - 3 &= x + 7 + 4\sqrt{x + 7} + 4 \\ x - 14 &= 4\sqrt{x + 7}. \end{aligned}$$

Now, square both sides and solve the resulting equation:

$$\begin{aligned} (x - 14)^2 &= \left(4\sqrt{x + 7}\right)^2 \\ x^2 - 28x + 196 &= 4^2 \cdot \left(\sqrt{x + 7}\right)^2 \\ x^2 - 28x + 196 &= 16(x + 7) \\ x^2 - 28x + 196 &= 16x + 112 \\ x^2 - 44x + 84 &= 0 \\ (x - 2)(x - 42) &= 0 \\ x - 2 = 0 \quad &\text{or} \quad x - 42 = 0 \\ x = 2 \qquad\qquad & \qquad\quad x = 42. \end{aligned}$$

Substituting 2 and 42 in the left side of the original equation shows that

$$\sqrt{2 \cdot 2 - 3} - \sqrt{2 + 7} = \sqrt{1} - \sqrt{9} = 1 - 3 = -2.$$
$$\sqrt{2 \cdot 42 - 3} - \sqrt{42 + 7} = \sqrt{81} - \sqrt{49} = 9 - 7 = 2.$$

Therefore, 42 is the only solution of the equation. ■

Example 6 Stella, who is standing at point A on the bank of a 2.5-km-wide river, wants to reach point B, 15 km downstream on the opposite bank. She plans to row to a point C on the opposite shore and then run to B, as shown in Figure 2–13. She can row at a rate of 4 km per hour and can run at 8 km per hour. If her trip is to take 3 hours, how far from B should she land?

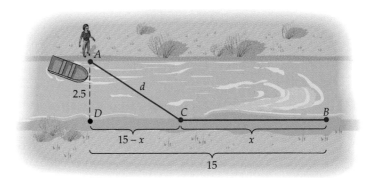

Figure 2–13

Solution Let x be the distance that Stella must run from C to B. Using the basic formula for distance, we have:

$$\text{Rate} \times \text{Time} = \text{Distance}$$

$$\text{Time} = \frac{\text{Distance}}{\text{Rate}} = \frac{x}{8}.$$

Similarly, the time required to row distance d is

$$\text{Time} = \frac{\text{Distance}}{\text{Rate}} = \frac{d}{4}.$$

Since $15 - x$ is the distance from D to C, the Pythagorean Theorem applied to right triangle ADC shows that

$$d^2 = (15 - x)^2 + 2.5^2 \qquad \text{or equivalently} \qquad d = \sqrt{(15 - x)^2 + 6.25}.$$

Therefore, the total time for the trip is given by

$$\text{Rowing time} + \text{Running time} = \frac{d}{4} + \frac{x}{8} = \frac{\sqrt{(x - 15)^2 + 6.25}}{4} + \frac{x}{8}.$$

If the trip is to take 3 hours, then we must solve the equation

$$\frac{\sqrt{(x - 15)^2 + 6.25}}{4} + \frac{x}{8} = 3.$$

GRAPHING EXPLORATION

Solve the equation by graphing

$$y = \frac{\sqrt{(x-15)^2 + 6.25}}{4} + \frac{x}{8} - 3$$

in the window with $0 \le x \le 15$ and $-2 \le y \le 2$ and finding its x-intercept.

This Exploration shows that Stella should land approximately 6.74 km from B in order to make the trip in 3 hours. ∎

Absolute Value Equations

If c is a real number, then by the definition of absolute value $|c|$ is either c or $-c$ (whichever one is positive). This fact can be used to solve absolute value equations algebraically.

Example 7 To solve $|3x - 4| = 8$, apply the fact stated above with $c = 3x - 4$. Then $|3x - 4|$ is either $3x - 4$ or $-(3x - 4)$, so that

$$3x - 4 = 8 \qquad \text{or} \qquad -(3x - 4) = 8$$
$$3x = 12 \qquad\qquad\qquad -3x + 4 = 8$$
$$x = 4 \qquad\qquad\qquad\qquad -3x = 4$$
$$x = -4/3.$$

So there are two possible solutions of the original equation $|3x - 4| = 8$. You can readily verify that both 4 and $-4/3$ actually are solutions. ∎

Example 8 Solve: $|x + 4| = 5x - 2$.

Solution The left side of the equation is either $x + 4$ or $-(x + 4)$ (why?). Hence,

$$x + 4 = 5x - 2 \qquad \text{or} \qquad -(x + 4) = 5x - 2$$
$$-4x + 4 = -2 \qquad\qquad\qquad -x - 4 = 5x - 2$$
$$-4x = -6 \qquad\qquad\qquad\qquad -6x = 2$$
$$x = \frac{-6}{-4} = \frac{3}{2} \qquad\qquad\qquad x = \frac{2}{-6} = -\frac{1}{3}.$$

Check each of these possible solutions in the original equation: $x = 3/2$ is a solution because

$$|x + 4| = \left|\frac{3}{2} + 4\right| = \left|\frac{11}{2}\right| = \frac{11}{2}$$

$$5x - 2 = 5\left(\frac{3}{2}\right) - 2 = \frac{15}{2} - \frac{4}{2} = \frac{11}{2},$$

but $x = -1/3$ is not a solution since

$$|x + 4| = \left|-\frac{1}{3} + 4\right| = \left|\frac{11}{3}\right| = \frac{11}{3}$$

$$5x - 2 = 5\left(-\frac{1}{3}\right) - 2 = -\frac{5}{3} - \frac{6}{3} = -\frac{11}{3}.$$

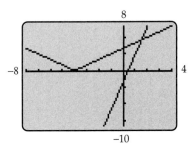

Figure 2–14

You can graphically confirm the fact that $x = -1/3$ is not a solution by graphing $y = |x + 4|$ and $y = 5x - 2$ on the same screen (Figure 2–14). Since there is no intersection point with a negative x-coordinate, $x = -1/3$ cannot be a solution. ■

Exercises 2.3

In Exercises 1–8, find all real solutions of each equation exactly.

1. $y^4 - 7y^2 + 6 = 0$ **2.** $x^4 - 2x^2 + 1 = 0$

3. $x^4 - 2x^2 - 35 = 0$ **4.** $x^4 - 2x^2 - 24 = 0$

5. $2y^4 - 9y^2 + 4 = 0$ **6.** $6z^4 - 7z^2 + 2 = 0$

7. $10x^4 + 3x^2 = 1$ **8.** $6x^4 - 7x^2 = 3$

In Exercises 9–28, find all real solutions of each equation. Find exact solutions when possible and approximate ones otherwise.

9. $\sqrt{x + 2} = 3$ **10.** $\sqrt{x - 7} = 4$

11. $\sqrt{4x + 9} = 5$ **12.** $\sqrt{3x - 2} = 7$

13. $\sqrt[3]{5 - 11x} = 3$ [*Hint:* Cube both sides.]

14. $\sqrt[3]{6x - 10} = 2$ **15.** $\sqrt[3]{x^2 - 1} = 2$

16. $(x + 1)^{2/3} = 4$ **17.** $\sqrt{x^2 - x - 1} = 1$

18. $\sqrt{x^2 - 5x + 4} = 2$ **19.** $\sqrt{x + 7} = x - 5$

20. $\sqrt{x + 5} = x - 1$ **21.** $\sqrt{3x^2 + 7x - 2} = x + 1$

22. $\sqrt{4x^2 - 10x + 5} = x - 3$

23. $\sqrt[3]{x^3 + x^2 - 4x + 5} = x + 1$

24. $\sqrt[3]{x^3 - 6x^2 + 2x + 3} = x - 1$

25. $\sqrt{5x + 6} = 3 + \sqrt{x + 3}$

26. $\sqrt{3y + 1} - 1 = \sqrt{y + 4}$

27. $\sqrt{2x - 5} = 1 + \sqrt{x - 3}$

28. $\sqrt{x - 3} + \sqrt{x + 5} = 4$

In Exercises 29–32, assume that all letters represent positive numbers and solve each equation for the required letter.

29. $A = \sqrt{1 + \dfrac{a^2}{b^2}}$ for b **30.** $T = 2\pi\sqrt{\dfrac{m}{g}}$ for g

31. $K = \sqrt{1 - \dfrac{x^2}{u^2}}$ for u **32.** $R = \sqrt{d^2 + k^2}$ for d

In Exercises 33–42, find all real solutions of each equation.

33. $|2x + 3| = 9$ **34.** $|3x - 5| = 7$

35. $|6x - 9| = 0$ **36.** $|4x - 5| = 9$

37. $|2x + 3| = 4x - 1$ **38.** $|3x - 2| = 5x + 4$

39. $|x - 3| = x$ **40.** $|2x - 1| = 2x + 1$

41. $|x^2 + 4x - 1| = 4$

42. In statistical quality control, one needs to find the proportion of the product that is not acceptable. The upper and lower control limits are found by solving the following equation (in which $\bar{p}$ is the mean percent defective and n is the sample size) for CL.

$$|CL - \bar{p}| = 3\sqrt{\frac{\bar{p}(1 - \bar{p})}{n}}.$$

Find the control limits when $\bar{p} = .02$ and $n = 200$.

43. What are the dimensions of a rectangle whose diagonal is 130 cm long and whose area is 6000 square centimeters?

44. The dimensions of a rectangular box are consecutive integers. If the box has volume 13,800 cubic centimeters, what are its dimensions?

45. The surface area S of the right circular cone in the figure is given by $S = \pi r\sqrt{r^2 + h^2}$. What radius should be used to produce a cone of height 5 inches and surface area 100 square inches?

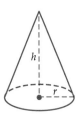

46. What is the radius of the base of a cone whose surface area is 18π square centimeters and whose height is 4 cm?

47. Find the radius of the base of a conical container whose height is 1/3 of the radius and whose volume is 180 cubic inches. [*Note:* The volume of a cone of radius r and height h is $\pi r^2 h/3$.]

48. The surface area of the right square pyramid in the figure is given by $S = b\sqrt{b^2 + 4h^2}$. If the pyramid has height 10 feet and surface area 100 square feet, what is the length of a side b of its base?

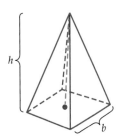

49. A rope is to be stretched at uniform height from a tree to a 35-foot-long fence, which is 20 feet from the tree, and then to the side of a building at a point 30 feet from the fence, as shown in the figure. If 63 feet of rope is to be used, how far from the building wall should the rope meet the fence?

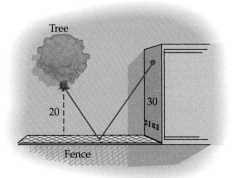

50. Anne is standing on a straight road and wants to reach her helicopter, which is located 2 miles down the road from her, a mile from the road in a field (see figure). She can run 5 miles per hour on the road and 3 miles per hour in the field. She plans to run down the road, then cut diagonally across the field to reach the helicopter. Where should she leave the road in order to reach the helicopter in exactly 42 minutes (.7 hour)?

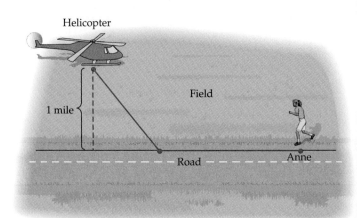

51. A power plant is located on the bank of a river that is $\frac{1}{2}$ mile wide. Wiring is to be laid across the river and then along the shore to a substation 8 miles downstream, as shown in the figure. It costs $12,000 per mile for underwater wiring and $8000 per mile for wiring on land. If $72,000 is to be spent on the project, how far from the substation should the wiring come to shore?

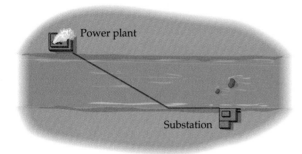

52. A homemade loaf of bread turns out to be a perfect cube. Five slices of bread, each .6 in. thick, are cut from one end of the loaf. The remainder of the loaf now has a volume of 235 cubic inches. What were the dimensions of the original loaf?

53. A rectangular bin with an open top and volume of 38.72 cubic feet is to be built. The length of its base must be twice the width and the bin must be at least 3 feet high. Material for the base of the bin costs $12 per square foot and material for the sides costs $8 per square foot. If it costs $538.56 to build the bin, what are its dimensions?

Thinker

54. One corner of an 8.5-by-11-inch piece of paper is folded over to the opposite side, as shown in the figure. The area of the darkly shaded triangle at the lower left is 6 square inches and we want to find the length x.
  (a) Take a piece of paper this size and experiment. Approximately, what is the largest value x could have (and still have the paper look like the figure)? With this value of x, what is the approximate area of the triangle? Try some other possibilities.
  (b) Now find an exact answer by constructing and solving a suitable equation. Explain why one of the solutions to the equation is not an answer to this problem.

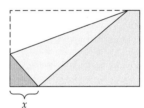

x

2.4 Linear Inequalities

Inequalities such as

$$5x + 3 \leq 7x + 6 \qquad \text{or} \qquad 4 < 3 - 5x < 18$$

are called **linear inequalities.** They may be solved by algebraic or graphical means, both of which are discussed here.

We say that two inequalities are **equivalent** if they have the same solutions. The usual strategy for solving inequalities algebraically is to use the following Basic Principles to transform a given inequality into an equivalent one whose solutions are known.

Basic Principles for Solving Inequalities

Performing any of the following operations on an inequality produces an equivalent inequality:

1. Add or subtract the same quantity on both sides of the inequality.
2. Multiply or divide both sides of the inequality by the same *positive* quantity.
3. Multiply or divide both sides of the inequality by the same *negative* quantity and *reverse the direction of the inequality.*

Note Principle 3 carefully. It says, for example, that when you multiply both sides of $-3 < 5$ by -2, you must reverse the direction of the inequality in order to have a true statement:

$$-3 < 5$$

Reverse direction: $(-2)(-3) > (-2)(5)$

$$6 > -10$$

Failure to do this is probably the most common algebraic error when solving inequalities.

Example 1   Solve: $3x + 1 < 5$.

Solution   *Algebraic Method:* We use the Basic Principles to transform this inequality into an equivalent one whose solutions are obvious.

$$3x + 1 < 5$$

Subtract 1 from both sides: $3x < 4$

Divide both sides by 3: $x < 4/3$

Therefore, the solutions are all real numbers that are less than $4/3$, as shown on the number line in Figure 2–15.

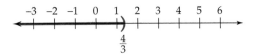

Figure 2–15

Graphical Method: One approach is to change the inequality to an equivalent one with 0 on one side:

$$3x + 1 < 5$$

Subtract 5 from both sides: $3x - 4 < 0.$

Now graph $y_1 = 3x - 4$ (Figure 2–16) and note how the graph is related to this last inequality:

When a point (x, y_1) lies below the x-axis, then its y-coordinate is negative, which means that

$$y_1 < 0, \quad \text{or equivalently,} \quad 3x - 4 < 0.$$

Thus, the solutions of the inequality are the x-coordinates of all points on the graph that lie below the x-axis. Figure 2–16 shows that these are the points to the left of the x-intercept of the graph. The calculator's root finder shows that the x-intercept is approximately 1.333 (see Figure 2–17). So the (approximate) solutions of this inequality are all x such that $x < 1.333$. Since the decimal expansion of 4/3 begins 1.333, this answer is consistent with the one obtained algebraically. ■

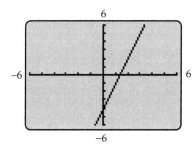

Figure 2–16

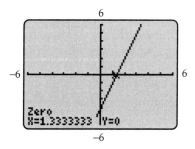

Figure 2–17

Example 2 Solve: $x + 2 \le 8x + 4$.

Solution *Algebraic Method:*

$$x + 2 \le 8x + 4$$

Subtract 2 from both sides: $$x \le 8x + 2$$

Subtract 8x from both sides: $$-7x \le 2$$

Divide both sides by -7 and reverse the direction of the inequality: $$x \ge -2/7.$$

Therefore, the solutions are all real numbers greater than or equal to $-2/7$. They are shown on the number line in Figure 2–18.

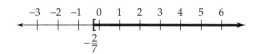

Figure 2–18

Graphical Method: The inequality can be solved by using the method of Example 1. A second method is to graph $y_1 = x + 2$ and $y_2 = 8x + 4$ on the same screen and note how points on the two graphs are related:

When the point (x, y_1) lies below the point (x, y_2), then the y-coordinate of (x, y_1) is smaller than the y-coordinate of (x, y_2), which means that

$$y_1 \leq y_2, \quad \text{or equivalently,} \quad x + 2 \leq 8x + 4.$$

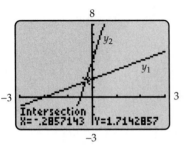

Figure 2–19

Thus, the solutions of the inequality are the x-coordinates of all points for which the graph of y_1 lies below or on the graph of y_2. Figure 2–19 shows that the graph of y_1 lies below the graph of y_2 to the right of the point where they intersect. The calculator's intersection finder shows that this point is approximately $(-.285714, 1.771429)$. Points to the right of the intersection point have larger x-coordinates. So the approximate solutions of the inequality are all x such that $x \geq -.285714$. Verify that this answer is consistent with the one obtained algebraically by finding the decimal expansion of $-2/7$. ∎

Example 3 Solve: $2 \leq 3x + 5 < 2x + 11$.

Solution By definition, the solutions of $2 \leq 3x + 5 < 2x + 11$ are the numbers that are solutions of *both* of these inequalities:

$$2 \leq 3x + 5 \quad \text{and} \quad 3x + 5 < 2x + 11.$$

Each of these inequalities can be solved by the methods used above. For the first one we have:

$$2 \leq 3x + 5$$

Subtract 5 from both sides: $\quad -3 \leq 3x$

Divide both sides by 3: $\quad -1 \leq x.$

The second inequality is solved similarly:

$$3x + 5 < 2x + 11$$

Subtract 5 from both sides: $\quad 3x < 2x + 6$

Subtract 2x from both sides: $\quad x < 6.$

> **CAUTION**
>
> All inequality signs in an inequality should point in the same direction. *Don't* write things like $4 < x > 2$ or $-3 \geq x < 5$.

The solutions of the original inequality are the numbers that satisfy *both* $-1 \leq x$ *and* $x < 6$, that is, all x with $-1 \leq x < 6$, as shown in Figure 2–20. ∎

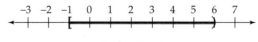

Figure 2–20

GRAPHING EXPLORATION

Graphically confirm the results in Example 3 as follows. Rewrite the first inequality $2 \le 3x + 5$ as $0 \le 3x + 3$ and graph $y_1 = 3x + 3$. The solutions are the x-coordinates of points that lie on or *above* the x-axis (why?). Solve the second inequality $3x + 5 < 2x + 11$ by finding the x-coordinates of the points where the graph of $y_2 = 3x + 5$ lies below the graph of $y_3 = 2x + 11$.

Example 4   Solve: $-1 < 4 - 3x < 7$.

Solution   When solving an inequality like this, in which the variable appears only in the middle part, you can proceed as follows:

$$-1 < 4 - 3x < 7$$

Subtract 4 from each part: $$-5 < -3x < 3$$

Divide each part by -3 *and* **reverse** *the direction of the inequalities:* $$\frac{5}{3} > x > -1.$$

Reading this last inequality from right to left, we see that the solutions are all real numbers x such that $-1 < x < 5/3$.

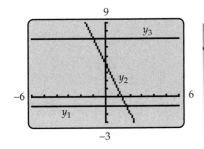

Figure 2–21

GRAPHING EXPLORATION

Confirm that these are the solutions by graphing $y_1 = -1$, $y_2 = 4 - 3x$, and $y_3 = 7$ on the same screen, as in Figure 2–21, and using an intersection finder to determine the x-coordinates of points on the graph of y_2 that lie above the graph of y_1 and below the graph of y_3. ■

Example 5   You can rent a car for $40 per day plus 5¢ per mile, or you can pay $10 per day plus 30¢ per mile. How many miles must you drive in one day for the first plan to be cheaper?

Solution   Let x be the number of miles driven in a day. We must solve this inequality:

Cost of driving x miles with plan 1 $<$ Cost of driving x miles with plan 2

$$\$40 + 5¢ \text{ per mile} < \$10 + 30¢ \text{ per mile}$$

$$40 + .05x < 10 + .30x$$

We'll solve it algebraically (it can also be solved graphically).

$$40 + .05x < 10 + .30x$$

Subtract 40 from both sides: $.05x < -30 + .30x$

Subtract .30x from both sides: $-.25x < -30$

Divide both sides by −.25 and reverse the direction of the inequality: $x > \dfrac{-30}{-.25}$

Simplify right side: $x > 120$

Therefore plan 1 is cheaper when you drive more than 120 miles.   ■

Absolute Value Inequalities

Recall that if r is a real number, then $|r|$ is the distance from r to 0 on the number line. Thus, the inequality $|r| \leq 5$ states that the distance from r to 0 is 5 units or less. A glance at the number line in Figure 2–22 shows that these are the numbers r such that $-5 \leq r \leq 5$.

Figure 2–22

Similarly, the numbers r such that $|r| \geq 5$ are those whose distance to 0 is 5 or more units, that is, the numbers r with $r \leq -5$ or $r \geq 5$. This argument works with any positive number k in place of 5 and proves the following facts (which are also true with $<$ and $>$ in place of $\leq$ and $\geq$):

Absolute Value Inequalities

> **Let k be a positive number and r any real number.**
>
> $|r| \leq k$ is equivalent to $-k \leq r \leq k.$
>
> $|r| \geq k$ is equivalent to $r \leq -k$ or $r \geq k.$

Example 6   Solve: $|3x - 5| \leq 4.$

Solution   *Algebraic Method:* Apply the first fact in the preceding box, with $3x - 5$ in place of r and 4 in place of k, to obtain this equivalent inequality:

$$-4 \leq 3x - 5 \leq 4$$

Add 5 to each part: $1 \leq 3x \leq 9$

Divide each part by 3: $\dfrac{1}{3} \leq x \leq 3.$

Therefore, the solutions of the original inequality are all real numbers x such that $1/3 \le x \le 3$.

Graphical Method: One way to solve the inequality is to graph $y_1 = |3x - 5|$ and $y_2 = 4$ on the same screen (Figure 2–23) and determine the x-coordinates of the points where the graph of y_1 lies on or below the graph of y_2. Alternatively, you can rewrite the original inequality as

$$|3x - 5| - 4 \le 0$$

and graph $y_3 = |3x - 5| - 4$ (Figure 2–24). The solutions of the inequality are the x-coordinates of points on the graph of y_3 that lie on or below the x-axis.

GRAPHING EXPLORATION

Use one of the graphical methods just described to solve the inequality. ■

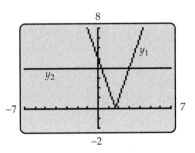

Figure 2–23

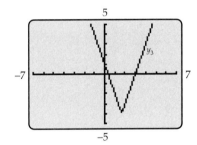

Figure 2–24

Example 7   Solve: $|5x + 3| > 3$.

Solution   Apply the second fact in the box before Example 6, with $5x + 2$ in place of r, 3 in place of k, and $>$ in place of $\ge$. This produces the equivalent statement

$$5x + 2 < -3 \quad \text{or} \quad 5x + 2 > 3$$
$$5x < -5 \qquad\qquad 5x > 1$$
$$x < -1 \quad \text{or} \quad x > \frac{1}{5}.$$

Therefore, the solution of the original inequality consists of the numbers that are either less than -1 or greater than $1/5$.

GRAPHING EXPLORATION

Graphically confirm that this solution is correct. Use either one of the methods described in Example 6. ■

Exercises 2.4

In Exercises 1–12, solve the inequality algebraically. Confirm your answer graphically.

1. $2x + 4 \leq 7$ **2.** $3x - 5 > -6$

3. $3 - 5x < 13$ **4.** $2 - 3x < 11$

5. $6x + 3 \leq x - 5$ **6.** $5x + 3 \leq 2x + 7$

7. $5 - 7x < 2x - 4$ **8.** $5 - 3x > 7x - 3$

9. $2 < 3x - 4 < 8$ **10.** $1 < 5x + 6 < 9$

11. $0 < 5 - 2x \leq 11$ **12.** $-4 \leq 7 - 3x < 0$

In Exercises 13–20, solve the inequality exactly (no decimal approximations) by any means.

13. $2x + 7(3x - 2) < 2(x - 1)$

14. $x + 3(x - 5) \geq 3x + 2(x + 1)$

15. $\dfrac{x + 1}{2} - 3x \leq \dfrac{x + 5}{3}$

16. $\dfrac{x - 1}{4} + 2x \geq \dfrac{2x - 1}{3} + 2$

17. $2x + 3 \leq 5x + 6 < -3x + 7$

18. $4x - 2 < x + 8 < 9x + 1$

19. $3 - x < 2x + 1 \leq 3x - 4$

20. $2x + 5 \leq 4 - 3x < 1 - 4x$

In Exercises 21–24, a, b, c, and d are positive constants. Solve the inequality for x.

21. $ax - b < c$ **22.** $d - cx > a$

23. $0 < x - c < a$ **24.** $-d < x - c < d$

25. The average temperature on the planet Mercury is 333°F. The temperature t on Mercury can be as much as 507°F above or 633°F below the average. Write the range of temperatures on Mercury as an inequality.

26. If you were single in 1999 and had a taxable income I between \$25,750 and \$62,450, then your federal income tax T was

    28% of (income − 25,750) + \$3862.50.

  (a) Write this income bracket as an inequality.

  (b) Write the tax range for this income bracket as an inequality.

27. A company has fixed overhead costs of \$350 per day and variable costs of \$7.50 per unit for every unit produced. During December, the total daily cost T varied from \$3150 to \$4575 per day.

  (a) Write an equation that expresses T in terms of the number x of units produced.

  (b) Write an inequality that expresses the variability of the total daily cost in December.

  (c) Write an inequality that gives the lowest and highest number of units produced daily during December.

28. Using statistical quality control methods, the management of a large hotel chain determines that the percentage of rooms that are not ready when the guest checks in ranges from 0.16% to 0.44%. When the room is not ready at check-in, a guest receives the first night free. The chain has 20,000 rooms, each of which costs \$85 per night. Let x be the number of rooms that are not ready at check-in and C the cost to the hotel chain for rooms not ready at check-in.

  (a) Write an equation that relates C and x.

  (b) Write an inequality that gives the range of values for x.

  (c) Write an inequality that gives the range of values of C.

In Exercises 29 and 30, read the solution of the inequality from the given graph.

29. $3 - 2x < .8x + 7$

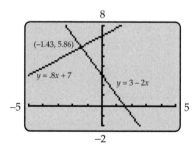

30. $8 - |7 - 5x| > 3$

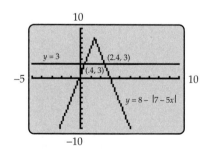

In Exercises 31–34, match the equation or inequality with the graph of its solution on the number line.

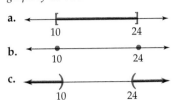

31. $|x - 17| = 7$ **32.** $|x - 17| \leq 7$

33. $|x - 17| \geq -7$ **34.** $|x - 17| > 7$

In Exercises 35–44, solve the inequality exactly (no decimal approximations).

35. $|3x + 2| \leq 2$ **36.** $|5x - 1| < 3$

37. $|3 - 2x| < 2/3$ **38.** $|4 - 5x| \leq 4$

39. $|2x + 3| > 1$ **40.** $|3x - 1| \geq 2$

41. $|5x + 2| \geq \dfrac{3}{4}$ **42.** $|2 - 3x| > 4$

43. $\left|\dfrac{12}{5} + 2x\right| > \dfrac{1}{4}$ **44.** $\left|\dfrac{5}{6} + 3x\right| < \dfrac{7}{6}$

45. A broker predicts that over the next six months, the price p of Gigantica stock will not vary from its current price of $\$25\frac{3}{4}$ by more than $\$4$. Use absolute value to write this prediction as an inequality.

46. About two-thirds of people taking a standard IQ test will score within 12 points of 100. Write an absolute value inequality that describes this situation. Be sure to state what the variable represents.

47. One freezer costs $\$623.95$ and uses 90 kilowatt-hours (kwh) of electricity each month. A second freezer costs $\$500$ and uses 100 kwh of electricity each month. The expected life of each freezer is 12 years. What is the minimum electric rate (in *cents* per kwh) for which the 12-year total cost (purchase price + electricity costs) will be less for the first freezer?

48. A business executive leases a car for $\$300$ per month. She decides to lease another brand for $\$250$ per month, but has to pay a penalty of $\$1000$ for breaking the first lease. How long must she keep the second car in order to come out ahead?

49. One salesperson is paid a salary of $\$1000$ per month plus a commission of 2% of her total sales. A second salesperson receives no salary, but is paid a commission of 10% of her total sales. What dollar amount of sales must the second salesperson have in order to earn more per month than the first?

50. A Gas Guzzler SUV has a 26-gallon gas tank and gets 12 miles per gallon. If it travels more than 210 miles and runs out of gas, what possible amounts of gas were in the tank at the beginning of the trip?

51. If $\$5000$ is invested at 8%, how much more should be invested at 10% in order to guarantee a total annual interest income between $\$800$ and $\$940$?

52. How many gallons of a 12% salt solution should be added to 10 gallons of an 18% salt solution in order to produce a solution whose salt content is between 14 and 16%?

2.5 Polynomial and Rational Inequalities

Polynomial and rational inequalities, such as

$$2x^3 - 15x < x^2, \quad \text{or} \quad 2x^2 + 3x - 4 \leq 0, \quad \text{or} \quad \frac{x}{x - 1} > -6,$$

may be solved by algebraic or graphical methods, or by a combination of the two. The Basic Principles for solving inequalities (page 164) are valid for all inequalities and will be used frequently. Whenever possible, we use algebra to obtain exact solutions. When algebraic methods are too difficult or unavailable, approximate graphical solutions will be found.

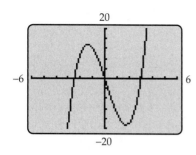

Figure 2–25

To illustrate the recommended graphical solution method, we consider these inequalities:

$$2x^3 - x^2 - 15x < 0 \quad \text{and} \quad 2x^3 - x^2 - 15x > 0.$$

To solve either one, we graph $y_1 = 2x^3 - x^2 - 15x$, as in Figure 2–25. A point (x, y_1) on the graph lies *below* the x-axis when its y-coordinate is negative, that is, when

$$y_1 < 0, \quad \text{or equivalently,} \quad 2x^3 - x^2 - 15x < 0.$$

Thus, the solutions of $2x^3 - x^2 - 15x < 0$ are the x-coordinates of all points on the graph that lie *below* the x-axis. Similarly, the point (x, y_1) lies above the x-axis when its y-coordinate is positive, that is, when

$$y_1 > 0, \quad \text{or equivalently,} \quad 2x^3 - x^2 - 15x > 0.$$

So the solutions of $2x^3 - x^2 - 15x > 0$ are the x-coordinates of all points on the graph that lie *above* the x-axis.

In order to determine precisely which points are above and which are below the x-axis, we must find the x-intercepts of the graph (the points where the graph intersects the x-axis). This should be done algebraically if possible and by graphical approximation otherwise.

Example 1 Solve these inequalities:

(a) $2x^3 - x^2 - 15x < 0$ (b) $2x^3 - x^2 - 15x > 0$.

Solution Figure 2–25 shows the graph of $y_1 = 2x^3 - x^2 - 15x$. To find its x-intercepts, we note that these are the points of the graph that are *on* the x-axis, that is, the points (x, y_1) with $y_1 = 0$. So we solve the equation $y_1 = 0$.

$$2x^3 - x^2 - 15x = 0$$

Factor out x: $$x(2x^2 - x - 15) = 0$$

Factor other term: $$x(2x + 5)(x - 3) = 0$$

$$x = 0 \quad \text{or} \quad 2x + 5 = 0 \quad \text{or} \quad x - 3 = 0$$
$$x = -5/2 \qquad\qquad x = 3.$$

Therefore, the x-intercepts are $-5/2$, 0, and 3. Now we can read the solutions of the two inequalities from the graph.

(a) Figure 2–25 shows that the graph lies below the x-axis when
$$x < -5/2 \quad \text{or} \quad 0 < x < 3.$$
Therefore, the numbers satisfying either of these conditions are the solutions of $2x^3 - x^2 - 15x < 0$.

(b) Figure 2–25 also shows that the graph lies above the x-axis when
$$-5/2 < x < 0 \quad \text{or} \quad x > 3.$$
The solutions of $2x^3 - x^2 - 15x > 0$ are all numbers x satisfying either of these conditions. ∎

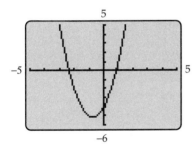

Figure 2–26

Example 2 The solutions of $2x^2 + 3x - 4 \leq 0$ are the numbers x for which the graph of $y = 2x^2 + 3x - 4$ lies on or below the x-axis (Figure 2–26). The x-intercepts of the graph can be found by using the quadratic formula to solve the equation $y = 0$.

$$2x^2 + 3x - 4 = 0$$

$$x = \frac{-3 \pm \sqrt{3^2 - 4 \cdot 2(-4)}}{2 \cdot 2} = \frac{-3 \pm \sqrt{41}}{4}$$

Figure 2–26 shows that the graph lies below the x-axis between the two x-intercepts. Therefore, the solutions to the inequality are all numbers x such that

$$\frac{-3 - \sqrt{41}}{4} \leq x \leq \frac{-3 + \sqrt{41}}{4}. \quad \blacksquare$$

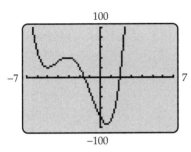

Figure 2–27

Example 3 Solve: $x^4 + 10x^3 + 21x^2 + 8 > 40x + 88$.

Solution We begin by rewriting the inequality in an equivalent form:

$$x^4 + 10x^3 + 21x^2 - 40x - 80 > 0.$$

The graph of $y = x^4 + 10x^3 + 21x^2 - 40x - 80$ in Figure 2–27 has two x-intercepts, one between -2 and -1, and the other near 2. Finding the x-intercepts algebraically would involve solving a fourth-degree equation. So we approximate them graphically.

GRAPHING EXPLORATION

Use a root finder to show that the approximate x-intercepts are -1.53 and 1.89.

Therefore, the approximate solutions of the inequality (the numbers x for which the graph is above the x-axis) are all numbers x such that $x < -1.53$ or $x > 1.89$. $\quad \blacksquare$

CAUTION

Do not attempt to write the solutions in Example 3, namely, $x < -1.53$ or $x > 1.89$, as a single inequality. If you do, the result will be a *nonsense* statement, such as $-1.53 > x > 1.89$ (which says, among other things, that $-1.53 > 1.89$).

Factorable Inequalities

The preceding examples show that solving a polynomial inequality depends only on knowing the x-intercepts of a graph and the places where it is above or below the x-axis. When the polynomial can be completely factored, a calculator isn't necessary to determine this information.

Example 4 Solve: $(x + 15)(x - 2)^6(x - 10) \le 0$.

Solution   The x-intercepts of the graph of $y = (x + 15)(x - 2)^6(x - 10)$ are the solutions of $y = 0$. They are easily read from the factored form.

$$y = 0$$
$$(x + 15)(x - 2)^6(x - 10) = 0$$

$$x + 15 = 0 \quad \text{or} \quad x - 2 = 0 \quad \text{or} \quad x - 10 = 0$$
$$x = -15 \qquad\qquad x = 2 \qquad\qquad x = 10.$$

We need only determine the places where the graph is below the x-axis. To do this without a calculator, note that the three x-intercepts divide the x-axis into four intervals:

$$x < -15, \qquad -15 < x < 2, \qquad 2 < x < 10, \qquad x > 10.$$

Consider, for example, the interval with $2 < x < 10$. As in previous examples, the graph is an unbroken curve that crosses the x-axis only at the x-intercepts. Since there are no x-intercepts between 2 and 10, the graph must lie *entirely above* or *entirely below* the x-axis on that interval. In order to determine which one, choose any number between 2 and 10, say $x = 4$. Then

$$y = (x + 15)(x - 2)^6(x - 10) = (4 + 15)(4 - 2)^6(4 - 10) = 19(2^6)(-6).$$

We don't need to finish the computation to see that y is negative when $x = 4$. Therefore, the point with x-coordinate 4 lies *below* the x-axis. Since one point between 2 and 10 lies below the x-axis, the entire graph must be below the x-axis between 2 and 10.

The location of the graph on the other intervals can also be determined algebraically by choosing a "test number" in each interval, as summarized in the following chart.

Interval	$x < -15$	$-15 < x < 2$	$2 < x < 10$	$x > 10$
Test number in this interval	-20	0	4	11
Value of y at test number	$(-5)(-22)^6(-30)$	$15(-2)^6(-10)$	$19(2^6)(-6)$	$26(9^6)(1)$
Sign of y at test number	$+$	$-$	$-$	$+$
Graph	Above x-axis	Below x-axis	Below x-axis	Above x-axis

The last line of the chart shows that the graph is below the x-axis when $-15 < x < 2$ and when $2 < x < 10$. Since the graph touches the x-axis at -15, 2, and 10, the solutions of the inequality (numbers for which the graph is *on* or below the x-axis) are all numbers x such that $-15 \le x \le 10$. ∎

The procedures used in Examples 1–4 may be summarized as follows.

Solving Polynomial Inequalities

1. To solve an inequality in one of these forms

$$y > 0, \qquad y \geq 0, \qquad y < 0, \qquad y \leq 0,$$

where y is a polynomial in x, consider the graph of y.

2. Determine the x-intercepts of the graph by solving the equation $y = 0$. Find exact solutions if possible (as in Examples 1, 2, and 4). Otherwise, approximate the solutions (as in Example 3).

3. Use a calculator graph (as in Examples 1–3) or a sign chart (as in Example 4) to determine the intervals where the graph is above or below the x-axis.

4. Use the preceding information to read off the solutions of the inequality.

Rational Inequalities

Inequalities involving rational expressions may be solved in much the same way as polynomial inequalities, with one difference. Unlike a polynomial graph that can cross the x-axis only at an x-intercept, the graph of a rational expression may change from one side of the x-axis to the other at its x-intercepts *or* at undefined points (where its denominator is 0).

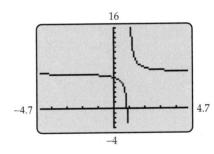

Figure 2–28

Example 5 Solve: $\dfrac{x}{x-1} > -6$.

Solution   There are three ways to solve this inequality.

Graphical Method: The fastest way to get approximate solutions is to replace the given inequality by an equivalent one

$$\frac{x}{x-1} + 6 > 0$$

and graph the equation $y = \dfrac{x}{x-1} + 6$, as in Figure 2–28.* The equation is undefined when $x = 1$ (why?), so there is no point on the graph there. The graph lies above the x-axis everywhere except between the x-intercept and the undefined point at $x = 1$. Using a root finder, we determine that the x-intercept is approximately .857. So the approximate solutions of the inequality are all numbers x such that

$$x < .857 \qquad \text{or} \qquad x > 1.$$

*If you don't use a decimal window, as we did, your graph may show an erroneous vertical line at $x = 1$.

Algebraic/Graphical Method: Proceed as before, but write the left side as a single fraction before graphing:

$$\frac{x}{x-1} + 6 > 0$$

$$\frac{x}{x-1} + \frac{6(x-1)}{x-1} > 0$$

$$\frac{x}{x-1} + \frac{6x-6}{x-1} > 0$$

$$\frac{7x-6}{x-1} > 0.$$

When the equation $y = \dfrac{x}{x-1} + 6$ is rewritten as $y = \dfrac{7x-6}{x-1}$, we obtain the same graph (Figure 2–28), but now its x-intercept can be determined exactly. The x-intercept occurs when $y = 0$, which happens when the numerator is 0:

$$7x - 6 = 0$$

$$x = 6/7.$$

So the x-intercept is 6/7 (whose decimal expansion begins .857) and the exact solutions of the original inequality are those numbers x such that

$$x < 6/7 \qquad \text{or} \qquad x > 1.$$

Algebraic Method: Begin as above by rewriting the given inequality as

$$\frac{7x-6}{x-1} > 0.$$

The numerator of the fraction is 0 when $x = 6/7$ and the denominator is 0 when $x = 1$. The numbers 6/7 and 1 divide the x-axis into three intervals. Use test numbers and a sign chart, instead of graphing, to determine the location of the graph of $y = \dfrac{7x-6}{x-1}$ on each interval.*

Interval	$x < 6/7$	$6/7 < x < 1$	$x > 1$
Test number in this interval	0	.9	2
Value of y at test number	$\dfrac{7 \cdot 0 - 6}{0 - 1}$	$\dfrac{7(.9) - 6}{.9 - 1}$	$\dfrac{7 \cdot 2 - 6}{2 - 1}$
Sign of y at test number	+	−	+
Graph	Above x-axis	Below x-axis	Above x-axis

*The justification for this approach is the same as in Example 4. The graph can change from one side of the x-axis to the other only at an x-intercept (where the numerator is 0) or at an undefined point (where the denominator is 0).

The last line of the chart shows that the solutions of the original inequality (the numbers x for which the graph is above the x-axis) are all such that $x < 6/7$ or $x > 1$. ■

CAUTION

Don't treat rational inequalities as if they are equations, as in this *incorrect* "solution" of the preceding example:

$$\frac{x}{x - 1} > -6$$

$$x > -6(x - 1) \quad \text{[Both } \textit{sides multiplied by } x - 1]$$

$$x > -6x + 6$$

$$7x > 6$$

$$x > \frac{6}{7}$$

According to this, the inequality has no negative solutions and $x = 1$ is a solution, but as we saw in Example 5, *every* negative number is a solution and $x = 1$ is not.*

The algebraic technique of writing the left side of an inequality as a single fraction is useful when the resulting numerator and denominator have low degree, so that setting them equal to 0 produces exact solutions (x-intercepts and undefined points). It can be omitted, however, when these solutions must be approximated.

Applications

Example 6 A computer store has determined that the cost C of ordering and storing x laser printers is given by

$$C = 2x + \frac{300{,}000}{x}.$$

If the delivery truck can bring at most 450 printers per order, how many printers should be ordered at a time to keep the cost below $1600?

Solution To find the values of x that make C less than 1600, we must solve the inequality

$$2x + \frac{300{,}000}{x} < 1600 \quad \text{or equivalently,} \quad 2x + \frac{300{,}000}{x} - 1600 < 0.$$

We shall solve this inequality graphically, although it can also be solved algebraically. In this context, the only solutions that make sense are those

---

*The source of the error is multiplying by $x - 1$, which is negative for some values of x and positive for others. So you must consider two cases and reverse the direction of the inequality when $x - 1$ is negative.

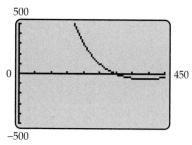

Figure 2–29

between 0 and 450. So we choose the viewing window in Figure 2–29 and graph

$$y = 2x + \frac{300{,}000}{x} - 1600.$$

Figure 2–29 shows that the desired solutions (numbers where the graph is below the x-axis) are all numbers between the x-intercept and 450. A root finder shows that the x-intercept is $x \approx 300$. In fact, this is the exact x-intercept because $y = 0$ when $x = 300$, as you can easily verify. Therefore, to keep costs under \$1600, x printers should be ordered each time, with $300 < x \le 450$. ∎

Exercises 2.5

In Exercises 1–22, solve the inequality exactly.

1. $x^2 - 4x + 3 \le 0$
2. $x^2 - 7x + 10 \le 0$
3. $x^2 + 9x + 14 \ge 0$
4. $x^2 + 8x + 15 \ge 0$
5. $x^2 \ge 9$
6. $x^2 \ge 16$
7. $6 + x - x^2 \le 0$
8. $4 - 3x - x^2 \ge 0$
9. $6x^2 - x \ge 2$
10. $4x^2 + 11x + 6 \ge 0$
11. $x^2 + 11x + 6 > 0$
12. $x^2 - 9x + 6 > 0$
13. $x^2 - 5x < 5$
14. $x^2 + 6x < 4$
15. $x^2 + 4x + 5 > 0$
16. $x^2 + 3x + 4 < 0$
17. $x^3 - x \ge 0$
18. $x^3 + 2x^2 + x > 0$
19. $x^3 - 2x^2 - 3x < 0$
20. $x^4 - 14x^3 + 48x^2 \ge 0$
21. $x^4 - 5x^2 + 4 < 0$
22. $x^4 - 10x^2 + 9 \le 0$

In Exercises 23–50, solve the inequality. Find exact solutions when possible and approximate ones otherwise.

23. $(x + 7)(x + 3)(x - 5)^2 \ge 0$
24. $(x - 5)(x + 4)(x - 3)(x + 2) \ge 0$
25. $(x + 4)^2(x + 1)(x - 3)^2 \le 0$
26. $(x - 6)^3(x + 2)^3(2x - 1) \le 0$
27. $x^3 + 6x^2 + 6x > 0$
28. $x^3 + 2x^2 < 4x$
29. $x^4 - 3x^3 \le 5x^2$
30. $x^4 + 3x^2 \ge 7x^3$
31. $x^3 - 2x^2 - 5x + 7 \ge 2x + 1$
32. $x^4 - 6x^3 + 2x^2 < 5x - 2$
33. $2x^4 + 3x^3 < 2x^2 + 4x - 2$
34. $x^5 + 5x^4 > 4x^3 - 3x^2 + 2$
35. $x^4 - 5x^3 \le 35x^2 + 4x - 9$

36. $x^4 + 5x^2 + 27x \ge 8x^3 + 6$
37. $4x^5 + 17x^4 - 12x^3 < 85x^2 + 60x - 4$
38. $x^5 + x^4 + 206x + 12 > 34x^3 + 69x^2$

39. $\dfrac{3x + 1}{2x - 4} > 0$
40. $\dfrac{2x - 1}{5x + 3} \ge 0$
41. $\dfrac{x^2 + x - 2}{x^2 - 2x - 3} < 0$
42. $\dfrac{2x^2 + x - 1}{x^2 - 4x + 4} \ge 0$
43. $\dfrac{x - 2}{x - 1} < 1$
44. $\dfrac{-x + 5}{2x + 3} \ge 2$
45. $\dfrac{x - 3}{x + 3} \le 5$
46. $\dfrac{2x + 1}{x - 4} > 3$
47. $\dfrac{2}{x + 3} \ge \dfrac{1}{x - 1}$
48. $\dfrac{1}{x - 1} < \dfrac{-1}{x + 2}$
49. $\dfrac{x^3 - 3x^2 + 5x - 29}{x^2 - 7} > 3$
50. $\dfrac{x^4 - 3x^3 + 2x^2 + 2}{x - 2} > 15$

51. Find all pairs of numbers that satisfy these two conditions: Their sum is 20 and the sum of their squares is less than 362.

52. The length of a rectangle is 6 inches longer than its width. What are the possible widths if the area of the rectangle is at least 667 square inches?

53. It costs a craftsman \$5 in materials to make a medallion. He has found that if he sells the medallions for $50 - x$ dollars each, where x is the number of medallions produced each week, then he can sell all that he makes. His fixed costs are \$350 per week. If he wants to sell all he makes and show a profit each week, what are the possible numbers of medallions he should make?

54. A retailer sells file cabinets for $80 - x$ dollars each, where x is the number of cabinets she receives from the supplier each week. She pays \$10 for each file cabinet and has fixed costs of \$600 per week. How many file cabinets should she order from the supplier each week in order to guarantee that she makes a profit?

In Exercises 55–58, you will need the formula for the height h of an object above the ground at time t seconds: $h = -16t^2 + v_0t + h_0$; this formula was explained on page 154.

55. A toy rocket is fired straight up from ground level with an initial velocity of 80 feet per second. During what time interval will it be at least 64 feet above the ground?

56. A projectile is fired straight up from ground level with an initial velocity of 72 feet per second. During what time interval is it at least 37 feet above the ground?

57. A ball is dropped from the roof of a 120-foot-high building. During what time period will it be strictly between 56 and 39 feet above the ground?

58. A ball is thrown straight up from a 40-foot-high tower with an initial velocity of 56 feet per second.
 (a) During what time interval is the ball at least 8 feet above the ground?
 (b) During what time interval is the ball between 53 feet and 80 feet above the ground?

Thinkers

In Exercises 59–61, solve the inequality.

59. $x^4 - 50x^3 + 125x^2 + 250x - 8 > 0$

60. $.1x^4 + 6x^3 + 22x^2 - 153x - 16 < 0$

61. $\dfrac{2x^2 + 6x - 8}{2x^2 + 5x - 3} < 1$

62. $\dfrac{2x^2}{x^2 + x - 2} > 1.8$

Chapter 2   *Review*

Important Concepts

Important Facts and Formulas

- *Quadratic Formula:* If $a \neq 0$, then the solutions of $ax^2 + bx + c = 0$ are
$$x = \frac{-b \pm \sqrt{b^2 - 4ac}}{2a}.$$

- If $a \neq 0$, then the number of real solutions of $ax^2 + bx + c = 0$ is 0, 1, or 2, depending on whether the discriminant $b^2 - 4ac$ is negative, zero, or positive.

- The solutions of $|x| \leq k$ are all numbers x such that $-k \leq x \leq k$.

- The solutions of $|x| \geq k$ are all numbers x such that $x \leq -k$ or $x \geq k$.

Review Questions

1. Solve for x: $2\left(\dfrac{x}{5} + 7\right) - 3x = \dfrac{x + 2}{5} - 4$

2. Solve for t: $\dfrac{t + 1}{t - 1} = \dfrac{t + 2}{t - 3}$

3. Solve for x: $\dfrac{3}{x} - \dfrac{2}{x - 1} = \dfrac{1}{2x}$

4. Solve for z: $2 - \dfrac{1}{z - 2} = \dfrac{z - 3}{z - 2}$

5. Solve for r: $Q = \dfrac{b - a}{2r}$

6. Solve for R: $\dfrac{v}{t} = 2\pi R h$

7. Solve for x in terms of y: $xy + 3 = x - 2y$

8. Bert weighs 10 times as much as the Thanksgiving turkey, and Sally weighs 7 times as much as the turkey. If Bert is 48 pounds heavier than Sally, how much does Bert weigh?

9. A jeweler wants to make a 1-ounce ring consisting of gold and silver, using $200 worth of metal. If gold costs $600 per ounce and silver $50 per ounce, how much of each metal should she use?

10. A calculator is on sale for 15% less than the list price. The sale price, plus a 5% shipping charge, totals $210. What is the list price?

11. Karen can do a job in 5 hours and Claire can do the same job in 4 hours. How long will it take them to do the job together?

12. A car leaves the city traveling at 54 mph. One-half hour later, a second car leaves from the same place and travels at 63 mph along the same road. How long will it take for the second car to catch up with the first?

13. A 12-foot-long rectangular board is cut in two pieces so that one piece is 4 times as long as the other. How long is the bigger piece?

14. George owns 200 shares of stock, 40% of which are in the computer industry. How many more shares must he buy in order to have 50% of his total shares in computers?

15. Solve for x: $3x^2 - 2x + 5 = 0$

16. Solve for x: $5x^2 + 6x = 7$

17. Solve for y: $3y^2 - 2y = 5$

18. Find the *number* of real solutions of the equation $20x^2 + 12 = 31x$.

19. Find the *number* of real solutions of the equation $\dfrac{1}{3x + 6} = \dfrac{2}{x + 2}$.

20. For what value of k does the equation $kt^2 + 5t + 2 = 0$ have exactly one real solution for t?

In Questions 21–24, find all real solutions of the equation. Do not approximate.

21. $x^4 - 11x^2 + 18 = 0$ 22. $x^6 - 4x^3 + 4 = 0$

23. $|3x - 1| = 4$ 24. $|2x - 1| = x + 4$

25. Do there exist two real numbers whose sum is 2 and whose product is 2? Justify your answer.

26. Find two consecutive integers, the sum of whose squares is 481.

27. A square region is changed into a rectangular one by making it 2 feet longer and twice as wide. If the area of the rectangular region is 3 times larger than the area of the original square region, what was the length of a side of the square before it was changed?

28. The radius of a circle is 10 inches. By how many inches should the radius be increased so that the area increases by 5π square inches?

In Questions 29–36, find all real solutions of the equation.

29. $x^4 - 2x^2 - 15 = 0$ 30. $x^4 - x^2 = 6$

31. $x^6 + 7x^3 - 8 = 0$ 32. $3y^7 - 3y^5 - 15y^3 = 0$

33. $\sqrt{x - 1} = 2 - x$ 34. $\sqrt[3]{1 - t^2} = -2$

35. $\sqrt{x + 1} + \sqrt{x - 1} = 1$ 36. $\sqrt{3x - 1} + \sqrt{x} = 2$

37. Solve for s: $t = \sqrt{\dfrac{2s}{g}}$

In Questions 38–42, solve the inequality.

38. $-3(x - 4) \le 5 + x$

39. $-4 < 2x + 5 < 9$

40. $-3 < 8x + 5 < 13$

41. $|3x + 2| \ge 2$

42. $\left|\dfrac{y + 2}{3}\right| \ge 5$

43. On which intervals is $\dfrac{2x - 1}{3x + 1} < 1$?

44. On which intervals is $\dfrac{2}{x + 1} < x$?

45. Let C = consumer spending and I = income. Suppose $C = 90 + \dfrac{3}{4}I$. Express the range of incomes (I) over which consumer spending exceeds income.

46. Solve for x: $x^2 + x > 12$.

47. Solve for x: $(x - 1)^2(x^2 - 1)x \le 0$.

48. If $0 < r \le s - t$, then which of these statements is *false*?

   (a) $s \ge r + t$

   (b) $t - s \le -r$

   (c) $-r \ge s - t$

   (d) $\dfrac{s - t}{r} > 0$

   (e) $s - r \ge t$

49. If $\dfrac{x + 3}{2x - 3} > 1$, then which of these statements is *true*?

   (a) $\dfrac{x - 3}{2x + 3} < -1$

   (b) $\dfrac{2x - 3}{x + 3} < -1$

   (c) $\dfrac{3 - 2x}{x + 3} > 1$

   (d) $2x + 3 < x - 3$

   (e) None of the preceding statements.

50. Solve:

$$2x - 3 \le 5x + 9 < -3x + 4.$$

In Questions 51–58, solve the inequality.

51. $|2 - 5x| \ge 2$

52. $x^2 + x - 20 > 0$

53. $\dfrac{x - 2}{x + 4} \le 3$

54. $(x + 1)^2(x - 3)^4(x + 2)^3(x - 7)^5 > 0$

55. $\dfrac{x^2 + x - 9}{x + 3} < 1$

56. $\dfrac{x^2 - x - 6}{x - 3} > 1$

57. $\dfrac{x^2 - x - 5}{x^2 + 2} > -2$

58. $\dfrac{x^4 - 3x^2 + 2x - 3}{x^2 - 4} < -1$

Discovery Project 2

Inequalities in Yes/No Decisions

In this chapter you learned techniques for determining the values for which an inequality is true. In many practical applications, it is not necessary to know what is happening for every value of the variable. In fact, in most situations we are only interested in a very limited range of values for the variable. Consider the construction of sports facilities, for example. An important consideration in the construction of the roofs of such facilities is whether or not the ball used is likely to hit the ceiling during play of the game.

1. A new field house is to be constructed on a college campus to support the very successful volleyball program. The roof design has the lights suspended from the ceiling at a height of 46 feet above the floor. When the ball is hit so that it rises high, its height h after t seconds is typically given by a formula such as $h = -16t^2 + 50t + 5$. Assuming this formula to be universally valid, can a shot hit the lights? [*Hint:* Are there any positive values of t for which $h > 46$?]

A similar question can be posed for more complicated paths.

2. A particularly well-hit baseball will leave the bat at 80 miles per hour with an upward angle of 30°. Its height h above the ground s feet from the bat is given by

$$h = \frac{-9s^2}{7744} + \frac{s}{2} + 3.$$

The height k of the new stadium roof along the right field line, s feet from home plate, is given by

$$k = \frac{2\sqrt{-5s^2 + 1500s + 145{,}000}}{5} - 420.$$

Will the ball hit the ceiling before crossing out of play 385 feet from home plate? In other words, is the height of the ball always less than the height of the ceiling?

3. A movie director wants to film a scene in which a speeding car evades a police car by crossing the state line first. The car passes a stopped police officer, who immediately begins the chase. The officer's position is given by $p = 2000\sqrt{\dfrac{t-4}{15}}$, where t is measured in seconds. The speeder drives at a constant 80 miles per hour, giving it a position of $s = \dfrac{352t}{3}$ after t seconds. If the speeder reaches the border in 45 seconds, will the police car catch up?

Looking for a house?

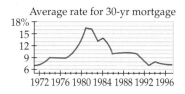

Average rate for 30-yr mortgage
18%
15
12
9
6
1972 1976 1980 1984 1988 1992 1996

If you buy a house, you'll probably need a mortgage. If you can get a low interest rate, your monthly payments are lower (or alternatively, you can afford a more expensive house). The timing of your purchase can make a difference because mortgage interest rates constantly fluctuate. In mathematical terms, rates are a function of time. The graph of this function provides a picture of how interest rates change. See Exercise 47 on page 215.

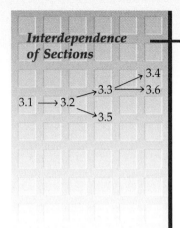

Chapter Outline

The concept of a function and functional notation are central to modern mathematics and its applications. In this chapter you will be introduced to functions and operations on functions, learn how to use functional notation, and develop skill in constructing and interpreting graphs of functions.

3.1 Functions

To understand the origin of the concept of function, it may help to consider some real-life situations in which one numerical quantity depends on, corresponds to, or determines another.

Example 1   The amount of income tax you pay depends on the amount of your income. The way in which the income determines the tax is given by the tax laws.   ■

Example 2   The weather bureau records the temperature over a 24-hour period in the form of a graph (Figure 3–1). The graph shows the temperature that corresponds to each given time.   ■

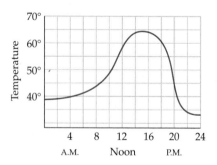

Figure 3–1

Example 3 Suppose a rock is dropped straight down from a high place. Physics tells us that the distance traveled by the rock in t seconds is $16t^2$ feet. So the distance depends on the time. ■

These examples share several common features. Each involves two sets of numbers, which we can think of as inputs and outputs. In each case there is a rule by which each input determines an output, as summarized here.

	Set of Inputs	Set of Outputs	Rule
Example 1	All incomes	All tax amounts	The tax laws
Example 2	Hours since midnight	Temperatures during the day	Time/temperature graph
Example 3	Seconds elapsed after dropping the rock	Distance rock travels	Distance $= 16t^2$

Each of these examples may be mentally represented by an idealized calculator that has a single operation key: a number is entered [*input*], the rule key is pressed [*rule*], and an answer is displayed [*output*]. The formal definition of function incorporates these common features (input/rule/output), with a slight change in terminology.

Functions

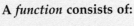

A *function* consists of:

A set of inputs (called the *domain*);

A *rule* by which each input determines one and only one output;

A set of outputs (called the *range*).

The phrase "one and only one" in the definition of the rule of a function may need some clarification. In Example 2, for each time of day (input), there is one and only one temperature (output). But it is quite possible to have the same temperature (output) at different times (inputs). In general,

For each input (number in the domain), the rule of a function determines exactly one output (number in the range). But different inputs may produce the same output.

Although real-world situations, such as Examples 1–3, are the motivation for functions, much of the emphasis in mathematics courses is on the functions themselves, independent of possible interpretations in specific situations, as illustrated in the following examples.

Example 4   Could either of the following be the table of values of a function?

(a)

Input	−4	−2	0	2	4
Output	21	7	1	3	7

(b)

Input	3	2	1	3	5
Output	4	0	2	6	9

Solution   In table (a), two different inputs (−2 and 4) produce the same output, but that's OK because each input produces exactly one output. So table (a) could be the table of values of a function. In table (b), however, the input 3 produces two different outputs (4 and 6), so this table cannot possibly represent a function.   ■

Technology Tip

The greatest integer function is denoted INT or FLOOR in the NUM submenu of the MATH menu of TI and Sharp 9600. It is denoted FLOOR in the REAL submenu of the HP-38 MATH menu, and INTG in the NUM submenu of the Casio 9850 OPTN menu.

Example 5   For each real number s that is not an integer, let $[s]$ denote the *integer* that is closest to s on the *left* side of s on the number line; if s is itself an integer, we define $[s] = s$. Here are some examples:

$$[-4.7] = -5, \quad [-3] = -3, \quad [-1.5] = -2,$$

$$[0] = 0, \quad \left[\frac{5}{3}\right] = 1, \quad [\pi] = 3.$$

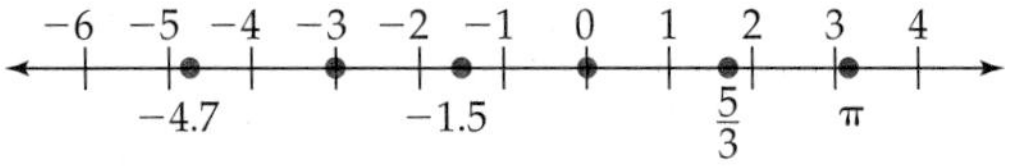

Figure 3–2

The **greatest integer function** is the function whose domain is the set of all real numbers, whose range is the set of integers, and whose rule is

     For each input (real number) x, the output is the integer $[x]$.   ■

Functions Defined by Equations

Equations in two variables are *not* the same things as functions. However, many equations can be used to define functions.

Example 6   The equation $4x - 2y^3 + 5 = 0$ can be solved uniquely for y:

$$2y^3 = 4x + 5$$

$$y^3 = 2x + \frac{5}{2}$$

$$y = \sqrt[3]{2x + \frac{5}{2}}.$$

If a number is substituted for x in this equation, then exactly one value of y is produced. So we can define a function whose domain is the set of all real numbers and whose rule is

$$\text{The input } x \text{ produces the output } \sqrt[3]{2x + 5/2}.$$

In this situation we say that the equation defines y **as a function of** x.

The original equation can also be solved for x:

$$4x = 2y^3 - 5$$
$$x = \frac{2y^3 - 5}{4}.$$

Now if a number is substituted for y, exactly one value of x is produced. So we can think of y as the input and the corresponding x as the output and say that the equation defines x **as a function of** y. ■

Example 7 If you solve the equation

$$y^2 - x + 1 = 0$$

for y, you obtain

$$y^2 = x - 1$$
$$y = \pm\sqrt{x - 1}.$$

This equation does *not* define y as a function of x because, for example, the input $x = 5$ produces two outputs: $y = \pm 2$. ■

Example 8 A group of students drive from Cleveland to Seattle, a distance of 2350 miles, at an average speed of 52 mph.

(a) Express their distance from Cleveland as a function of time.

(b) Express their distance from Seattle as a function of time.

Solution

(a) Let t denote the time traveled (in hours) after leaving Cleveland and D the distance from Cleveland at time t. Then the equation that expresses D as a function of t is

$$D = \text{distance traveled in } t \text{ hours at 52 mph} = 52t.$$

(b) At time t the car has traveled $52t$ miles of the 2350-mile journey, so the distance K remaining to Seattle is given by $K = 2350 - 52t$. This equation expresses K as a function of t. ■

Graphing calculators are designed to deal with equations that define y as a function of x. The table feature of a calculator is a convenient way to evaluate such functions (that is, to produce the outputs from various inputs).*

*TI-85 does not have a built-in table feature, but a program to provide one is in the Program Appendix.

Technology Tip

The table setup screen is labeled TBLSET on the TI-83/89 and Sharp 9600 keyboards and NUM SETUP on the HP-38 keyboard. It is labeled TBLSET in the TI-86 TABLE menu and RANG in the Casio 9850 TABLE menu.

The increment is labeled ΔTBL on TI, TBLSTEP on Sharp 9600, NUMSTEP on HP-38, and PITCH on Casio 9850.

The table type is labeled INDPNT on TI, INPUT on Sharp 9600, and NUMTYPE on HP-38.

Example 9   The equation $y = x^3 - 2x + 3$ defines y as a function of x. Find the outputs for each of the following inputs:

(a) $-3, -2, -1, 0, 1, 2, 3, 4, 5$ (b) $-5, -11, 8, 7.2, -.44$.

Solution

(a)   To use the table feature, we first enter $y = x^3 - 2x + 3$ in the equation memory, say as y_1. Then we call up the setup screen (see the Tip in the margin and Figure 3–3) and enter the *starting number* (-3), the *increment* (the amount the input changes for each subsequent entry, which is 1 here), and the *table type* (AUTO, which means the calculator will compute all the outputs at once).* Then press TABLE to obtain the table in Figure 3–4. To find values that don't appear on the screen in Figure 3–4, use the up and down arrow keys to scroll through the table.

(b)   With an apparently random list of inputs, as here, we change the table type to ASK (or USER or BUILD YOUR OWN).† Then key in each value of x and hit ENTER. This produces the table one line at a time, as in Figure 3–5.   ■

Figure 3–3

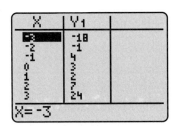

Figure 3–4

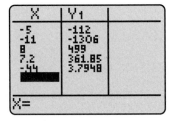

Figure 3–5

CALCULATOR EXPLORATION

Construct a table of values for the function in Example 9 that shows the outputs for these inputs: 2, 2.4, 2.8, 3.2, 3.6, and 4. What is the increment here?

Example 10   The number of public schools (K–12) that have computer networks can be approximated by the equation

$$y = .33x^3 - 6.32x^2 + 43.95x - 78.2 \quad (4 \le x \le 10),$$

*On Casio calculators, you must also enter a maximum value of x.
†This type of table is not available on Casio calculators.

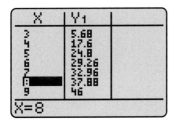

Figure 3–6

where $x = 4$ corresponds to 1994 and y is in thousands. In this case, the number y of computer networks is a function of the year x. What is the first year in which the number of networks exceeds 40,000?

Solution   We make a table of values for the function (Figure 3–6). Since y is in thousands, it shows that the number of networks is approximately 37,880 in 1998 ($x = 8$) and 46,000 in 1999 ($x = 9$). ∎

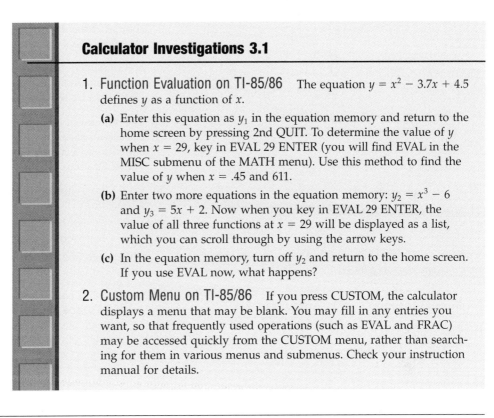

Calculator Investigations 3.1

1. **Function Evaluation on TI-85/86**   The equation $y = x^2 - 3.7x + 4.5$ defines y as a function of x.

   (a) Enter this equation as y_1 in the equation memory and return to the home screen by pressing 2nd QUIT. To determine the value of y when $x = 29$, key in EVAL 29 ENTER (you will find EVAL in the MISC submenu of the MATH menu). Use this method to find the value of y when $x = .45$ and 611.

   (b) Enter two more equations in the equation memory: $y_2 = x^3 - 6$ and $y_3 = 5x + 2$. Now when you key in EVAL 29 ENTER, the value of all three functions at $x = 29$ will be displayed as a list, which you can scroll through by using the arrow keys.

   (c) In the equation memory, turn off y_2 and return to the home screen. If you use EVAL now, what happens?

2. **Custom Menu on TI-85/86**   If you press CUSTOM, the calculator displays a menu that may be blank. You may fill in any entries you want, so that frequently used operations (such as EVAL and FRAC) may be accessed quickly from the CUSTOM menu, rather than searching for them in various menus and submenus. Check your instruction manual for details.

Exercises 3.1

In Exercises 1–4, determine whether or not the given table could possibly be a table of values of a function. Give reasons for your answer.

1.

Input	−2	0	3	1	−5
Output	2	3	−2.5	2	14

2.

Input	−5	3	0	−3	5
Output	7	3	0	5	−3

3.

Input	−5	1	3	−5	7
Output	0	2	4	6	8

4.

Input	1	−1	2	−2	3
Output	1	−2	±5	−6	8

Exercises 5–10 deal with the greatest integer function of Example 5, which is given by the equation $y = [x]$. Compute the following values of the function:

5. $[6.75]$ **6.** $[.75]$ **7.** $[-4/3]$ **8.** $[5/3]$ **9.** $[-16.0001]$

10. Does the equation $y = [x]$ define x as a function of y? Give reasons for your answer.

In Exercises 11–18, determine whether the equation defines y as a function of x or defines x as a function of y.

11. $y = 3x^2 - 12$

12. $y = 2x^4 + 3x^2 - 2$

13. $y^2 = 4x + 1$

14. $5x - 4y^4 + 64 = 0$

15. $3x + 2y = 12$

16. $y - 4x^3 - 14 = 0$

17. $x^2 + y^2 = 9$

18. $y^2 - 3x^4 + 8 = 0$

In Exercises 19–22, each equation defines y as a function of x. Create a table that shows the values of the function for the given values of x.

19. $y = x^2 + x - 4$; $x = -2, -1.5, -1, \ldots, 3, 3.5, 4$

20. $y = x^3 - x^2 + 4x + 1$; $x = 3, 3.1, 3.2, \ldots, 3.9, 4$

21. $y = \sqrt{4 - x^2}$; $x = -2, -1.2, -.04, .04, 1.2, 2$

22. $y = |x^2 - 5|$; $x = -8, -6, \ldots, 8, 10, 12$

Exercises 23–26 refer to Example 1. Assume that the state income tax law reads as follows.

Annual Income	Amount of Tax
Less than $2000	0
$2000–$6000	2% of income over $2000
More than $6000	$80 plus 5% of income over $6000

23. Find the output (tax amount) that is assigned to each of the following inputs (incomes):

$500, $1509, $3754,
$6783, $12,500, $55,342.

24. Find four different numbers in the domain of this function that produce the same output (number in the range).

25. Explain why your answer in Exercise 24 does *not* contradict the definition of a function (in the box on page 187).

26. Is it possible to do Exercise 24 if all four numbers in the domain are required to be greater than 2000? Why or why not?

27. The amount of postage required to mail a first-class letter is determined by its weight. In this situation, is weight a function of postage? Or vice versa? Or both?

28. Could the following statement ever be the rule of a function?

> For input x, the output is the number whose square is x.

Why or why not? If there is a function with this rule, what is its domain and range?

29. **(a)** Use the chart below to make two tables of values (one for an average man and one for an average woman) in which the inputs are the number of drinks per hour and the outputs are the corresponding blood alcohol contents.

   (b) Do each of these tables define a function? If so, what are the domain and range of each function? [Remember that you can have part of a drink.]

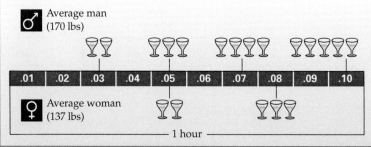

Blood alcohol content

A look at the number of drinks consumed and blood alcohol content in one hour under optimum conditions:

Source: National Highway Traffic Safety Administration   AP/Amy Kranz

30. The table (from Metropolitan Life Insurance Company) relates a large-framed woman's height to the weight at which the woman should live longest. In this situation, is weight a function of height? Is height a function of weight? Justify your answer.

Price p per pair	Demand D (number of pairs sold per day)
$34.99	70
37.50	67
40.00	65
40.00	62
49.99	59
58.00	53
79.00	45

Women (Large-Frame)	
Height (in shoes)	Weight (in pounds, in indoor clothing)
4'10"	118–131
4'11"	120–134
5'0"	122–137
5'1"	125–140
5'2"	128–143
5'3"	131–147
5'4"	134–151
5'5"	137–155
5'6"	140–159
5'7"	143–163
5'8"	146–167
5'9"	149–170
5'10"	152–173
5'11"	155–176
6'0"	158–179

32. The prime rate is the rate that large banks charge their best corporate customers for loans. The graph shows how the prime rate charged by a particular bank has varied in recent years. Answer the following questions by reading the graph as best you can.
   (a) What was prime rate in January 1996? In January 1992? In mid-1995?
   (b) In what time periods was the prime rate 6%?
   (c) Based on the data provided by this graph, can the prime rate be considered a function of time? Can time be considered a function of the prime rate?

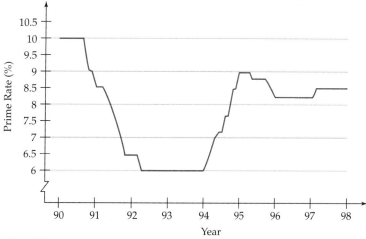

31. A chain of shoe stores collected the data in the table at the top of the next column over several days; it shows how the demand D for Nike Air sports shoes is related to the price p per pair on that day.
   (a) In this situation, is D a function of p? Why?
   (b) Is p a function of D? Why?

33. Find an equation that expresses the area A of a circle as a function of its
   (a) radius r (b) diameter d

34. Find an equation that express the area of a square as a function of its
    (a) side x (b) diagonal d

35. A box with a square base of side x is 4 times higher than it is wide. Express the volume V of the box as a function of x.

36. The surface area of a cylindrical can of radius r and height h is $2\pi r^2 + 2\pi rh$. If the can is twice as high as the diameter of its top, express its surface area S as a function of r.

37. Suppose you drop a rock from the top of a 400-ft-high building. Express the distance D from the rock to the ground as a function of time t. [*Hint:* See Example 3.] What is the range of this function?

38. A bicycle factory has weekly fixed costs of $26,000. In addition, the material and labor costs for each bicycle are $125. Express the total weekly cost C as a function of the number x of bicycles that are made.

39. The table shows the median sales price of existing single-family homes in the Midwest for selected years. (*Source:* National Association of Realtors)

Year	Median Price	Year	Median Price
1970	$20,100	1986	$63,500
1974	27,700	1990	74,000
1978	42,200	1994	87,900
1982	55,100		

(a) Use linear regression to find an equation that expresses the median sales price y (in thousands of dollars) as a function of the year x, with $x = 0$ corresponding to 1970.

(b) Use the function found in part (a) to estimate the median sales price in 1975, 1985, and 1995.

40. (a) Use linear regression on the data given in Exercise 31 to find an equation that expresses the approximate number y of pairs of shoes sold as a function of the price x.

(b) Use the function found in part (a) to estimate the number of pairs sold when the price is $36 and when it is $66.

Thinkers

41. Consider the function whose rule uses a calculator as follows: "press COS and then press LN; then enter a number in the domain and press ENTER."* Experiment with this function, then answer the following questions. You may not be able to prove your answers—just make the best estimate you can based on the evidence from your experiments.
    (a) What is the largest set of real numbers that could be used for the domain of this function? [If applying the rule to a number produces an error message or a complex number (on TI-85, TI-92, or HP-38), that number cannot be in the domain.]
    (b) Using the domain in part (a), what is the range of this function?

42. Do Exercise 41 for the function whose rule is "press 10^x and then press TAN; then enter a number in the domain and press ENTER."

43. The *integer part* function has the set of all real numbers (written as decimals) as its domain. The rule is "assign to a number in the domain the part of the number to the left of the decimal point." For instance, the input 37.986 produces the output 37 and the input -1.5 produces the output -1. On most calculators, the integer part function is denoted "iPart." On calculators that use "Intg" or "Floor" for the greatest integer function, the integer part function is denoted by "INT."
    (a) For each nonnegative real number input, explain why both the integer part function and the greatest integer function [Example 5] produce the same output.
    (b) For which negative numbers do the two functions produce the same output?
    (c) For which negative numbers do the two functions produce different outputs?

3.2 **Functional Notation**

Functional notation is a convenient shorthand language that facilitates the analysis of mathematical problems involving functions. It arises from real-life situations, such as the following.

*You don't need to know what these keys mean in order to do this exercise.

Example 1 According to the Connecticut Department of Revenue Services, the 1999 state income tax rates were as follows.

Taxable Income	Amount of Tax
$20,000 or less	3% of income
More than $20,000	$600 plus 4.5% of income over $20,000

Let I denote income and write $T(I)$ (read "T of I") to denote the amount of tax on income I. In this shorthand language, $T(7500)$ denotes "the tax on an income of $7500." The sentence "The tax on an income of $7500 is $225" is abbreviated as $T(7500) = 225$. Similarly, $T(25,000) = 825$ says that the tax on an income of $25,000 is $825. There is nothing that forces us to use the letters T and I here:

> **Any choice of letters will do, provided we make clear what is meant by these letters.** ■

Example 2 Recall that a falling rock travels $16t^2$ feet after t seconds. Let $d(t)$ stand for the phrase "the distance the rock has traveled after t seconds." Then the sentence "The distance the rock has traveled after t seconds is $16t^2$ feet" can be abbreviated as $d(t) = 16t^2$. For instance,

$$d(1) = 16 \cdot 1^2 = 16$$

means "the distance the rock has traveled after 1 second is 16 feet" and

$$d(4) = 16 \cdot 4^2 = 256$$

means "the distance the rock has traveled after 4 seconds is 256 feet." ■

CAUTION

In Example 2, the parentheses in $d(t)$ do *not* denote multiplication as in the algebraic equation $3(a + b) = 3a + 3b$. The entire symbol $d(t)$ is part of a *shorthand language*. In particular,

$$d(1 + 4) \text{ is } not \text{ equal to } d(1) + d(4).$$

For we saw earlier that $d(1) = 16$ and $d(4) = 256$, so that $d(1) + d(4) = 16 + 256 = 272$. But $d(1 + 4)$ is "the distance traveled after $1 + 4$ seconds," that is, the distance after 5 seconds, namely, $16 \cdot 5^2 = 400$. In general,

Functional notation is a convenient shorthand for phrases and sentences in the English language. It is *not* the same as ordinary algebraic notation.

Functional notation is easily adapted to mathematical settings, in which the particulars of time, distance, etc., are not mentioned. Suppose a function is given. Denote the function by f and let x denote a number in the domain. Then

$f(x)$ denotes the output produced by input x.

For example, $f(6)$ is the output produced by the input 6. The sentence

"y is the output produced by input x according
to the rule of the function f"

is abbreviated

$$y = f(x),$$

which is read "y equals f of x." The output $f(x)$ is sometimes called the **value** of the function f at x.

In actual practice, functions are seldom presented in the style of domain, rule, range, as they have been here. Usually, you will be given a phrase such as "the function $f(x) = \sqrt{x^2 + 1}$." This should be understood as a set of directions:

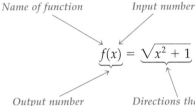

Name of function   *Input number*

$$f(x) = \sqrt{x^2 + 1}$$

Output number   *Directions that tell you what to do with input*
x in order to produce the corresponding output
f(x), namely, "square it, add 1, and take the
square root of the result."

Technology Tip

Functional notation can be used directly on TI-82/83/86/89, Sharp 9600, and HP-38. For example, if the function is entered in the function memory as y_1, then keying in $y_1(5)$ ENTER evaluates the function at $x = 5$.

You will find y_1 in the FUNCTION sub-menu of the TI-82/83 Y-VARS menu or in the EQVARS submenu of the Sharp 9600 VARS menu. On TI-86/89 and HP-38, type in y_1 on the keyboard.

Functional notation cannot be used this way on TI-85 or Casio 9850 (you'll get an answer, but it usually will be wrong).

For example, to find $f(3)$, the output of the function f for input 3, simply replace x by 3 in the formula:

$$f(x) = \sqrt{x^2 + 1}$$
$$f(3) = \sqrt{3^2 + 1} = \sqrt{10}.$$

Similarly, replacing x by -5 and 0 shows that

$$f(-5) = \sqrt{(-5)^2 + 1} = \sqrt{26} \quad \text{and} \quad f(0) = \sqrt{0^2 + 1} = 1.$$

Example 3   The expression $h(x) = \dfrac{x^2 + 5}{x - 1}$ defines the function h whose rule is:

For input x, the output is the number $\dfrac{x^2 + 5}{x - 1}$.

Find each of the following.

$$h(\sqrt{3}), \quad h(-2), \quad h(-a), \quad h(r^2 + 3), \quad h(\sqrt{c} + 2).$$

Solution   To find $h(\sqrt{3})$ and $h(-2)$, replace x by $\sqrt{3}$ and -2, respectively, in the rule of h:

$$h(\sqrt{3}) = \frac{(\sqrt{3})^2 + 5}{\sqrt{3} - 1} = \frac{8}{\sqrt{3} - 1} \qquad \text{and} \qquad h(-2) = \frac{(-2)^2 + 5}{-2 - 1} = -3.$$

The value of the function h at any quantity, such as $-a$, $r^2 + 3$, etc., can be found by using the same procedure: *replace x in the formula for $h(x)$ by that quantity*:

$$h(-a) = \frac{(-a)^2 + 5}{-a - 1} = \frac{a^2 + 5}{-a - 1}$$

$$h(r^2 + 3) = \frac{(r^2 + 3)^2 + 5}{(r^2 + 3) - 1} = \frac{r^4 + 6r^2 + 9 + 5}{r^2 + 2} = \frac{r^4 + 6r^2 + 14}{r^2 + 2}$$

$$h(\sqrt{c + 2}) = \frac{(\sqrt{c + 2})^2 + 5}{\sqrt{c + 2} - 1} = \frac{c + 2 + 5}{\sqrt{c + 2} - 1} = \frac{c + 7}{\sqrt{c + 2} - 1}. \quad \blacksquare$$

When functional notation is used in expressions such as $f(-x)$ or $f(x + h)$, the same basic rule applies: Replace x in the formula by the *entire* expression in parentheses.

Technology Tip

One way to evaluate a function $f(x)$ is to enter its rule as an equation $y = f(x)$ in the equation memory and use TABLE or (on TI-85/86) EVAL; see Example 9 or Calculator Investigation 1 in Section 3.1.

Example 4   If $f(x) = x^2 + x - 2$, then

$$f(-x) = (-x)^2 + (-x) - 2 = x^2 - x - 2.$$

Note that in this case $f(-x)$ is *not* the same as $-f(x)$, because $-f(x)$ is the negative of the number $f(x)$, that is,

$$-f(x) = -(x^2 + x - 2) = -x^2 - x + 2. \quad \blacksquare$$

Example 5   If $f(x) = x^2 - x + 2$ and $h \neq 0$, find

(a) $f(x + h)$ (b) $f(x + h) - f(x)$ (c) $\dfrac{f(x + h) - f(x)}{h}$.

Solution

(a) Replace x by $x + h$ in the rule of the function:

$$f(x) = x^2 - x + 2$$
$$f(x + h) = (x + h)^2 - (x + h) + 2 = x^2 + 2xh + h^2 - x - h + 2.$$

(b) By part (a),

$$f(x + h) - f(x) = [(x + h)^2 - (x + h) + 2] - [x^2 - x + 2]$$
$$= [x^2 + 2xh + h^2 - x - h + 2] - [x^2 - x + 2]$$
$$= x^2 + 2xh + h^2 - x - h + 2 - x^2 + x - 2$$
$$= 2xh + h^2 - h.$$

(c) By part (b), we have

$$\frac{f(x + h) - f(x)}{h} = \frac{2xh + h^2 - h}{h} = \frac{h(2x + h - 1)}{h} = 2x + h - 1. \quad \blacksquare$$

If f is a function, then the quantity $\dfrac{f(x + h) - f(x)}{h}$, as in Example 5(c), is called the **difference quotient** of f. Difference quotients, whose significance is explained in Excursion 3.6.A, play an important role in calculus.

As the preceding examples illustrate, functional notation is a specialized shorthand language. Treating it as ordinary algebraic notation may lead to mistakes.

C A U T I O N   Common Mistakes with Functional Notation

Each of the following statements may be FALSE:

1.	$f(a + b) = f(a) + f(b)$	**4.**	$f(ab) = af(b)$
2.	$f(a - b) = f(a) - f(b)$	**5.**	$f(ab) = f(a)b$
3.	$f(ab) = f(a)f(b)$		

Example 6   Here are examples of three of the errors listed in the Caution box.

1. If $f(x) = x^2$, then

$$f(3 + 2) = f(5) = 5^2 = 25.$$

But

$$f(3) + f(2) = 3^2 + 2^2 = 9 + 4 = 13.$$

So $f(3 + 2) \neq f(3) + f(2)$.

3. If $f(x) = x + 7$, then

$$f(3 \cdot 4) = f(12) = 12 + 7 = 19.$$

But

$$f(3)f(4) = (3 + 7)(4 + 7) = 10 \cdot 11 = 110.$$

So $f(3 \cdot 4) \neq f(3)f(4)$.

5. If $f(x) = x^2 + 1$, then

$$f(2 \cdot 3) = (2 \cdot 3)^2 + 1 = 36 + 1 = 37.$$

But

$$f(2) \cdot 3 = (2^2 + 1)3 = 5 \cdot 3 = 15.$$

So $f(2 \cdot 3) \neq f(2) \cdot 3$. ■

Domains

When the rule of a function is given by a formula, as in Examples 3–6, its domain (set of inputs) is determined by the following convention.

Domain Convention

Unless specific information to the contrary is given, the domain of a function f includes every real number (input) for which the rule of the function produces a real number as output.

Thus, the domain of a polynomial function such as $f(x) = x^3 - 4x + 1$ is the set of all real numbers, since $f(x)$ is defined for every value of x. In cases where applying the rule of a function leads to division by zero or to the square root of a negative number, however, the domain may not consist of all real numbers.

Example 7 Find the domain of the function given by

(a) $k(x) = \dfrac{x^2 - 6x}{x - 1}$

(b) $f(u) = \sqrt{u + 2}$.

Solution

(a) When $x = 1$, the denominator of $\dfrac{x^2 - 6x}{x - 1}$ is 0 and the fraction is not defined. When $x \neq 1$, however, the denominator is nonzero and the fraction *is* defined. Therefore, the domain of the function k consists of all real numbers *except* 1.

(b) Since negative numbers do not have real square roots, $\sqrt{u + 2}$ is a real number only when $u + 2 \geq 0$, that is, when $u \geq -2$. Therefore, the domain of f consists of all real numbers greater than or equal to -2. ■

Example 8 A **piecewise-defined** function is one whose rule includes several formulas, such as

$$f(x) = \begin{cases} 2x + 3 & \text{if } x < 4 \\ x^2 - 1 & \text{if } 4 \leq x \leq 10. \end{cases}$$

Find each of the following:

(a) $f(-5)$ (b) $f(8)$ (c) $f(c)$
(d) The domain of f.

Solution

(a) Since $-5 < 4$, the first part of the rule applies:
$$f(-5) = 2(-5) + 3 = -7.$$

(b) Since 8 is between 4 and 10, the second part of the rule applies:
$$f(8) = 8^2 - 1 = 63.$$

(c) We cannot find $f(c)$ unless we know whether $c < 4$ or $4 \leq c \leq 10$.

(d) The rule of f gives no directions when $x > 10$, so the domain of f consists of all real numbers x with $x \leq 10$. ■

Applications

The domain convention does not always apply when dealing with applications. Consider, for example, the distance function for falling objects, $d(t) = 16t^2$ (see Example 2). Since t represents time, only nonnegative values of t make sense here, even though the rule of the function is defined for all values of t. Analogous comments apply to other applications.

A real-life situation may lead to a function whose domain does not include all the numbers for which the rule of the function is defined.

Example 9   A glassware factory has fixed expenses (mortgage, taxes, machinery, etc.) of $12,000 per week. It costs 80 cents to make one cup (labor, materials, shipping). A cup sells for $1.95. At most 18,000 cups can be manufactured each week.

(a)   Express the weekly revenue as a function of the number x of cups made.

(b)   Express the weekly costs as a function of x.

(c)   Find the domain and the rule of the weekly profit function.

Solution

(a)   If $R(x)$ is the weekly revenue from selling x cups, then
$$R(x) = (\text{price per cup}) \times (\text{number sold})$$
$$R(x) = 1.95x.$$

(b)   If $C(x)$ is the weekly cost of manufacturing x cups, then
$$C(x) = (\text{cost per cup}) \times (\text{number sold}) + (\text{fixed expenses})$$
$$C(x) = .80x + 12,000.$$

(c)   If $P(x)$ is the weekly profit from selling x cups, then
$$P(x) = \text{revenue} - \text{cost}$$
$$P(x) = R(x) - C(x)$$
$$P(x) = 1.95x - (.80x + 12,000) = 1.95x - .80x - 12,000$$
$$P(x) = 1.15x - 12,000$$

Although this rule is defined for all real numbers x, the domain of the function P consists of the possible number of cups that can be made each week. Since you can only make whole cups and the maximum production is 18,000, the domain of P consists of all integers from 0 to 18,000.   ■

Example 10   Let P be the profit function in Example 9.

(a)   What is the profit from selling 5000 cups? From 14,000 cups?

(b)   What is the break-even point?

Solution

(a)   We evaluate the function $P(x) = 1.15x - 12,000$ at the required values of x:
$$P(5000) = 1.15(5000) - 12,000 = -\$6250$$
$$P(14,000) = 1.15(14,000) - 12,000 = \$4100.$$

Thus, sales of 5000 cups produce a loss of $6250, while sales of 14,000 produce a profit of $4100.

(b) The break-even point occurs when revenue equals costs (that is, when profit is 0). So we set $P(x) = 0$ and solve for x:

$$1.15x - 12,000 = 0$$
$$1.15x = 12,000$$
$$x = 12,000/1.15 \approx 10,434.78$$

Thus, the break-even point occurs between 10,434 and 10,435 cups. There is a slight loss from selling 10,434 cups and a slight profit from selling 10,435. ∎

Exercises 3.2

In Exercises 1 and 2, find the indicated values of the function by hand and by using the table feature of a calculator (or the EVAL key on TI-85/86). If your answers do not agree with each other or with those at the back of the book, you are either making algebraic mistakes or incorrectly entering the function in the equation memory.

1. $f(x) = \dfrac{x - 3}{x^2 + 4}$

(a) $f(-1)$ (b) $f(0)$ (c) $f(1)$ (d) $f(2)$ (e) $f(3)$

2. $g(x) = \sqrt{x + 4} - 2$

(a) $g(-2)$ (b) $g(0)$ (c) $g(4)$ (d) $g(5)$ (e) $g(12)$

Exercises 3–24 refer to these three functions:

$$f(x) = \sqrt{x + 3} - x + 1 \qquad g(t) = t^2 - 1$$
$$h(x) = x^2 + \frac{1}{x} + 2$$

In each case find the indicated value of the function.

3. $f(0)$ **4.** $f(1)$

5. $f(\sqrt{2})$ **6.** $f(\sqrt{2} - 1)$

7. $f(-2)$ **8.** $f(-3/2)$

9. $h(3)$ **10.** $h(-4)$

11. $h(3/2)$ **12.** $h(\pi + 1)$

13. $h(a + k)$ **14.** $h(-x)$

15. $h(2 - x)$ **16.** $h(x - 3)$

17. $g(3)$ **18.** $g(-2)$

19. $g(0)$ **20.** $g(x)$

21. $g(s + 1)$ **22.** $g(1 - r)$

23. $g(-t)$ **24.** $g(t + h)$

25. If $f(x) = x^3 + cx^2 + 4x - 1$ for some constant c and $f(1) = 2$, find c. [*Hint:* Use the rule of f to compute $f(1)$.]

26. If $g(x) = \sqrt{cx - 4}$ and $g(8) = 6$, find c.

27. If $f(x) = \dfrac{dx - 5}{x - 3}$ and $f(4) = 3$, find d.

28. If $g(x) = \dfrac{dx + 1}{x + 2}$ and $g(3) = .2$, find d.

In Exercises 29–32, find the values of x for which $f(x) = g(x)$.

29. $f(x) = 2x^2 + 4x - 3$; $g(x) = x^2 + 12x + 7$

30. $f(x) = 2x^2 + 12x - 14$; $g(x) = 7x - 2$

31. $f(x) = 3x^2 - x + 5$; $g(x) = x^2 - 2x + 26$

32. $f(x) = 2x^2 - x + 1$; $g(x) = x^2 - 4x + 4$

In Exercises 33–40, assume $h \neq 0$. Compute and simplify the difference quotient

$$\frac{f(x + h) - f(x)}{h}.$$

33. $f(x) = x + 1$ **34.** $f(x) = -10x$

35. $f(x) = 3x + 7$ **36.** $f(x) = x^2$

37. $f(x) = x - x^2$ **38.** $f(x) = x^3$

39. $f(x) = \sqrt{x}$ **40.** $f(x) = 1/x$

41. In each part compute $f(a)$, $f(b)$, and $f(a + b)$ and determine whether the statement "$f(a + b) = f(a) + f(b)$" is true or false for the given function.
(a) $f(x) = x^2$ (b) $f(x) = 3x$

42. In each part compute $g(a)$, $g(b)$, and $g(ab)$ and determine whether the statement "$g(ab) = g(a) \cdot g(b)$" is true or false for the given function.
(a) $g(x) = x^3$ (b) $g(x) = 5x$

43. If $f(x) = \begin{cases} x^2 + 2x & \text{if } x < 2 \\ 3x - 5 & \text{if } 2 \leq x \leq 20 \end{cases}$ find
(a) the domain of f
(b) $f(-3)$ (c) $f(-1)$ (d) $f(2)$ (e) $f(7/3)$

44. If $g(x) = \begin{cases} 2x - 3 & \text{if } x < -1 \\ |x| - 5 & \text{if } -1 \le x \le 2 \\ x^2 & \text{if } x > 2 \end{cases}$ find

(a) the domain of g

(b) $g(-2.5)$ **(c)** $g(-1)$ **(d)** $g(2)$ **(e)** $g(4)$

45. In a certain state the sales tax $T(p)$ on an item of price p dollars is 5% of p. Which of the following formulas give the correct sales tax in all cases?

  (i) $T(p) = p + 5$
  (ii) $T(p) = 1 + 5p$
  (iii) $T(p) = p/20$
  (iv) $T(p) = p + (5/100)p = p + .05p$
  (v) $T(p) = (5/100)p = .05p$

46. Let T be the sales tax function of Exercise 45 and find $T(3.60)$, $T(4.80)$, $T(.60)$, and $T(0)$.

In Exercises 47–60, determine the domain of the function according to the usual convention.

47. $f(x) = x^2$

48. $g(x) = \dfrac{1}{x^2} + 2$

49. $h(t) = |t| - 1$

50. $k(u) = \sqrt{u}$

51. $k(x) = |x| + \sqrt{x} - 1$

52. $h(x) = \sqrt{(x + 1)^2}$

53. $g(u) = \dfrac{|u|}{u}$

54. $h(x) = \dfrac{\sqrt{x - 1}}{x^2 - 1}$

55. $g(y) = [-y]$

56. $f(t) = \sqrt{-t}$

57. $g(u) = \dfrac{u^2 + 1}{u^2 - u - 6}$

58. $f(t) = \sqrt{4 - t^2}$

59. $f(x) = -\sqrt{9 - (x - 9)^2}$

60. $f(x) = \sqrt{-x} + \dfrac{2}{x + 1}$

61. Give an example of two different functions f and g that have all of the following properties:

$$f(-1) = 1 = g(-1) \quad \text{and} \quad f(0) = 0 = g(0)$$
$$\text{and} \quad f(1) = 1 = g(1).$$

62. Give an example of a function g with the property that $g(x) = g(-x)$ for every real number x.

In Exercises 63–66, the rule of a function f is given. Write an algebraic formula for $f(x)$.

63. Double the input, subtract 5, and take the square root of the result.

64. Square the input, multiply by 3, and subtract the result from 8.

65. Cube the input, add 6, and divide the result by 5.

66. Take the square root of the input, add 7, divide the result by 8, and add this result to the original input.

67. Suppose a car travels at a constant rate of 55 mph for two hours and travels 45 mph thereafter. Find the rule of the function that expresses the distance traveled as a function of time.

68. The table shows the 1999 federal income tax rates for a single person. Write the rule of a piecewise-defined function T such that $T(x)$ is the tax due on a 1999 taxable income of x dollars.

Taxable Income	Tax
Not over $25,750	15% of income
Over $25,750, but not over $62,450	$3862.50 + 28% of amount over $25,750
Over $62,450, but not over $130,250	$14,138.50 + 31% of amount over $62,450
Over $130,250, but not over $283,150	$35,156.50 + 36% of amount over $130,250
Over $283,150	$90,200.50 + 39.6% of amount over $283,150

69. According to the National Center for Health Statistics, the number of divorces and annulments in the United States was 393,000 in 1960; 1,190,000 in 1985; and 1,150,000 in 1996. Here are three functions that model this data (where $x = 0$ corresponds to 1960 and $f(x)$ is in thousands):

$$f(x) = 22.2x + 509$$
$$f(x) = -1.03x^2 + 60.8x + 302$$
$$f(x) = -.032x^3 + .74x^2 + 36.7x + 484.$$

(a) Make a table or otherwise evaluate each function at the values of x corresponding to 1960, 1985, and 1996. Determine which function provides the best model for the given data.

(b) Use the function determined in part (a) to estimate the number of divorces and annulments in 1993.

70. The Travel Industry Association of America estimates that the number of foreign tourists visiting the United States was 34.1 million in 1988, 47.3 million in 1992, and 46.3 million in 1996. This data can be modeled by the following functions (where $x = 0$ corresponds to 1986 and $g(x)$ is in millions):

$$g(x) = 2.05x + 29.5$$
$$g(x) = -.29x^2 + 4.97x + 25.6$$
$$g(x) = .1126x^3 - 2.4713x^2 + 17.213x + 8.658.$$

(a) Make a table or otherwise evaluate each function at the values of x corresponding to 1988, 1992, and 1996. Determine which function provides the best model for the given data.

(b) Use the function determined in part (a) to estimate the number of foreign tourists in 1997.

71. Jack and Jill are salespersons in the suit department of a clothing store. Jack is paid $200 per week plus $5 for each suit he sells, whereas Jill is paid $10 for every suit she sells.

(a) Let $f(x)$ denote Jack's weekly income and $g(x)$ Jill's weekly income from selling x suits. Find the rules of the functions f and g.

(b) Use algebra or a table to find: $f(20)$ and $g(20)$; $f(35)$ and $g(35)$; $f(50)$ and $g(50)$.

(c) If Jack sells 50 suits a week, how many must Jill sell to have the same income as Jack?

72. A potato chip factory has a daily overhead from salaries and building costs of $1800. The cost of ingredients and packaging to produce a pound of potato chips is 50¢. A pound of potato chips sells for $1.20. Show that the factory's daily profit is a function of the number of pounds of potato chips sold and find the rule of this function. (Assume that the factory sells all the potato chips it produces each day.)

73. A rectangular region of 6000 sq ft is to be fenced in on three sides with fencing costing $3.75 per foot and on the fourth side with fencing costing $2.00 per foot. Express the cost of the fence as a function of the length x of the fourth side.

74. A box with a square base measuring $t \times t$ ft is to be made of three kinds of wood. The cost of the wood for the base is 85¢ per sq ft; the wood for the sides costs 50¢ per sq ft and the wood for the top $1.15 per sq ft. The volume of the box is to be 10 cubic feet. Express the total cost of the box as a function of the length t.

75. Average tuition (and fees) in private four-year colleges in recent years were as follows. (*Source:* U.S. Department of Education and the College Board)

Year	Tuition	Year	Tuition
1992	$10,294	1995	$12,216
1993	$10,952	1996	$12,994
1994	$11,481	1997	$13,664

(a) Use linear regression to find the rule of a function f that gives the approximate average tuition in year x, where $x = 0$ corresponds to 1990.

(b) Find $f(3)$, $f(5)$, and $f(7)$. How do they compare with the actual figures?

(c) Use f to estimate average tuition in 1999.

76. The table shows the national debt (in billions of dollars) in selected years. (*Source:* U.S. Department of Treasury)

Year	Debt	Year	Debt
1980	907.7	1990	3233.3
1982	1142	1992	4064.6
1984	1572.3	1994	4692.8
1986	2125.3	1996	5224.8
1988	2602.3		

(a) Use linear regression to find the rule of a function f that gives the approximate national debt (in billions) in year x, where $x = 0$ corresponds to 1980.

(b) Find $f(6)$, $f(10)$, and $f(14)$. How do they compare with the actual figures?

(c) Use f to estimate the national debt in 2000.

3.3 Graphs of Functions

The graph of a function f is the graph of the *equation* $y = f(x)$. Hence

The graph of the function f consists of all points $(x, f(x))$, where x is any number in the domain of f.

The graphs of most functions whose rules are given by algebraic formulas are easily obtained on a calculator. However, there are situations in which a calculator-generated graph may be incomplete or misleading. So the emphasis here is on using your algebraic knowledge *before* reaching for a calculator. Doing this will often tell you that a calculator is inappropriate, or help you to interpret screen images when a calculator is used.

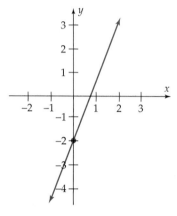

Figure 3–7

Example 1 The graph of the function $f(x) = 3x - 2$ is the graph of the equation $y = 3x - 2$. As we saw in Section 1.4, this graph is a straight line with slope 3 and y-intercept -2, which is easily graphed by hand (Figure 3–7). On a calculator, the graph of f will look a bit bumpy and jagged, rather than the smooth line in Figure 3–7 (try it!). ■

Example 2 The greatest integer function $f(x) = [x]$ was introduced in Example 5 of Section 3.1. It can easily be graphed by hand, by considering the values of the function between each two consecutive integers. For instance,

x	$-2 \leq x < -1$	$-1 \leq x < 0$	$0 \leq x < 1$	$1 \leq x < 2$	$2 \leq x < 3$
$[x]$	-2	-1	0	1	2

Figure 3–8

Thus, between $x = -2$ and $x = -1$, the value of $f(x) = [x]$ is always -2, so that the graph there is a horizontal line segment, all of whose points have second coordinate -2. The rest of the graph is obtained similarly (Figure 3–8). An open circle in Figure 3–8 indicates that the endpoint of the segment is *not* on the graph, whereas a closed circle indicates that the endpoint is on the graph. ■

A function whose graph consists of horizontal line segments, such as Figure 3–8, is called a **step function.** Graphing step functions accurately on a calculator requires some care.

Technology Tip

Directions for switching to dot graphing mode are in Calculator Investigation 5 at the end of Section 1.2.

GRAPHING EXPLORATION

Graph the greatest integer function $f(x) = [x]$ on your calculator (see the Technology Tip on page 188). Does your graph look like Figure 3–8 or does it include vertical segments? Now change the graphing mode of your calculator to "dot" rather than "connected" (see the Technology Tip in the margin), and graph again. How does this graph compare with Figure 3–8? Can you tell from the graph which endpoints are included?

Example 3 An overnight delivery service charges $18 for a package weighing less than 1 pound, $21 for one weighing at least 1 pound, but less than 2 pounds, $24 for one weighing at least 2 pounds, but less than 3 pounds, and so on. Verify that the cost $c(x)$ of shipping a package weighing x pounds is given by $c(x) = 18 + 3[x]$. For example,

If $2 \le x < 3$, then $[x] = 2$ and $c(x) = 18 + 3[x] = 18 + 3(2) = 24$.

Although this rule makes sense for all real numbers, the domain of this cost function consists of positive numbers (why?). The graph of c is in Figure 3–9. ■

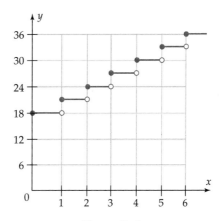

Figure 3–9

Example 4 The graph of the piecewise-defined function

$$y = \begin{cases} x^2 & \text{if } x \le 1 \\ x + 2 & \text{if } 1 < x \le 4 \end{cases}$$

is made up of *parts* of two graphs, corresponding to the different parts of the rule of the function:

$x \le 1$ For these values of x, the graph of f coincides with the graph of $y = x^2$, which was sketched in Figure 1–16 on page 72;

$1 < x \leq 4$ For these values of x, the graph of f coincides with the graph of $y = x + 2$, which is a straight line.

Therefore, we must graph

$y = x^2$ when $x \leq 1$ and $y = x + 2$ when $1 < x \leq 4$.

Combining these partial graphs produces the graph of f in Figure 3–10. ■

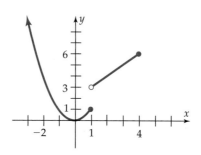

Figure 3–10

Piecewise-defined functions can be graphed on a calculator, provided that you use the correct syntax. Once again, however, the screen does not show which endpoints are included or excluded from the graph.

Technology Tip

Inequality symbols are in the TEST menu of TI-83/86, in the TESTS submenu of the HP-38 MATH menu, and in the INEQ submenu of the Sharp 9600 MATH menu. TI-89 has the symbols $<$, $>$, and $|$ on the keyboard; other inequality symbols and logical symbols (such as "and") are in the TEST submenu of the MATH menu.

GRAPHING EXPLORATION

Graph the function f of Example 4 on a calculator, as follows. On Sharp 9600 or HP-38 or TI-83/86 calculators, use the Tip in the margin to graph these two equations on the same screen:

$$y_1 = \frac{x^2}{(x \leq 1)}$$

$$y_2 = \frac{(x + 2)}{(x > 1)(x \leq 4)}.$$

On TI-89, use the Tip to graph these equations on the same screen:

$$y_1 = x^2 | x \leq 1$$

$$y_2 = x + 2 | x > 1 \text{ and } x \leq 4.$$

To graph f on Casio 9850, with the viewing window of Figure 3–10, graph these equations on the same screen (including commas and square brackets):

$$y_1 = x^2, [-6, 1]$$

$$y_2 = x + 2, [1, 4].$$

How does your graph compare with Figure 3–10?

Example 5 The absolute value function $f(x) = |x|$ is also a piecewise-defined function, since by definition

$$|x| = \begin{cases} x & \text{if } x \geq 0 \\ -x & \text{if } x < 0. \end{cases}$$

Its graph can be obtained by drawing the part of the line $y = x$ to the right of the origin and the part of the line $y = -x$ to the left of the origin (Figure 3–11) or by graphing $y = \text{ABS } x$ on a calculator (Figure 3–12). ∎

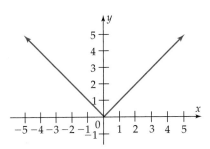

Figure 3–11   Figure 3–12

Local Maxima and Minima

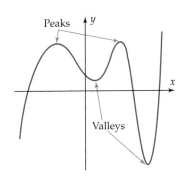

Figure 3–13

The graph of a function may include some peaks and valleys (Figure 3–13). A peak is not necessarily the highest point on the graph, but it is the highest point in its neighborhood. Similarly, a valley is the lowest point in the neighborhood, but not necessarily the lowest point on the graph.

More formally, we say that a function f has a **local maximum** at $x = c$ if the graph of f has a peak at the point $(c, f(c))$. This means that all nearby points $(x, f(x))$ have smaller y-coordinates, that is,

$$f(x) \leq f(c) \quad \text{for all } x \text{ near } c.$$

Similarly, a function has a **local minimum** at $x = d$ provided that

$$f(x) \geq f(d) \quad \text{for all } x \text{ near } d.$$

In other words, the graph of f has a valley at $(d, f(d))$ because all nearby points $(x, f(x))$ have larger y-coordinates.

Calculus is usually needed to find the exact location of local maxima and minima (the plural forms of maximum and minimum). However, they can be accurately approximated by the maximum finder or minimum finder of a calculator.

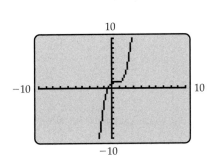

Figure 3–14

Example 6 The graph of $f(x) = x^3 - 1.8x^2 + x + 1$ in Figure 3–14 does not seem to have any local maxima or minima. However, if you use the trace feature to move along the flat segment to the right of the y-axis, you find that the y-coordinates increase, then decrease, then increase (try it!). To see what's really going on, we change viewing windows (Figure

3–15) and see that the function actually has a local maximum and a local minimum (Figure 3–16). The calculator's minimum finder shows that the local minimum occurs when $x \approx .7633$. ∎

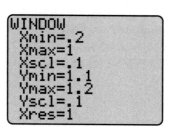

Figure 3–15

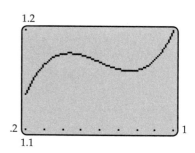

Figure 3–16

GRAPHING EXPLORATION

Graph the function in Example 6 in the viewing window of Figure 3–15. Use the maximum finder to approximate the location of the local maximum.

Example 7   The number of radio stations that primarily play country music is approximated by the function

$$g(x) = 1.2x^3 - 31.4x^2 + 192x + 2300,$$

where $x = 0$ corresponds to 1990. When was the number of country stations the largest during the 1990s?

Solution   We graph the function and use the maximum finder to determine that the local maximum occurs when $x \approx 3.95$ and $g(3.95) \approx 2642$, as shown in Figure 3–17. Therefore the maximum number of country stations was about 2642 and this occurred very late in 1993. ∎

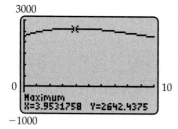

Figure 3–17

Increasing and Decreasing Functions

A function is said to be **increasing on an interval** if its graph always rises as you move from left to right over the interval, and **decreasing on an interval** if its graph always falls as you move from left to right over the interval. A function is said to be **constant on an interval** if its graph is horizontal over the interval.

Example 8   Figure 3–18 suggests that $f(x) = |x| + |x - 2|$ is decreasing on the interval where $x < 0$, increasing where $x > 2$, and constant on the interval where $0 \le x \le 2$. You can confirm that the function is actually constant between 0 and 2 by using the trace feature to move

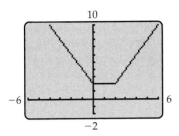

Figure 3–18

along the graph there (the y-coordinates remain the same, as they should on a horizontal segment). For an algebraic proof that f is constant when $0 \le x \le 2$, see Exercise 18. ■

CAUTION

A horizontal segment on a calculator graph does not always mean that the function is constant there. There may be **hidden behavior,** as was the case in Example 6. When in doubt, use the trace feature to see if the y-coordinates remain constant as you move along the "horizontal" segment, or change the viewing window.

Example 9 On what (approximate) intervals is the function $g(x) = .5x^3 - 3x$ increasing and decreasing?

Solution The (complete) graph of g in Figure 3–19 shows that g has a local maximum at P and a local minimum at Q. The maximum and minimum finders show that the approximate coordinates of P and Q are

$$P = (-1.4142, 2.8284) \quad \text{and} \quad Q = (1.4142, -3.8284).$$

Therefore, f is increasing when $x < -1.4142$ and when $x > 1.4142$. It is decreasing when $-1.4142 < x < 1.4142$. ■

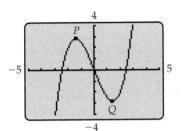

Figure 3–19

The Vertical Line Test

The following fact, which distinguishes graphs of functions from other graphs, can also be used to interpret some calculator-generated graphs.

Vertical Line Test

The graph of a function $y = f(x)$ has this property:

No vertical line intersects the graph more than once.

Conversely, any graph with this property is the graph of a function.

To see why this is true, consider Figure 3–20, in which the graph intersects the vertical line at two points. If this were the graph of a function f, then we would have $f(3) = 2$ [because $(3, 2)$ is on the graph] *and* $f(3) = -1$ [because $(3, -1)$ is on the graph]. This means that the input 3 produces two different outputs, which is impossible for a function. Therefore, Figure 3–20 is not the graph of a function. A similar argument works in the general case.

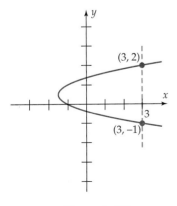

Figure 3–20

Example 10 The rule $g(x) = x^{15} + 2$ does give a function (each input produces exactly one output), but its graph in Figure 3–21 appears to be a vertical line near $x = 1$. To see why this is not actually true, change the viewing window. Figure 3–22 shows that the graph is not vertical between $x = 1$ and $x = 1.2$. This detail is lost in Figure 3–21 because all the x values in Figure 3–22 occupy only a single pixel width in Figure 3–21. ■

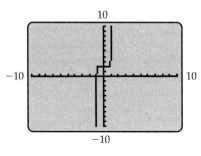

Figure 3–21   Figure 3–22

GRAPHING EXPLORATION

Find a viewing window that shows that the graph of the function g in Example 10 is not actually vertical near $x = -1$.

Graph Reading

Until now we have concentrated on translating statements into functional notation and functional notation into graphs. It is just as important, however, to be able to translate graphical information into equivalent statements in English or functional notation.

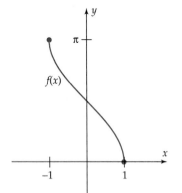

Figure 3–23

Example 11   The entire graph of a function f is shown in Figure 3–23. Find the domain and range of f.

Solution   The graph of f consists of all points of the form $(x, f(x))$. Thus, the first coordinates of points on the graph are the inputs (numbers in the domain of f) and the second coordinates are the outputs (the numbers in the range of f). Figure 3–23 shows that the first coordinates of points on the graph all satisfy $-1 \leq x \leq 1$, so these numbers are the domain of f. Similarly, the range of f consists of all numbers y such that $0 \leq y \leq \pi$, because these are the second coordinates of points on the graph. ■

Example 12   The consumer confidence level reflects people's feelings about their employment opportunities and income prospects. Let $C(t)$ be the consumer confidence level at time t (with $t = 0$ corresponding to 1970) and consider the graph of the function C in Figure 3–24.*

---

*The consumer confidence level is scaled to be 100 in 1985.

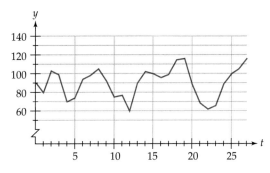

Figure 3–24

(a) How did the consumer confidence level vary in the 1980s?

(b) What was the lowest level of consumer confidence during the 1970s?

(c) During what time periods was the confidence level above 110?

Solution

(a) The 1980s correspond to the interval $10 \leq t \leq 20$, so we consider the part of the graph that lies between the vertical lines $t = 10$ and $t = 20$. The second coordinates of these points range from approximately 60 to 117. So the consumer confidence level varied from a low of 60 to a high of 117 during the 1980s.

(b) The 1970s correspond to the interval $0 \leq t \leq 10$. Figure 3–24 shows that the graph has local minimums at $t = 1$, $t = 4$, and $t = 10$. The lowest of these three points is the one at $t = 4$. Hence, the lowest level of consumer confidence before 1980 occurred at the beginning of 1974.

(c) We must find the values of t for which the graph lies above the horizontal line through 110. Figure 3–24 shows that this occurs approximately when $17.5 \leq t \leq 19.5$ and when $t \geq 26.5$. Thus, the confidence level was above 110 from the middle of 1987 to the middle of 1989 and after the middle of 1996. ■

Example 13   Use the graphs of the functions g and h in Figure 3–25 on the next page to find:

(a) All numbers x such that $h(x) < 0$;

(b) All numbers x with $-3 \leq x \leq 3$ such that $g(x) = 2$;

(c) The largest interval over which g is increasing and h is decreasing and $h(x) \geq g(x)$ for every x in the interval.

Solution

(a) The graph of h consists of all points $(x, h(x))$. The numbers such that $h(x) < 0$ correspond to the points on the graph with negative y-coordinates, that is, points that lie below the x-axis. The graph of h shows that these points are the ones with $-2 < x < 4$.

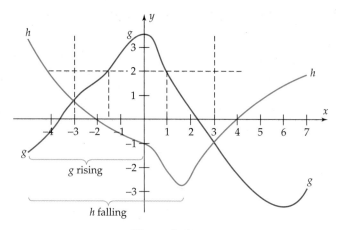

Figure 3–25

(b) The graph of g consists of the points $(x, g(x))$. The points on the graph that lie *between* the vertical lines through -3 and 3 and *on* the horizontal line through 2 have x-coordinate satisfying $-3 \le x \le 3$ and y-coordinate $g(x) = 2$. Figure 3–25 shows that the only such points are $(-1.5, 2)$ and $(2, 2)$. So the answer is $x = -1.5$ and $x = 1$.

(c) Figure 3–25 shows that the only place where the graph of h is falling *and* the graph of g is rising is where $-5 \le x \le 0$. Now $h(x) \ge g(x)$ when the point $(x, h(x))$ lies above the point $(x, g(x))$. This occurs when $-5 \le x \le -3$. ■

Exercises 3.3

In Exercises 1–4 sketch the graph of the function. You should be able to do this without using a calculator. Regardless of how you obtain these graphs, memorize their shapes.

1. $f(x) = x$ **2.** $g(x) = x^2$ **3.** $h(x) = x^3$

4. $p(x) = \sqrt{x}$

In Exercises 5–12, sketch the graph of the function, being sure to indicate which endpoints are included and which ones are excluded.

5. $f(x) = 2[x]$

6. $f(x) = -[x]$

7. $g(x) = [-x]$ [This is *not* the same function as Exercise 6.]

8. $h(x) = [x] + [-x]$

9. $f(x) = \begin{cases} x^2 & \text{if } x \ge -1 \\ 2x + 3 & \text{if } x < -1 \end{cases}$

10. $g(x) = \begin{cases} |x| & \text{if } x < 1 \\ -3x + 4 & \text{if } x \ge 1 \end{cases}$

11. $k(u) = \begin{cases} -2u - 2 & \text{if } u < -3 \\ u - [u] & \text{if } -3 \le u \le 1 \\ 2u^2 & \text{if } u > 1 \end{cases}$

12. $f(x) = \begin{cases} x^2 & \text{if } x < -2 \\ x & \text{if } -2 \le x < 4 \\ \sqrt{x} & \text{if } x \ge 4 \end{cases}$

13. At this writing first-class postage rates are 33¢ for the first ounce or fraction thereof, plus 22¢ for each additional ounce or fraction thereof. Assume that each first-class letter carries one 33¢ stamp and as many 22¢ stamps as are necessary. Then the *number* of stamps required for a first-class letter is a function of the weight of the letter in ounces. Call this function the *postage stamp function*.

   (a) Describe the rule of the postage stamp function algebraically.

   (b) Sketch the graph of the postage stamp function.

(c) Sketch the graph of the function whose rule is $f(x) = p(x) - [x]$, where p is the postage stamp function.

14. A common mistake is to graph the function f in Example 4 by graphing both $y = x^2$ and $y = x + 2$ on the same screen (with no restrictions on x). Explain why this graph could not possibly be the graph of a function.

In Exercises 15–17: (a) Use the fact that the absolute value function is piecewise-defined (see Example 5) to write the rule of the given function as a piecewise-defined function whose rule does not include any absolute value bars. (b) Graph the function.

15. $f(x) = |x| + 2$ 16. $g(x) = |x| - 4$

17. $h(x) = \dfrac{|x|}{2} - 2$

18. Show that the function $f(x) = |x| + |x - 2|$ is constant for $0 \le x \le 2$. [*Hint:* Use the definition of absolute value (see Example 5) to compute $f(x)$ when $0 \le x \le 2$.]

In Exercises 19–24, find the approximate location of all local maxima and minima of the function.

19. $f(x) = x^3 - x$ 20. $g(t) = -\sqrt{16 - t^2}$

21. $h(x) = \dfrac{x}{x^2 + 1}$ 22. $k(x) = x^3 - 3x + 1$

23. $f(x) = x^3 - 1.8x^2 + x + 2$

24. $g(x) = 2x^3 + x^2 + 1$

In Exercises 25–26, find the approximate intervals on which the function whose graph is shown is increasing and those on which it is decreasing.

25.

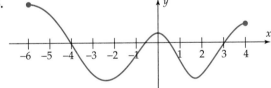

26.

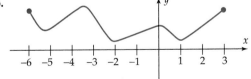

In Exercises 27–32, find the approximate intervals on which the function is increasing, those on which it is decreasing, and those on which it is constant.

27. $f(x) = |x - 1| - |x + 1|$

28. $g(x) = |x - 1| + |x + 2|$

29. $f(x) = -x^3 - 8x^2 + 8x + 5$

30. $f(x) = x^4 - .7x^3 - .6x^2 + 1$

31. $g(x) = .2x^4 - x^3 + x^2 - 2$

32. $g(x) = x^4 + x^3 - 4x^2 + x - 1$

In Exercises 33–34, sketch the graph of a function f that satisfies all of the given conditions. The function whose graph you sketch need not be given by an algebraic formula.

33. (i) The domain of f consists of all real numbers x with $-2 \le x \le 4$.
(ii) The range of f consists of all real numbers y with $-5 \le y \le 6$.
(iii) $f(-1) = f(3)$
(iv) $f(\tfrac{1}{2}) = 0$

34. (i) $f(-1) = 2$
(ii) $f(x) \ge 2$ when $-1 \le x < 1/2$
(iii) $f(x)$ starts decreasing when $x = 1$.
(iv) $f(0) = 3$ and $f(3) = 3$
(v) $f(x)$ starts increasing when $x = 5$.

35. Find the dimensions of the rectangle with perimeter 100 inches and largest possible area, as follows.

(a) Use the figure to write an equation in x and z that expresses the fact that the perimeter of the rectangle is 100.
(b) The area A of the rectangle is given by $A = xz$. (Why?) Write an equation that expresses A as a function of x. [*Hint:* Solve the equation in part (a) for z and substitute the result in the area equation.]
(c) Graph the function in part (b) and find the value of x that produces the largest possible value of A. What is z in this case?

36. Find the dimensions of the rectangle with area 240 square inches and smallest possible perimeter, as follows.
(a) Using the figure for Exercise 35, write an equation for the perimeter P of the rectangle in terms of x and z.
(b) Write an equation in x and z that expresses the fact that the area of the rectangle is 240.
(c) Write an equation that expresses P as a function of x. [*Hint:* Solve the equation in part (b) for z and substitute the result in the equation of part (a).]
(d) Graph the function in part (c) and find the value of x that produces the smallest possible value of P. What is z in this case?

37. Find the dimensions of a box with a square base that has a volume of 867 cubic inches and the smallest possible surface area, as follows.

(a) Write an equation for the surface area S of the box in terms of x and h. [Be sure to include all four sides, the top, and the bottom of the box.]

(b) Write an equation in x and h that expresses the fact that the volume of the box is 867.

(c) Write an equation that expresses S as a function of x. [*Hint:* Solve the equation in part (b) for h and substitute the result in the equation of part (a).]

(d) Graph the function in part (c) and find the value of x that produces the smallest possible value of S. What is h in this case?

38. Find the radius r and height h of a cylindrical can with a surface area of 60 square inches and the largest possible volume, as follows.

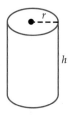

(a) Write an equation for the volume V of the can in terms of r and h.

(b) Write an equation in r and h that expresses the fact that the surface area of the can is 60. [*Hint:* Think of cutting the top and bottom off the can; then cut the side of the can lengthwise and roll it out flat; it's now a rectangle. The surface area is the area of the top and bottom plus the area of this rectangle. The length of the rectangle is the same as the circumference of the original can (why?).]

(c) Write an equation that expresses V as a function of r. [*Hint:* Solve the equation in part (b) for h and substitute the result in the equation of part (a).]

(d) Graph the function in part (c) and find the value of r that produces the largest possible value of V. What is h in this case?

39. Match each of the following functions with the graph that best fits the situation.

(a) The phases of the moon as a function of time;

(b) The demand for a product as a function of its price;

(c) The height of a ball thrown from the top of a building as a function of time;

(d) The distance a woman runs at constant speed as a function of time;

(e) The temperature of an oven turned on and set to 350° as a function of time.

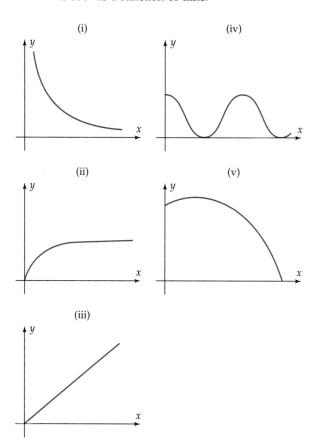

In Exercises 40–42, sketch a plausible graph of the given function. Label the axes and specify a reasonable domain and range.

40. The distance from the top of your head to the ground as you jump on a trampoline is a function of time.

41. The amount you spend on gas each week is a function of the number of gallons you put in your car.

42. The temperature of an oven that is turned on, set to 350°, and 45 minutes later turned off is a function of time.

43. A bacteria population in a laboratory culture contains about a million bacteria at 8 A.M. The culture grows very rapidly until noon, when a bactericide is introduced and the bacteria population plunges. By 4 P.M. the bacteria have adapted to the bactericide and the culture slowly increases in population until 9 P.M. when the culture is accidentally destroyed by the clean-up crew. Let $g(t)$ denote the bacteria population at time t and draw a plausible graph of the function g. [Many correct answers are possible.]

44. A plane flies from Austin, Texas, to Cleveland, Ohio, a distance of 1200 miles. Let f be the function whose rule is $f(t)$ = distance (in miles) from Austin at time t hours. Draw a plausible graph of f under the given circumstances. [There are many possible correct answers for each part.]
  (a) The flight is nonstop and takes less than 4 hours.
  (b) Bad weather forces the plane to land in Dallas (about 200 miles from Austin), remain overnight (for 8 hours), and continue the next day.
  (c) The flight is nonstop, but due to heavy traffic the plane must fly in a holding pattern over Cincinnati (about 200 miles from Cleveland) for an hour before going on to Cleveland.

In Exercises 45–46, the graph of a function f is shown. Find and label the given points on the graph.

45. (a) $(k, f(k))$
  (b) $(-k, f(-k))$
  (c) $(k, -f(k))$

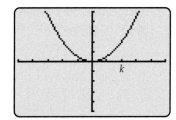

46. (a) $(k, f(k))$
  (b) $(k, .5f(k))$
  (c) $(.5k, f(.5k))$
  (d) $(2k, f(2k))$

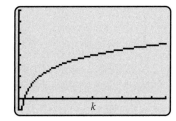

47. The figure shows the graph of the function f whose rule is $f(x)$ = average interest rate on a 30-year fixed rate mortgage in year x. Use it to answer the following questions (reasonable approximations are OK).

(a) $f(1973) = ?$ **(b)** $f(1982) = ?$ **(c)** $f(1995) = ?$
(d) In what year between 1990 and 1996 were rates the lowest? The highest?

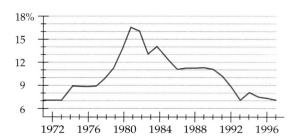

48. The figure shows the graph of the function g whose rule is $g(x)$ = the percentage of overdue bank credit card accounts in year x (where $x = 0$ corresponds to 1985). Use it to answer the following questions.
  (a) Over what time intervals was this function approximately constant?
  (b) Over what time intervals was this function decreasing?
  (c) When did the lowest rate of overdue accounts occur?

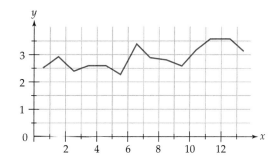

Exercises 49–58 deal with the function g whose entire graph is shown in the figure.

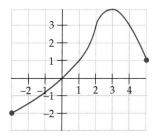

49. What is the domain of g?

50. What is the range of g?

51. If $t = 1.5$, then $g(2t) = ?$

52. If $t = 1.5$, then $2g(t) = ?$
53. If $y = 2$, then $g(y + 1.5) = ?$
54. If $y = 2$, then $g(y) + g(1.5) = ?$
55. If $y = 2$, then $g(y) + 1.5 = ?$
56. For what values of x is $g(x) < 0$?
57. For what values of z is $g(z) = 1$?
58. For what values of z is $g(z) = -1$?

Exercises 59–64 deal with the function f whose entire graph is shown in the figure.

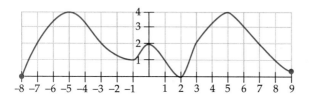

59. What is the domain of f?
60. What is the range of f?
61. Find all numbers x such that $f(x) = 2$.
62. Find all numbers x such that $f(x) > 2$.
63. Find all numbers x such that $f(x) = f(7)$.
64. Find two numbers x such that $f(x - 2) = 4$.

Exercises 65–70 deal with the functions f and g whose entire graphs are shown.

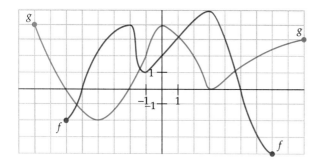

65. What is the domain of f? The domain of g?
66. What is the range of f? The range of g?
67. Find all numbers x, with $-3 \le x \le 1$, such that $f(x) = 2$.
68. For how many values of x is it true that $f(x) = g(x)$?
69. Find all intervals over which both functions are defined, f is decreasing, and g is increasing.
70. Find all intervals over which g is decreasing.

Exercises 71–74 deal with this situation: The owners of the Melville & Pluth Hammer Factory have determined that both their weekly manufacturing expenses and their weekly sales income are functions of the number of hammers manufactured each week. The figure shows the graphs of these two functions.

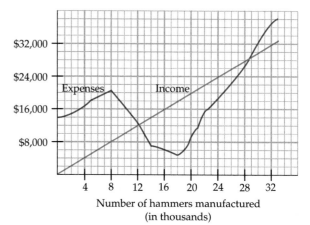

Number of hammers manufactured
(in thousands)

71. Use careful measurement on the graph and the fact that profit = income − expenses to determine the weekly profit if 5000 hammers are manufactured.

72. Do the same if 10,000, 14,000, 18,000, or 22,000 hammers are manufactured.

73. What is the smallest number of hammers that can be manufactured each week without losing money?

74. What is the largest number of hammers that can be manufactured without losing money?

75. As cell phone usage has grown, the average monthly bill for a cell phone has declined, as shown in the table. (*Source:* Cellular Telecommunications Industry Association)

Year	Average bill	Year	Average bill
1988	$98.02	1995	$51.00
1990	$80.90	1996	$47.70
1992	$68.68	1997	$42.78
1994	$56.21		

(a) Make a scatter plot of the data, with $x = 0$ corresponding to 1988.
(b) Use linear regression to find a function that models this data.

(c) Use your model to estimate the average bill in 1989 and 1993. How do your estimates compare with the actual figures of $89.30 and $61.48?

(d) Assuming the model remains accurate, when will the average bill be $35?

76. The table shows the number of people (in millions) receiving health care from health maintenance organizations (HMOs) in selected years. (*Source: American Association of Health Plans*)

Year	HMO members
1986	25.7
1988	32.7
1990	36.5
1992	41.4
1994	51.1
1996	61.8

(a) Make a scatter plot of the data, with $x = 0$ corresponding to 1986.

(b) Use linear regression to find a function that models this data.

(c) Use your model to estimate the number of HMO members in 1989 and 1995. How do your estimates compare with the actual figures of 34.7 and 59.1 million?

(d) Assuming the model remains accurate, how many people will be HMO members in 2001?

Thinkers

77. A jogger begins her daily run from her home. The graph shows her distance from home at time t minutes. The graph shows, for example, that she ran at a slow but steady pace for 10 minutes, then very quickly increased her pace for 5 minutes, all the time moving farther from home. Describe the rest of her run.

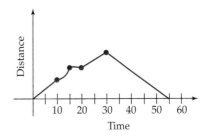

78. The graph shows the speed (in mph) at which a driver is going at time t minutes. Describe his journey.

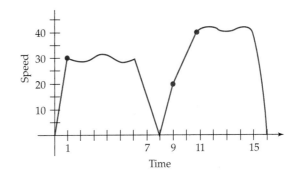

3.3.A *EXCURSION* **Parametric Graphing**

Most of the functions we have seen up to now can be described by equations in which y is a function of x, and hence, are easily graphed on a calculator. When an equation expresses x as a function of y, a different calculator graphing technique is needed.

Example 1 Graph: $x = y^3 - 3y^2 - 4y + 7$.

Solution This equation defines x as a function of y because each input (y value) leads to a single output (corresponding value of x). Let t be any real number. If $y = t$, then

$$x = y^3 - 3y^2 - 4y + 7 = t^3 - 3t^2 - 4t + 7.$$

Technology Tip

To change to parametric graphing mode, choose PAR (or PARAM or PARAMETRIC or PARM) in the TI MODE menu, or the HP-38 LIB menu, or the COORD submenu of the Sharp 9600 SETUP menu, or in the TYPE submenu of the Casio 9850 GRAPH menu (on the main menu).

Thus, the graph consists of all points (x, y) such that

$$x = t^3 - 3t^2 - 4t + 7 \qquad \text{and} \qquad y = t \quad (t \text{ any real number}).$$

When written in this form, the equation can be graphed as follows.

Change the graphing mode to **parametric mode** (see the Tip in the margin) and enter the equations

$$x_{1t} = t^3 - 3t^2 - 4t + 7$$

$$y_{1t} = t$$

in the equation memory.* Next set the viewing window so that

$$-6 \leq t \leq 6, \quad -10 \leq x \leq 10, \quad -6 \leq y \leq 6,$$

as partially shown in Figure 3–26 (scroll down to see the rest). We use the same range for t and y here because $y = t$. Note that we must also set "t-step" (or "t-pitch"), which determines how much t changes each time a point is plotted. A t-step between .05 and .15 usually produces a relatively smooth graph in a reasonable amount of time. Finally, pressing GRAPH (or PLOT or DRAW) produces the graph in Figure 3–27. ■

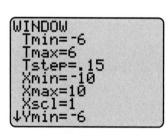

Figure 3–26

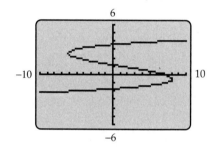

Figure 3–27

As illustrated in Example 1, the underlying idea of **parametric graphing** is to express both x and y as functions of a third variable t. The equations that define x and y are called **parametric equations** and the variable t is called the **parameter**. Example 1 illustrates just one of the many applications of parametric graphing. It can also be used to graph curves that are not graphs of a single equation in x and y.

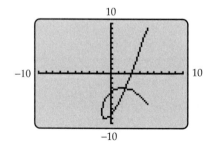

Figure 3–28

Example 2 Graph the curve given by

$$x = t^2 - t - 1 \qquad \text{and} \qquad y = t^3 - 4t - 6 \quad (-2 \leq t \leq 3).$$

Solution Using the standard viewing window we obtain the graph in Figure 3–28. Note that the graph crosses over itself at one point and that it does not extend forever to the left and right, but has "endpoints."

*On some calculators x_{1t} is denoted x_{t1} or $x_1(T)$, and similarly for y_{1t}.

Technology Tip

If you have trouble finding appropriate ranges for t, x, and y, it may help to use the TABLE feature to display a table of t-x-y values produced by the parametric equations.

GRAPHING EXPLORATION

Graph these same parametric equations, but set the range of t values so that $-4 \leq t \leq 4$. What happens to the graph? Now change the range of t values so that $-10 \leq t \leq 10$. Find a viewing window large enough to show the entire graph, including endpoints. ■

Any function of the form $y = f(x)$ can be expressed in terms of parametric equations and graphed that way. For instance, to graph $f(x) = x^2 + 1$, let $x = t$ and $y = f(t) = t^2 + 1$.

Exercises 3.3.A

In Exercises 1–6, each equation defines x as a function of y. Solve each equation for x and find a viewing window that shows a complete graph of the equation.

1. $x = y^3 + 5y^2 - 4y - 5$

2. $\sqrt[3]{y^2 - y + 1} - x + 2 = 0$

3. $xy^2 + xy + x = y^3 - 2y^2 + 4$

4. $2y = xy^2 + 180x$ 5. $x - \sqrt{y} + y^2 + 8 = 0$

6. $y^2 - x - \sqrt{y + 5} + 4 = 0$

In Exercises 7–12, find a viewing window that shows a complete graph of the curve determined by the parametric equations.

7. $x = 3t^2 - 5$ and $y = t^3$ ($-4 \leq t \leq 4$)

8. The Zorro curve: $x = .1t^3 - .2t^2 - 2t + 4$ and $y = 1 - t$ ($-5 \leq t \leq 6$)

9. $x = t^2 - 3t + 2$ and $y = 8 - t^3$ ($-4 \leq t \leq 4$)

10. $x = t^2 - 6t$ and $y = \sqrt{t + 7}$ ($-5 \leq t \leq 9$)

11. $x = 1 - t^2$ and $y = t^3 - t - 1$ ($-4 \leq t \leq 4$)

12. $x = t^2 - t - 1$ and $y = 1 - t - t^2$

Thinkers

13. Graph the curve given by
$$x = (t^2 - 1)(t^2 - 4)(t + 5) + t + 3$$
$$y = (t^2 - 1)(t^2 - 4)(t^3 + 4) + t - 1$$
$$(-2.5 \leq t \leq 2.5)$$
How many times does this curve cross itself?

14. Use parametric equations to describe a curve that crosses itself more times than the curve in Exercise 13. [Many correct answers are possible.]

3.4   **Graphs and Transformations**

In this section we shall see that when the rule of a function is algebraically changed in certain ways, so as to produce a new function, then the graph of the new function can be obtained from the graph of the original function by a simple geometric transformation. The same format will be used for each topic:

First,   you will be asked to assemble some evidence by doing a graphing exploration.

Next,   general conclusions deduced from the evidence will be summarized in the boxes.

Last,   there may be some additional examples or explorations.

Vertical Shifts

GRAPHING EXPLORATION

Using the standard viewing window, graph these three functions on the same screen:

$$f(x) = x^2 \qquad g(x) = x^2 + 5 \qquad h(x) = x^2 - 7$$

and answer these questions:

Do the graphs of g and h look very similar to the graph of f in *shape*?

How do their vertical positions differ?

Where would you predict that the graph of $k(x) = x^2 - 9$ is located relative to the graph of $f(x) = x^2$ and what is its shape?

Confirm your prediction by graphing k on the same screen as f, g, and h.

The results of this Exploration should make the following statement plausible:

Vertical Shifts

If $c > 0$, then the graph of $g(x) = f(x) + c$ is the graph of f shifted upward c units.

If $c > 0$, then the graph of $h(x) = f(x) - c$ is the graph of f shifted downward c units.

Example 1 A calculator was used to obtain a complete graph of $f(x) = .04x^3 - x - 3$ in Figure 3–29. The graph of

$$h(x) = f(x) - 4 = (.04x^3 - x - 3) - 4 = .04x^3 - x - 7$$

is the graph of f shifted 4 units downward, as shown in Figure 3–30.

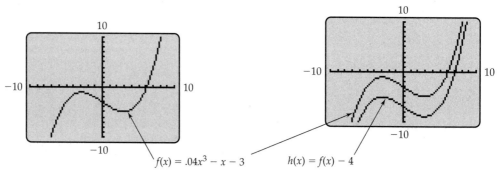

$f(x) = .04x^3 - x - 3$ $\qquad$ $h(x) = f(x) - 4$

Figure 3–29 $\qquad\qquad\qquad$ **Figure 3–30**

Although it may appear that the graph of h is closer to the graph of f at the outer edges of Figure 3–30 than in the center, this is an optical illusion:

The *vertical* distance between the graphs is always 4 units.

GRAPHING EXPLORATION

Use the trace feature of your calculator to confirm this fact.

Move the cursor to any point on the graph of f and note its coordinates.

Use the down arrow to drop the cursor to the graph of h and note the coordinates of the cursor in its new position.

The x-coordinates will be the same in both cases and the new y-coordinate will be 4 less than the original y-coordinate. ∎

Horizontal Shifts

GRAPHING EXPLORATION

Using the standard viewing window, graph these three functions on the same screen:

$$f(x) = 2x^3 \qquad g(x) = 2(x + 6)^3 \qquad h(x) = 2(x - 8)^3$$

and answer these questions:

Do the graphs of g and h look very similar to the graph of f in *shape*?

How do their horizontal positions differ?

Where would you predict that the graph of $k(x) = 2(x + 2)^3$ is located relative to the graph of $f(x) = 2x^3$ and what is its shape?

Confirm your prediction by graphing k on the same screen as $f, g,$ and h.

The results of this Exploration should make the following statement plausible:

Horizontal Shifts

Let f be a function and c a positive constant.

The graph of $g(x) = f(x + c)$ is the graph of f shifted horizontally c units to the left.

The graph of $h(x) = f(x - c)$ is the graph of f shifted horizontally c units to the right.

Technology Tip

If the function f of Example 2 is entered as $y_1 = x^2 - 7$, then the functions g and h can be entered as $y_2 = y_1(x + 5)$ and $y_3 = y_1(x - 4)$ on TI-82/83/86/89, Sharp 9600, and HP-38 (but *not* on TI-85 or Casio 9850). See the Tip on page 196 for how to enter y_1.

Example 2 In some cases, shifting the graph of a function f horizontally may produce a graph that overlaps the graph of f. For instance, a complete graph of $f(x) = x^2 - 7$ is shown in Figure 3–31. The graph of

$$g(x) = f(x + 5) = (x + 5)^2 - 7$$

is the graph of f shifted 5 units to the left and the graph of

$$h(x) = f(x - 4) = (x - 4)^2 - 7$$

is the graph of f shifted 4 units to the right, as shown in Figure 3–31. ∎

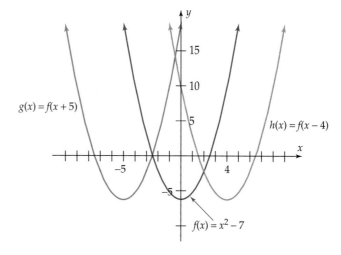

Figure 3–31

Expansions and Contractions

Technology Tip

On TI calculators (except TI-81) you can graph both functions in the Exploration at the same time by keying in

$$y = \{1, 3\}(x^2 - 4).$$

GRAPHING EXPLORATION

In the viewing window with $-5 \le x \le 5$ and $-15 \le y \le 15$, graph these functions on the same screen:

$$f(x) = x^2 - 4 \qquad g(x) = 3f(x) = 3(x^2 - 4).$$

One way to understand the relationship between the two graphs is to imagine that the graph of f is nailed to the x-axis at its intercepts (± 2) and that you can vertically "stretch" the graph by pulling from the top and bottom away from the x-axis (with the nails holding the x-intercepts in place) so that it fits onto the graph of g. In this process (Figure 3–32), the point $(0, -4)$ on the graph of f is stretched down to the point $(0, -12)$ on the graph of g—that is, it is stretched away from the x-axis by a factor of 3. Similarly, the point $(3, 5)$ on the graph of f is stretched up by a factor of 3 to the point $(3, 15)$ on the graph of g, shown in Figure 3–32.

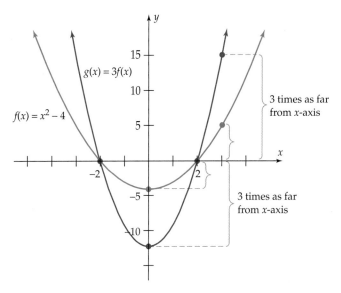

Figure 3–32

GRAPHING EXPLORATION

In the viewing window with $-5 \leq x \leq 5$ and $-5 \leq y \leq 10$, graph these functions on the same screen:

$$f(x) = x^2 - 4 \qquad h(x) = \tfrac{1}{4}(x^2 - 4).$$

Your screen should suggest that the graph of h is the graph of f "shrunk" vertically toward the x-axis by a factor of $1/4$.

Analogous facts are true in the general case:

Expansions and Contractions

If $c > 1$, then the graph of $g(x) = cf(x)$ is the graph of f stretched vertically away from the x-axis by a factor of c.

If $0 < c < 1$, then the graph of $h(x) = cf(x)$ is the graph of f shrunk vertically toward the x-axis by a factor of c.

Reflections

> ### GRAPHING EXPLORATION
>
> In the standard viewing window, graph these functions on the same screen:
>
> $$f(x) = .04x^3 - x \qquad g(x) = -f(x) = -(.04x^3 - x).$$
>
> Using your trace cursor to move from graph to graph, verify that for every point on the graph of f there is a point on the graph of g with the same first coordinate that is on the opposite side of the x-axis, the same distance from the x-axis.

This Exploration shows that the graph of g is the mirror image (reflection) of the graph of f, with the x-axis being the (two-way) mirror. The same thing is true in the general case:

Reflections

> Let f be a function.
>
> The graph of $g(x) = -f(x)$ is the graph of f reflected in the x-axis.

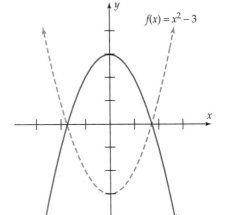

$f(x) = x^2 - 3$

$g(x) = -(x^2 - 3)$

Figure 3–33

Example 3 If $f(x) = x^2 - 3$, then the graph of

$$g(x) = -f(x) = -(x^2 - 3)$$

is the reflection of the graph of f in the x-axis, as shown in Figure 3–33. ■

> ### GRAPHING EXPLORATION
>
> In the standard viewing window graph these functions on the same screen:
>
> $$f(x) = \sqrt{5x + 10} \qquad \text{and} \qquad h(x) = f(-x) = \sqrt{5(-x) + 10}.$$
>
> Reflect carefully: How are the two graphs related to the y-axis? Now graph these two functions on the same screen:
>
> $$f(x) = x^2 + 3x - 3 \qquad \text{and}$$
> $$h(x) = f(-x) = (-x)^2 + 3(-x) - 3 = x^2 - 3x - 3.$$
>
> Are the graphs of f and h related in the same way as the first pair?

This Exploration shows that the graph of h in each case is the mirror image (reflection) of the graph of f, with the y-axis as the mirror. The same thing is true in the general case.

Reflections

Let f be a function.

The graph of $h(x) = f(-x)$ is the graph of f reflected in the y-axis.

For other types of algebraic operations on the rule of a function and their effects on its graph, see Exercises 43–50.

Combining Transformations

The transformations described above may be used in sequence to analyze the graphs of functions whose rules are algebraically complicated.

Example 4   To understand the graph of $g(x) = 2(x - 3)^2 - 1$, note that the rule of g may be obtained from the rule of $f(x) = x^2$ in three steps:

$$f(x) = x^2 \xrightarrow{\text{Step 1}} (x - 3)^2 \xrightarrow{\text{Step 2}} 2(x - 3)^2 \xrightarrow{\text{Step 3}} 2(x - 3)^2 - 1 = g(x)$$

Step 1 shifts the graph of f horizontally 3 units to the right; step 2 stretches the resulting graph away from the x-axis by a factor of 2; step 3 shifts this graph 1 unit downward, thus producing the graph of g in Figure 3–34.   ∎

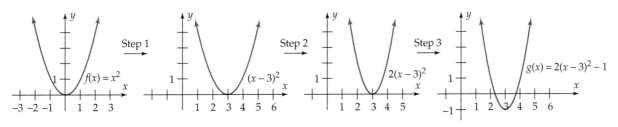

Figure 3–34

Calculator Investigations 3.4

Investigations 1 and 2 show that, because of its small screen size, a calculator may not always display clearly what it should.

1. Graph $f(x) = x^2 - x - 6$ in the standard viewing window. Describe verbally what the graph of $h(x) = f(x - 1000)$ should look like (see the box on page 222). Now find an appropriate viewing window and graph h. Can you find a viewing window that clearly displays both the graph of f and the graph of h?

2. Graph $f(x) = x^2 - x - 6$ in the standard viewing window. Describe verbally what the graph of $g(x) = 1000f(x)$ should look like (see the box on page 223). Now find an appropriate viewing window and graph g. Can you find a viewing window that clearly displays both the graph of f and the graph of g?

Exercises 3.4

If you have not already done Exercises 1–4 of Section 3.3, do them before doing Exercises 1–8 here. In Exercises 1–8, match the function with its graph, which is one of the ones shown here.

A.
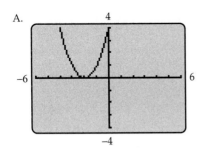

B.
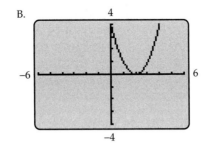

C.
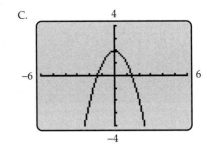

D.
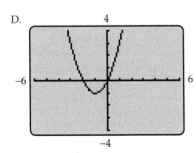

E.
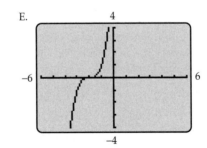

F.
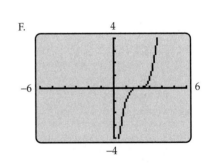

G.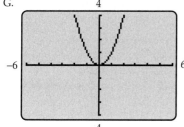

H.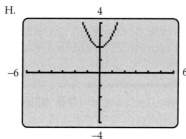

I.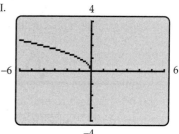

J.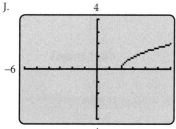

K.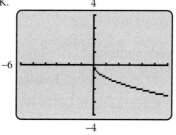

L.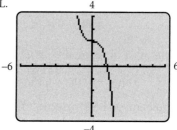

1. $f(x) = x^2 + 2$

2. $f(x) = \sqrt{x - 2}$

3. $g(x) = (x - 2)^3$

4. $g(x) = (x - 2)^2$

5. $f(x) = -\sqrt{x}$

6. $f(x) = (x + 1)^2 - 1$

7. $g(x) = -x^2 + 2$

8. $g(x) = -x^3 + 2$

In Exercises 9–12, use Example 5 of Section 3.3 and appropriate information from this section (but not a calculator) to sketch the graph of the function.

9. $f(x) = |x - 2|$

10. $g(x) = |x| - 2$

11. $g(x) = -|x|$

12. $f(x) = |x + 2| - 2$

In Exercises 13–16, find a single viewing window that shows complete graphs of the functions f, g, h.

13. $f(x) = .25x^3 - 9x + 5$; $g(x) = f(x) + 15$;
$h(x) = f(x) - 20$

14. $f(x) = \sqrt{x^2 - 9} - 5$; $g(x) = 3f(x)$; $h(x) = .5f(x)$

15. $f(x) = |x^2 - 5|$; $g(x) = f(x + 8)$; $h(x) = f(x - 6)$

16. $f(x) = .125x^3 - .25x^2 - 1.5x + 5$;
$g(x) = f(x) - 5$; $h(x) = 5 - f(x)$

In Exercises 17 and 18, find complete graphs of the functions f and g in the same viewing window.

17. $f(x) = \dfrac{4 - 5x^2}{x^2 + 1}$; $g(x) = -f(x)$

18. $f(x) = x^4 - 4x^3 + 2x^2 + 3$; $g(x) = f(-x)$

In Exercises 19–22, describe a sequence of transformations that will transform the graph of the function f into the graph of the function g.

19. $f(x) = x^2 + x$; $g(x) = (x - 3)^2 + (x - 3) + 2$

20. $f(x) = x^2 + 5$; $g(x) = (x + 2)^2 + 10$

21. $f(x) = \sqrt{x^3 + 5}$; $g(x) = -\dfrac{1}{2}\sqrt{x^3 + 5} - 6$

22. $f(x) = \sqrt{x^4 + x^2 + 1}$;
$g(x) = 10 - \sqrt{4x^4 + 4x^2 + 4}$

In Exercises 23–26, write the rule of a function g whose graph can be obtained from the graph of the function f by performing the transformations in the order given.

23. $f(x) = x^2 + 2$; shift the graph horizontally 5 units to the left and then vertically upward 4 units.

24. $f(x) = x^2 - x + 1$; reflect the graph in the x-axis, then shift it vertically upward 3 units.

25. $f(x) = \sqrt{x}$; shift the graph horizontally 6 units to the right, stretch it away from the x-axis by a factor of 2, and shift it vertically downward 3 units.

26. $f(x) = \sqrt{-x}$; shift the graph horizontally 3 units to the left, then reflect it in the x-axis, and shrink it toward the x-axis by a factor of 1/2.

In Exercises 27–30, use the graph of the function f in the figure to sketch the graph of the function g.

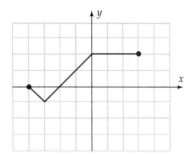

27. $g(x) = f(x) + 3$ **28.** $g(x) = f(x) - 1$

29. $g(x) = 3f(x)$ **30.** $g(x) = .25f(x)$

In Exercises 31–34, use the graph of the function f in the figure to sketch the graph of the function h.

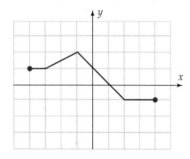

31. $h(x) = -f(x)$ **32.** $h(x) = -4f(x)$

33. $h(x) = f(-x)$ **34.** $h(x) = f(-x) + 2$

In Exercises 35–40, use the graph of the function f in the figure to sketch the graph of the function g.

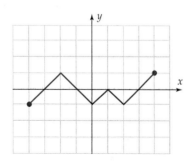

35. $g(x) = f(x + 3)$ **36.** $g(x) = f(x - 2)$

37. $g(x) = f(x - 2) + 3$ **38.** $g(x) = f(x + 1) - 3$

39. $g(x) = 2 - f(x)$ **40.** $g(x) = f(-x) + 2$

41. Graph $f(x) = -|x - 3| - |x - 17| + 20$ in the window with $0 \le x \le 20$ and $-2 \le y \le 12$. Think of the x-axis as a table and the graph as a side view of a fast-food carton placed upside down on the table (the flat part of the graph is the bottom of the carton). Find the rule of a function g whose graph (in this viewing window) looks like another fast-food carton, which has been placed right side up on top of the first one.

42. A factory has a linear cost function $c(x) = ax + b$, where b represents fixed costs and a represents the variable costs (labor and materials) of making one item, both in thousands of dollars.

  (a) If property taxes (part of the fixed costs) are increased by $28,000 per year, what effect does this have on the graph of the cost function?

  (b) If variable costs increase by 12 cents per item, what effect does this have on the graph of the cost function?

In Exercises 43–45, assume $f(x) = (.2x)^6 - 4$. Use the standard viewing window to graph the functions f and g on the same screen.

43. $g(x) = f(2x)$ **44.** $g(x) = f(3x)$

45. $g(x) = f(4x)$

46. Based on the results of Exercises 43–45, describe the transformation that transforms the graph of a function $f(x)$ into the graph of the function $f(cx)$, where c is a constant with $c > 1$. [*Hint:* How are the two graphs related to the y-axis? Stretch your mind.]

In Exercises 47–49, assume $f(x) = x^2 - 3$. Use the standard viewing window to graph the functions f and g on the same screen.

47. $g(x) = f\left(\dfrac{1}{2}x\right)$ **48.** $g(x) = f\left(\dfrac{1}{3}x\right)$

49. $g(x) = f\left(\dfrac{1}{4}x\right)$

50. Based on the results of Exercises 47–49, describe the transformation that transforms the graph of a function $f(x)$ into the graph of the function $f(cx)$, where c is a constant with $0 < c < 1$. [*Hint:* How are the two graphs related to the y-axis?]

3.4.A **ExCURSION**   **Symmetry**

A graph is **symmetric with respect to the y-axis** if the part of the graph on the right side of the y-axis is the mirror image of the part on the left side of the y-axis (with the y-axis being the mirror), as shown in Figure 3–35.

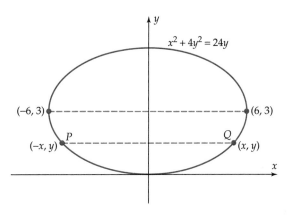

Figure 3–35

Each point P on the left side of the graph has a mirror image point Q on the right side of the graph, as indicated by the dashed lines. Note that

> Their second coordinates are the same.

> Their first coordinates are negatives of each other.

Thus, a graph is symmetric with respect to the y-axis provided that

> Whenever (x, y) is on the graph, then $(-x, y)$ is also on it.

In algebraic terms, this means that replacing x by $-x$ in the equation leads to the same number y. In other words, replacing x by $-x$ produces an equivalent equation.

Example 1   Replacing x by $-x$ in the equation $y = x^4 - 5x^2 + 3$ produces

$$(-x)^4 - 5(-x)^2 + 3,$$

which is the same equation because $(-x)^2 = x^2$ and $(-x)^4 = x^4$. Therefore, the graph is symmetric with respect to the y-axis.

GRAPHING EXPLORATION

Confirm this fact by graphing the equation.   ◼

x-Axis Symmetry

A graph is **symmetric with respect to the *x*-axis** if the part of the graph above the *x*-axis is the mirror image of the part below the *x*-axis (with the *x*-axis being the mirror), as shown in Figure 3–36.

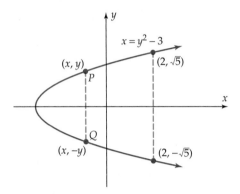

Figure 3–36

Using Figure 3–36 and argument analogous to the one preceding Example 1, we see that a graph is symmetric with respect to the *x*-axis provided that

Whenever (x, y) is on the graph, then $(x, -y)$ is also on it.

In algebraic terms, this means that replacing y by $-y$ in the equation leads to the same number x. In other words, replacing y by $-y$ produces an equivalent equation.

Example 2   Replacing y by $-y$ in the equation $y^2 = 4x - 12$ produces

$$(-y)^2 = 4x - 12,$$

which is the same equation, so the graph is symmetric with respect to the *x*-axis.

GRAPHING EXPLORATION

Confirm this fact by graphing the equation. In order to do this, note that every point on the graph of $y^2 = 4x - 12$ is also on the graph of either $y = \sqrt{4x - 12}$ or $y = -\sqrt{4x - 12}$. Each of these latter equations defines a function; graph them both on the same screen.   ∎

Origin Symmetry

A graph is **symmetric with respect to the origin** if a straight line through the origin and any point P on the graph also intersects the graph at a point Q such that the origin is the midpoint of segment PQ, as shown in Figure 3–37.

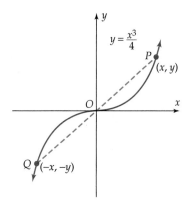

Figure 3–37

Using Figure 3–37, we can also describe symmetry with respect to the origin in terms of coordinates and equations (Exercise 34):

Whenever (x, y) is on the graph, then $(-x, -y)$ is also on it.

In algebraic terms, this means that replacing x by $-x$ and y by $-y$ in the equation produces an equivalent equation.

Example 3 Replacing x by $-x$ and y by $-y$ in the equation $y = x^3/10 - x$ yields

$$-y = \frac{(-x)^3}{10} - (-x), \qquad \text{that is,} \qquad -y = \frac{-x^3}{10} + x.$$

This equation is equivalent to $y = x^3/10 - x$ since it can be obtained from it by multiplying by -1. Therefore, the graph of $y = x^3/10 - x$ is symmetric with respect to the origin.

GRAPHING EXPLORATION

Confirm this fact by graphing the equation. ■

Here is a summary of the various tests for symmetry:

Symmetry Tests

Symmetry with Respect to	Coordinate Test for Symmetry	Algebraic Test for Symmetry
y-axis	(x, y) on graph implies $(-x, y)$ on graph.	Replacing x by $-x$ produces an equivalent equation.
x-axis	(x, y) on graph implies $(x, -y)$ on graph.	Replacing y by $-y$ produces an equivalent equation.
origin	(x, y) on graph implies $(-x, -y)$ on graph.	Replacing x by $-x$ and y by $-y$ produces an equivalent equation.

Even and Odd Functions

For *functions,* the algebraic description of symmetry takes a different form. A function f whose graph is symmetric with respect to the y-axis is called an **even function.** To say that the graph of $y = f(x)$ is symmetric with respect to the y-axis means that replacing x by $-x$ produces the same y value. In other words, the function takes the same value at both x and $-x$. Therefore,

Even Functions

> A function f is even provided that
>
> $f(x) = f(-x)$ for every number x in the domain of f.

For example, $f(x) = x^4 + x^2$ is even because

$$f(-x) = (-x)^4 + (-x)^2 = x^4 + x^2 = f(x).$$

Thus, the graph of f is symmetric with respect to the y-axis, as you can easily verify with your calculator (do it!).

Except for zero functions ($f(x) = 0$ for every x in the domain), *the graph of a function is never symmetric with respect to the x-axis.* The reason is the Vertical Line Test: The graph of a function never contains two points with the same first coordinate. If both $(5, 3)$ and $(5, -3)$, for instance, were on the graph, this would say that $f(5) = 3$ and $f(5) = -3$, which is impossible when f is a function.

A function whose graph is symmetric with respect to the origin is called an **odd function.** If both (x, y) and $(-x, -y)$ are on the graph of such a function f, then we must have both

$$y = f(x) \quad \text{and} \quad -y = f(-x)$$

so that $f(-x) = -y = -f(x)$. Therefore,

Odd Functions

A function f is **odd** provided that

$$f(-x) = -f(x) \text{ for every number } x \text{ in the domain of } f.$$

For example, $f(x) = x^3$ is an odd function because

$$f(-x) = (-x)^3 = -x^3 = -f(x).$$

Hence, the graph of f is symmetric with respect to the origin (verify this with your calculator).

Exercises 3.4.A

In Exercises 1–4, graph the equation. If the graph is symmetric with respect to the x-axis, the y-axis, or the origin, say so.

1. $y = x^2 + 2$ **2.** $x = (y - 3)^2$

3. $y = x^3 + 2$ **4.** $y = (x + 2)^3$

In Exercises 5–14, determine whether the given function is even, odd, or neither.

5. $f(x) = 4x$ **6.** $k(t) = -5t$

7. $f(x) = x^2 - |x|$ **8.** $h(u) = |3u|$

9. $k(t) = t^4 - 6t^2 + 5$ **10.** $f(x) = x(x^4 - x^2) + 4$

11. $f(t) = \sqrt{t^2 - 5}$ **12.** $h(x) = \sqrt{7 - 2x^2}$

13. $f(x) = \dfrac{x^2 + 2}{x - 7}$ **14.** $g(x) = \dfrac{x^2 + 1}{x^2 - 1}$

In Exercises 15–18, determine algebraically whether or not the graph of the given equation is symmetric with respect to the x-axis.

15. $x^2 - 6x + y^2 + 8 = 0$ **16.** $x^2 + 8x + y^2 = -15$

17. $x^2 - 2x + y^2 + 2y = 2$ **18.** $x^2 - x + y^2 - y = 0$

In Exercises 19–24, determine whether the given graph is symmetric with respect to the y-axis, the x-axis, or the origin.

19.

20.

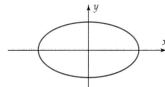

21.

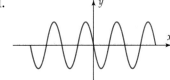

22.

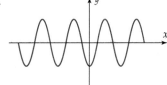

23.

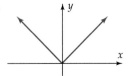

24.

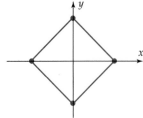

In Exercises 25–28, complete the graph of the given function, assuming that it satisfies the given symmetry condition.

25. Even

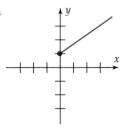

26. Even

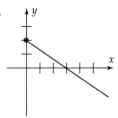

27. Odd

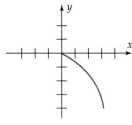

28. Odd

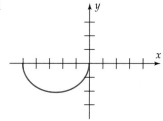

29. (a) Draw some coordinate axes and plot the points $(0, 1)$, $(1, -3)$, $(-5, 2)$, $(-3, 5)$, $(2, 3)$, and $(4, 1)$.

(b) Suppose the points in part (a) lie on the graph of an *even* function f. Plot the points $(0, f(0))$, $(-1, f(-1))$, $(5, f(5))$, $(3, f(3))$, $(-2, f(-2))$, and $(-4, f(-4))$.

30. Draw the graph of an *even* function that includes the points $(0, -3)$, $(-3, 0)$, $(2, 0)$, $(1, -4)$, $(2.5, -1)$, $(-4, 3)$, and $(-5, 3)$.*

31. (a) Plot the points $(0, 0)$, $(2, 3)$, $(3, 4)$, $(5, 0)$, $(7, -3)$, $(-1, -1)$, $(-4, -1)$, and $(-6, 1)$.

(b) Suppose the points in part (a) lie on the graph of an *odd* function f. Plot the points $(-2, f(-2))$, $(-3, f(-3))$, $(-5, f(-5))$, $(-7, f(-7))$, $(1, f(1))$, $(4, f(4))$, and $(6, f(6))$.

(c) Draw the graph of an odd function f that includes all the points plotted in parts (a) and (b).*

32. Draw the graph of an odd function that includes the points $(-3, 5)$, $(-1, 1)$, $(2, -6)$, $(4, -9)$, and $(5, -5)$.*

Thinkers

33. Show that any graph that has two of the three types of symmetry (x-axis, y-axis, origin) necessarily has the third type also.

34. Use the midpoint formula to show that $(0, 0)$ is the midpoint of the segment joining (x, y) and $(-x, -y)$. Conclude that the coordinate test for symmetry with respect to the origin (page 232) is correct.

*Many correct answers are possible.

3.5 **Operations on Functions**

We now examine ways in which two or more given functions can be used to create new functions. If f and g are functions, then their **sum** is the function h defined by the rule

$$h(x) = f(x) + g(x).$$

For example, if $f(x) = 3x^2 + x$ and $g(x) = 4x - 2$, then

$$h(x) = f(x) + g(x) = (3x^2 + x) + (4x - 2) = 3x^2 + 5x - 2.$$

Instead of using a different letter h for the sum function, we shall usually denote it by $f + g$. Thus, the sum $f + g$ is defined by the rule

$$(f + g)(x) = f(x) + g(x).$$

This rule is *not* just a formal manipulation of symbols. If x is a number, then so are $f(x)$ and $g(x)$. The plus sign in $f(x) + g(x)$ is addition of *numbers* and the result is a number. But the plus sign in $f + g$ is addition of *functions* and the result is a new function.

The **difference** $f - g$ is the function defined by the rule:

$$(f - g)(x) = f(x) - g(x).$$

The domain of the sum and difference functions is the set of all real numbers that are in both the domain of f and the domain of g.

Technology Tip

If you have two functions entered in the equation memory as y_1 and y_2 you can graph their sum by entering $y_1 + y_2$ as y_3 in the equation memory and graphing y_3. Differences, products, and quotients are graphed similarly.

   You can find y_1 and y_2 in the FUNCTION submenu of the TI-83 VARS or Y-VARS menu, in the EQVARS submenu of the Sharp 9600 VARS menu, and in the GRPH submenu of the Casio 9850 VARS menu. On other calculators, type them in from the keyboard.

Example 1 If $f(x) = \sqrt{9 - x^2}$ and $g(x) = \sqrt{x - 2}$, then

$$(f + g)(x) = \sqrt{9 - x^2} + \sqrt{x - 2}$$
$$(f - g)(x) = \sqrt{9 - x^2} - \sqrt{x - 2}.$$

The domain of f consists of all x such that $9 - x^2 \geq 0$ (so that the square root will be defined), that is, all x with $-3 \leq x \leq 3$. Similarly, the domain of g consists of all x such that $x \geq 2$. The domain of $f + g$ and $f - g$ consists of all real numbers in both the domain of f and the domain of g, namely, all x such that $2 \leq x \leq 3$. ∎

The product and quotient of functions f and g are the functions defined by the rules

$$(fg)(x) = f(x)g(x) \quad \text{and} \quad \left(\frac{f}{g}\right)(x) = \frac{f(x)}{g(x)}.$$

The domain of fg consists of all real numbers in both the domain of f and the domain of g. The domain of f/g consists of all real numbers x in both the domain of f and the domain of g such that $g(x) \neq 0$.

Example 2 If $f(x) = \sqrt{3x}$ and $g(x) = x^2 - 1$, then

$$(fg)(x) = \sqrt{3x}\,(x^2 - 1) = \sqrt{3x}\,x^2 - \sqrt{3x}$$

$$\left(\frac{f}{g}\right)(x) = \frac{\sqrt{3x}}{x^2 - 1}.$$

The domain of fg consists of all numbers x in both the domain of f (all nonnegative real numbers) and the domain of g (all real numbers), that is, all $x \geq 0$. The domain of f/g consists of all these x for which $g(x) \neq 0$, that is, all nonnegative real numbers *except $x = 1$*. ∎

If c is a real number and f is a function, then the product of f and the constant function $g(x) = c$ is usually denoted cf. For example, if the function $f(x) = x^3 - x + 2$, and $c = 5$, then $5f$ is the function given by

$$(5f)(x) = 5 \cdot f(x) = 5(x^3 - x + 2) = 5x^3 - 5x + 10.$$

Composition of Functions

Another way of combining functions is illustrated by the function $h(x) = \sqrt{x^3}$. To compute $h(4)$, for example, you first find $4^3 = 64$ and then take the square root $\sqrt{64} = 8$. So the rule of h may be rephrased as:

First apply the function $f(x) = x^3$,

Then apply the function $g(t) = \sqrt{t}$ to the result.

The same idea can be expressed in functional notation like this:

$$
\begin{array}{ccc}
\textit{first apply f} & \textit{then apply g to the result} & \\
x \xrightarrow{\hspace{2cm}} f(x) & \xrightarrow{\hspace{3cm}} & g(f(x)) \\
x & x^3 & \sqrt{x^3} \\
\end{array}
$$

apply h

So the rule of h may be written as $h(x) = g(f(x))$, where $f(x) = x^3$ and $g(t) = \sqrt{t}$. We can think of h as being made up of two simpler functions f and g, or we can think of f and g being "composed" to create the function h. Both viewpoints are useful.

Example 3 Suppose $f(x) = 4x^2 + 1$ and $g(t) = \dfrac{1}{t + 2}$. Define a new function h whose rule is "first apply f; then apply g to the result." In functional notation

$$
\begin{array}{ccc}
\textit{first apply f} & \textit{then apply g to the result} & \\
x \xrightarrow{\hspace{2cm}} f(x) & \xrightarrow{\hspace{3cm}} & g(f(x)) \\
\end{array}
$$

So the rule of the function h is $h(x) = g(f(x))$. Evaluating $g(f(x))$ means that whenever t appears in the formula for $g(t)$, we must replace it by $f(x) = 4x^2 + 1$:

$$h(x) = g(f(x)) = \frac{1}{f(x) + 2} = \frac{1}{(4x^2 + 1) + 2} = \frac{1}{4x^2 + 3}. \quad \blacksquare$$

The function h in Example 3 is an illustration of the following definition.

Composite Functions

> Let f and g be functions. The *composite function* of f and g is given by:
>
> **For input x, the output is $g(f(x))$.**
>
> This composite function is denoted $g \circ f$.

The symbol "$g \circ f$" is read "g circle f" or "f followed by g." (Note the order carefully; the functions are applied *right* to *left*.) So the rule of the composite function is:

$$(g \circ f)(x) = g(f(x)).$$

Example 4 If $f(x) = 2x + 5$ and $g(t) = 3t^2 + 2t + 4$, then

$$(f \circ g)(2) = f(g(2)) = f(3 \cdot 2^2 + 2 \cdot 2 + 4) = f(20) = 2 \cdot 20 + 5 = 45.$$

Similarly,

$$(g \circ f)(-1) = g(f(-1)) = g(2(-1) + 5) = g(3) = 3 \cdot 3^2 + 2 \cdot 3 + 4 = 37.$$

The value of a composite function can also be computed like this:

$$(g \circ f)(5) = g(f(5)) = 3(f(5)^2) + 2(f(5)) + 4 = 3(15^2) + 2(15) + 4 = 709.$$
$$\blacksquare$$

The domain of $g \circ f$ is determined by the usual convention:

Domains of Composite Functions

> The domain of the composite function $g \circ f$ is the set of all real numbers x such that x is in the domain of f and $f(x)$ is in the domain of g.

Example 5 If $f(x) = \sqrt{x}$ and $g(t) = t^2 - 5$, then

$$(g \circ f)(x) = g(f(x)) = (f(x))^2 - 5 = \left(\sqrt{x}\right)^2 - 5 = x - 5.$$

Although $x - 5$ is defined for every real number x, the domain of $g \circ f$ is *not* the set of all real numbers. The domain of g is the set of all real numbers, but the function $f(x) = \sqrt{x}$ is defined only when $x \geq 0$. So the domain of $g \circ f$ is the set of nonnegative real numbers. $\blacksquare$

Technology Tip

Evaluating composite functions is easy on TI-82/83/86/89, Sharp 9600, and HP-38. If the functions are entered in the equation memory as $y_1 = g(x)$ and $y_2 = h(x)$ (with f in place of y on HP-38), then keying in $y_2(y_1(5))$ ENTER produces the number $h(g(5))$.

On other calculators (including TI-85) this syntax does *not* produce $h(g(5))$; it produces $h(x) \cdot g(x) \cdot 5$ for whatever number is stored in the x-memory.

Example 6 If $h(x) = \sqrt{3x^2 + 1}$, then h may be considered as the composite $g \circ f$, where $f(x) = 3x^2 + 1$ and $g(u) = \sqrt{u}$ because

$$(g \circ f)(x) = g(f(x)) = g(3x^2 + 1) = \sqrt{3x^2 + 1} = h(x).$$

There are other ways to consider $h(x) = \sqrt{3x^2 + 1}$ as a composite function. For instance, h is also the composite $j \circ k$, where $j(x) = \sqrt{x + 1}$ and $k(x) = 3x^2$:

$$(j \circ k)(x) = j(k(x)) = j(3x^2) = \sqrt{3x^2 + 1} = h(x). \quad \blacksquare$$

Example 7 If $k(x) = (x^2 - 2x + \sqrt{x})^3$, then k is $g \circ f$, where $f(x) = x^2 - 2x + \sqrt{x}$ and $g(t) = t^3$ because

$$(g \circ f)(x) = g(f(x)) = g(x^2 - 2x + \sqrt{x}) = (x^2 - 2x + \sqrt{x})^3 = k(x). \quad \blacksquare$$

As you may have noticed, there are two possible ways to form a composite function from two given functions. If f and g are functions, we can consider either

or

$$(g \circ f)(x) = g(f(x)) \qquad \text{[The composite of } f \text{ and } g\text{]}$$

$$(f \circ g)(x) = f(g(x)) \qquad \text{[The composite of } g \text{ and } f\text{]}$$

The *order is important,* as we shall now see:

$g \circ f$ and $f \circ g$ usually are *not* the same function.

Example 8 If $f(x) = x^2$ and $g(x) = x + 3$,* then

$$(g \circ f)(x) = g(f(x)) = g(x^2) = x^2 + 3$$

but

$$(f \circ g)(x) = f(g(x)) = f(x + 3) = (x + 3)^2 = x^2 + 6x + 9.$$

Obviously, $g \circ f \neq f \circ g$ since, for example, they have different values at $x = 0$. $\quad \blacksquare$

CAUTION

Don't confuse the product function fg with the composite function $f \circ g$ (g followed by f). For instance, if $f(x) = 2x^2$ and $g(x) = x - 3$, then the product fg is given by:

$$(fg)(x) = f(x)g(x) = 2x^2(x - 3) = 2x^3 - 6x^2.$$

It is *not* the same as the composite $f \circ g$ because

$$(f \circ g)(x) = f(g(x)) = f(x - 3) = 2(x - 3)^2 = 2x^2 - 12x + 18.$$

---

*Now that you have the idea of composite functions, we'll use the same letter for the variable in both functions.

Applications

Composition of functions arises in applications involving several functional relationships simultaneously. In such cases one quantity may have to be expressed as a function of another.

Example 9 A circular puddle of liquid is evaporating and slowly shrinking in size. After t minutes, the radius r of the puddle measures $\dfrac{18}{2t+3}$ inches; in other words, the radius is a function of time. The area A of the puddle is given by $A = \pi r^2$, that is, area is a function of the radius r. We can express the area as a function of time by substituting $r = \dfrac{18}{2t+3}$ in the area equation:

$$A = \pi r^2 = \pi\left(\frac{18}{2t+3}\right)^2.$$

This amounts to forming the composite function $f \circ g$, where $f(r) = \pi r^2$ and $g(t) = \dfrac{18}{2t+3}$:

$$(f \circ g)(t) = f(g(t)) = f\left(\frac{18}{2t+3}\right) = \pi\left(\frac{18}{2t+3}\right)^2.$$

When area is expressed as a function of time, it is easy to compute the area of the puddle at any time. For instance, after 12 minutes the area of the puddle is

$$A = \pi\left(\frac{18}{2t+3}\right)^2 = \pi\left(\frac{18}{2\cdot 12 + 3}\right)^2 = \frac{4\pi}{9} \approx 1.396 \text{ sq in.} \quad \blacksquare$$

Exercises 3.5

In Exercises 1–4, find $(f+g)(x)$, $(f-g)(x)$, and $(g-f)(x)$.

1. $f(x) = -3x + 2, \quad g(x) = x^3$

2. $f(x) = x^2 + 2, \quad g(x) = -4x + 7$

3. $f(x) = 1/x, \quad g(x) = x^2 + 2x - 5$

4. $f(x) = \sqrt{x}, \quad g(x) = x^2 + 1$

In Exercises 5–8, find $(fg)(x)$, $(f/g)(x)$, and $(g/f)(x)$.

5. $f(x) = -3x + 2, \quad g(x) = x^3$

6. $f(x) = 4x^2 + x^4, \quad g(x) = \sqrt{x^2 + 4}$

7. $f(x) = x^2 - 3, \quad g(x) = \sqrt{x - 3}$

8. $f(x) = \sqrt{x^2 - 1}, \quad g(x) = \sqrt{x - 1}$

In Exercises 9–12, find the domains of fg and f/g.

9. $f(x) = x^2 + 1, \quad g(x) = 1/x$

10. $f(x) = x + 2, \quad g(x) = \dfrac{1}{x + 2}$

11. $f(x) = \sqrt{4 - x^2}, \quad g(x) = \sqrt{3x + 4}$

12. $f(x) = 3x^2 + x^4 + 2, \quad g(x) = 4x - 3$

In Exercises 13–16, find the indicated values, where $g(t) = t^2 - t$ and $f(x) = 1 + x$.

13. $g(f(0))$ **14.** $(f \circ g)(3)$

15. $g(f(2) + 3)$ **16.** $f(2g(1))$

In Exercises 17–20, find $(g \circ f)(3)$, $(f \circ g)(1)$, *and* $(f \circ f)(0)$.

17. $f(x) = 3x - 2,$ $g(x) = x^2$

18. $f(x) = |x + 2|,$ $g(x) = -x^2$

19. $f(x) = x,$ $g(x) = -3$

20. $f(x) = x^2 - 1,$ $g(x) = \sqrt{x}$

In Exercises 21–24, find the rule of the function $f \circ g$, *the domain of* $f \circ g$, *the rule of* $g \circ f$, *and the domain of* $g \circ f$.

21. $f(x) = x^2,$ $g(x) = x + 3$

22. $f(x) = -3x + 2,$ $g(x) = x^3$

23. $f(x) = 1/x,$ $g(x) = \sqrt{x}$

24. $f(x) = \dfrac{1}{2x + 1},$ $g(x) = x^2 - 1$

In Exercises 25–28, find the rules of the functions ff and $f \circ f$.

25. $f(x) = x^3$

26. $f(x) = (x - 1)^2$

27. $f(x) = 1/x$

28. $f(x) = \dfrac{1}{x - 1}$

In Exercises 29–32, verify that $(f \circ g)(x) = x$ *and* $(g \circ f)(x) = x$ *for every x.*

29. $f(x) = 9x + 2,$ $g(x) = \dfrac{x - 2}{9}$

30. $f(x) = \sqrt[3]{x - 1},$ $g(x) = x^3 + 1$

31. $f(x) = \sqrt[3]{x} + 2,$ $g(x) = (x - 2)^3$

32. $f(x) = 2x^3 - 5,$ $g(x) = \sqrt[3]{\dfrac{x + 5}{2}}$

Exercises 33 and 34 refer to the function f whose graph is shown in the figure.

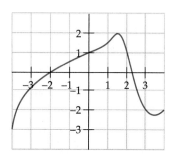

33. Let g be the composite function $f \circ f$ [that is, $g(x) = (f \circ f)(x) = f(f(x))$]. Use the graph of f to fill in the following table (approximate where necessary).

x	$f(x)$	$g(x) = f(f(x))$
-4		
-3		
-2	0	1
-1		
0		
1		
2		
3		
4		

34. Use the information obtained in Exercise 33 to sketch the graph of the function g.

In Exercises 35–38, fill the blanks in the given table. In each case the values of the functions f and g are given by these tables:

x	$f(x)$
1	3
2	5
3	1
4	2
5	3

t	$g(t)$
1	5
2	4
3	4
4	3
5	2

35.

x	$(g \circ f)(x)$
1	4
2	
3	5
4	
5	

36.

t	$(f \circ g)(t)$
1	
2	2
3	
4	
5	

37.

x	$(f \circ f)(x)$
1	
2	
3	3
4	
5	

38.

t	$(g \circ g)(t)$
1	
2	
3	
4	4
5	

In Exercises 39–44, write the given function as the composite of two functions, neither of which is the identity function, as in Examples 6 and 7. (There may be more than one way to do this.)

39. $f(x) = \sqrt[3]{x^2 + 2}$

40. $g(x) = \sqrt{x + 3} - \sqrt[3]{x + 3}$

41. $h(x) = (7x^3 - 10x + 17)^7$

42. $k(x) = \sqrt[3]{(7x - 3)^2}$

43. $f(x) = \dfrac{1}{3x^2 + 5x - 7}$

44. $g(t) = \dfrac{3}{\sqrt{t - 3}} + 7$

45. $f(x) = (x + 5)^2 + \dfrac{1}{x + 5}$

46. Give an example of a function f such that

$$f\left(\frac{1}{x}\right) \neq \frac{1}{f(x)}.$$

In Exercises 47 and 48, graph both $f \circ g$ and $g \circ f$ on the same screen. Use the graphs to determine whether $f \circ g$ is the same function as $g \circ f$.

47. $f(x) = x^5 - x^3 - x; \quad g(x) = x - 2$

48. $f(x) = x^3 + x; \quad g(x) = \sqrt[3]{x - 1}$

49. **(a)** What is the area of the puddle in Example 9 after one day? After a week? After a month?

(b) Does the puddle ever totally evaporate? Is this realistic? Under what circumstances might this area function be an accurate model of reality?

50. In a laboratory culture, the number $N(d)$ of bacteria (in thousands) at temperature d degrees Celsius is given by the function

$$N(d) = \frac{-90}{d + 1} + 20 \quad (4 \leq d \leq 32).$$

The temperature $D(t)$ at time t hours is given by the function $D(t) = 2t + 4 \quad (0 \leq t \leq 14)$.

(a) What does the composite function $N \circ D$ represent?

(b) How many bacteria are in the culture after 4 hours? After 10 hours?

51. A certain fungus grows in a circular shape. Its diameter after t weeks is $6 - \dfrac{50}{t^2 + 10}$ inches.

(a) Express the area covered by the fungus as a function of time.

(b) What is the area covered by the fungus when $t = 0$? What area does it cover at the end of 8 weeks?

(c) When is its area 25 square inches?

52. Tom left point P at 6 A.M. walking south at 4 miles per hour. Anne left point P at 8 A.M. walking west at 3.2 miles per hour.

(a) Express the distance between Tom and Anne as a function of the time t elapsed since 6 A.M.

(b) How far apart are Tom and Anne at noon?

(c) At what time are they 35 miles apart?

53. As a weather balloon is inflated its radius increases at the rate of 4 cm per second. Express the volume of the balloon as a function of time and determine the volume of the balloon after 4 seconds. [*Hint:* The volume of a sphere of radius r is $4\pi r^3/3$.]

54. Express the surface area of the weather balloon in Exercise 53 as a function of time. [*Hint:* The surface area of a sphere of radius r is $4\pi r^2$.]

3.6   Average Rates of Change

Rates of change play a central role in the analysis of many real-world situations. To understand the basic ideas involved in rates of change, we take another look at the falling rock from Sections 3.1 and 3.2. We saw that when the rock is dropped from a high place, then the distance it travels (ignoring wind resistance) is given by the function.

$$d(t) = 16t^2$$

with distance $d(t)$ measured in feet and time t in seconds. The following table shows the distance the rock has fallen at various times:

Time t	0	1	2	3	3.5	4	4.5	5
Distance $d(t)$	0	16	64	144	196	256	324	400

To find the distance the rock falls from time $t = 1$ to $t = 3$, we note that at the end of three seconds, the rock has fallen $d(3) = 144$ feet, whereas it had only fallen $d(1) = 16$ feet at the end of one second. So during this time interval the rock traveled

$$d(3) - d(1) = 144 - 16 = 128 \text{ feet.}$$

The distance traveled by the rock during other time intervals can be found similarly:

Time Interval	Distance Traveled
$t = 1$ to $t = 4$	$d(4) - d(1) = 256 - 16 = 240$
$t = 2$ to $t = 3.5$	$d(3.5) - d(2) = 196 - 64 = 132$
$t = 2$ to $t = 4.5$	$d(4.5) - d(2) = 324 - 64 = 260$

The same procedure works in general:

The distance traveled from time $t = a$ to time $t = b$ is $d(b) - d(a)$ feet.

In the preceding chart, the length of each time interval can be computed by taking the difference between the two times. For example, from $t = 1$ to $t = 4$ is a time interval of length $4 - 1 = 3$ seconds. Similarly, the interval from $t = 2$ to $t = 3.5$ is of length $3.5 - 2 = 1.5$ seconds and in general,

The time interval from $t = a$ to $t = b$ is an interval of $b - a$ seconds.

Since distance = average speed × time,

$$\text{average speed} = \frac{\text{distance traveled}}{\text{time interval}}.$$

Hence, the average speed over the time interval from $t = a$ to $t = b$ is

$$\text{average speed} = \frac{\text{distance traveled}}{\text{time interval}} = \frac{d(b) - d(a)}{b - a}.$$

For example, to find the average speed from $t = 1$ to $t = 4$, apply the preceding formula with $a = 1$ and $b = 4$:

$$\text{average speed} = \frac{d(4) - d(1)}{4 - 1} = \frac{256 - 16}{4 - 1} = \frac{240}{3} = 80 \text{ ft per sec.}$$

Similarly, the average speed from $t = 2$ to $t = 4.5$ is

$$\frac{d(4.5) - d(2)}{4.5 - 2} = \frac{324 - 64}{4.5 - 2} = \frac{260}{2.5} = 104 \text{ ft per sec.}$$

The units in which average speed is measured here (feet per second) indicate the number of units of distance traveled during each unit of time, that is, the *rate of change* of distance (feet) with respect to time (seconds). The preceding discussion can be summarized by saying that the average speed (rate of change of distance with respect to time) as time changes from $t = a$ to $t = b$ is given by

$$\text{average speed} = \text{average rate of change}$$

$$= \frac{\text{change in distance}}{\text{change in time}} = \frac{d(b) - d(a)}{b - a}.$$

Although speed is the most familiar example, rates of change play a role in many other situations as well, as illustrated in Examples 1–3 below. Consequently, we define the average rate of change of any function as follows.

Average Rate of Change

Let f be a function. The *average rate of change of* $f(x)$ **with respect to x as x changes from a to b is the number**

$$\frac{\text{change in } f(x)}{\text{change in } x} = \frac{f(b) - f(a)}{b - a}.$$

Example 1 A large heavy-duty balloon is being filled with water. Its approximate volume (in gallons) is given by

$$V(x) = \frac{x^3}{55},$$

where x is the radius of the balloon (in inches). Find the average rate of change of the volume of the balloon as the radius increases from 5 to 10 inches.

Solution

$$\frac{\text{change in volume}}{\text{change in radius}} = \frac{V(10) - V(5)}{10 - 5} \approx \frac{18.18 - 2.27}{10 - 5} = \frac{15.91}{5}$$

$$= 3.182 \text{ gallons per inch.} \qquad \blacksquare$$

Example 2   According to data from Dun & Bradstreet, the number of new businesses incorporated each year in the United States can be approximated by

$$f(x) = -.035x^4 + 1.51x^3 - 20.6x^2 + 102x + 531 \quad (0 \le x \le 17),$$

where $x = 0$ corresponds to 1980 and $f(x)$ is in thousands. Find the average rate of change in the number of business incorporations during the following time periods:

(a)   1980 to 1997   (b)   1984 to 1990.

Solution

(a)   Since 1980 corresponds to $x = 0$ and 1997 to $x = 17$, we have

$$\text{average rate of change} = \frac{f(17) - f(0)}{17 - 0} \approx \frac{807 - 531}{17} = \frac{276}{17} \approx 16.235.$$

Since $f(x)$ is measured in thousands, this means that incorporations were increasing at an average rate of about 16,235 per year during this period.

(b)   During this period ($x = 4$ to $x = 10$) the average rate of change was

$$\frac{f(10) - f(4)}{10 - 4} \approx \frac{651 - 697.08}{6} = \frac{-46.08}{6} = -7.68.$$

The rate is negative, which means that incorporations were *decreasing* at an average rate of about 7680 per year during this period. ∎

Example 3   Figure 3–38 is the graph of the temperature function f during a particular day; $f(x)$ is the temperature at x hours after midnight. What is the average rate of change of the temperature **(a)** from 4 A.M. to noon? **(b)** from 3 P.M. to 8 P.M.?

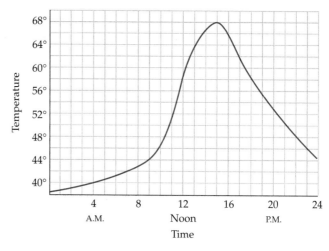

Figure 3–38

Solution

(a)   The graph shows that the temperature at 4 A.M. is $f(4) = 40°$ and the temperature at noon is $f(12) = 58°$. The average rate of change of temperature is

$$\frac{\text{change in temperature}}{\text{change in time}} = \frac{f(12) - f(4)}{12 - 4} = \frac{58 - 40}{12 - 4} = \frac{18}{8}$$

$$= 2.25° \text{ per hour.}$$

The rate of change is positive because the temperature is increasing at an average rate of 2.25° per hour.

(b)   Now 3 P.M. corresponds to $x = 15$ and 8 P.M. to $x = 20$. The graph shows that $f(15) = 68°$ and $f(20) = 53°$. Hence the average rate of change of temperature is:

$$\frac{\text{change in temperature}}{\text{change in time}} = \frac{f(20) - f(15)}{20 - 15} = \frac{53 - 68}{20 - 15} = \frac{-15}{5}$$

$$= -3° \text{ per hour.}$$

The rate of change is negative because the temperature is decreasing at an average rate of 3° per hour. ■

Geometric Interpretation of Average Rate of Change

If P and Q are points on the graph of a function f, then the straight line determined by P and Q is called a **secant line.** Figure 3–39 shows the secant line joining the points (4, 40) and (12, 58) on the graph of the temperature function f of Example 3.

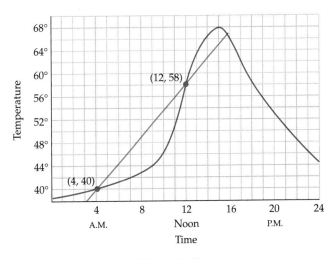

Figure 3–39

Using the points (4, 40) and (12, 58), we see that the slope of this secant line is $\dfrac{58 - 40}{12 - 4} = \dfrac{18}{8} = 2.25$. To say that (4, 40) and (12, 58) are on the graph of f means that $f(4) = 40$ and $f(12) = 58$. Thus,

$$\text{slope of secant line} = 2.25 = \frac{58 - 40}{12 - 4} = \frac{f(12) - f(4)}{12 - 4}$$

$$= \text{average rate of change as } x \text{ goes from 4 to 12.}$$

The same thing happens in the general case:

Secant Lines and Average Rates of Change

If f is a function, then the average rate of change of $f(x)$ with respect to x as x changes from $x = a$ to $x = b$ is the slope of the secant line joining the points $(a, f(a))$ and $(b, f(b))$ on the graph of f.

The fact in the box makes it easy to determine certain rates of change, as the next example illustrates.

Example 4 Current data and projections from the Office of Management and Budget show that between 1990 and 2004, the national debt (the amount the United States government owes) can be approximated by

$$f(x) = 199x + 3760 \quad (0 \le x \le 14),$$

where $x = 0$ corresponds to 1990 and $f(x)$ is in billions of dollars. At what average rate is the national debt changing from year to year during this period?

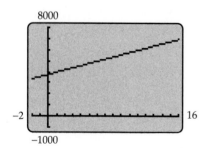

Figure 3–40

Solution   The graph of $f(x) = 199x + 3760$ is a straight line (Figure 3–40). The secant line joining any two points on the graph is just *the graph itself*, that is, the line $y = 199x + 3760$. As we saw in Section 1.4, the slope of this line is 199. Therefore, the average rate of change of the national debt function between any two values of x is 199. In other words, at any time during this period, the national debt is increasing at a constant rate of $199 billion per year. ■

The argument used in Example 4 works for any function whose graph is a straight line and leads to this conclusion:

The average rate of change of a linear function $f(x) = mx + b$, as x changes from one value to another, is the slope m of the line.

Exercises 3.6

1. A car moves along a straight test track. The distance traveled by the car at various times is shown in this table:

Time (sec)	0	5	10	15	20	25	30
Distance (ft)	0	20	140	400	680	1400	1800

Find the average speed of the car over the interval from
(a) 0 to 10 sec   (b) 10 to 20 sec
(c) 20 to 30 sec   (d) 15 to 30 sec

2. The yearly profit of a small manufacturing firm is shown in the following tables.

Year	1986	1987	1988	1989
Profit	$5000	$6000	$6500	$6800

Year	1990	1991	1992	1993
Profit	$7200	$6700	$6500	$7000

What is the average rate of change of profits over the given time span?
(a) 1986–1990   (b) 1986–1993
(c) 1989–1992   (d) 1988–1992

3. The graph shows the total amount spent on advertising (in millions of dollars) in the United States, as estimated by a leading advertising firm. Find the average rate of change in advertising expenditures over the following time periods:
(a) 1950–1970   (b) 1970–1980
(c) 1980–1997   (d) 1950–1997
(e) During which of these periods were expenditures increasing at the fastest rate?

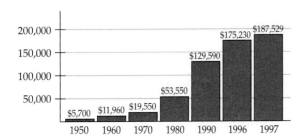

4. The table shows the total elementary and secondary school enrollment (in thousands) for selected years (projections used for 1996–2008). (*Source:* U.S. Department of Education)

Year	Enrollment
1964	47,716
1972	50,726
1980	46,208
1988	45,430
1996	51,413
2000	53,445
2008	54,268

Find the average rate of change of enrollment from
(a) 1964 to 1972   (b) 1972 to 1980
(c) 1980 to 1988   (d) 1996 to 2000
(e) During which of these periods was enrollment increasing at the fastest rate? At the slowest rate?
(f) During which of these periods was enrollment decreasing at the fastest rate? At the slowest rate?

5. The graph shows the hourly minimum wage (adjusted to 1997 dollars), with $x = 0$ corresponding to 1940. Find the average rate of change in the minimum wage from
(a) 1940 to 1968   (b) 1968 to 1992
(c) 1992 to 1996   (d) 1968 to 1996

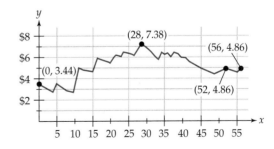

6. The table on the next page shows the total number of shares traded (in thousands) on the New York Stock Exchange in selected years. (*Source:* U.S. Securities and Exchange Commission) Find the average rate of change in share volume from

(a) 1960 to 1980   (b) 1980 to 1990
(c) 1990 to 1996   (d) 1980 to 1996
(e) During which of these periods did share volume increase at the fastest rate?

Year	Volume
1960	1,411,120
1970	4,834,887
1980	15,587,986
1990	53,746,087
1996	125,922,577

7. The graph in the figure shows the monthly sales of floral pattern ties (in thousands of ties) made by Neckwear, Inc., over a 48-month period. Sales are very low when the ties are first introduced, increase significantly, hold steady for a while, and then drop off as the ties go out of fashion. Find the average rate of change of sales (in ties per month) over the interval:
(a) 0 to 12   (b) 8 to 24   (c) 12 to 24
(d) 20 to 28   (e) 28 to 36   (f) 32 to 44
(g) 36 to 40   (h) 40 to 48

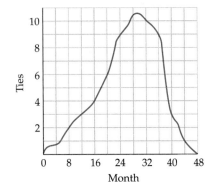

8. The XYZ Company has found that its sales are related to the amount of advertising it does in trade magazines. The graph in the figure shows the sales (in thousands of dollars) as a function of the amount of advertising (in number of magazine ad pages). Find the average rate of change of sales when the number of ad pages increases from
(a) 10 to 20   (b) 20 to 60
(c) 60 to 100   (d) 0 to 100
(e) Is it worthwhile to buy more than 70 pages of ads, if the cost of a one-page ad is $2000? If the cost is $5000? If the cost is $8000?

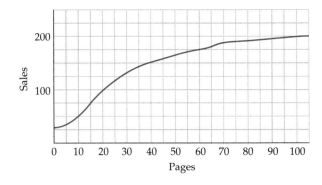

9. Find the average rate of change of the volume of the balloon in Example 1 as the radius increases from
(a) 2 to 5 inches   (b) 4 to 8 inches

10. When blood flows through an artery (which can be thought of as a cylindrical tube) its velocity is greatest at the center of the artery. Because of friction along the walls of the tube, the blood's velocity decreases as the distance r from the center of the artery increases, finally becoming 0 at the wall of the artery. The velocity (in centimeters per second) is given by the function $v = 18{,}500(.000065 - r^2)$, where r is measured in centimeters. Find the average rate of change of the velocity as the distance from the center changes from
(a) $r = .001$ to $r = .002$ (b) $r = .002$ to $r = .003$
(c) $r = 0$ to $r = .025$

11. A car is stopped at a traffic light and begins to move forward along a straight road when the light turns green. The distance (in feet) traveled by the car in t seconds is given by $s(t) = 2t^2$ $(0 \le t \le 30)$. What is the average speed of the car from
(a) $t = 0$ to $t = 5$?   (b) $t = 5$ to $t = 10$?
(c) $t = 10$ to $t = 30$?

In Exercises 12–20, find the average rate of change of the function f over the given interval.

12. $f(x) = 15x - 6$ from $x = 97$ to $x = 107$
13. $f(x) = -72x + 144$ from $x = 37$ to $x = 99$
14. $f(x) = 13x + 8$ from $x = -12$ to $x = 15$
15. $f(x) = 2 - x^2$ from $x = 0$ to $x = 2$
16. $f(x) = .25x^4 - x^2 - 2x + 4$ from $x = -1$ to $x = 4$
17. $f(x) = x^3 - 3x^2 - 2x + 6$ from $x = -1$ to $x = 3$
18. $f(x) = -\sqrt{x^4 - x^3 + 2x^2 - x + 4}$ from $x = 0$ to $x = 3$
19. $f(x) = \sqrt{x^3 + 2x^2 - 6x + 5}$ from $x = 1$ to $x = 2$
20. $f(x) = \dfrac{x^2 - 3}{2x - 4}$ from $x = 3$ to $x = 6$

21. Two cars race on a straight track, beginning from a dead stop. The distance (in feet) each car has covered at each time during the first 16 seconds is shown in the figure.

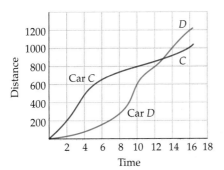

(a) What is the average speed of each car during this 16-second interval?

(b) Find an interval beginning at $t = 4$ during which the average speed of car D was approximately the same as the average speed of car C from $t = 2$ to $t = 10$.

(c) Use secant lines and slopes to justify the statement "car D traveled at a higher average speed than car C from $t = 4$ to $t = 10$."

22. The figure shows the profits earned by a certain company during the last quarters of three consecutive years.

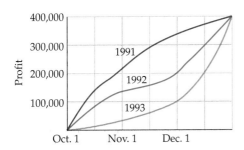

(a) Explain why the average rate of change of profits from October 1 to December 31 was the same in all three years.

(b) During what month in what year was the average rate of change of profits the greatest?

23. The graph shows the chipmunk population in a certain wilderness area. The population increases as the chipmunks reproduce, but then decreases sharply as predators move into the area.

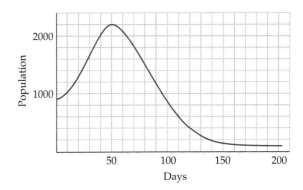

(a) During what approximate time period (beginning on day 0) is the average growth rate of the chipmunk population positive?

(b) During what approximate time period, beginning on day 0, is the average growth rate of the chipmunk population 0?

(c) What is the average growth rate of the chipmunk population from day 50 to day 100? What does this number mean?

(d) What is the average growth rate from day 45 to day 50? From day 50 to day 55? What is the approximate average growth rate from day 49 to day 51?

24. Lucy has a viral flu. How bad she feels depends primarily on how fast her temperature is rising at that time. The graph shows her temperature during the first day of the flu.

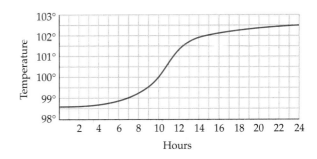

(a) At what average rate does her temperature rise during the entire day?

(b) During what 2-hour period during the day does she feel worst?

(c) Find two time intervals, one in the morning and one in the afternoon, during which she feels about the same (that is, during which her temperature is rising at the same average rate).

25. The table on the next page shows the median weekly earnings of full-time workers, ages 25 and over, who have had four or more years of college. (*Source:* U.S. Bureau of Labor Statistics)

Year	Earnings
1980	$376
1984	$486
1988	$585

Year	Earnings
1992	$697
1996	$758
1997	$779

(a) Make a scatter plot of the data, with $x = 0$ corresponding to 1980.
(b) Use linear regression to find a function that models this data.
(c) According to your function, what is the average rate of change in earnings over any time period between 1980 and 1997?
(d) Use the data to find the average rate of change in earnings from 1980 to 1997 and from 1992 to 1997. How do these rates compare with the rate given by the model?

26. The table shows the number of trips (in millions) for both business and pleasure taken by

Americans in selected years. (*Source:* Travel Industry Association of America)

Year	Trips
1985	808
1987	894
1989	945
1991	980

Year	Trips
1993	1058
1995	1173
1996	1161

(a) Make a scatter plot of the data, with $x = 0$ corresponding to 1985.
(b) Use linear regression to find a function that models this data.
(c) According to your function, what is the average rate of change in the number of trips over any time period between 1985 and 1996?
(d) Use the data to find the average rate of change in the number of trips from 1985 to 1991 and from 1991 to 1996. How do these rates compare with the rate given by the model?

 3.6.A

EXCURSION   Instantaneous Rates of Change

Recall that the distance (in feet) traveled by a falling rock in t seconds is given by $d(t) = 16t^2$. Consider the distances fallen by the rock over various time intervals:

Time Interval	Distance Traveled
$t = 0$ to $t = 1$	$d(1) - d(0) = 16 - 0 = 16$ feet
$t = 1$ to $t = 2$	$d(2) - d(1) = 64 - 16 = 48$ feet
$t = 2$ to $t = 3$	$d(3) - d(2) = 144 - 64 = 80$ feet

During each second, the rock falls farther than it did in the preceding second. So it must be traveling at faster and faster speeds as it falls. We know how to compute the *average* speed of the rock over any time interval, but what about its speed at a particular instant?

To make this discussion concrete, consider the exact speed of the rock at the instant when $t = 4$. We could approximate this speed by looking at a very small time interval near $t = 4$, say, from 4 to 4.01, or from 4 to 4.001. The average speed over such a short time interval cannot differ much from the exact speed at any time during the interval, so these average speeds should be reasonable approximations of the exact speed at $t = 4$. Note that

$$4.01 = 4 + .01 \qquad \text{and} \qquad 4.001 = 4 + .001.$$

Thus, we would do essentially the same thing in both cases: Compute the average speed (average rate of change of distance with respect to time) over an interval from 4 to $4 + h$, for some small nonzero quantity h.

Example 1 Find the average speed of the falling rock from $t = 4$ to $t = 4 + h$, where h is a small nonzero quantity. Use the result to compute the average speed of the rock from 4 to 4.01 seconds and from 4 to 4.001 seconds.

Solution Using the average speed (rate of change) formula from Section 3.6, we have:

$$\text{average speed} = \frac{d(4 + h) - d(4)}{(4 + h) - 4} = \frac{16(4 + h)^2 - 16 \cdot 4^2}{h}$$

$$= \frac{16(16 + 8h + h^2) - 256}{h} = \frac{256 + 128h + 16h^2 - 256}{h}$$

$$= \frac{128h + 16h^2}{h} = \frac{h(128 + 16h)}{h} = 128 + 16h.$$

Therefore, from 4 to 4.01 seconds (that is, 4 to $4 + h$, with $h = .01$), the rock's average speed is

$$128 + 16h = 128 + 16(.01) = 128.16 \text{ feet per second.}$$

Similarly, its average speed from 4 to 4.001 seconds (here $h = .001$) is

$$128 + 16h = 128 + 16(.001) = 128.016 \text{ feet per second.}$$

These results suggest that the exact speed of the rock when $t = 4$ is about 128 feet per second. ■

To approximate the speed of the rock at other times, we could perform a calculation similar to that in Example 1, with another number in place of 4. We can cover all the cases at once by considering a time interval from x to $x + h$.

Example 2 Find the average speed of the falling rock from time x to time $x + h$, where x is a fixed number and h is a small nonzero quantity.

Solution The average rate of change formula shows that*

$$\text{average speed} = \frac{d(x + h) - d(x)}{(x + h) - x} = \frac{16(x + h)^2 - 16x^2}{h}$$

$$= \frac{16(x^2 + 2xh + h^2) - 16x^2}{h} = \frac{16x^2 + 32xh + 16h^2 - 16x^2}{h}$$

$$= \frac{32xh + 16h^2}{h} = \frac{h(32x + 16h)}{h} = 32x + 16h. \quad ■$$

---

*Note that this calculation is the same as that in Example 1, except that 4 has been replaced by x.

When $x = 4$, then the formula in Example 2 states that the average speed from 4 to $4 + h$ is $32(4) + 16h = 128 + 16h$, which is exactly what we found in Example 1. To find the average speed from 3 to 3.1 seconds, apply the formula

$$\text{average speed} = 32x + 16h$$

with $x = 3$ and $h = .1$:

$$\text{average speed} = 32 \cdot 3 + 16(.1) = 96 + 1.6 = 97.6 \text{ ft per sec.}$$

More generally, we can compute the average rate of change of any function f over the interval from x to $x + h$ just as we did in Example 2: Apply the definition of average rate of change in the box on page 243 with x in place of a and $x + h$ in place of b:

$$\text{average rate of change} = \frac{f(b) - f(a)}{b - a} = \frac{f(x + h) - f(x)}{(x + h) - x}$$

$$= \frac{f(x + h) - f(x)}{h}.$$

This last quantity is just the difference quotient of f (see page 198). Therefore,

Difference Quotients and Rates of Change

▶

> If f is a function, then the average rate of change of f over the interval from x to $x + h$ is given by the difference quotient
>
> $$\frac{f(x + h) - f(x)}{h}.$$

Example 3 Find the difference quotient of $V(x) = x^3/55$ and use it to find the average rate of change of V as x changes from 8 to 8.01.

Solution Use the definition of the difference quotient and algebra:

$$\frac{V(x + h) - V(x)}{h} = \frac{\overbrace{\frac{(x + h)^3}{55}}^{V(x+h)} - \overbrace{\frac{x^3}{55}}^{V(x)}}{h} = \frac{\frac{1}{55}\left[(x + h)^3 - x^3\right]}{h}$$

$$= \frac{1}{55} \cdot \frac{(x + h)^3 - x^3}{h} = \frac{1}{55} \cdot \frac{x^3 + 3x^2h + 3xh^2 + h^3 - x^3}{h}$$

$$= \frac{1}{55} \cdot \frac{3x^2h + 3xh^2 + h^3}{h} = \frac{1}{55} \cdot \frac{h(3x^2 + 3xh + h^2)}{h}$$

$$= \frac{3x^2 + 3xh + h^2}{55}.$$

When x changes from 8 to $8.01 = 8 + .01$, we have $x = 8$ and $h = .01$. So the average rate of change is

$$\frac{3x^2 + 3xh + h^2}{55} = \frac{3 \cdot 8^2 + 3 \cdot 8(.01) + (.01)^2}{55} \approx 3.495. \quad ∎$$

Instantaneous Rates of Change

Calculus is needed to determine exactly the instantaneous rate of change of a function (that is, its rate of change at a particular instant). As the preceding discussion suggests, however, accurate approximations of instantaneous rates of change may be found by evaluating the difference quotient for very small values of h.

Example 4 A rock is dropped from a high place. What is its speed exactly 3 seconds after it is dropped?

Solution The distance the rock has fallen at time t is given by $d(t) = 16t^2$. Its speed at $t = 3$ can be approximated by determining the average rate of change of the function $d(t)$ from $t = 3$ to $3 + h$ for very small values of h. This average rate of change is given by the difference quotient $32x + 16h$, which was found in Example 2. When $x = 3$, the difference quotient is $32 \cdot 3 + 16h = 96 + 16h$ and we have:

Change in Time 3 to 3 + h	h	Average Speed [Difference Quotient at $x = 3$] 96 + 16h
3 to 3.1	.1	$96 + 16(.1) = 97.6$ ft per sec
3 to 3.01	.01	$96 + 16(.01) = 96.16$ ft per sec
3 to 3.005	.005	$96 + 16(.005) = 96.08$ ft per sec
3 to 3.00001	.00001	$96 + 16(.00001) = 96.00016$ ft per sec

The table suggests that exact speed of the rock at the instant $t = 3$ seconds is very close to 96 ft per sec. ∎

Example 5 A balloon is being filled with water in such a way that when its radius is x inches, then its volume is $V(x) = x^3/55$ gallons. What is the rate of change at the instant when the radius is 7 inches?

Solution The average rate of change when the radius goes from x to $x + h$ inches is given by the difference quotient of $V(x)$, which was found in Example 3:

$$\frac{V(x + h) - V(x)}{h} = \frac{3x^2 + 3xh + h^2}{55}.$$

Therefore, when $x = 7$ the difference quotient is

$$\frac{3 \cdot 7^2 + 3 \cdot 7 \cdot h + h^2}{55} = \frac{147 + 21h + h^2}{55}$$

and we have these average rates of change over small intervals near 7:

Change in Radius 7 to 7 + h	h	Average Rate of Change of Volume [Difference Quotient at x = 7] $\dfrac{147 + 21h + h^2}{55}$
7 to 7.01	.01	2.6765 gallons per inch
7 to 7.001	.001	2.6731 gallons per inch
7 to 7.0001	.0001	2.6728 gallons per inch
7 to 7.00001	.00001	2.6727 gallons per inch

The chart suggests that at the instant the radius is 7 inches, the volume is changing at a rate of approximately 2.673 gallons per inch.   ∎

Exercises 3.6.A

In Exercises 1–8, compute the difference quotient $\dfrac{f(x + h) - f(x)}{h}$.

1. $f(x) = x + 5$
2. $f(x) = 7x + 2$
3. $f(x) = x^2 + 3$
4. $f(x) = x^2 + 3x - 1$
5. $f(t) = 160{,}000 - 8000t + t^2$
6. $V(x) = x^3$
7. $A(r) = \pi r^2$
8. $V(p) = 5/p$

9. Water is draining from a large tank. After t minutes there are $160{,}000 - 8000t + t^2$ gallons of water in the tank.
   (a) Use the results of Exercise 5 to find the average rate at which the water runs out in the interval from 10 to 10.1 minutes.
   (b) Do the same for the interval from 10 to 10.01 minutes.
   (c) Estimate the rate at which the water runs out after exactly 10 minutes.

10. Use the results of Exercise 6 to find the average rate of change of the volume of a cube whose side has length x as x changes from

   (a) 4 to 4.1   (b) 4 to 4.01   (c) 4 to 4.001
   (d) Estimate the rate of change of the volume at the instant when $x = 4$.

11. Use the results of Exercise 7 to find the average rate of change of the area of a circle of radius r as r changes from
   (a) 3 to 3.5   (b) 3 to 3.2   (c) 3 to 3.1
   (d) Estimate the rate of change at the instant when $r = 3$.
   (e) How is your answer in part (d) related to the circumference of a circle of radius 3?

12. Under certain conditions, the volume V of a quantity of air is related to the pressure p (which is measured in kilopascals) by the equation $V = 5/p$. Use the results of Exercise 8 to estimate the rate at which the volume is changing at the instant when the pressure is 50 kilopascals.

Chapter 3 *Review*

Important Concepts

Important Facts and Formulas

· The average rate of change of a function f as x changes from a to b is the number

$$\frac{f(b) - f(a)}{b - a}$$

· The difference quotient of the function f is the quantity

$$\frac{f(x + h) - f(x)}{h}$$

· The average rate of change of a function f as x changes from a to b is the slope of the secant line joining the points $(a, f(a))$ and $(b, f(b))$.

Review Questions

1. Let $[x]$ denote the greatest integer function and evaluate
   (a) $[-5/2] = $ _____.   (b) $[1755] = $ _____.
   (c) $[18.7] + [-15.7] = $ _____.   (d) $[-7] - [7] = $ _____.

2. If $f(x) = x + |x| + [x]$, then find $f(0), f(-1), f(1/2)$, and $f(-3/2)$.

3. Let f be the function given by the rule $f(x) = 7 - 2x$. Complete this table:

x	0	1	2	-4	t	k	$b-1$	$1-b$	$6-2u$
$f(x)$	7								

4. What is the domain of the function g given by
   $$g(t) = \frac{\sqrt{t-2}}{t-3}?$$

5. In each case give a *specific* example of a function and numbers a, b to show that the given statement may be *false*.
   (a) $f(a+b) = f(a) + f(b)$ (b) $f(ab) = f(a)f(b)$

6. If $f(x) = |3 - x|\sqrt{x-3} + 7$, then $f(7) - f(4) = $ _____.

7. What is the domain of the function given by
   $$g(r) = \sqrt{r-4} + \sqrt{r-2}?$$

8. What is the domain of the function $f(x) = \sqrt{-x+2}$?

9. If $h(x) = x^2 - 3x$, then $h(t+2) = $ _____.

10. Which of the following statements about the greatest integer function $f(x) = [x]$ is true for *every* real number x?
    (a) $x - [x] = 0$ (b) $x - [x] \leq 0$
    (c) $[x] + [-x] \leq 0$ (d) $[-x] \geq [x]$
    (e) $3[x] = [3x]$

11. If $f(x) = 2x^3 + x + 1$, then $f(x/2) = $ _____.

12. If $g(x) = x^2 - 1$, then $g(x-1) - g(x+1) = $ _____.

13. The radius of an oil spill (in meters) is 50 times the square root of the time t (in hours).
    (a) Write the rule of a function f that gives the radius of the spill at time t.
    (b) Write the rule of a function g that gives the area of the spill at time t.
    (c) What are the radius and area of the spill after 9 hours?
    (d) When will the spill have an area of 100,000 square meters?

14. The cost of renting a limousine for 24 hours is given by
    $$C(x) = \begin{cases} 150 & \text{if } 0 < x \leq 25 \\ 1.75x + 150 & \text{if } x > 25 \end{cases},$$
    where x is the number of miles driven.
    (a) What is the cost if the limo is driven 20 miles? 30 miles?
    (b) If the cost is \$218.25, how many miles were driven?

15. Sketch the graph of the function f given by
    $$f(x) = \begin{cases} x^2 & \text{if } x \leq 0 \\ x+1 & \text{if } 0 < x < 4 \\ \sqrt{x} & \text{if } x \geq 4. \end{cases}$$

16. U.S. Express Mail rates in 1998 are shown in the following table. Sketch the graph of the function e, whose rule is $e(x) = $ cost of sending a package weighing x pounds by Express Mail.

Express Mail

Letter Rate—Post Office to Addressee Service

Up to 8 ounces	$10.75
Over 8 ounces to 2 pounds	15.00
Up to 3 pounds	17.25
Up to 4 pounds	19.40
Up to 5 pounds	21.55
Up to 6 pounds	25.40
Up to 7 pounds	26.45

17. Which of the following are graphs of functions of x?

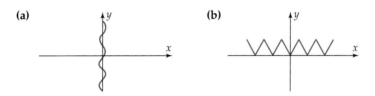

(a)   **(b)**

18. The function whose graph is shown gives the amount of money (in millions of dollars) spent on tickets for major concerts in selected years. (*Source:* Pollstar)

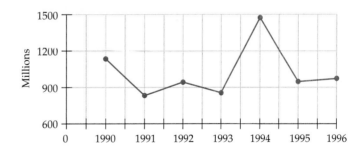

    (a) What is the domain of the function?
    (b) What is the approximate range of the function?
    (c) Over what two-year interval is the average rate of change the largest?

In Questions 19–22, determine the local maxima and minima of the function, the intervals on which the function is increasing, and the intervals on which it is decreasing.

19. $g(x) = \sqrt{x^2 + x + 1}$
20. $f(x) = 2x^3 - 5x^2 + 4x - 3$
21. $g(x) = x^3 + 8x^2 + 4x - 3$
22. $f(x) = .5x^4 + 2x^3 - 6x^2 - 16x + 2$

In Questions 23 and 24, sketch the graph of the curve given by the parametric equations.

23. $x = t^2 - 4$ and $y = 2t + 1$ $(-3 \le t \le 3)$

24. $x = t^3 + 3t^2 - 1$ and $y = t^2 + 1$ $(-3 \le t \le 2)$

25. Sketch a graph that is symmetric with respect to both the x-axis and the y-axis. [*Note:* There are many correct answers and your graph need not be the graph of an equation.]

26. Sketch the graph of a function that is symmetric with respect to the origin. [*Note:* There are many correct answers and you don't have to state the rule of your function.]

In Questions 27 and 28, determine algebraically whether the graph of the given equation is symmetric with respect to the x-axis, and y-axis, or the origin.

27. $x^2 = y^2 + 2$

28. $5y = 7x^2 - 2x$

In Questions 29–31, determine whether the given function is even, odd, or neither.

29. $g(x) = 9 - x^2$ **30.** $f(x) = |x|x + 1$

31. $h(x) = 3x^5 - x(x^4 - x^2)$

32. (a) Draw some coordinate axes and plot the points $(-2, 1)$, $(-1, 3)$, $(0, 1)$, $(3, 2)$, $(4, 1)$.
   (b) Suppose the points plotted in part (a) lie on the graph of an *even* function f. Plot these points: $(2, f(2))$, $(1, f(1))$, $(0, f(0))$, $(-3, f(-3))$, $(-4, f(-4))$.

33. Determine whether the circle with equation $x^2 + y^2 + 6y = -5$ is symmetric with respect to the x-axis, the y-axis, or the origin.

34. Sketch the graph of a function f that satisfies all of these conditions:
   (i) The domain of f consists of all x such that $-3 \le x \le 4$.
   (ii) The range of f consists of all y such that $-2 \le y \le 5$.
   (iii) $f(-2) = 0$ **(iv)** $f(1) > 2$
   [*Note:* There are many possible correct answers and the function whose graph you sketch need *not* have a simple algebraic rule.]

35. Sketch the graph of $g(x) = 5 + \dfrac{4}{x - 5}$.

Use the graph of the function f in the figure to answer Questions 36–39.

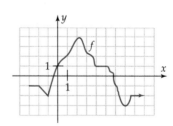

36. What is the domain of f?   **37.** What is the range of f?

38. Find all numbers x such that $f(x) = 1$.

39. Find a number x such that $f(x + 1) < f(x)$. [Many correct answers are possible.]

Use the graph of the function f in the figure to answer Questions 40–46.

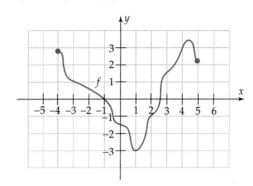

40. What is the domain of f?

41. $f(-3) = $ _____.

42. $f(2 + 2) = $ _____.

43. $f(-1) + f(1) = $ _____.

44. True or false: $2f(2) = f(4)$.

45. True or false: $3f(2) = -f(4)$.

46. True or false: $f(x) = 3$ for exactly one number x.

Use the graphs of the functions f and g in the figure to answer Questions 47–52.

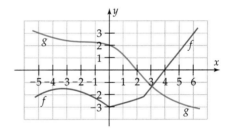

47. For which values of x is $f(x) = 0$?

48. True or false: If a and b are numbers such that $-5 \le a < b \le 6$, then $g(a) < g(b)$.

49. For which values of x is $g(x) \ge f(x)$?

50. Find $f(0) - g(0)$.

51. For which values of x is $f(x + 1) < 0$?

52. What is the distance from the point $(-5, g(-5))$ to the point $(6, g(6))$?

53. Fireball Bob and King Richard are two NASCAR racers. The following graph shows their distance traveled in a recent race as a function of time.
  (a) Which car made the most pit stops?
  (b) Which car started out the fastest?
  (c) Which car won the race?

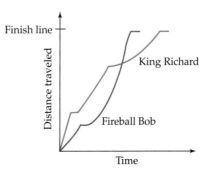

In Questions 54–57, list the transformations, in the order they should be performed on the graph of $g(x) = x^2$, so as to produce a complete graph of the function f.

54. $f(x) = (x - 2)^2$

55. $f(x) = .25x^2 + 2$

56. $f(x) = -(x + 4)^2 - 5$

57. $f(x) = -3(x - 7)^2 + 2$

58. The graph of a function f is shown in the figure. On the same coordinate plane, carefully draw the graphs of the functions g and h whose rules are:

$$g(x) = -f(x) \quad \text{and} \quad h(x) = 1 - f(x).$$

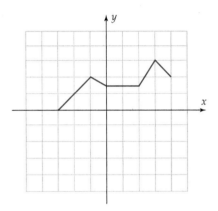

59. The figure shows the graph of a function f. If g is the function given by $g(x) = f(x + 2)$, then which of these statements about the graph of g is true?

(a) It does not cross the x-axis.

(b) It does not cross the y-axis.

(c) It crosses the y-axis at $y = 4$.

(d) It crosses the y-axis at the origin.

(e) It crosses the x-axis at $x = -3$.

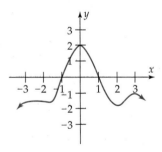

60. If $f(x) = 3x + 2$ and $g(x) = x^3 + 1$, find

(a) $(f + g)(-1)$ **(b)** $(f - g)(2)$ **(c)** $(fg)(0)$

61. If $f(x) = \dfrac{1}{x - 1}$ and $g(x) = \sqrt{x^2 + 5}$, find:

(a) $(f/g)(2)$ **(b)** $(g/f)(x)$

(c) $(fg)(c + 1)$ $(c \neq -1)$

62. Find two functions f and g such that neither is the identity function and

$$(f \circ g)(x) = (2x + 1)^2$$

63. Use the graph of the function g in the figure to fill in the table, in which h is the composite function $g \circ g$.

x	-4	-3	-2	-1	0	1	2	3	4
$g(x)$					-1				
$h(x) = g(g(x))$									

Questions 64–69 refer to the functions $f(x) = \dfrac{1}{x+1}$ and $g(t) = t^3 + 3$.

64. $(f \circ g)(1) =$ _____.

65. $(g \circ f)(2) =$ _____.

66. $g(f(-2)) =$ _____.

67. $(g \circ f)(x - 1) =$ _____.

68. $g(2 + f(0)) =$ _____.

69. $f(g(1) - 1) =$ _____.

70. Let f and g be the functions given by
$$f(x) = 4x + x^4 \quad \text{and} \quad g(x) = \sqrt{x^2 + 1}$$
   (a) $(f \circ g)(x) =$ _____.   **(b)** $(g - f)(x) =$ _____.

71. If $f(x) = \dfrac{1}{x}$ and $g(x) = x^2 - 1$, then
$$(f \circ g)(x) = \underline{\quad} \quad \text{and} \quad (g \circ f)(x) = \underline{\quad}.$$

72. Let $f(x) = x^2$. Give an example of a function g with domain all real numbers such that $g \circ f \neq f \circ g$.

73. If $f(x) = \dfrac{1}{1-x}$ and $g(x) = \sqrt{x}$, then find the domain of the composite function $f \circ g$.

74. These tables show the values of the functions f and g at certain numbers:

x	-1	0	1	2	3
$f(x)$	1	0	1	3	5

and

t	0	1	2	3	4
$g(t)$	-1	0	1	2	5

Which of the following statements are *true*?
   (a) $(g - f)(1) = 1$ **(b)** $(f \circ g)(2) = (f - g)(0)$
   (c) $f(1) + f(2) = f(3)$ **(d)** $(g \circ f)(2) = 1$
   (e) None of the above is true.

75. Find the average rate of change of the function $g(x) = \dfrac{x^3 - x + 1}{x + 2}$ as x changes from

   (a) -1 to 1   **(b)** 0 to 2

76. Find the average rate of change of the function $f(x) = \sqrt{x^2 - x + 1}$ as x changes from
   (a) -3 to 0   **(b)** -3 to 3.5   **(c)** -3 to 5

77. If $f(x) = 2x + 1$ and $g(x) = 3x - 2$, find the average rate of change of the composite function $f \circ g$ as x changes from 3 to 5.

78. If $f(x) = x^2 + 1$ and $g(x) = x - 2$, find the average rate of change of the composite function $f \circ g$ as x changes from -1 to 1.

In Questions 79–82, find the difference quotient of the function.

79. $f(x) = 3x + 4$ **80.** $g(x) = \sqrt{x}$

81. $g(x) = x^2 - 1$ **82.** $f(x) = x^2 + x$

83. The profit (in hundreds of dollars) from selling x tons of Wonderchem is given by $P(x) = .2x^2 + .5x - 1$. What is the average rate of change of profit when the number of tons of Wonderchem sold increases from
  (a) 4 to 8 tons?   **(b)** 4 to 5 tons?   **(c)** 4 to 4.1 tons?

84. On the planet Mars, the distance traveled by a falling rock (ignoring atmospheric resistance) in t seconds is $6.1t^2$ ft. How far must a rock fall in order to have an average speed of 25 ft per sec over that time interval?

85. The graph in the figure shows the population of fruit flies during a 50-day experiment in a controlled atmosphere.

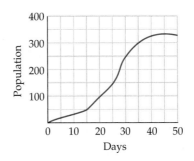

  (a) During what 5-day period is the average rate of population growth the slowest?
  (b) During what 10-day period is the average rate of population growth the fastest?
  (c) Find an interval beginning at the 30th day during which the average rate of population growth is the same as the average rate from day 10 to day 20.

86. The graph of the function g in the figure consists of straight line segments. Find an interval over which the average rate of change of g is
  (a) 0   **(b)** -3 **(c)** .5
  (d) Explain why the average rate of change of g is the same from -3 to -1 as it is from -2.5 to 0.

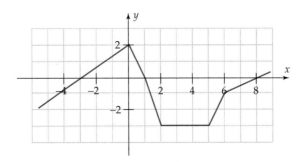

Discovery Project 3

Building an Odometer

Simple mathematical functions often occur in everyday life in ways that are transparent to the casual observer. Consider the construction of a device for measuring the distance a bicycle has traveled. In this device, a pin is placed on one of the spokes of the bicycle wheel and a device that counts the number of times the pin passes by is fixed to the frame of the bicycle. The counter device can be either mechanical, incrementing when the pin strokes the counter, or electronic, sensing the passage of the pin with some sort of electromagnetic radiation.

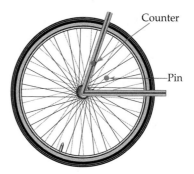

1. How many times will the pin pass by the counter over the course of 1 mile if the wheel has a diameter of 27 inches?

2. Write a function $d(x)$, where x is the number of times that the pin passes by the counter and $d(x)$ tells you the distance traveled in miles.

3. What is the domain of $d(x)$? What *kind* of numbers are in the domain?

4. The electronic version of the device described can be turned into a speedometer. Write a function $s_1(x)$, where x is the number of times that the pin passes by the counter *each second* and $s_1(x)$ is the speed of the bicycle in miles per hour.

5. What is the domain of $s_1(x)$? Calculate $s_1(1)$ and $s_1(2)$. Why is this a problem?

6. A more efficient method of using the device as a speedometer is to measure the time interval between clicks of the counter. Write a function $s_2(x)$, where x is the time interval between clicks and $s_2(x)$ is the speed of the bicycle in miles per hour. Why is $s_2(x)$ better than $s_1(x)$?

Chapter

4

Polynomial and Rational Functions

Can you afford to go to college?

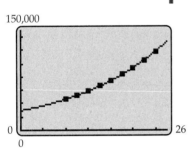

As you (and your parents) know, the cost of a four-year college education (tuition, fees, room, board, books) has steadily increased. It is expected to continue to do so in the foreseeable future, according to projections by a large insurance company. This growth can be modeled by a fourth-degree polynomial function. See Exercise 38 on page 342.

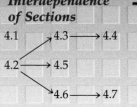
Polynomial functions arise naturally in many applications. Many complicated functions in applied mathematics can be approximated by polynomial functions or their quotients (rational functions).

4.1 Quadratic Functions and Models*

A **quadratic function** is a function whose rule can be written in the form

$$f(x) = ax^2 + bx + c$$

for some constants a, b, c, with $a \neq 0$. The graph of a quadratic function is called a **parabola.** As the following Exploration illustrates, all parabolas have a "cup shape." The "cup" may open upward or downward, broadly or narrowly.

GRAPHING EXPLORATION

Using the standard viewing window, graph the following quadratic functions on the same screen:

$$f(x) = x^2, \qquad f(x) = 3x^2 + 30x + 77, \qquad f(x) = -x^2 + 4x,$$
$$f(x) = -.2x^2 + 1.5x - 5.$$

The preceding Exploration also shows that the parabola opens upward when the coefficient of x^2 is positive and downward when this coefficient is negative.

---

*This section may be omitted or postponed. Section 3.4 (Graphs and Transformations) is a prerequisite.

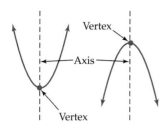

Figure 4–1

If a parabola opens upward, its **vertex** is the lowest point on the graph and if a parabola opens downward, its **vertex** is the highest point on the graph, as shown in Figure 4–1. Every parabola is symmetric with respect to the vertical line through its vertex; this line is called the **axis** of the parabola.

Parabolas are easily graphed on a calculator. The vertex can always be approximated by using the trace feature or a maximum/minimum finder. However, algebraic techniques can be used to find the vertex precisely.

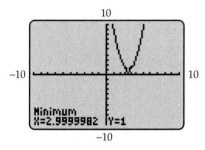

Figure 4–2

Example 1 Show that the function $g(x) = 2(x - 3)^2 + 1$ is quadratic, graph the function, and find its vertex.

Solution   The function g is quadratic because its rule can be written in the form $g(x) = ax^2 + bx + c$:

$$g(x) = 2(x - 3)^2 + 1 = 2(x^2 - 6x + 9) + 1 = 2x^2 - 12x + 19.$$

Graphing the function in the standard viewing window (Figure 4–2) and using the minimum finder, we see that the vertex is approximately $(2.999, 1)$. In order to find the vertex exactly, we use the techniques of Section 3.4. The graph of $g(x) = 2(x - 3)^2 + 1$ can be obtained from the graph of $y = x^2$ as follows:

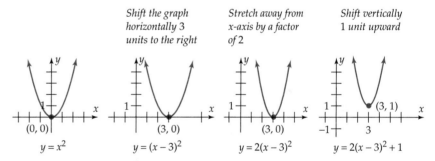

Figure 4–3

Figure 4–3 shows that when the vertex $(0, 0)$ of $y = x^2$ is shifted 3 units to the right and 1 unit upward, it moves to $(3, 1)$. Therefore, $(3, 1)$ is the vertex of $g(x) = 2(x - 3)^2 + 1$. Note how the coordinates of the vertex are related to the rule of the function g:

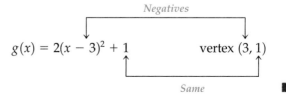

The vertex of the function g in Example 1 was easily determined because the rule of g had a special algebraic form. The vertex of the graph of any quadratic function can be determined in a similar fashion by first rewriting its rule.

Example 2   Find the vertex of the graph of $g(x) = 3x^2 + 30x + 77$ algebraically.

Solution   We rewrite the rule of g as follows, being careful at each step not to change its value:

$$g(x) = 3x^2 + 30x + 77$$

Factor out 3: $= 3(x^2 + 10x) + 77$

*Add $25 - 25$ inside parentheses:** $= 3(x^2 + 10x + 25 - 25) + 77$

Use the distributive law: $= 3(x^2 + 10x + 25) - 3 \cdot 25 + 77$

Simplify: $= 3(x^2 + 10x + 25) + 2$

Factor expression in parentheses: $= 3(x + 5)^2 + 2$

As we saw in Section 3.4, the graph of $g(x)$ is the graph of $f(x) = x^2$ shifted horizontally 5 units to the left, stretched by a factor of 3, and shifted 2 units upward, as shown in Figure 4–4. In this process, the vertex $(0, 0)$ of f moves to $(-5, 2)$. Therefore, $(-5, 2)$ is the vertex of $g(x) = 3(x + 5)^2 + 2$. Once again, note how the coordinates of the vertex are related to the rule of the function:

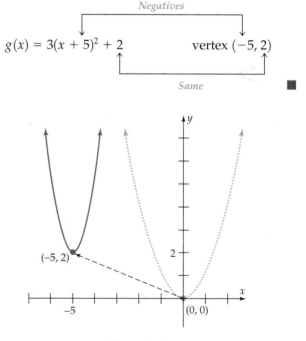

Figure 4–4

---

*Since $25 - 25 = 0$, we haven't changed the value of the function. The number 25, which is the square of half the coefficient of x, is chosen because it will enable us to factor an expression below. This technique is called *completing the square.*

The technique used to find the vertex in Example 2 works for any quadratic function $f(x) = ax^2 + bx + c$. In what follows, don't let all the letters scare you. In Example 2 we replace 3 by a, 30 by b, and 77 by c, and then use exactly the same five-step computation to rewrite the rule of f:

$$f(x) = ax^2 + bx + c$$

Factor out a:
$$= a\left(x^2 + \frac{b}{a}x\right) + c$$

Add $\dfrac{b^2}{4a^2} - \dfrac{b^2}{4a^2}$ inside parentheses:*
$$= a\left(x^2 + \frac{b}{a}x + \frac{b^2}{4a^2} - \frac{b^2}{4a^2}\right) + c$$

Use the distributive law:
$$= a\left(x^2 + \frac{b}{a}x + \frac{b^2}{4a^2}\right) - a\cdot\frac{b^2}{4a^2} + c$$

Simplify and rearrange:
$$= a\left(x^2 + \frac{b}{a}x + \frac{b^2}{4a^2}\right) + \left(c - \frac{b^2}{4a}\right)$$

Factor first expression in parentheses:†
$$= a\left(x + \frac{b}{2a}\right)^2 + \left(c - \frac{b^2}{4a}\right).$$

As in the preceding examples, the graph of f is just the graph of x^2 shifted horizontally, stretched by a factor of a, and shifted vertically. As above, the vertex of this parabola can be read from the rule of the function:

Negatives

$$f(x) = a\left(x + \frac{b}{2a}\right)^2 + \left(c - \frac{b^2}{4a}\right) \qquad \text{vertex}\left(\frac{-b}{2a}, c - \frac{b^2}{4a}\right)$$

Same

If we let $h = -\dfrac{b}{2a}$ and $k = c - \dfrac{b^2}{4a}$, then we have these useful facts:

Quadratic Functions

> The rule of the quadratic function $f(x) = ax^2 + bx + c$ can be rewritten in the form
>
> $$f(x) = a(x - h)^2 + k,$$
>
> where $h = -b/2a$. The graph of f is a parabola with vertex (h, k). It opens upward if $a > 0$ and downward if $a < 0$.

*Since $\dfrac{b^2}{4a^2} - \dfrac{b^2}{4a^2} = 0$, we haven't changed the value of the function. This is another example of "completing the square" (the number $b^2/4a^2$ is the square of half the coefficient of x). Note that when $a = 3$ and $b = 30$ (as in Example 2), then $\dfrac{b^2}{4a^2} = \dfrac{30^2}{4\cdot 3^2} = \dfrac{900}{36} = 25$.

†Verify that this factorization is correct by multiplying out $\left(x + \dfrac{b}{2a}\right)^2$.

Example 3   Describe the graph of

$$f(x) = -x^2 + 5x + 1$$

and find its vertex and x-intercepts exactly.

Solution   The graph is a downward-opening parabola (because f is a quadratic function and the coefficient of x^2 is negative). According to the preceding box, the x-coordinate of its vertex is

$$-\frac{b}{2a} = -\frac{5}{2(-1)} = \frac{-5}{-2} = \frac{5}{2} = 2.5.$$

To find the y-coordinate of the vertex we need only evaluate f at this number:

$$f(2.5) = -(2.5^2) + 5(2.5) + 1 = 7.25.$$

Therefore, the vertex is $(2.5, 7.25)$.

    The x-intercepts are the points on the graph with y-coordinate 0, that is, the points where $f(x) = 0$. So we must solve this equation:

$$-x^2 + 5x + 1 = 0$$

Multiply both sides by -1: $x^2 - 5x - 1 = 0$

Since the left side doesn't readily factor, we use the quadratic formula:

$$x = \frac{-b \pm \sqrt{b^2 - 4ac}}{2a} = \frac{-(-5) \pm \sqrt{(-5)^2 - 4(-1)(1)}}{2 \cdot 1} = \frac{5 \pm \sqrt{29}}{2}.$$

Therefore, the x intercepts are at $\dfrac{5 - \sqrt{29}}{2} \approx -.1923$ and $\dfrac{5 + \sqrt{29}}{2} \approx$ 5.1923.   ■

GRAPHING EXPLORATION

Graphically confirm the results of Example 3 by graphing $y = -x^2 + 5x + 1$ and using your maximum finder to approximate the vertex and your root finder to approximate the x-intercepts.

Example 4   Find the rule of the quadratic function whose graph is a parabola with vertex $(3, 4)$ that passes through the point $(-1, 36)$.

Solution   The rule of a quadratic function can be written in the form $f(x) = a(x - h)^2 + k$ and its graph is a parabola with vertex (h, k). In this case, the vertex is $(3, 4)$, so we have $h = 3$ and $k = 4$. Hence, the rule of f is $f(x) = a(x - 3)^2 + 4$. Since $(-1, 36)$ is on the graph, we have

$$f(-1) = 36$$

Substitute −1 for x in the rule of f: $a(-1 - 3)^2 + 4 = 36$

Simplify: $16a + 4 = 36$

Subtract 4 from both sides: $16a = 32$

Divide both sides by 16: $a = 2.$

Therefore, the rule of the function is $f(x) = 2(x - 3)^2 + 4.$ ■

Applications

Quadratic functions arise in the design of satellite dishes, microphones, and searchlights. They can also be used to track the path of a projectile, such as an artillery shell or a baseball.

Example 5 Sammy Sosa hits a baseball. The height of the ball is given by

$$h(x) = -.001x^2 + .45x + 3,$$

where x is the distance on the ground from the batter to the point directly under the ball.

(a) How high does the ball go?

(b) How far from the batter does it land?

Solution

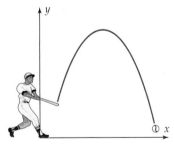

Figure 4–5

(a) Imagine a coordinate plane with the batter at the origin, the y-axis perpendicular to the ground, and the ball traveling in the direction of the x-axis, as shown in Figure 4–5 (which is not to scale). The ball follows a parabolic path, namely, the graph of h. The highest point it reaches (the vertex of the parabola) has x-coordinate

$$-\frac{b}{2a} = -\frac{.45}{2(-.001)} = 225$$

and y-coordinate

$$f(225) = -.001(225^2) + .45(225) + 3 = 53.625.$$

Therefore, the maximum height of the ball is 53.625 feet.

(b) The ball hits ground when its height is 0, that is, when

$$h(x) = 0$$

$$-.001x^2 + .45x + 3 = 0$$

This equation can be solved by the quadratic formula:

$$x = \frac{-b \pm \sqrt{b^2 - 4ac}}{2a} = \frac{-.45 \pm \sqrt{.45^2 - 4(-.001)(3)}}{2(-.001)}$$

$$= \frac{-.45 \pm \sqrt{.2145}}{-.002} \approx \begin{cases} -6.57 \\ \text{or} \\ 456.57 \end{cases}$$

The negative solution does not apply here, and we see that the ball hits ground 456.57 feet from the batter. ■

GRAPHING EXPLORATION

Graphically confirm the results of Example 5 by graphing $y = -.001x^2 + .45x + 3$ and using your maximum finder to approximate the vertex and your root finder to approximate the appropriate x-intercept.

Example 6 The owner of a 20-unit apartment complex has found that each $50 increase in monthly rent results in a vacant apartment. All units are now rented at $400 per month. How many $50 increases in rent will produce the largest possible income for the owner?

Solution Let x represent the number of $50 increases. Then the monthly rent will be $400 + 50x$ dollars. Since one apartment goes vacant for each increase, the number of occupied apartments will be $20 - x$. Then the owner's monthly income $R(x)$ is given by

$$R(x) = (\text{number of apartments rented}) \times (\text{rent per apartment})$$

$$R(x) = (20 - x)(400 + 50x)$$

$$R(x) = -50x^2 + 600x + 8000$$

There are three ways to find the maximum possible income.

Table Method. Make a table of values of $R(x)$ for $0 \leq x \leq 20$, as in Figure 4–6.* The table shows that the maximum income of $9800 occurs when $x = 6$. In this case, there will be $20 - 6 = 14$ apartments rented at a monthly rent of $400 + 6(50) = 700.

X	Y1
0	8000
1	8550
2	9000
3	9350
4	9600
5	9750
6	9800

X=0

X	Y1
7	9750
8	9600
9	9350
10	9000
11	8550
12	8000
13	7350

X=7

X	Y1
14	6600
15	5750
16	4800
17	3750
18	2600
19	1350
20	0

X=14

Figure 4–6

*This method is feasible here because there are only 20 apartments, but it cannot be used when the number of possibilities is very large or infinite.

Algebraic Method. The graph of $R(x) = -50x^2 + 600x + 8000$ is a downward-opening parabola (why?). Maximum income occurs at the vertex of this parabola, that is, when

$$x = \frac{-b}{2a} = \frac{-600}{2(-50)} = \frac{-600}{-100} = 6.$$

Therefore, six increases of $50 will produce maximum income.

Graphical Method. Graphing $R(x) = -50x^2 + 600x + 8000$ and using a maximum finder to determine the coordinates of the vertex, as in Figure 4–7, shows that maximum income of $9800 occurs when there are six rent increases. ■

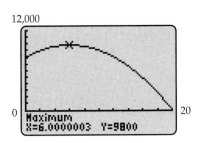

12,000

0   20

Maximum
X=6.0000003 Y=9800

Figure 4–7

Example 7 Find the area and dimensions of the largest rectangular field that can be enclosed with 3000 feet of fence.

Solution Let x denote the length and y the width of the field, as shown in Figure 4–8.

Perimeter $= x + y + x + y$

$= 2x + 2y$

Area $= xy$

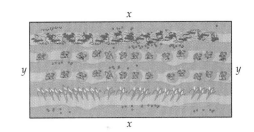

x

y y

x

Figure 4–8

Since the perimeter is the length of the fence, $2x + 2y = 3000$. Hence, $2y = 3000 - 2x$ and $y = 1500 - x$. Consequently, the area is

$$A = xy = x(1500 - x) = 1500x - x^2 = -x^2 + 1500x.$$

The largest possible area is just the maximum value of the quadratic function $A(x) = -x^2 + 1500x$. This maximum occurs at the vertex of the graph of $A(x)$ (which is a downward-opening parabola because the coefficient of x^2 is negative). The x-coordinate of the vertex is

$$\frac{-b}{2a} = \frac{1500}{2(-1)} = 750 \text{ ft.}$$

Hence, the y-coordinate of the vertex, the maximum value of $A(x)$, is

$$A(750) = -750^2 + 1500 \cdot 750 = 562{,}500 \text{ sq ft.}$$

It occurs when the length is $x = 750$. In this case the width is $y = 1500 - x = 1500 - 750 = 750$. ■

Exercises 4.1

In Exercises 1–8, without graphing, *determine the vertex of the parabola and state whether it opens upward or downward.*

1. $f(x) = 3(x - 5)^2 + 2$ **2.** $g(x) = -6(x - 2)^2 - 5$

3. $y = -(x - 1)^2 + 2$ **4.** $h(x) = -x^2 + 1$

5. $f(x) = x^2 - 6x + 3$ **6.** $g(x) = x^2 + 8x - 1$

7. $h(x) = x^2 + 3x + 6$ **8.** $f(x) = x^2 - 5x - 7$

In Exercises 9–16, match the function with its graph, which is one of those shown here.

A.
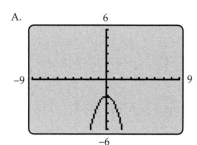

B.
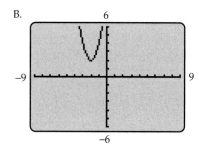

C.
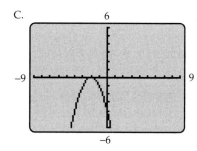

D.
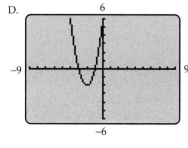

E.
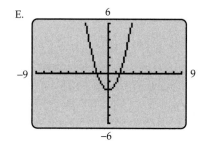

F.
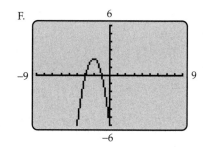

G.
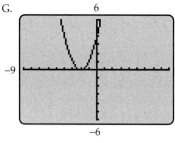

H.
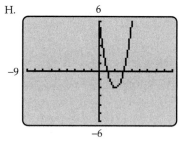

I.
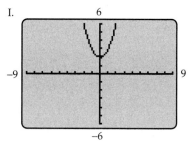

J.
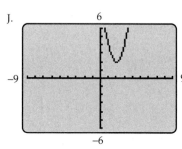

K.
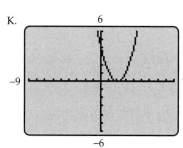

L.
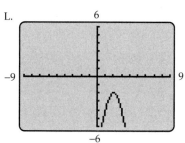

9. $f(x) = x^2 + 2$

10. $g(x) = x^2 - 2$

11. $g(x) = (x - 2)^2$

12. $f(x) = -(x + 2)^2$

13. $f(x) = 2(x - 2)^2 + 2$

14. $g(x) = -2(x - 2)^2 - 2$

15. $g(x) = -2(x + 2)^2 + 2$

16. $f(x) = 2(x + 2)^2 - 2$

In Exercises 17–24, find the rule of the quadratic function whose graph satisfies the given conditions.

17. Vertex at $(0, 0)$; passes through $(2, 12)$

18. Vertex at $(0, 1)$; passes through $(2, -7)$

19. Vertex at $(2, 5)$; passes through $(-3, 80)$

20. Vertex at $(4, 1)$; passes through $(2, -11)$

21. Vertex at $(-3, 4)$; passes through $(-2, 6)$

22. Vertex at $(-1, -2)$; passes through $(2, -29)$

23. Vertex at $(4, 3)$; passes through $(6, 5)$

24. Vertex at $(1/2, -3)$; passes through $(5, 3)$

In Exercises 25–32, sketch the graph of the parabola and determine its vertex and x-intercepts exactly.

25. $y = 2x^2 + 12x - 3$

26. $y = 3x^2 + 6x + 1$

27. $f(x) = -x^2 + 8x - 2$

28. $g(x) = -x^2 - 6x + 4$

29. $f(x) = -3x^2 + 4x + 5$

30. $g(x) = 2x^2 - x - 1$

31. $y = -x^2 + x$

32. $y = -2x^2 + 2x - 1$

33. The graph of the quadratic function g is obtained from the graph of $f(x) = x^2$ by vertically stretching it by a factor of 2 and then shifting vertically 5 units downward. What is the rule of the function g? What is the vertex of its graph?

34. The graph of the quadratic function g is obtained from the graph of $f(x) = x^2$ by shifting it horizontally 4 units to the left, then vertically stretching it by a factor of 3, and then shifting vertically 2 units upward. What is the rule of the function g? What is the vertex of its graph?

35. If the graph of the quadratic function h is shifted vertically 4 units downward, then shrunk by a factor of 1/2, and then shifted horizontally 3 units to the left, the resulting graph is the parabola $f(x) = x^2$. What is the rule of the function h? What is the vertex of its graph?

36. If the graph of the quadratic function h is shifted vertically 3 units upward, then reflected in the x-axis, and then shifted horizontally 5 units to the right, the resulting graph is the parabola $f(x) = x^2$. What is the rule of the function h? What is the vertex of its graph?

37. The Leslie Lahr Luggage Company has determined that its profit on its Luxury ensemble is given by $p(x) = 1600x - 4x^2 - 50,000$, where x is the number of units sold.

(a) What is the profit on 50 units? On 250 units?

(b) How many units should be sold to maximize profit? In that case, what will be the profit on each unit?

38. Based on data from past years, a consultant informs Bob's Bicycles that their profit from selling x bicycles is given by the function $p(x) = 250x - x^2/4 - 15,000$.

(a) How much profit do they make by selling 100 bicycles? By selling 400 bicycles?

(b) How many bicycles should be sold to maximize profit? In that case, what will be the profit per bicycle?

In Exercises 39–42, use the formula for the height h of an object (that is traveling vertically subject only to gravity) at time t: $h = -16t^2 + v_0 t + h_0$, where h_0 is the initial height and v_0 the initial velocity.

39. A ball is thrown upward from the top of a 96-foot-high tower with an initial velocity of 80 feet per second. When does the ball reach its maximum height and how high is it at that time?

40. A rocket is fired upward from ground level with an initial velocity of 1600 feet per second. When does it attain its maximum height, and what is that height?

41. A ball is thrown upward from a height of 6 feet with an initial velocity of 32 feet per second. Find its maximum height.

42. A bullet is fired upward from ground level with an initial velocity of 1500 feet per second. How high does it go?

43. A projectile is fired at an angle of 45° upward. Exactly t seconds after firing, its vertical height above the ground is $500t - 16t^2$ feet.

(a) What is the greatest height the projectile reaches, and at what time does that occur?

(b) When does the projectile hit the ground? [*Hint:* It's on the ground when its height is 0.]

44. Jack throws a baseball. Its height above the ground is given by $h(x) = -.0013x^2 + .26x + 5$, where x is the distance from Jack to a point on the ground directly below the ball.

(a) How far from Jack is the ball when it reaches the highest point on its flight? How high is the ball at that point?

(b) How far from Jack does the ball hit the ground?

45. During the Civil War, the standard heavy gun for coastal artillery was the 15-inch Rodman cannon, which fired a 330-pound shell. If one of these guns is fired from the top of a 50-foot-high shoreline embankment, then the height of the shell above the water (in feet) can be approximated by the function

$$p(x) = -.0000167x^2 + .23x + 50,$$

where x is the horizontal distance (in feet) from the foot of the embankment to a point directly under the shell. How high does the shell go, and how far away does it hit the water?

46. The Golden Gate Bridge is supported by two huge cables strung between the towers at each end of the bridge. The function

$$f(x) = .0001193x^2 - .50106x + 526.113$$

gives the approximate height of the cables above the roadway at a point on the road x feet from one of the towers. The cables touch the road halfway between the two towers. How far apart are the towers?

47. The braking distance (in meters) for a car with excellent brakes on a good road with an alert driver can be modeled by the quadratic function $B(s) = .01s^2 + .7s$, where s is the car's speed in kilometers per hour.
  (a) What is the braking distance for a car traveling 30 km/h? For one traveling 100 km/h?
  (b) If the car takes 60 meters to come to a complete stop, what was its speed?

48. The median sale price of existing single family homes in the United States from 1970 to 1996 can be approximated by the quadratic function

$$P(t) = -5.006t^2 + 3921.13t + 19{,}275.39,$$

where $t = 0$ corresponds to 1970.
  (a) Assuming the trend continues, estimate the median sale price in the year 2000.
  (b) Explain why this model needs a restricted domain.
  (c) In what year was the median sale price $100,000?

49. A potter can sell 120 bowls per week at $4 per bowl. For each 50¢ decrease in price 20 more bowls are sold. What price should be charged in order to maximize sales income?

50. A vendor can sell 200 souvenirs per day at a price of $2 each. Each 10¢ price increase decreases the number of sales by 25 per day. Souvenirs cost the vendor $1.50 each. What price should be charged to maximize the profit?

51. When a basketball team charges $4 per ticket, average attendance is 500 people. For each 20¢ decrease in ticket price, average attendance increases by 30 people. What should the ticket price be to ensure maximum income?

52. A ballpark concessions manager finds that each salesperson sells an average of 40 boxes of popcorn per game when 20 salespeople are working. When an additional salesperson is employed, each salesperson averages 1 less box

per game. How many salespeople should be hired to ensure maximum income?

53. Find two real numbers whose difference is 4 and whose product is as small as possible. [*Hint:* If the numbers are x and z, then $x - z = 4$. Use this fact to write the product xz as a function of x.]

54. Find two real numbers whose sum is -18 and whose product is as large as possible.

55. Find the dimensions of the rectangle with perimeter 250 and largest possible area. [*Hint:* Exercise 35 in Section 3.3.]

56. A gutter is to be made by bending up the edges of a 15-inch-wide piece of aluminum. What depth should the gutter be to have the maximum possible cross-sectional area?

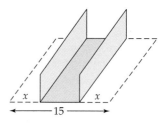

57. A gardener wants to use 130 feet of fencing to enclose a rectangular garden and divide it into two plots, as shown in the figure. What is the largest possible area for such a garden?

58. A rectangular box (with top) has a square base. The sum of the lengths of its 12 edges is 8 feet. What dimensions should the box have so that its surface area is as large as possible?

59. A field bounded on one side by a river is to be fenced on three sides so as to form a rectangular enclosure. If 200 ft of fencing is to be used, what dimensions will yield an enclosure of the largest possible area?

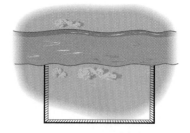

60. A rectangular garden next to a building is to be fenced on three sides. Fencing for the side parallel to the building costs $80 per foot and material for the other two sides costs $20 per foot. If $1800 is to be spent on fencing, what are the dimensions of the garden with the largest possible area?

61. At Middleton Place, a historic plantation near Charleston, South Carolina, there is a "joggling board" that was once used for courting. A young girl would sit at one end, her suitor at the other end, and her mother in the center. The mother would bounce on the board, thus causing the girl and her suitor to move closer together. A joggling board is 8 feet long, and an average mother sitting at its center causes the board to deflect 2 inches, as shown in the figure. The shape of the deflected board is parabolic.

(a) Find the equation of the parabola, assuming that the joggling board (with no one on it) lies on the *x*-axis with its center at the origin.

(b) How far from the center of the board is the deflection 1 inch?

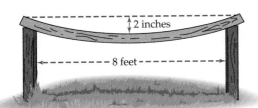

Thinker

62. The *discriminant* of a quadratic function $f(x) = ax^2 + bx + c$ is the number $b^2 - 4ac$. For each of the discriminants listed here, state which graphs could possibly be the graph of *f*.

(a) $b^2 - 4ac = 25$
(b) $b^2 - 4ac = 0$
(c) $b^2 - 4ac = -49$
(d) $b^2 - 4ac = 72$

(i)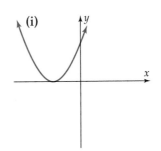

(ii)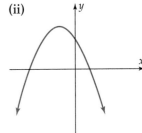

(iii)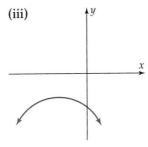

(iv)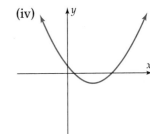

4.2 Polynomial Functions and Roots

A **polynomial function** is a function whose rule is given by a polynomial, such as

$$f(x) = x^4 - 2x^3 + 4x - 5 \quad \text{or} \quad g(x) = 2x^3 + x^2 - 3x + 1.$$

Long division of polynomials, which we now present, will play a key role in analyzing polynomial functions. The basic procedure, as illustrated in the following example, is quite similar to long division of numbers.

Example 1 Divide $8x^3 + 2x^2 + 1$ by $2x^2 - x$.

Solution We set up the division in the same way that is used for numbers:

Divisor → $\underline{2x^2 - x} \; | \; 8x^3 + 2x^2 + 1$ ← *Dividend*

Begin by dividing the first term of the divisor ($2x^2$) into the first term of the dividend ($8x^3$) and putting the result $\left(\text{namely, } \dfrac{8x^3}{2x^2} = 4x\right)$ on the top line, as shown in the following equation. Then multiply $4x$ times the entire divisor, put the result on the third line, and subtract.

$$
\begin{array}{r}
4x \qquad\qquad\quad \leftarrow \textit{Partial Quotient} \\
2x^2 - x \,\overline{\big)\, 8x^3 + 2x^2 + 1} \qquad\qquad \\
\underline{8x^3 - 4x^2} \qquad \leftarrow 4x(2x^2 - x) \\
6x^2 + 1 \quad \leftarrow \textit{Subtraction*}
\end{array}
$$

Now divide the first term of the divisor ($2x^2$) into $6x^2$ and put the result $\left(\dfrac{6x^2}{2x^2} = 3\right)$ on the top line as shown below. Then multiply 3 times the entire divisor, put the result in the fifth line, and subtract.

$$
\begin{array}{r}
4x \;\; + 3 \qquad\qquad\;\; \leftarrow \textit{Quotient} \\
2x^2 - x \,\overline{\big)\, 8x^3 + 2x^2 \qquad\; + 1} \qquad\qquad \\
\underline{8x^3 - 4x^2} \qquad\qquad\;\; \leftarrow 4x(2x^2 - x) \\
6x^2 \qquad + 1 \;\; \leftarrow \textit{Subtraction} \\
\underline{6x^2 - 3x} \qquad\;\; \leftarrow 3(2x^2 - x) \\
\textit{Remainder} \rightarrow 3x + 1 \;\; \leftarrow \textit{Subtraction}
\end{array}
$$

The division process stops when the remainder is 0 or has smaller degree than the divisor, which is the case here. ∎

Recall how you check a long division problem with numbers:

$$
\begin{array}{r}
145 \\
31 \,\overline{\big)\, 4509} \\
\underline{31} \\
140 \\
\underline{124} \\
169 \\
\underline{155} \\
14
\end{array}
\qquad
\textit{Check:}
\qquad
\begin{array}{r}
145 \;\; \leftarrow \textit{Quotient} \\
\underline{\times\; 31} \;\; \leftarrow \textit{Divisor} \\
145 \\
\underline{435} \\
4495 \\
\underline{+\; 14} \;\; \leftarrow \textit{Remainder} \\
4509 \;\; \leftarrow \textit{Dividend}
\end{array}
$$

We can summarize this process in one line:

$$\text{Divisor} \cdot \text{Quotient} + \text{Remainder} = \text{Dividend}.$$

The same thing works for division of polynomials, as you can see by examining the division problem from Example 1:

$$\overset{\textit{Divisor} \cdot \textit{Quotient} \;\; + \textit{Remainder}}{(2x^2 - x) \cdot (4x + 3) + (3x + 1)} = (8x^3 + 2x^2 - 3x) + (3x + 1)$$
$$= \underset{\textit{Dividend}}{8x^3 + 2x^2 + 1}$$

---

*If this subtraction is confusing, write it out horizontally and watch the signs:

$$(8x^3 + 2x^2 + 1) - (8x^3 - 4x^2) = 8x^3 + 2x^2 + 1 - 8x^3 + 4x^2 = 6x^2 + 1.$$

This fact is so important that it is given a special name and a formal statement.

The Division Algorithm

If a polynomial $f(x)$ is divided by a nonzero polynomial $h(x)$, then there is a quotient polynomial $q(x)$ and a remainder polynomial $r(x)$ such that

$$Dividend = Divisor \cdot Quotient + Remainder$$

$$f(x) = h(x)\, q(x) + r(x),$$

where either $r(x) = 0$ or $r(x)$ has degree less than the degree of the divisor $h(x)$.

Technology Tip

The TI-89 does polynomial division (use PROPFRAC in the ALGEBRA menu). It displays the answer as the sum of a fraction and a polynomial:

$$\frac{Remainder}{Divisor} + Quotient.$$

Example 2 Show that $2x^2 + 1$ is a factor of $6x^3 - 4x^2 + 3x - 2$.

Solution We divide $6x^3 - 4x^2 + 3x - 2$ by $2x^2 + 1$, and find that the remainder is 0:

$$
\begin{array}{r}
3x - 2 \\
2x^2 + 1 \overline{\smash{\big)}\, 6x^3 - 4x^2 + 3x - 2} \\
\underline{6x^3 \qquad\quad + 3x} \\
-4x^2 \qquad\quad - 2 \\
\underline{-4x^2 \qquad\quad - 2} \\
0.
\end{array}
$$

Since the remainder is 0, the Division Algorithm tells us that:

$$Dividend = Divisor \cdot Quotient + Remainder$$

$$6x^3 - 4x^2 + 3x - 2 = (2x^2 + 1)(3x - 2) + 0$$

$$= (2x^2 + 1)(3x - 2).$$

Therefore, $2x^2 + 1$ is a factor of $6x^3 - 4x^2 + 3x - 2$, and the other factor is the quotient $3x - 2$. ■

Example 2 illustrates this fact:

Remainders and Factors

The remainder in polynomial division is 0 exactly when the divisor is a factor of the dividend. In this case, the quotient is the other factor.

Example 3 Let $f(x) = x^3 - 2x^2 - 4x + 5$.

(a) Find the quotient and remainder when $f(x)$ is divided by $x - 3$.

(b) Find $f(3)$.

NOTE   When the divisor is a first-degree polynomial such as $x - 3$, there is a convenient shorthand method of division, called **synthetic division**. See Excursion 4.2.A for details.

Solution

(a) Using long division, we have:

$$
\begin{array}{r}
x^2 + x - 1 \\
x - 3 \overline{) x^3 - 2x^2 - 4x + 5} \\
\underline{x^3 - 3x^2} \\
x^2 - 4x + 5 \\
\underline{x^2 - 3x} \\
-x + 5 \\
\underline{-x + 3} \\
2
\end{array}
$$

Therefore, the quotient is $x^2 + x - 1$ and the remainder is 2.

(b) Using the Division Algorithm, we can write the dividend $f(x) = x^3 - 2x^2 - 4x + 5$ as

$$Dividend = Divisor \cdot Quotient + Remainder$$
$$f(x) = (x - 3)(x^2 + x - 1) + 2$$

so that

$$f(3) = (3 - 3)(3^2 + 3 - 1) + 2 = 0 + 2 = 2.$$

Note that the number $f(3)$ is the same as the remainder when $f(x)$ is divided by $x - 3$. ■

The argument used in Example 3 to show that $f(3)$ is the remainder when $f(x)$ is divided by $x - 3$ also works in the general case and proves this fact:

Remainder Theorem

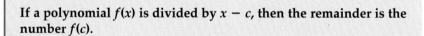

If a polynomial $f(x)$ is divided by $x - c$, then the remainder is the number $f(c)$.

Example 4   To find the remainder when $f(x) = x^{79} + 3x^{24} + 5$ is divided by $x - 1$, we apply the Remainder Theorem with $c = 1$. The remainder is

$$f(1) = 1^{79} + 3 \cdot 1^{24} + 5 = 1 + 3 + 5 = 9.$$ ■

Example 5   To find the remainder when $f(x) = 3x^4 - 8x^2 + 11x + 1$ is divided by $x + 2$, we must apply the Remainder Theorem *carefully*. The divisor in the theorem is $x - c$, not $x + c$. So we rewrite $x + 2$ as $x - (-2)$ and apply the theorem with $c = -2$. The remainder is

$$f(-2) = 3(-2)^4 - 8(-2)^2 + 11(-2) + 1 = 48 - 32 - 22 + 1 = -5.$$ ■

Remainders and Roots

If $f(x)$ is a polynomial, then a solution of the equation $f(x) = 0$ is called a **root** or **zero** of $f(x)$. Thus, a number c is a root of $f(x)$ if $f(c) = 0$. A root that is a real number is called a **real root**. For example, 4 is a real root of the polynomial $f(x) = 3x - 12$ because $f(4) = 3 \cdot 4 - 12 = 0$. There is an interesting connection between the roots of a polynomial and its factors.

Example 6 Let $f(x) = x^3 - 4x^2 + 2x + 3$.

(a) Show that 3 is a root of $f(x)$.

(b) Show that $x - 3$ is a factor of $f(x)$.

Solution

(a) Evaluating $f(x)$ at 3 shows that
$$f(3) = 3^3 - 4(3^2) + 2(3) + 3 = 0.$$
Therefore, 3 is a root of $f(x)$.

(b) If $f(x)$ is divided by $x - 3$, then by the Division Algorithm, there is a quotient polynomial $q(x)$ such that
$$f(x) = (x - 3)q(x) + \text{remainder}.$$
The Remainder Theorem shows that the remainder when $f(x)$ is divided by $x - 3$ is the number $f(3)$, which is 0, as we saw in part (a). Therefore,
$$f(x) = (x - 3)q(x) + 0 = f(x) = (x - 3)q(x).$$
Thus, $x - 3$ is a factor of $f(x)$. [To determine the other factor, the quotient $q(x)$, you have to perform the division.] ■

Example 6 illustrates this fact, which can be proved by the same argument used in the example:

Factor Theorem

The number c is a root of the polynomial $f(x)$ exactly when $x - c$ is a factor of $f(x)$.

The Factor Theorem and a calculator can sometimes be used to factor polynomials.

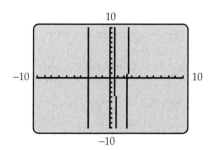

Figure 4–9

Example 7 The graph of $f(x) = 15x^3 - x^2 - 114x + 72$ in the standard viewing window (Figure 4–9) is obviously not complete, but it does suggest that -3 is an x-intercept, and hence a solution of the equation

$f(x) = 0$, that is, a root of $f(x)$. It is easy to verify that this is indeed the case:

$$f(-3) = 15(-3)^3 - (-3)^2 - 114(-3) + 72 = -405 - 9 + 342 + 72 = 0.$$

Since -3 is a root, $x - (-3) = x + 3$ is a factor of $f(x)$. Use division to verify that the other factor is $15x^2 - 46x + 24$. By factoring this quadratic, we obtain a complete factorization of $f(x)$:

$$f(x) = (x + 3)(15x^2 - 46x + 24) = (x + 3)(3x - 2)(5x - 12). \quad \blacksquare$$

Example 8 Find three polynomials of different degrees that have 1, 2, 3, and -5 as roots.

Solution A polynomial that has 1, 2, 3, and -5 as roots must have $x - 1, x - 2, x - 3$, and $x - (-5) = x + 5$ as factors. Many polynomials satisfy these conditions, such as

$$g(x) = (x - 1)(x - 2)(x - 3)(x + 5) = x^4 - x^3 - 19x^2 + 49x - 30$$

$$h(x) = 8(x - 1)(x - 2)(x - 3)^2(x + 5)$$

$$k(x) = 2(x + 4)^2(x - 1)(x - 2)(x - 3)(x + 5)(x^2 + x + 1).$$

Note that g has degree 4. When h is multiplied out, its leading term is $8x^5$, so h has degree 5. Similarly, k has degree 8 since its leading term is $2x^8$. $\quad \blacksquare$

If a polynomial $f(x)$ has four roots, say a, b, c, d, then by the same argument used in Example 8, it must have

$$(x - a)(x - b)(x - c)(x - d)$$

as a factor. Since $(x - a)(x - b)(x - c)(x - d)$ has degree 4 (multiply it out—its leading term is x^4), $f(x)$ must have degree at least 4. In particular, this means that no polynomial of degree 3 can have four or more roots. A similar argument works in the general case.

Number of Roots

A polynomial of degree n has at most n distinct roots.

Rational Roots

Finding the real roots of polynomials is the same as solving polynomial equations. So by using the techniques presented in Sections 2.1 and 2.2 we can always find the roots of first- and second-degree polynomials. The roots of higher degree polynomials are usually harder to find exactly. When a polynomial has integer coefficients, however, all of its **rational roots** (roots that are rational numbers) can be found exactly by using the following result.

Rational Root Test

> If a rational number r/s (in lowest terms) is a root of the polynomial
>
> $$a_n x^n + \cdots + a_1 x + a_0,$$
>
> where the coefficients $a_n, \ldots, a_1, a_0$ are integers with $a_n \neq 0, a_0 \neq 0$, then
>
> r is a factor of the constant term a_0 and
>
> s is a factor of the leading coefficient a_n.

The test states that every rational root must satisfy certain conditions.* By finding all the numbers that satisfy these conditions, we produce a list of *possible* rational roots. Then we must evaluate the polynomial at each number on the list to see if the number actually is a root. This testing process can be considerably shortened by using a calculator, as in the next example.

Example 9 Find the rational roots of

$$f(x) = 2x^4 + x^3 - 17x^2 - 4x + 6.$$

Solution If $f(x)$ has a rational root r/s, then by the Rational Root Test r must be a factor of the constant term 6. Therefore, r must be one of $\pm 1, \pm 2, \pm 3,$ or ± 6 (the only factors of 6). Similarly, s must be a factor of the leading coefficient 2, so s must be one of ± 1 or ± 2 (the only factors of 2). Consequently, the only *possibilities* for r/s are

$$\frac{\pm 1}{\pm 1}, \frac{\pm 2}{\pm 1}, \frac{\pm 3}{\pm 1}, \frac{\pm 6}{\pm 1}, \frac{\pm 1}{\pm 2}, \frac{\pm 2}{\pm 2}, \frac{\pm 3}{\pm 2}, \frac{\pm 6}{\pm 2}.$$

Eliminating duplications from this list, we see that the only *possible* rational roots are

$$1, -1, 2, -2, 3, -3, 6, -6, \frac{1}{2}, -\frac{1}{2}, \frac{3}{2}, -\frac{3}{2}.$$

Now graph $f(x)$ in a viewing window that includes all of these numbers on the x-axis, say $-7 \le x \le 7$ and $-5 \le y \le 5$ (Figure 4–10). A complete graph isn't necessary since we are interested only in the x-intercepts.

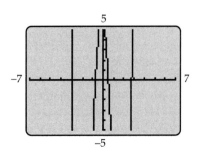

Figure 4–10

*Since the proof of the Rational Root Test sheds no light on how the test is actually used to solve equations, it will be omitted.

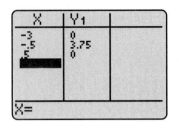

Figure 4–11

Figure 4–10 shows that the only numbers on our list that could possibly be roots (x-intercepts) are -3, $-1/2$, and $1/2$, so these are the only ones that need be tested. We use the table feature to evaluate $f(x)$ at these three numbers (Figure 4–11). The table shows that -3 and $1/2$ are the only rational roots of $f(x)$. Its other roots (x-intercepts) in Figure 4–10 must be irrational numbers. ■

Once some roots of a polynomial have been found, the Factor Theorem can be used to factor the polynomial, which may lead to additional roots.

Example 10 Find all the roots of $f(x) = 2x^4 + x^3 - 17x^2 - 4x + 6$.

Solution In Example 9 we saw that -3 and $1/2$ are the rational roots of $f(x)$. By the Factor Theorem $x - (-3) = x + 3$ and $x - 1/2$ are factors of $f(x)$. Using division twice, we have

$$2x^4 + x^3 - 17x^2 - 4x + 6 = (x + 3)(2x^3 - 5x^2 - 2x + 2)$$
$$= (x + 3)(x - .5)(2x^2 - 4x - 4)$$

The remaining roots of $f(x)$ are the roots of $2x^2 - 4x - 4$, that is, the solutions of

$$2x^2 - 4x - 4 = 0$$
$$x^2 - 2x - 2 = 0.$$

They are easily found by the quadratic formula.

$$x = \frac{-(-2) \pm \sqrt{(-2)^2 - 4 \cdot 1 \cdot (-2)}}{2 \cdot 1}$$
$$= \frac{2 \pm \sqrt{12}}{2} = \frac{2 \pm 2\sqrt{3}}{2} = 1 \pm \sqrt{3}.$$

Therefore, $f(x)$ has rational roots -3 and $1/2$, and irrational roots $1 + \sqrt{3}$ and $1 - \sqrt{3}$. ■

Example 11 Factor $f(x) = 2x^5 - 10x^4 + 7x^3 + 13x^2 + 3x + 9$ completely.

Solution We begin by finding as many roots of $f(x)$ as we can. By the Rational Root Test, every rational root is of the form r/s, where $r = \pm 1$, ± 3, or ± 9 and $s = \pm 1$ or ± 2. Thus, the possible rational roots are

$$\pm 1, \pm 3, \pm 9, \pm \frac{1}{2}, \pm \frac{3}{2}, \pm \frac{9}{2}.$$

The partial graph of $f(x)$ in Figure 4–12 shows that the only possible roots (x-intercepts) are -1 and 3. You can easily verify that both -1 and 3 are roots of $f(x)$.

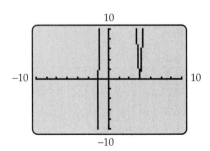

Figure 4–12

Since -1 and 3 are roots, $x - (-1) = x + 1$ and $x - 3$ are factors of $f(x)$ by the Factor Theorem. Division shows that

$$f(x) = 2x^5 - 10x^4 + 7x^3 + 13x^2 + 3x + 9$$
$$= (x + 1)(2x^4 - 12x^3 + 19x^2 + 3x + 9)$$
$$= (x + 1)(x - 3)(2x^3 - 6x^2 + x - 3).$$

The other roots of $f(x)$ are the roots of $g(x) = 2x^3 - 6x^2 + x - 3$. We first check for rational roots of $g(x)$. Since every root of $g(x)$ is also a root of $f(x)$ (why?), the only possible rational roots of $g(x)$ are -1 and 3 [the rational roots of $f(x)$]. We have

$$g(-1) = 2(-1)^3 - 6(-1)^2 + (-1) - 3 = -12;$$
$$g(3) = 2(3^3) - 6(3^2) + 3 - 3 = 0.$$

So -1 is not a root, but 3 is a root of $g(x)$. By the Factor Theorem, $x - 3$ is a factor of $g(x)$. Division shows that

$$f(x) = (x + 1)(x - 3)(2x^3 - 6x^2 + x - 3)$$
$$= (x + 1)(x - 3)(x - 3)(2x^2 + 1).$$

Since $2x^2 + 1$ has no real roots, it cannot be factored. So the factorization of $f(x)$ is complete. ■

Exercises 4.2

In Exercises 1–6, state the quotient and remainder when the first polynomial is divided by the second. Check your division by calculating (divisor) (quotient) + remainder.

1. $3x^4 + 2x^2 - 6x + 1;$ $\quad x + 1$

2. $x^5 - x^3 + x - 5;$ $\quad x - 2$

3. $x^5 + 2x^4 - 6x^3 + x^2 - 5x + 1;$ $\quad x^3 + 1$

4. $3x^4 - 3x^3 - 11x^2 + 6x - 1;$ $\quad x^3 + x^2 - 2$

5. $5x^4 + 5x^2 + 5;$ $\quad x^2 - x + 1$

6. $x^5 - 1;$ $\quad x - 1$

In Exercises 7–10, determine whether the first polynomial is a factor of the second.

7. $x^2 + 3x - 1;$ $\quad x^3 + 2x^2 - 5x - 6$

8. $x^2 + 9;$ $\quad x^5 + x^4 - 81x - 81$

9. $x^2 + 3x - 1;$ $\quad x^4 + 3x^3 - 2x^2 - 3x + 1$

10. $x^2 - 5x + 7;$ $\quad x^3 - 3x^2 - 3x + 9$

In Exercises 11–14, determine which of the given numbers are roots of the given polynomial.

11. $2, 3, 0, -1;$ $\quad g(x) = x^4 + 6x^3 - x^2 - 30x$

12. $1, 1/2, 2, -1/2, 3;$ $\quad f(x) = 6x^2 + x - 1$

13. $2\sqrt{2}, \sqrt{2}, -\sqrt{2}, 1, -1;$
  $h(x) = x^3 + x^2 - 8x - 8$

14. $\sqrt{3}, -\sqrt{3}, 1, -1;$ $\quad k(x) = 8x^3 - 12x^2 - 6x + 9$

In Exercises 15–22, find the remainder when $f(x)$ is divided by $g(x)$, without using division.

15. $f(x) = x^{10} + x^8;$ $\quad g(x) = x - 1$

16. $f(x) = x^6 - 10;$ $\quad g(x) = x - 2$

17. $f(x) = 3x^4 - 6x^3 + 2x - 1;$ $\quad g(x) = x + 1$

18. $f(x) = x^5 - 3x^2 + 2x - 1;$ $\quad g(x) = x - 2$

19. $f(x) = x^3 - 2x^2 + 5x - 4;$ $\quad g(x) = x + 2$

20. $f(x) = 10x^{75} - 8x^{65} + 6x^{45} + 4x^{32} - 2x^{15} + 5;$
  $g(x) = x - 1$

21. $f(x) = 2x^5 - 3x^4 + x^3 - 2x^2 + x - 8;$
$g(x) = x - 10$

22. $f(x) = x^3 + 8x^2 - 29x + 44;$ $\quad g(x) = x + 11$

In Exercises 23–26, use the Factor Theorem to determine whether or not $h(x)$ is a factor of $f(x)$.

23. $h(x) = x - 1;$ $\quad f(x) = x^5 + 1$

24. $h(x) = x + 2;$ $\quad f(x) = x^3 - 3x^2 - 4x - 12$

25. $h(x) = x + 1;$ $\quad f(x) = x^3 - 4x^2 + 3x + 8$

26. $h(x) = x - 1;$ $\quad f(x) = 14x^{99} - 65x^{56} + 51$

In Exercises 27–30, use the Factor Theorem and a calculator to factor the polynomial, as in Example 7.

27. $f(x) = 6x^3 - 7x^2 - 89x + 140$

28. $g(x) = x^3 - 5x^2 - 5x - 6$

29. $h(x) = 4x^4 + 4x^3 - 35x^2 - 36x - 9$

30. $f(x) = x^5 - 5x^4 - 5x^3 + 25x^2 + 6x - 30$

In Exercises 31–34, each graph is of a polynomial function $f(x)$ of degree 5 whose leading coefficient is 1. The graph is not drawn to scale. Use the Factor Theorem to find the polynomial. [Hint: What are the roots of $f(x)$? What does the Factor Theorem tell you?]

31.

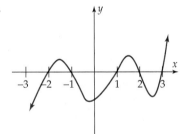

32.

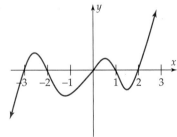

33.

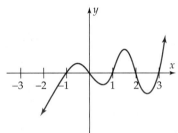

34.

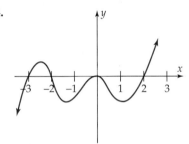

In Exercises 35–38, find a polynomial with the given degree n, the given roots, and no other roots.

35. $n = 3;$ roots $1, 7, -4$

36. $n = 3;$ roots $1, -1$

37. $n = 6;$ roots $1, 2, \pi$

38. $n = 5;$ root 2

39. Find a polynomial function f of degree 3 such that $f(10) = 17$ and the roots of $f(x)$ are 0, 5, and 8.

40. Find a polynomial function g of degree 4 such that the roots of g are $0, -1, 2, -3$ and $g(3) = 288$.

In Exercises 41–52, find all the rational roots of the polynomial.

41. $x^3 + 3x^2 - x - 3$ $\qquad$ **42.** $x^3 - x^2 - 3x + 3$

43. $x^3 + 5x^2 - x - 5$ $\qquad$ **44.** $3x^3 + 8x^2 - x - 20$

45. $f(x) = 2x^5 + 5x^4 - 11x^3 + 4x^2$ [*Hint:* The Rational Root Test can only be used on polynomials with nonzero constant terms. Factor $f(x)$ as a product of a power of x and a polynomial $g(x)$ with nonzero constant term. Then use the Rational Root Test on $g(x)$.]

46. $2x^6 - 3x^5 - 7x^4 - 6x^3$

47. $f(x) = \frac{1}{12}x^3 - \frac{1}{12}x^2 - \frac{2}{3}x + 1$ [*Hint:* The Rational Root Test can only be used on polynomials with integer coefficients. Note that $f(x)$ and $12\,f(x)$ have the same roots. (Why?)]

48. $\frac{2}{3}x^4 + \frac{1}{2}x^3 - \frac{5}{4}x^2 - x - \frac{1}{6}$

49. $\frac{1}{3}x^4 - x^3 - x^2 + \frac{13}{3}x - 2$

50. $\frac{1}{3}x^7 - \frac{1}{2}x^6 - \frac{1}{6}x^5 + \frac{1}{6}x^4$

51. $.1x^3 - 1.9x + 3$

52. $.05x^3 + .45x^2 - .4x + 1$

In Exercises 53–58, factor the polynomial as a product of linear factors and a factor $g(x)$ such that $g(x)$ is either a constant or a polynomial that has no rational roots.

53. $2x^3 - 4x^2 + x - 2$

54. $6x^3 - 5x^2 + 3x - 1$

55. $x^6 + 2x^5 + 3x^4 + 6x^3$

56. $x^5 - 2x^4 + 2x^3 - 3x + 2$

57. $x^5 - 4x^4 + 8x^3 - 14x^2 + 15x - 6$

58. $x^5 + 4x^3 + x^2 + 6x$

59. (a) Show that $\sqrt{2}$ is an irrational number. [*Hint:* $\sqrt{2}$ is a root of $x^2 - 2$. Does this polynomial have any rational roots?]
(b) Show that $\sqrt{3}$ is irrational.

60. Graph $f(x) = .001x^3 - .199x^2 - .23x + 6$ in the standard viewing window.
(a) How many roots does $f(x)$ appear to have? Without changing the viewing window, explain why $f(x)$ must have an additional root. [*Hint:* Each root corresponds to a factor of $f(x)$. What does the rest of the factorization consist of?]
(b) Find all the roots of $f(x)$.

61. According to data from the FBI, the number of people murdered each year per 100,000 population can be approximated by the polynomial function
$$f(x) = .0011x^4 - .0233x^3 + .1144x^2 + .0126x + 8.1104 \quad (0 \le x \le 10),$$
where $x = 0$ corresponds to 1987.
(a) What was the murder rate in 1990?
(b) In what year was the rate 8 people per 100,000?

62. During the first 150 hours of an experiment, the growth rate of a bacteria population at time t hours is $g(t) = -.0003t^3 + .04t^2 + .3t + .2$ bacteria per hour.
(a) What is the growth rate at 50 hours? At 100 hours?
(b) What is the growth rate at 145 hours? What does this mean?
(c) At what time is the growth rate 0?
(d) At what time is the growth rate -50 bacteria per hour?

63. In a sealed chamber where the temperature varies, the instantaneous rate of change of temperature with respect to time over an 11-day period is given by $F(t) = .0035t^4 - .4t^2 - .2t + 6$, where time is measured in days and temperature in degrees Fahrenheit (so that rate of change is in degrees per day).
(a) At what rate is the temperature changing at the beginning of the period ($t = 0$)? At the end of the period ($t = 11$)?
(b) When is the temperature increasing at a rate of 4°F per day?
(c) When is the temperature decreasing at a rate of 3°F per day?

4.2.A **EXCURSION** **Synthetic Division**

Synthetic division is a fast method of doing polynomial division, when the divisor is a first-degree polynomial of the form $x - c$ for some real number c. To see how it works, we first consider an example of ordinary long division:

$$
\begin{array}{r}
3x^3 + 6x^2 + 4x - 3 \quad \leftarrow \textit{Quotient} \\
x - 2 \,\big|\, \overline{3x^4 - 8x^2 - 11x + 1} \quad \leftarrow \textit{Dividend} \\
\underline{3x^4 - 6x^3} \\
6x^3 - 8x^2 \\
\underline{6x^3 - 12x^2} \\
4x^2 - 11x \\
\underline{4x^2 - 8x} \\
- 3x + 1 \\
\underline{- 3x + 6} \\
- 5 \quad \leftarrow \textit{Remainder}
\end{array}
$$

$\textit{Divisor} \rightarrow$ appears before $x - 2$ above.

This calculation obviously involves a lot of repetitions. If we insert 0 coefficients for terms that don't appear above and keep the various coefficients in the proper columns, we can eliminate the repetitions and all the x's:

$$
\begin{array}{r}
3 \quad 6 \quad 4 \quad -3 \qquad \leftarrow \textit{Quotient} \\
\textit{Divisor} \rightarrow 1 - 2 \,\big|\, \overline{3 \quad 0 \quad -8 \quad -11 \qquad 1} \quad \leftarrow \textit{Dividend} \\
\underline{-6} \\
6 \\
\underline{-12} \\
4 \\
\underline{-8} \\
-3 \\
\underline{+6} \\
-5 \quad \leftarrow \textit{Remainder}
\end{array}
$$

We can save space by moving the lower lines upward and writing 2 in the divisor position (since that's enough to remind us that the divisor is $x - 2$):

$$
\begin{array}{r}
3 \quad 6 \quad 4 \quad -3 \qquad \leftarrow \textit{Quotient} \\
\textit{Divisor} \rightarrow 2 \,\big|\, \overline{3 \quad 0 \quad -8 \quad -11 \qquad 1} \quad \leftarrow \textit{Dividend} \\
\underline{-6 \quad -12 \quad -8 \qquad 6} \\
6 \quad 4 \quad -3 \,\boxed{-5} \quad \leftarrow \textit{Remainder}
\end{array}
$$

Since the last line contains most of the quotient line, we can save more space and still preserve the essential information by inserting a 3 in the last line and omitting the top line:

$$
\begin{array}{r}
\textit{Divisor} \rightarrow 2 \,\big|\, \overline{3 \quad 0 \quad -8 \quad -11 \qquad 1} \quad \leftarrow \textit{Dividend} \\
\underline{-6 \quad -12 \quad -8 \qquad 6} \\
\underbrace{3 \quad 6 \quad 4 \quad -3}_{\textit{Quotient}} \,\boxed{-5} \quad \leftarrow \textit{Remainder}
\end{array}
$$

Synthetic division is a quick method for obtaining the last row of this array. Here is a step-by-step explanation of the division of $3x^4 - 8x^2 - 11x + 1$ by $x - 2$:

Step 1   In the first row list the 2 from the divisor and the coefficients of the dividend in order of decreasing powers of x (insert 0 coefficients for missing powers of x).

$$
2 \,\big|\, 3 \quad 0 \quad -8 \quad -11 \quad 1
$$

Step 2 Bring down the first dividend coefficient (namely, 3) to the third row.

$$2\underline{)}\ \ 3\quad 0\quad -8\quad -11\quad 1$$
$$\ 3\ \ \underline{}$$

Step 3 Multiply $2\cdot 3$ and insert the answer 6 in the second row, in the position shown here.

$$2\underline{)}\ \ 3\quad 0\quad -8\quad -11\quad 1$$
$$\ 6$$
$$\ 3\ \ \underline{}$$

Step 4 Add $0 + 6$ and write the answer 6 in the third row.

$$2\underline{)}\ \ 3\quad 0\quad -8\quad -11\quad 1$$
$$\ 6$$
$$\ 3\quad 6\ \ \underline{}$$

Step 5 Multiply $2\cdot 6$ and insert the answer 12 in the second row.

$$2\underline{)}\ \ 3\quad 0\quad -8\quad -11\quad 1$$
$$\ 6\quad 12$$
$$\ 3\quad 6\ \ \underline{}$$

Step 6 Add $-8 + 12$ and write the answer 4 in the third row.

$$2\underline{)}\ \ 3\quad 0\quad -8\quad -11\quad 1$$
$$\ 6\quad 12$$
$$\ 3\quad 6\quad 4\ \ \underline{}$$

Step 7 Multiply $2\cdot 4$ and insert the answer 8 in the second row.

$$2\underline{)}\ \ 3\quad 0\quad -8\quad -11\quad 1$$
$$\ 6\quad 12\quad 8$$
$$\ 3\quad 6\quad 4\ \ \underline{}$$

Step 8 Add $-11 + 8$ and write the answer -3 in the third row.

$$2\underline{)}\ \ 3\quad 0\quad -8\quad -11\quad 1$$
$$\ 6\quad 12\quad 8$$
$$\ 3\quad 6\quad 4\quad -3\ \ \underline{}$$

Step 9 Multiply $2\cdot(-3)$ and insert the answer -6 in the second row.

$$2\underline{)}\ \ 3\quad 0\quad -8\quad -11\quad 1$$
$$\ 6\quad 12\quad 8\quad -6$$
$$\ 3\quad 6\quad 4\quad -3\ \ \underline{}$$

Step 10 Add $1 + (-6)$ and write the answer -5 in the third row.

$$2\underline{)}\ \ 3\quad 0\quad -8\quad -11\quad 1$$
$$\ 6\quad 12\quad 8\quad -6$$
$$\ 3\quad 6\quad 4\quad -3\ \ \underline{|-5}$$

The last line of this array is the same as the last line of the array obtained from the long division process, and we can read off the quotient and remainder:

> The last number in the third row is the remainder.

> The other numbers in the third row are the coefficients of the quotient (arranged in order of decreasing powers of x).

Since we are dividing the *fourth*-degree polynomial $3x^4 - 8x^2 - 11x + 1$ by the *first*-degree polynomial $x - 2$, the quotient must be a polynomial of degree *three* with coefficients 3, 6, 4, -3, namely, $3x^3 + 6x^2 + 4x - 3$. The remainder is -5.

CAUTION

Synthetic division can be used *only* when the divisor is a first-degree polynomial of the form $x - c$. In the example above, $c = 2$. If you want to use synthetic division with a divisor such as $x + 3$, you must write it as $x - (-3)$, which is of the form $x - c$ with $c = -3$.

Technology Tip

Synthetic division
programs are in the
Program Appendix.

Example 1 To divide $x^5 + 5x^4 + 6x^3 - x^2 + 4x + 29$ by $x + 3$, we write the divisor as $x - (-3)$ and proceed as above:

$$
\begin{array}{r|rrrrrr}
-3 & 1 & 5 & 6 & -1 & 4 & 29 \\
   &   & -3 & -6 & 0 & 3 & -21 \\
\hline
   & 1 & 2 & 0 & -1 & 7 & \underline{8} \\
\end{array}
$$

The last row shows that the quotient is $x^4 + 2x^3 - x + 7$ and the remainder is 8. ∎

Example 2 Show that $x - 7$ is a factor of

$$8x^5 - 52x^4 + 2x^3 - 198x^2 - 86x + 14$$

and find the other factor.

Solution $x - 7$ is a factor exactly when division by $x - 7$ leaves remainder 0, in which case the quotient is the other factor. Using synthetic division we have:

$$
\begin{array}{r|rrrrrr}
7 & 8 & -52 & 2 & -198 & -86 & 14 \\
  &   & 56 & 28 & 210 & 84 & -14 \\
\hline
  & 8 & 4 & 30 & 12 & -2 & \underline{0} \\
\end{array}
$$

Since the remainder is 0, the divisor $x - 7$ and the quotient $8x^4 + 4x^3 + 30x^2 + 12x - 2$ are factors:

$$8x^5 - 52x^4 + 2x^3 - 198x^2 - 86x + 14$$
$$= (x - 7)(8x^4 + 4x^3 + 30x^2 + 12x - 2). \quad ∎$$

Exercises 4.2.A

In Exercises 1–8, use synthetic division to find the quotient and remainder.

1. $(3x^4 - 8x^3 + 9x + 5) \div (x - 2)$

2. $(4x^3 - 3x^2 + x + 7) \div (x - 2)$

3. $(2x^4 + 5x^3 - 2x - 8) \div (x + 3)$

4. $(3x^3 - 2x^2 - 8) \div (x + 5)$

5. $(5x^4 - 3x^2 - 4x + 6) \div (x - 7)$

6. $(3x^4 - 2x^3 + 7x - 4) \div (x - 3)$

7. $(x^4 - 6x^3 + 4x^2 + 2x - 7) \div (x - 2)$

8. $(x^6 - x^5 + x^4 - x^3 + x^2 - x + 1) \div (x + 3)$

In Exercises 9–12, use synthetic division to find the quotient and the remainder. In each divisor $x - c$, the number c is not an integer, but the same technique will work.

9. $(3x^4 - 2x^2 + 2) \div \left(x - \dfrac{1}{4} \right)$

10. $(2x^4 - 3x^2 + 1) \div \left(x - \dfrac{1}{2} \right)$

11. $(2x^4 - 5x^3 - x^2 + 3x + 2) \div \left(x + \dfrac{1}{2} \right)$

12. $\left(10x^5 - 3x^4 + 14x^3 + 13x^2 - \dfrac{4}{3}x + \dfrac{7}{3} \right) \div \left(x + \dfrac{1}{5} \right)$

In Exercises 13–16, use synthetic division to show that the first polynomial is a factor of the second and find the other factor.

13. $x + 4;$ $3x^3 + 9x^2 - 11x + 4$

14. $x - 5;$ $x^5 - 8x^4 + 17x^2 + 293x - 15$

15. $x - 1/2;$ $2x^5 - 7x^4 + 15x^3 - 6x^2 - 10x + 5$

16. $x + 1/3;$ $3x^6 + x^5 - 6x^4 + 7x^3 + 3x^2 - 15x - 5$

In Exercises 17 and 18, use a calculator and synthetic division to find the quotient and remainder.

17. $(x^3 - 5.27x^2 + 10.708x - 10.23) \div (x - 3.12)$

18. $(2.79x^4 + 4.8325x^3 - 6.73865x^2 + .9255x - 8.125)$
$\div (x - 1.35)$

Thinkers

19. When $x^3 + cx + 4$ is divided by $x + 2$, the remainder is 4. Find c.

20. If $x - d$ is a factor of $2x^3 - dx^2 + (1 - d^2)x + 5$, what is d?

4.3 Graphs of Polynomial Functions

The graphs of first- and second-degree polynomial functions are straight lines and parabolas, respectively (Sections 1.4 and 4.1). The emphasis here will be on higher degree polynomial functions. The simplest ones are those of the form $f(x) = ax^n$ (where a is a constant). Their graphs are of four types, as shown in the following chart.

Graph of f(x) = ax^n

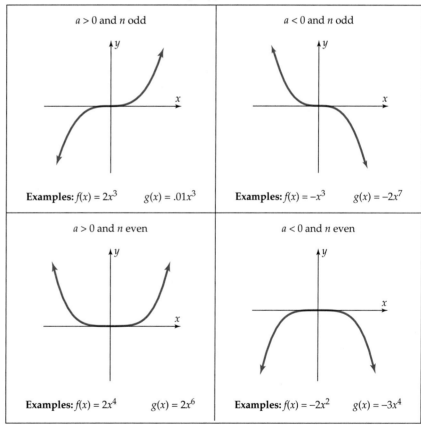

$a > 0$ and n odd	$a < 0$ and n odd
Examples: $f(x) = 2x^3$ $g(x) = .01x^3$	**Examples:** $f(x) = -x^3$ $g(x) = -2x^7$
$a > 0$ and n even	$a < 0$ and n even
Examples: $f(x) = 2x^4$ $g(x) = 2x^6$	**Examples:** $f(x) = -2x^2$ $g(x) = -3x^4$

GRAPHING EXPLORATION

Verify the accuracy of the preceding summary by graphing each of the examples in the window with $-5 \le x \le 5$ and $-30 \le y \le 30$.

Properties of Polynomial Graphs

The graphs of more complicated polynomial functions can vary considerably in shape. Understanding the properties discussed below should assist you to interpret screen images correctly and to determine when a polynomial graph is complete.

Continuity

Every polynomial graph is **continuous,** meaning that it is an unbroken curve, with no jumps, gaps, or holes. Furthermore, polynomial graphs have no sharp corners. Thus, neither of the graphs in Figure 4–13 is the graph of a polynomial function. On a calculator screen, however, a polynomial graph may look like a series of juxtaposed line segments, rather than a smooth, continuous curve.

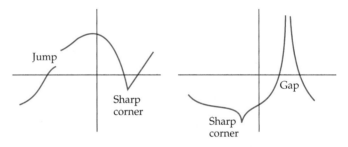

Figure 4–13

Shape of the Graph When |x| Is Large

The shape of a polynomial graph at the far left and far right of the coordinate plane is easily determined by using our knowledge of graphs of functions of the form $f(x) = ax^n$.

Example 1   Consider the function $f(x) = 2x^3 + x^2 - 6x$ and the function determined by its leading term $g(x) = 2x^3$.

> **GRAPHING EXPLORATION**
>
> Using the standard viewing window, graph f and g on the same screen.
>
> Do the graphs look different? Now graph f and g in the viewing window with $-20 \leq x \leq 20$ and $-10{,}000 \leq y \leq 10{,}000$. Do the graphs look almost the same?
>
> Finally, graph f and g in the viewing window with $-100 \leq x \leq 100$ and $-1{,}000{,}000 \leq y \leq 1{,}000{,}000$. Do the graphs look virtually identical?

The reason the answer to the last question is "yes" can be understood from this table:

x	-100	-50	70	100
$-6x$	600	300	-420	-600
x^2	10,000	2,500	4,900	10,000
$g(x) = 2x^3$	$-2,000,000$	$-250,000$	686,000	2,000,000
$f(x) = 2x^3 + x^2 - 6x$	$-1,989,400$	$-247,200$	690,480	2,009,400

It shows that when $|x|$ is large, the terms x^2 and $-6x$ are insignificant compared with $2x^3$ and play a very minor role in determining the value of $f(x)$. Hence the values of $f(x)$ and $g(x)$ are relatively close. ∎

Example 1 is typical of what happens in every case: When $|x|$ is very large, the highest power of x totally overwhelms all lower powers and plays the greatest role in determining the value of the function.

Behavior When $|x|$ Is Large

When $|x|$ is very large, the graph of a polynomial function closely resembles the graph of its highest degree term.

In particular, when the polynomial function has odd degree, one end of its graph shoots upward and the other end downward. When the polynomial function has even degree, both ends of its graph shoot upward or both ends shoot downward.

x-Intercepts

As we saw in Section 4.2 the x-intercepts of the graph of a polynomial function are the real roots of the polynomial. Since a polynomial of degree n has at most n distinct roots (page 282), we have:

x-Intercepts

The graph of a polynomial function of degree n meets the x-axis at most n times.

There is another connection between roots and graphs. For example, it is easy to see that the roots of $f(x) = (x + 3)^2(x + 1)(x - 1)^3$ are $-3, -1,$ and 1. We say that

-3 is a root of multiplicity 2;

-1 is a root of multiplicity 1;

1 is a root of multiplicity 3.

Observe that the graph of $f(x)$ in Figure 4–14 does not cross the x-axis at -3 (a root whose multiplicity is an *even* number), but does cross the

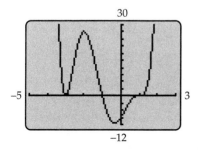

Figure 4–14

x-axis at -1 and 1 (roots of *odd* multiplicity). In the general case, a number c is a **root of multiplicity** k of a polynomial $f(x)$ if $(x - c)^k$ is a factor of $f(x)$ and no higher power of $(x - c)$ is a factor and we have this fact:

Multiplicity and Graphs

> Let c be a root of multiplicity k of a polynomial function f.
>
> If k is odd, the graph of f crosses the x-axis at c.
>
> If k is even, the graph of f touches, but does not cross, the x-axis at c.

Local Extrema

The term **local extremum** (plural, extrema) refers to either a local maximum or a local minimum, that is, a point where the graph has a peak or a valley.

> **GRAPHING EXPLORATION**
>
> Graph $f(x) = x^3 + 2x^2 - 4x - 3$ in the standard viewing window. What is the total number of peaks and valleys on the graph? What is the degree of $f(x)$?
>
> Now graph $g(x) = x^4 - 3x^3 - 2x^2 + 4x + 5$ in the standard viewing window. What is the total number of peaks and valleys on the graph? What is the degree of $g(x)$? ■

The two polynomials you have just graphed are illustrations of the following fact, which is proved in calculus:

Local Extrema

> A polynomial function of degree n has at most $n - 1$ local extrema. In other words, the total number of peaks and valleys on the graph is at most $n - 1$.

Complete Graphs of Polynomial Functions

By using the facts discussed earlier, you can often determine whether the graph of a polynomial function is complete (that is, shows all the important features).

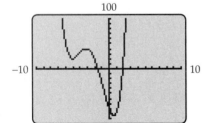

Figure 4–15

Example 2   Find a complete graph of
$$f(x) = x^4 + 10x^3 + 21x^2 - 40x - 80.$$

Solution   Since $f(0) = -80$, the standard viewing window probably won't show a complete graph, so we try the window with $-10 \le x \le 10$ and $-100 \le y \le 100$ and obtain Figure 4–15. The three peaks and valleys shown here are the only ones because a fourth-degree polynomial graph has at most three local extrema. There cannot be more x-intercepts than the two shown here because if the graph turned toward the x-axis farther out, there would be an additional peak, which is impossible. Finally, the outer ends of the graph resemble the graph of x^4, the highest degree term (see the chart on page 291). Hence, Figure 4–15 includes all the important features of the graph and is therefore complete. ■

Example 3 The graph of $f(x) = x^3 - 1.8x^2 + 2$ in Figure 4–16 is similar to the graph of its leading term $y = x^3$, but does not appear to have any local extrema. However, if you use the trace feature on the flat portion of the graph to the right of the x-axis, you see that the y-coordinates increase, then decrease, then increase (Try it!). Zooming in on the portion of the graph between 0 and 1 (Figure 4–17), we see that the graph actually has a tiny peak and valley (the maximum possible number of local extrema for a cubic). So Figures 4–16 and 4–17 together provide a complete graph of f. ■

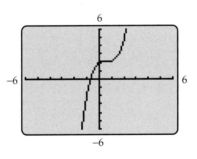

Figure 4–16

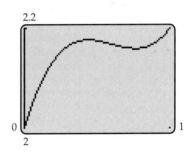

Figure 4–17

Exercise 49 shows that *no polynomial graph contains any horizontal line segments.* However, a calculator may erroneously show some, as in Figure 4–16. So always investigate such segments by using trace or zoom-in to determine any hidden behavior, as in Example 3.

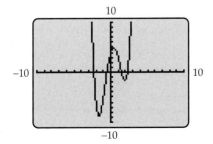

Figure 4–18

Example 4 In the standard viewing window, the graph of $f(x) = .01x^5 + x^4 - x^3 - 6x^2 + 5x + 4$ looks like Figure 4–18. This cannot be a complete graph because, when $|x|$ is large, the graph of $f(x)$ must resemble the graph of $g(x) = .01x^5$, whose left end goes downward (see the chart on page 291). So the graph of $f(x)$ must turn downward and cross the x-axis somewhere to the left of the origin. Even without graphing, we can see that there must be one more peak (where the graph turns downward), making a total of four local extrema (the most a fifth-degree polynomial can have), and another x-intercept for a total of five. When these additional features are shown, we will have a complete graph.

> ### GRAPHING EXPLORATION
>
> Find a viewing window that includes the local maximum and
> x-intercept not shown in Figure 4–18. When you do, the scale
> will be such that the local extrema and x-intercepts shown in
> Figure 4–18 will no longer be visible.

Consequently, a complete graph of $f(x)$ requires several viewing windows
in order to see all the important features. ■

The graphs obtained in Examples 2–4 were known to be complete because in each case they included the maximum possible number of local
extrema. In many cases, however, a graph may not have the largest possible number of peaks and valleys. In such cases, use any available information and try several viewing windows to obtain the most likely
complete graph.

Exercises 4.3

*In Exercises 1–6, decide whether the given graph could pos-
sibly be the graph of a polynomial function.*

1.

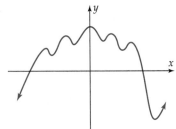

2.

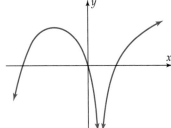

3.

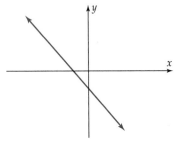

4.

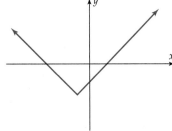

5.

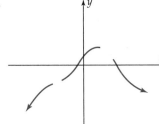

6.

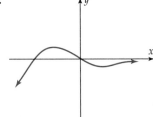

In Exercises 7–12, determine whether the given graph could possibly be the graph of a polynomial function of degree 3, of degree 4, or of degree 5.

7.

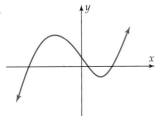

8.

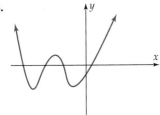

9.

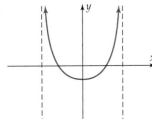

10.

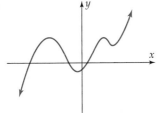

11.

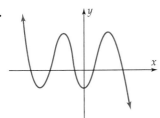

12.

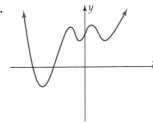

In Exercises 13 and 14, find a viewing window in which the graph of the given polynomial function f appears to have the same general shape as the graph of its leading term.

13. $f(x) = x^4 - 6x^3 + 9x^2 - 3$

14. $f(x) = x^3 - 5x^2 + 4x - 2$

In Exercises 15–18, the graph of a polynomial function is shown. List each root of the polynomial and state whether its multiplicity is even or odd.

15.

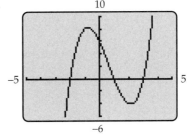

16.

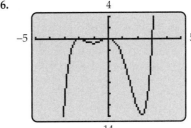

17.

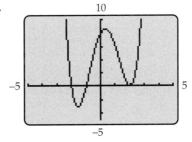

18.

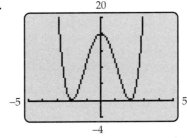

In Exercises 19–24, use your knowledge of polynomial graphs, not a calculator, to match the given function with its graph, which is one of those shown here.

A.
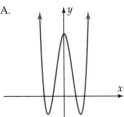

B.
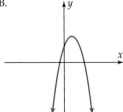

C.
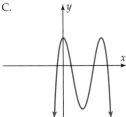

D.
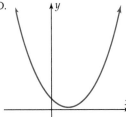

E.
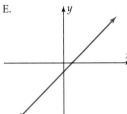

F.
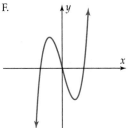

19. $f(x) = 2x - 3$

20. $g(x) = x^2 - 4x + 7$

21. $g(x) = x^3 - 4x$

22. $f(x) = x^4 - 5x^2 + 4$

23. $f(x) = -x^4 + 6x^3 - 9x^2 + 2$

24. $g(x) = -2x^2 + 3x + 1$

In Exercises 25–28, graph the function in the standard viewing window and explain why that graph cannot possibly be complete.

25. $f(x) = .01x^3 - .2x^2 - .4x + 7$

26. $g(x) = .01x^4 + .1x^3 - .8x^2 - .7x + 9$

27. $h(x) = .005x^4 - x^2 + 5$

28. $f(x) = .001x^5 - .01x^4 - .2x^3 + x^2 + x - 5$

In Exercises 29–34, find a complete graph of the function and list the viewing window(s) that show this graph.

29. $f(x) = x^3 + 8x^2 + 5x - 14$

30. $g(x) = x^3 - 3x^2 - 4x - 5$

31. $g(x) = -x^4 - 3x^3 + 24x^2 + 80x + 15$

32. $f(x) = x^4 - 10x^3 + 35x^2 - 50x + 24$

33. $f(x) = .1x^5 + 3x^4 - 4x^3 - 11x^2 + 3x + 2$

34. $g(x) = x^4 - 48x^3 - 101x^2 + 49x + 50$

35. (a) Explain why the graph of a cubic polynomial function has either two local extrema or none at all. [*Hint:* If it had only one, what would the graph look like when $|x|$ is very large?]

(b) Explain why the general shape of the graph of a cubic polynomial function must be one of the following.

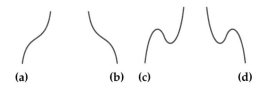

(a)   **(b)**   **(c)**   **(d)**

36. The figure shows an incomplete graph of a fourth-degree, even polynomial function f. (Even functions were defined in Excursion 3.4.A.)

(a) Find the roots of f.

(b) Explain why
$$f(x) = k(x - a)(x - b)(x - c)(x - d),$$
where a, b, c, d are the roots of f.

(c) Experiment with your calculator to find the value of k that produces the graph in the figure.

(d) Find all local extrema of f.

(e) List the approximate intervals on which f is increasing and those on which it is decreasing.

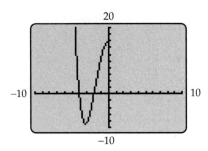

37. A complete graph of a polynomial function g is shown here.

(a) Is the degree of $g(x)$ even or odd?

(b) Is the leading coefficient of $g(x)$ positive or negative?

(c) What are the real roots of $g(x)$?

(d) What is the smallest possible degree of $g(x)$?

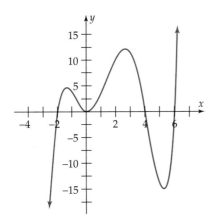

38. Do Exercise 37 for the polynomial function g whose complete graph is shown here.

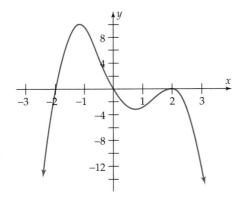

39. The figure is a partial view of the graph of a cubic polynomial whose leading coefficient is negative. Which of the patterns shown in Exercise 35 does this graph have?

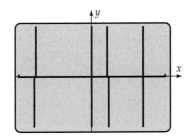

40. The figure is a partial view of the graph of a fourth-degree polynomial. Sketch the general shape of the graph and state whether the leading coefficient is positive or negative.

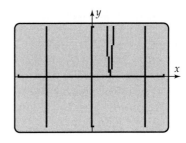

In Exercises 41–46, sketch a complete graph of the function. Label each x-intercept and the coordinates of each local extremum; find intercepts and coordinates exactly when possible and otherwise approximate them.

41. $f(x) = x^3 - 3x^2 + 4$

42. $g(x) = 4x - 4x^3/3$

43. $h(x) = .25x^4 - 2x^3 + 4x^2$

44. $f(x) = .25x^4 - 2x^3/3$

45. $g(x) = 3x^3 - 18.5x^2 - 4.5x - 45$

46. $h(x) = 2x^3 + x^2 - 4x - 2$

47. The number of short tons of carbon monoxide emissions (mostly from automobiles) can be approximated by the polynomial function
$$f(x) = .017x^4 - 2.675x^3 + 96.208x^2 - 45.167x + 94{,}621.034 \ (0 \le x \le 60),$$
where $x = 0$ corresponds to 1940. In what year between 1940 and 2000 were carbon monoxide emissions at their highest?

48. The number of divorces and annulments in the United States from 1945 to 2000 can be approximated by the polynomial function
$$g(x) = .0005x^4 - .1x^3 + 6.17x^2 - 114.25x + 953.5 \ (5 \le x \le 60),$$
where $x = 0$ corresponds to 1940 and $g(x)$ is in thousands.
(a) In what year between 1945 and 2000 was the number of divorces and annulments the highest?
(b) In what year was it the lowest?

Thinkers

49. (a) Graph $g(x) = .01x^3 - .06x^2 + .12x + 3.92$ in the viewing window with $-3 \le x \le 3$ and $0 \le y \le 6$ and verify that the graph appears to coincide with the horizontal line $y = 4$ between $x = 1$ and $x = 3$. In other words, it appears that every x with $1 \le x \le 3$ is a solution of the equation
$$.01x^3 - .06x^2 + .12x + 3.92 = 4.$$

Explain why this is impossible. Conclude that the actual graph is not horizontal between $x = 1$ and $x = 3$.

(b) Use the trace feature to verify that the graph is actually rising from left to right between $x = 1$ and $x = 3$. Find a viewing window that shows this.

(c) Show that it is not possible for the graph of a polynomial $f(x)$ to contain a horizontal segment. [*Hint:* A horizontal line segment is part of the horizontal line $y = k$ for some constant k. Adapt the argument in part (a), which is the case $k = 4$.]

50. For x-values in a particular interval, a nonpolynomial function $g(x)$ can often be approximated by a polynomial function, meaning that there is some polynomial $f(x)$ such that $g(x) \approx f(x)$ for every x in the interval.

(a) In the standard viewing window, graph both $g(x) = \sqrt{x}$ (which is not a polynomial function) and the polynomial function

$$f(x) = .26705x^3 - .78875x^2 + 1.3021x + .22033.$$

Are the graphs similar?

(b) Graph $f(x)$ and $g(x)$ in the viewing window with $0.26 \le x \le 1$ and $0 \le y \le 1$. Does it now appear that $f(x)$ is a good approximation of $g(x)$ over this interval?

(c) For any particular value of x, the error in this approximation is the difference between $f(x)$ and $g(x)$. In other words, $h(x) = f(x) - g(x)$ measures the error in the approximation. Graph the function $h(x)$ in the viewing

window with $.26 \le x \le 1$ and $-.001 \le y \le .001$ and use the trace feature to determine the maximum error in the approximation.

51. The graph of

$$f(x) = (x + 18)(x^2 - 20)(x - 2)^2(x - 10)$$

has x-intercepts at each of its roots, that is, at $x = -18, \pm\sqrt{20} \approx \pm4.472, 2,$ and 10. It is also true that $f(x)$ has a relative minimum at $x = 2$.

(a) Draw the x-axis and mark the roots of $f(x)$. Then use the fact that $f(x)$ has degree 6 (why?) to sketch the general shape of the graph (as was done for cubics in Exercise 35).

(b) Now graph $f(x)$ in the standard viewing window. Does the graph resemble your sketch? Does it even show all the x-intercepts between -10 and 10?

(c) Graph $f(x)$ in the viewing window with $-19 \le x \le 11$ and $-10 \le y \le 10$. Does this window include all the x-intercepts as it should?

(d) List viewing windows that give a complete graph of $f(x)$.

52. (a) Let $f(x)$ be a polynomial of odd degree. Explain why $f(x)$ must have at least one real root. [*Hint:* What does the graph of a cubic polynomial look like? Must it cross the x-axis?]

(b) Let $g(x)$ be a polynomial of even degree, with a negative leading coefficient and a positive constant term. Explain why $g(x)$ must have at least one positive and one negative root.

4.3.A   **EXCURSION**   **Optimization Applications**

Many real-life situations require you to find the largest or smallest quantity satisfying certain conditions. For instance, automotive engineers want to design engines with maximum fuel efficiency. A cereal manufacturer might want to know the dimensions of the box that requires the least amount of cardboard (and hence is cheapest). Many such optimization problems can be modeled by polynomial functions and solved by finding the local extrema of these functions. Although calculus is required for exact solutions, graphing technology can provide very accurate approximate solutions.

Example 1   A box with no top is to be made from a 22-by-30-inch sheet of cardboard by cutting squares of equal size from each corner and bending up the flaps, as shown in Figure 4–19. To the nearest hundredth of an inch, what size square should be cut from each corner in order to obtain a box with the largest possible volume and what is the volume of this box?

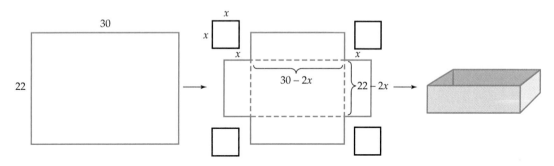

Figure 4–19

Solution Let x denote the length of the side of the square to be cut from each corner. Then,

$$\text{Volume of box} = \text{length} \times \text{width} \times \text{height}$$

$$= (30 - 2x) \cdot (22 - 2x) \cdot x$$

$$= 4x^3 - 104x^2 + 600x$$

Thus the equation $y = 4x^3 - 104x^2 + 660x$ gives the volume y of the box that results from cutting a square of side x from each corner. Since the shortest side of the cardboard is 22 inches, the length x of the side of the cut-out square must be less than 11 (why?).

Each point on the graph of the volume equation

$$y = 4x^3 - 104x^2 + 660x \quad (0 < x < 11)$$

in Figure 4–20 represents one of the possibilities:

The x-coordinate is the size of the square to be cut from each corner;

The y-coordinate is the volume of the resulting box.

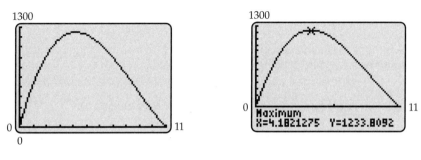

Figure 4–20   **Figure 4–21**

The box with the largest volume corresponds to the point with the largest y-coordinate, that is, the highest point in the viewing window. A maximum finder (Figure 4–21) shows that this point is approximately

(4.182, 1233.809). Therefore, a square measuring approximately 4.18 by 4.18 inches should be cut from each corner, producing a box of volume approximately 1233.81 cubic inches. ■

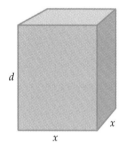

Figure 4–22

Example 2 A rectangular box with a square base (Figure 4–22) is to be mailed. The sum of the height of the box and the perimeter of the base is to be 84 inches, the maximum allowable under postal regulations. What are the dimensions of the box with largest possible volume that meets these conditions?

Solution   If the length of one side of the base is x, then the perimeter of the base (the sum of the lengths of its four sides) is $4x$. If the height of the box is d, then $4x + d = 84$, so that $d = 84 - 4x$ and hence the volume is

$$V = x \cdot x \cdot d = x \cdot x \cdot (84 - 4x) = 84x^2 - 4x^3.$$

The graph of the polynomial function $V(x) = 84x^2 - 4x^3$ in Figure 4–23 is complete (why?). However, the only relevant part of the graph in this situation is the portion with x and $V(x)$ positive (because x is a length and $V(x)$ is a volume). The graph of $V(x)$ has a local maximum between 10 and 20 and this local maximum value is the largest possible volume for the box.

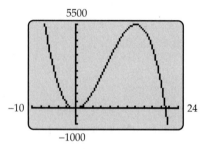

Figure 4–23

GRAPHING EXPLORATION

Use a maximum finder to find the x-value at which the local maximum occurs. State the dimensions of the box in this case. ■

Exercises 4.3.A

1. Name tags can be sold for $29 per thousand. The cost of manufacturing x thousand tags is $.001x^3 + .06x^2 - 1.5x$ dollars. Assume that all tags manufactured are sold.
   (a) Write the rule of the revenue function in this situation.
   (b) Write the rule of the profit function.
   (c) What number of tags should be made to guarantee a maximum profit? What will that profit be?

2. An auto parts manufacturer makes radiators that sell for $350 each. The cost of producing x radiators is approximated by the function $C(x) = 600,000 - 25x + .01x^2$.
   (a) What is the revenue function in this situation? What is the profit function?

   (b) What number of radiators will produce the largest possible profit?

3. (a) A company makes novelty bookmarks that sell for $142 per hundred. The cost (in dollars) of making x hundred bookmarks is $x^3 - 8x^2 + 20x + 40$. Because of other projects, a maximum of 600 bookmarks per day can be manufactured. Assuming that the company can sell all the bookmarks it makes, how many should it make each day to maximize profits?
   (b) Due to a change in other orders, as many as 1600 bookmarks can now be manufactured each day. How many should be made to maximize profits?

4. A manufacturer's revenue (in cents) from selling x items per week is given by $200x - .02x^2$. It costs $60x + 30,000$ cents to make x items.
   (a) Approximately how many items should be made each week to make a profit of $1100? [Don't confuse cents and dollars.]
   (b) How many items should be made each week to have the largest possible profit? What is that profit?

5. A 20-inch-square piece of metal is to be used to make an open-top box by cutting equal-sized squares from each corner and folding up the sides (as in Example 1). The length, width, and height of the box are each to be no more than 12 inches. What size squares should be cut out to produce a box with
   (a) volume 550 cubic inches?
   (b) largest possible volume?

6. A certain type of fencing comes in rigid 10-ft-long segments. Four uncut segments are to be used to fence in a garden on the side of a building, as shown in the figure. What value of x will result in a garden of the largest possible area and what is that area?

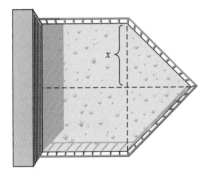

7. Find the dimensions of the rectangular box with a square base and no top that has volume 30,000 cubic centimeters and the smallest possible surface area. [*Hint:* See Example 3 in Section 2.3.]

8. An open-top box with a square base is to be constructed from 120 sq cm of material. What dimensions will produce a box
   (a) of volume 100 cubic centimeters?
   (b) with largest possible volume?

9. If $c(x)$ is the cost of producing x units, then $c(x)/x$ is the *average cost* per unit.* Suppose the cost of producing x units is given by $c(x) = .13x^3 - 70x^2 + 10,000x$ and that no more than 300 units can be produced per week.
   (a) If the average cost is $1100 per unit, how many units are being produced?
   (b) What production level should be used to minimize the average cost per unit? What is the minimum average cost?

10. A cylindrical waste container with no top, a diameter of at least 2 feet, and a volume of 25 cubic feet is to be constructed. What should its radius be if
   (a) 65 sq ft of material are to be used to construct it?
   (b) the smallest possible amount of material is to be used to construct it? In this case, how much material is needed?

*Depending on the situation, a unit of production might consist of a single item or several thousand items. Similarly, the cost of x units might be measured in thousands of dollars.

4.4 **Polynomial Models**†

Linear regression was used in Section 1.5 to construct a linear function that modeled a set of data points. When the scatter plot of the data points looks more like a higher degree polynomial graph than a straight line, similar least squares regression procedures are available on most calculators for constructing quadratic, cubic, and quartic (fourth-degree) polynomial functions to model the data.

†This section is optional; its prerequisite is Section 1.5. This material will be used in the optional Section 5.6 and in clearly identifiable exercises, but not elsewhere in the text.

Technology Tip

Quadratic, cubic, and quartic regression are denoted by QuadReg, CubicReg, QuartReg on the CALC submenu of the TI-83 STAT menu and TI-89 Data Editor (APPS menu); by P2Reg, P3Reg, P4Reg on the CALC menu of the TI-86 STAT menu; by x^2, x^3, x^4 on the CALC REG submenu of the Casio 9850 STAT menu; and by Rg_x^2, Rg_x^3, Rg_x^4 on the REG submenu of the Sharp 9600 STAT menu. On HP-38, select STAT in the LIB menu and choose quadratic or cubic on the SYMB SETUP menu; quartic regression is not available.

Example 1   Use the following data to construct a polynomial model for the population of San Francisco. Use the model to estimate the population in 1975 and 1985.

Year	1900	1950	1970	1980	1990	1996
Population	342,782	775,357	715,674	678,974	723,959	735,315

Solution   Let $x = 0$ correspond to 1900 and plot the data points, as in Figure 4–24. The points suggest the general shape of a cubic polynomial, so we use cubic regression on a calculator to obtain this function:

$$f(x) = 2.804x^3 - 507.220x^2 + 27,011.001x + 342,795.453.*$$

(The procedure is the same as for linear regression; see the Technology Tip on page 109 and the one in the margin here.) The graph of f in Figure 4–25 appears to fit the data well.

To estimate the population in 1975 and 1985, we evaluate the function f at $x = 75$ and $x = 85$, as shown in Figure 4–26. According to this model, the population was approximately 698,261 in 1975 and 695,803 in 1985. ∎

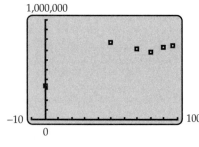

Figure 4–24

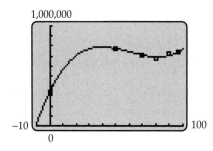

Figure 4–25

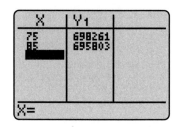

Figure 4–26

Example 2   The average yearly price of Pepsico Stock in selected years is given in the table.

Year	1981	1983	1985	1987	1989	1991	1993	1995	1997
Price	1.88	2.02	3.15	5.74	8.79	15.30	19.60	22.94	35.75

*Here and later, coefficients are rounded for convenient reading, but the full coefficients are used to produce the graphs and estimates.

Use this data to construct a polynomial model of the situation and to estimate the average price in 1990, 1996, and late 1998. How accurate are these estimates?

Solution   Let $x = 0$ correspond to 1980 and plot the points $(1, 1.88)$, $(3, 2.02)$, etc. The scatter plot in Figure 4–27 suggests a parabola. However, the data points climb quite steeply at the right side, so a fourth-degree polynomial graph might fit them better. Quartic regression produces this model for the data:

$$f(x) = .0018x^4 - .0626x^3 + .825x^2 - 3.0741x + 4.5459.$$

The graph of f in Figure 4–28 appears to be a reasonably good fit for the data.

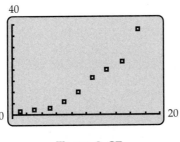

Figure 4–27

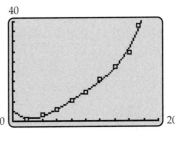

Figure 4–28

According to the function f,

$$f(10) \approx 11.89 \quad \text{and} \quad f(16) \approx 29.38,$$

meaning that the average price was approximately $11.89 in 1990 and $29.38 in 1996. These estimates compare favorably with the actual average prices of $11.75 and $31.01. However, a model may not be accurate when applied outside the range of points used to construct it. For instance, $f(18.75) \approx 49.19$, suggesting that the average price in late 1998 was about $49.19. In fact, Pepsico stock was selling at $29.44 on September 30, 1998. ■

Example 3   The table below, which is based on statistics from the Department of Health and Human Services, gives the cumulative number of reported cases of AIDS in the United States from 1982 through 1996. It shows, for example, that 41,662 cases were reported from 1982 through 1986.

Year	Cases	Year	Cases	Year	Cases
1982	1,563	1987	70,222	1992	278,189
1983	4,647	1988	105,489	1993	380,601
1984	10,845	1989	147,170	1994	457,789
1985	22,620	1990	188,872	1995	529,282
1986	41,662	1991	232,383	1996	598,433

Find a suitable polynomial model for this data.

Solution Letting $x = 0$ correspond to 1980 and plotting the data points (2, 1563), (3, 4647), etc., we obtain the scatter plot in Figure 4–29. The points are not in a straight line, but could be part of a polynomial graph of degree 2 or more. So we use the regression feature to find three possible models for this data:

Quadratic: $f(x) = 3362.1x^2 - 17{,}270.3x + 24{,}977.1$

Cubic: $g(x) = 2.18x^3 + 3303.21x^2 - 16{,}811.16x + 24{,}042.78$

Quartic: $h(x) = -13.17x^4 + 476.33x^3 - 2474.98x^2 + 10{,}384.48x - 15{,}338.41.$

The graph of the quadratic model f in Figure 4–30 appears to be a good fit. In fact the graphs of f, g, and h are virtually identical in this viewing window, as you can easily verify.

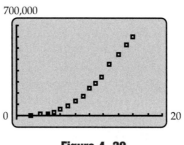

Figure 4–29

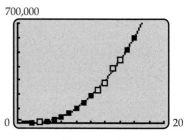

Figure 4–30

Although any one of f, g, or h provides a reasonable model for the given data, our knowledge of polynomial graphs suggests that the quartic model h should not be used for predicting future results. As x gets larger, the graph of h will resemble that of $y = -13.25x^4$, which turns downward. However, the cumulative number of cases can't decrease (even when there are no new cases, the cumulative total stays the same).

GRAPHING EXPLORATION

Graph the functions f, g, and h in the window with $0 \le x \le 26$ and $0 \le y \le 1{,}800{,}000$. In this window can you distinguish the graphs of f and g? Assuming no medical breakthroughs or changes in the current social situation, does the graph of f seem to be a plausible model for the next few years? What about the graph of h? ∎

NOTE You must have at least three data points for quadratic regression, at least four for cubic regression, and at least five for quartic regression. If you have exactly the required minimum number data points, no two of them can have the same first coordinate. In this case, the polynomial regression function will pass through all of the data points (an exact fit). When you have more than the minimum number of data points required, the fit will generally be approximate rather than exact.

Exercises 4.4

In Exercises 1–4, a scatter plot of data is shown. State the type of polynomial model that seems most appropriate for the data (linear, quadratic, cubic, or quartic). If none of them is likely to provide a reasonable model, say so.

1.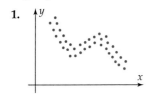

2.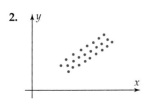

3.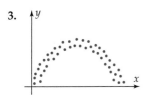

4.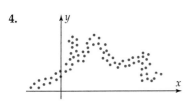

5. The table, which is based on the 1996 FBI Uniform Crime Report, shows the rate of property crimes per 100,000 population.

Year	Crimes	Year	Crimes
1980	5553.3	1990	5088.5
1982	5032.5	1992	4902.7
1984	4492.1	1994	4660
1986	4862.6	1996	4444.8
1988	5027.1		

   (a) Use cubic regression to find a polynomial function that models this data, with $x = 0$ corresponding to 1980.

   (b) According to this model, what was the property crime rate in 1987 and 1995?

   (c) The actual crime rate was about 4250 in 1997. What does the model predict?

   (d) For how many years in the future is this model likely to be a reasonable one?

6. The table, which is based on the FBI Uniform Crime Report, shows the rate of violent crimes per 100,000 population.

Year	Crimes	Year	Crimes
1980	596.6	1990	731.8
1982	571.1	1992	757.5
1984	539.2	1994	713.6
1986	617.7	1996	634.1
1988	637.2		

   (a) Use cubic regression to find a polynomial function that models this data, with $x = 0$ corresponding to 1980.

   (b) According to this model, what was the violent crime rate in 1983 and 1993?

   (c) According to the model, in what year were violent crimes at their highest level?

   (d) Do you think this model could accurately predict the crime rate in 2002?

7. The table shows the air temperature (in degrees Farenheit) at various times during a spring day in Gainesville, Florida.

Time	Temperature	Time	Temperature
6 A.M.	52	1 P.M.	82
7 A.M.	56	2 P.M.	86
8 A.M.	61	3 P.M.	85
9 A.M.	67	4 P.M.	83
10 A.M.	72	5 P.M.	78
11 A.M.	77	6 P.M.	72
noon	80		

   (a) Sketch a scatter plot of the data, with $x = 0$ corresponding to midnight.

   (b) Find a quadratic polynomial model for the data.

   (c) What is the predicted temperature for noon? For 9 A.M.? For 2 P.M.?

8. The table on the next page, which is based on data from the U.S. Education Department, Office of Special Education and Rehabilitation Services, shows the number of disabled children (in thousands) in federally supported programs.

School Year Beginning in Fall of	Number of Children
1976	3692
1980	4142
1985	4317
1987	4446
1988	4544
1989	4641
1990	4771
1991	4949
1992	5125
1993	5309
1994	5378
1995	5573

(a) Sketch a scatter plot of the data, with $x = 1$ corresponding to 1976.
(b) Find a cubic polynomial model for this data.

Use the following table for Exercises 9–10. It shows the median income of U.S. households in 1995 dollars. (Source: U.S. Census Bureau)

Year	Median Income	Year	Median Income
1980	$32,795	1988	$35,073
1981	32,263	1989	35,526
1982	32,155	1990	34,914
1983	31,957	1991	33,709
1984	32,878	1992	33,278
1985	33,452	1993	32,949
1986	34,620	1994	33,178
1987	34,962	1995	34,076

9. (a) Sketch a scatter plot of the data from 1980 to 1995, with $x = 0$ representing 1980.
   (b) Decide whether a quadratic or quartic model seems more appropriate.
   (c) Find an appropriate polynomial model.
   (d) Use the model to predict the median income in 2000. Does your answer seem reasonable?
   (e) According to this model, in what year will the median income exceed $50,000?

10. (a) Sketch a scatter plot of the data from 1989 to 1995, with $x = 0$ representing 1989.
   (b) Decide whether a quadratic or quartic model seems more appropriate.
   (c) Find an appropriate polynomial model.
   (d) Use the model to predict the median income in 2000. Does your answer seem reasonable?
   (e) According to this model, in what year will the median income exceed $50,000?
   (f) Which model [the one in part (c) or the one in Exercise 9] is the better predictor? Justify your answer.

11. The table shows the percentage of eighth graders who had smoked within 30 days of the time the data was collected each year.

Year	Percentage Who Smoked
1991	14.3
1992	15.5
1993	16.7
1994	18.6
1995	19.1
1996	21.0

(a) Sketch a scatter plot of the data, with $x = 0$ corresponding to 1990.
(b) Find a polynomial model of the data and justify your choice.
(c) Do you think your model will remain accurate into the future? For how long?

12. The table shows the U.S. public debt per person (in dollars). (*Source:* U.S. Department of Treasury, Bureau of Public Debt)

Year	Debt	Year	Debt
1980	$ 3,985	1989	$11,545
1981	4,338	1990	13,000
1982	4,913	1991	14,436
1983	5,870	1992	15,846
1984	6,640	1993	17,105
1985	7,598	1994	18,025
1986	8,774	1995	18,930
1987	9,615	1996	19,805
1988	10,534		

(a) Sketch a scatter plot of the data.
(b) Find a polynomial model of the data and justify your choice.

13. (a) Find both a cubic and a quartic model for the data on the number of unemployed people in the labor force in Example 4 of Section 1.5.
(b) Does either model seem likely to be accurate in the future?

14. The table shows the number of people murdered each year per 100,000 population. (*Source:* FBI)
(a) Sketch a scatter plot of the data, with $x = 0$ corresponding to 1987.

(b) Find a cubic model for the data.
(c) Compare the model in part (b) with the quartic model for the same data in Exercise 61 of Section 4.2. Which model seems likely to be accurate for years after 1996? For how long do you think it is likely to remain accurate?

Year	Number	Year	Number
1987	8.12	1993	8.74
1989	8.35	1994	8.47
1990	8.71	1995	8.12
1991	8.88	1996	7.71
1992	8.65		

4.5 Rational Functions

A **rational function** is a function whose rule is the quotient of two polynomials, such as

$$f(x) = \frac{1}{x}, \qquad t(x) = \frac{4x - 3}{2x + 1}, \qquad k(x) = \frac{2x^3 + 5x + 2}{x^2 - 7x + 6}.$$

A polynomial function is defined for every real number, but the rational function $f(x) = g(x)/h(x)$ is defined only when its denominator is non-zero. Hence,

Domain

> The domain of the rational function $f(x) = \dfrac{g(x)}{h(x)}$ is the set of all real numbers that are *not* roots of the denominator $h(x)$.

For instance, the domain of $f(x) = \dfrac{x^2 + 3x + 1}{x^2 - x - 6}$ is the set of all real numbers except -2 and 3 [the roots of $x^2 - x - 6 = (x + 2)(x - 3)$].

Calculators sometimes do a poor job of graphing rational functions. Hence, the emphasis here is on the algebraic analysis of rational functions, so that you will be able to interpret misleading screen images. The key to understanding the behavior of rational functions is this fact from arithmetic:

The Big–Little Principle

> If c is a number far from 0, then $1/c$ is a number close to 0. Conversely, if c is close to 0, then $1/c$ is far from 0. In less precise, but more suggestive terms:
>
> $$\frac{1}{\text{Big}} = \text{Little} \qquad \text{and} \qquad \frac{1}{\text{Little}} = \text{Big}.$$

For example, 5000 is big (far from 0) and 1/5000 is little (close to 0). Similarly, $-1/1000$ is very close to 0, but $\dfrac{1}{-1/1000} = -1000$ is far from 0. To see the role played by the Big–Little Principle, we consider a typical example.

Example 1 Without using a calculator, describe the graph of $f(x) = \dfrac{x + 1}{2x - 4}$ near $x = 2$ and far from $x = 2$. Then sketch the graph.

Solution The function is not defined when $x = 2$ because the denominator is 0 there. When $x > 2$ and is very close to 2, then

The numerator $x + 1$ is very close to $2 + 1 = 3$;

The denominator $2x - 4$ is positive and very close to $2 \cdot 2 - 4 = 0$.

Therefore,

$$f(x) = \frac{x + 1}{2x - 4} \approx \frac{3}{\text{little}} = 3 \cdot \frac{1}{\text{little}} = 3 \cdot \text{big} = \text{BIG!}$$

You can confirm this fact by making a table of values for $f(x)$ when $x = 2.1$, 2.01, 2.001, etc., as in Figure 4–31. In graphical terms, the points with x-coordinates slightly larger than 2 have gigantic y-coordinates, so that the graph shoots upward just to the right of $x = 2$ (Figure 4–32).

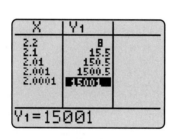

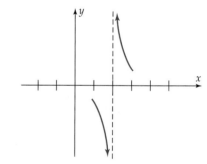

Figure 4–31   **Figure 4–32**

A similar analysis when $x < 2$ and very close to 2 shows that the numerator of $f(x)$ is very close to 3 and the denominator is negative and very close to 0, so that the quotient is a negative number far from 0. (If you doubt this, make a table.) Hence, the graph shoots downward just to the left of $x = 2$ (see Figure 4–32).

The dashed vertical line in Figure 4–32 is included for easier visualization; it is *not* part of the graph, but is called a **vertical asymptote** of the graph. The graph gets closer and closer to the vertical asymptote, but never crosses it because $f(x)$ is not defined when $x = 2$.

To see what the graph looks like far from $x = 2$, we rewrite the rule of f like this:

$$f(x) = \frac{x + 1}{2x - 4} = \frac{\dfrac{x + 1}{x}}{\dfrac{2x - 4}{x}} = \frac{1 + \dfrac{1}{x}}{2 - \dfrac{4}{x}}$$

As x gets larger in absolute value (far from 0), both $1/x$ and $4/x$ get very close to 0 by the Big–Little Principle. Consequently, $f(x) = \dfrac{1 + (1/x)}{2 - (4/x)}$ gets very close to $\dfrac{1+0}{2-0} = \dfrac{1}{2}$. So when $|x|$ is large, the graph gets closer and closer to the horizontal line $y = 1/2$, but never touches it, as shown in Figure 4–33. The line $y = 1/2$ is called a **horizontal asymptote** of the graph.

The preceding information, together with a few hand-plotted points, produces the graph in Figure 4–33. ■

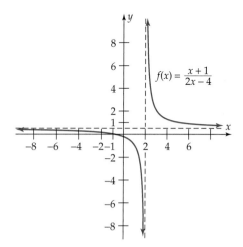

Figure 4–33

Technology Tip

On most calculators you can avoid an erroneous vertical line in a graph by choosing a window that has the vertical asymptote at its center. In Figure 4–34(c), for instance, the vertical asymptote is at $x = 2$, which is at the center of the window with $-8 \le x \le 12$.

Also, see the Tip on page 316.

Getting an accurate graph of a rational function on a calculator often depends on choosing an appropriate viewing window. For example, a TI-83 produced the following graphs of $f(x) = \dfrac{x + 1}{2x - 4}$ in Figure 4–34, two of which do not look like Figure 4–33 as they should.

The vertical segments in graphs (a) and (b) are not representations of the vertical asymptote. They are a result of the calculator evaluating $f(x)$ just to the left of $x = 2$ and just to the right of $x = 2$, but not at $x = 2$, and then erroneously connecting these points with a near vertical segment that looks like an asymptote.

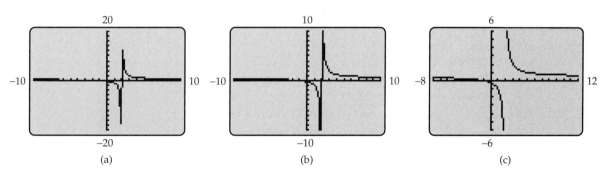

(a)   (b)   (c)

Figure 4–34

In the accurate graph (c) the calculator attempted to plot a point with $x = 2$ and when it found that $f(2)$ was not defined, skipped a pixel and did not join the points on either side of the skipped one.

GRAPHING EXPLORATION

Find a viewing window on your calculator (other than the one in Figure 4–29) that displays the graph of $f(x)$ without any erroneous vertical line segments being shown. The Tip on page 311 may be helpful.

The analysis in Example 1 works in the general case:

Linear Rational Functions ▶

The graph of $f(x) = \dfrac{ax + b}{cx + d}$ (with $c \neq 0$ and $ad \neq bc$) has two asymptotes:

The vertical asymptote occurs at the root of the denominator.

The horizontal asymptote is the line $y = a/c$.

Example 1 is the case where $a = 1, b = 1, c = 2,$ and $d = -4$. Figure 4–35 shows some additional examples, in which the asymptotes are indicated by dashed lines that are not part of the graph.

$$f(x) = \frac{-5x + 12}{2x - 4}$$

Vertical asymptote $x = 2$
Horizontal asymptote $y = -\frac{5}{2}$

$$k(x) = \frac{3x + 6}{x} = \frac{3x + 6}{1x + 0}$$

Vertical asymptote $x = 0$
Horizontal asymptote $y = \frac{3}{1} = 3$

$$f(x) = \frac{1}{x} = \frac{0x + 1}{1x + 0}$$

Vertical asymptote $x = 0$
Horizontal asymptote $y = \frac{0}{1} = 0$

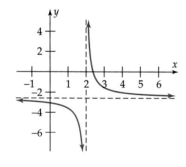

(a)

(b)

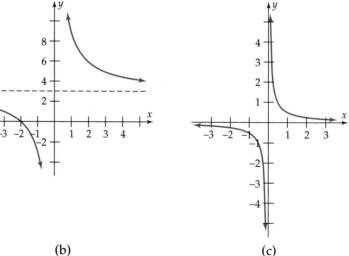

(c)

Figure 4–35

Properties of Rational Graphs

Here is a summary of the important characteristics of graphs of more complicated rational functions.

Continuity

There will be breaks in the graph of a rational function wherever the function is not defined. Except for breaks at these undefined points, the graph is a continuous unbroken curve. In addition, the graph has no sharp corners.

Local Maxima and Minima

The graph may have some local extrema (peaks and valleys) and calculus is needed to determine their exact location. There are no simple rules for the possible number of peaks and valleys as there were with polynomial functions.

Intercepts

As with any function, the y-intercept of the graph of a rational function f occurs at $f(0)$, provided that f is defined at $x = 0$. The x-intercepts of the graph of any function f occur at each number c for which $f(c) = 0$. Now a fraction is 0 only when its numerator is 0 and its denominator nonzero (since division by 0 is not defined). Thus,

Intercepts

The x-intercepts of the graph of the rational function $f(x) = \dfrac{g(x)}{h(x)}$ occur at the numbers that are roots of the numerator $g(x)$ but *not* of the denominator $h(x)$. If f has a y-intercept, it occurs at $f(0)$.

For example, the graph of $f(x) = \dfrac{x^2 - x - 2}{x - 5}$ has x-intercepts at $x = -1$ and $x = 2$ (which are the roots of $x^2 - x - 2 = (x + 1)(x - 2)$, but not of $x - 5$) and y-intercept at $y = 2/5$ (the value of f at $x = 0$).

Vertical Asymptotes

In Example 1 we saw that the graph of $f(x) = \dfrac{x + 1}{2x - 4}$ had a vertical asymptote at $x = 2$. Note that $x = 2$ is a root of the denominator $2x - 4$, but not of the numerator $x + 1$. The same thing occurs in the general case:

Vertical Asymptotes

The function $f(x) = \dfrac{g(x)}{h(x)}$ has a vertical asymptote at every number that is a root of the denominator $h(x)$, but *not* of the numerator $g(x)$.

Near a vertical asymptote, the graph of a rational function may look like the graph in Example 1, or like one of these graphs (Figure 4–36):

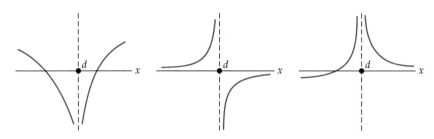

Figure 4–36

Behavior When |x| Is Large

The shape of a rational graph at the far left and far right (that is, when $|x|$ is large) can usually be found by algebraic analysis.

Example 2   Determine the shape of the graph when $|x|$ is large for the following functions.

(a) $f(x) = \dfrac{7x^4 - 6x^3 + 4}{2x^4 + x^2}$ **(b)** $g(x) = \dfrac{x^2 - 2}{x^3 - 3x^2 + x - 3}$

Solution

(a)   When $|x|$ is very large, a polynomial function behaves in essentially the same way as its highest degree term, as we saw on page 293. Consequently, we have this approximation

$$f(x) = \frac{7x^4 - 6x^3 + 4}{2x^4 + x^2} \approx \frac{7x^4}{2x^4} = \frac{7}{2} = 3.5.$$

Thus, when $|x|$ is large, the graph of $f(x)$ is very close to the horizontal line $y = 3.5$, which is a horizontal asymptote of the graph. This means that at the far left and far right, the graph of $f(x)$ is almost flat and very close to the line $y = 3.5$.

GRAPHING EXPLORATION

Graph $f(x)$ in the standard viewing window. Use the trace feature to determine how far the left and right ends of the graph are from the horizontal asymptote $y = 3.5$. Now find a wider viewing window in which the ends of the graph are within .1 of the horizontal asymptote.

(b)   When $|x|$ is large, the graph of g closely resembles the graph of $y = \dfrac{x^2}{x^3} = \dfrac{1}{x}$. By the Big–Little Principle, $1/x$ is very close to 0 when $|x|$ is large. So, the line $y = 0$ (that is, the x-axis) is the horizontal asymptote.

GRAPHING EXPLORATION

Find a viewing window that provides the answer to this question: As you move to the left from the origin does the graph of $g(x)$ approach its horizontal asymptote from above or from below? Does the graph ever cross the horizontal asymptote? ■

Arguments similar to those in the preceding example, using the highest degree terms in the numerator and denominator, carry over to the general case and lead to this conclusion:

Horizontal Asymptotes

Let $f(x) = \dfrac{ax^n + \cdots}{cx^k + \cdots}$ be a rational function whose numerator has degree n and whose denominator has degree k.

If $n = k$, then the line $y = a/c$ is a horizontal asymptote.

If $n < k$, then the x-axis (the line $y = 0$) is a horizontal asymptote.

The asymptotes of rational functions in which the denominator has smaller degree than the numerator are discussed in Excursion 4.5.A.

Graphs of Rational Functions

The facts presented above can be used in conjunction with a calculator to find accurate, complete graphs of rational functions whose numerators have degree less than or equal to the degree of their denominators. The basic procedure is as follows.

Graphing $f(x) = \dfrac{g(x)}{h(x)}$ When Degree $g(x) \leq$ Degree $h(x)$

1. Analyze the function algebraically to determine its vertical asymptotes and intercepts.

2. Determine the horizontal asymptote of the graph when $|x|$ is large by using the facts in the box above.

3. Use the preceding information to select an appropriate viewing window (or windows), to interpret the calculator's version of the graph (if necessary), and to sketch an accurate graph.

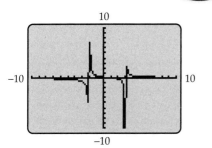

Figure 4–37

Example 3 If you ignore the preceding advice and simply graph $f(x) = \dfrac{x - 1}{x^2 - x - 6}$ in the standard viewing window, you get garbage (Figure 4–37). So let's try analyzing the function. We begin by factoring:

$$f(x) = \frac{x - 1}{x^2 - x - 6} = \frac{x - 1}{(x + 2)(x - 3)}.$$

The factored form allows us to read off the necessary information:

Vertical Asymptotes: $x = -2$ and $x = 3$ (roots of the denominator but not of the numerator).

Horizontal Asymptote: x-Axis (because denominator has larger degree than the numerator).

Intercepts: y-Intercept at $f(0) = \dfrac{0-1}{0^2 - 0 - 6} = \dfrac{1}{6}$; x-intercept at $x = 1$ (root of the numerator but not of the denominator).

Interpreting Figure 4–37 in the light of this information suggests that a complete graph of f looks something like Figure 4–38.

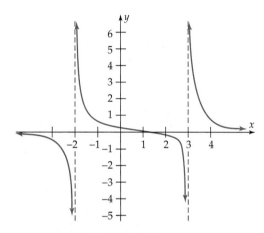

Figure 4–38

Technology Tip

When the vertical asymptotes of a rational function occur at numbers such as $-2.1, -2, -1.9, \ldots, 2.9,$ $3, 3.1$, a decimal window normally produces an accurate graph because the calculator actually evaluates the function at the asymptotes, finds that it is undefined, and skips a pixel.

GRAPHING EXPLORATION

Find a viewing window in which the graph of f looks similar to Figure 4–38. The Tip in the margin may be helpful.   ■

NOTE   The graph of a rational function never touches a horizontal asymptote when x is large in absolute value. For smaller values of x, however, the graph may cross the asymptote, as in Example 3.

Example 4   To graph $f(x) = \dfrac{2x^2}{x^2 + x - 2}$ we factor and then read off the necessary information:

$$f(x) = \frac{2x^2}{x^2 + x - 2} = \frac{2x^2}{(x + 2)(x - 1)}.$$

Vertical Asymptotes: $x = -2$ and $x = 1$ (roots of denominator).

Horizontal Asymptote: $y = 2/1 = 2$ (because numerator and denominator have the same degree; see the box on page 315).

Intercepts: x-Intercept at $x = 0$ (root of numerator); y-intercept at $f(0) = 0$.

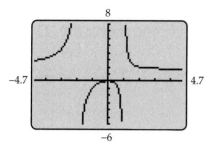

Figure 4–39

Using this information and selecting a viewing window that will accurately portray the graph near the vertical asymptotes, we obtain what seems to be a reasonably complete graph in Figure 4–39. The graph appears to be falling to the right of $x = 1$, but this is deceptive.

GRAPHING EXPLORATION

Graph f in this same viewing window and use the trace feature, beginning at approximately $x = 1.1$ and moving to the right. For what values of x is the graph above the horizontal asymptote $y = 2$? For what values of x is the graph below the horizontal asymptote?

This use of the trace feature indicates that there is some *hidden behavior* of the graph that is not visible in Figure 4–39.

GRAPHING EXPLORATION

To see this hidden behavior, graph both f and the line $y = 2$ in the viewing window with $1 \le x \le 50$ and $1.7 \le y \le 2.1$.

This Exploration shows that the graph has a local minimum near $x = 4$ and then stays below the asymptote, moving closer and closer to it as x takes larger values. ∎

Applications

Optimization applications of polynomial functions were considered in Excursion 4.3.A. Here is a similar situation, involving a rational function.

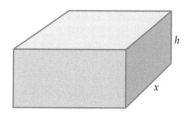

Figure 4–40

Example 5   A cardboard box with a square base and a volume of 1000 cubic inches is to be constructed (Figure 4–40). The box must be at least 2 inches in height.

(a)   What are the possible lengths for a side of the base if no more than 1100 square inches of cardboard can be used to construct the box?

(b)   What is the least possible amount of cardboard that can be used?

(c)   What are the dimensions of the box that uses the least possible amount of cardboard?

Solution   The amount of cardboard needed to construct the box is given by the surface area of the box. Since the top and bottom each have area x^2 (why?) and each of the four sides has area xh, the surface area S is given by

$$S = x^2 + x^2 + xh + xh + xh + xh = 2x^2 + 4xh$$

Since the volume of the box is given by:

$$\text{length} \times \text{width} \times \text{height} = \text{volume},$$

we have

$$x \cdot x \cdot h = 1000 \qquad \text{or equivalently,} \qquad h = \frac{1000}{x^2}.$$

Substituting this into the surface area formula allows us to express the surface area as a function of x:

$$S(x) = 2x^2 + 4xh = 2x^2 + 4x\left(\frac{1000}{x^2}\right) = 2x^2 + \frac{4000}{x} = \frac{2x^3 + 4000}{x}.$$

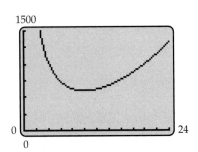

Figure 4–41

Although the rational function $S(x)$ is defined for all nonzero real numbers, x is a length here and must be positive. Furthermore, $x^2 \leq 500$ because if $x^2 > 500$, then $h = \dfrac{1000}{x^2}$ would be less than 2, contrary to specifications. Hence, the only values of x that make sense in this context are those with $0 < x \leq \sqrt{500}$. Since $\sqrt{500} \approx 22.4$, we choose the viewing window in Figure 4–41.

For each point (x, y) on the graph, x is a possible side length for the base of the box and y is the corresponding surface area.

(a)   The points on the graph corresponding to the requirement that no more than 1100 square inches of cardboard be used are those whose y-coordinates are less than or equal to 1100. The x-coordinates of these points are the possible side lengths. The x-coordinates of the points where the graph of S meets the horizontal line $y = 1100$ are the smallest and largest possible values for x, as indicated schematically in Figure 4–42.

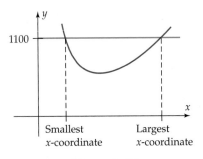

Figure 4–42

GRAPHING EXPLORATION

Graph $S(x)$ and $y = 1100$ on the same screen. Use an intersection finder to show that the possible side lengths that use no more than 1100 square inches of cardboard are those with $3.73 \leq x \leq 21.36$.

(b)   The least possible amount of cardboard corresponds to the point on the graph of $S(x)$ with the smallest y-coordinate.

> **GRAPHING EXPLORATION**
>
> Show that the graph of S has a local minimum at the point (10.00, 600.00). Consequently, the least possible amount of cardboard is 600 square inches and this occurs when $x = 10$.

(c) When $x = 10$, $h = 1000/10^2 = 10$. So the dimensions of the box using the least amount of cardboard are $10 \times 10 \times 10$. ∎

Exercises 4.5

In Exercises 1–6, find the domain of the function. You may need to use some of the techniques of Section 4.2.

1. $(x) = \dfrac{-3x}{2x + 5}$

2. $g(x) = \dfrac{x^3 + x + 1}{2x^2 - 5x - 3}$

3. $h(x) = \dfrac{6x - 5}{x^2 - 6x + 4}$

4. $g(x) = \dfrac{x^3 - x^2 - x - 1}{x^5 - 36x}$

5. $f(x) = \dfrac{x^5 - 2x^3 + 7}{x^3 - x^2 - 2x + 2}$

6. $h(x) = \dfrac{x^5 - 5}{x^4 + 12x^3 + 60x^2 + 50x - 125}$

In Exercises 7–12, use algebra to determine the location of the vertical asymptotes in the graph of the function.

7. $f(x) = \dfrac{x^2 + 4}{x^2 - 5x - 6}$

8. $g(x) = \dfrac{x - 5}{x^3 + 7x^2 + 2x}$

9. $f(x) = \dfrac{2}{x^3 + 2x^2 + x}$

10. $g(x) = \dfrac{1}{x^3 + 5x}$

11. $f(x) = \dfrac{x^2 - 4x + 4}{(x + 2)(x - 2)^3}$

12. $h(x) = \dfrac{x + 3}{x^2 - x - 6}$

In Exercises 13–18, find the horizontal asymptote of the graph of the function when $|x|$ is large and find a viewing window in which the ends of the graph are within .1 of this asymptote.

13. $f(x) = \dfrac{3x - 2}{x + 3}$

14. $g(x) = \dfrac{3x^2 + x}{2x^2 - 2x + 4}$

15. $h(x) = \dfrac{5 - x}{x - 2}$

16. $f(x) = \dfrac{4x^2 - 5}{2x^3 - 3x^2 + x}$

17. $g(x) = \dfrac{5x^3 - 8x^2 + 4}{2x^3 + 2x}$

18. $h(x) = \dfrac{8x^5 - 6x^3 + 2x - 1}{.5x^5 + x^4 + 3x^2 + x}$

In Exercises 19–36, analyze the function algebraically: List its vertical asymptotes and horizontal asymptote. Then, sketch a complete graph of the function.

19. $f(x) = \dfrac{1}{x + 5}$

20. $q(x) = \dfrac{-7}{x - 6}$

21. $k(x) = \dfrac{-3}{2x + 5}$

22. $g(x) = \dfrac{-4}{2 - x}$

23. $f(x) = \dfrac{3x}{x - 1}$

24. $p(x) = \dfrac{x - 2}{x}$

25. $f(x) = \dfrac{2 - x}{x - 3}$

26. $g(x) = \dfrac{3x - 2}{x + 3}$

27. $f(x) = \dfrac{1}{x(x + 1)^2}$

28. $g(x) = \dfrac{x}{2x^2 - 5x - 3}$

29. $f(x) = \dfrac{x - 3}{x^2 + x - 2}$

30. $g(x) = \dfrac{x + 2}{x^2 - 1}$

31. $h(x) - \dfrac{(x^2 + 6x + 5)(x + 5)}{(x + 5)^3(x - 1)}$

32. $f(x) = \dfrac{x^2 - 1}{x^3 - 2x^2 + x}$

33. $f(x) = \dfrac{-4x^2 + 1}{x^2}$

34. $k(x) = \dfrac{x^2 + 1}{x^2 - 1}$

35. $q(x) = \dfrac{x^2 + 2x}{x^2 - 4x - 5}$

36. $F(x) = \dfrac{x^2 + x}{x^2 - 2x + 4}$

In Exercises 37–42, find a viewing window, or windows, that shows a complete graph of the function (if possible, with no erroneous vertical line segments). Be alert for hidden behavior, such as that in Example 4.

37. $f(x) = \dfrac{x^3 + 4x^2 - 5x}{(x^2 - 4)(x^2 - 9)}$

38. $f(x) = \dfrac{x^3 - x + 1}{x^4 - 2x^3 - 2x^2 + x - 1}$

39. $h(x) = \dfrac{3x^2 + x - 4}{2x^2 - 5x}$ **40.** $f(x) = \dfrac{2x^2 - 1}{3x^3 + 2x + 1}$

41. $g(x) = \dfrac{x - 4}{2x^3 - 5x^2 - 4x + 12}$

42. $h(x) = \dfrac{x^2 - 9}{x^3 + 2x^2 - 23x - 60}$

43. The graph of $f(x) = \dfrac{2x^3 - 2x^2 - x + 1}{3x^3 - 3x^2 + 2x - 1}$ has a vertical asymptote. Find a viewing window that demonstrates this fact.

44. The percentage c of a drug in a person's bloodstream t hours after its injection is approximated by

$$c(t) = \frac{5t}{4t^2 + 5}.$$

(a) Approximately what percentage of the drug is in the person's bloodstream after four and a half hours?

(b) Graph the function c in an appropriate window for this situation.

(c) What is the horizontal asymptote of the graph? What does it tell you about the amount of the drug in the bloodstream?

(d) At what time is the percentage the highest? What is the percentage at that time?

45. It costs 2.5¢ per square inch to make the top and bottom of the box in Example 5. The sides cost 1.5¢ per square inch. What are the dimensions of the cheapest possible box?

46. A box with a square base and a volume of 1000 cubic inches is to be constructed. The material for the top and bottom of the box costs $3 per 100 square inches and the material for the sides costs $1.25 per 100 square inches.

(a) If x is the length of a side of the base, express the cost of constructing the box as a function of x.

(b) If the side of the base must be at least 6 inches long, for what value of x will the cost of the box be $7.50?

47. A truck traveling at a constant speed on a reasonably straight, level road burns fuel at the rate of $g(x)$ gallons per mile, where x is the speed of the truck (in miles per hour) and $g(x)$ is given by $g(x) = \dfrac{800 + x^2}{200x}$.

(a) If fuel costs $1.40 per gallon, find the rule of the cost function $c(x)$ that expresses the cost of fuel for a 500-mile trip as a function of the speed. [*Hint:* $500g(x)$ gallons of fuel are needed to go 500 miles (why?).]

(b) What driving speed will make the cost of fuel for the trip $250?

(c) What driving speed will minimize the cost of fuel for the trip?

48. Pure alcohol is being added to 50 gal of a coolant mixture that is 40% alcohol.

(a) Find the rule of the concentration function $c(x)$ that expresses the percentage of alcohol in the resulting mixture as a function of the number x of gallons of pure alcohol that are added. [*Hint:* The final mixture contains $50 + x$ gallons (why?). So $c(x)$ is the amount of alcohol in the final mixture divided by the total amount $50 + x$. How much alcohol is in the original 50-gal mixture? How much is in the final mixture?]

(b) How many gallons of pure alcohol should be added to produce a mixture that is at least 60% alcohol and no more than 80% alcohol?

(c) Determine algebraically the exact amount of pure alcohol that must be added to produce a mixture that is 70% alcohol.

49. A rectangular garden with an area of 250 sq m is to be located next to a building and fenced on three sides, with the building acting as a fence on the fourth side.

(a) If the side of the garden parallel to the building has length x meters, express the amount of fencing needed as a function of x.

(b) For what values of x will less than 60 m of fencing be needed?

(c) What value of x will result in the least possible amount of fencing being used? What are the dimensions of the garden in this case?

50. A certain company has fixed costs of $40,000 and variable costs of $2.60 per unit.

(a) Let x be the number of units produced. Find the rule of the average cost function. [*Hint:* The average cost is the cost of the units divided by the number of units.]

(b) Graph the average cost function in a window with $0 \le x \le 100{,}000$ and $0 \le y \le 20$.

(c) Find the horizontal asymptote of the average cost function. Explain what the asymptote means in this situation [how low can the average cost possibly be?].

51. Radioactive waste is stored in a cylindrical tank, whose exterior has radius r and height h as shown in the figure. The sides, top, and bottom of the tank are one foot thick and the tank has a volume of 150 cubic feet (including top, bottom, and walls).

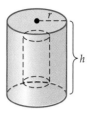

(a) Express the interior height h_1 (that is, the height of the storage area) as a function of h.

(b) Express the interior height as a function of r.

(c) Express the volume of the interior as a function of r.

(d) Explain why r must be greater than 1.

(e) What should the dimensions of the tank be for it to hold as much as possible?

52. The relationship between the fixed focal length F of a camera, the distance u from the object being photographed to the lens, and the distance v from the lens to the film is given by $\dfrac{1}{F} = \dfrac{1}{u} + \dfrac{1}{v}$.

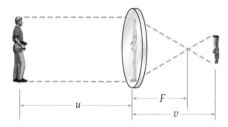

(a) If the focal length is 50 mm, express v as a function of u.

(b) What is the horizontal asymptote of the graph of the function in part (a)?

(c) Graph the function in part (a) when 50 mm $< u <$ 35,000 mm.

(d) When you focus the camera on an object, the distance between the lens and the film is changed. If the distance from the lens to the camera changes by less than .1 mm, the object will remain in focus. Explain why you have more latitude in focusing on distant objects than on very close ones.

53. The formula for the gravitational acceleration (in units of meters per second squared) of an object relative to the earth is

$$g(r) = \frac{3.987 \times 10^{14}}{(6.378 \times 10^6 + r)^2}$$

where r is the distance in meters above the earth's surface.

(a) What is the gravitational acceleration at the earth's surface?

(b) Graph the function $g(r)$ for $r \geq 0$.

(c) Can you ever escape the pull of gravity? [*Hint:* Does the graph have any r-intercepts?]

4.5.A **EXCURSION**   **Other Rational Functions**

We now examine the graphs of rational functions in which the degree of the denominator is smaller than the degree of the numerator. Such a graph has no horizontal asymptote. However, it does have some polynomial curve as an asymptote, which means that the graph will get very close to this curve when $|x|$ is very large.

Example 1 To graph $f(x) = \dfrac{x^3 + 3x^2 + x + 1}{x^2 + 2x - 1}$, we begin by finding the vertical asymptotes and the x- and y-intercepts. The quadratic formula can be used to find the roots of the denominator:

$$x = \frac{-2 \pm \sqrt{2^2 - 4 \cdot 1(-1)}}{2 \cdot 1} = \frac{-2 \pm \sqrt{8}}{2} = \frac{-2 \pm 2\sqrt{2}}{2} = -1 \pm \sqrt{2}.$$

It is easy to verify that neither of these numbers is a root of the numerator, so the graph has vertical asymptotes at $x = -1 - \sqrt{2}$ and $x = -1 + \sqrt{2}$.

The y-intercept is $f(0) = -1$. The x-intercepts are the roots of the numerator.

GRAPHING EXPLORATION

Use a calculator to verify that $x^3 + 3x^2 + x + 1$ has exactly one real root, located between -3 and -2.

Therefore, the graph of $f(x)$ has one x-intercept. Using this information and the calculator graph in Figure 4–43 (which erroneously shows some vertical segments), we conclude that the graph looks approximately like Figure 4–44.

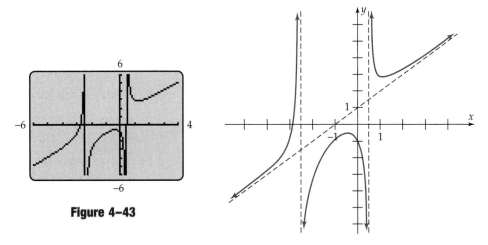

Figure 4–43

Figure 4–44

At the left and right ends, the graph moves away from the x-axis. To understand the behavior of the graph when $|x|$ is large, divide the numerator of $f(x)$ by its denominator:

$$
\begin{array}{r}
x + 1 \\
x^2 + 2x - 1 \overline{\smash{)}\, x^3 + 3x^2 + x + 1} \\
\underline{x^3 + 2x^2 - x } \\
x^2 + 2x + 1 \\
\underline{x^2 + 2x - 1} \\
2
\end{array}
$$

By the Division Algorithm

$$x^3 + 3x^2 + x + 1 = (x^2 + 2x - 1)(x + 1) + 2.$$

Dividing both sides by $x^2 + 2x - 1$, we have

$$\frac{x^3 + 3x^2 + x + 1}{x^2 + 2x - 1} = \frac{(x^2 + 2x - 1)(x + 1) + 2}{x^2 + 2x - 1}$$

$$f(x) = (x + 1) + \frac{2}{x^2 + 2x - 1}.$$

Now when x is very large in absolute value, so is $x^2 + 2x - 1$. Hence, $2/(x^2 + 2x - 1)$ is very close to 0 by the Big–Little Principle and $f(x)$ is very close to $(x + 1) + 0$. Therefore, as x gets larger in absolute value, the graph of $f(x)$ gets closer and closer to the line $y = x + 1$ (the dashed slanted line in Figure 4–44) and this line is an asymptote of the graph.* Note that $x + 1$ is just the quotient obtained in the long division above. ∎

It is instructive to examine the graph in Example 1 further to see that the asymptote accurately indicates the behavior of the function when $|x|$ is large.

> ### GRAPHING EXPLORATION
>
> Using the viewing window with $-20 \le x \le 20$ and $-20 \le y \le 20$, graph both $f(x)$ and $y = x + 1$ on the same screen.

Except near the vertical asymptotes of $f(x)$, the two graphs are virtually identical.

> ### GRAPHING EXPLORATION
>
> Now change the range values, so that the viewing window has $-100 \le x \le 100$ and $-100 \le y \le 100$.

In this viewing window, the vertical asymptotes of $f(x)$ are no longer visible and the graph is indistinguishable from the graph of the asymptote $y = x + 1$.

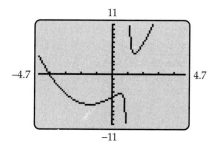

Figure 4–45

Example 2 To graph $g(x) = \dfrac{x^3 + 2x^2 - 7x + 5}{x - 1}$, we first note that there is a vertical asymptote at $x = 1$ (root of the denominator, but not the numerator). The y-intercept is at $g(0) = -5$. By carefully choosing a viewing window that accurately portrays the behavior of $g(x)$ near its vertical asymptote, we obtain Figure 4–45.

> ### GRAPHING EXPLORATION
>
> Verify that the x-intercept near $x = -4$ is the only one by showing graphically that the numerator of $g(x)$ has exactly one real root.

---

*An asymptote that is a nonvertical and nonhorizontal straight line is called an **oblique asymptote**.

To confirm that Figure 4–45 is a complete graph, we find its asymptote when $|x|$ is large. Divide the denominator by the numerator and use the Division Algorithm to rewrite $g(x)$:

$$\begin{array}{r} x^2 + 3x \;\; - 4 \\ x - 1 \overline{\smash{\big)}\, x^3 + 2x^2 - 7x + 5} \\ \underline{x^3 - x^2 } \\ 3x^2 - 7x + 5 \\ \underline{3x^2 - 3x } \\ - 4x + 5 \\ \underline{- 4x + 4} \\ 1 \end{array}$$

Hence, by the Division Algorithm,

$$x^3 + 2x^2 - 7x + 5 = (x - 1)(x^2 + 3x - 4) + 1$$

$$\frac{x^3 + 2x^2 - 7x + 5}{x - 1} = \frac{(x - 1)(x^2 + 3x - 4) + 1}{x - 1}$$

$$g(x) = (x^2 + 3x - 4) + \frac{1}{x - 1}.$$

When $|x|$ is large, $1/(x - 1)$ is very close to 0 (why?), so that $y = x^2 + 3x - 4$ is the asymptote. Once again, the asymptote is given by the quotient of the division.

GRAPHING EXPLORATION

Graph $g(x)$ and $y = x^2 + 3x - 4$ on the same screen to show that the graph of $g(x)$ does get very close to the asymptote when $|x|$ is large. Then find a large enough viewing window so that the two graphs appear to be identical. ■

The procedures used in the preceding examples may be summarized as follows.

Graphing

$$f(x) = \frac{g(x)}{h(x)}$$

When Degree g(x) > Degree h(x)

1. Analyze the function algebraically to determine its vertical asymptotes, and intercepts.

2. Divide the numerator $g(x)$ by the denominator $h(x)$. The quotient $q(x)$ is the nonvertical asymptote of the graph, which describes the behavior of the graph when $|x|$ is large.

3. Use the preceding information to select an appropriate viewing window (or windows), to interpret the calculator's version of the graph (if necessary), and to sketch an accurate graph.

Exercises 4.5.A

In Exercises 1–4, find the nonvertical asymptote of the graph of the function when $|x|$ is large and find a viewing window in which the ends of the graph are within .1 of this asymptote.

1. $f(x) = \dfrac{x^3 - 1}{x^2 - 4}$

2. $g(x) = \dfrac{x^3 - 4x^2 + 6x + 5}{x - 2}$

3. $h(x) = \dfrac{x^3 + 3x^2 - 4x + 1}{x + 4}$

4. $f(x) = \dfrac{x^3 + 3x^2 - 4x + 1}{x^2 - x}$

In Exercises 5–12, analyze the function algebraically: List its vertical asymptotes and holes, and determine its nonvertical asymptote. Then sketch a complete graph of the function.

5. $f(x) = \dfrac{x^2 - x - 6}{x - 2}$

6. $k(x) = \dfrac{x^2 + x - 2}{x}$

7. $Q(x) = \dfrac{4x^2 + 4x - 3}{2x - 5}$

8. $K(x) = \dfrac{3x^2 - 12x + 15}{3x + 6}$

9. $f(x) = \dfrac{x^3 - 2}{x - 1}$

10. $p(x) = \dfrac{x^3 + 8}{x + 1}$

11. $q(x) = \dfrac{x^3 - 1}{x - 2}$

12. $f(x) = \dfrac{x^4 - 1}{x^2}$

In Exercises 13–18, find a viewing window (or windows) that shows a complete graph of the function (if possible, with no erroneous vertical line segments). Be alert for hidden behavior.

13. $f(x) = \dfrac{2x^2 + 5x + 2}{2x + 7}$

14. $g(x) = \dfrac{2x^3 + 1}{x^2 - 1}$

15. $h(x) = \dfrac{x^3 - 2x^2 + x - 2}{x^2 - 1}$

16. $f(x) = \dfrac{3x^3 - 11x - 1}{x^2 - 4}$

17. $g(x) = \dfrac{2x^4 + 7x^3 + 7x^2 + 2x}{x^3 - x + 50}$

18. $h(x) = \dfrac{2x^3 + 7x^2 - 4}{x^2 + 2x - 3}$

19. (a) Show that when $0 < x < 4$, the rational function

$$r(x) = \dfrac{4096x^3 + 34{,}560x^2 + 19{,}440x + 729}{18{,}432x^2 + 34{,}560x + 5832}$$

is a good approximation of the function $s(x) = \sqrt{x}$ by graphing both functions in the viewing window with $0 \le x \le 4$ and $0 \le y \le 2$.

(b) For what values of x is $r(x)$ within .01 of $s(x)$?

20. Find a rational function f that has these properties:
  (i) The curve $y = x^3 - 8$ is an asymptote of the graph of f.
  (ii) $f(2) = 1$.
  (iii) The line $x = 1$ is a vertical asymptote of the graph of f.

4.6 Complex Numbers

If you are restricted to nonnegative integers, you can't solve the equation $x + 5 = 0$. Enlarging the number system to include negative integers makes it possible to solve this equation ($x = -5$). Enlarging it again, to include rational numbers, makes it possible to solve equations like $3x = 7$, which have no integer solutions. Similarly, the equation $x^2 = 2$ has no solutions in the rational number system, but has $\sqrt{2}$ and $-\sqrt{2}$ as solutions in the real number system. So the idea of enlarging a number system in order to solve an equation that can't be solved in the present system is a natural one.

Equations such as $x^2 = -1$ and $x^2 = -4$ have no solutions in the real number system because $\sqrt{-1}$ and $\sqrt{-4}$ are not real numbers. To solve such equations (or equivalently, to find square roots of negative numbers) we must enlarge the number system again. We claim that there is a number system, called the **complex number system,** with these properties:

Properties of the Complex Number System

1. The complex number system contains all real numbers.
2. Addition, subtraction, multiplication, and division of complex numbers obey the same rules of arithmetic that hold in the real number system, with one exception: the exponent laws hold for *integer* exponents, but not necessarily for fractional ones.
3. The complex number system contains a number (usually denoted by i) such that $i^2 = -1$.
4. Every complex number can be written in the *standard form* $a + bi$, where a and b are real numbers.*
5. Two complex numbers $a + bi$ and $c + di$ are equal exactly when $a = c$ and $b = d$.

In view of our past experience with enlarging the number system, this claim *ought* to appear plausible. But the mathematicians who invented the complex numbers in the 17th century were very uneasy about a number i such that $i^2 = -1$ $\left(\text{that is, } i = \sqrt{-1}\right)$. Consequently, they called numbers of the form bi (b any real number), such as $5i$ and $-\frac{1}{4}i$, **imaginary numbers.** The old familiar numbers (integers, rationals, irrationals) were called **real numbers.** Sums of real and imaginary numbers, numbers of the form $a + bi$, such as

$$5 + 2i, \qquad 7 - 4i, \qquad 18 + \frac{3}{2}i, \qquad \sqrt{3} - 12i$$

were called **complex numbers.**[†]

Every real number is a complex number; for instance, $7 = 7 + 0i$. Similarly, every imaginary number bi is a complex number since $bi = 0 + bi$. Since the usual laws of arithmetic still hold, it's easy to add, subtract, and multiply complex numbers. As the following examples demonstrate, *all symbols can be treated as if they were real numbers, provided that i^2 is replaced by -1*. Unless directed otherwise, express your answers in the standard form $a + bi$.

Example 1

(a) $(1 + i) + (3 - 7i) = 1 + i + 3 - 7i$
$$= (1 + 3) + (i - 7i) = 4 - 6i$$

---

*Hereafter whenever we write $a + bi$ or $c + di$, it is assumed that a, b, c, d are real numbers and $i^2 = -1$.

[†]This terminology is still used, even though there is nothing complicated, unreal, or imaginary about complex numbers—they are just as valid mathematically as are real numbers.

(b) $(4 + 3i) - (8 - 6i) = 4 + 3i - 8 - (-6i)$
$$= (4 - 8) + (3i + 6i) = -4 + 9i$$

(c) $4i\left(2 + \dfrac{1}{2}i\right) = 4i\cdot 2 + 4i\left(\dfrac{1}{2}i\right) = 8i + 4\cdot\dfrac{1}{2}\cdot i^2$
$$= 8i + 2i^2 = 8i + 2(-1) = -2 + 8i$$

(d) $(2 + i)(3 - 4i) = 2\cdot 3 + 2(-4i) + i\cdot 3 + i(-4i)$
$$= 6 - 8i + 3i - 4i^2 = 6 - 8i + 3i - 4(-1)$$
$$= (6 + 4) + (-8i + 3i) = 10 - 5i \quad\blacksquare$$

Technology Tip

You can do complex arithmetic directly by using the special i key on the TI-83/89 and Sharp 9600 keyboards and in the CPLX submenu of the Casio 9850 OPTN menu. On TI-86 and HP-38, enter $a + bi$ as (a, b). Other calculators can do complex arithmetic by means of matrices; see Calculator Investigation 1 on page 330.

The familiar multiplication patterns and exponent laws for integer exponents hold in the complex number system.

Example 2

(a) $(3 + 2i)(3 - 2i) = 3^2 - (2i)^2$
$$= 9 - 4i^2 = 9 - 4(-1) = 9 + 4 = 13$$

(b) $(4 + i)^2 = 4^2 + 2\cdot 4\cdot i + i^2 = 16 + 8i + (-1) = 15 + 8i$

(c) To find i^{54}, we first note that $i^4 = i^2 i^2 = (-1)(-1) = 1$ and that $54 = 52 + 2 = 4\cdot 13 + 2$. Consequently,
$$i^{54} = i^{52+2} = i^{52}i^2 = i^{4\cdot 13}i^2 = (i^4)^{13}i^2 = 1^{13}(-1) = -1. \quad\blacksquare$$

The **conjugate** of the complex number $a + bi$ is the number $a - bi$, and the conjugate of $a - bi$ is $a + bi$. For example, the conjugate of $3 + 4i$ is $3 - 4i$ and the conjugate of $-3i = 0 - 3i$ is $0 + 3i = 3i$. *Every real number is its own conjugate;* for instance, the conjugate of $17 = 17 + 0i$ is $17 - 0i = 17$.

For any complex number $a + bi$, we have
$$(a + bi)(a - bi) = a^2 - (bi)^2 = a^2 - b^2 i^2 = a^2 - b^2(-1) = a^2 + b^2.$$

Since a^2 and b^2 are nonnegative real numbers, so is $a^2 + b^2$. Therefore *the product of a complex number and its conjugate is a nonnegative real number.* This fact enables us to express quotients of complex numbers in standard form.

Example 3 To express $\dfrac{3 + 4i}{1 + 2i}$ in the form $a + bi$, *multiply both numerator and denominator by the conjugate of the denominator,* namely, $1 - 2i$:

$$\dfrac{3 + 4i}{1 + 2i} = \dfrac{3 + 4i}{1 + 2i}\cdot\dfrac{1 - 2i}{1 - 2i} = \dfrac{(3 + 4i)(1 - 2i)}{(1 + 2i)(1 - 2i)}$$

$$= \dfrac{3 + 4i - 6i - 8i^2}{1^2 - (2i)^2} = \dfrac{3 + 4i - 6i - 8(-1)}{1 - 4i^2} = \dfrac{11 - 2i}{1 - 4(-1)}$$

$$= \dfrac{11 - 2i}{5} = \dfrac{11}{5} - \dfrac{2}{5}i.$$

This is the form $a + bi$ with $a = 11/5$ and $b = -2/5$. $\quad\blacksquare$

Example 4 To express $\dfrac{1}{1-i}$ in standard form, note that the conjugate of the denominator is $1+i$ and therefore:

$$\frac{1}{1-i} = \frac{1(1+i)}{(1-i)(1+i)} = \frac{1+i}{1^2-i^2} = \frac{1+i}{1-(-1)} = \frac{1+i}{2} = \frac{1}{2} + \frac{1}{2}i.$$

We can check this result by multiplying $\dfrac{1}{2} + \dfrac{1}{2}i$ by $1-i$ to see if the product is 1 (which it should be if $\dfrac{1}{2} + \dfrac{1}{2}i = \dfrac{1}{1-i}$):

$$\left(\frac{1}{2} + \frac{1}{2}i\right)(1-i) = \frac{1}{2}\cdot 1 - \frac{1}{2}i + \frac{1}{2}i\cdot 1 - \frac{1}{2}i^2 = \frac{1}{2} - \frac{1}{2}(-1) = 1. \quad \blacksquare$$

Since $i^2 = -1$, we define $\sqrt{-1}$ to be the complex number i. Similarly, since $(5i)^2 = 5^2i^2 = 25(-1) = -25$, we define $\sqrt{-25}$ to be $5i$. In general,

Square Roots of Negative Numbers

For any positive real number b,

$$\sqrt{-b} \text{ is defined to be } \sqrt{b}\,i$$

because $\left(\sqrt{b}\,i\right)^2 = \left(\sqrt{b}\right)^2 i^2 = b(-1) = -b.$

CAUTION

$\sqrt{b}\,i$ is *not* the same as $\sqrt{bi}$. To avoid confusion it may help to write $\sqrt{b}\,i$ as $i\sqrt{b}$.

Example 5

(a) $\sqrt{-3} = \sqrt{3}i = i\sqrt{3}.$

(b) $\dfrac{1 - \sqrt{-7}}{3} = \dfrac{1 - \sqrt{7}i}{3} = \dfrac{1}{3} - \dfrac{\sqrt{7}}{3}i. \quad \blacksquare$

CAUTION

The property $\sqrt{cd} = \sqrt{c}\sqrt{d}$ (or equivalently in exponential notation, $(cd)^{1/2} = c^{1/2}d^{1/2}$), which is valid for positive real numbers, *does not hold* when both c and d are negative.

$$\sqrt{-20}\sqrt{-5} = \sqrt{20}i\cdot\sqrt{5}i = \sqrt{20}\sqrt{5}i^2\cdot\sqrt{20\cdot 5}(-1)$$
$$= \sqrt{100}(-1) = -10$$

But $\sqrt{(-20)(-5)} = \sqrt{100} = 10$, so that

$$\sqrt{(-20)(-5)} \ne \sqrt{-20}\sqrt{-5}.$$

To avoid difficulty, *always write square roots of negative numbers in terms of i before doing any simplification.*

Technology Tip

Most calculators that do complex number arithmetic automatically return a complex number when asked for the square root of a negative number. On TI-83/89, however, the MODE must be set to "rectangular" or "$a + bi$."

Example 6

$$(7 - \sqrt{-4})(5 + \sqrt{-9}) = (7 - \sqrt{4}i)(5 + \sqrt{9}i)$$
$$= (7 - 2i)(5 + 3i)$$
$$= 35 + 21i - 10i - 6i^2$$
$$= 35 + 11i - 6(-1) = 41 + 11i \quad \blacksquare$$

Since every negative real number has a square root in the complex number system, we can now find complex solutions for equations that have no real solutions. For example, the solutions of $x^2 = -25$ are $x = \pm\sqrt{-25} = \pm 5i$. In fact,

Every quadratic equation with real coefficients has solutions in the complex number system.

Example 7

To solve the equation $2x^2 + x + 3 = 0$, we apply the quadratic formula:

$$x = \frac{-1 \pm \sqrt{1^2 - 4\cdot2\cdot3}}{2\cdot2} = \frac{-1 \pm \sqrt{-23}}{4}.$$

Since $\sqrt{-23}$ is not a real number, this equation has no real number solutions. But $\sqrt{-23}$ is a complex number, namely, $\sqrt{-23} = \sqrt{23}i$. Thus the equation does have solutions in the complex number system:

$$x = \frac{-1 \pm \sqrt{-23}}{4} = \frac{-1 \pm \sqrt{23}i}{4} = -\frac{1}{4} \pm \frac{\sqrt{23}}{4}i.$$

Note that the two solutions, $-\frac{1}{4} + \frac{\sqrt{23}}{4}i$ and $-\frac{1}{4} - \frac{\sqrt{23}}{4}i$, are conjugates of each other. $\blacksquare$

Technology Tip

The polynomial solvers on TI-86, HP-38, Sharp 9600, and Casio 9850 and FX 2.0 produce all real and complex solutions of any polynomial equation that they can solve. See Calculator Investigation 1 on page 150 for details. On TI-89, use cSOLVE in the COMPLEX submenu of the ALGEBRA menu to find all solutions.

Example 8

To find *all* solutions of $x^3 = 1$, we rewrite the equation and use the Difference of Cubes pattern (see page 36) to factor:

$$x^3 = 1$$
$$x^3 - 1 = 0$$
$$(x - 1)(x^2 + x + 1) = 0$$
$$x - 1 = 0 \quad \text{or} \quad x^2 + x + 1 = 0.$$

The solution of the first equation is $x = 1$. The solutions of the second can be obtained from the quadratic formula:

$$x = \frac{-1 \pm \sqrt{1^2 - 4\cdot1\cdot1}}{2\cdot1} = \frac{-1 \pm \sqrt{-3}}{2} = \frac{-1 \pm \sqrt{3}i}{2} = -\frac{1}{2} \pm \frac{\sqrt{3}}{2}i.$$

Therefore, the equation $x^3 = 1$ has one real solution ($x = 1$) and two non-real complex solutions [$x = -1/2 + (\sqrt{3}/2)i$ and $x = -1/2 - (\sqrt{3}/2)i$]. Each of these solutions is said to be a **cube root of 1** or a **cube root of**

unity. Observe that the two nonreal complex cube roots of unity are conjugates of each other. ∎

The preceding examples illustrate this useful fact (whose proof is discussed in Section 4.7):

Conjugate Solutions

If $a + bi$ is a solution of a polynomial equation with *real* coefficients, then its conjugate $a - bi$ is also a solution of this equation.

Calculator Investigation 4.6

The following investigation of complex number arithmetic is for use with TI-81/82 and other calculators that do not do complex number arithmetic, but do have matrix capabilities. Before doing it, look up "matrix" or "matrices" in your instruction manual and learn how to enter and store 2 by 2 matrices and how to do addition, subtraction, and multiplication with them.

1. The complex number $a + bi$ is expressed in matrix notation as the matrix $\begin{pmatrix} a & b \\ -b & a \end{pmatrix}$. For example, $-3 + 6i$ is written as $\begin{pmatrix} -3 & 6 \\ -6 & -3 \end{pmatrix}$.

   (a) Write $3 + 4i$, $1 + 2i$, and $1 - i$ in matrix form and enter them in your calculator as $[A]$, $[B]$, $[C]$.

   (b) We know that $(3 + 4i) + (1 + 2i) = 4 + 6i$. Verify that $[A] + [B]$ is $\begin{pmatrix} 4 & 6 \\ -6 & 4 \end{pmatrix}$, which represents the complex number $4 + 6i$.

   (c) Use matrix addition, subtraction, and multiplication to find the following. Interpret the answers as complex numbers: $[A] - [C]$, $[B] + [C]$, $[A][B]$, $[B][C]$.

   (d) In Example 3 we saw that $\dfrac{3 + 4i}{1 + 2i} = \dfrac{11}{5} - \dfrac{2}{5}i = 2.2 - .4i$. Do this problem in matrix form by computing $[A] \cdot [B]^{-1}$ (use the x^{-1} key for the exponent).

   (e) Do each of the following calculations and interpret the problem in terms of complex numbers: $[A] \cdot [C]^{-1}$, $[B][A]^{-1}$, $[B][C]^{-1}$.

Exercises 4.6

In Exercises 1–54, perform the indicated operation and write the result in the form $a + bi$.

1. $(2 + 3i) + (6 - i)$

2. $(-5 + 7i) + (14 + 3i)$

3. $(2 - 8i) - (4 + 2i)$

4. $(3 + 5i) - (3 - 7i)$

5. $\dfrac{5}{4} - \left(\dfrac{7}{4} + 2i \right)$

6. $\left(\sqrt{3} + i \right) + \left(\sqrt{5} - 2i \right)$

7. $\left(\dfrac{\sqrt{2}}{2} + i \right) - \left(\dfrac{\sqrt{3}}{2} - i \right)$

8. $\left(\dfrac{1}{2} + \dfrac{\sqrt{3}i}{2}\right) + \left(\dfrac{3}{4} - \dfrac{5\sqrt{3}i}{2}\right)$

9. $(2 + i)(3 + 5i)$

10. $(2 - i)(5 + 2i)$

11. $(-3 + 2i)(4 - i)$

12. $(4 + 3i)(4 - 3i)$

13. $(2 - 5i)^2$

14. $(1 + i)(2 - i)i$

15. $\left(\sqrt{3} + 1\right)\left(\sqrt{3} - i\right)$

16. $\left(\dfrac{1}{2} - i\right)\left(\dfrac{1}{4} + 2i\right)$

17. i^{15} 18. i^{26} 19. i^{33} 20. $(-i)^{53}$

21. $(-i)^{107}$ 22. $(-i)^{213}$ 23. $\dfrac{1}{5 - 2i}$ 24. $\dfrac{1}{i}$

25. $\dfrac{1}{3i}$ 26. $\dfrac{i}{2 + i}$ 27. $\dfrac{3}{4 + 5i}$ 28. $\dfrac{2 + 3i}{i}$

29. $\dfrac{1}{i(4 + 5i)}$ 30. $\dfrac{1}{(2 - i)(2 + i)}$ 31. $\dfrac{2 + 3i}{i(4 + i)}$

32. $\dfrac{2}{(2 + 3i)(4 + i)}$ 33. $\dfrac{2 + i}{1 - i} + \dfrac{1}{1 + 2i}$

34. $\dfrac{1}{2 - i} + \dfrac{3 + i}{2 + 3i}$ 35. $\dfrac{i}{3 + i} - \dfrac{3 + i}{4 + i}$

36. $6 + \dfrac{2i}{3 + i}$ 37. $\sqrt{-36}$ 38. $\sqrt{-81}$

39. $\sqrt{-14}$ 40. $\sqrt{-50}$ 41. $-\sqrt{-16}$

42. $-\sqrt{-12}$ 43. $\sqrt{-16} + \sqrt{-49}$

44. $\sqrt{-25} - \sqrt{-9}$ 45. $\sqrt{-15} - \sqrt{-18}$

46. $\sqrt{-12}\sqrt{-3}$ 48. $\sqrt{-16}/\sqrt{-36}$

48. $-\sqrt{-64}/\sqrt{-4}$

49. $\left(\sqrt{-25} + 2\right)\left(\sqrt{-49} - 3\right)$

50. $\left(5 - \sqrt{-3}\right)\left(-1 + \sqrt{-9}\right)$

51. $\left(2 + \sqrt{-5}\right)\left(1 - \sqrt{-10}\right)$

52. $\sqrt{-3}\left(3 - \sqrt{-27}\right)$ 53. $1/\left(1 + \sqrt{-2}\right)$

54. $\left(1 + \sqrt{-4}\right)/\left(3 - \sqrt{-9}\right)$

In Exercises 55–58, find x and y. Remember that $a + bi = c + di$ exactly when $a = c$ and $b = d$.

55. $3x - 4i = 6 + 2yi$

56. $8 - 2yi = 4x + 12i$

57. $3 + 4xi = 2y - 3i$

58. $8 - xi = \dfrac{1}{2}y + 2i$

In Exercises 59–70, solve the equation and express each solution in the form $a + bi$.

59. $3x^2 - 2x + 5 = 0$ 60. $5x^2 + 2x + 1 = 0$

61. $x^2 + x + 2 = 0$ 62. $5x^2 - 6x + 2 = 0$

63. $2x^2 - x = -4$ 64. $x^2 + 1 = 4x$

65. $2x^2 + 3 = 6x$ 66. $3x^2 + 4 = -5x$

67. $x^3 - 8 = 0$ 68. $x^3 + 125 = 0$

69. $x^4 - 1 = 0$ 70. $x^4 - 81 = 0$

71. Simplify: $i + i^2 + i^3 + \cdots + i^{15}$

72. Simplify: $i - i^2 + i^3 - i^4 + i^5 - \cdots + i^{15}$

Thinkers

73. If $z = a + bi$ is a complex number, then its conjugate is usually denoted $\bar{z}$, that is $\bar{z} = a - bi$. Verify that
$$\bar{\bar{z}} = z$$

74. The **real part** of the complex number $a + bi$ is defined to be the real number a. The **imaginary part** of $a + bi$ is defined to the real number b (*not bi*).

(a) Show that the real part of $z = a + bi$ is $\dfrac{z + \bar{z}}{2}$.

(b) Show that the imaginary part of $z = a + bi$ is $\dfrac{z - \bar{z}}{2i}$.

75. If $z = a + bi$ (with a, b real numbers, not both 0), express $1/z$ in standard form.

4.7 The Fundamental Theorem of Algebra

The complex numbers were constructed in order to obtain a solution for the equation $x^2 = -1$, that is, a root of the polynomial $x^2 + 1$. In Section 4.6 we saw that *every* quadratic polynomial with real coefficients has roots in the complex number system. A natural question now arises: Do we have to enlarge the complex number system (perhaps many times) to find roots for higher degree polynomials? In this section we shall see that the somewhat surprising answer is no.

In order to give the full answer, we shall consider not just polynomials with real coefficients, but also those with complex coefficients, such as

$$x^3 - ix^2 + (4 - 3i)x + 1 \qquad \text{or} \qquad (-3 + 2i)x^6 - 3x + (5 - 4i).$$

The discussion of polynomial division in Section 4.2 can easily be extended to include polynomials with complex coefficients. In fact, *all of the results in Section 4.2 are valid for polynomials with complex coefficients.* For example, you can check that i is a root of $f(x) = x^2 + (i - 1)x + (2 + i)$ and that $x - i$ is a factor of $f(x)$:

$$f(x) = x^2 + (i - 1)x + (2 + i) = (x - i)[x - (1 - 2i)].$$

Since every real number is also a complex number, polynomials with real coefficients are just special cases of polynomials with complex coefficients. So in the rest of this section, "polynomial" means "polynomial with complex (possibly real) coefficients" unless specified otherwise. We can now answer the question posed in the first paragraph.

Fundamental Theorem of Algebra

> **Every nonconstant polynomial has a root in the complex number system.**

Although this is obviously a powerful result, neither the Fundamental Theorem nor its proof provides a practical method for *finding* a root of a given polynomial.* The proof of the Fundamental Theorem is beyond the scope of this book, but we shall explore some of the useful implications of the theorem, such as this one:

Factorization over the Complex Numbers

> **Let $f(x)$ be a polynomial of degree $n > 0$ with leading coefficients d. Then there are (not necessarily distinct) complex numbers $c_1, c_2, \ldots, c_n$ such that**
>
> $$f(x) = d(x - c_1)(x - c_2)(x - c_3) \cdots (x - c_n).$$
>
> **Furthermore, $c_1, c_2, \ldots, c_n$ are the only roots of $f(x)$.**

Proof By the Fundamental Theorem, $f(x)$ has a complex root c_1. The Factor Theorem shows that $x - c_1$ must be a factor of $f(x)$, say,

$$f(x) = (x - c_1)g(x),$$

---

*It may seem strange that you can prove that a root exists without actually exhibiting one. But such "existence theorems" are quite common. A rough analogy is the situation that occurs when someone is killed by a sniper's bullet. The police know that there *is* a killer, but *finding* the killer may be impossible.

where $g(x)$ has degree $n - 1$.* If $g(x)$ is nonconstant, then it has a complex root c_2 by the Fundamental Theorem. Hence $x - c_2$ is a factor of $g(x)$, so that

$$f(x) = (x - c_1)(x - c_2)h(x)$$

for some $h(x)$ of degree $n - 2$ [1 less than the degree of $g(x)$]. If $h(x)$ is nonconstant, then it has a complex root c_3 and the argument can be repeated. Continuing in this way, with the degree of the last factor going down by 1 at each step, we reach a factorization in which the last factor is a constant (degree 0 polynomial):

(∗) $$f(x) = (x - c_1)(x - c_2)(x - c_3)\cdots(x - c_n)d.$$

If the right side were multiplied out, it would look like

$$dx^n + \text{lower degree terms.}$$

Technology Tip

To find all the roots of a polynomial, see the Tip on page 329.

To factor a polynomial as a product of linear factors; use cFACTOR in the COMPLEX submenu of the TI-89 ALGEBRA menu.

So the constant factor d is the leading coefficient of $f(x)$.

It is easy to see from the factored form (∗) that the numbers $c_1, c_2, \ldots, c_n$ are roots of $f(x)$. If k is *any* root of $f(x)$, then

$$0 = f(k) = d(k - c_1)(k - c_2)(k - c_3)\cdots(k - c_n).$$

The product on the right is 0 only when one of the factors is 0. Since the leading coefficient d is nonzero, we must have

$$k - c_1 = 0 \quad \text{or} \quad k - c_2 = 0 \quad \text{or} \quad \cdots \quad \text{or} \quad k - c_n = 0$$
$$k = c_1 \quad \text{or} \quad k = c_2 \quad \text{or} \quad \cdots \quad k = c_n.$$

Therefore, k is one of the c's and $c_1, \ldots, c_n$ are the only roots of $f(x)$. This completes the proof. ∎

Since the n roots $c_1, \ldots, c_n$ of $f(x)$ may not all be distinct, we see that:

Number of Roots

▶ Every polynomial of degree $n > 0$ has at most n different roots in the complex number system.

Suppose $f(x)$ has repeated roots, meaning that some of the $c_1, \ldots, c_n$ are the same in factorization (∗). Recall that a root c is said to have multiplicity k if $(x - c)^k$ is a factor of $f(x)$, but no higher power of $(x - c)$ is a factor. Consequently, if every root is counted as many times as its multiplicity, then the statement in the preceding box implies that

A polynomial of degree n has exactly n roots.

Example 1 Find a polynomial $f(x)$ of degree 5 such that 1, -2, and 5 are roots, 1 is a root of multiplicity 3, and $f(2) = -24$.

---

*The degree of $g(x)$ is 1 less than the degree n of $f(x)$ because $f(x)$ is the product of $g(x)$ and $x - c_1$ (which has degree 1).

Solution Since 1 is a root of multiplicity 3, $(x - 1)^3$ must be a factor of $f(x)$. At least two other factors correspond to the roots -2 and 5: $x - (-2) = x + 2$ and $x - 5$. The product of these factors $(x - 1)^3(x + 2)(x - 5)$ has degree 5, as does $f(x)$, so $f(x)$ must look like this:

$$f(x) = d(x - 1)^3(x + 2)(x - 5),$$

where d is the leading coefficient. Since $f(2) = -24$ we have:

$$d(2 - 1)^3(2 + 2)(2 - 5) = f(2) = -24,$$

which reduces to $-12d = -24$. Therefore, $d = (-24)/(-12) = 2$ and

$$f(x) = 2(x - 1)^3(x + 2)(x - 5)$$
$$f(x) = 2x^5 - 12x^4 + 4x^3 + 40x^2 - 54x + 20. \quad \blacksquare$$

Polynomials with Real Coefficients

Recall that the **conjugate** of the complex number $a + bi$ is the number $a - bi$. We usually write a complex number as a single letter, say z, and indicate its conjugate by $\bar{z}$ (sometimes read "z bar"). For instance, if $z = 3 + 7i$, then $\bar{z} = 3 - 7i$. Conjugates play a role whenever a quadratic polynomial with real coefficients has complex roots.

Example 2 The quadratic formula shows that $x^2 - 6x + 13$ has two complex roots:

$$\frac{-(-6) \pm \sqrt{(-6)^2 - 4 \cdot 1 \cdot 13}}{2 \cdot 1} = \frac{6 \pm \sqrt{-16}}{2} = \frac{6 \pm 4i}{2} = 3 \pm 2i.$$

The complex roots are $z = 3 + 2i$ and its conjugate $\bar{z} = 3 - 2i$. $\quad \blacksquare$

Example 2 is a special case of a more general theorem, whose proof is outlined in Exercises 59 and 60:

Conjugate Roots Theorem

Let $f(x)$ be a polynomial with *real* coefficients. If the complex number z is a root of $f(x)$, then its conjugate $\bar{z}$ is also a root of $f(x)$.

Example 3 Find a polynomial with real coefficients whose roots include the numbers 2 and $3 + i$.

Solution Since $3 + i$ is a root, its conjugate $3 - i$ must also be a root. Consider the polynomial.

$$f(x) = (x - 2)[(x - (3 + i)][x - (3 - i)].$$

Obviously 2, $3 + i$, and $3 - i$ are roots of $f(x)$. Multiplying out this factored form shows that $f(x)$ *does* have real coefficients:

$$f(x) = (x - 2)[x^2 - (3 - i)x - (3 + i)x + (3 + i)(3 - i)]$$
$$= (x - 2)(x^2 - 3x + ix - 3x - ix + 9 - i^2)$$
$$= (x - 2)(x^2 - 6x + 10)$$
$$= x^3 - 8x^2 + 22x - 20.$$

The next-to-last line of this calculation also shows that $f(x)$ can be factored as a product of a linear and a quadratic polynomial, each with *real* coefficients. ∎

The technique in Example 3 works because the polynomial

$$[x - (3 + i)][x - (3 - i)]$$

turns out to have real coefficients. The proof of the following result shows why this must always be the case:

Factorization over the Real Numbers

> **Every nonconstant polynomial with real coefficients can be factored as a product of linear and quadratic polynomials with real coefficients in such a way that the quadratic factors, if any, have no real roots.**

Proof   The box on page 332 shows that

$$f(x) = d(x - c_1)(x - c_2)\cdots(x - c_n)$$

where $c_1, \ldots, c_n$ are the roots of $f(x)$. If some c_1 is a real number, then the factor $x - c_i$ is a linear polynomial with real coefficients.* If some c_j is a nonreal complex root, then its conjugate must also be a root. Thus some c_k is the conjugate of c_j, say, $c_j = a + bi$ (with a, b real) and $c_k = a - bi$.[†] In this case,

$$(x - c_j)(x - c_k) = [x - (a + bi)][x - (a - bi)]$$
$$= x^2 - (a - bi)x - (a + bi)x + (a + bi)(a - bi)$$
$$= x^2 - ax + bix - ax - bix + a^2 - (bi)^2$$
$$= x^2 - 2ax + (a^2 + b^2).$$

Therefore, the factor $(x - c_j)(x - c_k)$ of $f(x)$ is a quadratic with real coefficients (because a and b are real numbers). Its roots (c_j and c_k) are nonreal. By taking the real roots of $f(x)$ one at a time and the nonreal ones in conjugate pairs in this fashion, we obtain the desired factorization of $f(x)$. ∎

---

*In Example 3, for instance, 2 is a real root and $x - 2$ a linear factor.

[†]In Example 3, for instance, $c_j = 3 + i$ and $c_k = 3 - i$ are conjugate roots.

Example 4 Given that $1 + i$ is a root of $f(x) = x^4 - 2x^3 - x^2 + 6x - 6$, factor $f(x)$ completely over the real numbers.

Solution Since $1 + i$ is a root of $f(x)$, so is its conjugate $1 - i$, and hence $f(x)$ has this quadratic factor:

$$[x - (1 + i)][x - (1 - i)] = x^2 - 2x + 2.$$

Dividing $f(x)$ by $x^2 - 2x + 2$ shows that the other factor is $x^2 - 3$, which factors as $(x + \sqrt{3})(x - \sqrt{3})$. Therefore

$$f(x) = (x + \sqrt{3})(x - \sqrt{3})(x^2 - 2x + 2). \quad \blacksquare$$

Exercises 4.7

In Exercises 1–6, find the remainder when $f(x)$ is divided by $g(x)$ without using synthetic or long division.

1. $f(x) = x^{10} + x^8$; $g(x) = x - 1$

2. $f(x) = x^6 - 10$; $g(x) = x - 2$

3. $f(x) = 3x^4 - 6x^3 + 2x - 1$; $g(x) = x + 1$

4. $f(x) = x^5 - 3x^2 + 2x - 1$; $g(x) = x - 2$

5. $f(x) = x^3 - 2x^2 + 5x - 4$; $g(x) = x + 2$

6. $f(x) = 10x^{75} - 8x^{65} + 6x^{45} + 4x^{32} - 2x^{15} + 5$; $g(x) = x - 1$

In Exericses 7–10, list the roots of the polynomial and state the multiplicity of each root.

7. $f(x) = x^{54}\left(x + \dfrac{4}{5}\right)$

8. $g(x) = 3\left(x + \dfrac{1}{6}\right)\left(x - \dfrac{1}{5}\right)\left(x + \dfrac{1}{4}\right)$

9. $h(x) = 2x^{15}(x - \pi)^{14}[x - (\pi + 1)]^{13}$

10. $k(x) = (x - \sqrt{7})^7(x - \sqrt{5})^5(2x - 1)$

In Exercises 11–22, find all the roots of $f(x)$ in the complex number system; then write $f(x)$ as a product of linear factors.

11. $f(x) = x^2 - 2x + 5$ **12.** $f(x) = x^2 - 4x + 13$

13. $f(x) = 3x^2 + 2x + 7$ **14.** $f(x) = 3x^2 - 5x + 2$

15. $f(x) = x^3 - 27$ [*Hint:* Factor first.]

16. $f(x) = x^3 + 125$ **17.** $f(x) = x^3 + 8$

18. $f(x) = x^6 - 64$ [*Hint:* Let $u = x^3$ and factor $u^2 - 64$ first.]

19. $f(x) = x^4 - 1$ **20.** $f(x) = x^4 - x^2 - 6$

21. $f(x) = x^4 - 3x^2 - 10$ **22.** $f(x) = 2x^4 - 7x^2 - 4$

In Exercises 23–44, find a polynomial $f(x)$ with real coefficients that satisfies the given conditions. Some of these problems have many correct answers.

23. Degree 3; only roots are 1, 7, −4.

24. Degree 3; only roots are 1 and −1.

25. Degree 6; only roots are 1, 2, π.

26. Degree 5; only root is 2.

27. Degree 3; roots −3, 0, 4; $f(5) = 80$.

28. Degree 3; roots −1, 1/2, 2; $f(0) = 2$.

29. Roots include $2 + i$ and $2 - i$.

30. Roots include $1 + 3i$ and $1 - 3i$.

31. Roots include 2 and $2 + i$.

32. Roots include 3 and $4i - 1$.

33. Roots include −3, $1 - i$, $1 + 2i$.

34. Roots include 1, $2 + i$, $3i - 1$.

35. Degree 2; roots $1 + 2i$ and $1 - 2i$.

36. Degree 4; roots $3i$ and $-3i$, each of multiplicity 2.

37. Degree 4; only roots are 4, $3 + i$, and $3 - i$.

38. Degree 5; roots 2 (of multiplicity 3), i, and $-i$.

39. Degree 6; roots 0 (of multiplicity 3) and 3, $1 + i$, $1 - i$, each of multiplicity 1.

40. Degree 6; roots include i (of multiplicity 2) and 3.

41. Degree 2; roots include $1 + i$; $f(0) = 6$.

42. Degree 2; roots include $3 + i$; $f(2) = 3$.

43. Degree 3; roots include i and 1; $f(-1) = 8$.

44. Degree 3; roots include $2 + 3i$ and −2; $f(2) = -3$.

In Exercises 45–48, find a polynomial with complex coefficients that satisfies the given conditions.

45. Degree 2; roots i and $1 - 2i$.

46. Degree 2; roots $2i$ and $1 + i$.

47. Degree 3; roots 3, i, and $2 - i$.

48. Degree 4; roots $\sqrt{2}, -\sqrt{2}, 1 + i$, and $1 - i$.

In Exercises 49–56, one root of the polynomial is given; find all the roots.

49. $x^3 - 2x^2 - 2x - 3$; root 3.

50. $x^3 + x^2 + x + 1$; root i.

51. $x^4 + 3x^3 + 3x^2 + 3x + 2$; root i.

52. $x^4 - x^3 - 5x^2 - x - 6$; root i.

53. $x^4 - 2x^3 + 5x^2 - 8x + 4$; root 1 of multiplicity 2.

54. $x^4 - 6x^3 + 29x^2 - 76x + 68$; root 2 of multiplicity 2.

55. $x^4 - 4x^3 + 6x^2 - 4x + 5$; root $2 - i$.

56. $x^4 - 5x^3 + 10x^2 - 20x + 24$; root $2i$.

Thinkers

57. Let $z = a + bi$ and $w = c + di$ be complex numbers (a, b, c, d are real numbers). Prove the given equality by computing each side and comparing the results:
   (a) $\overline{z + w} = \overline{z} + \overline{w}$ (The left side says: First find $z + w$ and then take the conjugate. The right side says: First take the conjugates of z and w and then add.)
   (b) $\overline{z \cdot w} = \overline{z} \cdot \overline{w}$

58. Let $g(x)$ and $h(x)$ be polynomials of degree n and assume that there are $n + 1$ numbers $c_1, c_2, \ldots, c_n, c_{n+1}$ such that
$$g(c_i) = h(c_i) \quad \text{for every } i.$$
Prove that $g(x) = h(x)$. [*Hint:* Show that each c_i is a root of $f(x) = g(x) - h(x)$. If $f(x)$ is nonzero, what is its largest possible degree? To avoid a contradiction, conclude that $f(x) = 0$.]

59. Suppose $f(x) = ax^3 + bx^2 + cx + d$ has real coefficients and z is a complex root of $f(x)$.
   (a) Use Exercise 57 and the fact that $\overline{r} = r$, when r is a real number, to show that
$$\overline{f(z)} = \overline{az^3 + bz^2 + cz + d}$$
$$= a\overline{z}^3 + b\overline{z}^2 + c\overline{z} + d = f(\overline{z}).$$
   (b) Conclude that $\overline{z}$ is also a root of $f(x)$. [*Note:* $f(\overline{z}) = \overline{f(z)} = \overline{0} = 0$.]

60. Let $f(x)$ be a polynomial with real coefficients and z a complex root of $f(x)$. Prove that the conjugate $\overline{z}$ is also a root of $f(x)$. [*Hint:* Exercise 59 is the case when $f(x)$ has degree 3; the proof in the general case is similar.]

61. Use the statement in the box on page 335 to show that every polynomial with real coefficients and *odd* degree must have at least one real root.

62. Give an example of a polynomial $f(x)$ with complex, nonreal coefficients and a complex number z such that z is a root of $f(x)$, but its conjugate is not. Hence, the conclusion of the Conjugate Roots Theorem (page 334) may be *false* if $f(x)$ doesn't have real coefficients.

Chapter 4   *Review*

Important Concepts

Important Facts and Formulas

· The graph of $f(x) = ax^2 + bx + c$ is a parabola whose vertex has x-coordinate $-b/2a$.

Review Questions

In Questions 1–5, find the vertex of the graph of the quadratic function.

1. $f(x) = (x - 2)^2 + 3$ **2.** $f(x) = 2(x + 1)^2 - 1$

3. $f(x) = x^2 - 8x + 12$ **4.** $f(x) = x^2 - 7x + 6$ **5.** $f(x) = 3x^2 - 9x + 1$

6. Which of the following statements about the functions
$$f(x) = 3x^2 + 2 \quad \text{and} \quad g(x) = -3x^2 + 2$$
is *false*?
   (a) The graphs of f and g are parabolas.
   (b) The graphs of f and g have the same vertex.
   (c) The graphs of f and g open in opposite directions.
   (d) The graph of f is the graph of $y = 3x^2$ shifted 2 units to the right.

7. A preschool wants to construct a fenced playground. The fence will be attached to the building at two corners, as shown in the figure. There is 400 feet of fencing available, all of it to be used.
   (a) Write an equation in x and y that gives the amount of fencing to be used. Solve the equation for y.
   (b) Write the area of the playground as a function of x. [*Hint:* Part (a) may be helpful.]
   (c) What are the dimensions of the playground with the largest possible area?

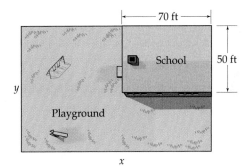

8. A model rocket is launched straight up from a platform at time $t = 0$ (where t is time measured in seconds). The altitude $h(t)$ of the rocket above the ground at given time (t) is given by $h(t) = 10 + 112t - 16t^2$ [where $h(t)$ is measured in feet].
   (a) What is the altitude of the rocket the instant it is launched?
   (b) What is the altitude of the rocket 2 seconds after launching?
   (c) What is the maximum *altitude* attained by the rocket?
   (d) At what *time* does the rocket return to the altitude at which it was launched?

9. A rectangular garden next to a building is to be fenced with 120 feet of fencing. The side against the building will not be fenced. What should the lengths of the other three sides be to ensure the largest possible area?

10. A factory offers 100 calculators to a retailer at a price of $20 each. The price per calculator on the entire order will be reduced 5¢ for each additional calculator over 100. What number of calculators will produce the largest possible sales revenue for the factory?

11. Which of the following are polynomials?
    (a) $2^3 + x^2$
    (b) $x + \dfrac{1}{x}$
    (c) $x^3 - \dfrac{1}{\sqrt{2}}$
    (d) $\sqrt[3]{x^4}$
    (e) $\pi^3 - x$
    (f) $\sqrt{2} + 2x^2$
    (g) $\sqrt{x} + 2x^2$
    (h) $|x|$

12. What is the remainder when $x^4 + 3x^3 + 1$ is divided by $x^2 + 1$?

13. What is the remainder when $x^{112} - 2x^8 + 9x^5 - 4x^4 + x - 5$ is divided by $x - 1$?

14. Is $x - 1$ a factor of $f(x) = 14x^{87} - 65x^{56} + 51$? Justify your answer.

15. Use synthetic division to show that $x - 2$ is a factor of
$$x^6 - 5x^5 + 8x^4 + x^3 - 17x^2 + 16x - 4,$$
and find the other factor.

16. List the roots of this polynomial and the multiplicity of each root:
$$f(x) = 5(x - 4)^3(x - 2)(x + 17)^3(x^2 - 4).$$

17. Find a polynomial f of degree 3 such that $f(-1) = 0, f(1) = 0$, and $f(0) = 5$.

18. Find the root(s) of $2\left(\dfrac{x}{5} + 7\right) - 3x - \dfrac{x + 2}{5} + 4$.

19. Find the roots of $3x^2 - 2x - 5$.

20. Factor the polynomial $x^3 - 8x^2 + 9x + 6$. [*Hint:* 2 is a root.]

21. Find all real roots of $9x^3 - 6x^2 - 35x + 26$. [*Hint:* Try $x = -2$.]

22. Find the rational roots of $x^4 - 2x^3 - 4x^2 + 1$.

23. Consider the polynomial $2x^3 - 8x^2 + 5x + 3$.
    (a) List the only *possible* rational roots.
    (b) Find one rational root.
    (c) Find all the roots of the polynomial.

24. (a) Find all rational roots of $x^3 + 2x^2 - 2x - 2$.
    (b) Find two consecutive integers such that an irrational root of $x^3 + 2x^2 - 2x - 2$ lies between them.

25. The polynomial $x^3 - 2x + 1$ has
    (a) no real roots.
    (b) only one real root.
    (c) three rational roots.
    (d) only one rational root.
    (e) None of the above.

26. Without using a calculator, explain why every fifth-degree polynomial function has at least one real root. [*Hint:* What must its graph look like?]

In Questions 27 and 28, compute and simplify the difference quotient of the function.

27. $f(x) = x^2 + x$

28. $g(x) = x^3 - x + 1$

29. Draw the graph of a function that could not possibly be the graph of a polynomial function and explain why.

30. Draw a graph that could be the graph of a polynomial function of degree 5. You need not list a specific polynomial, nor do any computation.

31. Which of the following statements is *not* true about the polynomial function f whose graph is shown in the figure?

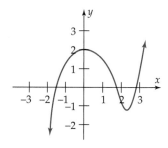

(a) f has three roots between -2 and 3.
(b) $f(x)$ could possibly be a fifth-degree polynomial.
(c) $(f \circ f)(0) > 0$.
(d) $f(2) - f(-1) < 3$.
(e) $f(x)$ is positive for all x such that $-1 \le x \le 0$.

32. Which of the statements (i)–(v) about the polynomial function f whose graph is shown in the figure are *false*?

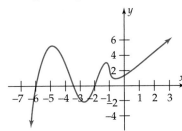

(i) f has 2 roots between -7 and -3.   **(iv)** $f(2) - 2 = 0$.
(ii) $f(-3) - f(-6) < 0$.   **(v)** f has degree ≤ 4.
(iii) $f(0) < f(1)$.

In Questions 33–36, find a viewing window (or windows) that shows a complete graph of the function. Be alert for hidden behavior.

33. $f(x) = .5x^3 - 4x^2 + x + 1$

34. $g(x) = .3x^5 - 4x^4 + x^3 - 4x^2 + 5x + 1$

35. $h(x) = 4x^3 - 100x^2 + 600x$

36. $f(x) = 32x^3 - 99x^2 + 100x + 2$

37. HomeArt makes plastic replicas of famous statues. Their total cost to produce copies of a particular statue is shown in the table.
(a) Sketch a scatter plot of the data.
(b) Use cubic regression to find a function $C(x)$ that models the data [that is, $C(x)$ is the cost of making x statues]. Assume C is reasonably accurate when $x \le 100$.
(c) Use C to estimate the cost of making the 71st statue.
(d) Use C to approximate the average cost per statue when 35 are made and when 75 are made. [Recall that the average cost of x statues is $C(x)/x$.]

Number of Statues	Total Cost
0	$2000
10	2519
20	2745
30	2938
40	3021
50	3117
60	3269
70	3425

38. The table gives the estimated cost of a college education at a public institution. Costs include tuition, fees, books, and room and board for four years. (*Source:* Teachers Insurance and Annuity Association College Retirement Equities Fund)

   (a) Sketch a scatter plot of the data (with $x = 0$ corresponding to 1990).
   (b) Use quartic regression to find a function C that models the data.
   (c) Estimate the cost of a college education in 2007 and in 2015.

Enrollment Year	Cost
1998	$ 46,691
2000	52,462
2002	58,946
2004	66,232
2006	74,418
2008	83,616
2010	93,951
2012	105,564
2014	118,611

In Questions 39–46, sketch a complete graph of the function. Label the x-intercepts, all local extrema, and asymptotes.

39. $f(x) = x^3 - 9x$

40. $g(x) = x^3 - 2x^2 + 3$

41. $h(x) = x^4 - x^3 - 4x^2 + 4x + 2$

42. $f(x) = x^4 - 3x - 2$

43. $g(x) = \dfrac{-2}{x + 4}$

44. $h(x) = \dfrac{3 - x}{x - 2}$

45. $k(x) = \dfrac{4x + 10}{3x - 9}$

46. $f(x) = \dfrac{x^2 + 1}{x^2 - 1}$

In Questions 47–50, find a viewing window (or windows) that shows a complete graph of the function. Be alert for hidden behavior.

47. $f(x) = \dfrac{x - 3}{x^2 + x - 2}$

48. $g(x) = \dfrac{x^2 - x - 6}{x^3 - 3x^2 + 3x - 1}$

49. $h(x) = \dfrac{x^4 + 4}{x^4 - 99x^2 - 100}$

50. $k(x) = \dfrac{x^3 - 2x^2 - 4x + 8}{x - 10}$

51. It costs the Junkfood Company 50¢ to produce a bag of Munchies. There are fixed costs of $500 per day for building, equipment, etc. The company has found that if the price of a bag of Munchies is set at $1.95 - \dfrac{x}{2000}$ dollars, where x is the number of bags produced per day, then all the bags that are produced will be sold. What number of bags can be produced each day if all are to be sold and the company is to make a profit? What are the possible retail prices?

52. Highway engineers have found that a good model of the relationship between the density of automobile traffic and the speed at which it moves along a particular section of highway is given by the function

$$s = \frac{100}{1 + d^2} \quad (0 \le d \le 3),$$

where the density of traffic d is measured in hundreds of cars per mile and the speed s at which traffic moves is measured in miles per hour. Then the traffic flow q is given by the product of the density and the speed, that is, $q = ds$, with q being measured in hundreds of cars per hour.
    (a) Express traffic flow as a function of traffic density.
    (b) For what densities will traffic flow be at least 3000 cars per hour?
    (c) What traffic density will maximize traffic flow? What is the maximum flow?

53. Charlie lives 150 miles from the city. He drives 40 miles to the station and catches a train to the city. The average speed of the train is 25 mph faster than the average speed of the car.
    (a) Express the total time for the journey as a function of the speed of the car. What speeds make sense in this context?
    (b) How fast should Charlie drive if the entire journey is to take no more than 2.5 hours?

54. The survival rate s of seedlings in the vicinity of a parent tree is given by

$$s = \frac{.5x}{1 + .4x^2},$$

where x is the distance from the seedling to the tree (in meters) and $0 < x \le 10$.
    (a) For what distances is the survival rate at least .21?
    (b) What distance produces the maximum survival rate?

In Questions 55 and 56, find the average rate of change of the function between x and $x + h$.

55. $f(x) = \dfrac{x}{x + 1}$

56. $f(x) = \dfrac{1}{x^2 + 1}$

In Questions 57–64, solve the equation in the complex number system.

57. $x^2 + 3x + 10 = 0$

58. $x^2 + 2x + 5 = 0$

59. $5x^2 + 2 = 3x$

60. $-3x^2 + 4x - 5 = 0$

61. $3x^4 + x^2 - 2 = 0$

62. $8x^4 + 10x^2 + 3 = 0$

63. $x^3 + 8 = 0$

64. $x^3 - 27 = 0$

65. One root of $x^4 - x^3 - x^2 - x - 2$ is i. Find all the roots.

66. One root of $x^4 + x^3 - 5x^2 + x - 6$ is i. Find all the roots.

67. Give an example of a fourth-degree polynomial with real coefficients whose roots include 0 and $1 + i$.

68. Find a fourth-degree polynomial f whose only roots are $2 + i$ and $2 - i$, such that $f(-1) = 50$.

Discovery Project 4

Architectural Arches

You can see arches almost everywhere you look—in windows, entryways, tunnels, and bridges. Common arch shapes are semicircles, semi-ellipses, and parabolas. When constructed on a level base, arches are symmetric left to right. This means that a mathematical function that describes an arch must be an even function.

A *semicircular arch* always has the property that its base is twice as wide as its height. This ratio can be modified by placing a rectangular area under the semicircle, giving a shape known as a *Norman arch*. This approach gives a tunnel or room a vaulted ceiling. *Parabolic arches* can also be created to give a more vaulted appearance.

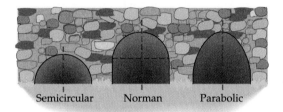

Semicircular   Norman   Parabolic

For the following exercises in arch modeling, you should always set the origin of your coordinate system to be the center of the base of the arch.

1. Show that the function that models a semicircular arch of radius r is $h(x) = \sqrt{r^2 - x^2}$.

2. Write a function $h(x)$ that models a semicircular arch that is 15 feet tall. How wide is the arch?

3. Write a function $n(x)$ that models a Norman arch that is 15 feet tall and 16 feet wide at the base.

4. Parabolic arches are typically modeled using the function $p(x) = H - ax^2$, where H is the height of the arch. Write a function $p(x)$ for an arch that is 15 feet tall and 16 feet wide at the base.

5. Would a truck that is 12 feet tall and 9 feet wide fit through all three arches? How could you fix an arch that is too small so that the truck would fit through?

Exponential and Logarithmic Functions

15,000

−6   16

−3000

How old is that dinosaur?

Population growth (of humans, fish, bacteria, etc.), compound interest, radioactive decay, and a host of other phenomena can be mathematically described by exponential functions (see Exercise 46 on page 356 and Exercise 41 on page 367). Archeologists sometimes use carbon-14 dating to determine the approximate age of an artifact (such as a dinosaur skeleton, a mummy, or a wooden statue). This involves using logarithms to solve an appropriate exponential equation. See Exercise 47 on page 398.

*Interdependence
of Sections*

5.2   5.5
5.1 → 5.3 → 5.4
         5.7   5.6

E xponential and logarithmic functions are essential for the mathematical description of a variety of phenomena in the physical sciences, economics, and engineering. Although a calculator is necessary to evaluate these functions at most numbers, you won't be able to use your calculator efficiently or interpret its answers unless you understand the properties of these functions. When calculations can readily be done by hand, you will be expected to do them without a calculator.

5.1 Exponential Functions

We now consider **exponential functions** such as

$$f(x) = 10^x, \quad g(x) = 2^x, \quad h(x) = \left(\frac{1}{2}\right)^x, \quad k(x) = \left(\frac{3}{2}\right)^x.$$

Before we can do this, however, we must clarify one point. We know how numbers such as 10^{-4} or $10^{3.7}$ are defined (see Sections 0.2 and 0.3), but what does 10^r mean when the exponent r is an irrational real number, say $10^{\sqrt{2}}$? We need the answer to this question in order for the rules of exponential functions to be defined.

Although we can't give a mathematically rigorous answer, we can illustrate the underlying idea. To define $10^{\sqrt{2}}$, we use the infinite decimal expansion $\sqrt{2} = 1.414213562\cdots$ (see Excursion 1.1.A). Each of

$$1.4, 1.41, 1.414, 1.4142, 1.41421, \ldots$$

is a rational number approximation of $\sqrt{2}$, and each is a more accurate approximation than the preceding one. We know how to raise 10 to each of these rational exponents:

$$10^{1.4} \approx 25.1189 \qquad 10^{1.4142} \approx 25.9537$$

$$10^{1.41} \approx 25.7040 \qquad 10^{1.41421} \approx 25.9543$$

$$10^{1.414} \approx 25.9418 \qquad 10^{1.414213} \approx 25.9545$$

As the exponent r gets closer and closer to $\sqrt{2}$, the number 10^r gets closer and closer and closer to a real number whose decimal expansion begins $25.954\ldots$. We define $10^{\sqrt{2}}$ to be this number.

Using similar arguments in the general case, we conclude that

For any real numbers a and r, with $a > 0$, a^r is a real number.

Furthermore, we shall assume this fact, whose proof is beyond the scope of this book:

The exponent laws (page 29) are valid for all real exponents.

Therefore, for any positive real number a, the function $f(x) = a^x$, which is called the **exponential function with base a,** is a well-defined function whose domain is the set of all real numbers.

Graphs of Exponential Functions

As the following explorations illustrate, the shape of the graph of $f(x) = a^x$ depends only on the size of the base a.

GRAPHING EXPLORATION

(a) Using the viewing window with $-3 \leq x \leq 7$ and $-2 \leq y \leq 18$, graph

$$f(x) = 1.3^x, \qquad g(x) = 2^x, \qquad h(x) = 10^x$$

on the same screen and observe their behavior to the *right* of the y-axis.

Which one rises least steeply? Which one most steeply?

How does the steepness of the graph of $f(x) = a^x$ seem to be related to the size of the base a?

(b) To see what's going on to the *left* of the y-axis, graph the same four functions in the viewing window with $-4 \leq x \leq 2$ and $-.5 \leq y \leq 2$.

As you move to the left, how does size of the base a seem to be related to how quickly the graph of $f(x) = a^x$ falls toward the x-axis?

The exploration illustrates this fact:

The exponential function $f(x) = a^x$ $(a > 1)$

When $a > 1$, then the graph of $f(x) = a^x$ has the following properties:

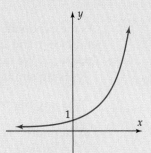

Graph is above x-axis.
y-intercept is 1.
$f(x)$ is increasing.
Negative x-axis is a horizontal asymptote.

When the number a is between 0 and 1, then the graph of $f(x) = a^x$ has a different shape.

GRAPHING EXPLORATION

Using the viewing window with $-4 \le x \le 4$ and $-1 \le y \le 4$, graph

$$f(x) = .2^x, \qquad g(x) = .4^x, \qquad h(x) = .6^x, \qquad k(x) = .8^x$$

on the same screen. Note that the bases of these exponential functions are increasing in size: $0 < .2 < .4 < .6 < .8 < 1$.

How is the size relationship of the bases related to the graphs? Which graph falls least steeply? Which one falls most steeply?

The exploration supports this conclusion:

The exponential function
$f(x) = a^x$
$(0 < a < 1)$

When $0 < a < 1$, then the graph of $f(x) = a^x$ has the following properties:

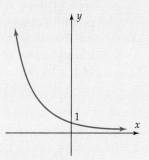

Graph is above x-axis.
y-intercept is 1.
$f(x)$ is decreasing.
Positive x-axis is a horizontal asymptote.

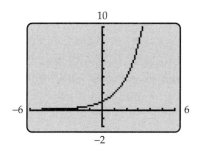

Figure 5–1

Example 1 The graph of $f(x) = 2^x$ is shown in Figure 5–1. Without graphing, describe the shape of each of the following graphs:

(a) $g(x) = 2^{x+3}$ (b) $h(x) = 2^{x-3} - 4$

Solution

(a) Since $f(x) = 2^x$, we have $f(x + 3) = 2^{x+3} = g(x)$. So the graph of $g(x)$ is just the graph of $f(x) = 2^x$ shifted horizontally 3 units to the left. [See page 222.]

GRAPHING EXPLORATION

Verify the preceding statement graphically by graphing $f(x)$ and $g(x)$ on the same screen.

(b) In this case $f(x - 3) - 4 = 2^{x-3} - 4 = h(x)$. So the graph of $h(x)$ is the graph of $f(x) = 2^x$ shifted horizontally 3 units to the right and vertically 4 units downward. [See pages 220 and 222.]

GRAPHING EXPLORATION

Verify the preceding statement graphically by graphing $f(x)$ and $h(x)$ on the same screen. ∎

Calculator screens don't really show how explosive exponential growth is. To see what this means, consider the graph of $f(x) = 2^x$ in Figure 5–1. Take a pencil and extend the x-axis to the right, keeping the same scale. Then $x = 50$ will be at the right edge of the page (try it). At this point the graph of $f(x) = 2^x$ is 2^{50} units high. Now the y-axis scale in Figure 5–1 is approximately 12 units per inch, which is equivalent to 144 units per foot or 760,320 units per mile. Therefore, the height of the graph at $x = 50$ is

$$\frac{2^{50}}{760,320} = 1,480,823,741 \text{ MILES,}$$

which would put that part of the graph well beyond the planet Saturn!

The exponential functions that model growth and decay are generally of the form $f(x) = Pa^{kx}$, such as

$$f(x) = 5 \cdot 2^{.45x}, \qquad g(x) = 3.5(10^{-.03x}), \qquad h(x) = (-6)(1.076^{2x}).$$

Their graphs have the same shape as the graph of $f(x) = a^x$. However, the graph may rise or fall more or less steeply, depending on the constants P, k, and a.

Example 2   Figure 5–2 shows the graphs of

$$f(x) = 3^x, \qquad g(x) = 3^{.15x}, \qquad h(x) = 3^{.35x}, \qquad k(x) = 3^{-x}, \qquad p(x) = 3^{-.4x}.$$

Note how the coefficient of x determines the steepness of the graph. When this coefficient is positive, the graph rises and when it is negative, the graph falls from left to right.

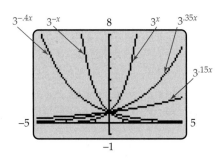

Figure 5–2

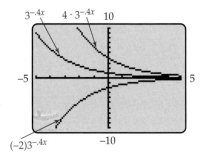

Figure 5–3

Figure 5–3 shows the graphs of

$$p(x) = 3^{-.4x}, \qquad q(x) = 4 \cdot 3^{-.4x}, \qquad r(x) = (-2) \cdot 3^{-.4x}.$$

As we saw in Section 3.4 the graph of $q(x) = 4 \cdot 3^{-.4x}$ is the graph of $p(x) = 3^{-.4x}$ stretched away from the x-axis by a factor of 4. The graph of $r(x) = (-2) \cdot 3^{-.4x}$ is the graph of $p(x) = 3^{-.4x}$ stretched away from the x-axis by a factor of 2 *and* reflected in the x-axis. ∎

Exponential Growth and Decay

Exponential functions have a number of practical applications, as illustrated in the following examples. We shall learn how to construct functions such as the ones in these examples in Section 5.2, but for now we concentrate on using them.

Example 3 (Finance)   If you invest $5000 in a high-flying stock that is increasing in value at the rate of 12% per year, then the value of your stock is given by the function $f(x) = 5000(1.12^x)$, where x is measured in years.

(a)   Assuming the value of your stock continues growing at this rate, how much will your investment be worth in four years?

(b)   When will your investment be worth $14,000?

Solution

(a)   In four years ($x = 4$), your stock is worth

$$f(4) = 5000(1.12^4) \approx \$7867.60.$$

(b)   To determine when your stock will be worth $14,000, we must find the value of x for which $f(x) = 14,000$. In other words, we must solve the equation

$$5000(1.12^x) = 14,000.$$

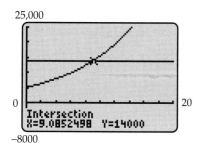

Figure 5–4

The solution is the x-coordinate of the intersection point of the graphs of $f(x) = 5000(1.12^x)$ and $y = 14,000$. Figure 5–4 shows that this point is approximately (9.085, 14,000). Therefore, the stock will be worth $14,000 in about 9.085 years.   ∎

Example 4 (Population Growth)   Based on figures from the past 50 years, the world population (in billions) can be approximated by the function $g(x) = 2.5(1.0185^x)$, where $x = 0$ corresponds to 1950.

(a)   Estimate the world population in 2000.

(b)   In what year will the world population be double what it is in 2000?

Solution

(a)   Since 2000 corresponds to $x = 50$, the population is

$$g(50) = 2.5(1.0185^{50}) \approx 6.25 \text{ billion people.}$$

(b)   Twice the population in 2000 is $2(6.25) = 12.5$ billion. We must find the value of x for which $g(x) = 12.5$, that is, solve the equation

$$2.5(1.0185^x) = 12.5.$$

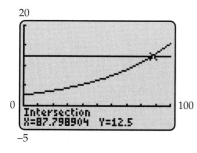

Figure 5–5

Figure 5–5 shows that the solution $x \approx 87.8$, or 88 when rounded to the nearest year. This corresponds to the year 2038. So the world population will double in just 38 years. This is what is meant by the "population explosion."   ∎

Example 5 (Radioactive Decay) To see why storage of nuclear waste is such a serious concern, consider the function

$$M(x) = .99997^x,$$

which gives the approximate amount remaining from one kilogram of plutonium (^{239}Pu) after x years. Since M is an exponential function with base smaller than 1, but very close to 1, its graph falls very slowly from left to right.

GRAPHING EXPLORATION

Verify that in a viewing window with $0 \le x \le 400$, the graph of M looks like a horizontal line. Find a viewing window in which you can actually see the graph falling to the right of the y-axis.

The fact that the graph falls so slowly as x gets large means that even after an extremely long time, a substantial amount of plutonium will remain. For instance, after 10 *thousand* years, almost three-fourths of it is still around because

$$M(10,000) \approx .74 \text{ kilogram.} \quad \blacksquare$$

The Number e and the Natural Exponential Function

There is an irrational number, denoted e, that arises naturally in a variety of phenomena and plays a central role in the mathematical description of the physical universe.* Its decimal expansion begins

$$e = 2.718281828459045 \cdots.$$

Your calculator has an e^x key that can be used to evaluate the **natural exponential function** $f(x) = e^x$. If you key in e^1, the calculator will display the first part of the decimal expansion of e. Every exponential function can be written in the form $f(x) = Pe^{kx}$ for suitable constants P and k. (See Exercise 80 of Section 5.3).

The graph of $f(x) = e^x$ has the same shape as the graph of $g(x) = 2^x$ in Figure 5–1, but climbs more steeply.

Technology Tip

On most calculators you use the e^x key but not the x^y or $\wedge$ keys to enter the function $f(x) = e^x$.

GRAPHING EXPLORATION

Graph $f(x) = e^x$, $g(x) = 2^x$, and $h(x) = 3^x$ on the same screen in a window with $-5 \le x \le 5$. The Tip in the margin may be helpful.

---

*For an example, see Example 5 in Section 5.2.

Example 6 (Population Growth)   If the population of the United States continues to grow as it has recently, then the approximate population of the United States (in millions) in year t will be given by the function

$$P(t) = 227e^{.0093t},$$

where 1980 corresponds to $t = 0$.

(a)   Estimate the population in 2015.

(b)   When will the population reach half a billion?

Solution

(a)   The population in 2015 (that is, $t = 35$) will be approximately
$$P(35) = 227e^{.0093(35)} \approx 314.3 \text{ million people.}$$

(b)   Half a billion is 500 million people. So we must find the value of t for which $P(t) = 500$, that is, we must solve the equation
$$227e^{.0093t} = 500.$$

This can be done graphically by finding the intersection of the graph of $P(t)$ and the horizontal line $y = 500$, which occurs when $t \approx 84.9$ (Figure 5–6). Therefore, the population will reach half a billion late in the year 2064. ∎

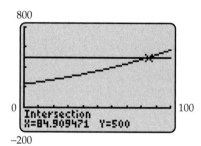

Figure 5–6

Other Exponential Functions

The population growth models in earlier examples do not take into account factors that may limit population growth in the future (wars, new diseases, etc.). Example 7 illustrates a function, called a **logistic model,** which is designed to model such situations more accurately.

Example 7 (Inhibited Population Growth)   There is an upper limit on the fish population in a certain lake due to the oxygen supply, available food, etc. The population of fish in this lake at time t months is given by the function

$$p(t) = \frac{20{,}000}{1 + 24e^{-t/4}} \quad (t \geq 0).$$

What is the upper limit on the fish population?

Solution   The graph of $p(t)$ in Figure 5–7 suggests that the horizontal line $y = 20{,}000$ is a horizontal asymptote of the graph.

In other words, the fish population never goes above 20,000. You can confirm this algebraically by rewriting the rule of p in this form:

$$p(t) = \frac{20{,}000}{1 + 24e^{-t/4}} = \frac{20{,}000}{1 + \dfrac{24}{e^{t/4}}}.$$

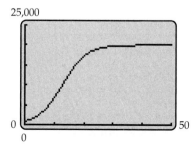

Figure 5–7

When t is very large, so is $t/4$, which means that $e^{t/4}$ is huge. Hence, by the Big–Little Principle (page 309), $\dfrac{24}{e^{t/4}}$ is very close to 0 and $p(t)$ is very close to $\dfrac{20{,}000}{1 + 0} = 20{,}000$. Since $e^{t/4}$ is positive, the denominator of $p(t)$ is slightly bigger than 1, so that $p(t)$ is always less than 20,000. ■

When a cable, such as a power line, is suspended between towers of equal height, it forms a curve called a **catenary,** which is the graph of a function of the form

$$f(x) = A(e^{kx} + e^{-kx})$$

for suitable constants A and k. The Gateway Arch in St. Louis (Figure 5–8) has the shape of an inverted catenary, which was chosen because it evenly distributes the internal structural forces.

Figure 5–8

GRAPHING EXPLORATION

Graph each of the following functions in the window with $-5 \le x \le 5$ and $-10 \le y \le 80$:

$$y_1 = 10(e^{.4x} + e^{-.4x}), \qquad y_2 = 10(e^{2x} + e^{-2x}), \qquad y_3 = 10(e^{3x} + e^{-3x}).$$

How does the coefficient of x affect the shape of the graph? Predict the shape of the graph of $y = -y_1 + 80$. Confirm your answer by graphing.

Exercises 5.1

In Exercises 1–6, sketch a complete graph of the function.

1. $f(x) = 4^{-x}$ **2.** $f(x) = (5/2)^{-x}$

3. $f(x) = 2^{3x}$ **4.** $g(x) = 3^{x/2}$

5. $f(x) = 2^{x^2}$ **6.** $g(x) = 2^{-x^2}$

In Exercises 7–12, list the transformations needed to transform the graph of $h(x) = 2^x$ into the graph of the given functions. [Section 3.4 may be helpful.]

7. $f(x) = 2^x - 5$ **8.** $g(x) = -(2^x)$

9. $k(x) = 3(2^x)$ **10.** $g(x) = 2^{x-1}$

11. $f(x) = 2^{x+2} - 5$ **12.** $g(x) = -5(2^{x-1}) + 7$

In Exercises 13 and 14, match the functions to the graphs. Assume $a > 1$ and $c > 1$.

13. $f(x) = a^x$

$g(x) = a^x + 3$

$h(x) = a^{x+5}$

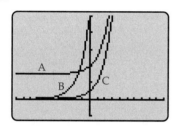

14. $f(x) = c^x$

$g(x) = -3c^x$

$h(x) = c^{x+5}$

$k(x) = -3c^x - 2$

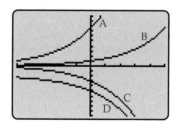

In Exercises 15–19, determine whether the function is even, odd, or neither (see Excursion 3.4.A).

15. $f(x) = 10^x$ **16.** $g(x) = 2^x - x$

17. $f(x) = \dfrac{e^x + e^{-x}}{2}$ **18.** $f(x) = \dfrac{e^x - e^{-x}}{2}$

19. $f(x) = e^{-x^2}$

20. Use the Big–Little Principle to explain why $e^x + e^{-x}$ is approximately equal to e^x when x is large.

In Exercises 21–24, find the average rate of change of the function.

21. $f(x) = x2^x$ as x goes from 1 to 3

22. $g(x) = 3^{x^2 - x}$ as x goes from -1 to 1

23. $h(x) = 5^{-x^2}$ as x goes from -1 to 0

24. $f(x) = e^x - e^{-x}$ as x goes from -3 to -1

In Exercises 25–28, find the difference quotient of the function.

25. $f(x) = 10^x$ **26.** $g(x) = 5^{x^2}$

27. $f(x) = 2^x + 2^{-x}$ **28.** $f(x) = e^x - e^{-x}$

In Exercises 29–36, find a viewing window (or windows) that shows a complete graph of the function.

29. $k(x) = e^{-x}$ **30.** $f(x) = e^{-x^2}$

31. $f(x) = \dfrac{e^x + e^{-x}}{2}$ **32.** $h(x) = \dfrac{e^x - e^{-x}}{2}$

33. $g(x) = 2^x - x$ **34.** $k(x) = \dfrac{2}{e^x + e^{-x}}$

35. $f(x) = \dfrac{5}{1 + e^{-x}}$ **36.** $g(x) = \dfrac{10}{1 + 9e^{-x/2}}$

In Exercises 37–42, list all asymptotes of the graph of the function and the approximate coordinates of each local extremum.

37. $f(x) = x2^x$ **38.** $g(x) = x2^{-x}$ **39.** $h(x) = e^{x^2/2}$

40. $k(x) = 2^{x^2 - 6x + 2}$ **41.** $f(x) = e^{-x^2}$

42. $g(x) = -xe^{x^2/20}$

43. (a) A genetic engineer is growing cells in a fermenter. The cells multiply by splitting in half every 15 minutes. The new cells have the same DNA as the original ones. Complete the following table.

Time (hours)	Number of Cells
0	1
.25	2
.5	4
.75	
1	

(b) Write the rule of the function that gives the number of cells at time t hours.

44. Do Exercise 43, using this table:

Time (hours)	Number of Cells
0	300
.25	600
.5	1200
.75	
1	

45. The Gateway Arch (see Figure 5–8) is 630 feet high and 630 feet wide at ground level. Suppose it were placed on a coordinate plane with the x-axis at ground level and the y-axis going through the center of the arch. Find a catenary function $g(x) = A(e^{kx} + e^{-kx})$ and a constant C such that the graph of the function $f(x) = g(x) + C$ provides a model of the arch. [*Hint:* Experiment with various values of A, k, C as in the Graphing Exploration on page 354.]

46. If you deposit \$750 at 2.2% interest, compounded annually and paid from the day of deposit to the day of withdrawal, your balance at time t is given by $B(t) = 750(1.022)^t$. How much will you have after two years? After three years and nine months?

47. The population of a colony of fruit flies t days from now is given by the function $p(t) = 100 \cdot 3^{t/10}$.
  (a) What will the population be in 15 days? In 25 days?
  (b) When will the population reach 2500?

48. A certain type of bacteria grows according to the function $f(x) = 5000e^{.4055x}$, where the time x is measured in hours.
  (a) What will the population be in 8 hours?
  (b) When will the population reach 1 million?

49. According to data from the National Center for Health Statistics, the life expectancy at birth for a person born in year x ($1900 \le x \le 2050$) is approximated by the function

$$D(x) = \frac{79.257}{1 + 9.7135 \cdot 10^{24} \cdot e^{-.0304x}}.$$

  (a) What is the life expectancy of someone born in 1980? In 2000?
  (b) In what year was life expectancy at birth 60 years?

50. Based on data from 1989 to 1994, the number of babies born each year in the United States through assisted reproductive technology (ART), such as in vitro fertilization, is approximated by the function

$$K(x) = \frac{12{,}439}{1 + 4.76 \cdot e^{-.4713x}},$$

where $x = 1$ corresponds to 1989. (*Source:* American Society for Reproductive Medicine)
  (a) In what year did the number of babies born through ART first exceed 8000?
  (b) If this model remains accurate in the future, will the number of babies born through ART ever reach 13,000 per year?

51. Take an ordinary piece of typing paper and fold it in half; then the folded sheet is twice as thick as the single sheet was. Fold it in half again, so that it is twice as thick as before. Keep folding it in half as long as you can. Soon the folded paper will be so thick and small that you will be unable to continue, but suppose you could keep folding the paper as long as you wanted. Assume the paper is .002 inches thick.
  (a) Make a table showing the thickness of the folded paper for the first four folds (with fold 0 being the thickness of the original unfolded paper).
  (b) Find a function of the form $f(x) = Pa^x$ that describes the thickness of the folded paper after x folds.
  (c) How thick would the paper be after 20 folds?
  (d) How many folds would it take to reach the moon (which is 243,000 miles from the earth)? [*Hint:* One mile is 5280 ft.]

52. An eccentric billionaire offers you a job for the month of September. She says that she will pay you 2¢ on the first day, 4¢ on the second day, 8¢ on the third day, and so on, doubling your pay on each successive day.
  (a) Let $P(x)$ denote your salary in *dollars* on day x. Find the rule of the function P.
  (b) Would you be better off financially if instead you were paid \$10,000 per day? [*Hint:* Consider $P(30)$.]

53. The estimated number of units that will be sold by a certain company t months from now is given by $N(t) = 100{,}000e^{-.09t}$.
  (a) What are current sales ($t = 0$)? What will sales be in two months? In six months?
  (b) Will sales ever start to increase again? [What does the graph of $N(t)$ look like?]

54. (a) The function $g(t) = 1 - e^{-.0479t}$ gives the percentage of the population (expressed as a decimal) that has seen a new TV show t weeks after it goes on the air. What percentage of people have seen the show after 24 weeks?

(b) Approximately when will 90% of the people have seen it?

55. (a) The beaver population near a certain lake in year t is approximately $p(t) = \dfrac{2000}{1 + 199e^{-.5544t}}$.

What is the population now $(t = 0)$ and what will it be in five years?

(b) Approximately when will there be 1000 beavers?

56. The figure is the graph of an exponential growth function $f(x) = Pa^x$.

(a) In this case, what is P? [*Hint:* What is $f(0)$?]

(b) Find the rule of the function f by finding a. [*Hint:* What is $f(2)$?]

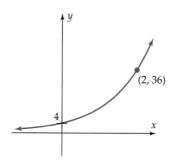

Thinkers

57. Find a function $f(x)$ with the property $f(r + s) = f(r)f(s)$ for all real numbers r and s. [*Hint:* Think exponential.]

58. Find a function $g(x)$ with the property $g(2x) = (g(x))^2$ for every real number x.

59. (a) Using the viewing window with $-4 \le x \le 4$ and $-1 \le y \le 8$, graph $f(x) = (\frac{1}{2})^x$ and $g(x) = 2^x$ on the same screen. If you think of the y-axis as a mirror, how would you describe the relationship between the two graphs?

(b) Without graphing, explain how the graphs of $g(x) = 2^x$ and $k(x) = 2^{-x}$ are related.

60. Approximating Exponential Functions by Polynomials. For each positive integer n, let f_n be the polynomial function whose rule is

$$f_n(x) = 1 + x + \frac{x^2}{2!} + \frac{x^3}{3!} + \frac{x^4}{4!} + \cdots + \frac{x^n}{n!},$$

where $k! = 1 \cdot 2 \cdot 3 \cdots k$.

(a) Using the viewing window with $-4 \le x \le 4$ and $-5 \le y \le 55$, graph $g(x) = e^x$ and $f_4(x)$ on the same screen. Do the graphs appear to coincide?

(b) Replace the graph of $f_4(x)$ by that of $f_5(x)$, then by $f_6(x)$, $f_7(x)$, and so on until you find a polynomial $f_n(x)$ whose graph appears to coincide with the graph of $g(x) = e^x$ in this viewing window. Use the trace feature to move from graph to graph at the same value of x to see how accurate this approximation is.

(c) Change the viewing window so that $-6 \le x \le 6$ and $-10 \le y \le 400$. Is the polynomial you found in part (b) a good approximation for $g(x)$ in this viewing window? What polynomial is?

5.2 Applications of Exponential Functions

In the previous section we encountered several exponential functions that modeled growth and decay. In this section we learn how to construct such exponential models in a variety of real-life situations.

Compound Interest

We begin with a subject that interests everyone: money and how it grows.

Example 1 If you invest $6000 at 8% interest, compounded annually, how much is in the account at the end of ten years?

Solution At the end of one year, the account balance is

$$6000 + 8\% \text{ of } 6000 = 6000 + .08 \cdot 6000$$

$$= 6000(1 + .08) = 6000(1.08) = \$6480.$$

Note that the ending balance is 1.08 times the original balance. During the second year with compound interest, you earn interest on your original $6000 *and* on the interest now in the account, that is, you earn interest on the entire $6480, so that at the end of the second year, the account balance is

$$6480 + 8\% \text{ of } 6480 = 6480 + .08 \cdot 6480$$

$$= 6480(1 + .08) = 6480(1.08) = \$6998.40.$$

Once again, the ending balance is 1.08 times the starting balance. The same pattern continues in subsequent years. Each year the previous balance is multiplied by 1.08, so the account grows like this:

| *start* | *year 1* | *year 2* | *year 3* |

$$6000 \rightarrow 6000 \cdot 1.08 \rightarrow 6000 \cdot 1.08 \cdot 1.08 \rightarrow 6000 \cdot 1.08 \cdot 1.08 \cdot 1.08 \rightarrow \ldots$$

that is,

$$6000 \rightarrow 6000 \cdot 1.08 \rightarrow 6000 \cdot 1.08^2 \rightarrow 6000 \cdot 1.08^3 \rightarrow \ldots$$

Thus, the account balance at the end of t years is given by

$$B(t) = 6000 \cdot 1.08^t.$$

The balance after ten years is

$$B(10) = 6000 \cdot 1.08^{10} = \$12,953.55.^* \quad \blacksquare$$

The argument used in Example 1 applies in the general case:

Compound Interest Formula

If P dollars is invested at interest rate r per time period (expressed as a decimal), then the amount A after t periods is

$$A = P(1 + r)^t.$$

In Example 1, for instance, we had $P = 6000$ and $r = .08$ (so that $1 + r = 1 + .08 = 1.08$); the number of periods (years) was $t = 10$.

Example 2 Suppose you borrow $50 from your friendly neighborhood loan shark, who charges 18% interest per week. How much do you owe after one year (assuming that he lets you wait that long to pay)?

Solution You use the compound interest formula with $P = 50, r = .18$, and $t = 52$ (because interest is compounded weekly and there are 52 weeks in a year). So you figure that you owe

$$A = P(1 + r)^t = 50(1 + .18)^{52} = 50 \cdot 1.18^{52} = \$273,422.58.$$

---

*Here and below, all financial answers are rounded to the nearest penny.

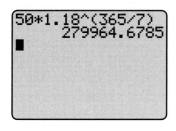

Figure 5–9

When you try to pay the shark this amount, however, he points out that a 365-day year has more than 52 weeks, namely, $\dfrac{365}{7} = 52\dfrac{1}{7}$ weeks. So you recalculate, with $t = 365/7$ (and careful use of parentheses, as shown in Figure 5–9) and find that you actually owe

$$A = P(1 + r)^t = 50(1 + .18)^{365/7} = 50 \cdot 1.18^{365/7} = \$279{,}964.68.$$

Ouch! ■

As Example 2 illustrates, the compound interest formula can be used even when the number of periods t is not an integer. You must also learn how to read "financial language" to apply the formula correctly, as shown in the following example.

Example 3 Determine the amount a \$3500 investment is worth after three and a half years at the following interest rates:

(a) 6.4% compounded annually;

(b) 6.4% compounded quarterly;

(c) 6.4% compounded monthly.

Solution

(a) Using the compound interest formula with $P = 3500$, $r = .064$, and $t = 3.5$, we have
$$A = 3500(1 + .064)^{3.5} = \$4348.74.$$

(b) "6.4% interest, compounded quarterly" means that the interest period is one-fourth of a year and the interest rate per period is $.064/4 = .016$. Since there are four interest periods per year, the number of periods in 3.5 years is $4(3.5) = 14$, so that
$$A = 3500\left(1 + \frac{.064}{4}\right)^{14} = 3500(1 + .016)^{14} = \$4370.99.$$

(c) Similarly, "6.4% compounded monthly" means that the interest period is one month (1/12 of a year) and the interest rate per period is $.064/12$. The number of periods (months) in 3.5 years is 42, so that
$$A = 3500\left(1 + \frac{.064}{12}\right)^{42} = \$4376.14.$$

Note that the more often interest is compounded, the larger the final amount. ■

Example 4 If \$5000 is invested at 7% interest, compounded daily, when will the investment be worth \$6800?

Solution The compound interest formula with $P = 5000$ and $r = .07/365$ (the daily interest rate) shows that after t days, the account is worth
$$5000(1 + .07/365)^t.$$

Technology Tip

The FINANCE menu on TI-83 and Sharp 9600 includes a solver for working compound interest problems such as those in Examples 1–3. Check your instruction manual for details. TI-86 users can download the same finance package from the TI website.

We must find the value of t for which

$$5000(1 + .07/365)^t = 6800.$$

GRAPHING EXPLORATION

Solve this equation by graphing $y_1 = 5000(1 + .07/365)^x$ and $y_2 = 6800$ in a viewing window with $0 \le x \le 3000$ and finding the intersection point.

The Exploration shows that the account will be worth $6800 when $t \approx 1603$ days (approximately 4.7 years). ■

Continuous Compounding and the Number e

Suppose you deposit money in a savings account. As we saw in Example 3, the more often your interest is compounded, the better off you are. But there is, alas, a limit.

Example 5 (**The Number** e)   You have $1 to invest for one year. The Exponential Bank offers to pay 100% annual interest, compounded n times per year and rounded to the nearest penny. You may pick any value you want for n. Can you choose n so large that your $1 will grow to some huge amount?

Solution   Since the interest rate 100% (= 1.00) is compounded n times per year, the interest rate per period is $r = 1/n$ and the number of periods in one year is n. According to the formula, the amount at the end of the year will be $A = \left(1 + \dfrac{1}{n}\right)^n$. Here's what happens for various values of n:

Interest Is Compounded	$n =$	$\left(1 + \dfrac{1}{n}\right)^n =$
Annually	1	$\left(1 + \frac{1}{1}\right)^1 = 2$
Semiannually	2	$\left(1 + \frac{1}{2}\right)^2 = 2.25$
Quarterly	4	$\left(1 + \frac{1}{4}\right)^4 \approx 2.4414$
Monthly	12	$\left(1 + \frac{1}{12}\right)^{12} \approx 2.6130$
Daily	365	$\left(1 + \frac{1}{365}\right)^{365} \approx 2.71457$
Hourly	8760	$\left(1 + \frac{1}{8760}\right)^{8760} \approx 2.718127$
Every minute	525,600	$\left(1 + \frac{1}{525,600}\right)^{525,600} \approx 2.7182792$
Every second	31,536,000	$\left(1 + \frac{1}{31,536,000}\right)^{31,536,000} \approx 2.7182818$

Since interest is rounded to the nearest penny, your dollar will grow no larger than $2.72, no matter how big n is. ∎

The last entry in the preceding table, 2.7182818, is the number e to seven decimal places. This is just one example of how e arises naturally in real-world situations. In calculus, it is proved that e is the *limit* of $\left(1 + \dfrac{1}{n}\right)^n$, meaning that as n gets larger and larger, $\left(1 + \dfrac{1}{n}\right)^n$ gets closer and closer to e.

GRAPHING EXPLORATION

Confirm this fact graphically by graphing the function $f(x) = \left(1 + \dfrac{1}{x}\right)^x$ and the horizontal line $y = e$ in the viewing window with $0 \leq x \leq 5000$ and $-1 \leq y \leq 4$ and noting that the two graphs appear identical.

When interest is compounded n times per year for larger and larger values of n, as in Example 5, we say that the interest is **continuously compounded.** In this terminology, Example 5 says that $1 will grow to $2.72 in one year at an interest rate of 100% compounded continuously. A similar argument with more realistic interest rates (see Exercise 38) produces the following result (Example 5 is the case when $P = 1$, $r = 1$, and $t = 1$).

Continuous Compounding

▶ | If P dollars is invested at interest rate r, compounded continuously, then the amount A after t years is
$$A = Pe^{rt}.$$

Example 6 Suppose $4000 is invested at 5% compounded continuously. How much will the investment be worth in three years?

Solution Applying the continuous compounding formula with $P = \$4000$, $r = .05$, and $t = 3$, we see that
$$A = 4000(e^{.05 \cdot 3}) = 4000e^{.15} = \$4647.34. \quad ∎$$

Exponential Growth

The rules of exponential growth functions are found in the same way that the formula for compound interest was developed.

Example 7   The world population in 1950 was about 2.5 billion people and has been increasing at approximately 1.85% per year. Find the rule of the function that gives the world population in year x, where $x = 0$ corresponds to 1950.

Solution   From one year to the next the population increases by 1.85%, so if P is the population at the start of a year, then the population at the end of that year is

$$P + 1.85\% \text{ of } P = P + .0185P = P(1 + .0185) = P(1.0185)$$

In other words, the population increases by a factor of 1.0185 every year:

$$2.5 \to 2.5(1.0185) \to 2.5(1.0185^2) \to 2.5(1.0185^3) \to \cdots$$

So the world population (in billions) in year x is given by

$$g(x) = 2.5(1.0185^x). \quad \blacksquare$$

Note the form of the function g in Example 7: The constant 2.5 is the initial population (when $t = 0$) and the number 1.0185 is the factor by which the population changes in one year, when population grows at a rate of 1.85% per year. This function and the compound interest formula are special cases of the following.

Exponential Growth

Exponential growth can be described by a function of the form

$$f(x) = Pa^x,$$

where $f(x)$ is the quantity at time x, P is the initial quantity (the amount when $x = 0$), and $a > 1$ is the factor by which the quantity changes when x increases by 1. If the quantity is growing at rate r per time period, then $a = 1 + r$ and

$$f(x) = Pa^x = P(1 + r)^x.$$

Example 8   At the beginning of an experiment a culture contains 1000 bacteria. Five hours later there are 7600 bacteria. Assuming that the bacteria grow exponentially, how many will there be after 24 hours?

Solution   The bacteria population is given by $f(x) = Pa^x$, where P is the initial population, a is the change factor, and x is the time in hours. We are given that $P = 1000$, so that $f(x) = 1000a^x$. The next step is to determine a. Since there are 7600 bacteria when $x = 5$, we have

$$7600 = f(5) = 1000a^5$$

so that

$$1000a^5 = 7600$$

$$a^5 = 7.6$$

$$a = \sqrt[5]{7.6} = (7.6)^{1/5} = (7.6)^{.2}.$$

Therefore, the population function is $f(x) = 1000(7.6^2)^x = 1000 \cdot (7.6)^{.2x}$. After 24 hours the bacteria population will be

$$f(24) = 1000 \cdot (7.6)^{.2(24)} \approx 16,900,721. \quad \blacksquare$$

Exponential Decay

In some situations a quantity decreases by a fixed factor as time goes on, as in the next example.

Example 9 When tap water is filtered through a layer of charcoal and other purifying agents, 30% of the chemical impurities in the water are removed and 70% remain. If the water is filtered through a second purifying layer, then the amount of impurities remaining is 70% of 70%, that is, $(.7)(.7) = .7^2 = .49$, or 49%. A third layer results in $.7^3$ of the impurities remaining. Thus, the function

$$f(x) = .7^x$$

gives the percentage of impurities remaining in the water after it passes through x layers of purifying material. How many layers are needed to ensure that 95% of the impurities are removed from the water?

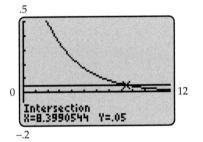

Figure 5–10

Solution   If 95% of the impurities are removed, then 5% will remain. Hence, we must find x such that $f(x) = .05$, that is, we must solve the equation $.7^x = .05$. This can be done numerically or graphically. Figure 5–10 shows that the solution is $x \approx 8.4$, so 8.4 layers of material are needed. $\blacksquare$

> ### GRAPHING EXPLORATION
>
> In Example 9, how many layers are needed to ensure that 99% of the impurities are removed?

Example 9 illustrates **exponential decay.** Note that the impurities were removed at a rate of $30\% = .3$ and that the amount of impurities remaining in the water was changing by a factor of $1 - .30 = .7$. The same thing is true in the general case:

Exponential Decay

> Exponential decay can be described by a function of the form
>
> $$f(x) = Pa^x,$$
>
> where $f(x)$ is the quantity at time x, P is the initial quantity (the amount when $x = 0$), and $0 < a < 1$; here a is the factor by which the quantity changes, when x increases by 1. If the quantity is decaying at rate r per period, then $a = 1 - r$ and
>
> $$f(x) = Pa^x = P(1 - r)^x.$$

The **half-life** of a radioactive substance is the time it takes a given quantity to decay to one-half of its original mass. The half-life depends only on the substance, not on the size of the sample. Since radioactive substances decay exponentially, their decay can be described by a function of the form $f(x) = Pa^x$, where x is measured in the same time units as the half-life (hours, days, years, etc.). The constant a can be determined from the half-life of the substance.

Suppose, for example, that the half-life is 25 years. Then after 25 years the initial amount P decays to $P/2$. This means that

$$f(25) = \frac{P}{2} = .5P$$

Use the rule of f: $\qquad\qquad Pa^{25} = .5P$

Divide both sides by P: $\qquad\qquad a^{25} = .5$

Raise both sides to the power 1/25: $\quad (a^{25})^{1/25} = .5^{1/25}$

Use exponent properties: $\qquad\qquad a = .5^{1/25}$

Therefore, the rule of the function is

$$f(x) = Pa^x = P(.5^{1/25})^x = P(.5^{x/25})$$

The same argument applies with any number h in place of 25 and we have:

Radioactive Decay

> Exponential decay of a radioactive substance is given by the function
>
> $$f(x) = P(.5^{x/h}),$$
>
> where P is the initial amount and h is the half-life of the substance.

Example 10   When a living organism dies, its carbon-14 decays exponentially. An archeologist determines that the skeleton of a mastodon has lost 64% of its carbon-14.* Use the fact that the half-life carbon-14 is 5730 years to estimate how long ago the mastodon died.

Solution   The amount of carbon-14 in the skeleton at time x is given by $f(x) = P(.5^{x/5730})$, where $x = 0$ corresponds to the death of the mastodon. Since the skeleton has lost 64% of its carbon-14, we know that the amount in the skeleton now is 36% of P, that is, $.36P$. So we must find the value of x for which

$$f(x) = .36P$$

Use the rule of f: $\qquad\qquad P(.5^{x/5730}) = .36P$

Divide both sides by P: $\qquad\qquad .5^{x/5730} = .36$

---

*The technique involves measuring the ratio of the radioactive isotope of carbon, carbon-14, to ordinary nonradioactive carbon-12 in the skeleton.

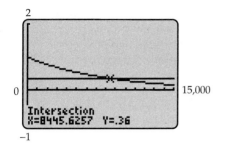

Figure 5–11

To solve this equation we graph $y_1 = .5^{x/5730}$ and $y_2 = .36$ on the same screen and find the x-coordinate of the intersection point. Figure 5–11 shows that $x \approx 8445.63$. Therefore, the mastodon died about 8446 years ago. ■

Exercises 5.2

1. If \$1000 is invested at 8%, find the value of the investment after five years if interest is compounded
   (a) annually (b) quarterly (c) monthly
   (d) weekly

2. If \$2500 is invested at 11.5%, what is the value of the investment after ten years if interest is compounded
   (a) annually (b) monthly (c) daily

In Exercises 3–12, determine how much money will be in a savings account if the initial deposit was \$500 and the interest rate is:

3. 2% compounded annually for 8 years.

4. 2% compounded annually for 10 years.

5. 2% compounded quarterly for 10 years.

6. 2.3% compounded monthly for 9 years.

7. 2.9% compounded daily for 8.5 years.

8. 3.5% compounded weekly for 7 years and 7 months.

9. 3% compounded continuously for 4 years.

10. 3.5% compounded continuously for 10 years.

11. 2.45% compounded continuously for 6.2 years.

12. 3.25% compounded continuously for 11.6 years.

*A sum of money P that can be deposited today to yield some larger amount A in the future is called the **present value** of A. In Exercises 13–18, find the present value of the given amount A. [Hint: Substitute A, the interest rate per period r, and the number t of periods in the compound interest formula and solve for P.]*

13. \$5000 at 6% compounded annually for 7 years

14. \$3500 at 5.5% compounded annually for 4 years

15. \$4800 at 7.2% compounded quarterly for 5 years

16. \$7400 at 5.9% compounded quarterly for 8 years

17. \$8900 at 11.3% compounded monthly for 3 years

18. \$9500 at 9.4% compounded monthly for 6 years

In Exercises 19–26, use the compound interest formula. In most cases, you will be given three of the quantities A, P, r, t and will have to solve an equation to find the remaining one.

19. A typical credit card company charges 18% annual interest, compounded monthly, on the unpaid balance. If your current balance is \$520 and you don't make any payments for 6 months, how much will you owe (assuming they don't sue you in the meantime)?

20. When his first child was born a father put \$3000 in a savings account that pays 4% annual interest, compounded quarterly. How much will be in the account on the child's 18th birthday?

21. You have \$10,000 to invest for two years. Fund A pays 13.2% interest, compounded annually. Fund B pays 12.7% interest, compounded quarterly, and Fund C pays 12.6% interest, compounded monthly. Which fund will return the most money?

22. If you invest \$7400 for five years, are you better off with an interest rate of 5% compounded quarterly or 4.8% compounded continuously?

23. If you borrow \$1200 at 14% interest, compounded monthly, and pay off the loan (principal and interest) at the end of two years, how much interest will you have paid?

24. A developer borrows \$150,000 at 6.5% interest, compounded quarterly, and agrees to pay off the loan in four years. How much interest will she owe?

25. A manufacturer has settled a lawsuit out of court by agreeing to pay $1.5 million four years from now. At this time how much should the company put in an account paying 6.4% annual interest, compounded monthly, in order to have $1.5 million in four years? [*Hint:* Exercises 13–18.]

26. Ellen Sklar wants to have $30,000 available in five years for a down payment on a house. She has inherited $25,000. How much of the inheritance should be invested at 5.7% annual interest, compounded quarterly, to accumulate the $30,000?

27. You win a contest and have a choice of prizes. You can take $3000 now or you can receive $4000 in four years. If money can be invested at 6% interest, compounded annually, which prize is more valuable in the long run?

28. If money can be invested at 7% compounded quarterly, which is worth more: $9000 now or $12,500 in five years?

29. If an investment of $1000 grows to $1407.10 in seven years with interest compounded annually, what is the interest rate?

30. If an investment of $2000 grows to $2700 in three and a half years, with an annual interest rate that is compounded quarterly, what is the annual interest rate?

31. If you put $3000 in a savings account today, what interest rate (compounded annually) must you receive in order to have $4000 after five years?

32. If interest is compounded continuously, what annual rate must you receive if your investment of $1500 is to grow to $2100 in six years?

33. At an interest rate of 8% compounded annually, how long will it take to double an investment of

  (a) $100   **(b)** $500   **(c)** $1200?

  (d) What conclusion about doubling time do parts (a)–(c) suggest?

34. At an interest rate of 6% compounded annually, how long will it take to double an investment of P dollars?

35. How long will it take to double an investment of $500 at 7% annual interest, compounded continuously?

36. How long will it take to triple an investment of $5000 at 8% annual interest, compounded continuously?

37. (a) Suppose P dollars is invested for one year at 12% interest compounded quarterly. What interest rate r would yield the same amount in one year with annual compounding? r is called the **effective rate of interest.** [*Hint:* Solve the equation $P(1 + .12/4)^4 = P(1 + r)$

for r. The left side of the equation is the yield after one year at 12% compounded quarterly and the right side is the yield after one year at r% compounded annually.]

(b) Fill the blanks in the following table.

12% Compounded	Effective Rate
Annually	12%
Quarterly	
Monthly	
Daily	

38. This exercise provides an illustration of why the continuous compounding formula (page 361) is valid, using a realistic interest rate. We shall determine the value of $4000 deposited for three years at 5% interest compounded n times per year for larger and larger values of n. In this case, the interest rate per period is $.05/n$ and the number of periods in three years is $3n$. So the amount in the account at the end of three years is:

$$A = 4000\left(1 + \frac{.05}{n}\right)^{3n} = 4000\left[\left(1 + \frac{.05}{n}\right)^n\right]^3.$$

(a) Fill in the missing entries in the following table.

n	$\left(1 + \dfrac{.05}{n}\right)^n$
1,000	
10,000	
500,000	
1,000,000	
5,000,000	
10,000,000	

(b) Compare the entries in the second column of the table with the number $e^{.05}$ and fill the blank in the following sentence:

As n gets larger and larger, the value of $\left(1 + \dfrac{.05}{n}\right)^n$ gets closer and closer to the number _____.

(c) Use your answer to part (b) to fill the blank in the following sentence:

As n gets larger and larger, the value of

$$A = 4000\left[\left(1 + \frac{.05}{n}\right)^n\right]^3$$ gets closer and closer

to _____.

(d) Compare your answer in part (c) to the value of the investment given by the continuous compounding formula.

39. A weekly census of the tree-frog population in Frog Hollow State Park produces the following results:

Week	1	2	3	4	5	6
Population	18	54	162	486	1458	4374

(a) Find a function of the form $f(x) = Pa^x$ that describes the frog population at time x weeks.

(b) What is the growth factor in this situation (that is, by what number must this week's population be multiplied to obtain next week's population)?

(c) Each tree frog requires 10 square feet of space and the park has an area of 6.2 square miles. Will the space required by the frog population exceed the size of the park in 12 weeks? In 14 weeks? [Remember: 1 square mile = 5280^2 square feet.]

40. The fruit fly population in a certain laboratory triples every day. Today there are 200 fruit flies.

(a) Make a table showing the number of fruit flies present for the first four days (today is day 0, tomorrow is day 1, etc.).

(b) Find a function of the form $f(x) = Pa^x$ that describes the fruit fly population at time x days.

(c) What is the growth factor here (that is, by what number must each day's population be multiplied to obtain the next day's population)?

(d) How many fruit flies will there be a week from now?

41. The population of Mexico was 67.4 million in 1980 and has been increasing by approximately 2.6% each year.

(a) If $g(x)$ is the population of Mexico (in millions) in year x (with $x = 0$ being 1980), find the rule of the function g. [See Example 7.]

(b) Estimate the population of Mexico in the year 2000.

42. The number of dandelions in your lawn increases by 5% a week and there are 75 dandelions now.

(a) If $f(x)$ is the number of dandelions in week x, find the rule of the function f.

(b) How many dandelions will there be in 16 weeks?

43. Average annual expenditure per pupil in public elementary and secondary schools was $5550 in 1989–1990 and has been increasing at about 3.68% a year.

(a) Write the rule of a function that gives the expenditure per pupil in year x, where $x = 0$ corresponds to the 1989–1990 school year.

(b) According to this model, what are the expenditures per pupil in 1999–2000?

(c) In what year did expenditures first exceed $7000 per pupil?

44. There are now 3.2 million people who play bridge and the number increases by 3.5% a year.

(a) Write the rule of a function that gives the number of bridge players in year x.

(b) How many people will be playing bridge in 15 years?

(c) When will there be 10 million bridge players?

45. At the beginning of an experiment a culture contains 200 h-$pylori$ bacteria. An hour later there are 205 bacteria. Assuming that the h-$pylori$ bacteria grow exponentially, how many will there be after 10 hours? After two days? [See Example 8.]

46. If the population of India was 650 million a decade ago and is now 790 million people and continues to grow exponentially at the same rate, what will the population be in five years?

47. Kerosene is passed through a pipe filled with clay to remove various pollutants. Each foot of pipe removes 25% of the pollutants.

(a) Write the rule of a function that gives the percentage of pollutants remaining in the kerosene after it has passed through x feet of pipe. [See Example 9.]

(b) How many feet of pipe are needed to ensure that 90% of the pollutants have been removed from the kerosene?

48. If inflation runs at a steady 3% per year, then the amount a dollar is worth decreases by 3% each year.

(a) Write the rule of a function that gives the value of a dollar in year x.

(b) How much will the dollar be worth in five years? In ten years?

(c) How many years will it take before today's dollar is worth only a dime?

49. (a) The half-life of radium is 1620 years. Find the rule of the function that gives the amount remaining from an initial quantity of 100 milligrams of radium after x years.

(b) How much radium is left after 800 years? After 1600 years? After 3200 years?

50. (a) The half-life of polonium-210 is 140 days. Find the rule of the function that gives the amount of polonium-210 remaining from an initial 20 milligrams after t days.

(b) How much polonium-210 is left after 15 weeks? After 52 weeks?

(c) How long will it take for the 20 milligrams to decay to 4 milligrams?

51. How old is a piece of ivory that has lost 58% of its carbon-14? [See Example 10.]

52. How old is a mummy that has lost 49% of its carbon-14?

5.3 Common and Natural Logarithmic Functions

Roadmap

We begin with the only logarithms that are in widespread use, common and natural logarithms. Those who prefer to begin with logarithms to an arbitrary base b should cover Excursion 5.4.A before reading this section.

From their invention in the 17th century until the development of computers and calculators, logarithms were the only effective tool for numerical computation in the natural sciences and engineering. Although they are no longer needed for computation, logarithmic functions still play an important role in science and engineering. In this section we examine the two most important types of logarithms, those to base 10 and those to base e.

Common Logarithms

The basic idea of logarithms can be seen from an example. We'll use the number 704. The *logarithm* of 704 is defined to be the solution of the equation $10^x = 704$. This number can be approximated in two ways:

Solve the equation $10^x = 704$ graphically (that is, find the intersection of $y = 10^x$ and $y = 704$), as in Figure 5–12,

or

Use the LOG key on your calculator, as in Figure 5–13.

Except for rounding, you get the same answer.

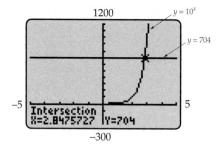

Figure 5–12

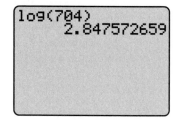

Figure 5–13

The fact that the number log 704 is the solution of $10^x = 704$ means that

log 704 is *the exponent to which* 10 *must be raised to produce* 704.

Figure 5–14

Similarly, the solution of $10^x = 56$ is the number log 56. Figure 5–14 shows that log $56 \approx 1.748188027$ and that 10 raised to this exponent is indeed 56.

More generally, whenever c is a positive real number, the horizontal line $y = c$ lies above the x-axis and hence intersects the graph of $y = 10^x$, so that the equation $10^x = c$ has a solution. Consequently, we have this definition:

Definition of Common Logarithms

> If $c > 0$, then the *common logarithm* of c, denoted log c, is the solution of the equation $10^x = c$.* In other words,
>
> log c is the exponent to which 10 must be raised to produce c.

Example 1 Without using a calculator, find

(a) log 1000   (b) log 1   (c) $\log \sqrt{10}$

Solution

(a) To find log 1000, ask yourself "what power of 10 equals 1000?" The answer is 3 because $10^3 = 1000$. Therefore, log 1000 = 3.

(b) To what power must 10 be raised in order to produce 1? Since $10^0 = 1$, we conclude that log 1 = 0.

(c) Log $\sqrt{10} = 1/2$ because 1/2 is the exponent to which 10 must be raised to produce $\sqrt{10}$, that is, $10^{1/2} = \sqrt{10}$. ■

A calculator is necessary to find most logarithms, but even then you should proceed as in Example 1 to get a rough estimate. For instance, log 795 is the exponent to which 10 must be raised to produce 795. Since $10^2 = 100$ and $10^3 = 1000$, this exponent must be between 2 and 3, that is, $2 < \log 795 < 3$.

Since logarithms are exponents, every statement about logarithms is equivalent to a statement about exponents. For instance, "log $v = u$" means "u is the exponent to which 10 must be raised to produce v," or in symbols, "$10^u = v$." In other words,

Logarithmic and Exponential Equivalence

> Let u and v be real numbers, with $v > 0$. Then
> $$\log v = u \qquad \text{exactly when} \qquad 10^u = v.$$

*The word "common" is usually omitted except when it is necessary to distinguish these logarithms from other types that are introduced below.

Example 2   Translate each of the following logarithmic statements into an equivalent exponential statement:

$$\log 29 = 1.4624 \qquad \log .47 = -.3279 \qquad \log (k + t) = d.$$

Solution   Using the preceding box, we have these translations:

Logarithmic Statement	Equivalent Exponential Statement
$\log 29 = 1.4624^*$	$10^{1.4624} = 29$
$\log .47 = -.3279$	$10^{-.3279} = .47$
$\log (k + t) = d$	$10^d = k + t$ ■

Example 3   Translate each of the following exponential statements into an equivalent logarithmic statement:

$$10^{5.5} = 316{,}227.766 \qquad 10^{.66} = 4.5708819 \qquad 10^{rs} = t$$

Solution   Translate as follows.

Exponential Statement	Equivalent Logarithmic Statement
$10^{5.5} = 316{,}227.766$	$\log 316{,}227.766 = 5.5$
$10^{.66} = 4.5708819$	$\log 4.5708819 = .66$
$10^{rs} = t$	$\log t = rs$ ■

Example 4   Solve the equation $\log x = 4$.

Solution $\log x = 4$ is equivalent to $10^4 = x$. So the solution is $x = 10{,}000$.   ■

Natural Logarithms

Common logarithms are closely related to the exponential function $f(x) = 10^x$. With the advent of calculus, however, it became clear that the most useful exponential function in science and engineering is $g(x) = e^x$. Consequently, a new type of logarithm, based on the number e instead of 10, was developed. This development is essentially a carbon copy of what was done above, with some minor changes in notation.

For example, the *natural logarithm* of 250 is defined to be the solution of the equation $e^x = 250$. This number can be approximated in two ways:

Solve the equation $e^x = 250$ graphically (that is, find the intersection of $y = e^x$ and $y = 250$), as in Figure 5–15,

or

Use the LN key on your calculator, as in Figure 5–16.

---

*Here and below all logarithms are rounded to four decimal places and an equal sign is used rather than the more correct "approximately equal."

Except for rounding, you get the same answer.

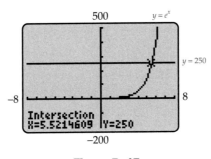

Figure 5–15   **Figure 5–16**

The fact that the number ln 250 is the solution of $e^x = 250$ means that

ln 250 is *the exponent to which e must be raised to produce 250.*

The same procedure works for any positive number.

CALCULATOR EXPLORATION

Find ln 87. Then raise e to this exponent. Except possibly for rounding, the answer will be 87.

More generally, we have this definition:

Definition of Natural Logarithms

> If $c > 0$, then the *natural logarithm* of c, denoted ln c, is the solution of the equation $e^x = c$. In other words,
>
> **ln c is the exponent to which e must be raised to produce c.**

Since natural logarithms are exponents, every statement about them is equivalent to a statement about exponents to the base e. For instance, "ln $v = u$" means "u is the exponent to which e must be raised to produce v," or in symbols, "$e^u = v$." In other words,

Logarithmic and Exponential Equivalence

> Let u and v be real numbers, with $v > 0$. Then
>
> $$\ln v = u \quad \text{exactly when} \quad e^u = v.$$

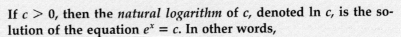

Example 5 Translate:

(a) $\ln 14 = 2.6391$ into an equivalent exponential statement.

(b) $e^{5.0626} = 158$ into an equivalent logarithmic statement.

Solution

(a) Using the preceding box, we see that $\ln 14 = 2.6391$ is equivalent to $e^{2.6391} = 14$.

(b) Similarly, $e^{5.0626} = 158$ is equivalent to $\ln 158 = 5.0626$. ■

Properties of Logarithms

Since common and natural logarithms have almost identical definitions (just replace 10 by e), it is not surprising that they share the same essential properties. You don't need a calculator to understand these properties. You need only use the definition of logarithms or translate logarithmic statements into equivalent exponential ones (or vice versa).

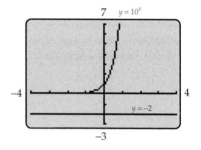

Figure 5–17

Example 6 What is $\log(-2)$?

Solution By definition $\log(-2)$ is the solution of the equation $10^x = -2$. But this equation has no solution, because the graph of $y_1 = 10^x$ lies entirely above the x-axis (every power of 10 is positive) and never intersects the horizontal line $y_2 = -2$, as you can see in Figure 5–17. So $\log(-2)$ is *not defined*.

> **GRAPHING EXPLORATION**
>
> Show that the equation $e^x = -2$ also has no solutions by graphing $y_1 = e^x$ and $y_2 = -2$ on the same screen. Conclude that $\ln(-2)$ is not defined.

Similar arguments apply with any negative number or 0 in place of -2. Hence,

$$\textbf{ln } v \textbf{ and log } v \textbf{ are defined only when } v > 0. \quad ■$$

Example 7 Find $\ln 1$.

Solution Ask yourself, to what power must e be raised to produce 1? We know that $e^0 = 1$, which means that $\ln 1 = 0$. Combining this fact with Example 1(b), we have:

$$\ln 1 = 0 \quad \text{and} \quad \log 1 = 0. \quad ■$$

Example 8 Find $\ln e^9$.

Solution To what power must e be raised to produce e^9? Obviously, the answer is 9. Hence, $\ln e^9 = 9$ and in general

$$\ln e^k = k \quad \text{for every real number } k.$$

Similarly,

$$\log 10^k = k \quad \text{for every real number } k$$

because k is the exponent to which 10 must be raised to produce 10^k. In particular, when $k = 1$, we have

$$\ln e = 1 \quad \text{and} \quad \log 10 = 1. \quad \blacksquare$$

Example 9 Find: $10^{\log 678}$ and $e^{\ln 678}$.

Solution By definition, $\log 678$ is the exponent to which 10 must be raised to produce 678. So if you raise 10 to this exponent, the answer will be 678, that is, $10^{\log 678} = 678$. Similarly, $\ln 678$ is the exponent to which e must be raised to produce 678, so that $e^{\ln 678} = 678$. The same argument works with any positive number v in place of 678:

$$e^{\ln v} = v \quad \text{and} \quad 10^{\log v} = v \quad \text{for every } v > 0. \quad \blacksquare$$

The facts presented in the preceding examples may be summarized as follows.

Properties of Logarithms

Natural Logarithms	**Common Logarithms**
1. $\ln v$ is defined only when $v > 0$;	$\log v$ is defined only when $v > 0$.
2. $\ln 1 = 0$ and $\ln e = 1$;	$\log 1 = 0$ and $\log 10 = 1$.
3. $\ln e^k = k$ for every real number k;	$\log 10^k = k$ for every real number k.
4. $e^{\ln v} = v$ for every $v > 0$;	$10^{\log v} = v$ for every $v > 0$.

The properties of logarithms can be used to simplify expressions and solve equations.

Example 10 Applying Property 3 with $k = 2x^2 + 7x + 9$ shows that $\ln e^{2x^2+7x+9} = 2x^2 + 7x + 9$. $\blacksquare$

Logarithmic Functions

The logarithmic functions, defined by

$$f(x) = \log x \quad \text{and} \quad g(x) = \ln x,$$

have as their domains all positive real numbers (the numbers for which logarithms are defined). Figure 5–18 shows that both functions are increasing and their graphs have the same basic shape. The only difference is how steeply they climb.

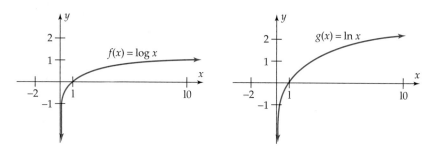

Figure 5–18

As indicated in Figure 5–18 the graphs drop sharply near the y-axis, which is a vertical asymptote of each graph. This fact is not visible in calculator-generated graphs (try it!), but can be confirmed by evaluating the functions when x is *very* small. For instance, we can evaluate $f(x) = \log x$ at $10^{-1} = .1$, $10^{-2} = .01$, $10^{-3} = .001$, and so on. We know that $\log 10^k = k$ for every k, which means that these points are on the graph of f:

$$(10^{-1}, -1), (10^{-2}, -2), \dots, (10^{-15}, -15), \dots, (10^{-50}, -50), \dots.$$

Thus, the graph goes lower and lower as x gets closer and closer to 0. A similar analysis applies to $g(x) = \ln x$.

Example 11   Sketch the graph of $f(x) = \ln (x - 2)$.

Solution   Using a calculator to graph $f(x) = \ln (x - 2)$, we obtain Figure 5–19, in which the graph appears to end abruptly near $x = 2$. Fortunately, however, we have read Section 3.4, so we know that this is *not* how the graph looks. From Section 3.4, we know that the graph of $f(x) = \ln (x - 2)$ is the graph of $g(x) = \ln x$ shifted horizontally 2 units to the right, as shown in Figure 5–20. In particular, the graph of f has a vertical asymptote at $x = 2$ and drops sharply downward there. ■

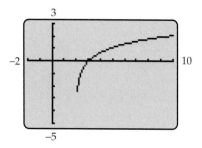

Figure 5–19

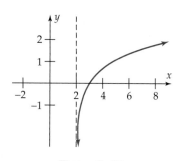

Figure 5–20

> ### GRAPHING EXPLORATION
>
> Graph $y_1 = \ln(x - 2)$ and $y_2 = -5$ in the same viewing window and verify that the graphs do not appear to intersect, as they should. Nevertheless, try to solve the equation $\ln(x - 2) = -5$ by finding the intersection point of y_1 and y_2. Some calculators will find the intersection point even though it does not show on the screen. Others produce an error message, in which case the SOLVER feature should be used instead of a graphical solution.

Although logarithms are only defined for positive numbers, many logarithmic functions include negative numbers in their domains.

Example 12 Find the domain of each of the following functions.

(a) $f(x) = \ln(x + 4)$ (b) $g(x) = \log x^2$

Solution

(a) $f(x) = \ln(x + 4)$ is defined only when $x + 4 > 0$, that is, when $x > -4$. So the domain of f consists of all real numbers greater than -4.

(b) Since $x^2 > 0$ for all nonzero x, the domain of $g(x) = \log x^2$ consists of all real numbers except 0. ∎

> ### GRAPHING EXPLORATION
>
> Verify the conclusions of Example 12 by graphing each of the functions. What is the vertical asymptote of each graph?

Example 13 If you invest money at interest rate r (expressed as a decimal and compounded annually), then the time it takes to double your investment is given by

$$D(r) = \frac{\ln 2}{\ln(1 + r)}.$$

(a) How long will it take to double an investment of $2500 at 6.5%?

(b) What interest rate is needed for the investment in part (a) to double in six years?

Solution

(a) The interest rate is $r = .065$, so the doubling time is

$$D(.065) = \frac{\ln 2}{\ln(1 + .065)} \approx 11 \text{ years.}$$

(b) For the investment to double in six years, we must have $D(r) = 6$. Thus, we must solve the equation

$$\frac{\ln 2}{\ln(1 + r)} = 6.$$

Algebraic solution techniques for such equations will be considered in Section 5.5. For now, we solve the equation graphically by finding the intersection of $y_1 = \dfrac{\ln 2}{\ln(1 + r)}$ and $y_2 = 6$. Figure 5–21 shows that $r \approx .1225$, so an interest rate of about 12.25% is needed to double the investment in six years. ∎

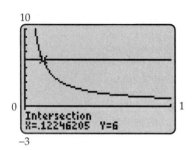

Figure 5–21

Exercises 5.3

Unless stated otherwise, all letters represent positive numbers.

In Exercises 1–4, find the logarithm, without using a calculator.

1. $\log 10{,}000$
2. $\log .001$
3. $\log \dfrac{\sqrt{10}}{1000}$
4. $\log \sqrt[3]{.01}$

In Exercises 5–14, translate the given logarithmic statement into an equivalent exponential statement.

5. $\log 1000 = 3$
6. $\log .001 = -3$
7. $\log 750 = 2.88$
8. $\log(.8) = -.097$
9. $\ln 3 = 1.0986$
10. $\ln 10 = 2.3026$
11. $\ln .01 = -4.6052$
12. $\ln s = r$
13. $\ln(x^2 + 2y) = z + w$
14. $\log(a + c) = d$

In Exercises 15–24, translate the given exponential statement into an equivalent logarithmic one.

15. $10^{-2} = .01$
16. $10^3 = 1000$
17. $10^{.4771} = 3$
18. $10^{7k} = r$
19. $e^{3.25} = 25.79$
20. $e^{-4} = .0183$
21. $e^{12/7} = 5.5527$
22. $e^k = t$
23. $e^{2/r} = w$
24. $e^{4uv} = m$

In Exercises 25–36, evaluate the given expression without using a calculator.

25. $\log 10^{\sqrt{43}}$
26. $\log 10^{\sqrt{x^2 + y^2}}$
27. $\ln e^{15}$

28. $\ln e^{3.78}$
29. $\ln \sqrt{e}$
30. $\ln \sqrt[5]{e}$
31. $e^{\ln 931}$
32. $e^{\ln 34.17}$
33. $\ln e^{x+y}$
34. $\ln e^{x^2 + 2y}$
35. $e^{\ln x^2}$
36. $e^{\ln \sqrt{x+3}}$

In Exercises 37–40, find the domain of the given function (that is, the largest set of real numbers for which the rule produces well-defined real numbers).

37. $f(x) = \ln(x + 1)$
38. $g(x) = \ln(x + 2)$
39. $h(x) = \log(-x)$
40. $k(x) = \log(2 - x)$

41. (a) Graph $y = x$ and $y = e^{\ln x}$ in separate viewing windows [or use a split-screen, if your calculator has that feature]. For what values of x are the graphs identical?
    (b) Use the properties of logarithms to explain your answer in part (a).

42. (a) Graph $y = x$ and $y = \ln(e^x)$ in separate viewing windows [or a split-screen, if your calculator has that feature]. For what values of x are the graphs identical?
    (b) Use the properties of logarithms to explain your answer in part (a).

43. Do the graphs of $f(x) = \log x^2$ and $g(x) = 2 \log x$ appear to be the same? How do they differ?

44. Do the graphs of $h(x) = \log x^3$ and $k(x) = 3 \log x$ appear to be the same?

In Exercises 45–50, list the transformations that will change the graph of $g(x) = \ln x$ into the graph of the given function. [Section 3.4 may be helpful.]

45. $f(x) = 2 \cdot \ln x$ **46.** $f(x) = \ln x - 7$

47. $h(x) = \ln(x - 4)$ **48.** $k(x) = \ln(x + 2)$

49. $h(x) = \ln(x + 3) - 4$ **50.** $k(x) = \ln(x - 2) + 2$

In Exercises 51–54, sketch the graph of the function.

51. $f(x) = \log(x - 3)$ **52.** $g(x) = 2 \ln x + 3$

53. $h(x) = -2 \log x$ **54.** $f(x) = \ln(-x) - 3$

In Exercises 55–60, find a viewing window (or windows) that shows a complete graph of the function.

55. $f(x) = \dfrac{x}{\ln x}$ **56.** $g(x) = \dfrac{\ln x}{x}$

57. $h(x) = \dfrac{\ln x^2}{x}$ **58.** $k(x) = e^{2/\ln x}$

59. $f(x) = 10 \log x - x$ **60.** $f(x) = \dfrac{\log x}{x}$

In Exercises 61–64, find the average rate of change of the function.

61. $f(x) = \ln(x - 2)$, as x goes from 3 to 5.

62. $g(x) = x - \ln x$, as x goes from .5 to 1.

63. $g(x) = \log(x^2 + x + 1)$, as x goes from -5 to -3.

64. $f(x) = x \log |x|$, as x goes from 1 to 4.

65. (a) What is the average rate of change of $f(x) = \ln x$, as x goes from 3 to $3 + h$?
   (b) What is the value of h when the average rate of $f(x) = \ln x$, as x goes from 3 to $3 + h$, is .25?

66. (a) Find the average rate of change of $f(x) = \ln x^2$, as x goes from .5 to 2.
   (b) Find the average rate of change of $g(x) = \ln(x - 3)^2$, as x goes from 3.5 to 5.
   (c) What is the relationship between your answers in parts (a) and (b) and why is this so?

67. The concentration of hydrogen ions in a given solution is denoted $[H^+]$ and is measured in moles per liter. For example, $[H^+] = .00008$ for beer and $[H^+] = .0004$ for wine. Chemists define the pH of the solution be the number $\text{pH} = -\log[H^+]$. The solution is said to be an *acid* if $\text{pH} < 7$ and a *base* if $\text{pH} > 7$.
   (a) Is beer an acid or a base? What about wine?
   (b) If a solution has a pH of 2, what is its $[H^+]$?
   (c) For hominy, $[H^+] = 5 \cdot 10^{-8}$. Is hominy a base?

68. Use the doubling function D of Example 13 for this exercise.
   (a) Find the time it takes to double your money at each of these interest rates: 4%, 6%, 8%, 12%, 18%, 24%, 36%.
   (b) Round the answers in part (a) to the nearest year and compare them with these numbers:

   $$72/4, 72/6, 72/8, 72/12, 72/18, 72/24, 72/36.$$

   Use this evidence to state a "rule of thumb" for determining approximate doubling time, without using the function D. This rule of thumb, which has long been used by bankers, is called the **rule of 72.**

69. Suppose $f(x) = A \ln x + B$, where A and B are constants. If $f(1) = 10$ and $f(e) = 1$, what are A and B?

70. (a) Graph all three of these function in a *square* viewing window:

   $$f(x) = e^x, \quad y = x, \quad g(x) = \ln x$$

   How are the graphs of f and g related to each other? [*Hint:* Think of the line $y = x$ as a mirror.]
   (b) Do part (a) for the functions

   $$f(x) = 10^x, \quad y = x, \quad g(x) = \log x.$$

   Is the answer the same?

71. The height h above sea level (in meters) is related to air temperature t (in degrees Celsius), the atmospheric pressure p (in centimeters of mercury at height h), and the atmosphere pressure c at sea level by

   $$h = (30t + 8000) \ln(c/p).$$

   If the pressure at the top of Mount Rainier is 44 centimeters on a day when sea level pressure is 75.126 centimeters and the temperature is 7°C, what is the height of Mount Rainier?

72. Mount Everest is 8850 meters high. What is the atmospheric pressure at the top of the mountain on a day when the temperature is -25°C and the atmospheric pressure at sea level is 75 centimeters? [See Exercise 71.]

73. A class in elementary Sanskrit is tested at the end of the semester and weekly thereafter on the same material. The average score on the exam taken after t weeks is given by the "forgetting function"

   $$g(t) = 77 - 10 \cdot \ln(t + 1).$$

   (a) What was the average score on the original exam?
   (b) What was the average score after two weeks? After five weeks?
   (c) When did the average score drop below 50?

74. Students in a precalculus class were given a final exam. Each month thereafter, they took an equivalent exam. The class average on the exam taken after t months is given by

$$F(t) = 82 - 8 \cdot \ln(t + 1).$$

(a) What was the class average after six months?
(b) After a year?
(c) When did the class average drop below 55?

75. One person with a flu virus visited the campus. The number T of days it took for the virus to infect x people was given by:

$$T = -.93 \ln \left[\frac{7000 - x}{6999x} \right].$$

(a) How many days did it take for 6000 people to become infected?
(b) After two weeks, how many people were infected?

76. **Approximating Logarithmic Functions by Polynomials.** For each positive integer n, let f_n be the polynomial function whose rule is

$$f_n(x) = x - \frac{x^2}{2} + \frac{x^3}{3} - \frac{x^4}{4} + \frac{x^5}{5} - \cdots \pm \frac{x^n}{n}$$

where the sign of the last term is $+$ if n is odd and $-$ if n is even. In the viewing window with $-1 \le x \le 1$ and $-4 \le y \le 1$, graph $g(x) = \ln(1 + x)$ and $f_4(x)$ on the same screen. For what values of x does f_4 appear to be a good approximation of g?

77. Using the viewing window in Exercise 76, find a value of n for which the graph of the function f_n (as defined in Exercise 76) appears to coincide with the graph of $g(x) = \ln(1 + x)$. Use the trace feature to move from graph to graph to see how good this approximation actually is.

78. The number N of days of training needed for a factory worker to produce x tools per day is given by

$$N = -25 \cdot \ln \left(1 - \frac{x}{60} \right).$$

(a) How many training days are needed for the worker to be able to produce 40 tools a day?
(b) It costs $135 to train one worker for one day. If the profit on one tool is $1.85, how many workdays does it take before the factory breaks even on the training costs for a worker who can produce 40 tools a day?

79. A bicycle store finds that the number N of bikes sold is related to the number d of dollars spent on advertising by $N = 51 + 100 \cdot \ln(d/100 + 2)$.
(a) How many bikes will be sold if nothing is spent on advertising? If $1000 is spent? If $10,000 is spent?
(b) If the average profit is $25 per bike, is it worthwhile to spend $1000 on advertising? What about $10,000?
(c) What are the answers in part (b) if the average profit per bike is $35?

Thinker

80. (a) Consider the function $f(x) = 4 \cdot 25^x$, which is a growth function since its base 25 is greater than 1. Find a constant k such that the rule of f can be written as $f(x) = 4e^{kx}$. [*Hint:* By Property 4 of logarithms, $e^{\ln 25} = 25$ and $\ln 25 \approx 3.2189$ is a constant.]
(b) Find a constant k such that the rule of $g(x) = 3(.05^x)$ (which is a decay function since its base .05 is less than 1) can be written in the form $g(x) = 3e^{kx}$.
(c) If $f(x) = Pa^x$, find a constant k such that the rule of f can be written in the form $f(x) = Pe^{kx}$.
(d) Explain why the constant k in part (c) is positive for a growth function and negative for a decay function. [*Hint:* f is a growth function if $a > 1$ and a decay function if $0 < a < 1$. For what values of a is $\ln a$ positive and for what values is it negative? See Figure 5–18.]

5.4 **Properties of Logarithms**

Logarithms have several important properties in addition to those presented in Section 5.3. These properties, which we shall call *logarithm laws*, arise from the fact that logarithms are exponents. Essentially, they are properties of exponents translated into logarithmic language.

The first law of exponents says that $b^m b^n = b^{m+n}$, or in words,

The exponent of a product is the sum of the exponents of the factors.

Since logarithms are just particular kinds of exponents, this statement translates as:

The logarithm of a product is the sum of the logarithms of the factors.

Here is the same statement in symbolic language:

Product Law for Logarithms

> For all $v, w > 0$,
>
> $$\ln(vw) = \ln v + \ln w$$
>
> and
>
> $$\log(vw) = \log v + \log w.$$

Proof   According to Property 4 of logarithms (in the box on page 373),

$$e^{\ln v} = v \quad \text{and} \quad e^{\ln w} = w.$$

Therefore, by the first law of exponents (with $m = \ln v$ and $n = \ln w$):

$$vw = e^{\ln v}e^{\ln w} = e^{\ln v + \ln w}.$$

So raising e to the exponent ($\ln v + \ln w$) produces vw. But the definition of logarithm says that $\ln vw$ is the exponent to which e must be raised to produce vw. Therefore, we must have $\ln vw = \ln v + \ln w$. A similar argument works for common logarithms (just replace e by 10 and "ln" by "log"). ■

Example 1   A calculator shows that $\ln 7 = 1.9459$ and $\ln 9 = 2.1972$. Therefore,

$$\ln 63 = \ln (7 \cdot 9) = \ln 7 + \ln 9 = 1.9459 + 2.1972 = 4.1341. \quad ■$$

CALCULATOR EXPLORATION

We know that $5 \cdot 7 = 35$. Key in LOG(35) ENTER. Then key in LOG(5) + LOG(7) ENTER. The answers are the same by the Product Law. Do you get the same answer if you key in LOG(5) $\times$ LOG(7) ENTER?

Example 2   Use the Product Law to write

(a) $\log (7xy)$ as a sum of three logarithms.

(b) $\log x^2 + \log y + 1$ as a single logarithm.

Solution

(a) $\log(7xy) = \log 7x + \log y = \log 7 + \log x + \log y$

(b) Note that $\log 10 = 1$ (why?). Hence,

$$\log x^2 + \log y + 1 = \log x^2 + \log y + \log 10$$
$$= \log(x^2 y) + \log 10$$
$$= \log(10x^2 y). \quad \blacksquare$$

CAUTION

A common error in applying the Product Law for Logarithms is to write the *false* statement

$$\ln 7 + \ln 9 = \ln(7 + 9)$$
$$= \ln 16$$

instead of the correct statement

$$\ln 7 + \ln 9 = \ln(7 \cdot 9)$$
$$= \ln 63.$$

GRAPHING EXPLORATION

Illustrate the Caution in the margin graphically by graphing both

$$f(x) = \ln x + \ln 9 \qquad \text{and} \qquad g(x) = \ln(x + 9)$$

in the standard viewing window and verifying that the graphs are not the same. In particular, the functions have different values at $x = 7$.

The second law of exponents, namely, $b^m / b^n = b^{m-n}$, may be roughly stated in words as

The exponent of the quotient is the difference of exponents.

When the exponents are logarithms, this says

The logarithm of a quotient is the difference of the logarithms.

In other words,

Quotient Law for Logarithms

For all $v, w > 0$,

$$\ln\left(\frac{v}{w}\right) = \ln v - \ln w$$

and

$$\log\left(\frac{v}{w}\right) = \log v - \log w.$$

The proof of the Quotient Law is very similar to the proof of the Product Law (see Exercise 26).

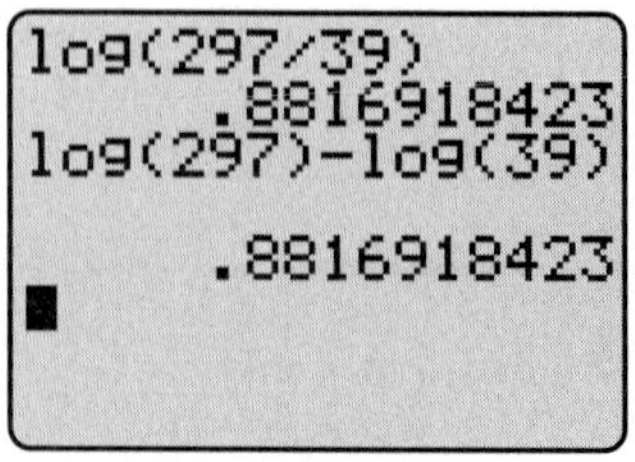

```
log(297/39)
       .8816918423
log(297)-log(39)
       .8816918423
■
```

Figure 5–22

Example 3 Figure 5–22 illustrates the Quotient Law by showing that

$$\log\left(\frac{297}{39}\right) = \log 297 - \log 39. \quad \blacksquare$$

Example 4 For any $w > 0$,

$$\ln\left(\frac{1}{w}\right) = \ln 1 - \ln w = 0 - \ln w = -\ln w$$

and

$$\log\left(\frac{1}{w}\right) = \log 1 - \log w = 0 - \log w = -\log w. \quad \blacksquare$$

CAUTION

Do not confuse $\ln\left(\dfrac{v}{w}\right)$ with the quotient $\dfrac{\ln v}{\ln w}$. They are *different* numbers. For example,

$$\ln\left(\frac{36}{3}\right) = \ln(12) = 2.4849, \quad \text{but} \quad \frac{\ln 36}{\ln 3} = \frac{3.5835}{1.0986} = 3.2619.$$

GRAPHING EXPLORATION

Illustrate the preceding Caution graphically by graphing both $f(x) = \ln(x/3)$ and $g(x) = (\ln x)/(\ln 3)$ and verifying that the graphs are not the same at $x = 36$ (or anywhere else, for that matter).

The third law of exponents, namely $(b^m)^k = b^{mk}$, can also be translated into logarithmic language:

Power Law for Logarithms

For all k and all $v > 0$,

$$\ln(v^k) = k(\ln v)$$

and

$$\log(v^k) = k(\log v).$$

Proof Since $v = 10^{\log v}$ (why?), the third law of exponents (with $b = 10$ and $m = \log v$) shows that

$$v^k = (10^{\log v})^k = 10^{(\log v)k} = 10^{k(\log v)}.$$

So raising 10 to the exponent $k(\log v)$ produces v^k. But the exponent to which 10 must be raised to produce v^k is, by definition, $\log(v^k)$. Therefore $\log(v^k) = k(\log v)$, and the proof is complete. A similar argument with e in place of 10 and "ln" in place of "log" works for natural logarithms. ■

Example 5   Express $\ln \sqrt{19}$ without radicals or exponents.

Solution   First write $\sqrt{19}$ in exponent notation, then use the Power Law:

$$\ln \sqrt{19} = \ln 19^{1/2}$$

$$= \frac{1}{2} \ln 19 \quad \text{or} \quad \frac{\ln 19}{2}. \quad ■$$

Example 6   Express as a single logarithm: $\dfrac{\log(x^2 + 1)}{3} - \log x$.

Solution

$$\frac{\log(x^2 + 1)}{3} - \log x = \frac{1}{3} \log (x^2 + 1) - \log x$$

$$= \log (x^2 + 1)^{1/3} - \log x \qquad \text{[Power Law]}$$

$$= \log \sqrt[3]{x^2 + 1} - \log x$$

$$= \log \left(\frac{\sqrt[3]{x^2 + 1}}{x} \right) \qquad \text{[Quotient Law]} \quad ■$$

Example 7   Express as a single logarithm: $\ln 3x + 4 \ln x - \ln 3xy$.

Solution

$$\ln 3x + 4 \cdot \ln x - \ln 3xy = \ln 3x + \ln x^4 - \ln 3xy \qquad \text{[Power Law]}$$

$$= \ln (3x \cdot x^4) - \ln 3xy \qquad \text{[Product Law]}$$

$$= \ln \frac{3x^5}{3xy} \qquad \text{[Quotient Law]}$$

$$= \ln \frac{x^4}{y} \qquad \text{[Cancel 3x]} \quad ■$$

Example 8   Simplify: $\ln \left(\dfrac{\sqrt{x}}{x} \right) + \ln \sqrt[4]{ex^2}$.

Solution   Begin by changing to exponential notation:

$$\ln\left(\frac{x^{1/2}}{x}\right) + \ln\,(ex^2)^{1/4} = \ln\,(x^{-1/2}) + \ln\,(ex^2)^{1/4}$$

$$= -\frac{1}{2}\cdot\ln x + \frac{1}{4}\cdot\ln ex^2 \qquad \textit{[Power Law]}$$

$$= -\frac{1}{2}\cdot\ln x + \frac{1}{4}\,(\ln e + \ln x^2) \qquad \textit{[Product Law]}$$

$$= -\frac{1}{2}\cdot\ln x + \frac{1}{4}\,(\ln e + 2\cdot\ln x) \qquad \textit{[Power Law]}$$

$$= -\frac{1}{2}\cdot\ln x + \frac{1}{4}\cdot\ln e + \frac{1}{2}\cdot\ln x$$

$$= \frac{1}{4}\cdot\ln e = \frac{1}{4} \qquad \textit{[ln e = 1]} \quad\blacksquare$$

Applications

Because logarithmic growth is slow, measurements on a logarithmic scale (that is, on a scale determined by a logarithmic function) can sometimes be deceptive.

Example 9 **(Earthquakes)** The magnitude $R(i)$ of an earthquake on the Richter scale is given by $R(i) = \log(i/i_0)$, where i is the amplitude of the ground motion of the earthquake and i_0 is the amplitude of the ground motion of the so-called zero earthquake.* A moderate earthquake might have 1000 times the ground motion of the zero earthquake (that is, $i = 1000i_0$). So its magnitude would be

$$\log\,(1000i_0/i_0) = \log 1000 = \log 10^3 = 3.$$

An earthquake with 10 times this ground motion (that is, $i = 10\cdot1000i_0 = 10,000i_0$) would have a magnitude of

$$\log\,(10,000i_0/i_0) = \log 10,000 = \log 10^4 = 4.$$

So a *tenfold* increase in ground motion produces only a 1-point change on the Richter scale. In general,

> **Increasing the ground motion by a factor of 10^k increases the Richter magnitude by k units.**[†]

For instance, the 1989 World Series earthquake in San Francisco measured 7.0 on the Richter scale, and the great earthquake of 1906 measured 8.3. The difference of 1.3 points means that the 1906 quake was $10^{1.3} \approx 20$ times more intense than the 1989 one in terms of ground motion. $\quad\blacksquare$

---

*The zero earthquake has ground motion amplitude of less than 1 micron on a standard seismograph 100 kilometers from the epicenter.

[†] *Proof:* If one quake has ground motion amplitude i and the other $10^k i$, then

$$R(10^k i) = \log\,(10^k i/i_0) = \log 10^k + \log\,(i/i_0)$$
$$= k + \log\,(i/i_0) = k + R(i).$$

Exercises 5.4

In Exercises 1–10, write the given expression as a single logarithm.

1. $\ln x^2 + 3 \ln y$

2. $\ln 2x + 2(\ln x) - \ln 3y$

3. $\log (x^2 - 9) - \log (x + 3)$

4. $\log 3x - 2[\log x - \log (2 + y)]$

5. $2(\ln x) - 3(\ln x^2 + \ln x)$

6. $\ln (e/\sqrt{x}) - \ln \sqrt{ex}$

7. $3 \ln (e^2 - e) - 3$ **8.** $2 - 2 \log (20)$

9. $\log (10x) + \log (20y) - 1$

10. $\ln (e^2 x) + \ln (ey) - 3$

In Exercises 11–16, let $u = \ln x$ and $v = \ln y$. Write the given expression in terms of u and v. For example, $\ln x^3 y = \ln x^3 + \ln y = 3 \ln x + \ln y = 3u + v$.

11. $\ln (x^2 y^5)$ **12.** $\ln (x^3 y^2)$ **13.** $\ln (\sqrt{x} \cdot y^2)$

14. $\ln \left(\dfrac{\sqrt{x}}{y} \right)$ **15.** $\ln \left(\sqrt[3]{x^2 \sqrt{y}} \right)$ **16.** $\ln \left(\dfrac{\sqrt{x^2 y}}{\sqrt[3]{y}} \right)$

In Exercises 17–22, use graphical or algebraic means to determine whether the statement is true or false.

17. $\ln |x| = |\ln x|$?   **18.** $\ln \left(\dfrac{1}{x} \right) = \dfrac{1}{\ln x}$?

19. $\log x^5 = 5(\log x)$?   **20.** $e^{x \ln x} = x^x \ (x > 0)$?

21. $\ln x^3 = (\ln x)^3$?   **22.** $\log \sqrt{x} = \sqrt{\log x}$?

In Exercises 23–24, find values of a and b for which the statement is false.

23. $\dfrac{\log a}{\log b} = \log \left(\dfrac{a}{b} \right)$ **24.** $\log (a + b) = \log a + \log b$

25. If $\ln b^7 = 7$, what is b?

26. Prove the Quotient Law for Logarithms: for v, $w > 0$, $\ln \left(\dfrac{v}{w} \right) = \ln v - \ln w$. (Use properties of exponents and the fact that $v = e^{\ln v}$ and $w = e^{\ln w}$.)

In Exercises 27–30, state the magnitude on the Richter scale of an earthquake that satisfies the given condition.

27. 100 times stronger than the zero quake.

28. $10^{4.7}$ times stronger than the zero quake.

29. 350 times stronger than the zero quake.

30. 2500 times stronger than the zero quake.

Exercises 31–34 deal with the energy intensity i of a sound, which is related to the loudness of the sound by the function $L(i) = 10 \cdot \log (i/i_0)$, where i_0 is the minimum intensity detectable by the human ear and $L(i)$ is measured in decibels. Find the decibel measure of the sound.

31. Ticking watch (intensity is 100 times i_0).

32. Soft music (intensity is 10,000 times i_0).

33. Loud conversation (intensity is 4 million times i_0).

34. Victoria Falls in Africa (intensity is 10 billion times i_0).

35. How much louder is the sound in Exercise 32 than the sound in Exercise 31?

36. The perceived loudness L of a sound of intensity I is given by $L = k \cdot \ln I$, where k is a certain constant. By how much must the intensity be increased to double the loudness? (That is, what must be done to I to produce $2L$?)

Thinkers

37. Compute each of the following pairs of numbers.

   (a) $\log 18$ and $\dfrac{\ln 18}{\ln 10}$

   (b) $\log 456$ and $\dfrac{\ln 456}{\ln 10}$

   (c) $\log 8950$ and $\dfrac{\ln 8950}{\ln 10}$

   (d) What do these results suggest?

38. Prove that for any positive number c, $\log c = \dfrac{\ln c}{\ln 10}$. [*Hint:* We know that $10^{\log c} = c$ (why?). Take natural logarithms on both sides and use a logarithm law to simplify and solve for $\log c$.]

39. Find each of the following logarithms.

   (a) $\log 8.753$ **(b)** $\log 87.53$ **(c)** $\log 875.3$

   (d) $\log 8753$ **(e)** $\log 87,530$

   (f) How are the numbers 8.753, 87.53, ... , 87,530 related to one another? How are their logarithms related? State a general conclusion that this evidence suggests.

40. Prove that for every positive number c, $\log c$ can be written in the form $k + \log b$, where $0 < b < 10$. [*Hint:* Write c in scientific notation and use logarithm laws to express $\log c$ in the required form.]

5.4.A

5.4.A **EXCURSION** **Logarithmic Functions to Other Bases**[*]

The same procedure used in Sections 5.3–5.4 can be carried out with any positive number b in place of 10 and e.

> **Throughout this excursion, b is a fixed positive number with $b > 1$.**[†]

The basic idea of logarithms can be seen from an example. The *logarithm of* 150 *to base* 7 is defined to be the solution of the equation $7^x = 150$ and is denoted $\log_7 150$. The solution of $7^x = 150$ (that is, $\log_7 150$) can be approximated graphically by finding the intersection of $y = 7^x$ and $y = 150$, as in Figure 5–23. Subject to rounding, Figure 5–24 shows that 7 raised to this power is 150. In other words, $\log_7 150$ is the exponent to which 7 must be raised to produce 150.

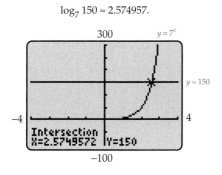

$\log_7 150 \approx 2.574957$.

Figure 5–23

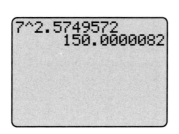

Figure 5–24

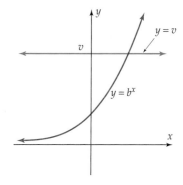

Figure 5–25

More generally, whenever b and v are positive numbers, then the horizontal line $y = v$ lies above the x-axis and hence intersects the graph of $y = b^x$ (Figure 5–25), so that the equation $b^x = v$ has a solution. Consequently, we have this definition.

Definition of Logarithms to Base b

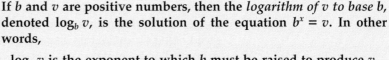

If b and v are positive numbers, then the *logarithm of v to base b,* denoted $\log_b v$, is the solution of the equation $b^x = v$. In other words,

$\log_b v$ is the exponent to which b must be raised to produce v.

---

[*]This section, which replicates the discussion of Sections 5.3 and 5.4 in a more general context, may be read before Section 5.3 if desired. It is not needed in the sequel.

[†]The discussion is also valid when $0 < b < 1$, but in that case the graphs have a different shape.

Example 1 To find $\log_2 16$, ask yourself "what power of 2 equals 16?" Since $2^4 = 16$, we see that $\log_2 16 = 4$. Similarly, $\log_2 (1/8) = -3$ because $2^{-3} = 1/8$. ∎

Example 2 Since logarithms are just exponents, every logarithmic statement can be translated into exponential language:

Logarithmic Statement	Equivalent Exponential Statement
$\log_3 81 = 4$	$3^4 = 81$
$\log_4 64 = 3$	$4^3 = 64$
$\log_{125} 5 = \dfrac{1}{3}$	$125^{1/3} = 5^*.$
$\log_8 \left(\dfrac{1}{4}\right) = -\dfrac{2}{3}$	$8^{-2/3} = \dfrac{1}{4}$ (verify!) ∎

Example 3 Solve: $\log_5 x = 3$.

Solution The equation $\log_5 x = 3$ is equivalent to the exponential statement $5^3 = x$, so the solution is $x = 125$. ∎

Example 4 Logarithms to the base 10 are called **common logarithms.** Is it customary to write $\log v$ instead of $\log_{10} v$. Then,

$$\log 100 = 2 \quad \text{because} \quad 10^2 = 100;$$

$$\log .001 = -3 \quad \text{because} \quad 10^{-3} = \frac{1}{10^3} = \frac{1}{1000} = .001.$$

Calculators have a LOG key for evaluating common logarithms. For instance,[†]

$$\log .4 = -0.3979, \quad \log 45.3 = 1.6561, \quad \log 685 = 2.8357. \quad ∎$$

Example 5 The most frequently used base for logarithms in modern applications is the number e ($\approx 2.71828\cdots$). Logarithms to the base e are called **natural logarithms** and use a different notation: We write $\ln v$ instead of $\log_e v$. Calculators also have an LN key for evaluating natural logarithms. For example,

$$\ln .5 = -0.6931, \quad \ln 65 = 4.1744, \quad \ln 158 = 5.0626. \quad ∎$$

---

[*]Because $125^{1/3} = \sqrt[3]{125} = 5.$

[†] Here and below, all logarithms are rounded to four decimal places. So strictly speaking, the equal sign should be replaced by an "approximately equal" sign ($\approx$).

You *don't* need a calculator to understand the essential properties of logarithms. You need only translate logarithmic statements into exponential ones (or vice versa).

Example 6   What is $\log(-25)$?

Translation: To what power must 10 be raised to produce -25? The graph of $f(x) = 10^x$ lies entirely above the x-axis (use your calculator), which means that *every* power of 10 is *positive*. So 10^x can *never* be -25, or any negative number, or zero. The same argument works for any base b:

$$\log_b v \text{ is defined only when } v > 0. \quad ■$$

Example 7   What is $\log_5 1$?

Translation: To what power must 5 be raised to produce 1? The answer, of course, is $5^0 = 1$. So $\log_5 1 = 0$. Similarly, $\log_5 5 = 1$ because 1 is the answer to "what power of 5 equals 5?" In general,

$$\log_b 1 = 0 \quad \text{and} \quad \log_b b = 1. \quad ■$$

Example 8   What is $\log_2 2^9$?

Translation: To what power must 2 be raised to produce 2^9? Obviously, the answer is 9. So $\log_2 2^9 = 9$ and, in general,

$$\log_b b^k = k \quad \text{for every real number } k.$$

This property holds even when k is a complicated expression. For instance, if x and y are positive, then

$$\log_6 6^{\sqrt{3x+y}} = \sqrt{3x+y} \quad \text{(here } k = \sqrt{3x+y}). \quad ■$$

Example 9   What is $10^{\log 439}$? Well, log 439 is the power to which 10 must be raised to produce 439, that is $10^{\log 439} = 439$. Similarly,

$$b^{\log_b v} = v \quad \text{for every } v > 0. \quad ■$$

Here is a summary of the facts illustrated in the preceding examples.

Properties of Logarithms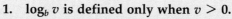

1. $\log_b v$ is defined only when $v > 0$.
2. $\log_b 1 = 0 \quad$ and $\quad \log_b b = 1.$
3. $\log_b (b^k) = k$ for every real number k.
4. $b^{\log_b v} = v$ for every $v > 0$.

Logarithm Laws

The first law of exponents states that $b^m b^n = b^{m+n}$, or in words,

> The exponent of a product is the sum of the exponents of the factors.

Since logarithms are just particular kinds of exponents, this statement translates as:

The logarithm of a product is the sum of the logarithms of the factors.

The second and third laws of exponents, namely, $b^m/b^n = b^{m-n}$ and $(b^m)^k = b^{mk}$, can also be translated into logarithmic language:

Logarithm Laws

Let b, v, w, k be real numbers, with b, v, w positive and $b \neq 1$.

Product Law: $\log_b (vw) = \log_b v + \log_b w.$

Quotient Law: $\log_b\left(\dfrac{v}{w}\right) = \log_b v - \log_b w.$

Power Law: $\log_b (v^k) = k(\log_b v).$

Proof   According to Property 4 in the box on page 387.

$$b^{\log_b v} = v \qquad \text{and} \qquad b^{\log_b w} = w.$$

Therefore, by the second law of exponents (with $m = \log_b v$ and $n = \log_b w$) we have:

$$\frac{v}{w} = \frac{b^{\log_b v}}{b^{\log_b w}} = b^{\log_b v - \log_b w}.$$

Since $\log_b (v/w)$ is the exponent to which b must be raised to produce v/w, we must have $\log_b (v/w) = \log_b v - \log_b w$. This proves the Quotient Law. The Product and Power Laws are proved in a similar fashion. ∎

Example 10   Simplify and write as a single logarithm:

(a) $\log_3 (x + 2) + \log_3 y - \log_3 (x^2 - 4)$
(b) $3 - \log_5 (125x)$

Solution

(a) $\log_3 (x + 2) + \log_3 y - \log_3 (x^2 - 4)$

$\qquad = \log_3[(x + 2)y] - \log_3 (x^2 - 4)$ *[Product Law]*

$\qquad = \log_3\left(\dfrac{(x + 2)y}{x^2 - 4}\right)$ *[Quotient Law]*

$\qquad = \log_3\left(\dfrac{(x + 2)y}{(x + 2)(x - 2)}\right)$ *[Factor denominator]*

$\qquad = \log_3\left(\dfrac{y}{x - 2}\right)$ *[Cancel common factor]*

(b) $3 - \log_5 (125x)$

$$= 3 - (\log_5 125 + \log_5 x) \qquad [\textit{Product Law}]$$

$$= 3 - \log_5 125 - \log_5 x$$

$$= 3 - 3 - \log_5 x \qquad [\log_5 125 = 3 \textit{ because } 5^3 = 125]$$

$$= -\log_5 x$$

$$= \log_5 x^{-1} = \log_5 \left(\frac{1}{x}\right) \qquad [\textit{Power Law}] \quad \blacksquare$$

CAUTION

1. A common error in using the Product Law is to write something like $\log 6 + \log 7 = \log (6 + 7) = \log 13$ instead of the correct statement $\log 6 + \log 7 = \log (6 \cdot 7) = \log 42$.

2. Do not confuse $\log_b \left(\dfrac{v}{w}\right)$ with the quotient $\dfrac{\log_b v}{\log_b w}$. They are *different* numbers. For example, when $b = 10$

$$\log \left(\frac{48}{4}\right) = \log 12 = 1.0792 \qquad \text{but} \qquad \frac{\log 48}{\log 4} = \frac{1.6812}{0.6021} = 2.7922.$$

For graphic illustrations of the errors mentioned in the Caution, see Exercises 82 and 83.

Example 11 Given that

$$\log_7 2 = .3562, \qquad \log_7 3 = .5646, \qquad \text{and} \qquad \log_7 5 = .8271,$$

find:

(a) $\log_7 10$; **(b)** $\log_7 2.5$; **(c)** $\log_7 48$.

Solution

(a) By the Product Law,
$$\log_7 10 = \log_7 (2 \cdot 5) = \log_7 2 + \log_7 5 = .3562 + .8271 = 1.1833.$$

(b) By the Quotient Law,
$$\log_7 2.5 = \log_7 \left(\frac{5}{2}\right) = \log_7 5 - \log_7 2 = .8271 - .3562 = .4709.$$

(c) By the Product and Power Laws,
$$\log_7 48 = \log_7 (3 \cdot 16) = \log_7 3 + \log_7 16 = \log_7 3 + \log_7 2^4$$

$$= \log_7 3 + 4 \cdot \log_7 2 = .5646 + 4(.3562)$$

$$= 1.9894. \quad \blacksquare$$

Example 11 worked because we were *given* several logarithms to base 7. But there's no $\log_7$ key on the calculator, so how do you find logarithms to base 7, or to any base other than e or 10? *Answer:* Use the LN key on the calculator and the following formula:

Change of Base Formula

For any positive numbers b and v,

$$\log_b v = \frac{\ln v}{\ln b}$$

Proof By Property 4 in the box on page 387, $b^{\log_b v} = v$. Take the natural logarithm of each side of this equation:

$$\ln (b^{\log_b v}) = \ln v.$$

Apply the Power Law for natural logarithms on the left side:

$$(\log_b v)(\ln b) = \ln v.$$

Dividing both sides by $\ln b$ finishes the proof:

$$\log_b v = \frac{\ln v}{\ln b}. \quad \blacksquare$$

Example 12 To find $\log_7 3$, apply the change of base formula with $b = 7$:

$$\log_7 3 = \frac{\ln 3}{\ln 7} = \frac{1.0986}{1.9459} = .5646. \quad \blacksquare$$

Exercises 5.4.A

Note: Unless stated otherwise, all letters represent positive numbers and $b \neq 1$.

In Exercises 1–8, fill in the missing entries in each table.

1.

x	0	1	2	4
$f(x) = \log_4 x$				

2.

x	1/25	5	25	$\sqrt{5}$
$g(x) = \log_5 x$				

3.

x		1/6	1	216
$h(x) = \log_6 x$	-2			

4.

x		10/3	4	6	12
$k(x) = \log_3 (x - 3)$					

5.

x	0	1/7	$\sqrt{7}$	49
$f(x) = 2 \log_7 x$				

6.

x			100	1000
$g(x) = 3 \log x$	6	3		

7.

x		-2.75	-1	1	29
$h(x) = 3 \log_2 (x + 3)$					

8.

x		1/e	1	e	e^2
$k(x) = 2 \ln x$					

In Exercises 9–18, translate the given exponential statement into an equivalent logarithmic one.

9. $10^{-2} - .01$

10. $10^3 = 1000$

11. $\sqrt[3]{10} = 10^{1/3}$

12. $10^{.4771} \approx 3$

13. $10^{7k} = r$

14. $10^{(a+b)} = c$

15. $7^8 = 5,764,801$

16. $2^{-3} = 1/8$

17. $3^{-2} = 1/9$

18. $b^{14} = 3379$

In Exercises 19–28, translate the given logarithmic statement into an equivalent exponential one.

19. $\log 10{,}000 = 4$

20. $\log .001 = -3$

21. $\log 750 \approx 2.88$

22. $\log (.8) = -.097$

23. $\log_5 125 = 3$

24. $\log_8 (1/4) = -2/3$

25. $\log_2 (1/4) = -2$

26. $\log_2 \sqrt{2} = 1/2$

27. $\log (x^2 + 2y) = z + w$

28. $\log (a + c) = d$

In Exercises 29–36, evaluate the given expression without using a calculator.

29. $\log 10^{\sqrt{97}}$

30. $\log_{17} (17^{17})$

31. $\log 10^{x^2 + y^2}$

32. $\log_{3.5} \left[3.5^{(x^2 - 1)} \right]$

33. $\log_{16} 4$

34. $\log_2 64$

35. $\log_{\sqrt{3}} (27)$

36. $\log_{\sqrt{3}} (1/9)$

In Exercises 27–40, a graph or a table of values for the function $f(x) = \log_b x$ is given. Find b.

37.

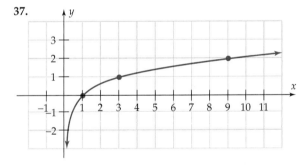

38.

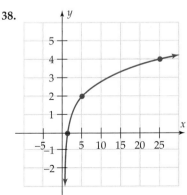

39.

x	.05	1	400	$2\sqrt{5}$
$f(x)$	-1	0	2	$1/2$

40.

x	1/25	1	5	125
$f(x)$	-4	0	2	6

In Exercises 41–46, find x.

41. $\log_3 243 = x$

42. $\log_{81} 27 = x$

43. $\log_{27} x = 1/3$

44. $\log_5 x = -4$

45. $\log_x 64 = 3$

46. $\log_x (1/9) = -2/3$

In Exercises 47–60, write the given expression as the logarithm of a single quantity, as in Example 10.

47. $2 \log x + 3 \log y - 6 \log z$

48. $5 \log_8 x - 3 \log_8 y + 2 \log_8 z$

49. $\log x - \log (x + 3) + \log (x^2 - 9)$

50. $\log_3 (y + 2) + \log_3 (y - 3) - \log_3 y$

51. $\dfrac{1}{2} \log_2 (25c^2)$

52. $\dfrac{1}{3} \log_2 (27b^6)$

53. $-2 \log_4 (7c)$

54. $\dfrac{1}{3} \log_5 (x + 1)$

55. $2 \ln (x + 1) - \ln (x + 2)$

56. $\ln (z - 3) + 2 \ln (z + 3)$

57. $\log_2 (2x) - 1$

58. $2 - \log_5 (25z)$

59. $2 \ln (e^2 - e) - 2$

60. $4 - 4 \log_5 (20)$

In Exercises 61–68 use a calculator and the change of base formula to find the logarithm.

61. $\log_2 10$

62. $\log_2 22$

63. $\log_7 5$

64. $\log_5 7$

65. $\log_{500} 1000$

66. $\log_{500} 250$

67. $\log_{12} 56$

68. $\log_{12} 725$

In Exercises 69–74, answer true *or* false *and give reasons for your answer.*

69. $\log_b (r/5) = \log_b r - \log_b 5$

70. $\dfrac{\log_b a}{\log_b c} = \log_b \left(\dfrac{a}{c} \right)$

71. $(\log_b r)/t = \log_b (r^{1/t})$

72. $\log_b (cd) = \log_b c + \log_b d$

73. $\log_5 (5x) = 5(\log_5 x)$

74. $\log_b (ab)^t = t(\log_b a) + t$

Thinkers

75. Which is larger: 397^{398} or 398^{397}? [*Hint:* $\log 397 \approx 2.5988$ and $\log 398 \approx 2.5999$ and $f(x) = 10^x$ is an increasing function.]

76. If $\log_b 9.21 = 7.4$ and $\log_b 359.62 = 19.61$, then what is $\log_b 359.62/\log_b 9.21$?

In Exercises 77–80, assume that a and b are positive, with $a \neq 1$ and $b \neq 1$.

77. Express $\log_b u$ in terms of logarithms to the base a.

78. Show that $\log_b a = 1/\log_a b$.

79. How are $\log_{10} u$ and $\log_{100} u$ related?

80. Show that $a^{\log b} = b^{\log a}$.

81. If $\log_b x = \dfrac{1}{2} \log_b v + 3$, show that $x = (b^3)\sqrt{v}$.

82. Graph the functions $f(x) = \log x + \log 7$ and $g(x) = \log(x + 7)$ on the same screen. For what values of x is it true that $f(x) = g(x)$? What do you conclude about the statement

$$\log 6 + \log 7 = \log(6 + 7)?$$

83. Graph the functions $f(x) = \log(x/4)$ and $g(x) = (\log x)/(\log 4)$. Are they the same? What does this say about a statement such as

$$\log\left(\frac{48}{4}\right) = \frac{\log 48}{\log 4}?$$

In Exercises 84–86, sketch a complete graph of the function, labeling any holes, asymptotes, or local extrema.

84. $f(x) = \log_5 x + 2$

85. $h(x) = x \log x^2$

86. $g(x) = \log_{20} x^2$

5.5 Algebraic Solutions of Exponential and Logarithmic Equations

Most of the exponential and logarithmic equations solved by graphical means earlier in this chapter could also have been solved algebraically. The algebraic techniques for solving such equations are based on the properties of logarithms.

The easiest exponential equations to solve are those in which both sides are powers of the same base.

Example 1   Solve: $8^x = 2^{x+1}$.

Solution   Using the fact that $8 = 2^3$, we rewrite the equation as follows.

$$8^x = 2^{x+1}$$

$$(2^3)^x = 2^{x+1}$$

$$2^{3x} = 2^{x+1}$$

Since the powers of 2 are equal, the exponents must be the same, that is,

$$3x = x + 1$$

$$2x = 1$$

$$x = \frac{1}{2}. \quad \blacksquare$$

When different bases are involved in an exponential equation, a new technique is needed.

Example 2 Solve: $5^x = 2$.

Solution

$$5^x = 2$$

Take logarithms on each side:* $\ln 5^x = \ln 2$

Use the Power Law: $x(\ln 5) = \ln 2$

Divide both sides by ln 5: $x = \dfrac{\ln 2}{\ln 5} \approx \dfrac{.6931}{1.6094} \approx .4307.$

Remember: $\dfrac{\ln 2}{\ln 5}$ is *not* $\ln \dfrac{2}{5}$ or $\ln 2 - \ln 5$. ■

Example 3 To solve $2^{4x-1} = 3^{1-x}$,

Take logarithms on each side: $\ln 2^{4x-1} = \ln 3^{1-x}$

Use the Power Law: $(4x - 1)(\ln 2) = (1 - x)(\ln 3)$

Multiply out both sides: $4x(\ln 2) - \ln 2 = \ln 3 - x(\ln 3)$

Rearrange terms: $4x(\ln 2) + x(\ln 3) = \ln 2 + \ln 3$

Factor left side: $(4 \cdot \ln 2 + \ln 3)x = \ln 2 + \ln 3$

Divide both sides by $(4 \cdot \ln 2 + \ln 3)$: $x = \dfrac{\ln 2 + \ln 3}{4 \cdot \ln 2 + \ln 3} \approx .4628.$ ■

Example 4 After 43 years, a 20-milligram sample of strontium-90 (^{90}Sr) decays to 6.071 mg. What is the half-life of strontium-90?

Solution As we saw in Section 5.2, the amount of strontium-90 at time x is given by

$$f(x) = 20(.5^{x/h}),$$

where h is the half-life of strontium-90. We know that $f(x) = 6.071$ when $x = 43$, that is, $6.071 = 20(.5^{43/h})$. We must solve this equation for h.

Divide both sides by 20: $\dfrac{6.071}{20} = .5^{43/h}$

Take logarithms on both sides: $\ln \dfrac{6.071}{20} = \ln .5^{43/h}$

---

*We shall use natural logarithms, but the same techniques are valid for logarithms to other bases (Exercise 24).

Use the Power Law: $$\ln \frac{6.071}{20} = \frac{43}{h}\ln .5$$

Multiply both sides by h: $$h \ln \frac{6.071}{20} = 43 \ln .5$$

Divide both sides by $\ln \frac{6.071}{20}$: $$h = \frac{43 \ln 5}{\ln(6.071/20)} \approx 25$$

Therefore, strontium-90 has a half-life of 25 years. ∎

Example 5 A certain bacteria is known to grow exponentially, with the population at time t given by a function of the form $g(t) = Pe^{kt}$, where P is the original population and k is the continuous growth rate. A culture shows 1000 bacteria present. Seven hours later there are 5000.

(a) Find the continuous growth rate k.

(b) Determine when the population will reach one billion.

Solution

(a) The original population is $P = 1000$, so the growth function is $g(t) = 1000e^{kt}$. We know that $g(7) = 5000$, that is,

$$1000e^{k \cdot 7} = 5000.$$

To determine the growth rate, we solve this equation for k.

Divide both sides by 1000: $$e^{7k} = 5$$

Take logarithms of both sides: $$\ln e^{7k} = \ln 5$$

Use the Power Law: $$7k \ln e = \ln 5$$

Since $\ln e = 1$ (why?), this equation becomes

$$7k = \ln 5$$

Divide both sides by 7: $$k = \frac{\ln 5}{7} \approx .22992.$$

Therefore, the growth function is $g(t) \approx 1000e^{.22992t}$.

(b) The population will reach one billion when $g(t) = 1,000,000,000$, that is, when

$$1000e^{.22992t} = 1,000,000,000.$$

So we solve this equation for t:

Divide both sides by 1000: $$e^{.22992t} = 1,000,000$$

Take logarithms on both sides: $$\ln e^{.22992t} = \ln 1,000,000$$

Use the Power Law: $$.22992t \ln e = \ln 1,000,000$$

Remember $\ln e = 1$: $$.22992t = \ln 1,000,000$$

Divide both sides by .22992: $$t = \frac{\ln 1,000,000}{.22992} \approx 60.09$$

Therefore, it will take a bit more than 60 hours for the culture to grow to one billion. ∎

Logarithmic Equations

Equations that involve only logarithmic terms may be solved by using this fact, which is proved in Exercise 23 (and is valid with log replaced by ln):

$$\text{If } \log u = \log v, \text{ then } u = v.$$

Example 6   Solve: $\log (3x + 2) + \log (x + 2) = \log (7x + 6)$.

Solution   First we write the left side as a single logarithm.

$$\log (3x + 2) + \log (x + 2) = \log (7x + 6)$$

Use the Product Law: $\log[(3x + 2)(x + 2)] = \log (7x + 6)$

Multiply out left side: $\log (3x^2 + 8x + 4) = \log (7x + 6).$

Since the logarithms are equal, we must have

$$3x^2 + 8x + 4 = 7x + 6$$

Subtract $7x + 6$ from both sides: $3x^2 + x - 2 = 0$

Factor: $(3x - 2)(x + 1) = 0$

$$3x - 2 = 0 \qquad \text{or} \qquad x + 1 = 0$$

$$3x = 2 \qquad\qquad\qquad x = -1$$

$$x = \frac{2}{3}$$

```
log(3*2/3+2)+log
(2/3+2)
          1.028028724
log(7*2/3+6)
          1.028028724
```

Figure 5–26

Thus, $x = 2/3$ and $x = -1$ are the *possible* solutions and must be checked in the *original* equation. When $x = 2/3$, both sides of the original equation have the same value, as shown in Figure 5–26. So $2/3$ is a solution. When $x = -1$, however, the right side of the equation is

$$\log (7x + 6) = \log [7(-1) + 6] = \log (-1),$$

which not defined. So -1 is not a solution.   ∎

Example 7   Solve: $\ln (x + 4) - \ln (x + 2) = \ln x$

Solution   First, write the left side as a single logarithm.

Use the Quotient Law: $\ln \left(\dfrac{x + 4}{x + 2} \right) = \ln x$

Therefore,

$$\frac{x + 4}{x + 2} = x$$

Multiply both sides by $x + 2$: $x + 4 = x(x + 2)$

Multiply out right side: $x + 4 = x^2 + 2x$

Rearrange terms: $x^2 + x - 4 = 0$

This equation does not readily factor, so we use the quadratic formula:

$$x = \frac{-1 \pm \sqrt{1^2 - 4 \cdot 1 \cdot (-4)}}{2 \cdot 1} = \frac{-1 \pm \sqrt{17}}{2}.$$

An easy way to verify that $\dfrac{-1 + \sqrt{17}}{2}$ is a solution is to store this number as A and then evaluate both sides of the original equation at $x = A$, as shown in Figure 5–27. The second possibility, however, is not a solution (because $x = \dfrac{-1 - \sqrt{17}}{2}$ is negative, so that $\ln x$ is not defined). ■

```
(-1+√(17))/2→A
          1.561552813
ln(A+4)-ln(A+2)
            .445680719
ln(A)
            .445680719
```

Figure 5–27

Equations that involve both logarithmic and constant terms may be solved by using this property of logarithms (see page 373):

(∗) $10^{\log v} = v$ and $e^{\ln v} = v.$

Example 8 Solve: $7 + 2 \log 5x = 11$.

Solution We start by getting all the logarithmic terms on one side and the constant on the other:

Subtract 7 from both sides: $2 \log 5x = 4$

Divide both sides by 2: $\log 5x = 2$

We know that if two quantities are equal, say $a = b$, then $10^a = 10^b$. We use this fact here, with the two sides of the preceding equation as a and b.

Exponentiate both sides: $10^{\log 5x} = 10^2$

Use the basic logarithm property (∗): $5x = 100$

Divide both sides by 5: $x = 20$

Verify that 20 is actually a solution of the original equation. ■

Example 9 Solve: $\ln (x - 3) = 5 - \ln (x - 3)$.

Solution We proceed as in Example 8, but since the base for these logarithms is e, we use e rather than 10 when we exponentiate.

$$\ln (x - 3) = 5 - \ln (x - 3)$$

Add ln (x − 3) to both sides: $2 \ln (x - 3) = 5$

Divide both sides by 2: $\ln (x - 3) = \dfrac{5}{2}$

Exponentiate both sides: $e^{\ln(x-3)} = e^{5/2}$

Use the basic property of logarithms (∗): $x - 3 = e^{5/2}$

Add 3 to both sides: $x = e^{5/2} + 3 \approx 15.1825$

This is the only possible solution. Use your calculator to verify that it actually is a solution of the original equation. ■

Example 10 Solve: $\log (x - 16) = 2 - \log (x - 1)$.

Solution

$$\log (x - 16) = 2 - \log (x - 1)$$

Add log (x − 1) to both sides: $\quad \log (x - 16) + \log (x - 1) = 2$

Use the Product Law: $\quad \log [(x - 16)(x - 1)] = 2$

Multiply out left side: $\quad \log (x^2 - 17x + 16) = 2$

Exponentiate both sides: $\quad 10^{\log (x^2 - 17x + 16)} = 10^2$

Use the basic logarithm property (∗): $\quad x^2 - 17x + 16 = 100$

Subtract 100 from both sides: $\quad x^2 - 17x - 84 = 0$

Factor: $\quad (x + 4)(x - 21) = 0$

$$x + 4 = 0 \quad \text{or} \quad x - 21 = 0$$
$$x = -4 \quad \text{or} \quad x = 21$$

You can easily verify that 21 is a solution of the original equation, but -4 is not [when $x = -4$, then $\log (x - 16) = \log (-20)$, which is not defined]. ■

Exercises 5.5

In Exercises 1–8, solve the equation without using logarithms.

1. $3^x = 81$ **2.** $3^x + 3 = 30$ **3.** $3^{x+1} = 9^{5x}$

4. $4^{5x} = 16^{2x-1}$ **5.** $3^{5x}9^{x^2} = 27$ **6.** $2^{x^2+5x} = \dfrac{1}{16}$

7. $9^{x^2} = 3^{-5x-2}$ **8.** $4^{x^2-1} = 8^x$

In Exercises 9–22, solve the equation. Express your answer in terms of natural logarithms (for instance, $x = (2 + \ln 5)/(\ln 3)$). Then use a calculator to find an approximation for the answer.

9. $3^x = 5$ **10.** $5^x = 4$ **11.** $2^x = 3^{x-1}$

12. $4^{x+2} = 2^{x-1}$ **13.** $3^{1-2x} = 5^{x+5}$

14. $4^{3x-1} = 3^{x-2}$ **15.** $2^{1-3x} = 3^{x+1}$

16. $3^{z+3} = 2^z$ **17.** $e^{2x} = 5$

18. $e^{-3x} = 2$ **19.** $6e^{-1.4x} = 21$

20. $3.4e^{-x/3} = 5.6$ **21.** $2.1e^{(x/2)\ln 3} = 5$

22. $7.8e^{(x/3)\ln 5} = 14$

23. Prove that if $\log u = \log v$, then $u = v$. [Hint: Property (∗) on page 396.]

24. (a) Solve $7^x = 3$, using natural logarithms. Leave your answer in logarithmic form; don't approximate with a calculator.

   (b) Solve $7^x = 3$, using common (base 10) logarithms. Leave your answer in logarithmic form.

   (c) Use the change of base formula in Excursion 5.4.A to show that your answers in parts (a) and (b) are the same.

In Exercises 25–34, solve the equation as in Examples 6 and 7.

25. $\ln (3x - 5) = \ln 11 + \ln 2$

26. $\log (4x - 1) = \log (x + 1) + \log 2$

27. $\log (3x - 1) + \log 2 = \log 4 + \log (x + 2)$

28. $\ln (x + 6) - \ln 10 = \ln (x - 1) - \ln 2$

29. $2 \ln x = \ln 36$

30. $2 \log x = 3 \log 4$

31. $\ln x + \ln (x + 1) = \ln 3 + \ln 4$

32. $\ln (6x - 1) + \ln x = \frac{1}{2}\ln 4$

33. $\ln x = \ln 3 - \ln (x + 5)$

34. $\ln (2x + 3) + \ln x = \ln e$

In Exercises 35–42, solve the equation, as in Examples 8–10.

35. $\ln (x + 9) - \ln x = 1$

36. $\ln (2x + 1) - 1 = \ln (x - 2)$

37. $\log x + \log (x - 3) = 1$

38. $\log (x - 1) + \log (x + 2) = 1$

39. $\log \sqrt{x^2 - 1} = 2$

40. $\log \sqrt[3]{x^2 + 21x} = \frac{2}{3}$

41. $\ln (x^2 + 1) - \ln (x - 1) = 1 + \ln (x + 1)$

42. $\dfrac{\ln (x + 1)}{\ln (x - 1)} = 2$

Exercises 43–48 deal with the half-life function $f(x) = P(.5^{x/h})$ that was introduced in Section 5.2 and used in Example 4 of this section.

43. A sample of 300 grams of uranium decays to 200 grams in .26 billion years. Find the half-life of uranium.

44. It takes 1000 years for a sample of 100 mg of radium-226 to decay to 65 mg. Find the half-life of radium-226.

45. A 3-gram sample of an isotope of sodium decays to 1 gram in 23.7 days. Find the half-life of the isotope of sodium.

46. After six days a sample of radon-222 decayed to 33.6% of its original mass. Find the half-life of radon-222. [*Hint:* When $x = 6$, then $f(x) = .336P$, where f is the half-life function.]

47. How old is a wooden statue that has lost two-thirds of its original carbon-14? [*Hint:* Remember that the half-life of carbon-14 is 5730 years and see Example 10 in Section 5.2.]

48. Krypton-85 loses 6.44% of its mass each year. What is its half-life?

Exercises 49–52 deal with the compound interest formula, $A = P(1 + r)^t$, which was discussed in Section 5.2.

49. How long will it take for a $1000 investment to grow to $1500 at an interest rate of 6%, compounded quarterly?

50. How long will it take for a $500 investment to grow to $1200 at an interest rate of 9%, compounded quarterly?

51. Is it possible to double an investment of $800 at an interest rate of 7.3%, compounded annually, in less than 10 years? Justify your answer.

52. Find a formula that gives the time needed for an investment of P dollars to triple at interest rate r, compounded annually. [*Hint:* Solve the compound interest formula for t when $A = 3P$.]

Exercises 53–56 deal with continuous growth that can be described by a function of the form $g(t) = Pe^{kt}$, where P is the original population and k is the continuous growth rate, as in Example 5.

53. The present concentration of carbon dioxide in the atmosphere is 364 parts per million (ppm) and is increasing exponentially at a continuous yearly rate of .4% (that is, $k = .004$). How many years will it take for the concentration to reach 500 ppm?

54. The amount P of ozone in the atmosphere is currently decaying exponentially each year at a continuous rate of $\frac{1}{4}$% (that is, $k = -.0025$). How long will it take for half the ozone to disappear (that is, when will the amount be $P/2$)? [Your answer is the half-life of ozone.]

55. The population of Brazil increased from 122 million in 1980 to 158 million in 1992.
   (a) At what continuous rate was the population growing during this period?
   (b) Assuming that Brazil's population continues to increase at this rate, when will it reach 250 million?

56. A colony of 1000 weevils grows exponentially to 1750 in one week.
   (a) At what continuous rate is the population growing?
   (b) How many weeks does it take for the weevil population to reach 3000?

57. The probability P percent of having an accident while driving a car is related to the alcohol level of the driver's blood by the formula $P = e^{kt}$, where k is a constant. Accident statistics show that the probability of an accident is 25% when the blood alcohol level is $t = .15$.
   (a) Find k. [Use $P = 25$, not .25.]
   (b) At what blood alcohol level is the probability of having an accident 50%?

58. Under normal conditions, the atmospheric pressure (in millibars) at height h feet above sea level is given by $P(h) = 1015e^{-kh}$, where k is a positive constant.
   (a) If the pressure at 18,000 feet is half the pressure at sea level, find k.
   (b) Using the information from part (a), find the atmospheric pressure at 1000 feet, 5000 feet, and 15,000 feet.

59. One hour after an experiment begins, the number of bacteria in a culture is 100. An hour later there are 500.

  (a) Find the number of bacteria at the beginning of the experiment and the number 3 hours later.

  (b) How long does it take the number of bacteria at any given time to double?

60. If the population at time t is given by $S(t) = ce^{kt}$, find a formula that gives the time it takes for the population to double.

Thinkers

61. According to one theory of learning, the number of words per minute N that a person can type after t weeks of practice is given by $N = c(1 - e^{-kt})$, where c is an upper limit that N cannot exceed and k is a constant that must be determined experimentally for each person.

  (a) If a person can type 50 wpm (words per minute) after four weeks of practice and 70 wpm after eight weeks, find the values of k and c for this person. According to the theory, this person will never type faster than c wpm.

  (b) Another person can type 50 wpm after four weeks of practice and 90 wpm after eight weeks. How many weeks must this person practice to be able to type 125 wpm?

62. Eileen has been offered two jobs, each with the same starting salary of $24,000 and identical benefits. Assuming satisfactory performance, she will receive a $1200 raise each year at the Great Gizmo Company, whereas the Wonder Widget Company will give her a 4% raise each year.

  (a) In what year (after the first year) would her salary be the same at either company? Until then, which company pays better? After that, which company pays better?

  (b) Answer the questions in part (a) assuming that the annual raise at Great Gizmo is $1800.

Exponential, Logarithmic, and Other Models*

Many data sets can be modeled by suitable exponential, logarithmic, and related functions. Most calculators have regression procedures for constructing the following models.

Model	Equation	Examples	
Power	$y = ax^r$	$y = 5x^{2.7}$	$y = 3.5x^{-.045}$
Exponential	$y = ab^x$ or $y = ae^{kx}$	$y = 2 \cdot (1.64)^x$	$y = 2 \cdot e^{.4947x}$
Logistic	$y = \dfrac{a}{1 + be^{-kx}}$	$y = \dfrac{20{,}000}{1 + 24e^{-.25x}}$	$y = \dfrac{650}{1 + 6e^{.3x}}$
Logarithmic	$y = a + b \ln x$	$y = 5 + 4.2 \ln x$	$y = 2 - 3 \ln x$

We begin by examining exponential models, such as $y = 3 \cdot 2^x$. A table of values for this model is shown on the next page. Look carefully at the ratio of successive y entries (that is, each entry divided by its predecessor).

*This section is optional; its prerequisites are Sections 2.5 and 4.4. It will be used in clearly identifiable exercises, but not elsewhere in the text.

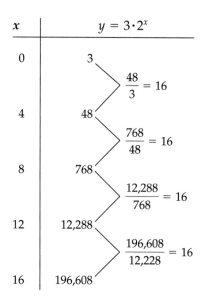

It should not be a surprise that the ratio of successive entries is constant. At each step, x changes from x to $x + 4$ (from 0 to 4, from 4 to 8, and so on) and y changes from $3 \cdot 2^x$ to $3 \cdot 2^{x+4}$. Hence, the ratio of successive terms is always

$$\frac{3 \cdot 2^{x+4}}{3 \cdot 2^x} = \frac{3 \cdot 2^x \cdot 2^4}{3 \cdot 2^x} = 2^4 = 16.$$

A similar argument applies to any exponential model $y = ab^x$ and shows that if x changes by a fixed amount k, then the ratio of the corresponding y values is the constant b^k (in our example b was 2 and k was 4). This suggests that

When the ratio of successive entries in a table of data is approximately constant, an exponential model is appropriate.

Example 1   In the years before the Civil War, the population of the United States grew rapidly, as shown in the following table from the U.S. Bureau of the Census. Find a model for this growth.

Year	Population (in millions)	Year	Population (in millions)
1790	3.93	1830	12.86
1800	5.31	1840	17.07
1810	7.24	1850	23.19
1820	9.64	1860	31.44

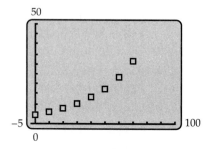

Figure 5–28

Solution   The data points (with $x = 0$ corresponding to 1790) are shown in Figure 5–28. Their shape suggests either a polynomial graph of even degree or an exponential graph. Since populations generally grow exponentially, an exponential model is likely to be a good choice. We can confirm this by looking at the ratios of successive entries in the table.

Year	Population		Year	Population
1790	3.93		1830	12.86
	$\dfrac{5.31}{3.93} \approx 1.351$			$\dfrac{17.07}{12.86} \approx 1.327$
1800	5.31		1840	17.07
	$\dfrac{7.24}{5.31} \approx 1.363$			$\dfrac{23.19}{17.07} \approx 1.359$
1810	7.24		1850	23.19
	$\dfrac{9.64}{7.24} \approx 1.331$			$\dfrac{31.44}{23.19} \approx 1.356$
1820	9.64		1860	31.44
	$\dfrac{12.86}{9.64} \approx 1.334$			
1830	12.86			

The ratios are almost constant, as they would be in an exponential model. So we use regression to find such a model. The process is the same as for linear and polynomial regression (see the Tips on pages 109 and 304). It produces this model:*

$$y = 3.9572 \cdot (1.0299^x).$$

The graph in Figure 5–29 appears to fit the data quite well. In fact, you can readily verify that the model has an error of less than 1% for each of the data points. Furthermore, as discussed before the example, when x changes by 10, the value of y changes by approximately $1.0299^{10} \approx 1.343$, which is very close to the successive ratios of the data that were computed above. ∎

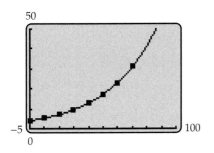

Figure 5–29

Example 2 After the Civil War, the U.S. population continued to increase, as shown in this table:

Year	Population (in millions)	Year	Population (in millions)	Year	Population (in millions)
1870	38.56	1920	106.02	1960	179.32
1880	50.19	1930	123.20	1970	202.30
1890	62.98	1940	132.16	1980	226.54
1900	76.21	1950	151.33	1990	248.72
1910	92.23				

However, the model from Example 1 does not remain valid, as can be seen in Figure 5–30, which shows its graph together with all the data points from 1790 through 1990 ($x = 0$ corresponds to 1790).

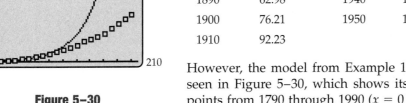

Figure 5–30

*Throughout this section, coefficients are rounded for convenient reading, but the full expansion is used for calculations and graphs.

The problem is that the *rate* of growth has steadily decreased since the Civil War. For instance, the ratio of the first two entries is $\frac{50.19}{38.56} \approx 1.302$ and the ratio of the last two is $\frac{248.72}{226.54} \approx 1.098$. So an exponential model may not be the best choice now. Other possibilities are polynomial models (which grow at a slower rate) or logistic models (in which the growth rate decreases with time). Figure 5–31 shows three possible models, each obtained by using the appropriate regression program on a calculator, with all the data points from 1790 through 1990.

Exponential Model

$$y = 5.5381 \cdot 1.02098^x$$

Polynomial Model

$$y = (5.61 \times 10^{-8})x^4 - (1.92 \times 10^{-5})x^3$$
$$+ .0084x^2 - .1276x + 5.247$$

Logistic Model

$$y = \frac{384.57}{1 + 54e^{-.0228x}}$$

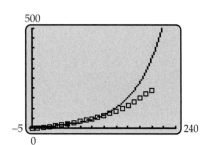
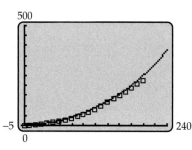
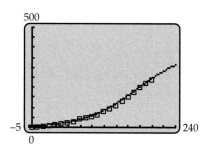

Figure 5–31

As expected, the polynomial and logistic models fit the data better than does the exponential model. The main difference between them is that the polynomial model indicates unlimited future growth, whereas the logistic model has the population growing more slowly in the future (and eventually leveling off—see Exercise 13). ■

In Example 1 we used the ratios of successive entries of the data table to determine that an exponential model was appropriate. Here is another way to make that determination. Consider the exponential function $y = ab^x$. Taking natural logarithms of both sides and using the logarithm laws on the right side shows that

$$\ln y = \ln (ab^x) = \ln a + \ln b^x = \ln a + x \ln b.$$

Now $\ln a$ and $\ln b$ are constants, say $k = \ln a$ and $m = \ln b$; so that

$$\ln y = mx + k.$$

Thus, the points $(x, \ln y)$ lie on the straight line with slope m and y-intercept k. Consequently, we have this guideline:

> **If (x, y) are data points and if the points $(x, \ln y)$ are approximately linear, then an exponential model may be appropriate for the data.**

Similarly, if $y = ax^r$ is a power function, then

$$\ln y = \ln (ax^r) = \ln a + r \ln x.$$

Since ln a is a constant, say $k = \ln a$, we have

$$\ln y = r \ln x + k,$$

which means that the points (ln x, ln y) lie on the straight line with slope r and y-intercept k. Consequently, we have this guideline:

> **If (x, y) are data points and if the points (ln x, ln y) are approximately linear, then a power model may be appropriate for the data.**

Example 3 The length of time that a planet takes to make one complete rotation around the sun is its year. The table shows the length (in earth years) of each planet's year and the distance of that planet from the sun (in millions of miles).* Find a model for this data in which x is the length of the year and y the distance from the sun.

Planet	Year	Distance		Planet	Year	Distance
Mercury	.24	36.0		Saturn	29.46	886.7
Venus	.62	67.2		Uranus	84.01	1783.0
Earth	1	92.9		Neptune	164.79	2794.0
Mars	1.88	141.6		Pluto	247.69	3674.5
Jupiter	11.86	483.6				

Solution Figure 5–32 shows the data points for the five planets with the shortest years. Figure 5–33 shows all the data points, but on this scale, the first four points look like a single large one near the origin.

Technology Tip

Suppose the x- and y-coordinates of the data points are stored in lists L_1 and L_2, respectively. On calculators other than TI-89, keying in

   ln L_2 STO $\rightarrow L_4$

produces the list L_4, whose entries are the natural logarithms of the numbers in list L_2, and stores it in the statistics editor. You can then use lists L_1 and L_4 to plot the points $(x, \ln y)$. For TI-89, check your instruction manual.

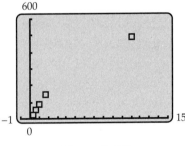

Figure 5–32

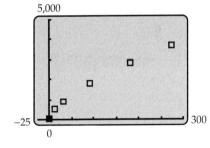

Figure 5–33

Plotting the point $(x, \ln y)$ for each data point (x, y) (see the Tip in the margin) produces Figure 5–34 on the next page. Its points do not form a linear pattern (four of them are almost vertical near the y-axis and the other five almost horizontal), so an exponential function is not an appropriate model. On the other hand, the points (ln x, ln y) in Figure 5–35 do form a linear pattern, which suggests that a power model will work.

---

*Since the orbit of a planet around the sun is not circular, its distance from the sun varies through the year. The number given here is the average of its maximum and minimum distances from the sun.

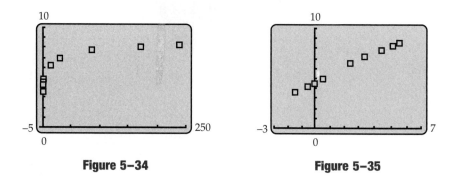

Figure 5-34   **Figure 5-35**

A calculator's power regression feature produces this model:

$$y = 92.8935 \, x^{.6669}.$$

Its graph in Figure 5-36 shows that it fits the original data points quite well. ▪

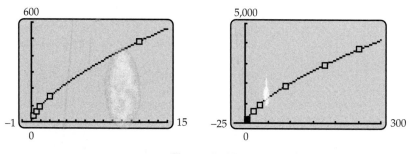

Figure 5-36

If $y = a + b \ln x$ is a logarithmic model, then the points $(\ln x, y)$ lie on the straight line with slope b and y-intercept a (why?). Thus we have this guideline:

> **If (x, y) are data points and if the points $(\ln x, y)$ are approximately linear, then a logarithmic model may be appropriate for the data.**

Example 4   Find a model for population growth in Anaheim, California, given the following data. (*Source:* U.S. Bureau of the Census)

Year	1950	1970	1980	1990	1994
Population	14,556	166,408	219,494	266,406	282,133

Solution The scatter plot of the data points (with $x = 50$ corresponding to 1950) in Figure 5–37 suggests a logarithmic curve with a very slight bend. So we plot the points (ln 50, 14,556), ... , (ln 94, 282,133) in Figure 5–38. Since these points appear to lie on a straight line, a logarithmic model is appropriate. Using logarithmic regression on a calculator, we obtain this model:

$$y = -1,643,983.42 + 424,768.97 \ln x.$$

Its graph is a good fit for the data (Figure 5–39). ∎

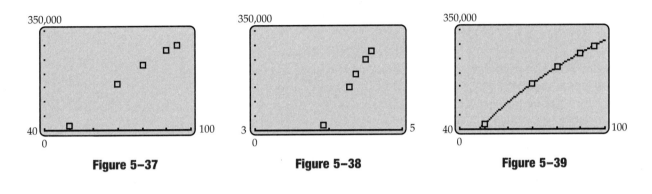

Figure 5–37   Figure 5–38   Figure 5–39

CAUTION

When using logarithmic models, you must have data points with positive first coordinates (since logarithms of negative numbers and 0 are not defined).

Exercises 5.6

In Exercises 1–10, state which of the following models might be appropriate for the given scatter plot of data. (More than one model may be appropriate.)

Model	Corresponding Function
A. Linear	$y = ax + b$
B. Quadratic	$y = ax^2 + bx + c$
C. Power	$y = ax^r$
D. Cubic	$y = ax^3 + bx^2 + cx + d$
E. Exponential	$y = ab^x$
F. Logarithmic	$y = a + b \ln x$
G. Logistic	$y = \dfrac{a}{1 + be^{-kx}}$

1.
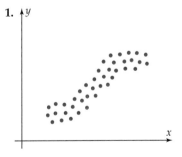

2.
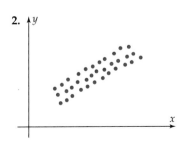

3.

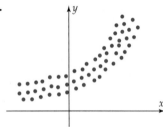

4.

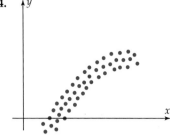

5.

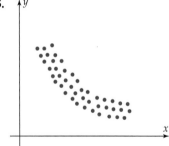

6.

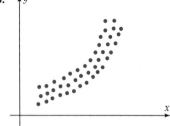

7.

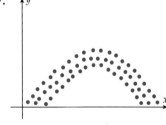

8.

9.

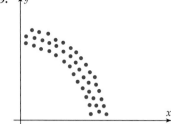

10.

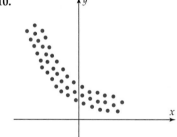

In Exercises 11 and 12, compute the ratios of successive entries in the table to determine whether or not an exponential model is appropriate for the data.

11.

x	0	2	4	6	8	10
y	3	15.2	76.9	389.2	1975.5	9975.8

12.

x	1	3	5	7	9	11
y	3	21	55	105	171	253

13. (a) Show algebraically that in the logistic model for the U.S. population in Example 2, the population can never exceed 384.57 million people.

(b) Confirm your answer in part (a) by graphing the logistic model in a window that includes the next three centuries.

14. According to estimates by the U.S. Bureau of the Census, the U.S. population was 267.2 million in 1997. Based on this information, which of the models in Example 2 appears to be the most accurate predictor?

15. Graph each of the following power functions in a window with $0 \le x \le 20$.
 (a) $f(x) = x^{-1.5}$ **(b)** $g(x) = x^{.75}$ **(c)** $h(x) = x^{2.4}$

16. Based on your graphs in Exercise 15, describe the general shape of the graph of $y = ax^r$, with $a > 0$ and
 (a) $r < 0$ **(b)** $0 < r < 1$ **(c)** $r > 1$

In Exercises 17–20, determine whether an exponential, power, or logarithmic model (or none or several of these) is appropriate for the data by determining which (if any) of the following sets of points is approximately linear:

$$\{(x, \ln y)\}, \qquad \{(\ln x, \ln y)\}, \qquad \{(\ln x, y)\},$$

where the given data set consists of the points $\{(x, y)\}$.

17.

x	1	3	5	7	9	11
y	2	25	81	175	310	497

18.

x	3	6	9	12	15	18
y	385	74	14	2.75	.5	.1

19.

x	5	10	15	20	25	30
y	17	27	35	40	43	48

20.

x	5	10	15	20	25	30
y	2	110	460	1200	2500	4525

21. The table shows the number of babies born as twins, triplets, quadruplets, etc. in recent years.

Year	Multiple Births
1989	92,916
1990	96,893
1991	98,125
1992	99,255
1993	100,613
1994	101,658
1995	101,709

(a) Sketch a scatter plot of the data, with $x = 1$ corresponding to 1989.

(b) Plot each of the following models on the same screen as the scatter plot:
$$f(x) = 93{,}201.973 + 4{,}545.977 \ln x;$$
$$g(x) = \frac{102{,}519.98}{1 + .1536e^{-.4263x}}$$

(c) Use the table feature to estimate the number of multiple births in 2000 and 2005.

(d) Over the long run which model do you think is the better predictor?

22. The graph shows the Census Bureau estimates of future U.S. population.
 (a) How well do the projections in the graph compare with those given by the logistic model in Example 2?
 (b) Find a logistic model of the U.S. population, using the data given in Example 2 for the years from 1900 to 1990 and the estimated 1997 population of 267.2 million.
 (c) How well do the projections in the graph compare with those given by the model in part (b)?

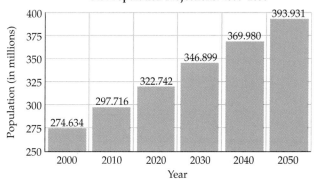

U.S. Population Projections: 2000–2050

23. Infant mortality rates in the United States are shown in the table.

Year	Infant Mortality Rate*	Year	Infant Mortality Rate*
1920	76.7	1980	12.6
1930	60.4	1985	10.6
1940	47.0	1990	9.2
1950	29.2	1995	7.6
1960	26.0	1996	7.2
1970	20.0		

*Rates are infant (under 1 year) deaths per 1000 live births.

(a) Sketch a scatter plot of the data, with $x = 0$ corresponding to 1900.

(b) Verify that the set of points $(x, \ln y)$, where (x, y) are the original data points, is approximately linear.

(c) Based on part (b), what type of model would be appropriate for this data? Find such a model.

24. The average number of students per computer in the U.S. public schools (elementary through high school) is shown in the table.

Fall of School Year	Students/Computer
1983	125
1984	75
1985	50
1986	37
1987	32
1988	25
1989	22
1990	20
1991	18
1992	16
1993	14
1994	10.5
1995	10

(a) Sketch a scatter plot of the data, with $x = 1$ corresponding to 1983.

(b) Find an exponential model for the data.

(c) Use the model to estimate the number of students per computer in 2000.

(d) In what year, according to this model, will each student have his or her own computer in school?

(e) What are the limitations of this model?

25. The number of children who were home schooled in the United States in selected years is shown in the table. [*Source:* National Home Education Research Institute]

Fall of School Year	Number of Children (in thousands)
1983	92.5
1985	183
1988	225
1990	301
1992	470
1993	588
1994	735
1995	800
1996	920
1997	1100

(a) Sketch a scatter plot of the data, with $x = 0$ corresponding to 1980.

(b) Find a quadratic model for the data.

(c) Find a logistic model for the data.

(d) What is the number of home-schooled children predicted by each model for the year 2000?

(e) What are the limitations of each model?

26. (a) Find an exponential model for the federal debt, based on the data in the table on the next page. Let $x = 0$ correspond to 1960.

(b) Use the model to predict the federal debt in 2000.

Accumulated Gross Federal Debt	
Year	Amount (in billions of dollars)
1960	291
1965	322
1970	381
1975	542
1980	909
1985	1818
1990	3207
1995	4921
1996	5182

27. Although the number of farms in the United States has been decreasing (see Exercise 28), the average size of farms (in acres) has increased, as shown in the table. [*Source:* U.S. Department of Agriculture]

Year	Average Farm Size (in acres)	Year	Average Farm Size (in acres)
1940	174	1980	426
1950	213	1990	460
1960	297	1997	470
1970	374		

(a) Sketch a scatter plot of the data, with $x = 0$ corresponding to 1940.
(b) Find a cubic model for the data. Explain why this is *not* a reasonable model for making predictions about the future.
(c) Find a logistic model for the data (which may be more reasonable because the amount of farming land is limited).
(d) Based on the logistic model, what is the estimated size of the average farm in 2000?
(e) According to the logistic model, when will the average farm size be 1000 acres?

28. (a) Find an exponential model for the number of farms in the United States, based on the table below. Let $x = 0$ correspond to 1950. [*Source:* U.S. Department of Agriculture]
(b) Use the model to predict the number of farms in 2000.
(c) Is this model realistic for the next millennium?

Year	Number of Farms (in millions)	Year	Number of Farms (in millions)
1950	5.647800	1980	2.439510
1955	4.653800	1985	2.292530
1960	3.962520	1990	2.145820
1965	3.356170	1995	2.071520
1970	2.949140	1996	2.063910
1975	2.521420	1997	2.058910

29. Worldwide production of computers has grown dramatically, as shown in the first two columns of the following table. [*Source:* Dataquest]

Year	Worldwide Shipments (thousands)	Predicted Number of Shipments (thousands)	Ratio
1985	14.7		
1986	15.1		
1987	16.7		
1988	18.1		
1989	21.3		
1990	23.7		
1991	27		
1992	32.4		
1993	38.9		
1994	47.9		
1995	60.2		
1996	70.9		
1987	84.3		

(a) Sketch a scatter plot of the data, with $x = 1$ corresponding to 1985.

(b) Find an exponential model for the data.

(c) Use the model to complete column 3 of the table.

(d) Fill in column 4 of the table by dividing each entry in column 2 by the preceding one.

(e) What does column 4 tell you about the appropriateness of the model?

30. A teacher invested $2000 in an individual retirement account (IRA) in 1986. Its value has increased since then, as shown in the table opposite.

(a) Find an exponential model for this data, with $x = 0$ corresponding to 1980.

(b) Based on your model, what is the effective rate of return on this investment?

(c) Assuming this rate of return continues, how much will be in the account in 2010?

Year	Value of Account	Year	Value of Account
1986	$2,000.00	1992	$ 6,687.36
1987	2,387.54	1993	7,410.43
1988	3,770.55	1994	7,383.54
1989	4,651.03	1995	9,513.97
1990	5,639.20	1996	10,980.12
1991	6,014.69	1997	13,424.73
		1998	15,207.44

5.7   Inverse Functions

According to the definition of logarithms, $\log v = u$ means that u is the exponent to which 10 must be raised to produce v. In other words,

$$\log v = u \quad \text{exactly when} \quad 10^u = v.$$

If $f(x) = 10^x$ and $g(x) = \log x$, then we can rewrite this statement in functional notation as

$$g(v) = u \quad \text{exactly when} \quad f(u) = v.$$

In other words, the logarithmic function g *reverses* the action of the exponential function f by taking each output v of f back to the input u that it came from. The inputs of g are the outputs of f and the output of g are the inputs of f. Consequently, we say that $g(x) = \log x$ is the *inverse function* of $f(x) = 10^x$. This is an example of the following definition.

Inverse Functions ▶

Let f be a function. If there is a function g such that

$$g(y) = x \quad \text{exactly when} \quad f(x) = y,$$

then g is said to be the *inverse function* of f. The domain of g is the range of f and the range of g is the domain of f.

Example 1   Find the inverse function of $f(x) = 3x - 2$, if it has one.

Solution   If f has an inverse function g, then g must satisfy

$$g(y) = x \quad \text{exactly when} \quad f(x) = y.$$

that is,

$$g(y) = x \qquad \text{exactly when} \qquad 3x - 2 = y.$$

To find the rule of g we need only solve this last equation for x:

$$3x - 2 = y$$

Add 2 to both sides: $\qquad 3x = y + 2$

Divide both sides by 3: $\qquad x = \dfrac{y + 2}{3}$

Since $g(y) = x$, we see that the rule of g is $g(y) = \dfrac{y + 2}{3}$. ■

Recall that the letter used for the variable of a function doesn't matter. For instance, $h(x) = x^2$ and $h(t) = t^2$ and $h(u) = u^2$ all describe the same function, whose rule is "square the input." When dealing with inverse functions, it is customary to use the same variable for both f and its inverse g. Consequently, the inverse function in Example 1 would normally be written as $g(x) = \dfrac{x + 2}{3}$.

Example 2 Find the inverse function of $f(x) = x^3 + 5$, if it has one.

Solution If f has an inverse function g, then g must satisfy

$$g(y) = x \qquad \text{exactly when} \qquad f(x) = y.$$

So its inverse can be found by solving for x in the equation $f(x) = y$, that is,

$$x^3 + 5 = y$$

Subtract 5 from both sides: $\qquad x^3 = y - 5$

Take cube roots on both sides: $\qquad x = \sqrt[3]{y - 5}$

Therefore, $g(y) = \sqrt[3]{y - 5}$ is the inverse function of f. Using the same variable as f, we write its rule as $g(x) = \sqrt[3]{x - 5}$. ■

Example 3 Find the inverse function of $f(x) = x^2$, if it has one.

Solution If there is an inverse function g, it must satisfy

$$g(y) = x \qquad \text{exactly when} \qquad f(x) = y.$$

As before, we solve for x in the equation $f(x) = y$, that is, $x^2 = y$. This equation, however, has two solutions:

$$x = \sqrt{y} \qquad \text{or} \qquad x = -\sqrt{y}.$$

This means that $g(y) = x = \pm\sqrt{y}$, which is *not* the rule of a function because it produces two outputs for a single input [for instance, $g(4)$ would be both 2 and -2]. Therefore, $f(x) = x^2$ does not have an inverse function. ■

The Round-Trip Properties

The inverse function g of a function f is designed to send each output of f back to the input it came from, that is,

$$g(d) = c \quad \text{exactly when} \quad f(c) = d.$$

Consequently, if you first apply f and then apply g to the result, you obtain the number you started with:

$$g(f(c)) = g(d) \quad \text{[because } f(c) = d\text{]}$$
$$= c \quad \text{[because } g(d) = c\text{]}.$$

A similar argument shows that $f(g(d)) = d$.

Example 4 As we saw in Example 1, the inverse function of $f(x) = 3x - 2$ is $g(x) = \dfrac{x + 2}{3}$. If we start with a number c and apply f we obtain $f(c) = 3c - 2$. If we now apply g to this result, we obtain

$$g(f(c)) = g(3c - 2) = \frac{(3c - 2) + 2}{3} = c.$$

So we are back where we started. Similarly, if we first apply g and then apply f to a number, we end up where we started:

$$f(g(c)) = f\left(\frac{c + 2}{3}\right) = 3\left(\frac{c + 2}{3}\right) - 2 = c.$$

The function $f(x) = x^3 + 5$ of Example 2 and its inverse function $g(x) = \sqrt[3]{x} - 5$ also have these "round-trip" properties. If you apply one function and then the other, you wind up at the number you started with:

$$g(f(x)) = g(x^3 + 5) = \sqrt[3]{(x^3 + 5) - 5} = \sqrt[3]{x^3} = x$$

and

$$f(g(x)) = f(\sqrt[3]{x} - 5) = (\sqrt[3]{x} - 5)^3 + 5 = (x - 5) + 5 = x. \quad \blacksquare$$

Not only do a function and its inverse have the round-trip properties illustrated in Example 4, but somewhat more is true (as is proved in Exercise 53).

Round-Trip Theorem

If f is a function with inverse function g, then

$$g(f(x)) = x \quad \text{for every } x \text{ in the domain of } f;$$
$$f(g(x)) = x \quad \text{for every } x \text{ in the domain of } g.$$

Conversely, if f and g are functions having these properties, then g is the inverse function of f.

Example 5 If $f(x) = e^x$ and $g(x) = \ln x$, then the domain of f consists of all real numbers and the domain of g of all positive real numbers. For every real number x (every number in the domain of f),

$$g(f(x)) = g(e^x) \qquad \textit{[Definition of f]}$$

$$= \ln e^x \qquad \textit{[Definition of g]}$$

$$= x. \qquad \textit{[Property 3 of logarithms (page 373)]}$$

Hence, $g(f(x)) = x$. Reversing the order of f and g, we have

$$f(g(x)) = f(\ln x) \qquad \textit{[Definition of g]}$$

$$= e^{\ln x} \qquad \textit{[Definition of f]}$$

$$= x. \qquad \textit{[Property 4 of logarithms (page 373)]}$$

Thus, $f(g(x)) = x$ for every $x > 0$ (every number in the domain of g). Therefore, by the Round-Trip Theorem, $g(x) = \ln x$ is the inverse function of $f(x) = e^x$. ∎

Example 6 If $f(x) = \dfrac{5}{2x - 4}$ and $g(x) = \dfrac{4x + 5}{2x}$, then for every x in the domain of f (that is, all $x \neq 2$),

$$g(f(x)) = g\left(\frac{5}{2x - 4}\right) = \frac{4\left(\dfrac{5}{2x-4}\right) + 5}{2\left(\dfrac{5}{2x-4}\right)} = \frac{\dfrac{20 + 5(2x-4)}{2x-4}}{\dfrac{10}{2x-4}}$$

$$= \frac{20 + 5(2x-4)}{10} = \frac{20 + 10x - 20}{10} = \frac{10x}{10} = x$$

and for every x in the domain of g (all $x \neq 0$),

$$f(g(x)) = f\left(\frac{4x+5}{2x}\right) = \frac{5}{2\left(\dfrac{4x+5}{2x}\right) - 4} = \frac{5}{\dfrac{4x+5}{x} - 4}$$

$$= \frac{5}{\dfrac{4x + 5 - 4x}{x}} = \frac{5}{\dfrac{5}{x}} = x.$$

Therefore, g is the inverse function of f by the Round-Trip Theorem. ∎

One-to-One Functions

We now examine a property that will enable us to identify those functions that have inverse functions. Recall that we were unable to find an inverse function for $f(x) = x^2$ in Example 3. The problem is that two different inputs of f can produce the same output. For instance,

$$f(2) = 2^3 = 4 \qquad \text{and} \qquad f(-2) = (-2)^2 = 4,$$

so there is no way to obtain a function g that sends 4 back to the input that produced it [since a function can't have $g(4) = 2$ and -2]. If we want to find a functions that do have inverses, we should look for those where this does not occur. In other words, we should look for a function f for which different inputs always produce different options, that is,

$$\text{if } a \neq b, \quad \text{then} \quad f(a) \neq f(b).$$

A function f with this property is said to be **one-to-one**.

If f is one-to-one, then every output of f comes from exactly one input (because different inputs produce different outputs). Consequently, we can define a function g that takes each output d of f back to the unique input c that it came from, that is,

$$g(d) = c \quad \text{exactly when} \quad f(c) = d.$$

Thus, the function g is the inverse function of f. Conversely, Exercise 54 shows that every function f that has an inverse function is one-to-one. In summary,

A function f has an inverse function exactly when f is one-to-one.

In graphical terms, the statement

$$\text{if } a \neq b, \quad \text{then} \quad f(a) \neq f(b)$$

means that two points on the graph, $(a, f(a))$ and $(b, f(b))$, that have different x-coordinates $[a \neq b]$ must also have different y-coordinates $[f(a) \neq f(b)]$. Consequently, these points cannot lie on the same horizontal line because all points on a horizontal line have the same y-coordinate. Therefore, we have this geometric test to determine if a function is one-to-one (and hence, has an inverse).

Horizontal Line Test

A function f is one-to-one exactly when it has this property:

No horizontal line intersects the graph of f more than once.

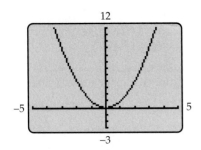

Figure 5–40

The graph of $f(x) = x^2$ (Figure 5–40) certainly fails the Horizontal Line Test because, for example, the horizontal lines $y = 1$, $y = 2$, etc., all intersect the graph more than once. This is geometric confirmation of what we saw above: $f(x) = x^2$ has no inverse function because it is not one-to-one.

Example 7 Which of the following functions are one-to-one and therefore have inverse functions?

(a) $f(x) = 7x^5 + 3x^4 - 2x^3 + 2x + 1$

(b) $g(x) = x^3 - 3x - 1$

(c) $h(x) = 1 - .2x^3$

Solution Complete graphs of each function are shown in Figure 5–41.

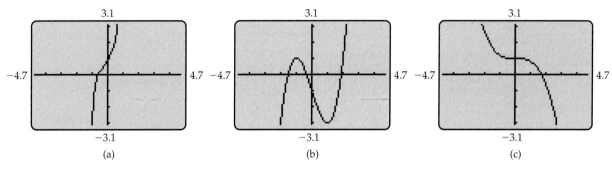

(a)   (b)   (c)

Figure 5–41

(a) The graph of f in Figure 5–41(a) passes the Horizontal Line Test since no horizontal line intersects the graph more than once. Hence, f is one-to-one.

(b) The graph of g in Figure 5–41(b) obviously fails the Horizontal Line Test. Therefore, g is not one-to-one.

(c) The graph of h in Figure 5–41(c) appears to contain a horizontal line segment. So, h appears to fail the Horizontal Line Test because the horizontal line through $(0, 1)$ seems to intersect the graph infinitely many times. But appearances are deceiving.

Technology Tip

Although a horizontal segment may appear on a calculator screen when the graph is actually rising or falling, there is another possibility. The graph may have a tiny wiggle (less than the height of a pixel) and thus fail the Horizontal Line Test:

You can usually detect such a wiggle by zooming in to magnify that portion of the graph, or by using the trace feature to see if the y-coordinates increase, then decrease (or vice versa) along the "horizontal" segment.

GRAPHING EXPLORATION

Graph $h(x) = 1 - .2x^3$ and use the trace feature to move from left to right along the "horizontal" segment. Do the y-coordinates stay the same, or do they decrease?

The Exploration shows that the graph is actually falling from left to right, so that each horizontal line intersects it only once. (It appears to have a horizontal segment because the amount the graph falls there is less than the height of a pixel on the screen.) Therefore, h is a one-to-one function. ■

The function f in Example 7 is an **increasing function** (its graph is always rising from left to right) and the function h is a **decreasing function** (its graph is always falling from left to right). Every increasing or decreasing function is necessarily one-to-one because its graph can never touch the same horizontal line twice (it would have to change from rising to falling, or vice versa, to do so). Therefore,

Every increasing function and every decreasing function has an inverse function.

Graphs of Inverse Functions

Finding the rule of the inverse function g of a one-to-one function f by solving the equation $y = f(x)$ for x, as in the preceding examples, is not always possible (some equations are hard to solve). But even if you don't know the rule of g, you can always find its graph, as shown below.

Suppose f is a one-to-one function and g is its inverse function. If (a, b) is on the graph of f, then by definition $f(a) = b$. Since the inverse function g takes each output of f back to its corresponding input, we know that $g(b) = a$. Hence, (b, a) is on the graph of g. A similar argument works in the other direction and leads to this conclusion:

(∗)   **(a, b) is on the graph of f exactly when (b, a) is on the graph of the inverse function g.**

This fact provides a method for graphing inverse functions. Exercise 51 shows that the line $y = x$ is the perpendicular bisector of the line through points (a, b) and (b, a). Thus, (a, b) and (b, a) lie on opposite sides of $y = x$, the same distance from it.* If you think of the line $y = x$ as a mirror, the points (a, b) and (b, a) are mirror images of each other. Consequently, the graph of the inverse function g is the mirror image of the graph of f, with the line $y = x$ being the mirror. In formal terms:

Inverse Function Graphs

> If g is the inverse function of f, then the graph of g is the reflection of the graph of f in the line $y = x$.

Example 8 We know that $g(x) = \ln x$ is the inverse function of $f(x) = e^x$. Figure 5–42 shows that the graph of g is the reflection (mirror image) of the graph of f in the line (mirror) $y = x$. ■

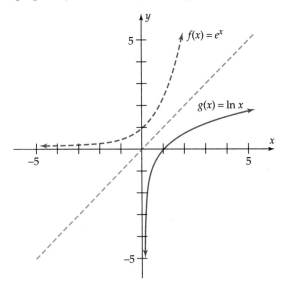

Figure 5–42

*In technical terms, (a, b) and (b, a) are said to be **symmetric** with respect to the line $y = x$.

Technology Tip

Inverse functions can be graphed directly on some calculators. When the function has been entered in the equation memory as y_1, use DRAWINV y_1 or DRINV y_1 in the TI-83 and Sharp 9600 DRAW menu or in the DRAW submenu of the TI-86/89 GRAPH menu.

GRAPHING EXPLORATION

Give another illustration of this property of inverse graphs by graphing the function $f(x) = x^3 + 5$ of Example 2, its inverse function $g(x) = \sqrt[3]{x - 5}$, and the line $y = x$ on the same screen. Use a square viewing window so that the mirror effect won't be distorted.

The easiest way to graph inverse functions is to use parametric graphing, which was explained in Excursion 3.3.A.*

Example 9 The function $f(x) = .7x^5 + .3x^4 - .2x^3 + 2x + .5$ can be graphed in parametric mode by letting

$$x = t \quad \text{and} \quad y = f(t) = .7t^5 + .3t^4 - .2t^3 + 2t + .5.$$

Its complete graph in Figure 5–43 shows that f has an inverse function g (why?). As we saw above, the graph of g can be obtained by reversing the coordinates of each point on the graph of f. In other words, g can be graphed parametrically by letting

$$x = .7t^5 + .3t^4 - .2t^3 + 2t + .5 \quad \text{and} \quad y = t.$$

Figure 5–44 shows the graphs of f and g on the same screen. ∎

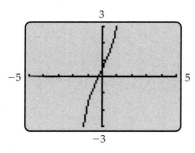

Figure 5–43

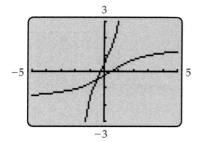

Figure 5–44

NOTE In many texts the inverse function of a function f is denoted f^{-1}. In this notation, for instance, the inverse of the function $f(x) = x^3 + 5$ in Example 2 would be written as $f^{-1}(x) = \sqrt[3]{x - 5}$. Similarly, the reversal properties of inverse functions become

$$f^{-1}(f(x)) = x \text{ for every } x \text{ in the domain of } f; \text{ and}$$

$$f(f^{-1}(x)) = x \text{ for every } x \text{ in the domain of } f^{-1}.$$

In this context, f^{-1} does *not* mean $1/f$ (see Exercise 47).

---

*If you are unfamiliar with parametric graphing, either read the Excursion or omit the next example.

Exercises 5.7

In Exercises 1–14, use algebra to find the inverse of the given one-to-one function.

1. $f(x) = -x$

2. $f(x) = -x + 1$

3. $f(x) = 5x - 4$

4. $f(x) = -3x + 5$

5. $f(x) = 5 - 2x^3$

6. $f(x) = (x^5 + 1)^3$

7. $f(x) = \sqrt{4x - 7}$

8. $f(x) = 5 + \sqrt{3x - 2}$

9. $f(x) = 1/x$

10. $f(x) = 1/\sqrt{x}$

11. $f(x) = \dfrac{1}{2x + 1}$

12. $f(x) = \dfrac{x}{x + 1}$

13. $f(x) = \dfrac{x^3 - 1}{x^3 + 5}$

14. $f(x) = \sqrt[5]{\dfrac{3x - 1}{x - 2}}$

In Exercises 15–20, use the Round-Trip Theorem to show that g is the inverse function of f.

15. $f(x) = x + 1; \qquad g(x) = x - 1$

16. $f(x) = 2x - 6; \qquad g(x) = \dfrac{x}{2} + 3$

17. $f(x) = \dfrac{1}{x + 1}; \qquad g(x) = \dfrac{1 - x}{x}$

18. $f(x) = \dfrac{-3}{2x + 5}; \qquad g(x) = \dfrac{-3 - 5x}{2x}$

19. $f(x) = x^5; \qquad g(x) = \sqrt[5]{x}$

20. $f(x) = 5^x; \qquad g(x) = \log_5 x$

In Exercises 21–28, use a calculator and the Horizontal Line Test to determine whether or not the function is one-to-one.

21. $f(x) = x^4 - 4x^2 + 3$

22. $f(x) = x^4 - 4x + 3$

23. $f(x) = x^3 + x - 5$

24. $f(x) = \begin{cases} x - 3 & \text{for } x \leq 3 \\ 2x - 6 & \text{for } x > 3 \end{cases}$

25. $f(x) = x^5 + 2x^4 - x^2 + 4x - 5$

26. $f(x) = x^3 - 4x^2 + x - 10$

27. $f(x) = .1x^3 - .1x^2 - .005x + 1$

28. $f(x) = .1x^3 + .005x + 1$

29. Show that the inverse function of the function f whose rule is $f(x) = \dfrac{2x + 1}{3x - 2}$ is f itself.

30. List three different functions (other than the one in Exercise 29), each of which is its own inverse. [Many correct answers are possible.]

In Exercises 31 and 32, the graph of a function f is given. Sketch the graph of the inverse function of f. [Reflect carefully.]

31.

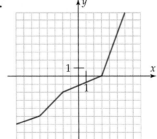

32.

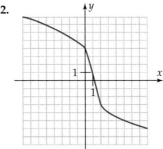

In Exercises 33–38, each given function has an inverse function. Sketch the graph of the inverse function.

33. $f(x) = \sqrt{x + 3}$

34. $f(x) = \sqrt{3x - 2}$

35. $f(x) = .3x^5 + 2$

36. $f(x) = \sqrt[3]{x + 3}$

37. $f(x) = \sqrt[5]{x^3 + x - 2}$

38. $f(x) = \begin{cases} x^2 - 1 & \text{for } x \leq 0 \\ -.5x - 1 & \text{for } x > 0 \end{cases}$

In Exercises 39–46, none of the functions has an inverse. State at least one way of restricting the domain of the function (that is, find a function with the same rule and a smaller domain) so that the restricted function has an inverse. Then find the rule of the inverse function.

Example: $f(x) = x^2$ has no inverse. But the function h with domain all $x \geq 0$ and rule $h(x) = x^2$ is increasing (its graph is the right half of the graph of f—see Figure 5–40) and therefore has an inverse.

39. $f(x) = |x|$

40. $f(x) = |x - 3|$

41. $f(x) = -x^2$

42. $f(x) = x^2 + 4$

43. $f(x) = \dfrac{x^2 + 6}{2}$

44. $f(x) = \sqrt{4 - x^2}$

45. $f(x) = \dfrac{1}{x^2 + 1}$

46. $f(x) = 3(x + 5)^2 + 2$

47. **(a)** Using the f^{-1} notation for inverse functions, find $f^{-1}(x)$ when $f(x) = 3x + 2$.

(b) Find $f^{-1}(1)$ and $1/f(1)$. Conclude that f^{-1} is not the same function as $1/f$.

48. Let C be the temperature in degrees Celsius. Then the temperature in degrees Fahrenheit is given by $f(C) = \frac{9}{5}C + 32$. Let g be the function that converts degrees Fahrenheit to degrees Celsius. Show that g is the inverse function of f and find the rule of g.

Thinkers

49. Let m and b be constants with $m \neq 0$. Show that the function $f(x) = mx + b$ has an inverse function g and find the rule of g.

50. Prove that the function $f(x) = 1 - .2x^3$ of Example 7(c) is one-to-one by showing that it satisfies the definition on page 414, namely, that

$$\text{If } a \neq b, \text{ then } f(a) \neq f(b).$$

[*Hint:* Use the rule of f to show that when $f(a) = f(b)$, then $a = b$. If this is the case, then it is impossible to have $f(a) = f(b)$ when $a \neq b$.]

51. Show that the points $P = (a, b)$ and $Q = (b, a)$ are symmetric with respect to the line $y = x$ as follows.

(a) Find the slope of the line through P and Q.

(b) Use slopes to show that the line through P and Q is perpendicular to $y = x$.

(c) Let R be the point where the line $y = x$ intersects line segment PQ. Since R is on $y = x$, it has coordinates (c, c) for some number c, as shown in the figure. Use the distance formula to show that segment PR has the same length as segment RQ. Conclude that the line $y = x$ is the perpendicular bisector of segment PQ. Therefore, P and Q are symmetric with respect to line $y = x$.

52. **(a)** Experiment with your calculator or use some of the preceding exercises to find four different increasing functions. For each function, sketch the graph of the function and the graphs of its inverse on the same set of axes.

(b) Based on the evidence in part (a), do you think the following statement true or false: The inverse function of every increasing function is also an increasing function.

(c) Do parts (a) and (b) with "increasing" replaced by "decreasing."

53. Suppose that functions f and g have these round-trip properties:

(1) $g(f(x)) = x$ for every x in the domain of f.

(2) $f(g(y)) = y$ for every y in the domain of g.

To complete the proof of the Round-Trip Theorem, we must show that g is the inverse function of f. Do this as follows.

(a) Prove that f is one-to-one by showing that

$$\text{if } \quad a \neq b, \quad \text{ then } \quad f(a) \neq f(b).$$

[*Hint:* If $f(a) = f(b)$, apply g to both sides and use **(1)** to show that $a = b$. Consequently, if $a \neq b$, it is impossible to have $f(a) = f(b)$.]

(b) If $g(y) = x$, show that $f(x) = y$. [*Hint:* Use **(2)**.]

(c) If $f(x) = y$, show that $g(y) = x$. [*Hint:* Use **(1)**.]

Parts (b) and (c) prove that

$$g(y) = x \quad \text{ exactly when } \quad f(x) = y.$$

Hence, g is the inverse function of f (see page 410).

54. Prove that every function f that has an inverse function g is one-to-one. [*Hint:* Page 412 shows that f and g have the round-trip properties; use Exercise 53(a).]

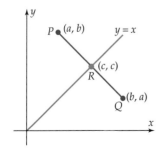

Chapter 5   *Review*

Important Concepts

**Important Facts
and Formulas**

· *Exponent Laws*

$$c^r c^s = c^{r+s} \qquad (cd)^r = c^r d^r$$

$$\frac{c^r}{c^s} = c^{r-s} \qquad \left(\frac{c}{d}\right)^r = \frac{c^r}{d^r}$$

$$(c^r)^s = c^{rs} \qquad c^{-r} = \frac{1}{c^r}$$

· *Logarithm Laws: For all $v, w > 0$ and any k:*

$$\ln(vw) = \ln v + \ln w \qquad \log_b(vw) = \log_b v + \log_b w$$

$$\ln\left(\frac{v}{w}\right) = \ln v - \ln w \qquad \log_b\left(\frac{v}{w}\right) = \log_b v - \log_b w$$

$$\ln(v^k) = k(\ln v) \qquad \log_b(v^k) = k(\log_b v)$$

· *Exponential Growth Functions*

$$f(x) = Pa^x \quad (a > 1) \qquad f(x) = Pe^{kx} \quad (k > 0)$$

$$f(x) = P(1 + r)^x \quad (0 < r < 1)$$

- Exponential Decay Functions
$$f(x) = Pa^x \quad (0 < a < 1) \qquad f(x) = Pe^{kx} \quad (k < 0)$$
$$f(x) = P(1 - r)^x \quad (0 < r < 1)$$
- Logistic Function: $f(x) = \dfrac{a}{1 + be^{-kx}}$
- Compound Interest Formula: $A = P(1 + r)^t$
- Radioactive Decay Function: $f(x) = P(.5^{x/h})$
- Change of Base Formula: $\log_b v = \dfrac{\ln v}{\ln b}$
- $g(x) = \log x$ is the inverse function of $f(x) = 10^x$.
- $g(x) = \ln x$ is the inverse function of $f(x) = e^x$.

Review Questions

In Questions 1–8, find a viewing window (or windows) that shows a complete graph of the function.

1. $f(x) = 3.7^x$

2. $g(x) = 2.5^{-x}$

3. $g(x) = 2^x - 1$

4. $f(x) = 2^{x-1}$

5. $f(x) = 2^{x^3 - x - 2}$

6. $g(x) = \dfrac{850}{1 + 5e^{-.4x}}$

7. $h(x) = \ln(x + 4) - 2$

8. $k(x) = \ln\left(\dfrac{x}{x - 2}\right)$

9. A computer software company claims the following function models the "learning curve" for their mathematical software:
$$P(t) = \dfrac{100}{1 + 48.2e^{-.52t}},$$
where t is measured in months and $P(t)$ is the average percent of the software program's capabilities mastered after t months.
   (a) Initially what percent of the program is mastered?
   (b) After six months what percent of the program is mastered?
   (c) Roughly, when can a person expect to "learn the most in the least amount of time"?
   (d) If the company's claim is true, how many months will it take to completely master the program?

10. Compunote has offered you a starting salary of $60,000 with $1000 yearly raises. Calcuplay offers you an initial salary of $30,000 and a guaranteed 6% raise each year.
   (a) Complete the following table for each company.

Year	Compunote
1	$60,000
2	$61,000
3	
4	
5	

Year	Calcuplay
1	$30,000
2	$31,800
3	
4	
5	

(b) For each company write a function that gives your salary in terms of years employed.

(c) If you plan on staying with the company for only five years, which job should you take to earn the most money?

(d) If you plan on staying with the company for 20 years, which is your best choice?

(e) In what year does the salary at Calcuplay exceed the salary at Compunote?

11. Phil borrows $800 at 9% interest, compounded annually.

(a) How much does he owe after six years?

(b) If he pays off the loan at the end of six years, how much interest will he owe?

12. If you invest $5000 for five years at 9% interest, how much more will you make if interest is compounded continuously than if it is compounded quarterly?

13. Mary Karen invests $2000 at 5.5% interest, compounded monthly.

(a) How much is her investment worth in three years?

(b) When will her investment be worth $12,000?

14. If a $2000 investment grows to $5000 in 14 years, with interest compounded annually, what is the interest rate?

15. Company sales are increasing at 6.5% per year. If sales this year are $56,000, write the rule of a function that gives the sales in year x (with $x = 0$ being the present year).

16. The population of Potterville is decreasing at an annual rate of 1.5%. If the population is now 38,500, what will the population be x years from now?

17. The half-life of carbon-14 is 5730 years. How much carbon-14 remains from an original 16-gram sample after 12,000 years?

18. How long will it take for 4 grams of carbon-14 to decay to 1 gram?

In Questions 19–24, translate the given exponential statement into an equivalent logarithmic one.

19. $e^{6.628} = 756$

20. $e^{5.8972} = 364$

21. $e^{r^2 - 1} = u + v$

22. $e^{a-b} = c$

23. $10^{2.8785} = 756$

24. $10^{c+d} = t$

In Questions 25–30, translate the given logarithmic statement into an equivalent exponential one.

25. $\ln 1234 = 7.118$

26. $\ln (ax + b) = y$

27. $\ln (rs) = t$

28. $\log 1234 = 3.0913$

29. $\log_5 (cd - k) = u$

30. $\log_d (uv) = w$

In Questions 31–34, evaluate the given expression without using a calculator.

31. $\ln e^3$

32. $\ln \sqrt[3]{e}$

33. $e^{\ln 3/4}$

34. $e^{\ln (x + 2y)}$

35. Simplify: $3 \ln \sqrt{x} + (1/2) \ln x$

36. Simplify: $\ln (e^{4e})^{-1} + 4e$

In Questions 37–39, write the given expression as a single logarithm.

37. $\ln 3x - 3 \ln x + \ln 3y$

38. $\log_7 7x + \log_7 y - 1$

39. $4 \ln x - 2(\ln x^3 + 4 \ln x)$

40. $\log (-.01) = ?$

41. $\log_{20} 400 = ?$

42. You are conducting an experiment about memory. The people who participate agree to take a test at the end of your course and every month thereafter for a period of two years. The average score for the group is given by the model $M(t) = 91 - 14 \ln (t + 1)$, $0 \le t \le 24$, where t is time in months after the first test.

(a) What is the average score on the initial exam?

(b) What is the average score after three months?

(c) When will the average drop below 50%?

(d) Is the magnitude of the rate of memory loss greater in the first month after the course (from $t = 0$ to $t = 1$) or after the first year (from $t = 12$ to $t = 13$)?

(e) Hypothetically, if the model could be extended past $t = 24$ months, would it be possible for the average score to be 0%?

43. Which of the following statements is *true*?

(a) $\ln 10 = (\ln 2)(\ln 5)$ (b) $\ln (e/6) = \ln e + \ln 6$

(c) $\ln (1/7) + \ln 7 = 0$ (d) $\ln (-e) = -1$

(e) None of the above are true.

44. Which of the following statements are *false*?

(a) $10 (\log 5) = \log 50$ (b) $\log 100 + 3 = \log 10^5$

(c) $\log 1 = \ln 1$ (d) $\log 6/\log 3 = \log 2$

(e) All of the above are false.

Use these graphs to answer Questions 45 and 46.

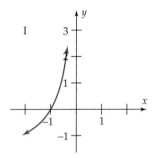

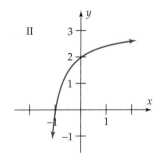

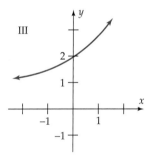

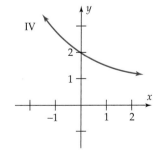

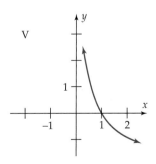

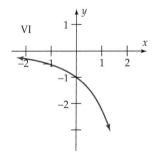

45. If $b > 1$, then the graph of $f(x) = -\log_b x$ could possibly be:

(a) I   (b) IV

(c) V   (d) VI

(e) None of these

46. If $0 < b < 1$, then the graph of $g(x) = b^x + 1$ could possibly be:
  (a) II   **(b)** III   **(c)** IV
  (d) VI   **(e)** None of these

47. If $\log_3 9^{x^2} = 4$, what is x?

48. What is the domain of the function $f(x) = \ln\left(\dfrac{x}{x-1}\right)$?

In Questions 49–57, solve the equation for x.

49. $8^x = 4^{x^2-3}$

50. $e^{3x} = 4$

51. $2 \cdot 4^x - 5 = -4$

52. $725e^{-4x} = 1500$

53. $u = c + d \ln x$

54. $2^x = 3^{x+3}$

55. $\ln x + \ln(3x - 5) = \ln 2$

56. $\ln(x + 8) - \ln x = 1$

57. $\log(x^3 - 1) = 2 + \log(x + 1)$

58. At a small community college the spread of a rumor for the population of 500 faculty and students can be modeled by:

$$\ln(n) - \ln(1000 - 2n) = .65t - \ln 998,$$

where n is the number of people who have heard the rumor after t days.
  (a) How many people know the rumor initially? (at $t = 0$)
  (b) How many people have heard the rumor after four days?
  (c) Roughly, in how many weeks will the entire population have heard the rumor?
  (d) Use the properties of logarithms to write n as a function of t; in other words, solve the model above for n in terms of t.
  (e) Enter the function you found in part (d) into your calculator and use the table feature to check your answers to parts (a), (b), and (c). Do they agree?
  (f) Now graph the function. Roughly over what time interval does the rumor seem to "spread" the fastest?

59. The half-life of polonium (^{210}Po) is 140 days. If you start with 10 milligrams, how much will be left at the end of a year?

60. An insect colony grows exponentially from 200 to 2000 in three months' time. How long will it take for the insect population to reach 50,000?

61. Hydrogen-3 decays at a rate of 5.59% per year. Find its half-life.

62. The half-life of radium-88 is 1590 years. How long will it take for 10 grams to decay to 1 gram?

63. How much money should be invested at 8% per year, compounded quarterly, to have $1000 in 10 years?

64. At what annual interest rate should you invest your money if you want to double it in six years?

65. One earthquake measures 4.6 on the Richter scale. A second earthquake is 1000 times more intense than the first. What does it measure on the Richter scale?

66. The table gives the population of Austin, TX.

Year	1950	1970	1980	1990	1994
Population	132,459	253,539	345,890	465,648	514,013

(a) Sketch a scatter plot of the data, with $x = 0$ corresponding to 1950.
(b) Find an exponential model for the data.
(c) Use the model to estimate the population of Austin in 1960 and 1985.

67. The wind-chill factor is the temperature that would produce the same cooling effect on a person's skin if there were no wind. The table shows the wind-chill factors for various wind speeds when the temperature is 35°F. [*Source:* National Weather Service]

Wind Speed (mph)	5	10	15	20	25	30	35	40	45
Wind-Chill Temperature (in °F)	33	22	16	12	8	6	4	3	2

(a) What does a 20 mph wind make 35°F feel like?
(b) Sketch a scatter plot of the data.
(c) Explain why an exponential model would be appropriate.

68. Find the inverse of the function $f(x) = 2x + 1$.

69. Find the inverse of the function $f(x) = \sqrt{5 - x} + 7$.

70. Find the inverse of the function $f(x) = \sqrt[5]{x^3 + 1}$.

71. The graph of a function f is shown in the figure. Sketch the graph of the inverse function of f.

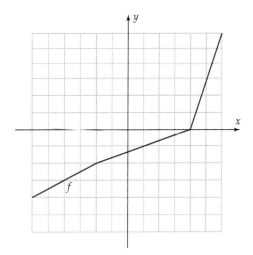

72. Which of the following functions have inverse functions (give reasons for your answers):
(a) $f(x) = x^3$ (b) $f(x) = 1 - x^2, \quad x \le 0$
(c) $f(x) = |x|$

In Exercises 73–75, determine whether or not the given function has an inverse function [give reasons for your answer]. If it does, find the graph of the inverse function.

73. $f(x) = 1/x$ 74. $f(x) = .02x^3 - .04x^2 + .6x - 4$

75. $f(x) = .2x^3 - 4x^2 + 6x - 15$

Discovery Project 5

Exponential and Logistic Modeling of Diseases

Diseases that are contagious and are transmitted homogeneously through a population often appear to be spreading exponentially. That is, the rate of spread is proportional to the number of people in the population who are already infected. This is a reasonable model as long as the number of infected people is relatively small compared with the number of people in the population who can be infected. The standard exponential model looks like this:

$$f(t) = Y_0 e^{rt}.$$

Y_0 is the initial number of infected people (the number on the arbitrarily decided day 0) and r is the rate by which the disease spreads through the population. If the time t is measured in days, then r is the ratio of new infections to current infections each day.

Suppose that in Big City, population 3855, there is an outbreak of dingbat disease. On the first Monday after the outbreak was discovered (day 0), 72 people have dingbat disease. On the following Monday (day 7), 193 people have dingbat disease.

1. Using the exponential model, Y_0 is clearly 72. Calculate the value of r.

2. Using your values of Y_0 and r, predict the number of cases of dingbat disease that will be reported on day 14.

It turns out that eventually the spread of disease must slow as the number of infected people approaches the number of susceptible people. What happens is that some of the people to whom the disease would spread are already infected. As time goes on, the spread of the disease becomes proportional to the number of susceptible and uninfected people. The disease then follows the logistic model:

$$g(t) = \frac{rY_0}{aY_0 + (r - aY_0)e^{-rt}}.$$

Y_0 is still the initial value, and r serves the same function as before, at least at the initial time. The extra parameter a is not so obvious, but it is inversely related to the number of people susceptible to the disease. Unfortunately, the algebra to solve for a is quite complicated. It is much easier to approximate a using the same r from the exponential model.

3. On day 14 in Big City, 481 people have dingbat disease. Using the values of Y_0 and r from Exercise 1 in the rule of the function g, determine the value of a. [*Hint:* $g(14) = 481$.] Does g overestimate or underestimate the number of people with dingbat disease on day 7?

4. Use the function g from Exercise 3 to approximate the number of people in Big City who are susceptible to the disease. Does this model make sense? [Remember, as time goes on, the number of people infected approaches the number of people susceptible.]

5. In the logistic model, the rate at which the disease spreads tends to fall over time. This means that the value of r you calculated in Exercise 1 is a little low. Raise the value of r and find the new value of a as in Exercise 3. Experiment until you find a value of r for which $g(7) = 193$ (meaning that the model g matches the data on day 7).

6. Using the function g from Exercise 5, repeat Exercise 4.

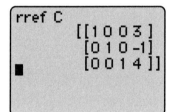

[[1 0 0 3]
[0 1 0 -1]
[0 0 1 4]]

Chapter 6

Systems of Equations and Inequalities

Is this a diamond in the rough?

The structure of certain crystals can be described by a large system of linear equations (more than a hundred equations and variables). A variety of resource allocation problems involving many variables can be handled by solving an appropriate system of equations. The fastest solution methods involve matrices and are easily implemented on a computer or calculator. See Exercise 53 on page 463.

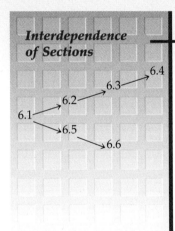

Interdependence of Sections

6.1 → 6.2 → 6.3 → 6.4
6.1 → 6.5 → 6.6

Chapter Outline

This chapter deals primarily with *systems* of equations and inequalities, such as

$$2x - 5y + 3z = 1 \qquad\qquad 2x + 5y + z + w = 0 \qquad\qquad 2x + y > 0$$
$$x + 2y - z = 2 \qquad\qquad\quad 2y - 4z + 41w = 5 \qquad\qquad x - y < -3$$
$$3x + y + 2z = 11 \qquad\quad 3x + 7y + 5z - 8w = -6$$

| *Three equations in* | *Three equations in* | *Two inequalities in* |
| *three variables* | *four variables* | *two variables* |

A *solution of a system* is a solution that satisfies all the equations or inequalities in the system. For instance, in the first system of equations above, $x = 1$, $y = 2$, $z = 3$ is a solution of all three equations (check it!) and hence is a solution for the system. On the other hand, $x = 0$, $y = 7$, $z = 12$ is a solution of the first two equations, but not the third (verify!), and hence, is not a solution of the system.

We also consider matrices, which are used for solving systems of equations and other purposes, and linear programming, which is a widely used method for dealing with optimization problems involving several variables.

6.1 Systems of Linear Equations in Two Variables

Systems of linear equations in two variables may be solved graphically or algebraically. The graphical method is similar to what we have done previously.

Example 1 Solve the following system graphically.

$$2x - y = 1$$
$$3x + 2y = 4$$

Solution First, we solve each equation for y:

$$2x - y = 1 \qquad\qquad 3x + 2y = 4$$
$$-y = -2x + 1 \qquad\qquad 2y = -3x + 4$$
$$y = 2x - 1 \qquad\qquad y = \frac{-3x + 4}{2}.$$

Next, we graph both equations on the same screen (Figure 6–1). As we saw in Section 1.4, each graph is a straight line and every point on the graph represents a solution of the equation. Therefore, the solution of the system is given by the coordinates of the point that lies on both lines. An intersection finder (Figure 6–2) shows that the approximate coordinates of this point are

$$x \approx .85714286 \qquad \text{and} \qquad y \approx .71428571. \quad \blacksquare$$

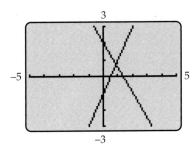

Figure 6–1   **Figure 6–2**

As shown in Example 1, the solutions of a system of linear equations are determined by the points where their graphs intersect. There are exactly three geometric possibilities for two lines in the plane: They are parallel, they intersect at a single point, or they coincide, as illustrated in Figure 6–3. Each of these possibilities leads to a different number of solutions for the system, as summarized here and illustrated on the next page.

Number of Solutions of a System

A system of two linear equations in two variables must have

No solutions *or*

Exactly one solution *or*

An infinite number of solutions.

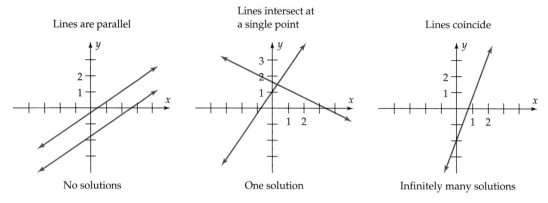

Figure 6–3

The Substitution Method

When you use a calculator to solve systems graphically, you may have to settle for an approximate solution, as we did in Example 1. Algebraic methods, however, produce exact solutions. Furthermore, algebraic methods are often as easy to implement as graphical ones, so we shall use them, whenever practical, to obtain exact solutions. One algebraic method is **substitution,** which is explained in the next example.

Example 2   Use substitution to find the exact solution of the system in Example 1:

$$2x - y = 1$$
$$3x + 2y = 4.$$

Solution   Any solution of this system must satisfy the first equation: $2x - y = 1$. Solving this equation for y, as in Example 1, shows that

$$y = 2x - 1.$$

Substituting this expression for y in the second equation, we have

$$3x + 2y = 4$$
$$3x + 2(2x - 1) = 4$$
$$3x + 4x - 2 = 4$$
$$7x = 6$$
$$x = 6/7.$$

Therefore, every solution of the original system must have $x = 6/7$. But when $x = 6/7$, we see from the first equation that

$$2x - y = 1$$
$$2\left(\frac{6}{7}\right) - y = 1$$

$$\frac{12}{7} - y = 1$$

$$-y = -\frac{12}{7} + 1$$

$$y = \frac{12}{7} - 1 = \frac{5}{7}.$$

(We would also have found that $y = 5/7$ if we had substituted $x = 6/7$ in the second equation.) Consequently, the exact solution of the original system is $x = 6/7$, $y = 5/7$. ∎

When using the substitution method, you may solve either of the given equations for either one of the variables and then substitute that result in the other equation. In Example 2, we solved for y in the first equation because that avoided the fractional expressions that would have occurred if we had solved for x or had solved the second equation for x or y.

CAUTION

To guard against arithmetic mistakes, you should always *check your answers* by substituting them into *all* the equations of the original system. We have in fact checked the answers in all the examples, but these checks are omitted to save space.

The Elimination Method

The **elimination method** of solving systems of linear equations is often more convenient than substitution. It depends on this fact:

> **Multiplying both sides of an equation by a nonzero constant does not change the solutions of the equation.**

For example, the equation $x + 3 = 5$ has the same solution as $2x + 6 = 10$ (the first equation multiplied by 2). The elimination method also uses this fact from basic algebra:

If $A = B$ and $C = D$, then $A + C = B + D$ and $A - C = B - D$.

Example 3 To solve the system

$$x - 3y = 4$$
$$2x + y = 1$$

we first replace the first equation by an equivalent one (that is, one with the same solutions):

$$-2x + 6y = -8 \qquad \textit{[First equation multiplied by -2]}$$
$$2x + y = 1.$$

The multiplier -2 was chosen so that the coefficients of x in the two equations would be negatives of each other. Any solution of this last system must also be a solution of the sum of the two equations:

$$-2x + 6y = -8$$
$$\underline{2x + y = 1}$$
$$7y = -7. \qquad \textit{[The first variable has been eliminated.]}$$

Solving this last equation we see that $y = -1$. Substituting this value in the first of the original equations shows that

$$x - 3(-1) = 4$$

$$x = 1.$$

Therefore, $x = 1$, $y = -1$ is the solution of the original system.   ■

Example 4   Any solution of the system

$$5x - 3y = 3$$

$$3x - 2y = 1$$

must also be a solution of this system:

$$10x - 6y = 6 \qquad \textit{[First equation multiplied by 2]}$$

$$-9x + 6y = -3. \qquad \textit{[Second equation multiplied by } -3\textit{]}$$

The multipliers 2 and -3 were chosen so that the coefficients of y in the new equations would be negatives of each other. Any solution of this last system must also be a solution of the equation obtained by adding these two equations:

$$\begin{array}{rcr} 10x - 6y = & 6 \\ -9x + 6y = & -3 \\ \hline x = & 3. \end{array} \qquad \textit{[The second variable has been eliminated.]}$$

Substituting $x = 3$ in the first of the original equations shows that

$$5(3) - 3y = 3$$

$$-3y = -12$$

$$y = 4.$$

Therefore the solution of the original system is $x = 3$, $y = 4$.   ■

Example 5   To solve the system

$$2x - 3y = 5$$

$$4x - 6y = 1$$

we multiply the first equation by -2 and add:

$$\begin{array}{rcr} -4x + 6y = & -10 \\ 4x - 6y = & 1 \\ \hline 0 = & -9. \end{array}$$

Since $0 = -9$ is always false, the original system cannot possibly have any solutions. A system with no solutions is said to be **inconsistent.**   ■

> **GRAPHING EXPLORATION**
>
> Confirm the result of Example 5 geometrically by graphing the two equations in the system. Do these lines intersect or are they parallel?

Example 6 To solve the system

$$3x - y = 2$$
$$6x - 2y = 4$$

we multiply the first equation by 2 to obtain the system:

$$6x - 2y = 4$$
$$6x - 2y = 4.$$

The two equations are identical. So the solutions of this system are the same as the solutions of the single equation $6x - 2y = 4$, which can be rewritten as:

$$2y = 6x - 4$$
$$y = 3x - 2.$$

This equation, and hence the original system, has infinitely many solutions. They can be described as follows: Choose any real number for x, say $x = b$. Then $y = 3x - 2 = 3b - 2$. So the solutions of the system are all pairs of numbers of the form

$$x = b, \quad y = 3b - 2 \quad \text{where } b \text{ is any real number.}$$

A system such as this is said to be **dependent.** ■

Applications

Producers are happy to supply items at a high price, but if they do, the demand for them by consumers may be low. At lower prices, more items will be demanded, but if the price is too low, producers may not want to supply the items. The **equilibrium price** is the price at which the number of items demanded by consumers is the same as the number supplied by producers. The number of items demanded and supplied at the equilibrium price is the **equilibrium quantity.**

Example 7 The consumer demand for a certain type of floor tile is related to its price by the equation $p = 60 - .75x$, where p is the price (in dollars) at which x thousand boxes of tile will be demanded. The supply of these tiles is related to their price by the equation $p = .8x + 5.75$, where p is the price at which x thousand boxes will be supplied by the producer. Find the equilibrium quantity and the equilibrium price.

Solution We must find the values of x and p that satisfy both the supply and demand equations. In other words, we must solve this system:

$$p = 60 - .75x$$
$$p = .8x + 5.75.$$

Algebraic Method. Since both equations are already solved for p, we use substitution. Substituting the value of p given by the first equation into the second we obtain

$$60 - .75x = .8x + 5.75$$

Subtract .8x from both sides: $\qquad 60 - 1.55x = 5.75$

Subtract 60 from both sides: $\qquad -1.55x = -54.25$

Divide both sides by -1.55: $\qquad x = \dfrac{-54.25}{-1.55} = 35.$

We can determine p by substituting $x = 35$ in either equation, say the first one.

$$p = 60 - .75x$$
$$p = 60 - .75(35) = 33.75$$

Therefore, the equilibrium quantity is 35,000 boxes (x is measured in thousands) and the equilibrium price is $33.75 per box.

Graphical Method. We graph the two equations in the form $y = 60 - .75x$ and $y = .8x + 5.75$ on the same screen and find their intersection point (Figure 6–4). The intersection point (35, 33.75) is called the **equilibrium point**. Its first coordinate is the equilibrium quantity and its second coordinate, the equilibrium price. ∎

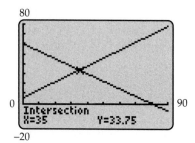

Figure 6–4

Example 8 575 people attend a ball game and total ticket sales are $2575. If adult tickets cost $5 and children's tickets $3, how many adults attended the game? How many children?

Solution Let x be the number of adults and y the number of children. Then,

Number of adults + Number of children = Total attendance

$$x + y = 575.$$

We can obtain a second equation by using the information about ticket sales:

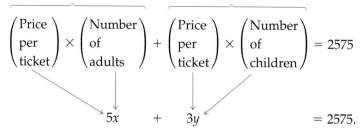

To find x and y we need only solve this system of equations:

$$x + y = 575$$
$$5x + 3y = 2575$$

Multiplying the first equation by -3 and adding we have:

$$-3x - 3y = -1725$$
$$\underline{5x + 3y = 2575}$$
$$2x = 850$$
$$x = 425$$

So, 425 adults attended the game. The number of children was

$$y = 575 - x = 575 - 425 = 150. \quad \blacksquare$$

Example 9 A plane flies 3000 miles from San Francisco to Boston in 5 hours, with a tailwind all the way. The return trip on the same route, now with a headwind, takes 6 hours. Assuming both remain constant, find the speed of the plane and the speed of the wind.

Solution Let x be the plane's speed and y the wind speed (both in miles per hour). Then on the trip to Boston with a tailwind,

$x + y$ = actual speed of the plane (wind and plane go in same direction).

On the return trip against a headwind,

$x - y$ = actual speed of plane (wind and plane go in opposite directions).

Using the basic rate/distance equation we have:

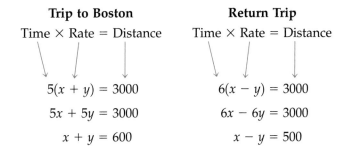

Trip to Boston	**Return Trip**
Time $\times$ Rate = Distance	Time $\times$ Rate = Distance
$5(x + y) = 3000$	$6(x - y) = 3000$
$5x + 5y = 3000$	$6x - 6y = 3000$
$x + y = 600$	$x - y = 500$

Thus, we need only solve this system of equations:

$$x + y = 600$$
$$x - y = 500.$$

Adding the two equations shows that

$$2x = 1100$$
$$x = 550$$

Substituting this result in the first equation, we have

$$550 + y = 600$$

$$y = 50$$

Thus, the plane's speed is 550 mph and the wind speed is 50 mph. ∎

Exercises 6.1

In Exercises 1–6, determine whether the given values of x, y, and z are a solution of the system of equations.

1. $x = -1, y = 3$

$$2x + y = 1$$
$$-3x + 2y = 9$$

2. $x = 3, y = 4$

$$2x + 6y = 30$$
$$x + 2y = 11$$

3. $x = 2, y = -1$

$$\frac{1}{3}x + \frac{1}{2}y = \frac{1}{6}$$
$$\frac{1}{2}x + \frac{1}{3}y = \frac{2}{3}$$

4. $x = .4, \quad y = .7$

$$3.1x - 2y = -.16$$
$$5x - 3.5y = -.48$$

5. $x = \frac{1}{2}, y = 3, z = -1$

$$2x - y + 4z = -6$$
$$3y + 3z = 6$$
$$2z = 2$$

6. $x = 2, y = \frac{3}{2}, z = -\frac{1}{2}$

$$3x + 4y - 2z = 13$$
$$\frac{1}{2}x + 8z = -3$$
$$x - 3y + 5z = -5$$

In Exercises 7–14, use substitution to solve the system.

7. $x - 2y = 5$
  $2x + y = 3$

8. $3x - y = 1$
  $-x + 2y = 4$

9. $3x - 2y = 4$
  $2x + y = -1$

10. $5x - 3y = -2$
  $-x + 2y = 3$

11. $r + s = 0$
  $r - s = 5$

12. $t = 3u + 5$
  $t = u + 5$

13. $x + y = c + d$ (where c, d are constants)
  $x - y = 2c - d$

14. $x + 3y = c - d$ (where c, d are constants)
  $2x - y = c + d$

In Exercises 15–32, use the elimination method to solve the system.

15. $2x - 2y = 12$
  $-2x + 3y = 10$

16. $3x + 2y = -4$
  $4x - 2y = -10$

17. $x + 3y = -1$
  $2x - y = 5$

18. $4x - 3y = -1$
  $x + 2y = 19$

19. $2x + 3y = 15$
  $8x + 12y = 40$

20. $2x + 5y = 8$
  $6x + 15y = 18$

21. $3x - 2y = 4$
  $6x - 4y = 8$

22. $2x - 8y = 2$
  $3x - 12y = 3$

23. $12x - 16y = 8$
  $42x - 56y = 28$

24. $\frac{1}{3}x + \frac{2}{5}y = \frac{1}{6}$
  $20x + 24y = 10$

[*Hint:* First, eliminate fractions by multiplying both sides of the first equation by 30.]

25. $9x - 3y = 1$
  $6x - 2y = -5$

26. $8x + 4y = 3$
  $10x + 5y = 1$

27. $\frac{x}{3} - \frac{y}{2} = -3$
  $\frac{2x}{5} + \frac{y}{5} = -2$

28. $\frac{x}{3} + \frac{3y}{5} = 4$
  $\frac{x}{6} - \frac{y}{2} = -3$

29. $\frac{x + y}{4} - \frac{x - y}{3} = 1$
  $\frac{x + y}{4} + \frac{x - y}{2} = 9$

30. $\frac{x - y}{4} + \frac{x + y}{3} = 1$
  $\frac{x + 2y}{3} + \frac{3x - y}{2} = -2$

31. $ax + by = r$ (where a, b, c, d, r, s are
  $cx + dy = s$ constants and $ad - bc \neq 0$)

32. $ax + by = ab$ (where a, b are nonzero
  $bx - ay = ab$ constants)

In Exercises 33–36, find the equilibrium quantity and the equilibrium price. In the supply and demand equations, p is price (in dollars) and x is quantity (in thousands).

33. Supply: $p = .75x$
  Demand: $p = 16 - 1.25x$

34. Supply: $p = 1.4x - .6$
  Demand: $p = -2x + 3.2$

35. Supply: $p = 300 - 30x$
Demand: $p = 80 + 25x$

36. Supply: $p = 181 - .01x$
Demand: $p = 52 + .02x$

37. Let c be any real number. Show that the following system has exactly one solution.

$$x + 2y = c$$
$$6x - 3y = 4$$

[*Hint:* The graph of each equation is a straight line. What are their slopes? See Figure 6–3.]

38. (a) Find the values of c for which this system has an infinite number of solutions.

$$2x - 4y = 6$$
$$-3x + 6y = c$$

(b) Find the values of c for which the system in part (a) has no solutions.

In Exercises 39 and 40, find the values of c and d for which both given points lie on the given straight line.

39. $cx + dy = 2$; $(0, 4)$ and $(2, 16)$

40. $cx + dy = -6$; $(1, 3)$ and $(-2, 12)$

*In Exercises 41–42, R is the revenue (in dollars) from the sale of x units and C is the cost (in dollars) of producing x units. Find the **break-even point,** that is, the point where revenue equals cost.*

41. $R = 35x$ and $C = 28x + 840$

42. $R = 200x$ and $C = 150x + 18,000$

43. The cost C of making x hedge trimmers is given by $C = 45x + 6000$. Each hedge trimmer can be sold for $60.
(a) Find an equation that expresses the revenue R from selling x hedge trimmers.
(b) How many hedge trimmers must be sold for the company to break even?

44. The cost C of making x cases of pet food is given by $C = 100x + 2600$. Each case of 100 boxes sells for $120.
(a) Find an equation that expresses the revenue from selling x cases.
(b) How many cases must be sold for the company to break even?

In Exercises 45–48, you are the manager of a firm that must determine what it will take for a new product to break even and decide whether or not to produce the product.

45. The marketing department estimates that no more than 450 units can be sold. Costs are expected to be given by $C = 80x + 7500$ and revenues by $R = 95x$.

46. Costs are given by $C = 1750x + 95,175$ and revenue by $R = 1975x$; no more than 800 units can be sold.

47. Revenue is given by $R = 155x$ and costs by $C = 125x + 42,000$; no more than 2000 units can be sold.

48. Costs are given by $C = 140x + 3000$ and revenue by $R = 150x$; no more than 250 units can be sold.

49. A 200-seat children's theater charges $3 for adults and $1.50 for children. At a recent performance, all seats were filled and the total ticket income was $510. How many adults and how many children were in the audience?

50. A theater charges $4 for main-floor seats and $2.50 for balcony seats. If all seats are sold, the ticket income is $2100. At one show, 25% of the main-floor seats and 40% of the balcony seats were sold and ticket income was $600. How many seats are on the main floor and how many in the balcony?

51. A boat made a 4-mile trip upstream against a constant current in 15 minutes. The return trip at the same constant speed with the same current took 12 minutes. What is the speed of the boat and of the current?

52. A plane flying into a headwind travels 2000 miles in 4 hours and 24 minutes. The return flight along the same route with a tailwind takes 4 hours. Find the wind speed and the plane's speed (assuming both are constant).

53. Bill and Ann plan to install a heating system for their swimming pool. Since gas is not available, they have a choice of electric or solar heat. They have gathered the following cost information.

System	Installation Costs	Monthly Operational Cost
Electric	$2,000	$80
Solar	$14,000	$ 9.50

(a) Ignoring changes in fuel prices, write a linear equation for each heating system that expresses its total cost y in terms of the number of *years* x of operation.
(b) What is the five-year total cost of electric heat? Of solar heat?
(c) In what year will the total cost of the two heating systems be the same? Which is the cheapest system before that time? After that time?

54. One parcel of land is worth $100,000 now and is increasing in value at the rate of $3000 per year. A second parcel is now worth $60,000 and is increasing in value at the rate of $7500 per year.
  (a) For each parcel of land write an equation that expresses the value y of the land in year x.
  (b) When will the two parcels be worth the same amount?

55. A toy company makes Boomie Babies, as well as collector cases for each Boomie Baby. To make x cases costs the company $5000 in fixed overhead, plus $7.50 per case. An outside supplier has offered to produce any desired volume of cases for $8.20 per case.
  (a) Write an equation that expresses the company's cost to make x cases itself.
  (b) Write an equation that expresses the cost of buying x cases from the outside supplier.
  (c) Graph both equations on the same axes and determine when the two costs are the same.
  (d) When should the company make the cases themselves and when should they buy them from the outside supplier?

56. The sum of two numbers is 40. The difference between twice the first number and the second is 11. What are the numbers? [*Hint:* If the numbers are x and y, then $x + y = 40$. Use the other information to obtain a second equation and solve the resulting system.]

57. An investor has part of her money in an account that pays 9% annual interest, and the rest in an account that pays 11% annual interest. If she has $8000 less in the higher paying account than in the lower paying one and her total annual interest income is $2010, how much does she have invested in each account?

58. Joyce has money in two investment funds. Last year the first fund paid a dividend of 8% and the second a dividend of 2% and Joyce received a total of $780. This year the first fund paid a 10% dividend and the second only 1% and Joyce received $810. How much money does she have invested in each fund?

59. At a certain store, cashews cost $4.40/lb and peanuts $1.20/lb. If you want to buy exactly 3 lb of nuts for $6.00, how many pounds of each kind of nut should you buy? [*Hint:* If you buy x pounds of cashews and y pounds of peanuts, then $x + y = 3$. Find a second equation by considering cost and solve the resulting system.]

60. A store sells deluxe tape recorders for $150. The regular model costs $120. The total tape recorder inventory would sell for $43,800. But during a recent month the store actually sold half of its deluxe models and two-thirds of the regular models and took in a total of $26,700. How many of each kind of recorder did they have at the beginning of the month?

61. A winemaker has two large casks of wine. One wine is 8% alcohol and the other is 18% alcohol. How many liters of each wine should be mixed to produce 30 liters of wine that is 12% alcohol? [*Hint:* If x is the number of liters of the 8% wine and y the number of liters of the 18% wine, then $x + y = ?$ Use the fact that the amount of alcohol in the mixture is "8% of $x + 18\%$ of y" to find a second equation.]

62. How many cubic centimeters of a solution that is 20% acid and of another solution that is 45% acid should be mixed to produce 100 cubic centimeters of a solution that is 30% acid?

63. How many grams of a 50%-silver alloy should be mixed with a 75%-silver alloy to obtain 40 grams of a 60%-silver alloy?

64. A machine in a pottery factory takes 3 minutes to form a bowl and 2 minutes to form a plate. The material for a bowl costs .25 and the material for a plate costs .20. If the machine runs for 8 hours straight and exactly $44 is spent for material, how many bowls and plates can be produced?

Thinker

65. Because Chevrolet and Saturn produce cars in the same price range, Chevrolet's sales are not only a function of Chevy prices (x), but of Saturn prices (y) as well. Saturn prices are related similarly to both Saturn and Chevy prices. Suppose General Motors forecasts the demand z_1 for Chevrolets and the demand z_2 for Saturns to be given by

$$z_1 = 68{,}000 - 6x + 4y \quad \text{and} \quad z_2 = 42{,}000 + 3x - 3y.$$

Solve this system of equations and express
  (a) the price x of Chevrolets in terms of z_1 and z_2.
  (b) the price y of Saturns in terms of z_1 and z_2.

EXCURSION **Systems of Nonlinear Equations**

Some systems that include nonlinear equations can be solved algebraically by using substitution.

Example 1 Solve the system

$$-2x + y = -1$$
$$xy = 3.$$

Solution We solve the first for y

$$y = 2x - 1$$

and substitute this into the second equation:

$$xy = 3$$
$$x(2x - 1) = 3$$
$$2x^2 - x = 3$$
$$2x^2 - x - 3 = 0$$
$$(2x - 3)(x + 1) = 0$$

$$2x - 3 = 0 \qquad \text{or} \qquad x + 1 = 0$$
$$x = 3/2 \qquad\qquad x = -1$$

Using the equation $y = 2x - 1$ to find the corresponding values of y, we see that:

$$\text{If } x = \frac{3}{2}, \qquad \text{then } y = 2\left(\frac{3}{2}\right) - 1 = 2.$$

$$\text{If } x = -1, \qquad \text{then } y = 2(-1) - 1 = -3.$$

Therefore, the solutions of the system are $x = 3/2$, $y = 2$ and $x = -1$, $y = -3$. ∎

Example 2 Solve the system

$$x^2 + y^2 = 8$$
$$x^2 - y = 6.$$

Solution Solve the second equation for y, obtaining $y = x^2 - 6$, and substitute this into the first equation:

$$x^2 + y^2 = 8$$
$$x^2 + (x^2 - 6)^2 = 8$$
$$x^2 + x^4 - 12x^2 + 36 = 8$$
$$x^4 - 11x^2 + 28 = 0$$
$$(x^2 - 4)(x^2 - 7) = 0$$

$$x^2 - 4 = 0 \qquad \text{or} \qquad x^2 - 7 = 0$$
$$x^2 = 4 \qquad\qquad x^2 = 7$$
$$x = \pm 2 \qquad\qquad x = \pm\sqrt{7}$$

Using the equation $y = x^2 - 6$ to find the corresponding values of y, we find that the solutions of the system are

$$x = 2, y = -2; \qquad x = -2, y = -2; \qquad x = \sqrt{7}, y = 1;$$
$$x = -\sqrt{7}, y = 1. \quad \blacksquare$$

Algebraic techniques were successful in Examples 1 and 2 because substitution led to equations whose solutions could be found exactly. When this is not the case, graphical methods are needed.

Example 3 Solve the system

$$y = x^4 - 4x^3 + 9x - 1$$
$$y = 3x^2 - 3x - 7.$$

Solution Graph both equations on the same screen. In the viewing window of Figure 6–5, the graphs intersect at three points. However, the graphs seem to be getting closer together as they run off the screen at the top right, which suggests that there may be another intersection point.

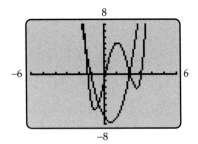

Figure 6–5

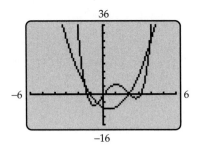

Figure 6–6

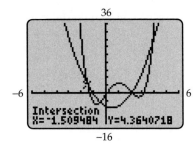

Figure 6–7

The larger window in Figure 6–6 shows four intersection points. An intersection finder (Figure 6–7) shows that one of the approximate solutions of the system is

$$x = -1.509484, \qquad y = 4.3640718.$$

> ### GRAPHING EXPLORATION
>
> Graph the two equations in the viewing window of Figure 6–6 and use your intersection finder to approximate the other three solutions of this system. ■

Example 4 Solve this system graphically:

$$x^2 - 4x - y + 1 = 0$$
$$10x^2 + 25y^2 = 100.$$

Solution It's easy to graph the first equation since it can be rewritten as $y = x^2 - 4x + 1$. To graph the second equation, we must first solve for y:

$$10x^2 + 25y^2 = 100$$
$$25y^2 = 100 - 10x^2$$
$$y^2 = \frac{100 - 10x^2}{25}$$
$$y = \sqrt{\frac{100 - 10x^2}{25}} \qquad \text{or} \qquad y = -\sqrt{\frac{100 - 10x^2}{25}}$$

Technology Tip

On Sharp 9600 and TI calculators, you can graph both of these equations at once by keying in

$$y = \{1, -1\}\frac{\sqrt{100 - 10x^2}}{25}.$$

Each of these preceding equations can be graphed on a calculator (see the Tip in the margin). By graphing both equations on the same screen, we obtain the complete graph of $10x^2 + 25y^2 = 100$, as shown in Figure 6–8. The graphs of both equations of the original system are shown in Figure 6–9. An intersection finder shows that the approximate intersection points (solutions of the system) are

$$x = -.2348, y = 1.9945 \qquad \text{and} \qquad x = .9544, y = -1.9067. \quad ■$$

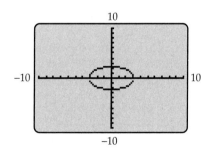

Figure 6–8

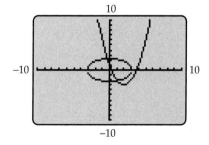

Figure 6–9

Example 5 A 52-ft-long piece of wire is to be cut into three pieces, two of which are the same length. The two equal pieces are to be bent into squares and the third piece into a circle. What should the length of each piece be if the total area enclosed by the two squares and the circle is 100 square feet?

Solution   Let x be the length of each piece of wire that is to be bent into a square and y the length of the piece that is to be bent into a circle. Since the original wire is 52 ft long,

$$x + x + y = 52, \qquad \text{or equivalently,} \qquad y = 52 - 2x.$$

If a piece of wire of length x is bent into a square, the side of the square will have length $x/4$ and hence the area of the square will be $(x/4)^2 = x^2/16$. The remaining piece of wire will be made into a circle of circumference (length) y. Since the circumference is 2π times the radius (that is, $y = 2\pi r$), the circle has radius $r = y/2\pi$. Therefore, the area of the circle is

$$\pi r^2 = \pi\left(\frac{y}{2\pi}\right)^2 = \frac{\pi y^2}{4\pi^2} = \frac{y^2}{4\pi}.$$

The sum of the areas of the two squares and the circle is 100, that is,

$$\frac{x^2}{16} + \frac{x^2}{16} + \frac{y^2}{4\pi} = 100$$

$$\frac{y^2}{4\pi} = 100 - \frac{2x^2}{16}$$

$$y^2 = 4\pi\left(100 - \frac{x^2}{8}\right).$$

$$y = \sqrt{4\pi\left(100 - \frac{x^2}{8}\right)} \qquad \text{or} \qquad y = -\sqrt{4\pi\left(100 - \frac{x^2}{8}\right)}$$

Since x and y must both be positive (because they are lengths), we must solve this system of equations:

$$y = 52 - 2x$$

$$y = \sqrt{4\pi\left(100 - \frac{x^2}{8}\right)}$$

We solve it by graphing both equations on the same screen and finding their intersection point. Figure 6–10 shows that its coordinates are $x \approx 9.25$ and $y \approx 33.50$. Therefore, the wire should be cut into two 9.25-ft pieces and one 33.5-ft piece. ■

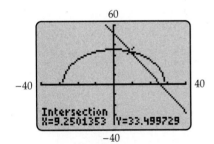

60

−40   40

Intersection
X=9.2501353 Y=33.499729

−40

Figure 6–10

Exercises 6.1.A

In Exercises 1–12, solve the system algebraically.

1. $x^2 - y = 0$
   $-2x + y = 3$

2. $x^2 - y = 0$
   $-3x + y = -2$

3. $x^2 - y = 0$
   $x + 3y = 6$

4. $x^2 - y = 0$
   $x + 4y = 4$

5. $x + y = 10$
   $xy = 21$

6. $2x + y = 4$
   $xy = 2$

7. $xy + 2y^2 = 8$
   $x - 2y = 4$

8. $xy + 4x^2 = 3$
   $3x + y = 2$

9. $x^2 + y^2 - 4x - 4y = -4$
$x - y = 2$

10. $x^2 + y^2 - 4x - 2y = -1$
$x + 2y = 2$

11. $x^2 + y^2 = 25$
$x^2 + y = 19$

12. $x^2 + y^2 = 1$
$x^2 - y = 5$

In Exercises 13–20, solve the system by any means.

13. $y = x^3 - 3x^2 + 4$
$y = -.5x^2 + 3x - 2$

14. $y = -x^3 + 3x^2 + x - 3$
$y = -2x^2 + 5$

15. $y = x^3 - 3x + 2$
$y = \dfrac{3}{x^2 + 3}$

16. $y = .25x^4 - 2x^2 + 4$
$y = x^3 - x^2 - 2x + 1$

17. $25x^2 - 16y^2 = 400$
$-9x^2 + 4y^2 = -36$

18. $9x^2 + 16y^2 = 140$
$-x^2 + 4y^2 = -4$

19. $5x^2 + 3y^2 - 20x + 6y = -8$
$x - y = 2$

20. $4x^2 + 9y^2 = 36$
$2x - y = -1$

21. What would be the answer in Example 5 if the original piece of wire were 70 ft long? Explain.

22. A 52-ft-long piece of wire is to be cut into three pieces, two of which are the same length. The two

equal pieces are to be bent into circles and the third piece into a square. What should the length of each piece be if the total area enclosed by the two circles and the square is 100 feet?

23. A rectangular box (including top) with square ends and a volume of 16 cubic meters is to be constructed from 40 square meters of cardboard. What should its dimensions be?

24. A rectangular sheet of metal is to be rolled into a circular tube. If the tube is to have a surface area (excluding ends) of 210 square inches and a volume of 252 cubic inches, what size sheet of metal should be used?

25. Find two real numbers whose sum is -16 and whose product is 48.

26. Find two real numbers whose sum is 34.5 and whose product is 297.

27. Find two positive real numbers whose difference is 1 and whose product is 4.16.

28. Find two real numbers whose difference is 25.75 and whose product is 127.5.

29. Find two real numbers whose sum is 3 such that the sum of their squares is 369.

30. Find two real numbers whose sum is 2 such that the difference of their squares is 60.

31. Find the dimensions of a rectangular room whose perimeter is 58 ft and whose area is 204 sq ft.

32. Find the dimensions of a rectangular room whose perimeter is 53 ft and whose area is 165 sq ft.

33. A rectangle has area 120 square inches and a diagonal of length 17 inches. What are its dimensions?

34. A right triangle has area 225 square centimeters and a hypotenuse of length 35 centimeters. To the nearest tenth of a centimeter, how long are the legs of the triangle?

6.2 Large Systems of Linear Equations

Systems of linear equations in three variables can be interpreted geometrically as the intersection of planes. However, algebraic methods are the only practical means to solve such systems or ones with more variables. You need to understand the basic algebraic method presented here to implement it effectively on a calculator (which will be done in the next section).

Large systems of linear equations can be solved by **Gaussian elimination**,* which is an extension of the elimination method used in Section 6.1. Two systems of equations are said to be **equivalent** if they have the same solutions. The basic idea of Gaussian elimination is to transform a given system into an equivalent system that can easily be solved.

There are several operations on a system of equations that leave the solutions to the system unchanged, and hence, produce an equivalent system. The first one is to

Interchange any two equations in the system,

which obviously won't affect the solutions of the system. The second is to

Multiply an equation in the system by a nonzero constant.

This does not change the solutions of the equation, and therefore, does not change the solutions of the system. To understand how the next operation works, we shall examine an earlier example from a different viewpoint.

Example 1 In Example 3 of Section 6.1, we solved the system

$$x - 3y = 4$$
$$2x + y = 1$$

by multiplying the first equation by -2 and adding it to the second to eliminate the variable x:

$$-2x + 6y = -8 \qquad \text{[-2 times first equation]}$$
$$\underline{2x + y = 1} \qquad \text{[Second equation]}$$
$$7y = -7. \qquad \text{[Sum of second equation and -2 times first equation]}$$

We then solved this last equation for y and substituted the answer, $y = -1$, in the original first equation to find that $x = 1$. What we did, in effect, was

Replace the original system by the following system, in which x has been eliminated from the second equation; then solve this new system.

$$\text{(∗)} \qquad \begin{aligned} x - 3y &= 4 \qquad \text{[First equation]} \\ 7y &= -7. \qquad \text{[Sum of second equation and -2 times first equation]} \end{aligned}$$

The solution of system (∗) is easily seen to be $y = -1$, $x = 1$, and you can readily verify that this is the solution of the original system. So the two systems are equivalent. [*Note:* We are not claiming that the second equations in the two systems have the same solutions—they don't—but only that the two *systems* have the same solutions.] ■

Example 1 is an illustration of the third of the following operations.

---

*Named after the German mathematician K. F. Gauss (1777–1855).

Elementary Operations

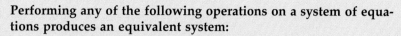

Performing any of the following operations on a system of equations produces an equivalent system:

1. Interchange any two equations in the system.
2. Replace an equation in the system by a nonzero constant multiple of itself.
3. Replace an equation in the system by the sum of itself and a constant multiple of another equation in the system.

The next example shows how elementary operations can be used to transform a system into an equivalent system that can be solved.

Example 2 Solve the system

$$x + 4y - 3z = 1 \quad \text{[Equation A]}$$
$$-3x - 6y + z = 3 \quad \text{[Equation B]}$$
$$2x + 11y - 5z = 0 \quad \text{[Equation C]}$$

Solution We first use elementary operations to produce an equivalent system in which the variable x has been eliminated from the second and third equations.

To eliminate x from equation B, replace equation B by the sum of itself and 3 times equation A:

$$
\begin{array}{lr}
\text{[3 times A]} & 3x + 12y - 9z = 3 \\
\text{[B]} & -3x - 6y + z = 3 \\
\hline
& 6y - 8z = 6
\end{array}
$$

$$x + 4y - 3z = 1 \quad \text{[A]}$$
$$6y - 8z = 6 \quad \text{[Sum of B and 3 times A]} \leftarrow$$
$$2x + 11y - 5z = 0 \quad \text{[C]}$$

To eliminate x from equation C we replace equation C by the sum of itself and -2 times equation A:

$$
\begin{array}{lr}
\text{[-2 times A]} & -2x - 8y + 6z = -2 \\
\text{[C]} & 2x + 11y - 5z = 0 \\
\hline
& 3y + z = -2
\end{array}
$$

$$x + 4y - 3z = 1$$
$$6y - 8z = 6$$
$$3y + z = -2 \quad \text{[Sum of C and -2 times A]} \leftarrow$$

The next step is to eliminate the y term in one of the last two equations. This can be done by replacing the second equation by the sum of itself and -2 times the third equation.

$$
\begin{array}{ll}
\textit{[Second equation]} & 6y - 8z = 6 \\
\textit{[-2 times third equation]} & \underline{-6y - 2z = 4} \\
& -10z = 10
\end{array}
$$

$$
\begin{aligned}
x + 4y - 3z &= 1 \\
-10z &= 10 \qquad \textit{[Sum of second equation and} \\
3y + z &= -2. \qquad \textit{-2 times third equation]}
\end{aligned}
$$

Finally, interchange the last two equations:

$$
\begin{aligned}
x + 4y - 3z &= 1 \\
3y + z &= -2 \\
-10z &= 10.
\end{aligned}
$$

We now have a system that is equivalent to the original system. To solve this system, begin with the last equation, which shows that

$$
-10z = 10, \qquad \text{or equivalently,} \qquad z = -1.
$$

Substituting $z = -1$ in the second equation shows that

$$
\begin{aligned}
3y + z &= -2 \\
3y + (-1) &= -2 \\
3y &= -1 \\
y &= -\frac{1}{3}.
\end{aligned}
$$

Substituting $y = -1/3$ and $z = -1$ in the first equation yields:

$$
\begin{aligned}
x + 4y - 3z &= 1 \\
x + 4\left(-\frac{1}{3}\right) - 3(-1) &= 1 \\
x = 1 + \frac{4}{3} - 3 &= -\frac{2}{3}.
\end{aligned}
$$

Therefore, the original system has just one solution: $x = -2/3$, $y = -1/3$, $z = -1$. ∎

     The solution process used at the end of Example 2 is called **back substitution** because you begin with the last equation and work back to the first. It works because the system in Example 2 is in **triangular form:** The first variable in the first equation, x, does not appear in any subsequent equation. The first variable in the second equation, y, does not appear in any subsequent equation, and so on.

Example 2 illustrates the following fact.

Gaussian Elimination

> Any system of linear equations can be transformed into an equivalent system in triangular form by using a finite number of elementary operations. If the system has solutions, they can be found by back substitution in the triangular form system.

In the next example of Gaussian elimination, we won't show the calculation of the sum of one equation and a constant multiple of another, as we did in Example 2. If you can't do them mentally by looking at the equations involved, write them out on scratch paper.

Example 3 Solve the system

$$x + 2y + 4z = 6$$
$$x + 3y + 3z = 6$$
$$-2x + 6y - 12z = -24$$
$$2x + 5y + 7z = 12.$$

Solution We begin by eliminating the x terms from the last three equations:

$$x + 2y + 4z = 6$$
$$y - z = 0 \quad \text{[Sum of second equation and } -1 \text{ times first equation]}$$
$$-2x + 6y - 12z = -24$$
$$2x + 5y + 7z = 12$$

$$x + 2y + 4z = 6$$
$$y - z = 0$$
$$10y - 4z = -12 \quad \text{[Sum of third equation and 2 times first equation]}$$
$$2x + 5y + 7z = 12$$

$$x + 2y + 4z = 6$$
$$y - z = 0$$
$$10y - 4z = -12$$
$$y - z = 0 \quad \text{[Sum of fourth equation and } -2 \text{ times first equation]}$$

Next, we use the second equation to eliminate the y term from the third and fourth equations.

$$x + 2y + 4z = 6$$
$$y - z = 0$$
$$6z = -12 \quad \text{[Sum of third equation and } -10 \text{ times second equation]}$$
$$y - z = 0$$

$$x + 2y + 4z = 6$$
$$y - z = 0$$
$$6z = -12$$
$$0 = 0 \quad \text{[Sum of fourth equation and } -1 \text{ times second equation]}$$

We now have a triangular form system that is easily solved. Use back substitution to verify that the solution is

$$x = 18, \qquad y = -2, \qquad z = -2. \quad \blacksquare$$

If your calculator has a solver for systems of equations (see the Tip in the margin), check your instruction manual to see how to use it. However, you should be aware of its limitations. It only works when the number of equations and number of variables are the same (so you can't use it in Example 3). Furthermore, when the solver gives an error message, it may mean that there is no solution or it may mean that there are infinitely many solutions. In the next section, we introduce calculator solution methods that can be used with all systems.

Technology Tip

Some (but not all) systems with the same number of linear equations as variables can be solved directly by using SIMULT on TI-86, SYSTEM in the TOOL menu of SHARP 9600, or SIMULTANEOUS in the Casio 9850 EQUATION menu (on the main menu).

Example 4 An investor wants to invest $30,000 in corporate bonds. The bonds are rated AAA, AA, A, B, C, and those with lower ratings (more risk) pay a higher rate of interest. Currently, the average yield is 5% on AAA bonds, 6% on A bonds, and 10% on B bonds. Being conservative, the investor wants to have twice as much in AAA bonds as in B bonds. How much should she invest in each type of bond to have an interest income of $2000?

Solution Let x be the amount invested in AAA bonds, y the amount in A bonds, and z the amount in B bonds. Since the total amount to be invested is $30,000,

$$x + y + z = 30,000.$$

Since x must be twice as large as z, we have

$$x = 2z, \qquad \text{or equivalently,} \qquad x - 2z = 0.$$

Finally, the income from the bonds is

$$\frac{\text{Interest on}}{\text{AAA bonds}} + \frac{\text{Interest on}}{\text{A bonds}} + \frac{\text{Interest on}}{\text{B bonds}} = \text{Total Interest}$$

$$(5\% \text{ of } x) + (6\% \text{ of } y) + (10\% \text{ of } z) = 2000.$$

$$.05x + .06y + .10z = 2000.$$

Therefore, we must solve this system of equations:

$$x + y + z = 30{,}000$$
$$x \qquad - 2z = 0$$
$$.05x + .06y + .10z = 2000$$

$$x + y + z = 30{,}000$$
$$-y - 3z = -30{,}000 \qquad \text{[\textit{Sum of second equation and}}$$
$$.05x + .06y + .10z = 2000 \qquad \text{\textit{−1 times first equation}]}$$

$$x + y + z = 30{,}000$$
$$-y - 3z = -30{,}000$$
$$.01y + .05z = 500 \qquad \text{[\textit{Sum of third equation and}}$$
$$\text{\textit{−.05 times first equation}]}$$

$$x + y + z = 30{,}000$$
$$-y - 3z = -30{,}000$$
$$.02z = 200 \qquad \text{[\textit{Sum of third equation and}}$$
$$\text{\textit{.01 times second equation}]}$$

Back substitution now shows that

$$z = \frac{200}{.02} = 10{,}000, \qquad y = 0, \qquad x = 20{,}000.$$

Therefore, \$20,000 should be invested in AAA bonds, \$10,000 in B bonds, and nothing should be invested in A bonds. ∎

Exercises 6.2

In Exercises 1–4, solve the system by back substitution.

1. $x + 3y - 4z + 2w = 1$
$y + z - w = 4$
$2z + 2w = -6$
$3w = 9$

2. $x + 5z + 6w = 10$
$y + 3z - 2w = 4$
$z - 4w = -6$
$2w = 4$

3. $2x + 2y - 4z + w = -5$
$3y + 4z - w = 0$
$2z - 7w = -6$
$5w = 15$

4. $3x - 2y - 4z + 2w = 6$
$2y + 5z - 3w = 7$
$3z + 4w = 0$
$3w = 15$

In Exercises 5–12, obtain an equivalent system by performing the stated elementary operation on the system.

5. Interchange equations 1 and 2.
$2x - 4y + 5z = 1$
$x \qquad - 3z = 2$
$5x - 8y + 7z = 6$
$3x - 4y + 2z = 3$

6. Interchange equations 1 and 3.

$$2x - 2y + z = -6$$
$$3x + y + 2z = 2$$
$$x + y - 2z = 0$$

7. Multiply the second equation by -1.

$$3x + z + 2w + 18v = 0$$
$$4x - y + w + 24v = 0$$
$$7x - y + z + 3w + 42v = 0$$
$$4x + z + 2w + 24v = 0$$

8. Multiply the third equation by $1/2$.

$$x + 2y + 4z = 3$$
$$x + 2z = 0$$
$$2x + 4y + z = 3$$

9. Replace the second equation by the sum of itself and -2 times the first equation.

$$x + y + 2z + 3w = 1$$
$$2x + y + 3z + 4w = 1$$
$$3x + y + 4z + 5w = 2$$

10. Replace the third equation by the sum of itself and -1 times the first equation.

$$x + 2y + 4z = 6$$
$$y + z = 1$$
$$x + 3y + 5z = 10$$

11. Replace the third equation by the sum of itself and -2 times the second equation.

$$x + 12y - 3z + 4w = 10$$
$$2y + 3z + w = 4$$
$$4y + 5z + 2w = 1$$
$$6y - 2z - 3w = 0$$

12. Replace the third equation by the sum of itself and 3 times the second equation.

$$2x + 2y - 4z + w = -5$$
$$2y + 4z - w = 2$$
$$-6y - 4z + 2w = 6$$
$$2y + 5z - 3w = 7$$

In Exercises 13–20, use Gaussian elimination to solve the system.

13. $-x + 3y + 2z = 0$
$$2x - y - z = 3$$
$$x + 2y + 3z = 0$$

14. $3x + 7y + 9z = 0$
$$x + 2y + 3z = 2$$
$$x + 4y + z = 2$$

15. $x - 2y + 4z = 6$
$$x + y + 13z = 6$$
$$-2x + 6y - z = -10$$

16. $x - y + 5z = -6$
$$3x + 3y - z = 10$$
$$x + 3y + 2z = 5$$

17. $x + y + z = 200$
$$x - 2y = 0$$
$$2x + 3y + 5z = 600$$
$$2x - y + z = 200$$

18. $2x - y + 2z = 3$
$$-x + 2y - z = 0$$
$$3y - 2z = 1$$
$$x + y - z = 1$$

19. $x + y = 3$
$$5x - y = 3$$
$$9x - 4y = 1$$

20. $3x - y + z = 6$
$$x + 2y - z = 0$$
$$2x - 2y + z = -2$$
$$3x - 7y + 4z = 4$$

In Exercises 21–24, solve the system graphically by finding the point that lies on the graphs of all equations in the system, if there is one. Since your intersection finder only works for two equations at a time, you may have to use it several times to verify that there is a solution.

21. $x + 2y = 6$
$$2x - y = 3$$
$$15x - 5y = 27$$

22. $x - y = 2$
$$2x - y = 3$$
$$2x + y = 1$$

23. $6x + 2y = 8$
$$x + y = 2$$
$$10x - 5y = 6$$

24. $2x - y = 1$
$$x + 2y = 4$$
$$20x - 5y = 16$$

25. An investor want to invest $35,000 in the AAA, A, and B bonds of Example 4. If he wants to have twice as much in AAA bonds as in B bonds and wants to get annual interest of $2200, how much should he invest in each type of bond?

26. An investor wants to invest $30,000 in bonds. She wants to have twice as much in AAA bonds as in B bonds and wants to receive $2000 in annual interest. If AAA bonds yield an average of 4%, A bonds an average of 7%, and B bonds an average of 9%, how much should she invest in each type of bond?

27. The Snack Company wants to make a 100-pound mixture of corn chips, nuts, and pretzels that will cost $4 per pound. Corn chips cost $2 per pound, nuts cost $6 per pound, and pretzels cost

$3 per pound. If the mixture is to have three times as many corn chips as pretzels (by weight), how many pounds of each ingredient should be used?

28. An animal feed is to be made from alfalfa, cornmeal, and meat by-products. Each unit of alfalfa provides 10 units of fiber, 30 units of fat, and 20 units of protein. Each unit of cornmeal provides 20 units of fiber, 20 units of fat, and 40 units of protein. Each unit of by-products provides 30 units of fiber, 40 units of fat, and 25 units of protein. The animal feed must provide 1800 units of fiber, 2800 units of fat, and 2200 units of protein. How many units of alfalfa, cornmeal, and by-products should the feed contain?

29. A minor league baseball park has 7000 seats. Box seats cost $6, grandstand seats cost $4, and bleacher seats cost $2. When all seats are sold, the revenue is $26,400. If the number of box seats is one-third the number of bleacher seats, how many of each type of seat are there?

30. Shipping charges at the online bookstore Hercules.com are $4 for one book, $6 for two books, and $7 for three to five books. Last week, there were 6400 orders of five or fewer books and total shipping charges for these orders were $33,600. The number of shipments with $7 charges was 1000 less than the number with $6 charges. How many shipments were made in each category (one book, two books, three to five books)?

6.3 Matrix Solution Methods

We now develop two versions of Gaussian elimination that can be implemented on a calculator. When solving systems by hand, a lot of time is wasted copying the x's, y's, z's, and so on. This fact suggests a shorthand system for representing a system of equations. For example, the system

$$(*)\qquad\begin{aligned} x + 2y + 3z &= -2 \\ 2x + 6y + z &= 2 \\ 3x + 3y + 10z &= -2 \end{aligned}$$

can be represented by the following rectangular array of numbers, consisting of the coefficients of the variables and the constants on the right of the equal sign, arranged in the same order they appear in the system:

$$\begin{pmatrix} 1 & 2 & 3 & -2 \\ 2 & 6 & 1 & 2 \\ 3 & 3 & 10 & -2 \end{pmatrix}$$

This array is called the **augmented matrix** of the system.* It has three horizontal rows and four vertical columns.

Example 1 Use the matrix form of system (*) above to solve the system.

Solution To solve the system in its original equation form, we would use elementary operations to eliminate the x terms from the last two equations, and then eliminate the y term from the last equation. With matrices we do essentially the same thing, with the elementary operations

---

*The plural of "matrix" is "matrices."

on equations being replaced by the corresponding **row operations** on the augmented matrix to make certain entries in the first and second columns 0. Here is a side-by-side development of the two solution methods.

Equation Method	**Matrix Method**

Equation Method

Replace the second equation by the sum of itself and -2 times the first equation:

$$\begin{aligned} x + 2y + 3z &= -2 \\ 2y - 5z &= 6 \\ 3x + 3y + 10z &= -2. \end{aligned}$$

Matrix Method

Replace the second row by the sum of itself and -2 times the first row:

$$\left(\begin{array}{ccc|c} 1 & 2 & 3 & -2 \\ 0 & 2 & -5 & 6 \\ 3 & 3 & 10 & -2 \end{array} \right)$$

Technology Tip

Check your instruction manual to learn how to enter and store matrices in the matrix memory.

Equation Method

Replace the third equation by the sum of itself and -3 times the first equation:

$$\begin{aligned} x + 2y + 3z &= -2 \\ 2y - 5z &= 6 \\ -3y + z &= 4. \end{aligned}$$

Matrix Method

Replace the third row by the sum of itself and -3 times the first row:

$$\left(\begin{array}{ccc|c} 1 & 2 & 3 & -2 \\ 0 & 2 & -5 & 6 \\ 0 & -3 & 1 & 4 \end{array} \right)$$

Multiply the second equation by $1/2$ (so that y has coefficient 1):

$$\begin{aligned} x + 2y + 3z &= -2 \\ y - \frac{5}{2}z &= 3 \\ -3y + z &= 4. \end{aligned}$$

Multiply the second row by $1/2$:

$$\left(\begin{array}{ccc|c} 1 & 2 & 3 & -2 \\ 0 & 1 & -\dfrac{5}{2} & 3 \\ 0 & -3 & 1 & 4 \end{array} \right)$$

Replace the third equation by the sum of itself and 3 times the second equation:

$$\begin{aligned} x + 2y + 3z &= -2 \\ y - \frac{5}{2}z &= 3 \\ -\frac{13}{2}z &= 13. \end{aligned}$$

Replace the third row by the sum of itself and 3 times the second row:

$$\left(\begin{array}{ccc|c} 1 & 2 & 3 & -2 \\ 0 & 1 & -\dfrac{5}{2} & 3 \\ 0 & 0 & -\dfrac{13}{2} & 13 \end{array} \right)$$

Finally, multiply the last equation by $-2/13$:*

$$\begin{aligned} x + 2y + 3z &= -2 \\ (**) \qquad y - \frac{5}{2}z &= 3 \\ z &= -2. \end{aligned}$$

Finally, multiply the last row by $-2/13$:

$$\left(\begin{array}{ccc|c} 1 & 2 & 3 & -2 \\ 0 & 1 & -\dfrac{5}{2} & 3 \\ 0 & 0 & 1 & -2 \end{array} \right)$$

---

*This step isn't necessary, but it is often convenient to have 1 as the coefficient of the first variable in each equation.

System (**) is easily solved. The third equation shows that $z = -2$ and substituting this in the second equation shows that

$$y - \frac{5}{2}(-2) = 3$$

$$y = 3 - 5 = -2.$$

Substituting $y = -2$ and $z = -2$ in the first equation yields

$$x + 2(-2) + 3(-2) = -2$$

$$x = -2 + 4 + 6 = 8.$$

Therefore, the only solution of the original system is $x = 8$, $y = -2$, $z = -2$. ■

When using matrix notation, row operations replace elementary operations on equations, as shown in Example 1. The solution process ends when you reach a matrix (and corresponding system), such as the final matrix in Example 1:

(**)
$$\begin{pmatrix} 1 & 2 & 3 & -2 \\ 0 & 1 & -\frac{5}{2} & 3 \\ 0 & 0 & 1 & -2 \end{pmatrix}$$

$$x + 2y + 3z = -2$$
$$y - \frac{5}{2}z = 3$$
$$z = -2.$$

The final matrix should satisfy these conditions:

All rows consisting entirely of zeros (if any) are at the bottom.

The first nonzero entry in each nonzero row is a 1 (called a **leading 1**).

Each leading 1 appears to the right of leading 1's in any preceding rows.

Such a matrix is said to be in **row echelon form.**

Most calculators have a key that uses row operations to put a given matrix into row echelon form (see the Tip in the margin). For example, using the TI-83 REF key on the first matrix in Example 1 produced this row echelon matrix and corresponding system of equations:

$$\begin{pmatrix} 1 & 1 & \frac{10}{3} & -\frac{2}{3} \\ 0 & 1 & -\frac{17}{12} & \frac{5}{6} \\ 0 & 0 & 1 & -2 \end{pmatrix}$$

$$x + y + \frac{10}{3}z = -\frac{2}{3}$$
$$y - \frac{17}{12}z = \frac{5}{6}$$
$$z = -2.$$

Because the calculator used a different sequence of row operations than was used in Example 1, it produced a row echelon matrix (and corresponding system) that differs from matrix (**) above. You can easily verify, however that the preceding system has the same solutions as system (**), namely, $x = 8$, $y = -2$, $z = -2$.

Technology Tip

To put a matrix in row echelon form, use REF in the MATH or OPS submenu of the TI-83/86 MATRIX menu, or in the MATRIX submenu of the TI-89 MATH menu; or use rowEF in the MATH submenu of the Sharp 9600 MATRIX menu.

Example 2 Solve the system

$$x + y + 2z = 1$$
$$2x + 4y + 5z = 2$$
$$3x + 5y + 7z = 4.$$

Solution If you try to use a systems equation solver on a calculator, you will get an error message. So we form the augmented matrix and reduce it to row echelon form by using the REF key. A TI-86 produced this row echelon matrix:

Technology Tip

On TI calculators, using the FRAC key in conjunction with the RREF key, as in Figure 6–11, usually eliminates long decimals and makes the matrix easier to read.

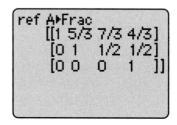

Figure 6–11

Look at the last row of the matrix in Figure 6–11; it represents the equation

$$0x + 0y + 0z = 1.$$

Since this equation has no solutions (the left side is always 0 and the right side 1), neither does the original system. Such a system is said to be **inconsistent.** ■

The Gauss-Jordan Method

Gaussian elimination on a calculator is an efficient method of solving systems of equations, but may involve some messy calculations when you solve the final triangular form system by hand. Most hand computations can be eliminated by using a slight variation, known as the **Gauss-Jordan method*,** which is illustrated in the next example.

Example 3 Use the Gauss-Jordan method to solve this system:

$$x - y + 5z = -6$$
$$3x + 3y - z = 10$$
$$x - 5y + 8z = -17$$
$$x + 3y + 2z = 5.$$

---

*This method was developed by the German engineer Wilhelm Jordan (1842–1899).

Solution   The augmented matrix of the system is shown in Figure 6–12 and an equivalent row echelon matrix (obtained by using the REF and FRAC keys) is shown in Figure 6–13.

$$\begin{pmatrix} 1 & -1 & 5 & -6 \\ 3 & 3 & -1 & 10 \\ 1 & -5 & 8 & -17 \\ 1 & 3 & 2 & 5 \end{pmatrix}$$

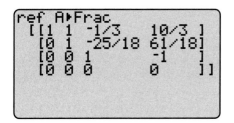

Figure 6–12   **Figure 6–13**

At this point in Gaussian elimination we would use back substitution to solve the triangular form system represented by the matrix in Figure 6–13. In the Gauss-Jordan method, however, additional elimination of variables replaces back substitution. Look at the leading 1 in the third row (shown in color):

$$\begin{pmatrix} 1 & 1 & -\dfrac{1}{3} & \dfrac{10}{3} \\ 0 & 1 & -\dfrac{25}{18} & \dfrac{61}{18} \\ 0 & 0 & 1 & -1 \\ 0 & 0 & 0 & 0 \end{pmatrix}.$$

In Gauss-Jordan elimination, we make the entries above this leading 1 into 0's:

Replace the second row by the sum of itself and 25/18 times the third row:
$$\begin{pmatrix} 1 & 1 & -\dfrac{1}{3} & \dfrac{10}{3} \\ 0 & 1 & 0 & 2 \\ 0 & 0 & 1 & -1 \\ 0 & 0 & 0 & 0 \end{pmatrix}$$

Replace the first row by the sum of itself and 1/3 times the third row:
$$\begin{pmatrix} 1 & 1 & 0 & 3 \\ 0 & 1 & 0 & 2 \\ 0 & 0 & 1 & -1 \\ 0 & 0 & 0 & 0 \end{pmatrix}.$$

Now consider the leading 1 in the second row (shown in color), and make the entry above it 0:

Replace the first row by the sum of itself and −1 times the second row:
$$\begin{pmatrix} 1 & 0 & 0 & 1 \\ 0 & 1 & 0 & 2 \\ 0 & 0 & 1 & -1 \\ 0 & 0 & 0 & 0 \end{pmatrix}.$$

This last matrix represents the following system, whose solution is obvious.

$$x \qquad\qquad = 1$$
$$y \qquad = 2$$
$$z = -1.$$ ∎

A row echelon form matrix, such as the last one in Example 3, in which any column containing a leading 1 has 0's in all other positions, is said to be in **reduced row echelon form.** The goal in the Gauss-Jordan method is to use row operations to put a given augmented matrix into reduced row echelon form (from which the solutions can be read immediately, as in Example 3).

As a general rule, Gaussian elimination (matrix version) is the method of choice when working by hand (the additional row operations needed to put a matrix in reduced row echelon form are usually more time consuming—and error prone—than back substitution). With a calculator or computer, however, it's better to find a reduced row echelon matrix for the system. You can do this in one step on a calculator by using the RREF key (see the Tip in the margin).

Technology Tip

To put a matrix in reduced row echelon form, use RREF in the MATH or OPS submenu of the TI-83/86 MATRIX menu, or rowEF in the MATH submenu of the Sharp 9600 MATRIX menu, or RREF in the MATRIX submenu of the HP-38 or TI-89 MATH menu. A RREF program for TI-82 is in the Program Appendix.

Example 4   Solve this system:

$$2x + 5y + z + 3w = 0$$
$$2y - 4z + 6w = 0$$
$$2x + 17y - 23z + 40w = 0$$

Solution   A system such as this, in which all the constants on the right side are zero, is called a **homogeneous system.** Every homogeneous system has at least one solution, namely, $x = 0$, $y = 0$, $z = 0$, $w = 0$, which is called the **trivial solution.** The issue with homogeneous systems is whether or not they have any nonzero solutions. The augmented matrix of the system is shown in Figure 6–14 and an equivalent reduced row echelon form matrix in Figure 6–15.*

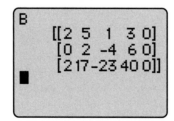

Figure 6–14

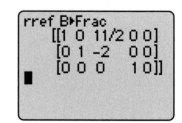

Figure 6–15

*When dealing with homogeneous systems, it's not really necessary to include the last column of zeros, as is done here, because row operations do not change this column.

The system corresponding to the reduced echelon form matrix in Figure 6–15 is

$$x + \frac{11}{2}z = 0$$

$$y - 2z = 0$$

$$w = 0.$$

The second equation shows that

$$y = 2z.$$

This equation has an infinite number of solutions, for instance,

$$z = 1, y = 2 \quad \text{or} \quad z = 3, y = 6 \quad \text{or} \quad z = -2.5, y = -5.$$

In fact, for each real number t, there is a solution: $z = t$, $y = 2t$. Substituting $z = t$ into the first equation shows that

$$x + \frac{11}{2}t = 0$$

$$x = -\frac{11}{2}t.$$

NOTE Every system that has more variables than equations (as in Example 4) is dependent (or inconsistent), but other systems may be dependent as well.

Therefore, this system, and hence the original one, has an infinite number of solutions, one for each real number t:

$$x = -\frac{11}{2}t, \quad y = 2t, \quad z = t, \quad w = 0.$$

A system with infinitely many solutions, such as this one, is said to be **dependent.** ∎

Applications

In calculus it is sometimes necessary to write a complicated rational expression as the sum of simpler ones. One technique for doing this involves systems of equations.

Example 5 Find constants A, B, and C such that

$$\frac{2x^2 + 15x + 10}{(x - 1)(x + 2)^2} = \frac{A}{x - 1} + \frac{B}{x + 2} + \frac{C}{(x + 2)^2}.$$

Solution Multiply both sides of the equation by the common denominator $(x - 1)(x + 2)^2$ and collect like terms on the right side:

$$2x^2 + 15x + 10 = A(x + 2)^2 + B(x - 1)(x + 2) + C(x - 1)$$

$$= A(x^2 + 4x + 4) + B(x^2 + x - 2) + C(x - 1)$$

$$= Ax^2 + 4Ax + 4A + Bx^2 + Bx - 2B + Cx - C$$

$$= (A + B)x^2 + (4A + B + C)x + (4A - 2B - C).$$

Since the polynomials on the left and right sides of the last equation are equal, their coefficients must be equal term by term, that is,

$$A + B \quad\quad = 2 \quad\quad [\textit{Coefficients of } x^2]$$
$$4A + B + C = 15 \quad\quad [\textit{Coefficients of } x]$$
$$4A - 2B - C = 10 \quad\quad [\textit{Constant terms}]$$

We can consider this as a system of equations with unknowns A, B, C. The augmented matrix of the system is shown in Figure 6–16 and an equivalent reduced row echelon form matrix in Figure 6–17.

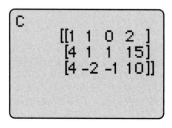

Figure 6–16

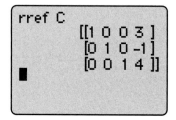

Figure 6–17

The solutions of the system can be read from the reduced row echelon form matrix in Figure 6–17:

$$A = 3, \quad B = -1, \quad C = 4.$$

Therefore,

$$\frac{2x^2 + 15x + 10}{(x - 1)(x + 2)^2} = \frac{3}{x - 1} + \frac{-1}{x + 2} + \frac{4}{(x + 2)^2}.$$

The right side of this equation is called the **partial fraction decomposition** of the fraction on the left side. ■

Example 6   Charlie is starting a small business and borrows $10,000 on three different credit cards, with annual interest rates of 18%, 15%, and 9%, respectively. He borrows three times as much on the 15% card as on the 18% card, and his total annual interest on all three cards is $1244.25. How much did he borrow on each credit card?

Solution   Let x be the amount on the 18% card, y the amount on the 15% card, and z the amount on the 9% card. Then, $x + y + z = 10,000$. Furthermore,

$$\begin{array}{ccccccc}
\text{Interest on} & + & \text{Interest on} & + & \text{Interest on} & = & \text{Total} \\
\text{18\% card} & & \text{15\% card} & & \text{9\% card} & & \text{interest} \\
\downarrow & & \downarrow & & \downarrow & & \downarrow \\
.18x & + & .15y & + & .09z & & = 1244.25.
\end{array}$$

Finally, we have

$$\text{Amount on } 15\% \text{ card} = 3 \text{ times amount on } 18\% \text{ card}$$

$$y = 3x,$$

which is equivalent to $3x - y = 0$. Therefore, we must solve this system of equations:

$$x + y + z = 10{,}000$$
$$.18x + .15y + .09z = 1244.25$$
$$3x - y = 0$$

whose augmented matrix is

$$\begin{pmatrix} 1 & 1 & 1 & 10{,}000 \\ .18 & .15 & .09 & 1244.25 \\ 3 & -1 & 0 & 0 \end{pmatrix}.$$

CALCULATOR EXPLORATION

Enter this matrix in your calculator. Use the RREF key to put it in reduced row echelon form. Read the solutions of the system from this last matrix.

The Exploration shows that Charlie borrowed $1275 on the 18% card, $3825 on the 15% card, and $4900 on the 9% card. ■

The preceding examples illustrate the following fact, whose proof is omitted.

Number of Solutions of a System

Any system of linear equations must have

No solutions (an inconsistent system) *or*

Exactly one solution *or*

An infinite number of solutions (a dependent system).

Exercises 6.3

In Exercises 1–4, write the augmented matrix of the system.

1. $2x - 3y + 4z = 1$
$x + 2y - 6z = 0$
$3x - 7y + 4z = -3$

2. $x + 2y - 3w + 7z = -5$
$2x - y + 2z = 4$
$3x + 7w - 6z = 0$

3. $x - \dfrac{1}{2}y + \dfrac{7}{4}z = 0$

$2x - \dfrac{3}{2}y + 5z = 0$

$- 2y + \dfrac{1}{3}z = 0$

4. $2x - \dfrac{1}{2}y + \dfrac{7}{2}w - 6z = 1$

$\dfrac{1}{4}x - 6y + 2w - z = 2$

$4y - \dfrac{1}{2}w + z = 3$

$2x + 3y + \dfrac{1}{2}z = 4$

In Exercises 5–8, the augmented matrix of a system of equations is given. Express the system in equation notation.

5. $\begin{pmatrix} 2 & -3 & 1 \\ 4 & 7 & 2 \end{pmatrix}$
6. $\begin{pmatrix} 2 & 3 & 5 & 2 \\ 1 & 6 & 9 & 0 \end{pmatrix}$

7. $\begin{pmatrix} 1 & 0 & 1 & 0 & 1 \\ 1 & -1 & 4 & -2 & 3 \\ 4 & 2 & 5 & 0 & 2 \end{pmatrix}$
8. $\begin{pmatrix} 1 & 7 & 0 & 4 \\ 2 & 3 & 1 & 6 \\ -1 & 0 & 2 & 3 \end{pmatrix}$

In Exercises 9–12, the reduced row echelon form of the augmented matrix of a system of equations is given. Find the solutions of the system.

9. $\begin{pmatrix} 1 & 0 & 0 & 0 & 3/2 \\ 0 & 1 & 0 & 0 & 5 \\ 0 & 0 & 1 & 0 & -2 \\ 0 & 0 & 0 & 1 & 0 \end{pmatrix}$
10. $\begin{pmatrix} 1 & 0 & 0 & 0 & 0 & 5 \\ 0 & 1 & 0 & 0 & 0 & 4 \\ 0 & 0 & 1 & 0 & 0 & 3 \\ 0 & 0 & 0 & 0 & 1 & 2 \\ 0 & 0 & 0 & 0 & 0 & 1 \end{pmatrix}$

11. $\begin{pmatrix} 1 & 0 & 0 & 1 & 2 \\ 0 & 1 & 0 & 2 & -3 \\ 0 & 0 & 1 & 0 & 4 \\ 0 & 0 & 0 & 0 & 0 \end{pmatrix}$
12. $\begin{pmatrix} 1 & 0 & 0 & 0 & 7 \\ 0 & 1 & 0 & 0 & 1 \\ 0 & 0 & 1 & 0 & -5 \\ 0 & 0 & 0 & 1 & 4 \\ 0 & 0 & 0 & 0 & 0 \\ 0 & 0 & 0 & 0 & 0 \end{pmatrix}$

In Exercises 13–16, use the matrix form of Gaussian elimination to solve the system.

13. $x + y + z = 1$
$x - 2y + 2z = 4$
$2x - y + 3z = 5$

14. $2x - y + z = 1$
$3x + y + z = 0$
$7x - y + 3z = 2$

15. $x + y = 5$
$x + 2z = 0$
$2x + y - z = 7$
$4x + 3y - z = 17$

16. $2x + y - z = 4$
$x + 2z = 9$
$3x + y + z = 13$
$-3x - y + 2z = 9$

In Exercises 17–20, use the Gauss-Jordan method to solve the system.

17. $2x - y + z = 1$
$x + y - z = 2$
$-x - y + z = 0$

18. $x + 2y + 3z = 4$
$2x - y + z = 3$
$3x + y + 4z = 7$

19. $3x + y - 2z = 4$
$-5x + 2z = 5$
$-7x - y + 3z = -2$

20. $3x - y + z = 6$
$x + 2y - z = 0$

In Exercises 21–34, solve the system by any means.

21. $11x + 10y + 9z = 5$
$x + 2y + 3z = 1$
$3x + 2y + z = 1$

22. $-x + 2y - 3z + 4w = 8$
$2x - 4y + z + 2w = -3$
$5x - 4y + z + 2w = -3$

23. $x - 4y - 13z = 4$
$x - 2y - 3z = 2$
$-3x + 5y + 4z = 2$

24. $2x - 4y + z = 3$
$x + 3y - 7z = 1$
$-2x + 4y - z = 10$

25. $4x + y + 3z = 7$
$x - y + 2z = 3$
$3x + 2y + z = 4$

26. $x + 4y + z = 3$
$-x + 2y + 2z = 0$
$2x + 2y - z = 3$

27. $x + y + z + w = 10$
$x + y + 2z = 11$
$x - 3y + w = -14$
$y + 3z - w = 7$
$2x - 2y + z + 2w = -4$

28. $2x + y + z = 3$
$y + z + w = 5$
$4x + z + w = 0$
$3y - 2z - w = 6$
$2x + 4y - z - w = 9$

29. $x + 2y + 3z = 1$
$3x + 2y + 4z = -1$
$2x + 6y + 8z + w = 3$
$2x + 2z - 2w = 3$

30. $x + y + z = 0$
$x - y - z = 0$
$x - y + z = 0$

31. $2x + y + 3z - 2w = -6$
$4x + 3y + z - w = -2$
$x + y + z + w = -5$
$-2x - 2y + 2z + 2w = -10$

32.
$$\begin{aligned} x + y + z + w &= -1 \\ -x + 4y + z - w &= 0 \\ x - 2y + z - 2w &= 11 \\ -x - 2y + z + 2w &= -3 \end{aligned}$$

33.
$$\begin{aligned} x + 2y + z + 4w &= 1 \\ y + 3z - w &= 2 \\ x + 4y + 7z - 2w &= 5 \\ 3x + 7y + 6z + 11w &= 5 \end{aligned}$$

34.
$$\begin{aligned} 3x - y + 2z &= 0 \\ -x + 3y + 2z + 5w &= 0 \\ x + 2y + 5z - 4w &= 0 \\ 2x - y + 3w &= 0 \end{aligned}$$

In Exercises 35–38, solve the system. [Note: The REF and RREF keys on some calculators produce an error message when there are more rows than columns in a matrix, in which case you will have to solve the system by some other means.]

35.
$$\begin{aligned} 2x - y &= 1 \\ 3x + y &= 2 \\ 4x - 2y &= 2 \\ 5x + 5y &= 4 \end{aligned}$$

36.
$$\begin{aligned} x + y &= 3 \\ -x + 2y &= 3 \\ 5x - y &= 3 \\ -7x + 5y &= 3 \end{aligned}$$

37.
$$\begin{aligned} x + 2y &= 3 \\ 2x + 3y &= 4 \\ 3x + 4y &= 5 \\ 4x + 5y &= 6 \end{aligned}$$

38.
$$\begin{aligned} x - y &= 2 \\ x + y &= 4 \\ 2x + 3y &= 9 \\ 3x - 2y &= 6 \end{aligned}$$

In Exercises 39–44, find the constants A, B, C.

39. $\dfrac{x}{(x + 1)(x + 2)} = \dfrac{A}{x + 1} + \dfrac{B}{x + 2}$

40. $\dfrac{1}{(x + 1)(x - 1)} = \dfrac{A}{x + 1} + \dfrac{B}{x - 1}$

41. $\dfrac{2x + 1}{(x + 2)(x - 3)^2} = \dfrac{A}{x + 2} + \dfrac{B}{x - 3} + \dfrac{C}{(x - 3)^2}$

42. $\dfrac{x^2 - x - 21}{(2x - 1)(x^2 + 4)} = \dfrac{A}{2x - 1} + \dfrac{Bx + C}{x^2 + 4}$

43. $\dfrac{5x^2 + 1}{(x + 1)(x^2 - x + 1)} = \dfrac{A}{x + 1} + \dfrac{Bx + C}{x^2 - x + 1}$

44. $\dfrac{x - 2}{(x + 4)(x^2 + 2x + 2)} = \dfrac{A}{x + 4} + \dfrac{Bx + C}{x^2 + 2x + 2}$

In Exercises 45 and 46, solve the system.

45.
$$\begin{aligned} \frac{3}{x} - \frac{1}{y} + \frac{4}{z} &= -13 \\ \frac{1}{x} + \frac{2}{y} - \frac{1}{z} &= 12 \\ \frac{4}{x} - \frac{1}{y} + \frac{3}{z} &= -7 \end{aligned}$$

[*Hint:* Let $u = 1/x$, $v = 1/y$, $w = 1/z$.]

46.
$$\begin{aligned} \frac{1}{x + 1} - \frac{2}{y - 3} + \frac{3}{z - 2} &= 4 \\ \frac{5}{y - 3} - \frac{10}{z - 2} &= -5 \\ \frac{-3}{x + 1} + \frac{4}{y - 3} - \frac{1}{z - 2} &= -2 \end{aligned}$$

[*Hint:* Let $u = 1/(x + 1)$, $v = 1/(y - 3)$, $w = 1/(z - 2)$.]

47. A collection of nickels, dimes, and quarters totals $6.00. If there are 52 coins altogether and twice as many dimes as nickels, how many of each kind of coin are there?

48. A collection of nickels, dimes, and quarters totals $8.20. The number of nickels and dimes together is twice the number of quarters. The value of the nickels is one-third the value of the dimes. How many of each kind of coin are there?

49. Lillian borrows $10,000. She borrows some from her friend at 8% annual interest, twice as much as that from her bank at 9%, and the remainder from her insurance company at 5%. She pays a total of $830 in interest for the first year. How much did she borrow from each source?

50. An investor puts a total of $25,000 into three very speculative stocks. She invests some of it in Crystalcomp and $2000 more than one-half that amount in Flyboys. The remainder is invested in Zumcorp. Crystalcomp rises 16% in value, Flyboys 20%, and Zumcorp 18%. Her investment in the three stocks is now worth $29,440. How much was originally invested in each stock?

51. An investor has $70,000 invested in a mutual fund, bonds, and a fast-food franchise. She has twice as much invested in bonds as in the mutual fund. Last year the mutual fund paid a 2% dividend, the bonds 10%, and the fast-food franchise 6%; her dividend income was $4800. How much is invested in each of the three investments?

52. Tickets to a band concert cost $2 for children, $3 for teenagers, and $5 for adults. 570 people attended the concert and total ticket receipts were $1950. Three-fourths as many teenagers as children attended. How many children, adults, and teenagers attended?

53. Comfort Systems, Inc., sells three models of humidifiers: the bedroom model weighs 10 pounds and comes in an 8-cubic-ft box; the living room model weighs 20 pounds and comes in an 8-cubic-ft box; the whole-house model weighs 60 pounds and comes in a 28-cubic-ft box. Each of their delivery vans has 248 cubic ft of space and can hold a maximum of 440 pounds. For a van to be as fully loaded as possible, how many of each model should it carry?

54. Peanuts cost $3 per pound, almonds $4 per pound, and cashews $8 per pound. How many pounds of each should be used to produce 140 pounds of a mixture costing $6 per pound, in which there are twice as many peanuts as almonds?

55. If Tom, Dick, and Harry work together, they can paint a large room in 4 hours. When only Dick and Harry work together, it takes 8 hours to paint the room. Tom and Dick, working together, take 6 hours to paint the room. How long would it take each of them to paint the room alone? [*Hint:* If x is the amount of the room painted in 1 hour by Tom, y the amount painted by Dick, and z the amount painted by Harry, then $x + y + z = 1/4$.]

56. Pipes R, S, T are connected to the same tank. When all three pipes are running, they can fill the tank in 2 hours. When only pipes S and T are running, they can fill the tank in 4 hours. When only R and T are running, they can fill the tank in 2.4 hours. How long would it take each pipe running alone to fill the tank?

57. A furniture manufacturer has 1950 machine hours available each week in the cutting department, 1490 hours in the assembly department, and 2160 in the finishing department. Manufacturing a chair requires 0.2 hours of cutting, 0.3 hours of assembly, and 0.1 hours of finishing. A chest requires 0.5 hours of cutting, 0.4 hours of assembly, and 0.6 hours of finishing. A table requires 0.3 hours of cutting, 0.1 hours of assembly, and 0.4 hours of finishing. How many chairs, chests, and tables should be produced to use all the available production capacity?

58. A stereo equipment manufacturer produces three models of speakers, R, S, and T, and has three kinds of delivery vehicles, trucks, vans, and station wagons. A truck holds two boxes of model R, one of model S, and three of model T. A van holds one box of model R, three of model S, and two of model T. A station wagon holds one box of model R, three of model S, and one of model T. If 15 boxes of model R, 20 boxes of model S, and 22 boxes of model T are to be delivered, how many vehicles of each type should be used so that all operate at full capacity?

6.4 Matrix Methods for Square Systems

Matrices were used in Section 6.3 as a convenient shorthand for solving systems of linear equations. We now consider matrices in a more general setting and show how the algebra of matrices provides an alternative method for solving systems of equations that are not dependent and have the same number of equations as variables.

Let m and n be positive integers. An $m \times n$ **matrix** (read "m by n matrix") is a rectangular array of numbers, with m horizontal rows and n vertical columns. For example,

$$\begin{pmatrix} 3 & 2 & -5 \\ 6 & 1 & 7 \\ -2 & 5 & 0 \end{pmatrix} \qquad \begin{pmatrix} -3 & 4 \\ 2 & 0 \\ 0 & 1 \\ 7 & 3 \\ 1 & -6 \end{pmatrix} \qquad \begin{pmatrix} 3 & 0 & 1 & 0 \\ \sqrt{2} & -\frac{1}{2} & 4 & \frac{8}{3} \\ 10 & 2 & -\frac{3}{4} & 12 \end{pmatrix} \qquad \begin{pmatrix} \sqrt{3} \\ 2 \\ 0 \\ 11 \end{pmatrix}$$

3×3 matrix	5×2 matrix	3×4 matrix	4×1 matrix
3 rows	5 rows	3 rows	4 rows
3 columns	2 columns	4 columns	1 column

In a matrix, the *rows* are horizontal and are numbered from top to bottom. The *columns* are vertical and are numbered from left to right. For example,

$$
\begin{array}{c}
\text{Row 1} \longrightarrow \\
\text{Row 2} \longrightarrow \\
\text{Row 3} \longrightarrow
\end{array}
\begin{pmatrix}
11 & 3 & 14 \\
-2 & 0 & -5 \\
\frac{1}{3} & 6 & 7
\end{pmatrix}
$$

$$
\begin{array}{ccc}
\uparrow & \uparrow & \uparrow \\
\textit{Column 1} & \textit{Column 2} & \textit{Column 3}
\end{array}
$$

Each entry in a matrix can be located by stating the row and column in which it appears. For instance, in the preceding 3×3 matrix, 14 is the entry in row 1, column 3, and 0 is the entry in row 2, column 2. When you enter a matrix on a calculator, the words "row" and "column" won't be displayed, but the row numbers will always be listed before the column number. Thus a display such as "$A[3, 2]$," or simply "3, 2," indicates the entry in row 3, column 2.

Two matrices are said to be **equal** if they have the same size (same number of rows and columns) and the corresponding entries are equal. For example,

$$
\begin{pmatrix} 3 & (-1)^2 \\ 6 & 12 \end{pmatrix} = \begin{pmatrix} 3 & 1 \\ \sqrt{36} & 12 \end{pmatrix}, \quad \text{but} \quad \begin{pmatrix} 6 & 4 \\ 5 & 1 \end{pmatrix} \neq \begin{pmatrix} 6 & 5 \\ 4 & 1 \end{pmatrix}.
$$

Matrix Multiplication

There is an extensive arithmetic of matrices, which is explored in Excursion 6.4.A. Here we shall need only matrix multiplication. The simplest case is the product of a matrix with a single row and a matrix with a single column, where the row and column have the same number of entries. This is done by multiplying corresponding entries (first by first, second by second, and so on) and then adding the results. An example is shown in Figure 6–18.

$$
(3 \quad 1 \quad 2)\begin{pmatrix} 2 \\ 0 \\ 1 \end{pmatrix} = 3 \cdot 2 + 1 \cdot 0 + 2 \cdot 1 = 8
$$

$$
\begin{array}{ccc}
\uparrow & \uparrow & \uparrow \\
\textit{First} & \textit{Second} & \textit{Third} \\
\textit{Terms} & \textit{Terms} & \textit{Terms}
\end{array}
$$

Figure 6–18

Note that the product of a row and a column is a single number.

Now let A be an $m \times n$ matrix and B an $n \times p$ matrix, so that the number of columns of A is the same as the number of rows of B (namely, n). The product matrix AB is defined to be an $m \times p$ matrix (same number of rows as A and same number of columns as B). The product AB is defined as follows.

Matrix Multiplication

> If A is an $m \times n$ and B is an $n \times p$ matrix, then AB is the $m \times p$ matrix whose entry in the ith row and jth column is
>
> the product of the ith row of A and the jth column of B.

Example 1 If it is defined, find the product AB, where

$$A = \begin{pmatrix} 3 & 1 & 2 \\ -1 & 0 & 4 \end{pmatrix} \quad \text{and} \quad B = \begin{pmatrix} 2 & -3 & 0 & 1 \\ 0 & 5 & 2 & 7 \\ 1 & 8 & -4 & 1 \end{pmatrix}.$$

Solution A has three columns and B has three rows. So the product matrix AB is defined. AB has two rows (same as A) and four columns (same as B). Its entries are calculated as follows. The entry in row 1, column 1 of AB is the product of row 1 of A and column 1 of B, which is the number 8, as shown in Figure 6–18 and indicated at the right here:

$$\text{row 1, column 1} \begin{pmatrix} 3 & 1 & 2 \\ -1 & 0 & 4 \end{pmatrix} \begin{pmatrix} 2 & -3 & 0 & 1 \\ 0 & 5 & 2 & 7 \\ 1 & 8 & -4 & 1 \end{pmatrix} = \begin{pmatrix} 8 & & & \end{pmatrix} \qquad 3 \cdot 2 + 1 \cdot 0 + 2 \cdot 1 = 8$$

The other entries in AB are obtained similarly:

$$\text{row 1, column 2} \begin{pmatrix} 3 & 1 & 2 \\ -1 & 0 & 4 \end{pmatrix} \begin{pmatrix} 2 & -3 & 0 & 1 \\ 0 & 5 & 2 & 7 \\ 1 & 8 & -4 & 1 \end{pmatrix} = \begin{pmatrix} 8 & 12 & & \end{pmatrix} \qquad 3(-3) + 1 \cdot 5 + 2 \cdot 8 = 12$$

$$\text{row 1, column 3} \begin{pmatrix} 3 & 1 & 2 \\ -1 & 0 & 4 \end{pmatrix} \begin{pmatrix} 2 & -3 & 0 & 1 \\ 0 & 5 & 2 & 7 \\ 1 & 8 & -4 & 1 \end{pmatrix} = \begin{pmatrix} 8 & 12 & -6 & \end{pmatrix} \qquad 3 \cdot 0 + 1 \cdot 2 + 2(-4) = -6$$

$$\text{row 1, column 4} \begin{pmatrix} 3 & 1 & 2 \\ -1 & 0 & 4 \end{pmatrix} \begin{pmatrix} 2 & -3 & 0 & 1 \\ 0 & 5 & 2 & 7 \\ 1 & 8 & -4 & 1 \end{pmatrix} = \begin{pmatrix} 8 & 12 & -6 & 12 \end{pmatrix} \qquad 3 \cdot 1 + 1 \cdot 7 + 2 \cdot 1 = 12$$

$$\text{row 2, column 1} \begin{pmatrix} 3 & 1 & 2 \\ -1 & 0 & 4 \end{pmatrix} \begin{pmatrix} 2 & -3 & 0 & 1 \\ 0 & 5 & 2 & 7 \\ 1 & 8 & -4 & 1 \end{pmatrix} = \begin{pmatrix} 8 & 12 & -6 & 12 \\ 2 & & & \end{pmatrix} \qquad (-1)2 + 0 \cdot 0 + 4 \cdot 1 = 2$$

$$\text{row 2, column 2} \begin{pmatrix} 3 & 1 & 2 \\ -1 & 0 & 4 \end{pmatrix} \begin{pmatrix} 2 & -3 & 0 & 1 \\ 0 & 5 & 2 & 7 \\ 1 & 8 & -4 & 1 \end{pmatrix} = \begin{pmatrix} 8 & 12 & -6 & 12 \\ 2 & 35 & & \end{pmatrix} \qquad (-1)(-3) + 0 \cdot 5 + 4 \cdot 8 = 35$$

$$\text{row 2, column 3} \begin{pmatrix} 3 & 1 & 2 \\ -1 & 0 & 4 \end{pmatrix} \begin{pmatrix} 2 & -3 & 0 & 1 \\ 0 & 5 & 2 & 7 \\ 1 & 8 & -4 & 1 \end{pmatrix} = \begin{pmatrix} 8 & 12 & -6 & 12 \\ 2 & 35 & -16 & \end{pmatrix} \qquad (-1)0 + 0 \cdot 2 + 4(-4) = -16$$

row 2, column 4 $\begin{pmatrix} 3 & 1 & 2 \\ -1 & 0 & 4 \end{pmatrix} \begin{pmatrix} 2 & -3 & 0 & 1 \\ 0 & 5 & 2 & 7 \\ 1 & 8 & -4 & 1 \end{pmatrix} = \begin{pmatrix} 8 & 12 & -6 & 12 \\ 2 & 35 & -16 & 3 \end{pmatrix}$ $(-1)1 + 0 \cdot 7 + 4 \cdot 1 = 3$

The last matrix on the right is the product AB. ∎

Example 2 Let A, B, C, D be the following matrices.

$$A = \begin{pmatrix} 3 & 2 & 1 \\ -2 & 0 & 4 \\ 1 & -2 & 5 \end{pmatrix}, \qquad B = \begin{pmatrix} 5 & -2 \\ 1 & -1 \\ 4 & 2 \end{pmatrix}, \qquad C = \begin{pmatrix} 4 & -2 & 7 \\ 6 & 3 & 1 \\ 2 & 1 & 4 \end{pmatrix},$$

$$D = \begin{pmatrix} 1 & 0 & -5 \\ 2 & 3 & -4 \\ 3 & 7 & 2 \end{pmatrix}$$

Find each of the following matrices, if possible.

(a) AB (b) BC (c) CD and DC

Solution

(a) Following the same procedure as in Example 1 we have:

$$AB = \begin{pmatrix} 3 & 2 & 1 \\ -2 & 0 & 4 \\ 1 & -2 & 5 \end{pmatrix} \begin{pmatrix} 5 & -2 \\ 1 & -1 \\ 4 & 2 \end{pmatrix}$$

$$= \begin{pmatrix} 3 \cdot 5 + 2 \cdot 1 + 1 \cdot 4 & 3(-2) + 2(-1) + 1 \cdot 2 \\ (-2) \cdot 5 + 0 \cdot 1 + 4 \cdot 4 & (-2)(-2) + 0(-1) + 4 \cdot 2 \\ 1 \cdot 5 + (-2) \cdot 1 + 5 \cdot 4 & 1(-2) + (-2)(-1) + 5 \cdot 2 \end{pmatrix}$$

$$= \begin{pmatrix} 21 & -6 \\ 6 & 12 \\ 23 & 10 \end{pmatrix}.$$

(b) Matrix B is 3×2 and C is 3×3. Since the number of columns in B is different from the number of rows in C, the product is not defined.

(c) We use the matrix editor of a calculator to enter the matrices (Figure 6–19).

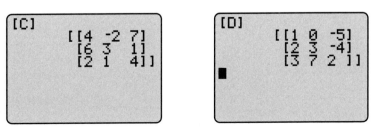

Figure 6–19

We then use the calculator to compute both products (Figure 6–20).

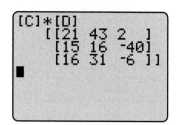
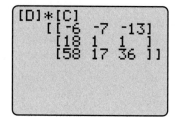

Figure 6–20

Note that DC is not equal to CD. ∎

It can be shown that matrix multiplication is associative, meaning that $A(BC) = (AB)C$ for all matrices A, B, C for which the products are defined. As we saw in Example 2, however, matrix multiplication is *not* commutative, that is AB may not be equal to BA, even when both products are defined.

Identity Matrices and Inverses

The $n \times n$ **identity matrix** I_n is the matrix with 1's on the diagonal from upper left to the lower right and 0's everywhere else; for example,

$$I_2 = \begin{pmatrix} 1 & 0 \\ 0 & 1 \end{pmatrix} \qquad I_3 = \begin{pmatrix} 1 & 0 & 0 \\ 0 & 1 & 0 \\ 0 & 0 & 1 \end{pmatrix} \qquad I_4 = \begin{pmatrix} 1 & 0 & 0 & 0 \\ 0 & 1 & 0 & 0 \\ 0 & 0 & 1 & 0 \\ 0 & 0 & 0 & 1 \end{pmatrix}.$$

The number 1 is the multiplicative identity of the number system because $a \cdot 1 = a = 1 \cdot a$ for every number a. The identity matrix I_n is the multiplicative identity for $n \times n$ matrices:

Identity Matrix

For any $n \times n$ matrix A,

$$AI_n = A = I_nA.$$

For example, in the 2 × 2 case

$$\begin{pmatrix} a & b \\ c & d \end{pmatrix}\begin{pmatrix} 1 & 0 \\ 0 & 1 \end{pmatrix} = \begin{pmatrix} a \cdot 1 + b \cdot 0 & a \cdot 0 + b \cdot 1 \\ c \cdot 1 + d \cdot 0 & c \cdot 0 + d \cdot 1 \end{pmatrix} = \begin{pmatrix} a & b \\ c & d \end{pmatrix}.$$

Verify that the same answer results if you reverse the order of multiplication.

Every nonzero number c has a multiplicative inverse $c^{-1} = 1/c$ with the property that $cc^{-1} = 1$. The analogous statement for matrix multipli-

Technology Tip

To display an $n \times n$ identity matrix, use IDENT(ITY) n in the OPS or MATH submenu of the TI MATRIX menu, in the OPE submenu of the Sharp 9600 MATRIX menu, or in the MAT submenu of the Casio 9850 OPTN menu. On HP-38 and TI-89, use IDENMAT n or IDENTITY in the MATRIX submenu of the MATH menu.

cation does not always hold, and special terminology is used when it does. An $n \times n$ matrix A is said to be **invertible** (or **nonsingular**) if there is an $n \times n$ matrix B such that $AB = I_n$. In this case it can be proved that $BA = I_n$ also. The matrix B is called the **inverse** of A and is sometimes denoted A^{-1}.

Example 3 You can readily verify that

$$\begin{pmatrix} 2 & 1 \\ 3 & 1 \end{pmatrix}\begin{pmatrix} -1 & 1 \\ 3 & -2 \end{pmatrix} = \begin{pmatrix} 1 & 0 \\ 0 & 1 \end{pmatrix} = \begin{pmatrix} -1 & 1 \\ 3 & -2 \end{pmatrix}\begin{pmatrix} 2 & 1 \\ 3 & 1 \end{pmatrix}.$$

Therefore, $A = \begin{pmatrix} 2 & 1 \\ 3 & 1 \end{pmatrix}$ is an invertible matrix with inverse $A^{-1} = \begin{pmatrix} -1 & 1 \\ 3 & -2 \end{pmatrix}$. ∎

Example 4 Find the inverse of the matrix $\begin{pmatrix} 2 & 6 \\ 1 & 4 \end{pmatrix}$.

Solution We must find numbers x, y, u, v such that

$$\begin{pmatrix} 2 & 6 \\ 1 & 4 \end{pmatrix}\begin{pmatrix} x & u \\ y & v \end{pmatrix} = \begin{pmatrix} 1 & 0 \\ 0 & 1 \end{pmatrix}$$

which is the same as

$$\begin{pmatrix} 2x + 6y & 2u + 6v \\ x + 4y & u + 4v \end{pmatrix} = \begin{pmatrix} 1 & 0 \\ 0 & 1 \end{pmatrix}.$$

Since corresponding entries in these last two matrices are equal, finding x, y, u, v amounts to solving these systems of equations:

$$\begin{array}{ccc} 2x + 6y = 1 & & 2u + 6v = 0 \\ x + 4y = 0 & \text{and} & u + 4v = 1. \end{array}$$

We shall solve the systems by the Gauss-Jordan method of Section 6.3. The augmented matrices for the two systems are

$$A = \begin{pmatrix} 2 & 6 & \vdots & 1 \\ 1 & 4 & \vdots & 0 \end{pmatrix} \quad \text{and} \quad B = \begin{pmatrix} 2 & 6 & \vdots & 0 \\ 1 & 4 & \vdots & 1 \end{pmatrix}.$$

Note that the row operations needed for both A and B will be the same (because the first two columns are the same in both A and B). Consequently, we can save space and time by combining both of these matrix into this single matrix:

$$\begin{pmatrix} 2 & 6 & 1 & \vdots & 0 \\ 1 & 4 & 0 & \vdots & 1 \end{pmatrix}.$$

The first three columns of this matrix are matrix A and the first two and last columns are matrix B. Performing row operations on this matrix amounts to doing the operations simultaneously on both A and B.

Multiply row 1 by 1/2:

$$\begin{pmatrix} 1 & 3 & \vdots & \frac{1}{2} & 0 \\ 1 & 4 & \vdots & 0 & 1 \end{pmatrix}$$

Replace row 2 by the sum of itself and −1 times row 1:

$$\begin{pmatrix} 1 & 3 & \vdots & \frac{1}{2} & 0 \\ 0 & 1 & \vdots & -\frac{1}{2} & 1 \end{pmatrix}$$

Replace row 1 by the sum of itself and −3 times row 2:

$$\begin{pmatrix} 1 & 0 & \vdots & 2 & -3 \\ 0 & 1 & \vdots & -\frac{1}{2} & 1 \end{pmatrix}$$

The first three columns of the last matrix show that $x = 2$ and $y = -1/2$. Similarly, the first two and last columns show that $u = -3$ and $v = 1$. Therefore,

$$A^{-1} = \begin{pmatrix} 2 & -3 \\ -\frac{1}{2} & 1 \end{pmatrix}.$$

Observe that A^{-1} is just the right half of the final form of the preceding augmented matrix and that the left half is the identity matrix I_2. ∎

Although the technique in Example 4 can be used to find the inverse of any matrix that has one, it's quicker to use a calculator. Any calculator with matrix capabilities can find the inverse of an invertible matrix (see the Tip in the margin).

Technology Tip

On calculators other than TI-89 you can find the inverse A^{-1} of matrix A by keying in A (or Mat A) and using the x^{-1} key. Using the $\wedge$ key and -1 produces A^{-1} on TI-89 and HP-38, but leads to an error message on other calculators.

CAUTION

A calculator should produce an error message when asked for the inverse of a matrix A that does not have one. However, because of round-off errors, it may sometimes display a matrix which it says is A^{-1}. As an accuracy check when finding inverses, multiply A by A^{-1} to see if the product is the identity matrix. If it isn't, A does not have an inverse.

Inverse Matrices and Systems of Equations

Any system of linear equations can be expressed in matrix form, as shown in the next example.

Example 5   Use matrix multiplication to express this system of equations in matrix form:

$$\begin{aligned} x + y + z &= 2 \\ 2x + 3y &= 5 \\ x + 2y + z &= -1. \end{aligned}$$

Solution   Let A be the 3×3 matrix of coefficients on the left side of the equations, B the column matrix of constants on the right side, and X the column matrix of unknowns:

$$A = \begin{pmatrix} 1 & 1 & 1 \\ 2 & 3 & 0 \\ 1 & 2 & 1 \end{pmatrix}, \qquad X = \begin{pmatrix} x \\ y \\ z \end{pmatrix} \qquad B = \begin{pmatrix} 2 \\ 5 \\ -1 \end{pmatrix}.$$

Then AX is a matrix with three rows and one column, as is B:

$$AX = \begin{pmatrix} 1 & 1 & 1 \\ 2 & 3 & 0 \\ 1 & 2 & 1 \end{pmatrix} \begin{pmatrix} x \\ y \\ z \end{pmatrix} = \begin{pmatrix} x + & y + & z \\ 2x + & 3y + & 0z \\ x + & 2y + & z \end{pmatrix} \quad \text{and} \quad B = \begin{pmatrix} 2 \\ 5 \\ -1 \end{pmatrix}.$$

The entries in AX are just the left sides of the equations of the system and the entries in B are the constants on the right sides. Therefore, the system can be expressed as the matrix equation $AX = B$. ■

Suppose a system of equations is written in matrix form $AX = B$ and that the matrix A has an inverse. Then we can solve $AX = B$ by multiplying both sides of A^{-1}:

$$A^{-1}(AX) = A^{-1}B$$
$$(A^{-1}A)X = A^{-1}B \qquad \textit{[Matrix multiplication is associative]}$$
$$I_nX = A^{-1}B \qquad \textit{[$A^{-1}A$ is the identity matrix]}$$
$$X = A^{-1}B \qquad \textit{[Product of identity matrix and X is X]}$$

The next example shows how this works in practice.

Example 6 Solve the system

$$x + y + z = 2$$
$$2x + 3y = 5$$
$$x + 2y + z = -1.$$

Solution As we saw in Example 5, this system is equivalent to the matrix equation

$$AX = B$$

$$\begin{pmatrix} 1 & 1 & 1 \\ 2 & 3 & 0 \\ 1 & 2 & 1 \end{pmatrix} \begin{pmatrix} x \\ y \\ z \end{pmatrix} = \begin{pmatrix} 2 \\ 5 \\ -1 \end{pmatrix}.$$

Use a calculator to find the inverse of the coefficient matrix A and multiply both sides of the equation by A^{-1}:

$$A^{-1}AX = A^{-1}B$$

$$\begin{pmatrix} 1.5 & .5 & -1.5 \\ -1 & 0 & 1 \\ .5 & -.5 & .5 \end{pmatrix} \begin{pmatrix} 1 & 1 & 1 \\ 2 & 3 & 0 \\ 1 & 2 & 1 \end{pmatrix} \begin{pmatrix} x \\ y \\ z \end{pmatrix} = \begin{pmatrix} 1.5 & .5 & -1.5 \\ -1 & 0 & 1 \\ .5 & -.5 & .5 \end{pmatrix} \begin{pmatrix} 2 \\ 5 \\ -1 \end{pmatrix}$$

$$\begin{pmatrix} 1 & 0 & 0 \\ 0 & 1 & 0 \\ 0 & 0 & 1 \end{pmatrix} \begin{pmatrix} x \\ y \\ z \end{pmatrix} = \begin{pmatrix} 1.5 & .5 & -1.5 \\ -1 & 0 & 1 \\ .5 & -.5 & .5 \end{pmatrix} \begin{pmatrix} 2 \\ 5 \\ -1 \end{pmatrix} \qquad \textit{[Since $A^{-1}A = I_3$]}$$

$$\begin{pmatrix} x \\ y \\ z \end{pmatrix} = \begin{pmatrix} 7 \\ -3 \\ -2 \end{pmatrix} \qquad \textit{[Since $I_3X = X$]}$$

Therefore, the solution of the original system is $x = 7, y = -3, z = -2$. ■

Only a matrix with the same number of rows as columns can possibly have an inverse. Consequently, the method of Example 6 can be tried only when the system has the same number of equations as unknowns. In this case, you should use your calculator to verify that the coefficient matrix actually has an inverse (see the caution on page 470). If it does not, other methods must be used. Here is a summary of the possibilities:

Matrix Solution of a System of Equations

Suppose a system of equations is written in matrix form as $AX = B$.

If the matrix A has an inverse, then the unique solution of the system is

$$X = A^{-1}B.$$

If A does not have an inverse (which is always the case when the number of equations differs from the number of unknowns), then the system either has no solutions or has infinitely many solutions. Its solutions (if any) may be found using Gauss-Jordan elimination (Section 6.3).

Example 7 Solve the system

$$2x + \ y \ - \ z = 2$$
$$x + 3y + 2z = 1$$
$$x + \ y + \ z = 2.$$

Solution   Since there are the same number of equations as unknowns, we can try the method of Example 6. In this case we have

$$A = \begin{pmatrix} 2 & 1 & -1 \\ 1 & 3 & 2 \\ 1 & 1 & 1 \end{pmatrix} \quad \text{and} \quad B = \begin{pmatrix} 2 \\ 1 \\ 2 \end{pmatrix}.$$

CALCULATOR EXPLORATION

Verify that the matrix A does have an inverse. Show that the solutions of the system are $x = 2, y = -1, z = 1$ by computing $A^{-1}B$. ■

Applications

Just as two points determine a unique line, three points (that aren't on the same line) determine a unique parabola, as the next example demonstrates.

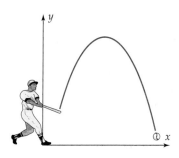

Figure 6–21

Example 8   A batter hits a baseball, and special measuring devices locate its position at various times during its flight. If the path of the ball is drawn on a coordinate plane, with the batter at the origin, it looks like Figure 6–21. According to the measuring devices, the ball passes through the points (7, 9), (47, 38), and (136, 70).

(a)   What is the equation of the path of the ball?

(b)   How far from the batter does the ball hit the ground?

Solution

(a)   The path of the ball appears to be part of a parabola (a fact that will not be proved here) and hence has an equation of the form

$$y = ax^2 + bx + c$$

for some constants a, b, c. Since (7, 9) is on the graph, we know that

when $x = 7$, then $y = 9$,

so that

$$9 = a(7^2) + b(7) + c, \qquad \text{or equivalently,} \qquad 49a + 7b + c = 9.$$

Similarly, since (47, 38) and (136, 70) are on the graph, we have

$$38 = a(47^2) + b(47) + c, \qquad \text{or equivalently,} \qquad 2209a + 47b + c = 38$$

and

$$70 = a(136^2) + b(136) + c, \qquad \text{or equivalently,} \qquad 18{,}496a + 136b + c = 70.$$

We can determine $a, b,$ and c by solving this system of equations:

$$49a + \quad 7b + c = 9$$
$$2209a + \quad 47b + c = 38$$
$$18{,}496a + 136b + c = 70,$$

or in matrix form,

$$\begin{pmatrix} 49 & 7 & 1 \\ 2209 & 47 & 1 \\ 18{,}496 & 136 & 1 \end{pmatrix} \begin{pmatrix} a \\ b \\ c \end{pmatrix} = \begin{pmatrix} 9 \\ 38 \\ 70 \end{pmatrix}.$$

A calculator shows that the solution is

$$\begin{pmatrix} a \\ b \\ c \end{pmatrix} = \begin{pmatrix} 49 & 7 & 1 \\ 2209 & 47 & 1 \\ 18{,}496 & 136 & 1 \end{pmatrix}^{-1} \begin{pmatrix} 9 \\ 38 \\ 70 \end{pmatrix} \approx \begin{pmatrix} -.00283 \\ .87798 \\ 2.99296 \end{pmatrix}.$$

Therefore the approximate equation for the ball's path is

$$y = -.00283x^2 + .87798x + 2.99296.$$

(b)   The ball hits the ground at a point where its height y is 0, that is, when

$$-.00283x^2 + .87798x + 2.99296 = 0.$$

Use the quadratic formula or an equation solver to verify that the solutions of this equation are $x \approx -3.37$ (which is not applicable here) and $x \approx 313.61$. Therefore, the ball hits the ground approximately 314 feet from the batter.   ∎

Given any three points not on a straight line, the method in Example 8 can be used to find the unique parabola that passes through these points. This parabola can also be found by quadratic regression.*

CALCULATOR EXPLORATION

Use quadratic regression on the three points in Example 8 and verify that the equation obtained is the same as the one in Example 8.

To see why regression produces the parabola that actually passes through the three points, recall that the error in a quadratic model is measured by taking the difference between the y-coordinate of each data point and the y-coordinate of the corresponding point on the model and summing the squares of these errors. If the data points actually lie on a parabola (which is always the case with three points not on a line), then the error for that parabola will be 0 (smallest possible error). Hence, it will be the parabola produced by the least squares quadratic regression procedure.

Exercises 6.4

In Exercises 1–6, determine if the product AB or BA is defined. If a product is defined, state its size (number of rows and columns). Do not calculate any products.

1. $A = \begin{pmatrix} 3 & 6 & 7 \\ 8 & 0 & 1 \end{pmatrix}$, $B = \begin{pmatrix} 2 & 5 & 9 & 1 \\ 7 & 0 & 0 & 6 \\ -1 & 3 & 8 & 7 \end{pmatrix}$

2. $A = \begin{pmatrix} -1 & -2 & -5 \\ 9 & 2 & -1 \\ 10 & 34 & 5 \end{pmatrix}$, $B = \begin{pmatrix} 17 & -9 \\ -6 & 12 \\ 3 & 5 \end{pmatrix}$

3. $A = \begin{pmatrix} 1 & 0 \\ 1 & 1 \\ 0 & 1 \end{pmatrix}$, $B = \begin{pmatrix} 5 & 6 & 11 \\ 7 & 8 & 15 \end{pmatrix}$

4. $A = \begin{pmatrix} 1 & -5 & 7 \\ 2 & 4 & 8 \\ 1 & -1 & 2 \end{pmatrix}$, $B = \begin{pmatrix} -2 & 4 & 9 \\ 13 & -2 & 1 \\ 5 & 25 & 0 \end{pmatrix}$

5. $A = \begin{pmatrix} -4 & 15 \\ 3 & -7 \\ 2 & 10 \end{pmatrix}$, $B = \begin{pmatrix} 1 & 2 \\ 3 & 4 \end{pmatrix}$

6. $A = \begin{pmatrix} 10 & 12 \\ -6 & 0 \\ 1 & 23 \\ -4 & 0 \end{pmatrix}$, $B = \begin{pmatrix} 1 & 2 & 3 \\ 3 & 2 & 1 \end{pmatrix}$

In Exercises 7–12, find AB.

7. $A = \begin{pmatrix} 3 & 2 \\ 2 & 4 \end{pmatrix}$, $B = \begin{pmatrix} 1 & -2 & 3 \\ 0 & 3 & 1 \end{pmatrix}$

8. $A = \begin{pmatrix} -1 & 2 & 3 \\ 0 & -1 & 2 \\ 1 & 2 & 0 \end{pmatrix}$, $B = \begin{pmatrix} 3 & -2 & -1 \\ 1 & 0 & 5 \\ 1 & -1 & -1 \end{pmatrix}$

9. $A = \begin{pmatrix} 1 & 0 & -4 \\ 0 & 2 & -1 \\ 2 & 3 & 4 \end{pmatrix}$, $B = \begin{pmatrix} 1 & 1 \\ 1 & 0 \\ 0 & 1 \end{pmatrix}$

10. $A = \begin{pmatrix} 1 & -2 \\ 3 & 0 \\ 0 & -1 \\ 2 & 1 \end{pmatrix}$, $B = \begin{pmatrix} -1 & 3 & -2 & 0 \\ 6 & 1 & 0 & -2 \end{pmatrix}$

11. $A = \begin{pmatrix} 2 & 0 & -1 \\ 1 & 1 & 2 \\ 0 & 2 & -3 \\ 2 & 3 & 0 \end{pmatrix}$, $B = \begin{pmatrix} 1 & 0 & 1 & 1 \\ 1 & 1 & 0 & 1 \\ 1 & 1 & 1 & 0 \end{pmatrix}$

12. $A = \begin{pmatrix} 10 & 0 & 1 & 0 \\ -1 & 1 & 0 & 1 \end{pmatrix}$, $B = \begin{pmatrix} 2 & -1 & 0 & 1 \\ -2 & 3 & 1 & -4 \\ 3 & 5 & 2 & -5 \end{pmatrix}$

---

*If you have not read Sections 1.5 and 4.4, you may skip this discussion.

In Exercises 13–16, show that AB is not equal to BA by computing both products.

13. $A = \begin{pmatrix} 3 & 2 \\ 5 & 1 \end{pmatrix}$, $B = \begin{pmatrix} 7 & -5 \\ -2 & 6 \end{pmatrix}$

14. $A = \begin{pmatrix} 3/2 & 2 \\ 4 & 7/2 \end{pmatrix}$, $B = \begin{pmatrix} 1 & -3/2 \\ 5/2 & 1 \end{pmatrix}$

15. $A = \begin{pmatrix} 4 & 2 & -1 \\ 0 & 1 & 2 \\ -3 & 0 & 1 \end{pmatrix}$, $B = \begin{pmatrix} 1 & 7 & -5 \\ 2 & -2 & 6 \\ 0 & 0 & 0 \end{pmatrix}$

16. $A = \begin{pmatrix} 1 & 1 & -1 & 1 \\ 2 & 0 & 3 & 2 \\ -3 & 0 & 0 & 1 \\ 1 & -1 & 1 & 2 \end{pmatrix}$, $B = \begin{pmatrix} 0 & 1 & 7 & 7 \\ 2 & 3 & -2 & 1 \\ 5 & 0 & 1 & 0 \\ -1 & 0 & 1 & 0 \end{pmatrix}$

In Exercises 17–24, find the inverse of the matrix, if it exists.

17. $\begin{pmatrix} 1 & 2 \\ 3 & 4 \end{pmatrix}$

18. $\begin{pmatrix} 3 & 5 \\ 1 & 4 \end{pmatrix}$

19. $\begin{pmatrix} 3 & -1 \\ -6 & 2 \end{pmatrix}$

20. $\begin{pmatrix} 1 & -1 & 0 \\ 1 & 0 & -1 \\ 6 & -2 & -3 \end{pmatrix}$

21. $\begin{pmatrix} 1 & 2 & 0 \\ 3 & -1 & 2 \\ -2 & 3 & -2 \end{pmatrix}$

22. $\begin{pmatrix} 1 & -3 & 4 \\ 2 & -5 & 7 \\ 0 & -1 & 1 \end{pmatrix}$

23. $\begin{pmatrix} 5 & 0 & 2 \\ 2 & 2 & 1 \\ -3 & 1 & -1 \end{pmatrix}$

24. $\begin{pmatrix} -1 & 3 & 1 \\ 2 & 5 & 0 \\ 3 & 1 & -2 \end{pmatrix}$

In Exercises 25–28, solve the system of equations by using the method of Example 6.

25.
$$-x + y \quad\quad = 1$$
$$-x \quad\quad + z = -2$$
$$6x - 2y - 3z = 3$$

26.
$$x + 2y + 3z = 1$$
$$2x + 5y + 3z = 0$$
$$x + \quad\quad 8z = -1$$

27.
$$2x + y \quad\quad = 0$$
$$-4x - y - 3z = 1$$
$$3x + y + 2z = 2$$

28.
$$-3x - 3y - 4z = 2$$
$$y + z = 1$$
$$4x + 3y + 4z = 3$$

In Exercises 29–39, solve the system by any method.

29.
$$x + y + \quad\quad 2w = 3$$
$$2x - y + z - w = 5$$
$$3x + 3y + 2z - 2w = 0$$
$$x + 2y + z \quad\quad = 2$$

30.
$$x - 2y + 3z \quad\quad = 1$$
$$y - z + w = -2$$
$$-2x + 2y - 2z + 4w = 5$$
$$2y - 3z + w = 8$$

31.
$$x + y + 6z + 2v \quad\quad = 1.5$$
$$x + \quad\quad 5z + 2v - 3w = 2$$
$$3x + 2y + 17z + 6v - 4w = 2.5$$
$$4x + 3y + 21z + 7v - 2w = 3$$
$$-6x - 5y - 36z - 12v + 3w = 3.5$$

32.
$$x - 1.5y + \quad\quad 1.5v - 4w = 3$$
$$-1.5y + \quad\quad .5w = 0$$
$$-x + 2y + .5z - 2v + 4.5w = 2$$
$$-.5y - 2.5z + .5v - .75w = 8$$
$$y - .5z + \quad\quad .5w = 4$$

33.
$$x + 2y + 2z - \quad\quad 2w = -23$$
$$4x + 4y - z + 5w = 7$$
$$-2x + 5y + 6z + 4w = 0$$
$$5x + 13y + 7z + 12w = -7$$

34.
$$x + 4y + 5z + 2w = 0$$
$$2x + y + 4z - 2w = 0$$
$$-x + 7y + 10z + 5w = 0$$
$$-4x + 2y + z + 5w = 0$$

35.
$$x + 2y + 5z - 2v + 4w = 0$$
$$2x - 4y + 6z + v + 4w = 0$$
$$5x + 2y - 3z + 2v + 3w = 0$$
$$6x - 5y - 2z + 5v + 3w = 0$$
$$x + 2y - z - 2v + 4w = 0$$

36.
$$4x + 2y + 3z + \quad\quad 3v + 2w = 1$$
$$2x + \quad\quad 2z + 2u + v - w = 2$$
$$10x + 2y + 10z + 10u + 3v + 4w = 5$$
$$16x + 4y + 16z + 18u + 7v - 2w = -3$$
$$x + 2y + 4z - 6u + 2v + w = -2$$
$$6x + 2y + 6z + 6u + 3v \quad\quad = 1$$

37.
$$x + 2y + 3z \quad\quad = 1$$
$$3x + 2y + 4z \quad\quad = -1$$
$$2x + 6y + 8z + w = 3$$
$$2x + \quad\quad 2z - 2w = 3$$

38.
$$x + 2y + 4z = 6$$
$$y + z = 1$$
$$x + 3y + 5z = 10$$

39.
$$x \quad\quad\quad + 3w = -2$$
$$x - 4y - z + 3w = -7$$
$$4y + z \quad\quad = 5$$
$$-x + 12y + 3z - 3w = 17$$

In Exercises 40–46, find constants a, b, c such that the three given points lie on the parabola $y = ax^2 + bx + c$. See Example 8.

40. $(-3, 2), (1, 1), (2, -1)$

41. $(-3, 15), (1, -7), (5, 111)$

42. $(1, -2), (3, 1), (4, -1)$

43. $(1, 0), (-1, 6), (2, 3)$

44. $(1, 1), (0, 0), (-1, 2)$

45. $(-1, 6), (-2, 16), (1, 4)$

46. $(-1, -6), (2, -3), (4, -25)$

47. Find constants a, b, c such that the points $(0, -2)$, $(\ln 2, 1)$, and $(\ln 4, 4)$ lie on the graph of $f(x) = ae^x + be^{-x} + c$. [*Hint:* Proceed as in Example 8.]

48. Find constants a, b, c such that the points $(0, -1)$, $(\ln 2, 4)$, and $(\ln 3, 7)$ lie on the graph of $f(x) = ae^x + be^{-x} + c$.

49. New Army Stores, a national chain of casual clothing stores, recently sent shipments of jeans, jackets, sweaters, and shirts to its stores in various cities. The number of items shipped to each city and their total wholesale cost are shown in the table below. Find the wholesale price of one pair of jeans, one jacket, one sweater, and one shirt.

City	Jeans	Jackets	Sweaters	Shirts	Total Cost
Cleveland	3000	3000	2200	4200	$507,650
St. Louis	2700	2500	2100	4300	459,075
Seattle	5000	2000	1400	7500	541,225
Phoenix	7000	1800	600	8000	571,500

50. A 100-bed nursing home provides two levels of long-term care: regular and maximum. Patients at each level have a choice of a private room or a less expensive semiprivate room. The table below shows the number of patients in each category at various times last year and the total daily cost for these patients. Find the daily cost of each of the following: a private room (regular care), a private room (maximum care), a semiprivate room (regular care), and a semiprivate room (maximum care).

51. A candy company produces three types of gift boxes: A, B, and C. A box of variety A contains .6 pound of chocolates and .4 pound of mints. A box of variety B contains .3 pound of chocolates, .4 pounds of mints, and .3 pound of caramels. A box of variety C contains .5 pound of chocolates, .3 pound of mints, and .2 pound of caramels. The company has 41,400 pounds of chocolates, 29,400 pounds of mints, and 16,200 pounds of caramels in stock. How many boxes of each kind should be made in order to use up all their stock?

52. Certain circus animals are fed the same three food mixes: R, S, and T. Lions receive 1.1 units of R, 2.4 units of S, and 3.7 units of T each day. Horses receive 8.1 units of R, 2.9 units of S, and 5.1 units of T each day. Bears receive 1.3 units of R, 1.3 units of S, and 2.3 units of T each day. If 16,000 units of R, 28,000 units of S, and 44,000 units of T are available each day, how many of each type of animal can be supported?

Month	Regular Care Patients Semiprivate	Regular Care Patients Private	Maximum Care Patients Semiprivate	Maximum Care Patients Private	Daily Cost
January	22	8	60	10	$18,824
April	26	8	54	12	$18,738
July	24	14	56	6	$18,606
October	20	10	62	8	$18,824

EXCURSION **Matrix Algebra**

Matrices and matrix multiplication were introduced in Section 6.4. In this section, we consider other aspects of matrix algebra. Recall that an ***m × n*** **matrix** is a rectangular array of numbers, with m horizontal **rows** (numbered from top to bottom) and n vertical **columns** (numbered from left to right), such as

$$\begin{pmatrix} 3 & 2 & -5 \\ 6 & 1 & 7 \\ -2 & 0 & 5 \end{pmatrix} \qquad \begin{pmatrix} \frac{3}{2} \\ -5 \\ 0 \\ 12 \end{pmatrix}$$

3 × 3 matrix   4 × 1 matrix

When discussing matrices in a general context, a typical matrix may be denoted by a capital letter, such as A, or by an array like this

$$\begin{pmatrix} a_{11} & a_{12} & a_{13} & \cdots & a_{1n} \\ a_{21} & a_{22} & a_{23} & \cdots & a_{2n} \\ a_{31} & a_{32} & a_{33} & \cdots & a_{3n} \\ \vdots & \vdots & \vdots & & \vdots \\ a_{m1} & a_{m2} & a_{m3} & \cdots & a_{mn} \end{pmatrix}$$

in which a_{ij} denotes the entry in row i and column j. This array notation is often abbreviated as (a_{ij}).

We have already seen how matrices can be used in the solution of systems of linear equations. They also provide a convenient way to display data.

Example 1 At the beginning of a laboratory experiment, five baby rats measured 5.5, 6.1, 6.8, 7.4, and 6.2 centimeters in length and weighed 139, 141, 148, 153, and 145 grams, respectively. We can summarize this data with the following 2 × 5 matrix:

$$\begin{pmatrix} 5.5 & 6.1 & 6.8 & 7.4 & 6.2 \\ 139 & 141 & 148 & 153 & 145 \end{pmatrix},$$

in which row 1 represents length and row 2 represents weight, with each column corresponding to one of the rats. ■

Recall that two matrices are said to be **equal** if they have the same size and the corresponding entries are equal. For example,

$$\begin{pmatrix} -3 & (-1)^2 & \frac{6}{8} \\ 6 & 12 & 0 \end{pmatrix} = \begin{pmatrix} -3 & 1 & \frac{3}{4} \\ \sqrt{36} & 3 \cdot 4 & 0 \end{pmatrix}$$

But

$$\begin{pmatrix} 6 & 4 \\ 5 & 1 \end{pmatrix} \neq \begin{pmatrix} 6 & 5 \\ 4 & 1 \end{pmatrix}$$

because the corresponding entries in row 1, column 2, aren't equal $(4 \neq 5)$, and similarly in row 2, column 1. More formally,

Equality of Matrices

If $A = (a_{ij})$ and $B = (b_{ij})$ are $m \times n$ matrices,

then $A = B$ means $a_{ij} = b_{ij}$ for every i and j.

Matrix Addition and Subtraction

Unlike the situation with multiplication, matrix addition and subtraction is defined only when the two matrices have the same size. In that case, the rule is: add (or subtract) the corresponding entries. In formal terms,

Matrix Addition

The sum of the $m \times n$ matrices $A = (a_{ij})$ and $B = (b_{ij})$ is the $m \times n$ matrix $A + B$ whose entry in row i, column j, is

$$a_{ij} + b_{ij}$$

Example 2   Let A, B, C, D be the following matrices.

$$A = \begin{pmatrix} 1 & 2 \\ 3 & 0 \\ -7 & 4 \end{pmatrix}, \quad B = \begin{pmatrix} 5 & 8 \\ -3 & 7 \\ 4 & 5 \end{pmatrix}, \quad C = \begin{pmatrix} 6 & -3 \\ -2 & 8 \end{pmatrix}, \quad D = \begin{pmatrix} 1 & 4 \\ 5 & -4 \end{pmatrix}$$

Find each of these:

(a) $A + B$ (b) $B + A$ (c) $B + C$ (d) $C - D$.

Solution

(a) $A + B = \begin{pmatrix} 1 & 2 \\ 3 & 0 \\ -7 & 4 \end{pmatrix} + \begin{pmatrix} 5 & 8 \\ -3 & 7 \\ 4 & 5 \end{pmatrix} = \begin{pmatrix} 1+5 & 2+8 \\ 3+(-3) & 0+7 \\ -7+4 & 4+5 \end{pmatrix} = \begin{pmatrix} 6 & 10 \\ 0 & 7 \\ -3 & 9 \end{pmatrix}.$

(b) $B + A = \begin{pmatrix} 5 & 8 \\ -3 & 7 \\ 4 & 5 \end{pmatrix} + \begin{pmatrix} 1 & 2 \\ 3 & 0 \\ -7 & 4 \end{pmatrix} = \begin{pmatrix} 5+1 & 8+2 \\ -3+3 & 7+0 \\ 4+(-7) & 5+4 \end{pmatrix} = \begin{pmatrix} 6 & 10 \\ 0 & 7 \\ -3 & 9 \end{pmatrix}.$

Note that $A + B = B + A$.

(c)   The sum $B + C$ is not defined because B and C are different sizes.

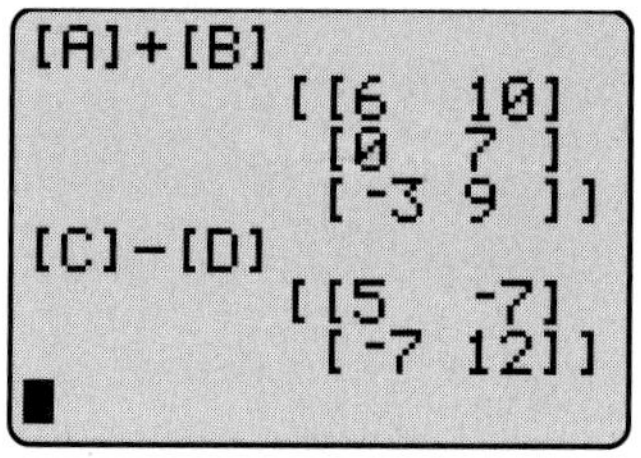

Figure 6–22

(d)　$C - D = \begin{pmatrix} 6 & -3 \\ -2 & 8 \end{pmatrix} - \begin{pmatrix} 1 & 4 \\ 5 & -4 \end{pmatrix} = \begin{pmatrix} 6-1 & -3-4 \\ -2-5 & 8-(-4) \end{pmatrix} = \begin{pmatrix} 5 & -7 \\ -7 & 12 \end{pmatrix}.$

The sums and differences in this example can also be found on a calculator (Figure 6–22).　■

Example 3　Back to the laboratory. As we saw in Example 1, the matrix

$$A = \begin{pmatrix} 5.5 & 6.1 & 6.8 & 7.4 & 6.2 \\ 139 & 141 & 148 & 153 & 145 \end{pmatrix}$$

summarizes the length (in centimeters) and weight (in grams) of five rats. Two weeks later, their lengths and weights are given by the matrix

$$B = \begin{pmatrix} 10.1 & 11.3 & 11.3 & 12.5 & 10.9 \\ 195 & 197 & 220 & 252 & 234 \end{pmatrix}.$$

The amount that the length and weight of each rat has changed during this two-week period can be represented by the matrix

$$B - A = \begin{pmatrix} 10.1 & 11.3 & 11.3 & 12.5 & 10.9 \\ 195 & 197 & 220 & 252 & 234 \end{pmatrix} - \begin{pmatrix} 5.5 & 6.1 & 6.8 & 7.4 & 6.2 \\ 139 & 141 & 148 & 153 & 145 \end{pmatrix}$$

$$= \begin{pmatrix} 4.6 & 5.2 & 4.5 & 5.1 & 4.7 \\ 56 & 56 & 72 & 99 & 89 \end{pmatrix}.$$

The matrix shows, for example that the second rat had the largest gain in length (5.2 cm) and the fourth rat had the largest weight gain (99 g).　■

Scalar Multiplication

Matrix arithmetic also has an operation that has no analogue in the arithmetic of real numbers. The product of a single number c and a matrix is defined by this rule: Multiply every entry in the matrix by c. For example,

$$2\begin{pmatrix} 3 & -4 & 0 \\ 1 & 5 & 2 \end{pmatrix} = \begin{pmatrix} 2 \cdot 3 & 2(-4) & 2 \cdot 0 \\ 2 \cdot 1 & 2 \cdot 5 & 2 \cdot 2 \end{pmatrix} = \begin{pmatrix} 6 & -8 & 0 \\ 2 & 10 & 4 \end{pmatrix}.$$

The process of multiplying a number by a matrix is called **scalar multiplication,** and the number is sometimes called a **scalar.** Here is the formal definition:

Scalar Multiplication

The product of an $m \times n$ matrix $A = (a_{ij})$ and a scalar c is the $m \times n$ matrix cA whose entry in row i, column j, is ca_{ij}.

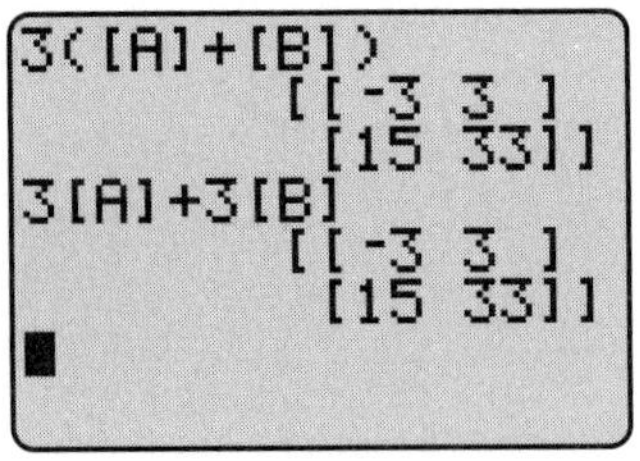

```
3([A]+[B])
        [[-3  3 ]
         [15 33]]
3[A]+3[B]
        [[-3  3 ]
         [15 33]]
■
```

Figure 6–23

Example 4 Show that $3(A + B) = 3A + 3B$, where

$$A = \begin{pmatrix} 2 & -1 \\ 0 & 4 \end{pmatrix} \quad \text{and} \quad B = \begin{pmatrix} -3 & 2 \\ 5 & 7 \end{pmatrix}.$$

Solution Figure 6–23 shows that the matrix $3(A + B)$ is the same as the matrix $3A + 3B$ and hence, $3(A + B) = 3A + 3B$. Here is how the same calculation is done by hand:

$$A + B = \begin{pmatrix} 2 & -1 \\ 0 & 4 \end{pmatrix} + \begin{pmatrix} -3 & 2 \\ 5 & 7 \end{pmatrix} = \begin{pmatrix} 2 + (-3) & -1 + 2 \\ 0 + 5 & 4 + 7 \end{pmatrix} = \begin{pmatrix} -1 & 1 \\ 5 & 11 \end{pmatrix}$$

so that

$$3(A + B) = 3\begin{pmatrix} -1 & 1 \\ 5 & 11 \end{pmatrix} = \begin{pmatrix} 3(-1) & 3 \cdot 1 \\ 3 \cdot 5 & 3 \cdot 11 \end{pmatrix} = \begin{pmatrix} -3 & 3 \\ 15 & 33 \end{pmatrix}.$$

On the other hand,

$$3A = 3\begin{pmatrix} 2 & -1 \\ 0 & 4 \end{pmatrix} = \begin{pmatrix} 3 \cdot 2 & 3(-1) \\ 3 \cdot 0 & 3 \cdot 4 \end{pmatrix} = \begin{pmatrix} 6 & -3 \\ 0 & 12 \end{pmatrix}$$

and

$$3B = 3\begin{pmatrix} -3 & 2 \\ 5 & 7 \end{pmatrix} = \begin{pmatrix} 3(-3) & 3 \cdot 2 \\ 3 \cdot 5 & 3 \cdot 7 \end{pmatrix} = \begin{pmatrix} -9 & 6 \\ 15 & 21 \end{pmatrix}$$

so that

$$3A + 3B = \begin{pmatrix} 6 & -3 \\ 0 & 12 \end{pmatrix} + \begin{pmatrix} -9 & 6 \\ 15 & 21 \end{pmatrix} = \begin{pmatrix} -3 & 3 \\ 15 & 33 \end{pmatrix}.$$

Therefore $3(A + B) = 3A + 3B$. ■

Here is a summary of the properties of matrix addition and subtraction and scalar multiplication. Property 1 was illustrated in Example 2 and Property 3 in Example 4. The others are illustrated in Exercises 23–27.

Properties of Matrix Addition and Scalar Multiplication ▶

For any $m \times n$ matrices A, B, C and numbers c, d:

1. $A + B = B + A$
2. $A + (B + C) = (A + B) + C$
3. $c(A + B) = cA + cB$
4. $(cd)A = c(dA)$
5. $(c + d)A = cA + dA$

Matrix Multiplication

In Example 2(c) of Section 6.4, we saw that matrix multiplication, unlike multiplication of numbers, is not commutative, that is, AB may not be equal to BA. Furthermore, the product of two nonzero matrices may be the zero matrix, unlike the situation with numbers (where a product is zero only if one of the factors is zero). For instance, if

$$A = \begin{pmatrix} 4 & 6 \\ 2 & 3 \end{pmatrix} \quad \text{and} \quad B = \begin{pmatrix} -3 & -9 \\ 2 & 6 \end{pmatrix},$$

then

$$AB = \begin{pmatrix} 4(-3) + 6 \cdot 2 & 4(-9) + 6 \cdot 6 \\ 2(-3) + 3 \cdot 2 & 2(-9) + 3 \cdot 6 \end{pmatrix} = \begin{pmatrix} 0 & 0 \\ 0 & 0 \end{pmatrix}.$$

However, matrix multiplication does have some other familiar properties in common with real-number arithmetic.

Properties of Matrix Multiplication

For any matrices *A*, *B*, *C* for which each of the following sums and products are defined,

1. $A(BC) = (AB)C$
2. $A(B + C) = AB + AC$
3. $(B + C)A = BA + CA$

CALCULATOR EXPLORATION

Enter the following matrices in your calculator:

$$A = \begin{pmatrix} 2 & 0 & -2 \\ 4 & 1 & -5 \\ 7 & 3 & -6 \end{pmatrix}, \quad B = \begin{pmatrix} -2 & 1.5 & 4 \\ 1 & 2 & 0 \\ 0 & 3.2 & -1 \end{pmatrix}, \quad C = \begin{pmatrix} 9 & 8 & 7 \\ 4 & 5 & 6 \\ 3 & 2 & 1 \end{pmatrix}.$$

Verify Property 2 in the preceding box by showing that $A(B + C)$ and $AB + AC$ are the same matrix.

Applications

Matrix arithmetic can sometimes be used with data that is presented in matrix form.

Example 5 McChicken Fast Food has three locations in Roanoke, VA, each of which sells chicken nuggets at $2.35, Buffalo wings at $2.95, and filet of chicken sandwiches at $1.95. The number of each item sold daily at each location is given in the following matrix:

	Nuggets	Wings	Sandwiches
Store 1	220	400	280
Store 2	360	300	225
Store 3	180	380	300

Use matrix multiplication to find the daily revenue at each store.

Solution The daily revenue per item can be represented by the 3×1 matrix

$$
\begin{array}{c} \\ Nuggets \\ Wings \\ Sandwiches \end{array}
\begin{array}{c} Revenue \\ \begin{pmatrix} 2.35 \\ 2.95 \\ 1.95 \end{pmatrix} \end{array}.
$$

The daily revenue per store is the product:

$$(\text{Store} \times \text{Item}) \cdot (\text{Item} \times \text{Revenue}) = \quad \text{Store} \times \text{Revenue}$$

$$
\begin{pmatrix} 220 & 400 & 280 \\ 360 & 300 & 225 \\ 180 & 380 & 300 \end{pmatrix} \cdot \begin{pmatrix} 2.35 \\ 2.95 \\ 1.95 \end{pmatrix} = \begin{pmatrix} 220(2.35) + 400(2.95) + 280(1.95) \\ 360(2.35) + 300(2.95) + 225(1.95) \\ 180(2.35) + 380(2.95) + 300(1.95) \end{pmatrix}
$$

$$
= \begin{pmatrix} \$2243 \\ \$2169.75 \\ \$2129 \end{pmatrix}.
$$

Therefore, the daily revenue is $2243 at Store 1, $2169.75 at Store 2, and $2129 at Store 3. ■

Matrices also have applications in cryptography (the study of secret codes).

Example 6 Use the matrix

$$
A = \begin{pmatrix} 3 & -1 & 2 & 6 \\ 1 & 3 & -1 & 0 \\ 5 & -2 & 1 & -3 \\ 3 & -7 & 4 & 4 \end{pmatrix}
$$

to encode the message "Attack at dawn."

Solution The first step is to translate the message into numerical form by assigning a number to each letter. For simplicity, we use:

0 = space	7 = G	14 = N	21 = U
1 = A	8 = H	15 = O	22 = V
2 = B	9 = I	16 = P	23 = W
3 = C	10 = J	17 = Q	24 = X
4 = D	11 = K	18 = R	25 = Y
5 = E	12 = L	19 = S	26 = Z
6 = F	13 = M	20 = T	

Since the encoding matrix A is 4×4, we break our message into blocks of four letters (using # for space) and write them in numerical form:

$$\begin{array}{cccc} A\ T\ T\ A & C\ K\ \#\ A & T\ \#\ D\ A & W\ N\ \#\ \# \\ (1\ 20\ 20\ 1) & (3\ 11\ 0\ 1) & (20\ 0\ 4\ 1) & (23\ 1\ 0\ 0) \end{array}$$

Each message block is a 4×1 matrix; to encode it, we multiply by the matrix A:

Plain Text	**Encoding Matrix**	**Coded Text**

$$(1\ \ 20\ \ 20\ \ 1)\begin{pmatrix} 3 & -1 & 2 & 6 \\ 1 & 3 & -1 & 0 \\ 5 & -2 & 1 & -3 \\ 3 & -7 & 4 & 4 \end{pmatrix} = (126\ \ 12\ \ 6\ \ -50)$$

$$(3\ \ 11\ \ 0\ \ 1)\begin{pmatrix} 3 & -1 & 2 & 6 \\ 1 & 3 & -1 & 0 \\ 5 & -2 & 1 & -3 \\ 3 & -7 & 4 & 4 \end{pmatrix} = (23\ \ 23\ \ -1\ \ 22)$$

CALCULATOR EXPLORATION

Enter the matrix A in your calculator and encode the rest of the message by computing $(20\ \ 0\ \ 4\ \ 1)A$ and $(23\ \ 1\ \ 0\ \ 0)A$.

Using the results of the Exploration, we send the encoded message as a single list of numbers:

$$126, 12, 6, -50, 23, 23, -1, 22, 83, -35, 48, 112, 70, -20, 45, 138.$$

The story continues in the next example. ■

Example 7 After the message in Example 6 was sent, the following reply was received in the same code:

$$122, -126, 75, 32, 68, 38, 14, 94, 94, 17, 33, 129,$$
$$180, -151, 99, 59, 129, -17, 43, 84, 174, -97, 84, 95.$$

Decode this message.

Solution We first break the message into blocks of four numbers each, which we think of as 1×4 matrices:

$(122\ \ -126\ \ 75\ \ 32),\quad (68\ \ 38\ \ 14\ \ 94),\quad (94\ \ 17\ \ 33\ \ 129),$

$(180\ \ -151\ \ 99\ \ 59),\quad (129\ \ -17\ \ 43\ \ 84),\quad (174\ \ -97\ \ 84\ \ 95).$

Recall how this message was encoded. Each four-letter block is written numerically as a 4×1 matrix E and the coded text is the product EA. To obtain the original matrix E from the coded matrix EA we need only multiply by A^{-1} and use the fact that AA^{-1} is the identity matrix I_4:

$$(EA)A^{-1} = E(AA^{-1}) = EI_4 = E.$$

Using a calculator (Figures 6–24 and 6–25), we see that the decoded form of the first four blocks of text is

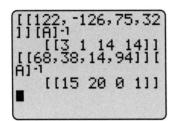

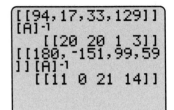

Figure 6–24   **Figure 6–25**

| 3 1 14 14 | 15 20 0 1 | 20 20 1 3 | 11 0 21 14 |
| C A N N | O T # A | T T A C | K # U N |

So the message begins "Cannot attack un . . . ."

CALCULATOR EXPLORATION

Use your calculator to decode the last two blocks of the message by computing $(129 \quad -17 \quad 43 \quad 84)A^{-1}$ and $(174 \quad -97 \quad 84 \quad 95)A^{-1}$. What is the entire message? ■

Exercises 6.4.A

In Exercises 1–8, compute $A + B$, AB, BA, and $2A - 3B$.

1. $A = \begin{pmatrix} 3 & 2 \\ 5 & 1 \end{pmatrix}$, $\quad B = \begin{pmatrix} 7 & -5 \\ -2 & 6 \end{pmatrix}$

2. $A = \begin{pmatrix} -6 & 2 \\ 7 & -1 \end{pmatrix}$, $\quad B = \begin{pmatrix} -8 & 4 \\ 2 & 7 \end{pmatrix}$

3. $A = \begin{pmatrix} \frac{3}{2} & 2 \\ 4 & \frac{7}{2} \end{pmatrix}$, $\quad B = \begin{pmatrix} \frac{1}{2} & -\frac{3}{2} \\ \frac{5}{2} & 1 \end{pmatrix}$

4. $A = \begin{pmatrix} \frac{3}{4} & 7 \\ 6 & -\frac{5}{4} \end{pmatrix}$, $\quad B = \begin{pmatrix} \frac{1}{2} & 3 \\ -5 & \frac{3}{2} \end{pmatrix}$

5. $A = \begin{pmatrix} 0 & 1 \\ 1 & 0 \end{pmatrix}$, $\quad B = \begin{pmatrix} 3 & 5 \\ 7 & 9 \end{pmatrix}$

6. $A = \begin{pmatrix} 1 & -2 \\ -3 & 5 \end{pmatrix}$, $\quad B = \begin{pmatrix} -1 & 2 \\ 3 & -5 \end{pmatrix}$

7. $A = \begin{pmatrix} 5 & 2 \\ 3 & -1 \end{pmatrix}$, $\quad B = \begin{pmatrix} -1 & 2 \\ 3 & -4 \end{pmatrix}$

8. $A = \begin{pmatrix} 1 & -1 \\ 2 & -2 \end{pmatrix}$, $\quad B = \begin{pmatrix} 3 & -1 \\ 3 & -1 \end{pmatrix}$

In Exercises 9–18, perform the indicated operations, where

$$A = \begin{pmatrix} 1 & -2 \\ 3 & 4 \end{pmatrix}, \quad B = \begin{pmatrix} -1 & 0 \\ 1 & 2 \end{pmatrix}, \quad C = \begin{pmatrix} 1 & 0 \\ 0 & 1 \end{pmatrix},$$

$$D = \begin{pmatrix} 1 & -2 & 3 \\ 0 & 4 & 1 \end{pmatrix}, \quad E = \begin{pmatrix} 1 & 2 \\ -2 & 1 \\ 0 & 5 \end{pmatrix},$$

$$F = \begin{pmatrix} 0 & 0 \\ 0 & 0 \\ 0 & 0 \end{pmatrix}, \quad G = \begin{pmatrix} 2 & -4 \\ 3 & 0 \\ -1 & 5 \end{pmatrix}.$$

9. $B - C$

10. $2A + C$

11. $-3B + 4C$

12. $A - 2B$

13. $AD + BD$

14. $EA + EB$

15. $2E + 3G$

16. $B^2 - AB$

17. $DG - B$

18. $DE + 2C$

In Exercises 19–22, find a matrix X satisfying the given equation, where

$$A = \begin{pmatrix} 1 & -2 \\ 4 & 3 \end{pmatrix} \quad \text{and} \quad B = \begin{pmatrix} 2 & -1 \\ 0 & 5 \end{pmatrix}.$$

19. $2X = 2A + 3B$

20. $3X = A - 3B$

21. $2X + 3A = 4B$

22. $3X + 2B = 3A$

In Exercises 23–30, verify that the given statement is true when $c = 2$, $d = 3$, and

$$A = \begin{pmatrix} 1 & 2 \\ 3 & 0 \end{pmatrix}, \quad B = \begin{pmatrix} -1 & 2 \\ 3 & 4 \end{pmatrix}, \quad C = \begin{pmatrix} 2 & 3 \\ 1 & 2 \end{pmatrix}.$$

23. $A + B = B + A$

24. $A + (B + C) = (A + B) + C$

25. $c(A + B) = cA + cB$

26. $(c + d)A = cA + dA$

27. $(cd)A = c(dA)$

28. $A(BC) = (AB)C$

29. $A(B + C) = AB + AC$

30. $(B + C)A = BA + CA$

31. If A is an $n \times k$ matrix and B is an $r \times t$ matrix, what conditions must n, k, r, t satisfy so that both AB and BA are defined?

In Exercises 32–36, verify that the statement is false for the given matrices.

32. $AB = BA$; $A = \begin{pmatrix} 1 & 2 \\ 3 & 4 \end{pmatrix}$ and $B = \begin{pmatrix} -1 & 4 \\ 5 & 2 \end{pmatrix}$

33. If $AB = AC$, then $B = C$; $A = \begin{pmatrix} 1 & 2 \\ 2 & 4 \end{pmatrix}$,

$B = \begin{pmatrix} 3 & 6 \\ -\frac{3}{2} & -3 \end{pmatrix}, C = \begin{pmatrix} 0 & 0 \\ 0 & 0 \end{pmatrix}$

34. $A^2 = \begin{pmatrix} 0 & 0 \\ 0 & 0 \end{pmatrix}$ only if $A = \begin{pmatrix} 0 & 0 \\ 0 & 0 \end{pmatrix}$; $A = \begin{pmatrix} 0 & 2 \\ 0 & 0 \end{pmatrix}$

35. $(A + B)(A - B) = A^2 - B^2$; $A = \begin{pmatrix} 3 & 1 \\ 2 & -4 \end{pmatrix}$ and

$B = \begin{pmatrix} 2 & -1 \\ 5 & 3 \end{pmatrix}$

36. $(A + B)(A + B) = A^2 + 2AB + B^2$;

$A = \begin{pmatrix} 2 & -1 \\ 3 & 5 \end{pmatrix}$ and $B = \begin{pmatrix} 1 & 2 \\ -3 & 4 \end{pmatrix}$

In Exercises 37–40, the transpose *of the matrix A is denoted by A^t and is defined by this rule: Row 1 of A is column 1 of A^t; row 2 of A is column 2 of A^t, etc.*

37. Find A^t and B^t, when

$$A = \begin{pmatrix} a & b \\ c & d \end{pmatrix} \quad \text{and} \quad B = \begin{pmatrix} r & s \\ u & v \end{pmatrix}$$

In Exercises 38–40, assume A and B are as in Exercise 37 and show that:

38. $(A + B)^t = A^t + B^t$ 39. $(A^t)^t = A$

40. $(AB)^t = B^t A^t$ (note order)

41. McChicken Fast Foods has raised its prices. Chicken nuggets now cost \$2.75, Buffalo wings cost \$3.25, and filet of chicken sandwiches cost \$2.20. Assuming the same number of sales at each location as in Example 5, find a matrix that represents the daily revenue per store.

42. (a) The LaPointe Lounger Company manufactures loungers and couches. Each is available in three different styles: classic, art deco, and modern. In September, the company shipped 12 classic loungers, 10 art deco loungers, 7 modern loungers, 22 classic couches, 14 art deco couches, and 9 modern couches to its Charleston distributor. Write 2×3 matrix A that summarizes this information, in which the rows represent loungers and couches (in that order) and the columns represent classic, art deco, and modern (in that order).

(b) The company also shipped loungers and couches to its distributors in St. Louis and Seattle in September. The number of each model is given by the following matrices:

$$B = \overset{St.\ Louis}{\begin{pmatrix} 24 & 26 & 40 \\ 44 & 48 & 29 \end{pmatrix}} \quad C = \overset{Seattle}{\begin{pmatrix} 30 & 33 & 34 \\ 31 & 33 & 30 \end{pmatrix}}.$$

Express the matrix that represents the total number of each model shipped to all distributors in September in terms of matrices A, B, C. Then find the matrix.

(c) The sales of the various models by the Seattle distributor are given by

$$D = \begin{pmatrix} 18 & 12 & 11 \\ 9 & 21 & 16 \end{pmatrix}.$$

Express the matrix that represents the inventory of the Seattle distributor at the end of September, in terms of matrices C and D. Then find the matrix.

43. A drug company is testing 400 patients to see if Lopress (a new blood-pressure medicine) is effective. Half the patients take Lopress for six

months and the other half take a placebo. The results are summarized in this matrix:

Effective?

Yes No

Patient took Lopress $\begin{pmatrix} 88 & 12 \\ 32 & 68 \end{pmatrix}$
Patient took placebo

The test was repeated for three more groups of 200, with the results given by

$$\begin{pmatrix} 92 & 8 \\ 12 & 88 \end{pmatrix}, \quad \begin{pmatrix} 84 & 16 \\ 24 & 76 \end{pmatrix}, \quad \begin{pmatrix} 76 & 24 \\ 40 & 60 \end{pmatrix}.$$

Use matrix addition to construct a matrix that gives total results of the test for all 800 patients.

44. A new 300-home subdivision is to have two styles of homes: traditional and contemporary. Each style comes in three different models: 2 bedrooms, 2 baths, or 3 bedrooms, 3 baths, or 4 bedrooms, 3.5 baths (the half-bath is a powder room). The numbers of each model are given by this matrix:

Traditional   Contemporary

$$\begin{array}{c} 2/2 \\ 3/3 \\ 4/3.5 \end{array} \begin{pmatrix} 0 & 90 \\ 30 & 60 \\ 60 & 60 \end{pmatrix} = A.$$

The materials needed for the exterior of each style of house are given by

Concrete Lumber Bricks Shingles

$$\begin{array}{c} \text{Traditional} \\ \text{Contemporary} \end{array} \begin{pmatrix} 30 & 6 & 0 & 6 \\ 150 & 3 & 60 & 6 \end{pmatrix} = B,$$

where concrete is measured in cubic yards, lumber in thousands of board feet, bricks in thousands, and shingles in units of 100 square feet. The cost per unit of each kind of material is given by

Cost

$$\begin{array}{c} \text{Concrete} \\ \text{Lumber} \\ \text{Bricks} \\ \text{Shingles} \end{array} \begin{pmatrix} 25 \\ 200 \\ 70 \\ 30 \end{pmatrix} = C.$$

(a) Explain why the matrix AB shows the amount of each material needed for each type of house. Find AB.

(b) Explain why the matrix $(AB)C$ gives the total cost for each model.

(c) Use matrix AB to find a 1×4 matrix D that represents the total amount of each type of material needed for the entire subdivision.

[*Hint:* What do the columns of matrix AB represent?]

(d) What matrix product represents the total cost of materials for the entire subdivision? Find this product.

In Exercises 45 and 46, use matrix A in Example 6 to encode the given message.

45. Shipment arrives Monday.

46. Follow the fourth car.

In Exercises 47 and 48, decode the given message (which was encoded by using matrix A in Example 6).

47. 140, −1, 38, 81, 120, −43, 34, −9, 88, −64, 36, −16, 164, −148, 116, 168, 6, −2, 4, 12.

48. 102, −22, 22, −16, 104, −11, 10, −57, 116, −47, −40, 15, 67, −35, 45, 108, 123, −101, 92, 170, 128, −110, 73, 50, 175, −170, 128, 175, 76, 19, 11, 33

In Exercises 49 and 50, use this matrix:

$$\begin{pmatrix} 1 & 2 & 3 & 4 & 5 \\ 9 & 8 & 7 & 6 & -5 \\ 2 & 0 & -3 & 0 & 4 \\ 1 & -2 & 5 & -6 & 0 \\ 6 & 0 & -3 & 0 & 2 \end{pmatrix}.$$

49. Encode this message: "The trumpet sounds loudly."

50. Decode this message: 100, 18, 11, 10, 150, 89, 16, 22, 4, 146, 56, 0, 125, −38, 131, 250, 214, 241, 166, −58, 322, 146, 102, 124, 147, 76, −32, 57, −106, 105, 81, 20, 156, −50, 37.

51. (a) Encode the message "Rats!" using the matrix

$$\begin{pmatrix} 1 & 3 & 5 & 3 \\ 2 & 4 & -1 & 0 \\ 4 & -2 & 1 & 2 \\ -1 & 9 & 3 & 1 \end{pmatrix}.$$

(b) Now decode the message you sent in part (a). What goes wrong? What condition must a matrix satisfy if it is to be used for encoding and decoding messages?

52. Tom uses matrix A to encode a message and sends it to Anne. Anne, who does not know what matrix A is, uses matrix B to encode the message she receives from Tom and sends it on to Laura. Laura has both matrices A and B. What matrix should she use to decode the message?

6.5 Systems of Linear Inequalities

Recall that a linear equation in two variables, such as $y = 2x + 1$, has infinitely many solutions, which we usually picture by drawing the graph of the equation. The graph consists of all points (c, d) such that $x = c$ and $y = d$ is a solution of the equation. Similarly, a linear inequality, such as

$$y \le 2x + 1 \qquad \text{or} \qquad x - y < -3$$

has infinitely many solutions and its graph consists of all points (c, d) such that $x = c$, $y = d$ is a solution of the inequality.

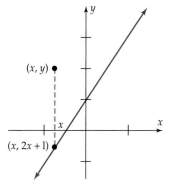

Figure 6–26

Example 1 Graph the inequality $y \le 2x + 1$.

Solution We first graph the line $y = 2x + 1$, all of whose points are solutions of the inequality. Now consider any point (x, y) that lies *above* this line, as in Figure 6–26. It lies directly above the point $(x, 2x + 1)$ and, hence, must have a larger y-coordinate; that is, $y > 2x + 1$. Thus, *none of the points above the line is a solution of $y \le 2x + 1$.*
  Now consider any point (x, y) that lies *below* the line $y = 2x + 1$, as in Figure 6–27. It lies directly below the point $(x, 2x + 1)$ on the line and has a smaller y-coordinate, that is, $y < 2x + 1$. Therefore, every point below the line is a solution of the inequality $y \le 2x + 1$. Therefore, the graph of $y \le 2x + 1$ is the half-plane consisting of all points *on or below the line* $y = 2x + 1$, as shown by the shading in Figure 6–28. ∎

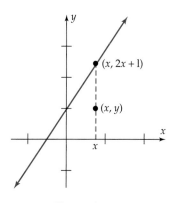

Figure 6–27

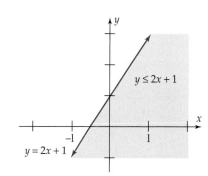

Figure 6–28

Example 2 Graph the inequality $y > 2x + 1$.

Solution Example 1 shows that the points on or below the line $y = 2x + 1$ are the solutions of $y \le 2x + 1$. Furthermore, the first paragraph of Example 1 shows that every point *above the line $y = 2x + 1$ is a* solution of $y > 2x + 1$. Thus, the graph of $y > 2x + 1$ is the shaded half-plane in Figure 6–29. The line $y = 2x + 1$ is dashed to indicate that it is *not* part of the graph of $y > 2x + 1$.   ■

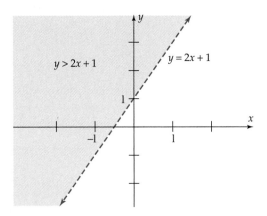

Figure 6–29

Example 3   Graph $3x + 2y \geq 6$.

Solution   First, solve the inequality for y:

$$2y \geq -3x + 6$$

$$y \geq -\frac{3}{2}x + 3$$

$$y \geq -1.5x + 3.$$

The $\geq$ symbol tells us that the solutions are all points whose y-coordinates lie *on or above* the line $y = -1.5x + 3$. The graph of the line can be sketched by hand and the area above it shaded (Figure 6–30), or the entire graph can be obtained on a calculator. See Figure 6–31 and the Tip in the margin.   ■

Technology Tip

To shade the area above the graph of Y_1 on TI-83, use the left arrow key to move to the left of Y_1 in the Y = list and repeatedly press ENTER until ◥ appears. On TI-86 repeatedly press STYLE in the GRAPH Y = menu until ◥ appears. On TI-89, choose "above" in the STYLE submenu (F6) of the Y = menu. For TI-82/85 and Sharp 9600, read the directions for using SHADE in the DRAW menu (or DRAW submenu of the GRAPH menu). For Casio 9850, read the directions for graphing inequalities.

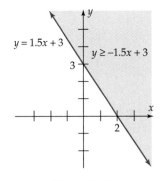

Figure 6–30

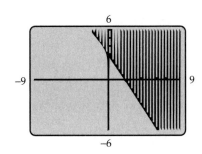

Figure 6–31

Example 4 Graph each of these inequalities:

(a) $x > 3$

(b) $y \leq 4.*$

Solution

(a) The graph of $x > 3$ consists of all points (x, y) with first coordinate greater than 3. So the graph is the half-plane to the right of the vertical line $x = 3$, as shown in Figure 6–32. The vertical line $x = 3$ is dashed because it is not part of the graph.

(b) The graph of $y \leq 4$ consists of all points (x, y) whose second coordinate is less than or equal to 4, that is, all points that are on or below the horizontal line $y = 4$, as shown in Figure 6–33. The horizontal line $y = 4$ is solid because it is part of the graph. ■

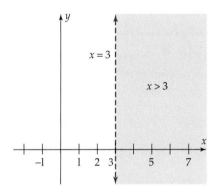

Figure 6–32

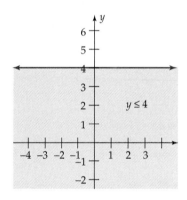

Figure 6–33

Systems of Inequalities

The graph of a system of inequalities consists of all points that are solutions of *every* inequality in the system.

Example 5 Sketch the graph of the system

$$2x + y > 0$$

$$x - y > -3.$$

Solution By rearranging each inequality the system becomes:

$$y > -2x$$

$$y < x + 3.$$

The solutions of the first inequality are the points *above* the line $y = -2x$ (Figure 6–34). The solutions of the second are the points *below* the line

*All inequalities here are assumed to have two variables, one of which may have a zero coefficient. For instance, $y \leq 4$ is $0x + y \leq 4$.

$y = x + 3$ (Figure 6–35). So the solutions of the *system* are the points that satisfy both of these conditions (Figure 6–36).   ■

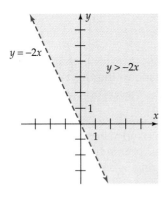

Figure 6–34

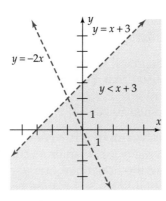

Figure 6–35

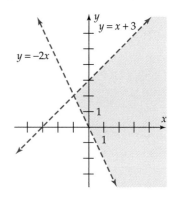

Figure 6–36

Example 6   Graph the system

$$x + y > 11$$
$$2x - y < 10$$
$$x - 2y > -16.$$

Solution   Begin by rewriting the system as

$$y > -x + 11$$
$$y > 2x - 10$$
$$y < \frac{1}{2}x + 8.$$

The solutions of the system consist of all points that lie *above both* of the lines $y = -x + 11$ and $y = 2x - 10$, and also *below* the line $y = \frac{1}{2}x + 8$, as shown in Figure 6–37.   ■

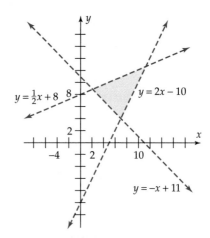

Figure 6–37

Example 7 Graph the system

$$6x + 5y \leq 100$$

$$15x + 6y \leq 168$$

$$x \geq 0, \qquad y \geq 0.$$

Solution We solve each inequality for y and rewrite the system as

$$y \leq -\frac{6}{5}x + \frac{100}{5} \qquad\qquad\qquad y \leq -1.2x + 20$$

$$y \leq -\frac{15}{6}x + \frac{168}{6} \qquad \text{or equivalently,} \qquad y \leq -2.5x + 28$$

$$x \geq 0, \qquad y \geq 0 \qquad\qquad\qquad x \geq 0, \qquad y \geq 0.$$

A point (x, y) that is a solution of $x \geq 0$ and $y \geq 0$ must have both coordinates nonnegative, that is, it must lie in the first quadrant. Therefore, the solutions of the system are all points in the first quadrant that lie on or below both of the lines

$$y = -1.2x + 20 \qquad \text{and} \qquad y = -2.5x + 28,$$

as shown in Figure 6–38. ■

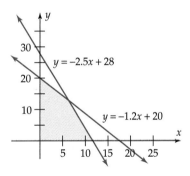

Figure 6–38

Example 8 A candy store sells two kinds of 4-pound gift boxes. The Regular Mix contains 2 pounds of assorted chocolates and 2 pounds of assorted caramels. The Chewy Mix has 1 pound of chocolates and 3 pounds of caramels. If 100 pounds of chocolates and 180 pounds of caramels are available, what possible numbers of both mixes can be made up?

Solution Let x be the number of Regular Mixes to be made and y the number of Chewy Mixes. We know that

$$\left(\begin{array}{c} \text{Pounds of chocolates} \\ \text{in } x \text{ boxes of Regular} \\ \text{(2 pounds per box)} \end{array} \right) + \left(\begin{array}{c} \text{Pounds of chocolates} \\ \text{in } y \text{ boxes of Chewy} \\ \text{(1 pound per box)} \end{array} \right) \leq 100$$

$$2x \qquad + \qquad 1y \qquad\qquad\qquad \leq 100$$

Similarly,

$$\left(\begin{array}{c} \text{Pounds of caramels} \\ \text{in } x \text{ boxes of Regular} \\ \text{(2 pounds per box)} \end{array} \right) + \left(\begin{array}{c} \text{Pounds of caramels} \\ \text{in } y \text{ boxes of Chewy} \\ \text{(1 pound per box)} \end{array} \right) \le 180$$

$$2x \quad + \quad 3y \quad \le 180$$

Furthermore, both x and y must be nonnegative (you can't have a negative number of boxes). So the possible numbers of the two mixes must be solutions of this system of inequalities:

$$2x + y \le 100$$
$$2x + 3y \le 180$$
$$x \ge 0$$
$$y \ge 0$$

We can picture the possibilities by graphing this system. To do this we rewrite it as:

$$y \le -2x + 100$$
$$y \le -\frac{2}{3}x + 60$$
$$x \ge 0, \qquad y \ge 0$$

So the graph (Figure 6–39) consists of the points that satisfy *all* of these conditions:

On or below the lines $y = -2x + 100$ and $y = -\frac{2}{3}x + 60$;

On or above the line $y = 0$ (the x-axis);

On or to the right of the vertical line $x = 0$ (the y-axis):

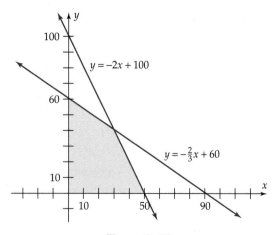

Figure 6–39

Since we can't have fractional boxes, the points on the graph with integer coordinates represent the possible numbers of Regular and Chewy Mixes that can be made from the available supplies. ■

Exercises 6.5

In Exercises 1–14, sketch the graph of the inequality.

1. $y < x + 3$

2. $y > 5 - 2x$

3. $y \geq -1.5x - 2$

4. $y \leq 1.5x + 2$

5. $y \geq -3$

6. $x \leq -1$

7. $x > 2$

8. $y < 3$

9. $2x - y \leq 4$

10. $4x - 3y \leq 24$

11. $3x - 2y \geq 18$

12. $2x + 5y \geq 10$

13. $3x + 4y > 12$

14. $4x - 3y > 9$

In Exercises 15–20, match the inequality with its graph, which is one of those shown here.

15. $y \geq -x - 2$ **16.** $y \leq 2x - 2$ **17.** $y \leq x + 2$

18. $y \geq x + 1$ **19.** $6x + 4y \geq -12$ **20.** $3x - 2y \geq -4$

In Exercises 21–42, sketch the graph of the system of inequalities.

21. $y \geq 3x - 6$

$\quad\ y \geq -x + 1$

22. $x + y \leq 4$

$\quad\ x - y \geq 2$

23. $2x + \ y \leq 5$

$\quad\ x + 2y \leq 5$

24. $4x + \ y \geq 9$

$\quad\ 2x + 3y \leq 7$

25. $2x + y > 8$

$\quad\ 4x - y < 3$

26. $x + \ y > 5$

$\quad\ x - 2y < 2$

A.
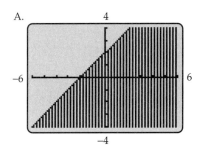

B.
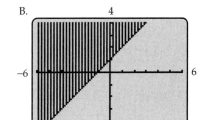

C.
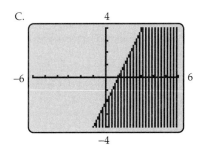

D.
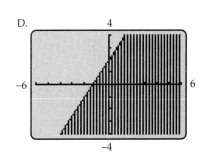

E.
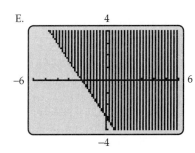

F.
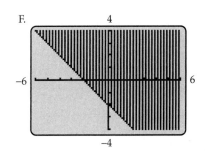

27. $3x + y \geq 6$
$x + 2y \geq 7$
$x \geq 0, \quad y \geq 0$

28. $2x + 3y \geq 12$
$x + y \geq 4$
$x \geq 0, \quad y \geq 0$

29. $3x + 2y < 18$
$x + 2y < 10$
$x \geq 0, \quad y \geq 0$

30. $x + 2y > 5$
$3x + 4y > 11$
$x \geq 0, \quad y \geq 0$

31. $2x + y \geq 8$
$2x - y \geq 0$
$x \leq 10, \quad y \geq 2$

32. $3x + y \leq -4$
$2x - y \leq 6$
$x \geq -5, \quad y \geq -8$

33. $3x + 2y \geq 12$
$3x + 2y \leq 18$
$3x - y \geq 0$

34. $4x + 5y \leq 20$
$x + 4y \geq 8$
$6x + y \geq 6$

35. $3x - y \leq 4$
$y - 3 \leq 0$
$x + y \geq 0$

36. $5x + 10y \leq 600$
$3x + 4y \leq 240$
$x \geq 0, \quad y \geq 0$

37. $x + y \geq 2$
$2x - y \leq 4$
$x \geq 0, \quad y \geq 0$

38. $x + y \geq 3$
$2x + 4y \geq 4$
$x \geq 0, \quad y \geq 0$

39. $x + y \leq 25$
$x \geq 15$
$0 \leq y \leq 5$

40. $x + y \leq 1200$
$2y \leq x$
$x - 3y \leq 600$
$x \geq 0, \quad y \geq 0$

41. $x + 2y \geq 6$
$3x + y \geq 8$
$2y \leq x + 8$
$x \geq 0, \quad y \geq 0$

42. $2x + 5y \leq 80$
$x + y \leq 25$
$x - y \leq -11$
$x \geq 0, \quad y \geq 0$

In Exercises 43–46, find a system of inequalities that has the given graph.

43.

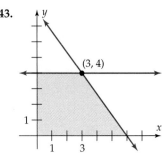

44.

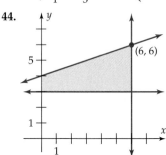

45.

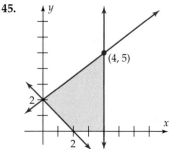

46.

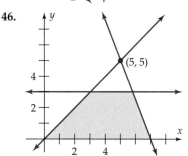

In Exercises 47–50, find a system of inequalities whose graph is the interior of the given figure.

47. Rectangle with vertices $(2, 3)$, $(2, -1)$, $(7, 3)$, $(7, -1)$.

48. Parallelogram with vertices $(2, 4)$, $(2, -2)$, $(5, 6)$, $(5, 0)$.

49. Triangle with vertices $(2, 4)$, $(-4, 0)$, $(2, -1)$.

50. Quadrilateral with vertices $(-2, 3)$, $(2, 4)$, $(5, 0)$, $(0, -3)$.

In Exercises 51–56, find a system of inequalities that describes all the possible answers and sketch its graph.

51. For an oil company to make 1 storage tank of regular gasoline, it takes 2 hours of processing and 3 hours of refining. One tank of premium gas takes 3 hours of processing and 3.5 hours of refining. The processing plant can be operated for at most 12 hours a day and the refinery for at most 16 hours a day. What are the possible outputs of regular and premium gas?

52. A can opener manufacturer makes both manual and electric models. The demand for manual can openers is never more than half of that for electric ones. The factory's production cannot exceed 1200 can openers per month. What are the possibilities for making x manual and y electric can openers each month?

53. Jack takes 4 hours to assemble a toy chest, and Jill takes 3 hours to decorate it. To make a silverware case takes Jack 2 hours for assembly and Jill 4 hours for decorating. Neither Jack nor Jill wants to work more than 20 hours a week. What are the possible production levels for toy and silverware chests?

54. A builder plans to construct a building containing both one-bedroom apartments, each with 1000 square feet of space, and two-bedroom apartments, each with 1500 square feet. The maximum building size is 45,000 square feet. If there can be at most 40 apartments, what are the possibilities for the number of one- and two-bedroom units?

55. A pet food is to be made of grain and meat by-products. One serving must provide at most 12 units of fat, at least 2 units of carbohydrates, and at least 1 unit of protein. Each gram of grain contains 2 units of fat, 2 units of carbohydrates,

and no units of protein. A gram of meat by-products has 3 units of fat, 1 unit of carbohydrate, and 1 unit of protein. What are the possible ways of combining x grams of grain and y grams of meat by-products in one serving?

56. An airline dietician is planning a snack package of candy and nuts. Each ounce of candy will supply 1 unit of protein, 2 units of carbohydrates, and 1 unit of fat. Each ounce of nuts will supply 1 unit of protein, 1 unit of carbohydrates, and 1 unit of fat. Every package must provide at least 7 units of protein, at least 10 units of carbohydrates, and no more than 9 units of fat. What are the possible ways of mixing x ounces of candy and y ounces of nuts in each package?

6.6 Introduction to Linear Programming

Many problems in business, economics, and the sciences are concerned with finding the optimal value of a function (for instance, the maximum value of a profit function or the minimum value of a cost function), subject to various constraints (such as transportation costs, environmental protection laws, interest rates, availability of parts, etc.). The function to be optimized is usually a function of several variables and the constraints are expressed as linear inequalities.

Example 1 Find the maximum and minimum values of $F = 2x + 5y$, subject to these constraints:

$$x + y \geq 1$$
$$-2x + 4y \leq 8$$
$$3x + 2y \leq 6$$
$$x \geq 0, \quad y \geq 0.$$

Solution The possible solutions of this problem are the points (x, y) that satisfy all of the constraint inequalities. We call them the *feasible solutions*. By graphing the system of constraint inequalities, we see that the feasible solutions are the points in the shaded area of Figure 6–40. We must find the feasible solution that makes the value of $F = 2x + 5y$ smallest and the one that makes it largest.

Consider different possible values of the function F, for instance,

$F = 0$	$F = 5$	$F = 10$	$F = 15$
$2x + 5y = 0$	$2x + 5y = 5$	$2x + 5y = 10$	$2x + 5y = 15$

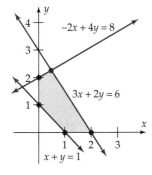

Figure 6–40

Each value of F leads to an equation, such as $2x + 5y = 10$, whose graph is a straight line. The points where this line intersects the shaded region are the feasible solutions for which $F = 10$ (Figure 6–41). Similarly, the points on the line $2x + 5y = 5$ that lie in the shaded region are the feasible solutions for which $F = 5$. The lines $2x + 5y = 0$ and $2x + 5y = 15$ do not intersect the shaded region, so there are no feasible solutions for which $F = 0$ or $F = 15$.

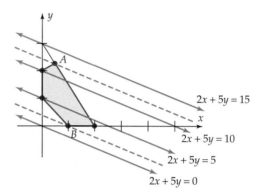

Figure 6–41

Look at the pattern in Figure 6–41. The lines

$$2x + 5y = 0, \qquad 2x + 5y = 5, \qquad 2x + 5y = 10, \qquad \text{etc.}$$

are all parallel (because they all have the same slope $-2/5$) and the constant on the right side of the equal sign determines their positions: As this number (the value of F) gets larger, the lines move upward; as it gets smaller, the lines move downward.

The minimum value of F corresponds to the lowest line that contains feasible solutions. As Figure 6–41 shows, this is the dashed line through the vertex B. All lines below this one (corresponding to smaller values of F) do not contain any feasible solutions. Therefore, the coordinates of B will produce the minimum value of F. As Figure 6–40 shows, B has coordinates $(1, 0)$. Therefore, the minimum value of F occurs when $x = 1$ and $y = 0$. This mininum value is

$$F = 2x + 5y = 2(1) + 5(0) = 2.$$

Similarly, the maximum value of F corresponds to the highest line that contains feasible solutions, namely, the dashed line through vertex A. So the coordinates of A will produce the maximum value of F. Since A is the intersection of the lines

$$2x + 4y = 8 \qquad \text{and} \qquad 3x + 2y = 6,$$

its coordinates can be found

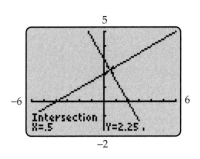

Figure 6–42

| *Algebraically* | or | *Graphically* |

Algebraically

Solve the system

$$-2x + 4y = 8$$
$$3x + 2y = 6$$

as in Section 6.1 and find that $x = .5$ and $y = 2.25$.

Graphically

Solve each equation for y:

$$y = .5x + 2$$
$$y = -1.5x + 3.$$

Graph them and use an intersection finder, as in Figure 6–42.

Thus, the maximum value of F occurs when $x = .5$ and $y = 2.25$. This maximum value is

$$F = 2x + 5y = 2(.5) + 5(2.25) = 12.25. \quad ■$$

Example 1 is a **linear programming problem** in two variables. Such problems consist of a linear function of the form $F = ax + by$ (called the **objective function**), subject to certain **constraints** that are expressed as a system of linear inequalities. Any pair (x, y) that satisfies all the constraints is a **feasible solution**. The goal is to find an **optimal solution,** that is, a feasible solution that produces the maximum or minimum value of the objective function.

As Example 1 illustrates, the set of feasible solutions (the graph of the system of constraint inequalities) is a region of the plane (called the **feasible region**) whose edges are straight-line segments. The feasible region may be either **bounded** as in Example 1, or **unbounded,** as in Figure 6–43 on the next page. The points where the edges of the feasible region meet are the **vertices.** The analysis in Example 1 showed that the minimum F occurred at the vertex $(1, 0)$ of the feasible region and that the maximum value of F occurred at the vertex $(.5, 2.25)$. The same kind of analysis works in the general case and proves this fact:

Linear Programming Theorem

> If the feasible region is bounded, then the objective function has both a maximum and a minimum value and each occurs at a vertex.
>
> If the feasible region is an unbounded region in the first quadrant and both coefficients of the objective function are positive*, then the objective function has no maximum value, but does have a minimum value and this occurs at a vertex.

Example 2 Find the maximum and minimum values of $F = 3x + 2y$, subject to the constraints:

$$2x + y \geq 9$$
$$x + y \geq 5$$
$$y \geq 0$$
$$x \geq 1.5.$$

*This is the only case of an unbounded feasible region that occurs in most applications.

Solution   By solving the first two constraints for y, we can easily find the graph of the system of constraint inequalities (the feasible region), as shown in Figure 6–43. It is an unbounded region that continues forever as you move to the right or upward.

$$y \geq -2x + 9$$
$$y \geq -x + 5$$
$$y \geq 0$$
$$x \geq 1.5$$

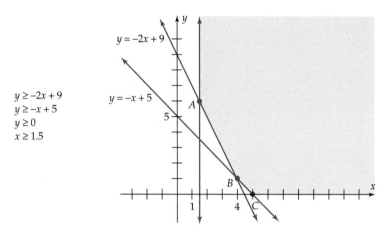

Figure 6–43

Each vertex is the intersection of two edges. Its coordinates can be found by solving the system of equations given by the two edges (either algebraically, which is easiest here, or graphically):

Vertex:	A	B	C
Edges:	$y = -2x + 9$	$y = -2x + 9$	$y = -x + 5$
	$x = 1.5$	$y = -x + 5$	$y = 0$
Coordinates:	$(1.5, 6)$	$(4, 1)$	$(5, 0)$

Since the feasible region is unbounded, F has no maximum value. Its minimum value must occur at one of the vertices, so we test each one of them:

Vertex	$F = 3x + 2y$
$(1.5, 6)$	16.5
$(4, 1)$	14
$(5, 0)$	15

Therefore, the minimum value of F is 14 and it occurs when $x = 4$ and $y = 1$. ■

Applications

Example 3   A candy store sells two kinds of 4-pound gift boxes: Regular and Chewy. A box of Regular has 2 pounds of chocolates and 2 pounds of caramels. A box of Chewy has 1 pound of chocolates and 3 pounds of caramels. 100 pounds of chocolates and 180 pounds of caramels are available. If the profit is $3 on each box of Regular and $4 on each box of Chewy, how many boxes of each kind should be made to maximize the profit?

Solution First, we find the rule of the profit function. Let x be the number of boxes of Regular and y the number of boxes of Chewy. Then the profit P is given by

$$P = \text{Profit on Regular} + \text{Profit on Chewy}$$

$$P = \underset{\text{box}}{\text{Price per}} \quad \underset{\text{of boxes}}{\text{Number}} \qquad \underset{\text{box}}{\text{Price per}} \quad \underset{\text{of boxes}}{\text{Number}}$$

$$P = \qquad\qquad 3x \quad + \quad 4y$$

So the profit function is $P = 3x + 4y$.

In Example 8 on page 491, we saw that the available amounts of chocolates and caramels led to this system of constraints:

$$2x + y \le 100 \qquad\qquad\qquad y \le -2x + 100$$

$$2x + 3y \le 180 \qquad \text{or equivalently,} \qquad y \le -\frac{2}{3}x + 60$$

$$x \ge 0, \quad y \ge 0 \qquad\qquad\qquad x \ge 0, \quad y \ge 0.$$

We must find the maximum value of $P = 3x + 4y$, subject to these constraints. The graph of the feasible region is shown in Figure 6–44.

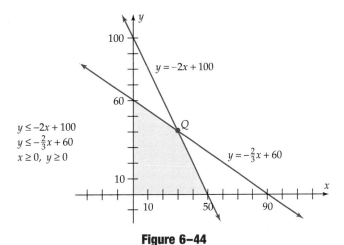

Figure 6–44

We need only evaluate $P = 3x + 4y$ at the four vertices of the graph, namely $(0, 0)$, $(0, 60)$, $(50, 0)$ and Q. To find the coordinates of Q, we solve this system of equations

$$y = -2x + 100$$

$$y = -\frac{2}{3}x + 60$$

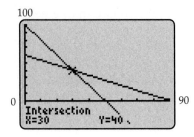

Figure 6–45

or use an intersection finder (Figure 6–45) to find that Q is (30, 40). So we have:

Vertex	Profit $= 3x + 4y$
(0, 0)	$P = 3 \cdot 0 + 4 \cdot 0 \quad = 0$
(0, 60)	$P = 3 \cdot 0 + 4 \cdot 60 = 240$
(50, 0)	$P = 3 \cdot 50 + 4 \cdot 0 = 150$
(30, 40)	$P = 3 \cdot 30 + 4 \cdot 40 = 250$

Therefore, making 30 boxes of Regular and 40 of Chewy will produce the maximum profit. ■

Example 4   An animal food is to be made from grain and silage. Each pound of grain provides 2400 calories and 20 grams of protein and costs 18¢. Each pound of silage provides 600 calories and 2 grams of protein and costs 3¢. An animal must have at least 18,000 calories and 120 grams of protein each day. It is unhealthy for an animal to have more than 12 pounds of grain or 20 pounds of silage per day. What combination of grain and silage should be used to keep costs as low as possible?

Solution   Let x be the number of pounds of grain an animal eats per day, and y the number of pounds of silage. Then the cost function to be minimized is given by $C = .18x + .03y$. The health limits on the amount of food imply these constraints:

$$0 \le x \le 12 \quad \text{and} \quad 0 \le y \le 20.$$

The minimal calorie requirement also leads to a constraint:

$$\left(\begin{matrix}\text{Calories in } x \\ \text{pounds of grain}\end{matrix}\right) + \left(\begin{matrix}\text{Calories in } y \\ \text{pounds of silage}\end{matrix}\right) \ge 18{,}000$$

$$2400x \quad + \quad 600y \qquad \ge 18{,}000$$

A final constraint comes from the minimal protein requirement:

$$\left(\begin{matrix}\text{Protein in } x \\ \text{pounds of grain}\end{matrix}\right) + \left(\begin{matrix}\text{Protein in } y \\ \text{pounds of silage}\end{matrix}\right) \ge 120$$

$$20x \quad + \quad 2y \qquad \ge 120$$

So the entire system of constraints is given by:

$$2400x + 600y \ge 18{,}000$$
$$20x + 2y \ge 120$$
$$x \le 12, \quad y \le 20$$
$$x \ge 0, \quad y \ge 0$$

which is equivalent to

$$4x + y \ge 30$$
$$10x + y \ge 60$$
$$x \le 12, \quad y \le 20$$
$$x \ge 0, \quad y \ge 0$$

The graph (Figure 6–46) of the feasible region and its vertices are found as above:

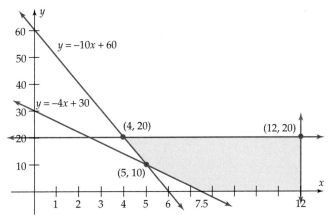

Figure 6–46

Since the set of feasible solutions is bounded, the cost function has both a maximum and a minimum value, each of which occurs at a vertex. The following table shows that the minimum is $1.20 and occurs when 5 pounds of grain and 10 pounds of silage are used.

Vertex	$C = .18x + .03y$
(4, 20)	$C = .18(4) + .03(20)\ \ = \$1.32$
(5, 10)	$C = .18(5) + .03(10)\ \ = \$1.20$
(7.5, 0)	$C = .18(7.5) + .03(0)\ \ = \$1.35$
(12, 0)	$C = .18(12) + .03(0)\ \ = \$2.16$
(12, 20)	$C = .18(12) + .03(20) = \$2.76$

Exercises 6.6

In Exercises 1–4, use the graph of the feasible region in the figure. Find the minimum and maximum values of the objective function F.

1. $F = 5x + 2y$

2. $F = x + 4y$

3. $F = 3x + y$

4. $F = 8x + 12y$

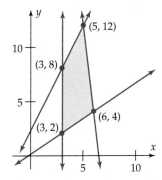

In Exercises 5–8, the graph of the feasible region is given. Find the minimum and maximum values of the objective function F.

5. $F = 3x + 2y$

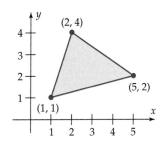

6. $F = 2x - y$

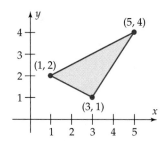

7. $F = 3x + 4y$

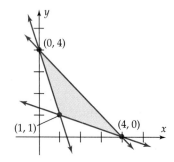

8. $F = 4x + 2y$

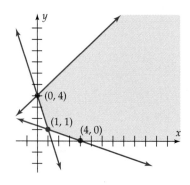

In Exercises 9–12, the graph of the feasible region is given. Determine the coordinates of each vertex. Then find minimum and maximum values of the objective function F.

9. $F = 12x + 8y$

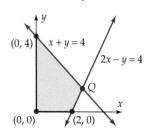

10. $F = 1.5x + 2.5y$

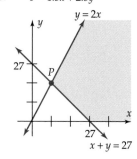

11. $F = 3.5x + 2y$

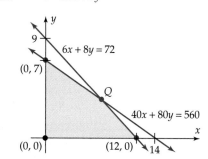

12. $F = 4x + 2.2y$

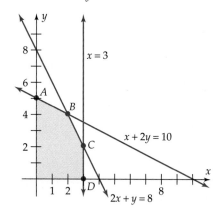

In Exercises 13–24, find the minimum and maximum values of the function F, subject to the given constraints.

13. $F = 4x + 3y$; constraints:
$$3x + y \leq 6$$
$$2x + y \leq 5$$
$$x \geq 0, \qquad y \geq 0$$

14. $F = 3x + 7y$; constraints:
$$6x + 5y \leq 100$$
$$8x + 5y \leq 120$$
$$x \geq 0, \qquad y \geq 0$$

15. $F = 3x + 2y$; constraints:

$$11x + 5y \geq 75$$
$$4x + 15y \geq 80$$
$$x \geq 0, \quad y \geq 0$$

16. $F = 6x + 3y$; constraints:

$$-2x + 3y \leq 9$$
$$-x + 3y \leq 12$$
$$x \geq 0, \quad y \geq 0$$

17. $F = 3x + y$; constraints:

$$2x + y \leq 20$$
$$10x + y \geq 36$$
$$2x + 5y \geq 36$$

18. $F = 2x + 6y$; constraints:

$$x + y \geq 6$$
$$-x + y \leq 2$$
$$2x - y \leq 8$$

19. $F = x + 3y$; constraints:

$$2x + 3y \leq 30$$
$$-x + y \leq 5$$
$$x + y \geq 5$$
$$0 \leq x \leq 10$$
$$y \geq 0$$

20. $F = 1100 - 2x - y$; constraints:

$$x + y \leq 80$$
$$x + y \geq 30$$
$$0 \leq x \leq 40$$
$$0 \leq y \leq 70$$

21. $F = 3x + 4y$; constraints:

$$x + y \leq 6$$
$$-x + y \leq 2$$
$$2x - y \leq 8$$
$$x \geq 0, \quad y \geq 0$$

22. $F = 4x + 3y$; constraints:

$$x + 2y \geq 10$$
$$2x + y \geq 12$$
$$x - y \leq 8$$
$$x \geq 0, \quad y \geq 0$$

23. $F = 2x + 3y$; constraints:

$$2x + 5y \leq 22$$
$$4x + 3y \geq 28$$
$$x - y \geq 20$$
$$x \geq 0, \quad y \geq 0$$

24. $F = 1.5x + 4.5y$; constraints:

$$10x + 7y \leq 252$$

$$4x + 10y \geq 85$$
$$-x + 4y \leq 48$$
$$x \geq 0, \quad y \geq 0$$

25. Wonderboat Company makes four-person and two-person inflatable rafts. A four-person raft requires 6 hours of fabrication and 1 hour of finishing. A two-person raft requires 4 hours of fabrication and 1 hour of finishing. Each week 108 hours of fabrication time and 24 hours of finishing time are available. If $40 is the profit on a four-person raft and $30 on a two-person one, how many of each kind should be made to maximize weekly profit?

26. The Coop sells summer-weight and winter-weight sleeping bags. A summer bag requires .9 hour of labor for cutting and .8 hour for assembly. A winter bag requires 1.8 hours for cutting and 1.2 hours for assembly. There are at most 864 hours of labor available in the cutting department each month and at most 672 hours in the assembly department. If the profit is $25 on a summer bag and $40 on a winter bag, how many of each kind should be made to maximize profit?

27. Sixty pounds of chocolates and 100 pounds of mints are available to make up 5-pound boxes of candy. A box of Choco-Mix has 4 pounds of chocolates and 1 pound of mints and sells for $10. A box of Minto-Mix has 2 pounds of chocolates and 3 pounds of mints and sells for $16. How many boxes of each kind should be made to maximize revenue?

28. CalenCo makes large wall calendars and small desk calendars. Printing a wall calendar takes 2 minutes and binding it 1 minute. A desk calendar requires 1 minute for printing and 3 minutes for binding. The printing machine is available for 3 hours a day and the binding machine for 5 hours. The profit is $1.00 on a wall calendar and $1.20 on a desk calendar. How many of each type should be made to maximize daily profit?

29. A person who is suffering from nutritional deficiencies is told to take at least 2400 mg of iron, 2100 mg of vitamin B_1, and 1500 mg of vitamin B_2. One Supervites pill contains 40 mg of iron, 10 mg of B_1, and 5 mg of B_2, and costs 6¢. One Vitahealth pill provides 10 mg of iron, 15 mg of B_1, and 15 mg of B_2, and costs 8¢. What combination of Supervites and Vitahealth pills will meet the requirements at lowest cost?

30. A fertilizer is to contain two major ingredients. Each pound of ingredient I contains 4 oz of nitrogen and 2 oz of phosphates and costs $1.00.

Each pound of ingredient II contains 3 oz of nitrogen and 5 oz of phosphates and costs $2.50. A bag of fertilizer must contain at least 70 oz of nitrogen and 110 oz of phosphates. How much of each ingredient should be used to produce a bag of fertilizer at minimum cost? How many pounds of each ingredient does it contain?

31. A pound of chicken feed A contains 3000 units of Nutrigood and 1000 units of Fasgrow, and costs 20¢. A pound of chicken feed B contains 4000 units of Nutrigood and 4000 units of Fasgrow, and costs 40¢. The minimum daily requirement for one chicken farm is 36,000 units of Nutrigood and 20,000 units of Fasgrow. How many pounds of each chicken feed should be used each day to minimize food costs while meeting or exceeding the daily nutritional requirements?

32. A company has warehouses in Titusville and Rockland. It has 80 stereo systems stored in Titusville and 70 in Rockland. Superstore orders 35 systems and Giantval orders 60. It costs $8 to ship a system from Titusville to Superstore and $12 to ship one to Giantval. It costs $10 to ship a system from Rockland to Superstore and $13 to ship one to Giantval. How should the orders be filled to keep shipping costs as low as possible?

33. A cereal company plans to combine bran and oats to manufacture at least 28 tons of a new cereal. The cereal must contain at least twice as much oats as bran. If bran costs $350 per ton and oats cost $225 per ton, how much of each should be used to minimize costs?

34. Newsmag publishes a U.S. and a Canadian edition each month. There are 30,000 subscribers in the United States and 20,000 in Canada. Other copies are sold at newsstands. Shipping costs average $80 per thousand copies for U.S. newsstands and $60 per thousand copies for Canadian newsstands. Surveys show that no more than 120,000 copies of each issue can be sold (including subscriptions) and that the number of copies of the Canadian edition should not exceed twice the number of copies of the U.S. edition. The publisher can spend at most $8400 a month on shipping costs to newsstands. If the profit is $200 for each thousand copies of the U.S. edition and $150 for each thousand copies of the Canadian edition, how many copies of each version should be printed to earn as large a profit as possible?

35. Students at Zonker College are required to take at least three humanities and four science courses. The maximum allowable number of science courses is 12. Each humanities course carries 4 credits and each science course 5 credits. The total number of credits in humanities and science cannot exceed 80. Quality points for each course are assigned in the usual way: the number of credit hours times 4 for an A grade; times 3 for a B grade; times 2 for a C grade. A student expects to get B's in all her science courses. She expects to get C's in half her humanities courses, B's in one-fourth of them, and A's in the rest. Under these assumptions how many courses of each kind should she take to earn the maximum possible number of quality points?

36. An investor has $20,000 to invest in bonds. He expects an annual yield of 10% for grade B bonds and 6% for grade AAA bonds. He decides that he shouldn't invest more than half his money in B bonds, and should invest at least one-fourth in AAA bonds. He wants the amount invested in B bonds to be at least two-thirds of the amount invested in AAA bonds. How much should he invest in each type of bond to maximize his annual yield?

Thinkers

37. Explain why it is impossible to maximize the function $P = 3x + 4y$, subject to the constraints: $x + y \geq 8$; $2x + y \leq 10$; $x + 2y \leq 8$; $x \geq 0$; $y \geq 0$.

38. The Linear Programming Theorem on page 497 guarantees that there is at least one vertex at which the optimal solution occurs (when it exists). This allows the possibility that the optimal solution might also occur at other points.
    (a) Show that this is indeed the case here:

    Maximize $P = 3x + 2y$, subject to the constraints:
    $$3x + 2y \leq 22; x \leq 6; y \leq 5; x \geq 0; y \geq 0$$

    (b) Show that the optimal value of P in part (a) also occurs at every point on the line segment joining the two vertices where the optimal value occurs. [*Hint:* The maximum value of P is 22; so any optimal solution satisfies $3x + 2y = 22$. Where does the line $3x + 2y = 22$ intersect the set of feasible solutions?]

39. In Exercise 35, find the student's grade point average (total number of quality points divided by total number of credit hours) at each vertex of the set of feasible solutions. Does the distribution of courses that produces the highest number of quality points also yield the highest grade point average? Is this a contradiction?

Chapter 6   *Review*

Important Concepts

Review Questions

In Questions 1–10, solve the system of linear equations by any means.

1. $-5x + 3y = 4$
   $2x - y = -3$

2. $3x - y = 6$
   $2x + 3y = 7$

3. $3x - 5y = 10$
   $4x - 3y = 6$

4. $\dfrac{1}{4}x - \dfrac{1}{3}y = -\dfrac{1}{4}$

   $\dfrac{1}{10}x + \dfrac{2}{5}y = \dfrac{2}{5}$

5. $3x + y - z = 13$
   $x + 2z = 9$
   $-3x - y + 2z = 9$

6. $x + 2y + 3z = 1$
   $4x + 4y + 4z = 2$
   $10x + 8y + 6z = 4$

7. $4x + 3y - 3z = 2$
   $5x - 3y + 2z = 10$
   $2x - 2y + 3z = 14$

8. $x + y - 4z = 0$
   $2x + y - 3z = 2$
   $-3x - y + 2z = -4$

9. $x - 2y - 3z = 1$
   $5y + 10z = 0$
   $8x - 6y - 4z = 8$

10. $4x - y - 2z = 4$
    $x - y - \dfrac{1}{2}z = 1$
    $2x - y - z = 8$

11. The sum of one number and three times a second number is -20. The sum of the second number and two times the first number is 55. Find the two numbers.

12. You are given \$144 in \$1, \$5, and \$10 bills. There are 35 bills. There are two more \$10 bills than \$5 bills. How many bills of each type do you have?

13. Let L be the line with equation $4x - 2y = 6$ and M the line with equation $-10x + 5y = -15$. Which of the following statements is true?
    (a) L and M do not intersect. (b) L and M intersect at a single point.
    (c) L and M are the same line. (d) All of the above are true.
    (d) None of the above are true.

14. Which of the following statements about this system of equations are *false*?
    $$x + z = 2$$
    $$6x + 4y + 14z = 24$$
    $$2x + y + 4z = 7$$
    (a) $x = 2, y = 3, z = 0$ is a solution.
    (b) $x = 1, y = 1, z = 1$ is a solution.
    (c) $x = 1, y = -3, z = 3$ is a solution.
    (d) The system has an infinite number of solutions.
    (e) $x = 2, y = 5, z = -1$ is not a solution.

15. Tickets to a lecture cost \$1 for students, \$1.50 for faculty, and \$2 for others. Total attendance at the lecture was 460, and the total income from tickets was \$570. Three times as many students as faculty attended. How many faculty members attended the lecture?

16. An alloy containing 40% gold and an alloy containing 70% gold are to be mixed to produce 50 pounds of an alloy containing 60% gold. How much of each alloy is needed?

In Questions 17–22, solve the system.

17. $x^2 - y = 0$
    $y - 2x = 3$

18. $x^2 + y^2 = 25$
    $x^2 + y = 19$

19. $x^2 + y^2 = 16$
  $x + y = 2$

20. $6x^2 + 4xy + 3y^2 = 36$
  $x^2 - xy + y^2 = 9$

21. $x^3 + y^3 = 26$
  $x^2 + y = 6$

22. $x^2 - 3xy + 2y^2 - y + x = 0$
  $5x^2 - 10xy + 5y^2 = 8$

23. Write the augmented matrix of the system
$$x - 2y + 3z = 4$$
$$2x + y - 4z = 3$$
$$-3x + 4y - z = -2$$

24. Use matrix methods to solve the system in Question 23.

25. Write the coefficient matrix of the system
$$2x - y - 2z + 2u = 0$$
$$x + 3y - 2z + u = 0$$
$$-x + 4y + 2z - 3u = 0$$

26. Use matrix methods to solve the system in Question 25.

In Questions 27 and 28, find the constants A, B, C that make the statement true.

27. $\dfrac{4x - 7}{x^2 - x - 6} = \dfrac{A}{x - 3} + \dfrac{B}{x + 2}$

28. $\dfrac{6x^2 + 6x - 6}{(x^2 - 1)(x + 2)} = \dfrac{A}{x + 1} + \dfrac{B}{x - 1} + \dfrac{C}{x + 2}$

In Questions 29–32, perform the indicated matrix multiplication or state that the product is not defined. Use these matrices:

$$A = \begin{pmatrix} -1 & 0 \\ 0 & -1 \end{pmatrix}, \quad B = \begin{pmatrix} 2 & -3 \\ 4 & 1 \end{pmatrix}, \quad C = \begin{pmatrix} 3 & 2 \\ 2 & 4 \end{pmatrix},$$

$$D = \begin{pmatrix} -3 & 1 & 2 \\ 1 & 0 & 4 \end{pmatrix}, \quad E = \begin{pmatrix} 1 & 2 \\ -3 & 4 \\ 0 & 5 \end{pmatrix}, \quad F = \begin{pmatrix} 2 & 3 \\ 6 & 3 \\ 6 & 1 \end{pmatrix}$$

29. AB **30.** CD **31.** AE **32.** DF

In Questions 33–36, find the inverse of the matrix, if it exists.

33. $\begin{pmatrix} 3 & -7 \\ 4 & -9 \end{pmatrix}$

34. $\begin{pmatrix} 2 & 6 \\ 1 & 3 \end{pmatrix}$

35. $\begin{pmatrix} 3 & 2 & 6 \\ 1 & 1 & 2 \\ 2 & 2 & 5 \end{pmatrix}$

36. $\begin{pmatrix} 1 & -1 & 1 \\ 2 & -3 & 2 \\ -4 & 6 & 1 \end{pmatrix}$

In Questions 37 and 38, use matrix inverses to solve the system.

37. $x + 2z + 6w = 2$
  $3x + 4y - 2z - w = 0$
  $5x + 2z - 5w = -4$
  $4x - 4y + 2z + 3w = 1$

38. $2x + y + 2z + u = 2$
  $x + 3y - 4z - 2u + 2v = -2$
  $2x + 3y + 5z - 4u + v = 1$
  $x - 2z + 4v = 4$
  $2x + 6z - 5v = 0$

In Questions 39 and 40, find the equation of the parabola passing through the given points.

39. $(-3, 52), (2, 17), (8, 305)$ **40.** $(-2, -18), (2, 6), (4, -12)$

41. An animal feed is to be made from corn, soybeans, and meat by-products. One bag is to supply 1800 units of fiber, 2800 units of fat, and 2200 units of protein. Each pound of corn has 10 units of fiber, 30 units of fat, and 20 units of protein. Each pound of soybeans has 20 units of fiber, 20 units of fat, and 40 units of protein. Each pound of by-products has 30 units of fiber, 40 units of fat, and 25 units of protein. How many pounds of corn, soybeans, and by-products should each bag contain?

42. A company produces three camera models: *A*, *B*, and *C*. Each model *A* requires 3 hours of lens polishing, 2 hours of assembly time, and 2 hours of finishing time. Each model *B* requires 2 hours of lens polishing, 2 hours of assembly time, and 1 hour of finishing time. Each model *C* requires 1, 3, and 1 hours of lens polishing, assembly, and finishing time, respectively. There are 100 hours available for lens polishing, 100 hours for assembly, and 65 hours for finishing each week. How many of each model should be produced if all available time is to be used?

In Questions 43–48, use the matrices A, B, C, D, E, F given before Question 29. Perform the indicated matrix operations (if they are defined) or state which operation is not defined.

43. $3F - 2E$ **44.** $3A + B - C$ **45.** $AB + AC$

46. $DA - A$ **47.** $DF + FE$ **48.** $DF - DE$

49. A recent survey on campaign reform legislation produced the following results. Among men, 227 favored the legislation, 124 opposed it, and 49 had no opinion. Among women, 266 favored the legislation, 40 opposed it, and 94 had no opinion. Write a 2×3 matrix that expresses this information.

50. Use the matrix *A* to encode this message: "Fire when ready."

$$A = \begin{pmatrix} 2 & -5 & 4 \\ 1 & 6 & -2 \\ 3 & 4 & 5 \end{pmatrix}$$

In Questions 51–53, sketch the graph of the system:

51. $x + y \geq 0$ **52.** $2x + y \geq 10$
       $-x + y \leq 6$ $-x + y \leq 1$
            $x \leq 5$ $x \geq 0, \quad y \geq 0$

53. $x + y \leq 4$
    $x - y \geq 1$
       $y \leq 0$

In Questions 54–56, find the minimum and maximum values (if any) of F subject to the given constraints. State the values of x and y that produce each optimal value of F.

54. $F = 5x + 7y$; constraints: **55.** $F = .08x + .05y$; constraints:
     $10x + 20y \leq 480$ $3x + y \geq 20$
     $10x + 30y \leq 570$ $3x + 6y \geq 60$
     $x \geq 0, \quad y \geq 0$ $x \geq 0, \quad y \geq 0$

56. $F = 2x - 3y$; constraints:

$2x + 3y \geq 12$

$3x + 2y \leq 18$

$3x - y \leq 0$

$x \geq 0, \qquad y \geq 0$

57. A private trash collection firm charges $260 for each container it removes. A container weighs 20 pounds and is 5 cubic feet in volume. The firm charges $200 for each barrel it removes. A barrel weighs 20 pounds and is 3 cubic feet in volume. If the truck can carry at most 1000 pounds and 180 cubic feet of material, what combination of containers and barrels will produce the largest possible revenue? Will there be any unused space in the truck?

58. A grass seed mixture contains bluegrass seeds costing 30¢ an ounce and rye seeds costing 15¢ an ounce. What is the cheapest way to obtain a mixture of 500 pounds that is at least 60% bluegrass?

Discovery Project 6

Input-Output Analysis

Wassily Leontief won the Nobel Prize in economics in 1973 for his method of input-output analysis of the economies of industrialized nations. This method has become a permanent part of production planning and forecasting by both national governments and private corporations. During the Arab oil boycott in 1973, for example, General Electric used input-output analysis on 184 sectors of the economy (such as energy, agriculture, and transportation) to predict the effect of the energy crisis on public demand for its products. The key to Leontief's method is knowing how much each sector of the economy needs from other sectors in order to do its job.

A simple economic model will illustrate the basic ideas behind input-output analysis. Suppose the country Hypothetica has just two sectors in its economy: agriculture and manufacturing. The production of a ton of agricultural products requires the use of .1 ton of agricultural products and .1 ton of manufactured products. Similarly, the production of a ton of manufactured products consumes .1 ton of agricultural products and .3 ton of manufactured products. The key economic question is: How many tons of each sector must Hypothetica produce to have enough surplus to export 10,000 tons of agricultural products and 10,000 tons of manufactured goods?

Let A be the total amount of agricultural goods and M the total amount of manufactured goods produced. The total agricultural production A is the sum of the amount needed for producing the agricultural products, plus the amount needed for producing manufactured products, plus the amount targeted for export. In other words,

Agricultural needs		Manufacturing needs		Export needs		Total agricultural production
$.1A$	$+$	$.1M$	$+$	$100{,}000$	$=$	$A.$

1. Write a similar equation for manufactured goods.

2. Solve the system of equations given by the agricultural equation above and the manufacturing equation of Problem 1. What level of production of agricultural and manufactured goods should Hypothetica work toward to reach its export goals?

3. The same kind of analysis can be used to determine the surplus available for exports in an economy. Suppose that Hypothetica can produce a total of 120,000 tons of agricultural products and 28,000 tons of manufactured goods. How much of each is available for export?

4. Input-output analysis can also be used for allocating resources in smaller scale situations. Suppose a horse outfitter is hired by the State of Washington Department of Fish and Wildlife to haul 20 salt blocks into a game area for winter feeding. Each horse can carry a 200-pound load. A single person can handle three horses. Each person requires one horse to ride, and one-half horse to carry his or her personal gear. A single horse can carry the feed and equipment for five horses. How many people and horses must go on the trip? [*Note:* Each horseload of salt must consist of an even number of blocks. Each salt block weighs 50 pounds. Also, fractional numbers of horses and people are not allowed. Make sure to adjust your answer to integer values and be sure that the answer still provides sufficient carrying capacity for the salt.]

Chapter

7

Discrete Algebra

What's next?

CD and CD-ROM players, fax machines, cameras, and other devices incorporate digital technology, which uses sequences of 0's and 1's to send signals. Determining the monthly payment on a car loan involves the sum of a geometric sequence. Problems involving the action of bouncing balls, vacuum pumps, and other devices can sometimes be solved by using sequences. See Exercises 37, 41, and 42 on page 533.

Chapter Outline

Interdependence of Sections

Sections 7.1, 7.4, 7.5, and 7.7 are independent of one another and may be read in any order. The interdependence of the other sections is as follows.

7.1 → 7.2
7.1 → 7.3
7.4
7.5 → 7.6
7.7

This chapter deals with a variety of subjects involving counting processes and the nonnegative integers $0, 1, 2, 3, \ldots$.

7.1 Sequences and Sums

A **sequence** is an ordered list of numbers. We usually write it horizontally, with the ordering understood to be from left to right. The same number may appear several times on the list. Each number on the list is called a **term** of the sequence. We are primarily interested in infinite sequences, such as

$$2, 4, 6, 8, 10, 12, \ldots$$

$$1, -3, 5, -7, 9, -11, 13, \ldots$$

$$2, 1, \frac{2}{3}, \frac{2}{4}, \frac{2}{5}, \frac{2}{6}, \frac{2}{7}, \ldots$$

where the dots indicate that the same pattern continues forever.*

When the pattern in an ordered list of numbers isn't obvious, the sequence is usually described as follows: The first term is denoted a_1, the second term a_2, and so on. Then a formula is given for the nth term a_n.

---

*Such a list defines a function f whose domain is the set of positive integers. The rule is $f(1) =$ first number on the list, $f(2) =$ second number on the list, and so on. Conversely, any function g whose domain is the set of positive integers leads to an ordered list of numbers, namely, $g(1), g(2), g(3), \ldots$. So a sequence is formally defined to be a function whose domain is the set of positive integers.

Example 1 Consider the sequence $a_1, a_2, a_3, \ldots, a_n, \ldots$, where a_n is given by the formula

$$a_n = \frac{n^2 - 3n + 1}{2n + 5}.$$

To find a_1 we substitute $n = 1$ in the formula for a_n; to find a_2 we substitute $n = 2$ in the formula; and so on:

$$a_1 = \frac{1^2 - 3 \cdot 1 + 1}{2 \cdot 1 + 5} = -\frac{1}{7}$$

$$a_2 = \frac{2^2 - 3 \cdot 2 + 1}{2 \cdot 2 + 5} = -\frac{1}{9}$$

$$a_3 = \frac{3^2 - 3 \cdot 3 + 1}{2 \cdot 3 + 5} = \frac{1}{11}.$$

Thus, the sequence begins $-1/7, -1/9, 1/11, \ldots$. The 39th term is

$$a_{39} = \frac{39^2 - 3 \cdot 39 + 1}{2 \cdot 39 + 5} = \frac{1405}{83}. \quad \blacksquare$$

Example 2 It is easy to list the first few terms of the sequence

$$a_1, a_2, a_3, \ldots \qquad \text{where} \qquad a_n = \frac{(-1)^n}{n + 2}.$$

Substituting $n = 1$, $n = 2$, and so on, in the formula for a_n shows that:

$$a_1 = \frac{(-1)^1}{1 + 2} = -\frac{1}{3}, \qquad a_2 = \frac{(-1)^2}{2 + 2} = \frac{1}{4}, \qquad a_3 = \frac{(-1)^3}{3 + 2} = -\frac{1}{5}.$$

Similarly,

$$a_{41} = \frac{(-1)^{41}}{41 + 2} = -\frac{1}{43} \qquad \text{and} \qquad a_{206} = \frac{(-1)^{206}}{206 + 2} = \frac{1}{208}. \quad \blacksquare$$

Example 3 Here are some other sequences whose nth term can be described by a formula:

Sequence	nth Term	First 5 Terms
$a_1, a_2, a_3, \ldots$	$a_n = n^2 + 1$	$2, 5, 10, 17, 26$
$b_1, b_2, b_3, \ldots$	$b_n = \dfrac{1}{n}$	$1, \dfrac{1}{2}, \dfrac{1}{3}, \dfrac{1}{4}, \dfrac{1}{5}$
$c_1, c_2, c_3, \ldots$	$c_n = \dfrac{(-1)^{n+1}2n}{(n + 1)(n + 2)}$	$\dfrac{1}{3}, -\dfrac{1}{3}, \dfrac{3}{10}, -\dfrac{4}{15}, \dfrac{5}{21}$
$x_1, x_2, x_3, \ldots$	$x_n = 3 + \dfrac{1}{10^n}$	$3.1, 3.01, 3.001, 3.0001, 3.00001. \quad \blacksquare$

A **constant sequence** is a sequence in which every term is the same number, such as the sequence 7, 7, 7, 7, ... or the sequence $a_1, a_2, a_3, \ldots$, where $a_n = -18$ for every $n \geq 1$.

The subscript notation for sequences is sometimes abbreviated by writing $\{a_n\}$ in place of $a_1, a_2, a_3, \ldots$.

Example 4 $\{1/2^n\}$ denotes the sequence whose first four terms are

$$a_1 = \frac{1}{2^1}, \qquad a_2 = \frac{1}{2^2} = \frac{1}{4}, \qquad a_3 = \frac{1}{2^3} = \frac{1}{8}, \qquad a_4 = \frac{1}{2^4} = \frac{1}{16}.$$

Similarly, $\{(-1)^n n^2\}$ denotes the sequence with first three terms

$$a_1 = (-1)^1 \cdot 1^2 = -1, \qquad a_2 = (-1)^2 \cdot 2^2 = 4, \qquad a_3 = (-1)^3 \cdot 3^2 = -9$$

and 23rd term $a_{23} = (-1)^{23} \cdot 23^2 = -529.$ ■

There are several ways to display the terms of a sequence on a calculator, as we now see.

Example 5

(a)   Display the first five terms of the sequence $\{a_n = n^2 - n - 3\}$ on your calculator screen.

(b)   Display the first, fifth, ninth, and thirteenth terms of this sequence.

Solution

Method 1. Enter the sequence in the function memory as $y_1 = x^2 - x - 3$. In the table set-up screen, begin the table at $x = 1$ and set the increment at 1 and display a table of values (Figure 7–1). To display the first, fifth, ninth, and thirteenth terms, set the table increment at 4 (Figure 7–2).

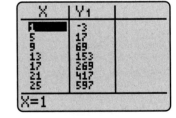

Figure 7–1   **Figure 7–2**

Method 2. Using the tip in the margin, enter the following, which produces Figure 7–3.

$$\text{SEQ}(x^2 - x - 3, x, 1, 5, 1).^*$$

To display the first, fifth, ninth, and thirteenth terms, enter

$$\text{SEQ}(x^2 - x - 3, x, 1, 13, 4),$$

*The final 1 is optional on TI and Sharp 9600; omit "x" on Sharp 9600 and use MAKELIST in place of SEQ on HP-38.

Technology Tip

SEQ is in the OPS submenu of the TI-83/86 LIST menu, in the LIST submenu of the TI-89 MATH menu, in the OPE submenu of the Sharp 9600 LIST menu, in the LIST submenu of the Casio 9850 OPTN menu. MAKELIST is in the LIST submenu of the HP-38 MATH menu.

which tells the calculator to look at every fourth term from 1 to 13, and produces Figure 7–4. ■

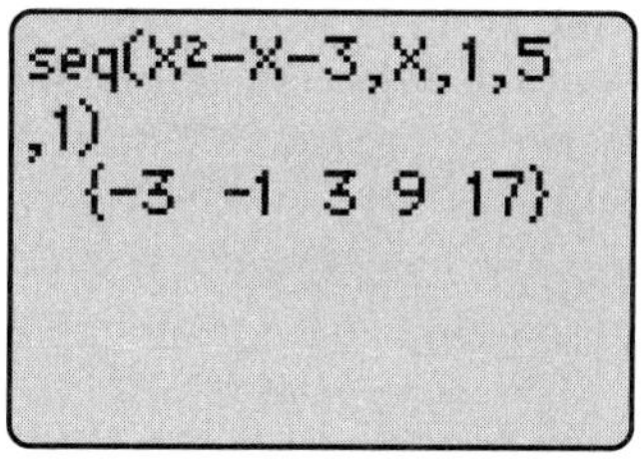

```
seq(X²-X-3,X,1,5
,1)
  {-3 -1 3 9 17}
```

Figure 7–3

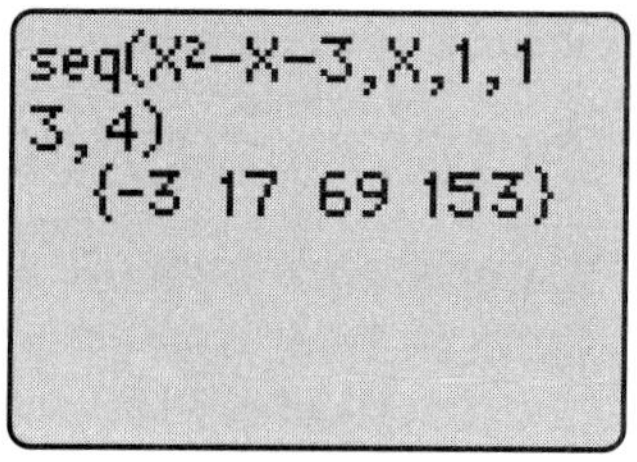

```
seq(X²-X-3,X,1,1
3,4)
  {-3 17 69 153}
```

Figure 7–4

Example 6 Display the first ten terms of the sequence $\left\{ a_n = \dfrac{n}{n+1} \right\}$ on your calculator screen in fractional form, if possible.

Solution Creating a table always produces decimal approximations, as you can easily verify. The same is usually true of the SEQ key, unless you take special steps. On HP-38, change the number format mode to "fraction" (MODE menu). On TI calculators (other than TI-89), either use the FRAC key after obtaining the sequence (Figure 7–5) or use parentheses and the FRAC key as part of the function: Entering

$$\text{SEQ}(x/(x+1) \quad \blacktriangleright\text{FRAC}, x, 1, 10, 1)$$

produces Figure 7–6. In each figure, you must use the arrow key to scroll to the right to see all the terms. ■

Technology Tip

Some calculators have a sequence graphing mode. It can be chosen in the TI-83/89 MODE menu, the HP-38 LIB menu, the COORD submenu of the Sharp 9600 SETUP menu, or the RECUR submenu of the Casio 9850 main menu. On such calculators, recursively defined functions may be entered into the sequence memory (Y = list). Check your instruction manual for the correct syntax.

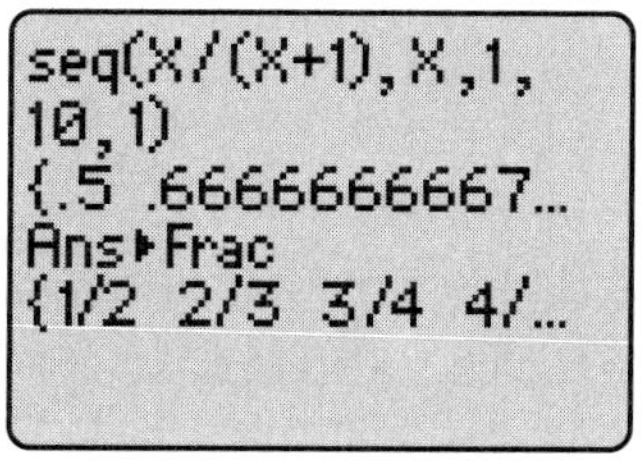

```
seq(X/(X+1),X,1,
10,1)
{.5 .6666666667…
Ans▶Frac
{1/2 2/3 3/4 4/…
```

Figure 7–5

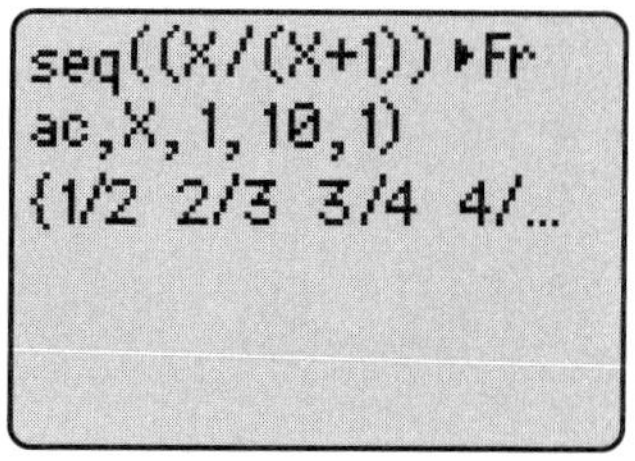

```
seq((X/(X+1))▶Fr
ac,X,1,10,1)
{1/2 2/3 3/4 4/…
```

Figure 7–6

A sequence is said to be defined **recursively** (or **inductively**) if the first term is given (or the first several terms) and there is a method of determining the nth term by using the terms that precede it.

Example 7 Consider the sequence whose first two terms are

$$a_1 = 1 \qquad \text{and} \qquad a_2 = 1$$

and whose nth term (for $n \geq 3$) is the sum of the two preceding terms:

$$a_3 = a_2 + a_1 = 1 + 1 = 2$$
$$a_4 = a_3 + a_2 = 2 + 1 = 3$$
$$a_5 = a_4 + a_3 = 3 + 2 = 5.$$

For each integer n, the two preceding integers are $n - 1$ and $n - 2$. So

$$a_n = a_{n-1} + a_{n-2} \quad (n \geq 3).$$

This sequence $1, 1, 2, 3, 5, 8, 13, \ldots$ is called the **Fibonacci sequence,** and the numbers that appear in it are called **Fibonacci numbers.** Fibonacci numbers have many surprising and interesting properties. See Exercises 58–64 for details. ∎

Example 8 The sequence given by

$$a_1 = -7 \quad \text{and} \quad a_n = a_{n-1} + 3 \quad \text{for } n \geq 2$$

is defined recursively. Its first three terms are

$$a_1 = -7, \qquad a_2 = a_1 + 3 = -7 + 3 = -4,$$
$$a_3 = a_2 + 3 = -4 + 3 = -1. \quad ∎$$

Sometimes it is convenient or more natural to begin numbering the terms of a sequence with a number other than 1. So we may consider sequences such as

$$b_4, b_5, b_6, \ldots \qquad \text{or} \qquad c_0, c_1, c_2, \ldots .$$

Example 9 The sequence $4, 5, 6, 7, \ldots$ can be conveniently described by saying $b_n = n$, with $n \geq 4$. In the brackets notation, we write $\{n\}_{n \geq 4}$. Similarly, the sequence

$$2^0, 2^1, 2^2, 2^3, \ldots$$

may be described as $\{2^n\}_{n \geq 0}$ or by saying $c_n = 2^n$, with $n \geq 0$. ∎

Summation Notation

It is sometimes necessary to find the sum of various terms in a sequence. For instance, we might want to find the sum of the first nine terms of the sequence $\{a_n\}$. Mathematicians often use the Greek letter sigma (Σ) to abbreviate such a sum:*

$$\sum_{k=1}^{9} a_k = a_1 + a_2 + a_3 + a_4 + a_5 + a_6 + a_7 + a_8 + a_9.$$

Similarly, for any positive integer m and numbers $c_1, c_2, \ldots, c_m$

---

*Σ is the letter S in the Greek alphabet, the first letter in *Sum.*

Summation Notation

$$\sum_{k=1}^{m} c_k \quad \text{means} \quad c_1 + c_2 + c_3 + \cdots + c_m.$$

Example 10 Compute each of these sums:

(a) $\displaystyle\sum_{k=1}^{5} k^2$ (b) $\displaystyle\sum_{k=1}^{4} k^2(k-2)$ (c) $\displaystyle\sum_{k=1}^{6} (-1)^k k.$

Solution

(a) We successively substitute 1, 2, 3, 4, 5 for k in the expression k^2 and add the results:

$$\sum_{k=1}^{5} k^2 = 1^2 + 2^2 + 3^2 + 4^2 + 5^2 = 55.$$

(b) Successively substituting 1, 2, 3, 4 for k in $k^2(k-2)$ and adding the results, we have:

$$\sum_{k=1}^{4} k^2(k-2) = 1^2(1-2) + 2^2(2-2) + 3^2(3-2) + 4^2(4-2)$$

$$= 1(-1) + 4(0) + 9(1) + 16(2) = 40.$$

(c) $\displaystyle\sum_{k=1}^{6} (-1)^k k =$

$$(-1)^1 \cdot 1 + (-1)^2 \cdot 2 + (-1)^3 \cdot 3 + (-1)^4 \cdot 4 + (-1)^5 \cdot 5 + (-1)^6 \cdot 6$$

$$= -1 + 2 - 3 + 4 - 5 + 6 = 3. \quad\blacksquare$$

In sums such as $\displaystyle\sum_{k=1}^{5} k^2$ and $\displaystyle\sum_{k=1}^{6} (-1)^k k$, the letter k is called the **summation index.** Any letter may be used for the summation index, just as the rule of a function f may be denoted by $f(x)$ or $f(t)$ or $f(k)$, etc. For example, $\displaystyle\sum_{n=1}^{5} n^2$ means: Take the sum of the terms n^2 as n takes values from 1 to 5. In other words, $\displaystyle\sum_{n=1}^{5} n^2 = \sum_{k=1}^{5} k^2$. Similarly,

$$\sum_{k=1}^{4} k^2(k-2) = \sum_{j=1}^{4} j^2(j-2) = \sum_{n=1}^{4} n^2(n-2).$$

The Σ notation for sums can also be used for sums that don't begin with $k = 1$. For instance,

$$\sum_{k=4}^{10} k^2 = 4^2 + 5^2 + 6^2 + 7^2 + 8^2 + 9^2 + 10^2 = 371$$

$$\sum_{j=0}^{3} j^2(2j + 5) = 0^2(2 \cdot 0 + 5) + 1^2(2 \cdot 1 + 5) + 2^2(2 \cdot 2 + 5) + 3^2(2 \cdot 3 + 5).$$

Technology Tip

On TI-86/89, Casio 9850, and HP-38, SUM (or ΣLIST) is in the same submenu as SEQ (or MAKELIST). On TI-83 and Sharp 9600, SUM is in the MATH submenu of the LIST menu.

Example 11 Use a calculator to compute these sums:

(a) $\displaystyle\sum_{k=1}^{50} k^2$ (b) $\displaystyle\sum_{k=38}^{75} k^2.$

Solution In each case, use SUM together with SEQ (or ΣLIST and MAKELIST on HP-38), with the same syntax for SEQ as in Example 5:

$$\sum_{k=1}^{50} k^2 = \text{SUM SEQ}(x^2, x, 1, 50, 1) = 42{,}925$$

and

$$\sum_{k=38}^{75} k^2 = \text{SUM SEQ}(x^2, x, 38, 75, 1) = 125{,}875$$

as shown in Figure 7–7. ■

```
sum(seq(X², X, 1, 5
0,1))
              42925
```

```
sum(seq(X², X, 38,
75,1))
             125875
```

(a)   (b)

Figure 7–7

Partial Sums

Suppose $\{a_n\}$ is a sequence and k is a positive integer. The sum of the first k terms of the sequence is called the **kth partial sum** of the sequence. Thus,

Partial Sums

The kth partial sum of $\{a_n\} = \displaystyle\sum_{n=1}^{k} a_n = a_1 + a_2 + a_3 + \cdots + a_k.$

Example 12 Here are some partial sums of the sequence $\{n^3\}$:

First partial sum: $\displaystyle\sum_{n=1}^{1} n^3 = 1^3 = 1$

Second partial sum: $\displaystyle\sum_{n=1}^{2} n^3 = 1^3 + 2^3 = 9$

Sixth partial sum: $\displaystyle\sum_{n=1}^{6} n^3 = 1^3 + 2^3 + 3^3 + 4^3 + 5^3 + 6^3 = 441.$ ■

Example 13 The sequence $\{2^n\}_{n \ge 0}$ begins with the 0th term, so the fourth partial sum (the sum of the first four terms) is

$$2^0 + 2^1 + 2^2 + 2^3 = \sum_{n=0}^{3} 2^n.$$

Similarly, the fifth partial sum of the sequence $\left\{ \dfrac{1}{n(n-2)} \right\}_{n \ge 3}$ is the sum of the first five terms:

$$\frac{1}{3(3-2)} + \frac{1}{4(4-2)} + \frac{1}{5(5-2)} + \frac{1}{6(6-2)} + \frac{1}{7(7-2)} = \sum_{n=3}^{7} \frac{1}{n(n-2)}. \ \blacksquare$$

Certain calculations can be written very compactly in summation notation. For example, the distributive law shows that

$$ca_1 + ca_2 + ca_3 + \cdots + ca_r = c(a_1 + a_2 + a_3 + \cdots + a_r).$$

In summation notation this becomes

$$\sum_{n=1}^{r} ca_n = c\left(\sum_{n=1}^{r} a_n\right).$$

This proves the first of the following statements.

Properties of Sums

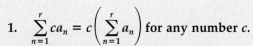

1. $\displaystyle\sum_{n=1}^{r} ca_n = c\left(\sum_{n=1}^{r} a_n\right)$ for any number c.

2. $\displaystyle\sum_{n=1}^{r} (a_n + b_n) = \sum_{n=1}^{r} a_n + \sum_{n=1}^{r} b_n$

3. $\displaystyle\sum_{n=1}^{r} (a_n - b_n) = \sum_{n=1}^{r} a_n - \sum_{n=1}^{r} b_n$

To prove statement 2, use the commutative and associative laws repeatedly to show that

$$(a_1 + b_1) + (a_2 + b_2) + (a_3 + b_3) + \cdots + (a_r + b_r)$$
$$= (a_1 + a_2 + a_3 + \cdots + a_r) + (b_1 + b_2 + b_3 + \cdots + b_r),$$

which can be written in summation notation as

$$\sum_{n=1}^{r} (a_n + b_n) = \sum_{n=1}^{r} a_n + \sum_{n=1}^{r} b_n.$$

The last statement is proved similarly.

Exercises 7.1

In Exercises 1–10, find the first five terms of the sequence $\{a_n\}$.

1. $a_n = 2n + 6$

2. $a_n = 2^n - 7$

3. $a_n = \dfrac{1}{n^3}$

4. $a_n = \dfrac{1}{(n+3)(n+1)}$

5. $a_n = (-1)^n\sqrt{n+2}$

6. $a_n = (-1)^{n+1}n(n-1)$

7. $a_n = 4 + (-.1)^n$

8. $a_n = 5 - (.1)^n$

9. $a_n = (-1)^n + 3n$

10. $a_n = (-1)^{n+2} - (n+1)$

In Exercises 11–14, express the sum in Σ notation.

11. $1 + 2 + 3 + 4 + 5 + 6 + 7 + 8 + 9 + 10 + 11$

12. $1^1 + 2^2 + 3^3 + 4^4 + 5^5$

13. $\dfrac{1}{2^7} + \dfrac{1}{2^8} + \dfrac{1}{2^9} + \dfrac{1}{2^{10}} + \dfrac{1}{2^{11}} + \dfrac{1}{2^{12}} + \dfrac{1}{2^{13}}$

14. $(-6)^{11} + (-6)^{12} + (-6)^{13} + (-6)^{14} + (-6)^{15}$

In Exercises 15–20, find the sum.

15. $\displaystyle\sum_{i=1}^{5} 3i$

16. $\displaystyle\sum_{i=1}^{4} \dfrac{1}{2^i}$

17. $\displaystyle\sum_{n=1}^{16} (2n - 3)$

18. $\displaystyle\sum_{n=1}^{75} (-1)^n(3n + 1)$

19. $\displaystyle\sum_{n=15}^{36} (n^2 - 8)$

20. $\displaystyle\sum_{k=0}^{25} (2k^2 - 5k + 1)$

In Exercises 21–26, find a formula for the nth term of the sequence whose first few terms are given.

21. $-1, 1, -1, 1, -1, 1, \ldots$

22. $2, -2, 2, -2, 2, -2, \ldots$

23. $\dfrac{1}{2}, \dfrac{2}{3}, \dfrac{3}{4}, \dfrac{4}{5}, \dfrac{5}{6}, \ldots$

24. $\dfrac{1}{2 \cdot 3}, \dfrac{1}{3 \cdot 4}, \dfrac{1}{4 \cdot 5}, \dfrac{1}{5 \cdot 6}, \dfrac{1}{6 \cdot 7}, \ldots$

25. $2, 7, 12, 17, 22, 27, \ldots$

26. $8, -5, 2, -11, -4, -17, -10, \ldots$

In Exercises 27–34, find the first five terms of the given sequence.

27. $a_1 = 4$ and $a_n = 2a_{n-1} + 3$ for $n \geq 2$

28. $a_1 = -3$ and $a_n = (-1)^n 4a_{n-1} - 5$ for $n \geq 2$

29. $a_1 = 1, a_2 = -2, a_3 = 3,$ and $a_n = a_{n-1} + a_{n-2} + a_{n-3}$ for $n \geq 4$

30. $a_1 = 1, a_2 = 3,$ and $a_n = 2a_{n-1} + 3a_{n-2}$ for $n \geq 3$

31. $a_0 = 2, a_1 = 3,$ and $a_n = (a_{n-1})\left(\dfrac{1}{2}a_{n-2}\right)$ for $n \geq 2$

32. $a_0 = 1, a_1 = 1,$ and $a_n = na_{n-1}$ for $n \geq 2$

33. a_n is the nth digit in the decimal expansion of π.

34. a_n is the nth digit in the decimal expansion of $1/13$.

In Exercises 35–38, find the third and the sixth partial sums of the sequence.

35. $\{n^2 - 5n + 2\}$

36. $\{(2n - 3n^2)^2\}$

37. $\{(-1)^{n+1}5\}$

38. $\{2^n(2 - n^2)\}_{n \geq 0}$

In Exercises 39–42, express the given sum in Σ notation.

39. $\dfrac{1}{3} + \dfrac{1}{5} + \dfrac{1}{7} + \dfrac{1}{9} + \dfrac{1}{11} + \dfrac{1}{13}$

40. $2 + 1 + \dfrac{4}{5} + \dfrac{5}{7} + \dfrac{2}{3} + \dfrac{7}{11} + \dfrac{8}{13}$

41. $\dfrac{1}{8} - \dfrac{2}{9} + \dfrac{3}{10} - \dfrac{4}{11} + \dfrac{5}{12}$

42. $\dfrac{2}{3 \cdot 5} + \dfrac{4}{5 \cdot 7} + \dfrac{8}{7 \cdot 9} + \dfrac{16}{9 \cdot 11} + \dfrac{32}{11 \cdot 13} + \dfrac{64}{13 \cdot 15} + \dfrac{128}{15 \cdot 17}$

In Exercises 43–48, use a calculator to approximate the required term or sum.

43. a_{12} where $a_n = \left(1 + \dfrac{1}{n}\right)^n$

44. a_{50} where $a_n = \dfrac{\ln n}{n^2}$

45. a_{102} where $a_n = \dfrac{n^3 - n^2 + 5n}{3n^2 + 2n - 1}$

46. a_{125} where $a_n = \sqrt[n]{n}$

47. $\displaystyle\sum_{k=1}^{14} \dfrac{1}{k^2}$

48. $\displaystyle\sum_{n=8}^{22} \dfrac{1}{n}$

49. From 1991 through 1998, the annual dividends per share of Coca-Cola stock were approximated by the sequence $\{a_n\}$, where $n = 1$ corresponds to 1991 and $a_n = .175 + .055n$.

(a) What were the approximate dividends per share in 1993, 1995, and 1998?

(b) If you owned one share from 1991 through 1998, how much would you have collected in dividends?

50. Calcuplay Company offers you a job, with salary b_n in year n, where $\{b_n\}$ is the sequence defined by $b_n = (1.06)^{n-1}30{,}000$.
    (a) What is your starting salary $(n = 0)$?
    (b) What is your salary two years from now? Ten years from now?
    (c) If you work for Calcuplay for 25 years, how much will you earn?

51. Book sales in the United States (in billions of dollars) in year n are projected to be given by the sequence $\{a_n\}$, where $n = 0$ corresponds to 1990 and $a_n = .6n + 15.2$.
    (a) What were book sales in 1993? In 2000?
    (b) Find the projected total book sales from 1995 through 2003.

52. Annual revenue (in billions of dollars) from home video sales and rentals is approximated by the sequence $\{b_n\}$, where $n = 1$ corresponds to 1980 and $b_n = .972n - .768$. What will be the total revenue for the period from 1995 through 2005?

Thinkers

Exercises 53–57 deal with prime numbers. A positive integer greater than 1 is prime *if its only positive integer factors are itself and 1. For example, 7 is prime because its only factors are 7 and 1, but 15 is not prime because it has factors other than 15 and 1 (namely, 3 and 5).*

53. (a) Let $\{a_n\}$ be the sequence of prime integers in their usual ordering. Verify that the first ten terms are 2, 3, 5, 7, 11, 13, 17, 19, 23, 29.
    (b) Find $a_{17}, a_{18}, a_{19}, a_{20}$.

In Exercises 54–57, find the first five terms of the sequence.

54. a_n is the nth prime integer larger than 10. [*Hint:* $a_1 = 11$.]

55. a_n is the square of the nth prime integer.

56. a_n is the number of prime integers less than n.

57. a_n is the largest prime integer less than $5n$.

Exercises 58–64 deal with the Fibonacci sequence $\{a_n\}$ that was discussed in Example 7.

58. Leonardo Fibonacci discovered the sequence in the 13th century in connection with this problem: A rabbit colony begins with one pair of adult rabbits (one male, one female). Each adult pair produces one pair of babies (one male, one female) every month. Each pair of baby rabbits becomes adult and produces the first offspring at age two months. Assuming that no rabbits die, how many adult pairs of rabbits are in the colony at the end of n months $(n = 1, 2, 3, \ldots)$? [*Hint:* It may be helpful to make up a chart listing for each month the number of adult pairs, the number of one-month-old pairs, and the number of baby pairs.]

59. (a) List the first ten terms of the Fibonacci sequence.
    (b) List the first ten partial sums of the sequence.
    (c) Do the partial sums follow an identifiable pattern?

60. Verify that every positive integer less than or equal to 15 can be written as a sum of Fibonacci numbers, with none used more than once.

61. Verify that $5(a_n)^2 + 4(-1)^n$ is always a perfect square for $n = 1, 2, \ldots, 10$.

62. Verify that $(a_n)^2 = a_{n+1}a_{n-1} + (-1)^{n-1}$ for $n = 2, \ldots, 10$.

63. Show that $\displaystyle\sum_{n=1}^{k} a_n = a_{k+2} - 1$. [*Hint:* $a_1 = a_3 - a_2$; $a_2 = a_4 - a_3$; etc.]

64. Show that $\displaystyle\sum_{n=1}^{k} a_{2n-1} = a_{2k}$, that is, the sum of the first k odd-numbered terms is the kth even-numbered term. [*Hint:* $a_3 = a_4 - a_2$; $a_5 = a_6 - a_4$; etc.]

7.2 Arithmetic Sequences

An **arithmetic sequence** (sometimes called an **arithmetic progression**) is a sequence in which the difference between each term and the preceding one is always the same constant.

Example 1 In the sequence, 3, 8, 13, 18, 23, 28, … , the difference between each term and the preceding one is always 5. So this is an arithmetic sequence. ■

Example 2 In the sequence 14, 10, 6, 2, −2, −6, −10, −14, … the difference between each term and the preceding one is −4 (for example, $10 - 14 = -4$ and $-6 - (-2) = -4$). Hence, this sequence is arithmetic. ■

If $\{a_n\}$ is an arithmetic sequence, then for each $n \geq 2$, the term preceding a_n is a_{n-1} and the difference $a_n - a_{n-1}$ is some constant—call it d. Therefore, $a_n - a_{n-1} = d$, or equivalently.

Arithmetic Sequences

> In an arithmetic sequence $\{a_n\}$
>
> $$a_n = a_{n-1} + d$$
>
> for some constant d and all $n \geq 2$.

The number d is called the **common difference** of the arithmetic sequence.

Example 3 If $\{a_n\}$ is an arithmetic sequence with $a_1 = 3$ and $a_2 = 4.5$, then the common difference is $d = a_2 - a_1 = 4.5 - 3 = 1.5$. So the sequence begins with 3, 4.5, 6, 7.5, 9, 10.5, 12, 13.5, … . ■

Example 4 The sequence $\{-7 + 4n\}$ is an arithmetic sequence because for each $n \geq 2$,

$$a_n - a_{n-1} = (-7 + 4n) - [-7 + 4(n - 1)]$$
$$= (-7 + 4n) - (-7 + 4n - 4) = 4.$$

Therefore, the common difference is $d = 4$. ■

If $\{a_n\}$ is an arithmetic sequence with common difference d, then for each $n \geq 2$ we know that $a_n = a_{n-1} + d$. Applying this fact repeatedly shows that

$$a_2 = a_1 + d$$
$$a_3 = a_2 + d = (a_1 + d) + d = a_1 + 2d$$
$$a_4 = a_3 + d = (a_1 + 2d) + d = a_1 + 3d$$
$$a_5 = a_4 + d = (a_1 + 3d) + d = a_1 + 4d$$

and in general,

nth Term of an Arithmetic Sequence

If an arithmetic sequence $\{a_n\}$ with common difference d

$$a_n = a_1 + (n - 1)d \quad \text{for every } n \geq 1.$$

Example 5 Find the nth term of the arithmetic sequence with first term -5 and common difference 3. Since $a_1 = -5$ and $d = 3$, the formula in the box shows that

$$a_n = a_1 + (n - 1)d = -5 + (n - 1)3 = 3n - 8. \quad \blacksquare$$

Example 6 What is the 45th term of the arithmetic sequence whose first three terms are 5, 9, 13?

Solution   The first three terms show that $a_1 = 5$ and that the common difference d is 4. Applying the formula in the box with $n = 45$, we have

$$a_{45} = a_1 + (45 - 1)d = 5 + (44)4 = 181. \quad \blacksquare$$

Example 7   If $\{a_n\}$ is an arithmetic sequence with $a_6 = 57$ and $a_{10} = 93$, find a_1 and a formula for a_n.

Solution   Apply the formula $a_n = a_1 + (n - 1)d$ with $n = 6$ and $n = 10$:

$$a_6 = a_1 + (6 - 1)d \quad \text{and} \quad a_{10} = a_1 + (10 - 1)d$$
$$57 = a_1 + 5d \qquad\qquad\quad 93 = a_1 + 9d$$

We can find a_1 and d by solving this system:

$$a_1 + 9d = 93$$
$$a_1 + 5d = 57.$$

Subtracting the second equation from the first shows that $4d = 36$, and hence $d = 9$. Substituting $d = 9$ in the second equation shows that $a_1 = 12$. So the formula for a_n is

$$a_n = a_1 + (n - 1)d = 12 + (n - 1)9 = 9n + 3. \quad \blacksquare$$

Partial Sums

It's easy to compute partial sums of arithmetic sequences by using the following formulas.

Partial Sums of an Arithmetic Sequence

> If $\{a_n\}$ is an arithmetic sequence with common difference d, then for each positive integer k the kth partial sum can be found by using *either* of these formulas:
>
> 1. $\displaystyle\sum_{n=1}^{k} a_n = \frac{k}{2}(a_1 + a_k)$ or
>
> 2. $\displaystyle\sum_{n=1}^{k} a_n = ka_1 + \frac{k(k-1)}{2}d$

Proof   Let S denote the kth partial sum $a_1 + a_2 + \cdots + a_k$. For reasons that will become apparent later we shall calculate the number $2S$:

$$2S = S + S = (a_1 + a_2 + \cdots + a_k) + (a_1 + a_2 + \cdots + a_k).$$

Now we rearrange the terms on the right by grouping the first and last terms together, then the first and last of the remaining terms, and so on:

$$2S = (a_1 + a_k) + (a_2 + a_{k-1}) + (a_3 + a_{k-2}) + \cdots + (a_k + a_1).$$

Since adjacent terms of the sequence differ by d we have:

$$a_2 + a_{k-1} = (a_1 + d) + (a_k - d) = a_1 + a_k.$$

Using this fact,

$$a_3 + a_{k-2} = (a_2 + d) + (a_{k-1} - d) = a_2 + a_{k-1} = a_1 + a_k.$$

Continuing in this manner we see that every pair in the sum for $2S$ is equal to $a_1 + a_k$. Therefore,

$$2S = (a_1 + a_k) + (a_2 + a_{k-1}) + (a_3 + a_{k-2}) + \cdots + (a_k + a_1)$$
$$= (a_1 + a_k) + (a_1 + a_k) + (a_1 + a_k) + \cdots + (a_1 + a_k) \quad (k \text{ terms})$$
$$= k(a_1 + a_k).$$

Dividing both sides of this last equation by 2 shows that $S = \dfrac{k}{2}(a_1 + a_k)$. This proves the first formula. To obtain the second one, note that

$$a_1 + a_k = a_1 + [a_1 + (k-1)d] = 2a_1 + (k-1)d.$$

Substituting the right side of this equation in the first formula for S shows that

$$S = \frac{k}{2}(a_1 + a_k) = \frac{k}{2}[2a_1 + (k-1)d] = ka_1 + \frac{k(k-1)}{2}d.$$

This proves the second formula.   ∎

Example 8   Find the 12th partial sum of the arithmetic sequence that begins $-8, -3, 2, 7, \ldots$.

Solution   We first note that the common difference d is 5. Since $a_1 = -8$ and $d = 5$, the second formula in the box with $k = 12$ shows that

$$\sum_{n=1}^{12} a_n = 12(-8) + \frac{12(11)}{2}5 = -96 + 330 = 234. \quad \blacksquare$$

Example 9   Find the sum of all multiples of 3 from 3 to 333.

Solution   Note that this sum is just a partial sum of the arithmetic sequence $3, 6, 9, 12, \ldots$. Since this sequence can be written in the form

$$3 \cdot 1, 3 \cdot 2, 3 \cdot 3, 3 \cdot 4, 3 \cdot 5, 3 \cdot 6, \ldots$$

we see that $333 = 3 \cdot 111$ is the 111th term. The 111th partial sum of this sequence can be found by using the first formula in the box with $k = 111$, $a_1 = 3$, and $a_{111} = 333$:

$$\sum_{n=1}^{111} a_n = \frac{111}{2}(3 + 333) = \frac{111}{2}(336) = 18{,}648. \quad \blacksquare$$

Example 10   If the starting salary for a job is $20,000 and you get a $2000 raise at the beginning of each subsequent year, what will your salary be during the tenth year? How much will you earn during the first ten years?

Solution   Your yearly salary rates form a sequence: 20,000, 22,000, 24,000, 26,000, and so on. It is an arithmetic sequence with $a_1 = 20{,}000$ and $d = 2000$. Your tenth-year salary is

$$a_{10} = a_1 + (10 - 1)d = 20{,}000 + 9 \cdot 2000 = \$38{,}000.$$

Your ten-year total earnings are the tenth partial sum of the sequence:

$$\frac{10}{2}(a_1 + a_{10}) + \frac{10}{2}(20{,}000 + 38{,}000) = 5(58{,}000) = \$290{,}000. \quad \blacksquare$$

Exercises 7.2

In Exercises 1–6, the first term a_1 and the common difference d of an arithmetic sequence are given. Find the fifth term and the formula for the nth term.

1. $a_1 = 5, d = 2$

2. $a_1 = -4, d = 5$

3. $a_1 = 4, d = \dfrac{1}{4}$

4. $a_1 = -6, d = \dfrac{2}{3}$

5. $a_1 = 10, d = -\dfrac{1}{2}$

6. $a_1 = \pi, d = \dfrac{1}{5}$

In Exercises 7–12, find the kth partial sum of the arithmetic sequence $\{a_n\}$ with common difference d.

7. $k = 6, a_1 = 2, d = 5$

8. $k = 8, a_1 = \dfrac{2}{3}, d = -\dfrac{4}{3}$

9. $k = 7, a_1 = \dfrac{3}{4}, d = -\dfrac{1}{2}$

10. $k = 9, a_1 = 6, a_9 = -24$

11. $k = 6, a_1 = -4, a_6 = 14$

12. $k = 10, a_1 = 0, a_{10} = 30$

In Exercises 13–18, show that the sequence is arithmetic and find its common difference.

13. $\{3 - 2n\}$

14. $\left\{4 + \dfrac{n}{3}\right\}$

15. $\left\{\dfrac{5 + 3n}{2}\right\}$

16. $\left\{\dfrac{\pi - n}{2}\right\}$

17. $\{c + 2n\}$ (*c* constant)

18. $\{2b + 3nc\}$ (*b, c* constants)

In Exercises 19–24, use the given information about the arithmetic sequence with common difference d to find a_5 and a formula for a_n.

19. $a_4 = 12, d = 2$

20. $a_7 = -8, d = 3$

21. $a_2 = 4, a_6 = 32$

22. $a_7 = 6, a_{12} = -4$

23. $a_5 = 0, a_9 = 6$

24. $a_5 = -3, a_9 = -18$

In Exercises 25–28, find the sum.

25. $\displaystyle\sum_{n=1}^{20} (3n + 4)$

26. $\displaystyle\sum_{n=1}^{25} \left(\dfrac{n}{4} + 5\right)$

27. $\displaystyle\sum_{n=1}^{40} \dfrac{n + 3}{6}$

28. $\displaystyle\sum_{n=1}^{30} \dfrac{4 - 6n}{3}$

29. Find the sum of all the even integers from 2 to 100.

30. Find the sum of all the integer multiples of 7 from 7 to 700.

31. Find the sum of the first 200 positive integers.

32. Find the sum of the positive integers from 101 to 200 (inclusive). [*Hint:* What's the sum from 1 to 100? Use it and Exercise 31.]

33. A business makes a $10,000 profit during its first year. If the yearly profit increases by $7500 in each subsequent year, what will the profit be in the tenth year and what will the total profit for the first ten years be?

34. If a man's starting salary is $15,000 and he receives a $1000 increase every six months, what will his salary be during the last six months of the sixth year? How much will he earn during the first six years?

35. A lecture hall has six seats in the first row, eight in the second, ten in the third, and so on, through row 12. Rows 12 through 20 (the last row) all have the same number of seats. Find the number of seats in the lecture hall.

36. A monument is constructed by laying a row of 60 bricks at ground level. A second row, with two fewer bricks, is centered on that; a third row, with two fewer bricks, is centered on the second; and so on. The top row contains ten bricks. How many bricks are in the monument?

37. A ladder with nine rungs is to be built, with the bottom rung 24 inches wide and the top rung 18 inches wide. If the lengths of the rungs decrease uniformly from bottom to top, how long should each of the seven intermediate rungs be?

38. Find the first eight numbers in an arithmetic sequence in which the sum of the first and seventh term is 40 and the product of the first and fourth terms is 160.

7.3 Geometric Sequences

A **geometric sequence** (sometimes called a **geometric progression**) is a sequence in which the quotient of each term and the preceding one is the same constant *r*. This constant *r* is called the **common ratio** of the geometric sequence.

Example 1 The sequence $3, 9, 27, \ldots, 3^n, \ldots$ is geometric with common ratio 3. For instance, $a_2/a_1 = 9/3 = 3$ and $a_3/a_2 = 27/9 = 3$. If 3^n is any term ($n \geq 2$), then the preceding term is 3^{n-1} and

$$\frac{3^n}{3^{n-1}} = \frac{3 \cdot 3^{n-1}}{3^{n-1}} = 3. \quad \blacksquare$$

Example 2   The sequence $\{5/2^n\}$ which begins $5/2, 5/4, 5/8, 5/16, \ldots$ is geometric with common ratio $r = 1/2$ because for each $n \geq 1$

$$\frac{5/2^n}{5/2^{n-1}} = \frac{5}{2^n} \cdot \frac{2^{n-1}}{5} = \frac{2^{n-1}}{2^n} = \frac{2^{n-1}}{2^{n-1} \cdot 2} = \frac{1}{2}. \quad \blacksquare$$

If $\{a_n\}$ is a geometric sequence with common ratio r, then for each $n \geq 2$ the term preceding a_n is a_{n-1} and

$$\frac{a_n}{a_{n-1}} = r, \quad \text{or equivalently,} \quad a_n = ra_{n-1}.$$

Applying this last formula for $n = 2, 3, 4, \ldots$ we have

$$a_2 = ra_1$$
$$a_3 = ra_2 = r(ra_1) = r^2 a_1$$
$$a_4 = ra_3 = r(r^2 a_1) = r^3 a_1$$
$$a_5 = ra_4 = r(r^3 a_1) = r^4 a_1$$

and in general

nth Term of a Geometric Sequence

> If $\{a_n\}$ is a geometric sequence with common ratio r, then for all $n \geq 1$,
>
> $$a_n = r^{n-1} a_1.$$

Example 3   To find a formula for the nth term of the geometric sequence $\{a_n\}$ where $a_1 = 7$ and $r = 2$, we use the equation in the box:

$$a_n = r^{n-1} a_1 = 2^{n-1} \cdot 7.$$

So the sequence is $\{7 \cdot 2^{n-1}\}$. $\blacksquare$

Example 4   If the first two terms of a geometric sequence are 2 and $-2/5$, then the common ratio must be

$$r = \frac{a_2}{a_1} = \frac{-2/5}{2} = \frac{-2}{5} \cdot \frac{1}{2} = -\frac{1}{5}.$$

Using the equation in the box, we now see that the formula for the nth term is

$$a_n = r^{n-1} a_1 = \left(-\frac{1}{5}\right)^{n-1} (2) = \frac{(1)^{n-1}}{(-5)^{n-1}} (2) = \frac{2}{(-5)^{n-1}}.$$

So, the sequence begins $2, -2/5, 2/5^2, -2/5^3, 2/5^4, \ldots$. $\blacksquare$

Example 5 If $\{a_n\}$ is a geometric sequence with $a_2 = 20/9$ and $a_5 = 160/243$, then by the equation in the preceding box

$$\frac{160/243}{20/9} = \frac{a_5}{a_2} = \frac{r^4 a_1}{r a_1} = r^3.$$

Consequently,

$$r = \sqrt[3]{\frac{160/243}{20/9}} = \sqrt[3]{\frac{160}{243} \cdot \frac{9}{20}} = \sqrt[3]{\frac{8 \cdot 9}{243}} = \sqrt[3]{\frac{8}{27}} = \frac{2}{3}.$$

Since $a_2 = r a_1$ we see that

$$a_1 = \frac{a_2}{r} = \frac{20/9}{2/3} = \frac{20}{9} \cdot \frac{3}{2} = \frac{10}{3}.$$

Therefore,

$$a_n = r^{n-1} a_1 = \left(\frac{2}{3}\right)^{n-1} \cdot \frac{10}{3} = \frac{2^{n-1} \cdot 2 \cdot 5}{3^{n-1} \cdot 3} = \frac{2^n \cdot 5}{3^n} = 5 \left(\frac{2}{3}\right)^n. \quad \blacksquare$$

Partial Sums

If the common ratio r of a geometric sequence is the number 1, then we have

$$a_n = 1^{n-1} a_1 \quad \text{for every } n \geq 1.$$

Therefore, the sequence is just the constant sequence $a_1, a_1, a_1, \ldots$. For any positive integer k, the kth partial sum of this constant sequence is

$$\underbrace{a_1 + a_2 + \cdots + a_1}_{k \text{ terms}} = k a_1.$$

In other words, the kth partial sum of a constant sequence is just k times the constant. If a geometric sequence is not constant (that is, $r \neq 1$), then its partial sums are given by the following formula.

Partial Sums of a Geometric Sequence ▶

The kth partial sum of the geometric sequence $\{a_n\}$ with common ratio $r \neq 1$ is

$$\sum_{n=1}^{k} a_n = a_1 \left(\frac{1 - r^k}{1 - r}\right).$$

Proof If S denotes the kth partial sum, then the formula for the nth term of a geometric sequence shows that

$$S = a_1 + a_2 + \cdots + a_k = a_1 + a_1 r + a_1 r^2 + a_1 r^3 + \cdots + a_1 r^{k-1}.$$

Use this equation to compute $S - rS$:

$$S = a_1 + a_1r + a_1r^2 + a_1r^3 + \cdots + a_1r^{k-1}$$
$$rS = \qquad a_1r + a_1r^2 + a_1r^3 + \cdots + a_1r^{k-1} + a_1r^k$$
$$\overline{}$$
$$S - rS = a_1 \qquad\qquad\qquad\qquad\qquad\qquad\qquad\qquad - a_1r^k$$
$$(1 - r)S = a_1(1 - r^k)$$

Since $r \neq 1$, we can divide both sides of this last equation by $1 - r$ to complete the proof:

$$S = \frac{a_1(1 - r^k)}{1 - r} = a_1\left(\frac{1 - r^k}{1 - r}\right). \quad \blacksquare$$

Example 6 Find the sum

$$-\frac{3}{2} + \frac{3}{4} - \frac{3}{8} + \frac{3}{16} - \frac{3}{32} + \frac{3}{64} - \frac{3}{128} + \frac{3}{256} - \frac{3}{512}.$$

Solution   Note that this is the ninth partial sum of the geometric sequence $\left\{3\left(\dfrac{-1}{2}\right)^n\right\}$. The common ratio is $r = -\dfrac{1}{2}$. The formula in the box shows that

$$\sum_{n=1}^{9} 3\left(\frac{-1}{2}\right)^n = a_1\left(\frac{1 - r^9}{1 - r}\right) = \left(\frac{-3}{2}\right)\left[\frac{1 - (-1/2)^9}{1 - (-1/2)}\right]$$

$$= \left(\frac{-3}{2}\right)\left(\frac{1 + 1/2^9}{3/2}\right) = \left(\frac{-3}{2}\right)\left(\frac{2}{3}\right)\left(1 + \frac{1}{2^9}\right)$$

$$= -1 - \frac{1}{2^9} = -1 - \frac{1}{512} = -\frac{513}{512}. \quad \blacksquare$$

Example 7   A superball is dropped from a height of 9 feet. It hits the ground and bounces to a height of 6 feet. It continues to bounce up and down. On each bounce it rises to 2/3 of the height of the previous bounce. How far has the ball traveled (both up and down) when it hits the ground for the seventh time?

Solution   We first consider how far the ball travels on each bounce. On the first bounce it rises 6 feet and falls 6 feet for a total of 12 feet. On the second bounce it rises and falls 2/3 of the previous height, and hence travels 2/3 of 12 feet. If a_n denotes the distance traveled on the nth bounce, then

$$a_1 = 12 \qquad a_2 = \left(\frac{2}{3}\right)a_1 \qquad a_3 = \left(\frac{2}{3}\right)a_2 = \left(\frac{2}{3}\right)^2 a_1$$

and in general

$$a_n = \left(\frac{2}{3}\right)a_{n-1} = \left(\frac{2}{3}\right)^{n-1} a_1.$$

So $\{a_n\}$ is a geometric sequence with common ratio $r = 2/3$. When the ball hits the ground for the seventh time it has completed six bounces. Therefore, the total distance it has traveled is the distance it was originally dropped (9 feet) plus the distance traveled in six bounces, namely,

$$9 + a_1 + a_2 + a_3 + a_4 + a_5 + a_6 = 9 + \sum_{n=1}^{6} a_n = 9 + a_1\left(\frac{1 - r^6}{1 - r}\right)$$

$$= 9 + 12\left[\frac{1 - (2/3)^6}{1 - (2/3)}\right] \approx 41.84 \text{ feet} \quad \blacksquare$$

Exercises 7.3

In Exercises 1–8, determine whether the sequence is arithmetic, geometric, or neither.

1. 2, 7, 12, 17, 22, ...   **2.** 2, 6, 18, 54, 162, ...

3. 13, 13/2, 13/4, 13/8, ...

4. $-1, -1/2, 0, 1/2, ...$

5. 50, 48, 46, 44, ...

6. 2, -3, 9/2, $-27/4$, $-81/8$, ...

7. 3, $-3/2$, 3/4, $-3/8$, 3/16, ...

8. $-6, -3.7, -1.4, .9, 3.2, ...$

In Exercises 9–14, the first term a_1 and the common ratio r of a geometric sequence are given. Find the sixth term and a formula for the nth term.

9. $a_1 = 5, r = 2$ **10.** $a_1 = 1, r = -2$

11. $a_1 = 4, r = \dfrac{1}{4}$ **12.** $a_1 = -6, r = \dfrac{2}{3}$

13. $a_1 = 10, r = -\dfrac{1}{2}$ **14.** $a_1 = \pi, r = \dfrac{1}{5}$

In Exercises 15–18, find the kth partial sum of the geometric sequence $\{a_n\}$ with common ratio r.

15. $k = 6, a_1 = 5, r = \dfrac{1}{2}$ **16.** $k = 8, a_1 = 9, r = \dfrac{1}{3}$

17. $k = 7, a_2 = 6, r = 2$ **18.** $k = 9, a_2 = 6, r = \dfrac{1}{4}$

In Exercises 19–22, show that the given sequence is geometric and find the common ratio.

19. $\left\{\left(-\dfrac{1}{2}\right)^n\right\}$ **20.** $\{2^{3n}\}$

21. $\{5^{n+2}\}$ **22.** $\{3^{n/2}\}$

In Exercises 23–28, use the given information about the geometric sequence $\{a_n\}$ to find a_5 and a formula for a_n.

23. $a_1 = 256, a_2 = -64$ **24.** $a_1 = 1/6, a_2 = -1/18$

25. $a_2 = 4, a_5 = 1/16$ **26.** $a_3 = 4, a_6 = -32$

27. $a_4 = -4/5, r = 2/5$ **28.** $a_2 = 6, a_7 = 192$

In Exercises 29–34, find the sum.

29. $\displaystyle\sum_{n=1}^{7} 2^n$ **30.** $\displaystyle\sum_{k=1}^{6} 3\left(\frac{1}{2}\right)^k$

31. $\displaystyle\sum_{n=1}^{9} \left(-\frac{1}{3}\right)^n$ **32.** $\displaystyle\sum_{n=1}^{5} 5 \cdot 3^{n-1}$

33. $\displaystyle\sum_{j=1}^{6} 4\left(\frac{3}{2}\right)^{j-1}$ **34.** $\displaystyle\sum_{t=1}^{8} 6(.9)^{t-1}$

35. For 1987–1998, the annual revenue per share in year n of Walt Disney stock is approximated by $a_n = 1.71(1.191)^n$, where $n = 7$ represents 1987.
   (a) Show that the sequence $\{a_n\}$ is a geometric sequence.
   (b) Approximate the total revenues per share for the period 1987–1998.

36. The annual dividends per share of Walt Disney stock from 1989 through 1998 are approximated by the sequence $\{b_n\}$, where $n = 9$ corresponds to 1989 and $b_n = .0228(1.1999)^n$.
   (a) Show that the sequence $\{b_n\}$ is a geometric sequence.
   (b) Approximate the total dividends per share for the period 1989–1998.

37. A ball is dropped from a height of 8 feet. On each bounce it rises to half its previous height. When the ball hits the ground for the seventh time, how far has it traveled?

38. A ball is dropped from a height of 10 feet. On each bounce it rises to 45% of its previous height. When it hits the ground for the tenth time, how far has it traveled?

39. If you are paid a salary of 1¢ on the first day of March, 2¢ on the second day, and your salary continues to double each day, how much will you earn in the month of March?

40. Starting with your parents, how many ancestors do you have for the preceding ten generations?

41. A car that sold for $8000 depreciates in value 25% each year. What is it worth after five years?

42. A vacuum pump removes 60% of the air in a container at each stroke. What percentage of the original amount of air remains after six strokes?

Thinkers

43. Suppose $\{a_n\}$ is a geometric sequence with common ratio $r > 0$ and each $a_n > 0$. Show that the sequence $\{\log a_n\}$ is an arithmetic sequence with common difference $\log r$.

44. Suppose $\{a_n\}$ is an arithmetic sequence with common difference d. Let C be any positive number. Show that the sequence $\{C^{a_n}\}$ is a geometric sequence with common ratio C^d.

45. In the geometric sequence $1, 2, 4, 8, 16, \ldots$, show that each term is 1 plus the sum of all preceding terms.

46. In the geometric sequence $2, 6, 18, 54, \ldots$, show that each term is twice the sum of 1 and all preceding terms.

47. The minimum monthly payment for a certain bank credit card is the larger of 1/25 of the outstanding balance or $5. If the balance is less than $5, the entire balance is due. If you make only the minimum payment each month, how long will it take to pay off a balance of $200 (excluding any interest that might be due)?

7.3.A *EXCURSION* **Infinite Series**

We now introduce a topic that is closely related to infinite sequences and has some very useful applications. We can give only a few highlights here; complete coverage requires calculus.

Consider the sequence $\{3/10^n\}$ and let S_k denote its kth partial sum; then

$$S_1 = \frac{3}{10}$$

$$S_2 = \frac{3}{10} + \frac{3}{10^2} = \frac{33}{100}$$

$$S_3 = \frac{3}{10} + \frac{3}{10^2} + \frac{3}{10^3} = \frac{333}{1000}$$

$$S_4 = \frac{3}{10} + \frac{3}{10^2} + \frac{3}{10^3} + \frac{3}{10^4} = \frac{3333}{10,000}.$$

These partial sums $S_1, S_2, S_3, S_4, \ldots$ themselves form a sequence:

$$\frac{3}{10}, \frac{33}{100}, \frac{333}{1000}, \frac{3333}{10,000}, \ldots$$

The terms in the sequence of partial sums appear to be getting closer and closer to $1/3$. In other words, as k gets larger and larger, the corresponding partial sum S_k gets closer and closer to $1/3$. Consequently, we write

$$\frac{3}{10} + \frac{3}{10^2} + \frac{3}{10^3} + \frac{3}{10^4} + \cdots = \frac{1}{3}$$

and say that $1/3$ is the *sum* of the *infinite series*

$$\frac{3}{10} + \frac{3}{10^2} + \frac{3}{10^3} + \frac{3}{10^4} + \cdots .$$

In the general case, an **infinite series** (or simply **series**) is defined to be an expression of the form

$$a_1 + a_2 + a_3 + a_4 + a_5 + \cdots$$

in which each a_n is a real number. This series is also denoted by the symbol $\sum_{n=1}^{\infty} a_n$.

Example 1

(a) $\displaystyle\sum_{n=1}^{\infty} 2(.6)^n$ denotes the series

$$2(.6) + 2(.6)^2 + 2(.6)^3 + 2(.6)^4 + \cdots .$$

(b) $\displaystyle\sum_{n=1}^{\infty} \left(\frac{-1}{2}\right)^n$ denotes the series

$$-\frac{1}{2} + \left(\frac{-1}{2}\right)^2 + \left(\frac{-1}{2}\right)^3 + \left(\frac{-1}{2}\right)^4 + \cdots = -\frac{1}{2} + \frac{1}{4} - \frac{1}{8} + \frac{1}{16} + \cdots . \quad \blacksquare$$

The **partial sums** of the series $a_1 + a_2 + a_3 + a_4 + \cdots$ are

$$S_1 = a_1$$
$$S_2 = a_1 + a_2$$
$$S_3 = a_1 + a_2 + a_3$$

and in general, for any $k \geq 1$

$$S_k = a_1 + a_2 + a_3 + a_4 + \cdots + a_k.$$

If it happens that the terms $S_1, S_2, S_3, S_4, \ldots$ of the *sequence* of partial sums get closer and closer to a particular real number S in such a way that the partial sum S_k is arbitrarily close to S when k is large enough, then we say that the series **converges** and that S is the **sum of the convergent series.** For example, we just saw that the series

$$\frac{3}{10} + \frac{3}{10^2} + \frac{3}{10^3} + \frac{3}{10^4} + \cdots$$

converges and that its sum is $1/3$.

A sequence is a *list* of numbers $a_1, a_2, a_3, \ldots$. Intuitively, you can think of a convergent series $a_1 + a_2 + a_3 + \cdots$ as an "infinite sum" of numbers. But be careful: Not every series has a sum. For instance, the partial sums of the series

$$1 + 2 + 3 + 4 + \cdots$$

get larger and larger (compute some) and do not get closer and closer to a single real number. So this series is not convergent.

Example 2 Although no proof will be given here, it is intuitively clear that every infinite decimal may be thought of as the sum of a convergent series. For instance,

$$\pi = 3.1415926 \cdots = 3 + .1 + .04 + .001 + .0005 + .00009 + \cdots.$$

Note that the third partial sum is $3 + .1 + .04 = 3.14$, which is π to two decimal places. Similarly, the kth partial sum of this series is just π to $k - 1$ decimal places. ■

Infinite Geometric Series

If $\{a_n\}$ is a geometric sequence with common ratio r, then the corresponding infinite series

$$a_1 + a_2 + a_3 + a_4 + a_5 + \cdots$$

is called an **infinite geometric series.** By using the formula for the nth term of a geometric sequence, we can also express the corresponding geometric series in the form

$$a_1 + ra_1 + r^2a_1 + r^3a_1 + r^4a_1 + \cdots.$$

Under certain circumstances, an infinite geometric series is convergent and has a sum:

Sum of an Infinite Geometric Series

If $|r| < 1$, then the infinite geometric series

$$a_1 + ra_1 + r^2a_1 + r^3a_1 + r^4a_1 + \cdots$$

converges and its sum is

$$\frac{a_1}{1 - r}.$$

Although we cannot prove this fact rigorously here, we can make it highly plausible both geometrically and algebraically.

Example 3 $\displaystyle\sum_{n=1}^{\infty} \frac{8}{5^n} = \frac{8}{5} + \frac{8}{5^2} + \frac{8}{5^3} + \cdots$ is an infinite geometric series with $a_1 = 8/5$ and $r = 1/5$. The kth partial sum of this series is the same as the kth partial sum of the sequence $\{8/5^n\}$ and hence from the box on page 530 we know that

$$S_k = a_1\left(\frac{1 - r^k}{1 - r}\right) = \frac{8}{5}\left[\frac{1 - \left(\frac{1}{5}\right)^k}{1 - \frac{1}{5}}\right]$$

$$= \frac{8}{5}\left(\frac{1 - \frac{1}{5^k}}{\frac{4}{5}}\right) = \frac{8}{5}\cdot\frac{5}{4}\left(1 - \frac{1}{5^k}\right)$$

$$= 2\left(1 - \frac{1}{5^k}\right) = 2 - \frac{2}{5^k}.$$

The function $f(x) = 2 - 2/5^x$ is defined for all real numbers. When $x = k$ is a positive integer, then $f(k)$ is S_k, the kth partial sum of the series. Using a calculator we obtain the graph of $f(x)$ in Figure 7–8.

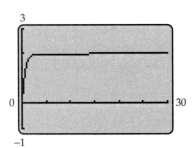

Figure 7–8

GRAPHING EXPLORATION

Graph $f(x)$ in the same viewing window as in Figure 7–8. Use the trace feature to move the cursor along the graph. As x gets larger, what is the apparent value of $f(x)$ (that is, the value of the partial sum)?

Your calculator will probably tell you that every partial sum is 2, once you move beyond approximately $x = 15$. Actually, the partial sums are slightly smaller than 2, but are rounded to 2 by the calculator. In any case, the horizontal line through 2 is a horizontal asymptote of the graph (meaning that the graph gets very close to the line as x gets larger), so it is very plausible that the sequence converges to the number 2. But 2 is exactly what the preceding box says the sum should be:

$$\frac{a_1}{1 - r} = \frac{\frac{8}{5}}{1 - \frac{1}{5}} = \frac{\frac{8}{5}}{\frac{4}{5}} = \frac{8}{4} = 2. \quad \blacksquare$$

Example 3 is typical of the general case, as can be seen algebraically. Consider the geometric series $a_1 + a_2 + a_3 + \cdots$ with common ratio r such that $|r| < 1$. The kth partial sum S_k is the same as the kth partial sum of the geometric sequence $\{a_n\}$ and hence

$$S_k = a_1\left(\frac{1 - r^k}{1 - r}\right).$$

As k gets larger and larger, the number r^k gets very close to 0 because $|r| < 1$ (for instance, $(-.6)^{20} \approx .0000366$ and $.2^9 \approx .000000512$). Consequently, when k is very large, $1 - r^k$ is very close to $1 - 0$ so that

$$S_k = a_1\left(\frac{1 - r^k}{1 - r}\right) \qquad \text{is very close to} \qquad a_1\left(\frac{1 - 0}{1 - r}\right) = \frac{a_1}{1 - r}.$$

Example 4 $\displaystyle\sum_{n=1}^{\infty}\left(\frac{-1}{2}\right)^n = -\frac{1}{2} + \frac{1}{4} - \frac{1}{8} + \frac{1}{16} + \cdots$ is an infinite geometric series with $a_1 = -1/2$ and $r = -1/2$. Since $|r| < 1$, this series converges and its sum is

$$\frac{a_1}{1 - r} = \frac{-\dfrac{1}{2}}{1 - \left(-\dfrac{1}{2}\right)} = \frac{-\dfrac{1}{2}}{\dfrac{3}{2}} = -\frac{1}{3}. \qquad \blacksquare$$

Infinite geometric series provide another way of writing an infinite repeating decimal as a rational number.

Example 5 To express $6.8573573573\cdots$ as a rational number, we first write it as $6.8 + 0.573573573\cdots$. Consider $0.573573573\cdots$ as an infinite series:

$$.0573 + .0000573 + .0000000573 + .0000000000573 + \cdots$$

which is the same as

$$.0573 + (.001)(.0573) + (.001)^2(.0573) + (.001)^3(.0573) + \cdots.$$

This is a convergent geometric series with $a_1 = .0573$ and $r = .001$. Its sum is

$$\frac{a_1}{1 - r} = \frac{.0573}{1 - .001} = \frac{.0573}{.999} = \frac{573}{9990}.$$

Therefore,

$$6.8573573573\cdots = 6.8 + [.0573 + .0000573 + \cdots]$$

$$= 6.8 + \frac{573}{9990}$$

$$= \frac{68}{10} + \frac{573}{9990}$$

$$= \frac{68{,}505}{9990} = \frac{4567}{666}. \qquad \blacksquare$$

Exercises 7.3.A

In Exercises 1–8, find the sum of the infinite series, if it has one.

1. $\sum_{n=1}^{\infty} \dfrac{1}{2^n}$ 2. $\sum_{n=1}^{\infty} \left(-\dfrac{3}{4}\right)^n$ 3. $\sum_{n=1}^{\infty} (.06)^n$

4. $1 - .5 + .25 - .125 + .0625 - \cdots$

5. $500 + 200 + 80 + 32 + \cdots$

6. $9 - 3\sqrt{3} + 3 - \sqrt{3} + 1 - \dfrac{1}{\sqrt{3}} + \cdots$

7. $2 + \sqrt{2} + 1 + \dfrac{1}{\sqrt{2}} + \dfrac{1}{2} + \cdots$

8. $\sum_{n=1}^{\infty} \left(\dfrac{1}{2^n} - \dfrac{1}{3^n}\right)$

In Exercises 9–15, express the repeating decimal as a rational number.

9. $.22222\cdots$ 10. $.37373737\cdots$

11. $5.4272727\cdots$ 12. $85.131313\cdots$

13. $2.1425425425\cdots$ 14. $3.7165165165\cdots$

15. $1.74241241241\cdots$

16. If $\{a_n\}$ is an arithmetic sequence with common difference $d > 0$ and each $a_i > 0$, explain why the infinite series $a_1 + a_2 + a_3 + a_4 + \cdots$ is not convergent.

17. (a) Verify that $\sum_{n=1}^{\infty} 2(1.5)^n$ is a geometric series with $a_1 = 3$ and $r = 1.5$.

    (b) Find the kth partial sum of the series and use this expression to define a function f, as in Example 3.

    (c) Graph the function f in a viewing window with $0 \leq x \leq 30$. As x gets very large, what happens to the corresponding value of $f(x)$? Does the graph get closer and closer to some horizontal line, as in Example 3? What does this say about the convergence of the series?

18. Use the graphical approach illustrated in Example 3 to find the sum of the series in Example 4. Does the graph get very close to the horizontal line through $-1/3$? What's going on?

7.4 The Binomial Theorem

The Binomial Theorem provides a formula for calculating the product $(x + y)^n$ for any positive integer n. Before we state the theorem, some preliminaries are needed.

Let n be a positive integer. The symbol $n!$ (read **n factorial**) denotes the product of all the integers from 1 to n. For example,

$$2! = 1\cdot 2 = 2, \qquad 3! = 1\cdot 2\cdot 3 = 6, \qquad 4! = 1\cdot 2\cdot 3\cdot 4 = 24,$$

$$5! = 1\cdot 2\cdot 3\cdot 4\cdot 5 = 120,$$

$$10! = 1\cdot 2\cdot 3\cdot 4\cdot 5\cdot 6\cdot 7\cdot 8\cdot 9\cdot 10 = 3{,}628{,}800.$$

In general, we have this result:

n Factorial

Let n be a positive integer. Then

$$n! = 1\cdot 2\cdot 3\cdot 4 \cdots (n - 2)(n - 1)n.$$

0! is defined to be the number 1.

Learn to use your calculator to compute factorials. You will find ! in the PROB (or PRB) submenu of the MATH or OPTN menu.

CALCULATOR EXPLORATION

15! is such a large number your calculator will switch to scientific notation to express it. What is this approximation? Many calculators cannot compute factorials larger than 69! If yours does compute larger ones, how large a one can you compute without getting an error message [or on HP-38, getting the number $9.9999 \cdots E499$]?

If r and n are integers with $0 \leq r \leq n$, then

Binomial Coefficients

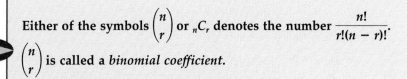

Either of the symbols $\binom{n}{r}$ or $_nC_r$ denotes the number $\dfrac{n!}{r!(n-r)!}$.

$\binom{n}{r}$ is called a *binomial coefficient.*

For example,

$$_5C_3 = \binom{5}{3} = \frac{5!}{3!(5-3)!} = \frac{5}{3!2!} = \frac{1 \cdot 2 \cdot 3 \cdot 4 \cdot 5}{(1 \cdot 2 \cdot 3)(1 \cdot 2)} = \frac{4 \cdot 5}{2} = 10$$

$$_4C_2 = \binom{4}{2} = \frac{4!}{2!(4-2)!} = \frac{4!}{2!2!} = \frac{1 \cdot 2 \cdot 3 \cdot 4}{(1 \cdot 2)(1 \cdot 2)} = \frac{3 \cdot 4}{2} = 6.$$

Binomial coefficients can be computed on a calculator by using $_nC_r$ or Comb in the PROB (or PRB) submenu of the MATH or OPTN menu.

CALCULATOR EXPLORATION

Compute $_{56}C_{47} = \binom{56}{47}$. Although calculators cannot compute 475!, they can compute many binomial coefficients, such as $\binom{475}{400}$, because most of the factors cancel out (as in the previous example). Check yours. Will it also compute $\binom{475}{50}$?

The preceding examples illustrate a fact whose proof will be omitted: *Every binomial coefficient is an integer.* Furthermore, for every nonnegative integer n,

$$\binom{n}{0} = 1 \qquad \text{and} \qquad \binom{n}{n} = 1$$

because

$$\binom{n}{0} = \frac{n!}{0!(n-0)!} + \frac{n!}{0!n!} = \frac{n!}{n!} = 1 \quad \text{and}$$

$$\binom{n}{n} = \frac{n!}{n!(n-n)!} = \frac{n!}{n!0!} = \frac{n!}{n!} = 1.$$

If we list the binomial coefficients for each value of n in this manner:

$n = 0$ $\qquad\qquad\qquad\qquad\qquad \binom{0}{0}$

$n = 1$ $\qquad\qquad\qquad\qquad \binom{1}{0} \qquad\qquad \binom{1}{1}$

$n = 2$ $\qquad\qquad\qquad \binom{2}{0} \qquad\quad \binom{2}{1} \qquad\quad \binom{2}{2}$

$n = 3$ $\qquad\qquad \binom{3}{0} \qquad\quad \binom{3}{1} \qquad\quad \binom{3}{2} \qquad\quad \binom{3}{3}$

$n = 4$ $\qquad \binom{4}{0} \qquad\quad \binom{4}{1} \qquad\quad \binom{4}{2} \qquad\quad \binom{4}{3} \qquad\quad \binom{4}{4}$

$\vdots$ $\qquad\cdot\cdot$ $\qquad\qquad\qquad\qquad\qquad\qquad\qquad\qquad\qquad \ddots$

and then calculate each of them, we obtain the following array of numbers:

row 0					1				
row 1				1		1			
row 2			1		2		1		
row 3		1		3		3		1	
row 4	1		4		6		4		1

$\qquad\qquad \vdots \quad \cdot\cdot \qquad\qquad\qquad\qquad\qquad \ddots$

This array is called **Pascal's triangle.** Its pattern is easy to remember: Each entry (except the 1's at the beginning or end of a row) is the sum of the two closest entries in the row above it. In the fourth row, for instance, 6 is the sum of the two 3's above it, and each 4 is the sum of the 1 and 3 above it. See Exercise 47 for a proof.

To develop a formula for calculating $(x + y)^n$, we first calculate these products for small values of n to see if we can find some kind of pattern:

$n = 0 \qquad (x + y)^0 = \qquad\qquad\qquad 1$

$n = 1 \qquad (x + y)^1 = \qquad\qquad 1x + 1y$

(*) $\qquad n = 2 \qquad (x + y)^2 = \qquad 1x^2 + 2xy + 1y^2$

$n = 3 \qquad (x + y)^3 = \quad 1x^3 + 3x^2y + 3xy^2 + 1y^3$

$n = 4 \qquad (x + y)^4 = 1x^4 + 4x^3y + 6x^2y^2 + 4xy^3 + 1y^4$

One pattern is immediately obvious: the coefficients here (shown in color) are the top part of Pascal's triangle! In the case $n = 4$, for example, this means that the coefficients are the numbers

$$\begin{array}{ccccc} 1 & 4 & 6 & 4 & 1 \\ \binom{4}{0}, & \binom{4}{1}, & \binom{4}{2}, & \binom{4}{3}, & \binom{4}{4}. \end{array}$$

If this pattern holds for larger n, then the coefficients in the expansion of $(x + y)^n$ are

$$\binom{n}{0}, \binom{n}{1}, \binom{n}{2}, \binom{n}{3}, \ldots, \binom{n}{n-1}, \binom{n}{n}.$$

As for the xy-terms associated with each of these coefficients, look at the pattern in (∗) above: the exponent of x goes down by 1 and the exponent of y goes up by 1 as you go from term to term, which suggests that the terms of the expansion of $(x + y)^n$ (without the coefficients) are:

$$x^n, \qquad x^{n-1}y, \qquad x^{n-2}y^2, \qquad x^{n-3}y^3, \ldots, xy^{n-1}, \quad y^n.$$

Combining the patterns of coefficients and xy-terms and using the fact that $\binom{n}{0} = 1$ and $\binom{n}{n} = 1$ suggests that the following result is true about the expansion of $(x + y)^n$.

The Binomial Theorem

For each positive integer n,

$$(x + y)^n = x^n + \binom{n}{1}x^{n-1}y + \binom{n}{2}x^{n-2}y^2 +$$

$$\binom{n}{3}x^{n-3}y^3 + \cdots + \binom{n}{n-1}xy^{n-1} + y^n.$$

Using summation notation and the fact that $\binom{n}{0} = 1 = \binom{n}{n}$, we can write the Binomial Theorem compactly as

$$(x + y)^n = \sum_{j=0}^{n} \binom{n}{j}x^{n-j}y^j.$$

The Binomial Theorem will be proved in Section 7.7 by means of mathematical induction. We shall assume its truth for now and illustrate some of its uses.

Example 1   Expand $(x + y)^8$.

Solution   We apply the Binomial Theorem in the case $n = 8$:

$$(x + y)^8 = x^8 + \binom{8}{1}x^7y + \binom{8}{2}x^6y^2 + \binom{8}{3}x^5y^3$$

$$+ \binom{8}{4}x^4y^4 + \binom{8}{5}x^3y^5 + \binom{8}{6}x^2y^6 + \binom{8}{7}xy^7 + y^8.$$

The coefficients can be computed individually by hand or by using $_nC_r$ (or COMB) on a calculator; for instance,

$$_8C_2 = \binom{8}{2} = \frac{8!}{2!6!} = 28 \quad \text{or} \quad _8C_3 = \binom{8}{3} = \frac{8!}{3!5!} = 56.$$

Alternatively, you can display all the coefficients at once by making a table of values for the function $f(x) = {}_8C_x$, as shown in Figure 7–9.*

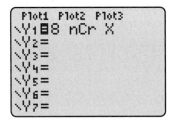

Figure 7–9

Substituting these values in the preceding expansion, we have

$$(x + y)^8 = x^8 + 8x^7y + 28x^6y^2 + 56x^5y^3$$

$$+ 70x^4y^4 + 56x^3y^5 + 28x^2y^6 + 8xy^7 + y^8. \quad \blacksquare$$

Example 2   Expand $(1 - z)^6$.

Solution   Note that $1 - z = 1 + (-z)$ and apply the Binomial Theorem with $x = 1$, $y = -z$, and $n = 6$:

$$(1 - z)^6 = 1^6 + \binom{6}{1}1^5(-z) + \binom{6}{2}1^4(-z)^2 + \binom{6}{3}1^3(-z)^3 + \binom{6}{4}1^2(-z)^4 + \binom{6}{5}1(-z)^5 + (-z)^6$$

$$= 1 - \binom{6}{1}z + \binom{6}{3}z^2 - \binom{6}{3}z^3 + \binom{6}{4}z^4 - \binom{6}{5}z^5 + z^6$$

$$= 1 - 6z + 15z^2 - 20z^3 + 15z^4 - 6z^5 + z^6. \quad \blacksquare$$

*Thanks to Nick Goodbody for suggesting this.

Technology Tip

Binomial expansions, such as those in Examples 1–3, can be done on TI-89 by using EXPAND in the ALGEBRA menu.

Example 3 Expand $(x^2 + x^{-1})^4$.

Solution Use the Binomial Theorem with x^2 in place of x and x^{-1} in place of y:

$$(x^2 + x^{-1})^4 = (x^2)^4 + \binom{4}{1}(x^2)^3(x^{-1}) + \binom{4}{2}(x^2)^2(x^{-1})^2$$

$$+ \binom{4}{3}(x^2)(x^{-1})^3 + (x^{-1})^4$$

$$= x^8 + 4x^6x^{-1} + 6x^4x^{-2} + 4x^2x^{-3} + x^{-4}$$

$$= x^8 + 4x^5 + 6x^2 + 4x^{-1} + x^{-4}. \quad \blacksquare$$

Example 4 Show that $(1.001)^{1000} > 2$ without using a calculator.

Solution We write 1.001 as $1 + .001$ and apply the Binomial Theorem with $x = 1$, $y = .001$, and $n = 1000$:

$$(1.001)^{1000} = (1 + .001)^{1000}$$

$$= 1^{1000} + \binom{1000}{1}1^{999}(.001) + \text{other positive terms}$$

$$= 1 + \binom{1000}{1}(.001) + \text{other positive terms.}$$

But $\binom{1000}{1} = \dfrac{1000!}{1!999!} = \dfrac{1000 \cdot 999!}{999!} = 1000$. Therefore, $\binom{1000}{1}(.001) = 1,000(.001) = 1$ and

$$(1.001)^{1000} = 1 + 1 + \text{other positive terms} = 2 + \text{other positive terms.}$$

Hence, $(1.001)^{1000} > 2$. $\quad \blacksquare$

Sometimes we need to know only one term in the expansion of $(x + y)^n$. If you examine the expansion given by the Binomial Theorem, you will see that in the second term y has exponent 1, in the third term y has exponent 2, and so on. Thus,

Properties of the Binomial Expansion

In the binomial expansion of $(x + y)^n$,

The exponent of y is always one less than the number of the term.

Furthermore, in each of the middle terms of the expansion,

The coefficient of the term containing y^r is $\binom{n}{r}$.

The sum of the x exponent and the y exponent is n.

For instance, in the *ninth* term of the expansion of $(x + y)^{13}$, y has exponent 8, the coefficient is $\binom{13}{8}$, and x must have exponent 5 (since $8 + 5 = 13$). Thus, the ninth term is $\binom{13}{8} x^5 y^8$.

Exercises 7.4

In Exercises 1–10, evaluate the expression.

1. $6!$ **2.** $\dfrac{11!}{8!}$ **3.** $\dfrac{12!}{9!3!}$ **4.** $\dfrac{9! - 8!}{7!}$

5. $\dbinom{5}{3} + \dbinom{5}{2} - \dbinom{6}{3}$ **6.** $\dbinom{12}{11} - \dbinom{11}{10} + \dbinom{7}{0}$

7. $\dbinom{6}{0} + \dbinom{6}{1} + \dbinom{6}{2} + \dbinom{6}{3} + \dbinom{6}{4} + \dbinom{6}{5} + \dbinom{6}{6}$

8. $\dbinom{6}{0} - \dbinom{6}{1} + \dbinom{6}{2} - \dbinom{6}{3} + \dbinom{6}{4} - \dbinom{6}{5} + \dbinom{6}{6}$

9. $\dbinom{100}{96}$ **10.** $\dbinom{75}{72}$

In Exercises 11–16, expand the expression.

11. $(x + y)^5$ **12.** $(a + b)^7$ **13.** $(a - b)^5$

14. $(c - d)^8$ **15.** $(2x + y^2)^5$ **16.** $(3u - v^3)^6$

In Exercises 17–26, use the Binomial Theorem to expand and (where possible) simplify the expression.

17. $\left(\sqrt{x} + 1\right)^6$ **18.** $\left(2 - \sqrt{y}\right)^5$

19. $(1 - c)^{10}$ **20.** $\left(\sqrt{c} + \dfrac{1}{\sqrt{c}}\right)^7$

21. $(x^{-3} + x)^4$ **22.** $(3x^{-2} - x^2)^6$

23. $\left(1 + \sqrt{3}\right)^4 + \left(1 - \sqrt{3}\right)^4$

24. $\left(\sqrt{3} + 1\right)^6 - \left(\sqrt{3} - 1\right)^6$

25. $(1 + i)^6$, where $i^2 = -1$

26. $\left(\sqrt{2} - i\right)^4$, where $i^2 = -1$

In Exercises 27–32, find the indicated term of the expansion of the given expression.

27. third, $(x + y)^5$ **28.** fourth, $(a + b)^6$

29. fifth, $(c - d)^7$ **30.** third, $(a + 2)^8$

31. fourth, $\left(u^{-2} + \dfrac{u}{2}\right)^7$ **32.** fifth, $\left(\sqrt{x} - \sqrt{2}\right)^7$

33. Find the coefficient of $x^5 y^8$ in the expansion of $(2x - y^2)^9$.

34. Find the coefficient of $x^{12} y^6$ in the expansion of $(x^3 - 3y)^{10}$.

35. Find the coefficient of $1/x^3$ in the expansion of $\left(2x + \dfrac{1}{x^2}\right)^6$.

36. Find the constant term in the expansion of $\left(y - \dfrac{1}{2y}\right)^{10}$.

37. **(a)** Verify that $\dbinom{9}{1} = 9$ and $\dbinom{9}{8} = 9$.

     (b) Prove that for each positive integer n, $\dbinom{n}{1} = n$ and $\dbinom{n}{n - 1} = n$. [*Note:* Part (a) is just the case when $n = 9$ and $n - 1 = 8$.]

38. **(a)** Verify that $\dbinom{7}{2} = \dbinom{7}{5}$.

     (b) Let r and n be integers with $0 \le r \le n$. Prove that $\dbinom{n}{r} = \dbinom{n}{n - r}$. [*Note:* Part (a) is just the case when $n = 7$ and $r = 2$.]

39. Prove that for any positive integer n,

$$2^n = \binom{n}{0} + \binom{n}{1} + \binom{n}{2} + \cdots + \binom{n}{n}.$$

[*Hint:* $2 = 1 + 1$.]

40. Prove that for any positive integer n,

$$\binom{n}{0} - \binom{n}{1} + \binom{n}{2} - \binom{n}{3} + \binom{n}{4} - \cdots$$
$$+ (-1)^k \binom{n}{k} + \cdots + (-1)^n \binom{n}{n} = 0.$$

41. **(a)** Let f be the function given by $f(x) = x^5$. Let h be a nonzero number and compute $f(x + h) - f(x)$ (but leave all binomial coefficients in the form $\dbinom{5}{r}$ here and below).

(b) Use part (a) to show that h is a factor of
$$f(x + h) - f(x) \text{ and find } \frac{f(x + h) - f(x)}{h}.$$

(c) If h is *very* close to 0, find a simple approximation of the quantity $\dfrac{f(x + h) - f(x)}{h}$.
[See part (b).]

42. Do Exercise 41 with $f(x) = x^8$ in place of $f(x) = x^5$.

43. Do Exercise 41 with $f(x) = x^{12}$ in place of $f(x) = x^5$.

44. Let n be a fixed positive integer. Do Exercise 41 with $f(x) = x^n$ in place of $f(x) = x^5$.

Thinkers

45. Let r and n be integers such that $0 \le r \le n$.
  (a) Verify that $(n - r)! = (n - r)[n - (r + 1)]!$
  (b) Verify that $(n - r)! = [(n + 1) - (r + 1)]!$
  (c) Prove that $\dbinom{n}{r + 1} + \dbinom{n}{r} = \dbinom{n + 1}{r + 1}$ for any

$r \le n - 1$. [*Hint:* Write out the terms on the left side and use parts (a) and (b) to express each of them as a fraction with denominator $(r + 1)!(n - r)!$. Then add these two fractions, simplify the numerator, and compare the result with $\dbinom{n + 1}{r + 1}$.]

(d) Use part (c) to explain why each entry in Pascal's triangle (except the 1's at the beginning or end of a row) is the sum of the two closest entries in the row above it.

46. (a) Find these numbers and write them one *below* the next: 11^0, 11^1, 11^2, 11^3, 11^4.

  (b) Compare the list in part (a) with rows 0 to 4 of Pascal's triangle. What's the explanation?

  (c) What can be said about 11^5 and row 5 of Pascal's triangle?

  (d) Calculate all integer powers of 101 from 101^0 to 101^8, list the results one under the other, and compare the list with rows 0 to 8 of Pascal's triangle. What's the explanation? What happens with 101^9?

7.5 **Permutations and Combinations**

How many different choices are there in a lottery in which you select six numbers from 1 to 44? How many seven-digit phone numbers can there be in any one area code? The answers to these and many other questions require systematic counting techniques that are introduced here.

Example 1   Anne, Bill, Charlie, Dana, Elsie, and Fred enter a short-story competition in which three prizes will be awarded (and there will be no ties). If they are the only contestants, in how many possible ways can the prizes be awarded?

Solution   For convenience we use initials. There are six possibilities for first place: *A, B, C, D, E, F.* For each possible first-place winner there are five possible second-place winners:

1. 2.	1. 2.	1. 2.	1. 2.	1. 2.	1. 2.
A B	B A	C A	D A	E A	F A
A C	B C	C B	D B	E B	F B
A D	B D	C D	D C	E C	F C
A E	B E	C E	D E	E D	F D
A F	B F	C F	D F	E F	F E

So the total number of ways first and second prize can be awarded is $6 \cdot 5 = 30$. If, for instance, A and B take first and second, then there are four possibilities for third:

$$A, B, C; \quad A, B, D; \quad A, B, E; \quad A, B, F$$

Similarly, for *each* of the 30 ways that first and second prizes can be awarded, there are four ways to award third prize. So there are $30 \cdot 4 = 120$ possible ways to award all three prizes.   ■

The argument in Example 1 may be summarized like this:

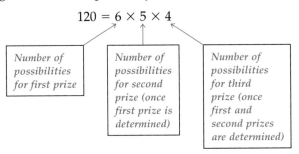

When thought of this way, the example is an illustration of the following principle.

Fundamental Counting Principle

Consider a list of k events. Suppose the first event can occur in n_1 ways; the second event can occur in n_2 ways (once the first event has occurred); the third event can occur in n_3 ways (once the first two events have occurred); and so on. Then the total number of ways that all k events can occur is the product $n_1 n_2 n_3 \cdots n_k$.

Example 2   How many license plates can be made consisting of three letters followed by 3 one-digit numbers, subject to these conditions: No plate may begin with I or O, and the first number may not be a zero?

Solution   Consider the filling of each of the six positions on a license plate as an event. Since O and I are excluded, there are 24 possible letters for the first position ($n_1 = 24$). There are 26 possible letters for each of the second and third positions ($n_2 = 26$, $n_3 = 26$). Since 0 cannot appear in the fourth position (the first position for a number), there are nine possibilities ($n_4 = 9$). Any digit from 0 through 9 may appear in the last two positions, so there are ten possible digits for each position ($n_5 = 10$, $n_6 = 10$). By the Fundamental Counting Principle, the total number of license plates is the product

$$24 \cdot 26 \cdot 26 \cdot 9 \cdot 10 \cdot 10 = 14,601,600. ■$$

Example 3 A committee consisting of eight Democrats and six Republicans must choose a chairperson, a vice-chairperson, a secretary, and a treasurer. The chairperson must be a Democrat and the vice-chairperson a Republican. The secretary and treasurer may belong to either party. If no person can hold more than one office, in how many different ways can these four officers be chosen?

Solution Consider the choice of a chairperson as the first event. Since there are eight Democrats, there are eight possible ways for this event to occur ($n_1 = 8$). The second event is the choice of a vice-chairperson. It can occur in six ways because there are six Republicans ($n_2 = 6$). Once the chairperson and vice-chairperson are determined, any of the remaining 12 people may be chosen secretary. So that event can occur in 12 ways ($n_3 = 12$). The fourth and last event ($k = 4$) is the choice of a treasurer, who can be any of the 11 people remaining ($n_4 = 11$). By the Fundamental Counting Principle the number of ways that all four officers can be chosen is the product $n_1 n_2 n_3 n_4 = 8 \cdot 6 \cdot 12 \cdot 11 = 6336$. ■

Permutations

A **permutation** of a set of n elements is an ordering of the elements of the set. We can use the Fundamental Counting Principle to determine the number of possible permutations of a set.

Example 4 The number of possible batting orders for a nine-person baseball team is the number of permutations of a set of nine elements. There are nine possibilities for first batter, eight possibilities for second (after the first one is determined), seven possibilities for third (after the first two are determined), and so on, down to the eighth (two possibilities) and ninth batter (one possibility). So the number of possible batting orders is

$$9 \cdot 8 \cdot 7 \cdot 6 \cdot 5 \cdot 4 \cdot 3 \cdot 2 \cdot 1 = 362{,}880. ■$$

The argument in the example carries over to the general case with n in place of 9: To determine a permutation of n elements, there are n possible choices for first position, $n - 1$ choices for second position, and so on. Therefore

Permutations of n Elements

The total number of permutations of a set of n elements is

$$n! = n(n - 1)(n - 2) \cdots 3 \cdot 2 \cdot 1^*$$

*Factorial notation ($n!$) is explained on page 538.

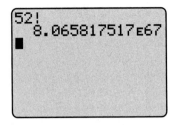

Figure 7–10

Example 5 How many ways can a 52-card deck of cards be shuffled (that is, how many different orderings of the deck are possible)?

Solution The number of orderings is the number of permutations of a 52-element set, namely 52!. Using the factorial key on a calculator (located in the MATH menu or its PROB submenu), we see that this is *enormous* (approximately 8.066×10^{67}, as shown in Figure 7–10). ∎

In many cases we are interested in the order of some (but not necessarily all) of the elements of a set. For instance, in Example 1 (the short-story contest), we considered the possible ways of ordering three of the six contestants (the prize winners). Such an ordering is called a permutation of six things taken three at a time. More generally, if $r \leq n$, an ordering of r elements of an n-element set is called a **permutation of n elements taken r at a time.**

The number of permutations of n elements taken r at a time is denoted $_nP_r$. Example 1 shows that $_6P_3$, the number of permutations of six elements taken three at a time, is

$$_6P_3 = 6 \times 5 \times 4 = 120.$$

Note that $_6P_3$ is the product of three consecutive integers, beginning with 6 and working downward. A similar pattern holds in other cases. For example, to compute $_8P_5$, the number of permutations of eight elements taken five at a time, there are eight choices for the first position, seven choices for the second, six choices for third, five choices for fourth, and four choices for the fifth position. By the Fundamental Counting Principle,

$$_8P_5 = 8 \cdot 7 \cdot 6 \cdot 5 \cdot 4 = 6720.$$

Thus, $_8P_5$ is the product of five consecutive integers, beginning with 8 and working downward. The same argument works in the general case:

Permutations of n Elements Taken r at a Time

▶ Let r and n be positive integers with $r \leq n$. Then

$_nP_r$ **is the product of r consecutive integers, beginning with n and working downward.**

Example 6 Find $_{20}P_4$ and $_9P_6$.

Solution For $_{20}P_4$, we start with 20 and go down 4 factors:

$$_{20}P_4 = \underbrace{20 \cdot 19 \cdot 18 \cdot 17}_{4 \text{ factors}} = 116,280.$$

For $_9P_6$ we start with 9 and go down 6 times:

$$_9P_6 = \underbrace{9 \cdot 8 \cdot 7 \cdot 6 \cdot 5 \cdot 4}_{6 \text{ factors}} = 60{,}480. \quad \blacksquare$$

By using an algebraic trick, we can find another useful formula for computing $_nP_r$. For example, since $\dfrac{3!}{3!} = 1$,

$$_8P_5 = 8 \cdot 7 \cdot 6 \cdot 5 \cdot 4 = \frac{8 \cdot 7 \cdot 6 \cdot 5 \cdot 4}{1} \cdot \frac{3!}{3!} = \frac{8 \cdot 7 \cdot 6 \cdot 5 \cdot 4 \cdot 3 \cdot 2 \cdot 1}{3!} = \frac{8!}{3!} = \frac{8!}{(8-5)!}.$$

A similar computation works in the general case (with n in place of 8 and r in place of 5) and leads to this description:

$${}_nP_r$$

$$_nP_r = \frac{n!}{(n-r)!}.$$

Although it is sometimes necessary to use the two formulas for $_nP_r$ directly, most numerical computations can be readily done on a calculator by using the *nPr* key (which is in the MATH menu or its PROB submenu).

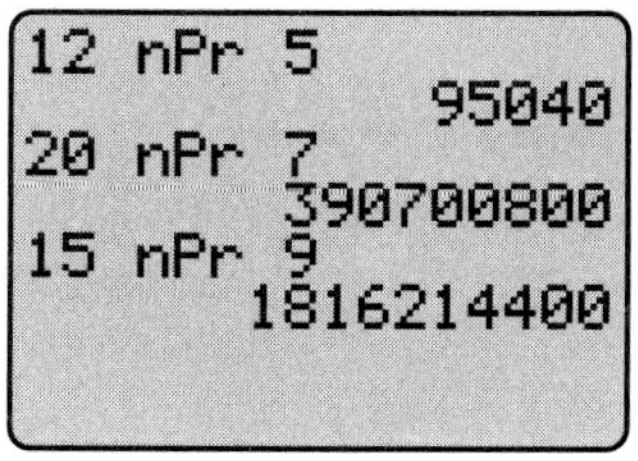

```
12 nPr 5
              95040
20 nPr 7
          390700800
15 nPr 9
        1816214400
```

Figure 7–11

Example 7 Use a calculator to find $_{12}P_5$, $_{20}P_7$, and $_{15}P_9$.

Solution See Figure 7–11. ∎

Combinations

To play the Ohio state lottery, you select six different integers from 1 to 44. The order in which you list them doesn't matter. If these six numbers are drawn in *any* order, you win. Each possible choice of six numbers is called a combination of 44 elements taken 6 at a time. More generally, a **combination of n elements taken r at a time** is any collection of r distinct objects in a set of n objects, without regard to the order in which the r objects might be chosen.

Example 8 The alphabet has 26 letters. So the letters B, Q, T form a combination of 26 elements taken 3 at a time. If we list these three letters as B, T, Q, or Q, B, T, or Q, T, B, or T, B, Q, or T, Q, B, we still have the *same combination* because the order doesn't matter. The situation with permutations is quite different. There are six different permutations of these three letters (BQT, BTQ, QBT, QTB, TBQ, TQB) because order *does* matter for permutations. ∎

The number of combinations of n elements taken r at a time is denoted $_nC_r$. We can compute this number by considering the relationship between combinations and permutations. A *permutation* of n elements taken r at a time is determined by two events:

1. A choice of r elements in the n-element set, that is, a combination of n elements taken r at a time; and

2. A specific ordering of these r elements, that is, a permutation of r elements.

So the number $_nP_r$ of permutations of n elements taken r at a time is the same as the total number of ways events 1 and 2 can occur. By the Fundamental Counting Principle,

$$_nP_r = \begin{pmatrix}\text{Number of ways} \\ \text{event 1 can occur}\end{pmatrix} \times \begin{pmatrix}\text{Number of ways} \\ \text{event 2 can occur}\end{pmatrix}$$

$$_nP_r = \begin{pmatrix}\text{Number of combinations of} \\ n \text{ elements taken } r \text{ at a time}\end{pmatrix} \times \begin{pmatrix}\text{Number of permutations} \\ \text{of } r \text{ elements}\end{pmatrix}$$

$$\frac{n!}{(n - r)!} = {}_nC_r \times r!$$

Dividing both sides of this last equation by $r!$ we have this result:

Combinations of n Elements Taken r at a Time

The number of combinations of n elements taken r at a time is

$$_nC_r = \frac{n!}{r!(n - r)!}.*$$

Example 9 The number of three-person committees that can be selected from a group of ten people is the number of combinations of ten elements taken three at a time:

$$_{10}C_3 = \frac{10!}{3!(10 - 3)!} = \frac{10!}{3! \cdot 7!} = 120. \quad ■$$

Example 10 The number of possible five-card poker hands is the number of combinations of 52 elements taken 5 at a time:

---

*If you have read Section 7.4, you see that the number of combinations of n elements taken r at a time is precisely the binomial coefficient, which is why the same notation $_nC_r$ is used for both.

$$_{52}C_5 = \frac{52!}{5!(52 - 5)!} = \frac{52!}{5! \cdot 47!} = 2{,}598{,}960. \quad \blacksquare$$

The easiest way to evaluate the number $_nC_r$ is to use the *nCr* key on a calculator (in the MATH menu or its PROB submenu).

Example 11 To play the Ohio state lottery, which has a minimum prize of $4 million, you select six different integers from 1 to 44. At $1 per ticket (one choice), how much would you have to spend to guarantee that one of your tickets would be a winner?

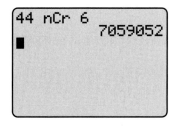

Figure 7–12

Solution To guarantee that you have a winning ticket, you must have a ticket for every possible six-number choice. The number of choices is the number of combinations of 44 elements taken 6 at a time, namely, $_{44}C_6$. At $1 per ticket, this would cost $7,059,052, as shown in Figure 7–12. Not a very good investment if the prize is $4 million (even worse if someone else also has a winning ticket and you have to split the prize). $\blacksquare$

Example 12 The city council consists of 18 Democrats, 20 Republicans, and 7 Independents. In how many ways can you select a committee of three Democrats, four Republicans, and two Independents?

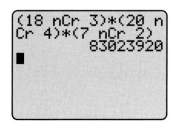

Figure 7–13

Solution The number of ways of choosing 3 of the 18 Democrats is $_{18}C_3$, the number of combinations of 18 elements taken 3 at a time. Similarly, the number of ways of choosing four Republicans is $_{20}C_4$, and the number of ways of choosing two Independents is $_7C_2$. By the Fundamental Counting Principle, the number of ways all these choices can be made is the product is

$$_{18}C_3 \cdot {}_{20}C_4 \cdot {}_7C_2 = 83{,}023{,}920,$$

as shown in Figure 7–13. $\blacksquare$

Exercises 7.5

In Exercises 1–16, compute the number.

1. $_4P_3$ **2.** $_7P_5$ **3.** $_8P_8$ **4.** $_5P_1$

5. $_{12}P_2$ **6.** $_{11}P_3$ **7.** $_{14}C_2$ **8.** $_{20}C_3$

9. $_{99}C_{95}$ **10.** $_{65}C_{60}$ **11.** $_nP_2$ **12.** $_nP_{n-1}$

13. $_nC_{n-1}$ **14.** $_nC_1$ **15.** $_nC_2$ **16.** $_nP_1$

In Exercises 17–22, find the number n that makes the statement true.

17. $_nP_4 = 8(_nP_3)$ **18.** $_nP_5 = 7(_nP_4)$

19. $_nP_6 = 9(_{n-1}P_5)$ **20.** $_nP_7 = 11(_{n-1}P_6)$

21. $_nP_5 = 21(_{n-1}P_3)$ **22.** $_nP_6 = 30(_{n-2}P_4)$

In Exercises 23–32, use the Fundamental Counting Principle to answer the question.

23. Every Social Security number has nine digits. How many possible Social Security numbers are there?

24. (a) How many five-digit zip codes are possible?
(b) How many nine-digit zip codes are possible?

25. How many four-letter radio station call letters are possible, if each must start with either W or K?

26. How many seven-digit phone numbers can be formed from the numerals 0 through 9 if none of the first three digits can be 0?

27. An early-bird restaurant menu offers 6 appetizers, 3 salads, 12 entrees, and 4 desserts. How many different meals can be ordered by choosing one item from each category?

28. A man has four pairs of pants, six sports coats, and eight ties. How many different outfits can he wear?

29. How many five-letter identification codes can be formed from the first 20 letters of the alphabet if no repetitions are allowed?

30. How many five-digit numbers can be formed from the numerals 2, 3, 4, 5, 6 if
    (a) no repeated digits are allowed?
    (b) repeated digits are allowed?

31. In how many different ways can an eight-question true-false test be answered?

32. An exam consists of ten multiple-choice questions with four choices for each question. In how many ways can this exam be answered?

In Exercises 33–38, use permutations to answer the question.

33. In how many different orders can eight people be seated in a row?

34. In how many orders can six pictures be arranged in a vertical line?

35. How many outcomes are possible in a four-team tournament (in which there are no ties)?

36. In how many ways can six mathematics texts be stacked in one pile on your desk?

37. In how many ways can the first four people in the batting order of a nine-player baseball team be chosen?

38. In how many different ways can first through third prizes be awarded in a ten-person race (assuming no ties)?

In Exercises 39–44, use combinations to answer the question.

39. A student must answer any five questions on an eight-question exam. In how many ways can this be done?

40. How many different straight lines are determined by eight points in the plane, no three of which lie on the same straight line? [*Hint:* A straight line is determined by two points.]

41. How many different six-card cribbage hands are possible from a 52-card deck?

42. Six pens are to be randomly selected from 100 newly manufactured pens to estimate the defect rate. How many six-pen samples are possible?

43. A committee of five students is to be chosen from a class of 22. How many different committees are possible?

44. An ice-cream store has 31 flavors. How many different two-scoop cones can they offer (assuming two different flavors for each cone)?

In the remaining exercises, use any appropriate technique (or several of them) to answer the question.

45. In how many ways can four men and eight women be seated in a row of 12 chairs if men sit at both ends of the row?

46. In how many ways can six men and six women be seated in a row of 12 chairs if the women sit in even-numbered seats?

47. How many different positive integers can be formed from the digits 3, 5, 7, 9 if no repeated digits are allowed? [*Hint:* How many one-digit numbers can be formed? How many two-digit ones?]

48. How many games must be played in an eight-team league if each team is to play every other team exactly twice?

49. How many batting orders are possible for a nine-player baseball team if the four infielders bat first and the pitcher bats last?

50. An electronic lock has five buttons. To unlock it you press two buttons simultaneously, then two more, then two more. For instance, you might have to press 1-5, then 2-3, then 4-2. How many codes are possible for this lock?

51. (a) A three-digit code is needed to open the combination lock in the figure. To open it, rotate the dial to the right until the first digit of the code is at the marker at the top of the dial; then rotate the dial to the left until the second digit of the code is at the marker; then rotate the dial to the right until the third digit is at the marker. How many possible "combinations" are there for this lock?

    (b) Explain why combination locks should really be called permutation locks.

52. A dish holds six yellow, eight red, and four brown M&Ms and you randomly pick four of them. How many such samples are possible in which the M&Ms are

(a) all red?   (b) all yellow?

(c) 2 yellow, 2 brown?

(d) 2 red, 1 yellow, 1 brown?

53. In the Michigan state lottery, you must pick 6 integers from 1 to 49. If a ticket (one choice) costs $1, how much would it cost to guarantee that you held a winning ticket?

54. Do Exercise 53 for the Powerball lottery that is played in 15 different states. You must select five distinct integers from 1 to 45 and one "Powerball number," also an integer from 1 to 45 (possibly the same as one of the five selected first). You win if your first five numbers and your Powerball number match the ones drawn in the lottery.

In Exercises 55 and 56, you may want to refer to this figure, which shows a standard deck of cards. It consists of four suits (spades, hearts, diamonds, clubs), each with 13 cards (Ace, 2, 3, 4, 5, 6, 7, 8, 9, 10, Jack, Queen, King).

55. How many 5-card poker hands consist of three aces and two kings?

56. How many 5-card poker hands are flushes (5 cards of the same suit)? [*Hint:* The number of club flushes is the number of ways 5 cards can be chosen from the 13 clubs.]

57. How many 8-digit numbers can be formed using three 6's and five 7's?

58. How many different 5-person committees consisting of 3 women and 2 men can be chosen from a group of 10 women and 8 men?

59. How many ways can two committees, one of 4 people and one of 3 people, be chosen from a group of 11 people if no person serves on both committees?

60. How many 6-digit numbers can be formed using only the numerals 5 and 9?

61. (a) A basketball squad consists of 5 people who can play center or forward and 7 people who can play only guard. In how many ways can a team consisting of a center, 2 forwards, and 2 guards be chosen?

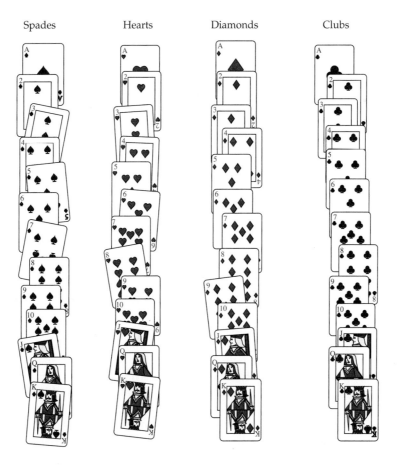

Spades   Hearts   Diamonds   Clubs

(b) How many teams are possible if the squad consists of 2 centers, 4 forwards, and 6 people who can play either guard or forward?

62. **(a)** How many different pairs are possible from a standard 52-card deck? [*Hint:* How many pairs can be formed from the 4 aces? How many from the 4 kings?]

(b) How many 5-card poker hands consist of a pair of aces and 3 other cards, none of which are aces?

(c) How many 5-card poker hands contain at least 2 aces?

(d) How many 5-card poker hands are full houses (3 of a kind and a pair of another kind)?

Thinkers

63. **(a)** In how many different orders can three objects be placed around a circle? (Note that ABC, BCA, and CAB are different orders when the three letters are in a line, but the *same* order in a circular arrangement. However, ABC is not the same circular arrangement as ACB, because B is to the left of A in one case and to the right of A in the other.)

(b) Do part (a) for 4 objects.

(c) Do part (a) for 5 objects.

(d) Let *n* be a positive integer and do part (a) for *n* objects.

64. In how many ways can 6 different foods be arranged around the edge of a circular table? (See Exercise 63.)

65. In how many different orders could King Arthur and 12 knights be seated at the Round Table? (See Exercise 63.)

66. All students at a certain college are required to take year-long courses in economics, mathematics, and history during their first year. The economics classes meet at 9, 11, 1, and 3 o'clock; mathematics classes meet at 10, 12, 2, and 4 o'clock; history classes meet at 8, 10, 12, and 5 o'clock. How many different schedules are possible for a student? Assume each class lasts 50 minutes and that a student has the same schedule for the entire year.

67. A promoter wants to make up a program consisting of ten acts. He has seven singing acts and nine instrumental acts available. If singing and instrumental acts are to be alternated, how many different program orders are possible? [*Hint:* First consider the possibilities when a singing act begins the program.]

7.5.A   ***EXCURSION*** **Distinguishable Permutations**

Figure 7–14

Suppose you have three identical red marbles, three identical white marbles, and two identical blue ones as in Figure 7–14. How many different color patterns can be formed by placing the eight marbles in a row? Each way of placing the marbles in a row is a permutation, and we know that there are 8! permutations of eight marbles. But the *same* color pattern may result from *different* permutations. To see how this can happen, mentally label the red marbles R_1, R_2, R_3, the white ones W_1, W_2, W_3, and the blue ones B_1, B_2. Here are several different permutations of the marbles that all produce the *same* color pattern:

$$R_3\ W_1\ B_1\ R_1\ W_3\ B_2\ R_2\ W_2; \qquad R_2\ W_2\ B_1\ R_3\ W_1\ B_2\ R_1\ W_3;$$

$$R_1\ W_3\ B_2\ R_3\ W_2\ B_1\ R_2\ W_1.$$

Permutations that produce the same color pattern, such as the three just listed, are said to be *indistinguishable.* Observe that any one of these three indistinguishable permutations may be obtained from any of the others simply by rearranging the red marbles among themselves, the white marbles among themselves, and the blue marbles among themselves.

Permutations that produce different color patterns are said to be *distinguishable.* If two permutations are distinguishable, then one *cannot* be obtained from the other simply by rearranging the marbles of the same color.

Finding the total number of different color patterns is the same as finding the total number of distinguishable permutations of the eight marbles.

To find the total number of different color patterns (distinguishable permutations), we shall proceed indirectly. We shall examine the way in which a specific permutation of the marbles is determined. Any *permutation* of the marbles (for instance, $R_1 W_3 B_2 R_3 W_2 B_1 R_2 W_1$) is determined by these four things:

    (i) A color pattern (in the example, the pattern R W B R W B R W).

    (ii) The order in which the red marbles appear in the red positions of the pattern (in the example, $R_1 R_3 R_2$), that is, a permutation of the three red marbles.

    (iii) The order in which the white marbles appear in the white positions of the pattern (in the example, $W_3 W_2 W_1$), that is, a permutation of the three white marbles.

    (iv) The order in which the blue marbles appear in the blue positions of the pattern (in the example, $B_2 B_1$), that is, a permutation of the two blue marbles.

A specific choice of one possibility for *each* of items (i) to (iv) will lead to exactly one permutation of the eight marbles. Conversely, every permutation of the eight marbles uniquely determines a choice in each one of items (i) to (iv), as shown by the example above. So the total number of permutations of the eight marbles, namely, 8!, is the same as the total number of ways that items (i) to (iv) can all occur. According to the Fundamental Counting Principle, this number is the product of the numbers of ways each of the four items can occur. Therefore

$$8! = \begin{pmatrix} \text{Number of} \\ \text{color} \\ \text{patterns} \end{pmatrix} \times \begin{pmatrix} \text{Number of} \\ \text{permutations of} \\ \text{three red marbles,} \\ \text{namely, 3!} \end{pmatrix} \times \begin{pmatrix} \text{Number of} \\ \text{permutations of} \\ \text{three white marbles,} \\ \text{namely, 3!} \end{pmatrix} \times \begin{pmatrix} \text{Number of} \\ \text{permutations of} \\ \text{two blue marbles,} \\ \text{namely, 2!} \end{pmatrix}.$$

If we let N denote the number of color patterns, this statement becomes:

$$8! = N \cdot 3! \cdot 3! \cdot 2!.$$

Solving this equation for N we see that

$$N = \frac{8!}{3! \cdot 3! \cdot 2!} = 560.$$

More generally, suppose that n and $k_1, k_2, \ldots, k_t$ are positive integers such that $n = k_1 + k_2 + \cdots + k_t$. Suppose that a set consists of n objects and that k_1 of these objects are all of one kind (such as red marbles), that k_2 of the objects are all of another kind (such as white marbles), that k_3 of the objects are all of a third kind, and so on. We say that two permutations of this set are **distinguishable** if one cannot be obtained from the other simply by rearranging objects of the same kind. The total number of distinguishable permutations can be found by using the same method that was used to find the total number of color patterns in the marble example (where we had $n = 8$ and $k_1 = 3$, $k_2 = 3$, $k_3 = 2$):

Distinguishable Permutations

Given a set of n objects in which k_1 are of one kind, k_2 are of a second kind, k_3 are of a third kind, and so on, then the number of distinguishable permutations of the set is

$$\frac{n!}{k_1! \cdot k_2! \cdot k_3! \cdots k_t!}$$

Example 1 The number of distinguishable ways that the letters in the word TENNESSEE can be arranged is just the number of distinguishable permutations of the set consisting of the nine symbols T, E, N, N, E, S, S, E, E. There are four E's, two N's, two S's, and one T. So we apply the formula in the preceding box with $n = 9$; $k_1 = 4$; $k_2 = 2$; $k_3 = 2$; $k_4 = 1$; and find that the number of distinguishable permutations is

$$\frac{9!}{4!\,2!\,2!\,1!} = \frac{9 \cdot 8 \cdot 7 \cdot 6 \cdot 5 \cdot 4 \cdot 3 \cdot 2 \cdot 1}{4 \cdot 3 \cdot 2 \cdot 1 \cdot 2 \cdot 1 \cdot 2 \cdot 1 \cdot 1} = 3780. \quad \blacksquare$$

Exercises 7.5.A

In Exercises 1–4, determine the number of distinguishable ways the letters in the word can be arranged.

1. LOOK
2. MISSISSIPPI
3. CINCINNATI
4. BOOKKEEPER

5. How many color patterns can be obtained by placing five red, seven black, three white, and four orange disks in a row?

6. In how many different ways can you write the algebraic expression $x^4 y^2 z^3$ without using exponents or fractions?

7.6 Introduction to Probability

In business and the sciences, people often must deal with factors that cannot be known with absolute certainty:

The chance of rain tomorrow is 40%.

There is a high probability of another serious earthquake in California.

The vaccine is effective with 87% of the patients.

Probability theory, which was first developed in the 17th century to analyze games of chance, provides the mathematical tools for dealing with situations like these that involve uncertainty.

In the study of probability, an **experiment** is any activity or occurrence with an observable result; this result is called an **outcome** of the experiment. For instance, flipping a coin is an experiment with two possible outcomes: heads or tails. Drawing a card from a standard deck is an experiment with 52 possible outcomes. An experiment with the outcomes "satisfactory" or "defective" occurs when an item coming off a factory assembly line is tested by the quality control department.

We first consider experiments with **equally likely** outcomes. Flipping a fair coin (one that is not weighted or shaved to favor one side over the other) is an example of such an experiment, as is rolling a fair die (one that isn't loaded so that one number comes up more often than the others).* A well-run factory, however, will produce far more satisfactory than defective items, so that the two outcomes of a quality control test will not be equally likely.

The set of all possible outcomes of an experiment is called the **sample space** of the experiment.

Example 1

(a) In the experiment of rolling a die, the sample space is

$$(1, 2, 3, 4, 5, 6)$$

since these numbers are the possible outcomes of rolling the die.

(b) If the experiment consists of flipping two coins, one after the other, then there are four possible outcomes:

Heads on both coins (HH);

Heads on the first coin and tails on the second (HT);

Tails on the first coin and heads on the second (TH);

Tails on both coins (TT).

Using the preceding labels, the sample space is the set

$$\{HH, HT, TH, TT\}. \quad \blacksquare$$

Any subset of the sample space (that is, any collection of some of the possible outcomes) is called an **event.** In the experiment of rolling a die, the subset $\{2, 4, 6\}$ is the event "rolling an even number" and the subset $\{3, 4, 5, 6\}$ is the event "rolling a number greater than 2." The event "rolling a number less than 7" is the entire sample space $\{1, 2, 3, 4, 5, 6\}$.

The basic idea of probability theory is to assign to each event a number between 0 and 1 that indicates its likelihood:

**Probability of
an Event**

> If S is the sample space of an experiment with equally likely outcomes and E is an event, then the *probability* of E is denoted $P(E)$ and defined by
>
> $$P(E) = \frac{\text{Number of outcomes in } E}{\text{Number of outcomes in } S}$$

*Hereafter all dice and coins in the discussion are assumed to be fair.

Example 2 In the experiment of rolling a single die, find the probability of rolling

(a) 5;   (b) an even number;

(c) a number greater than 2;   (d) a number less than 7;

(e) a number greater than 6.

Solution

(a) The sample space $S = \{1, 2, 3, 4, 5, 6\}$ contains six possible outcomes. The event E of "rolling a 5" has just one outcome. So the probability of rolling a 5 is

$$P(E) = \frac{\text{Number of outcomes in } E}{\text{Number of outcomes in } S} = \frac{1}{6}.$$

(b) The event F of "rolling an even number" has three possible outcomes (2, 4, and 6). Therefore, the probability of rolling an even number is

$$P(F) = \frac{\text{Number of outcomes in } F}{\text{Number of outcomes in } S} = \frac{3}{6} = \frac{1}{2}.$$

(c) The event G of "rolling a number greater than 2" has four possible outcomes (3, 4, 5, and 6). Hence,

$$P(G) = \frac{\text{Number of outcomes in } G}{\text{Number of outcomes in } S} = \frac{4}{6} = \frac{2}{3}.$$

(d) Rolling a number less than 7 is an event that *always* occurs because the set of possible outcomes (1, 2, 3, 4, 5, and 6) is the entire sample space S. Its probability is

$$P(S) = \frac{\text{Number of outcomes in } S}{\text{Number of outcomes in } S} = \frac{6}{6} = 1.$$

(e) The event T of rolling a number greater than 6 *never* occurs, so there are no outcomes for this event and its probability is

$$P(T) = \frac{\text{Number of outcomes in } T}{\text{Number of outcomes in } S} = \frac{0}{6} = 0. \quad \blacksquare$$

As Example 2 illustrates, the numerator of the fraction

$$P(E) = \frac{\text{Number of outcomes in } E}{\text{Number of outcomes in } S}$$

is always less than or equal to the denominator (because the event E cannot have more outcomes than the entire sample space). Therefore,

Properties of Probability

For any event E,

$$0 \le P(E) \le 1.$$

If an event E must always occur, then $P(E) = 1$.

If an event E can never occur, then $P(E) = 0$.

Example 3   Assuming that the probability of a girl being born is the same as that of a boy, find the probability that a family with three children has

(a)   at least two girls;   (b) exactly two girls.

Solution   The possible ways for three children to be born lead to the following sample space, in which b stands for boy and g for girl:

$$S = \{bbb, bbg, bgb, bgg, gbb, gbg, ggb, ggg\}.$$

(a)   The event of having at least two girls is

$$E = \{bgg, gbg, ggb, ggg\}$$

and

$$P(E) = \frac{\text{Number of outcomes in } E}{\text{Number of outcomes in } S} = \frac{4}{8} = \frac{1}{2}.$$

(b)   The event of having exactly two girls is

$$F = \{bgg, gbg, ggb\}$$

whose probability is

$$P(F) = \frac{\text{Number of outcomes in } F}{\text{Number of outcomes in } S} = \frac{3}{8}. \blacksquare$$

Figure 7–15

Example 4   If a pair of dice are rolled as in Figure 7–15, find the probability that the total is   (a)   7;   (b)   11.

Solution   Each outcome in the sample space can be denoted by an ordered pair of numbers: The first coordinate is the number showing on the first die and the second coordinate is the number showing on the second die. The sample space consists of all 36 such pairs, as shown here:

(1, 6)	(2, 6)	(3, 6)	(4, 6)	(5, 6)	(6, 6)
(1, 5)	(2, 5)	(3, 5)	(4, 5)	(5, 5)	(6, 5)
(1, 4)	(2, 4)	(3, 4)	(4, 4)	(5, 4)	(6, 4)
(1, 3)	(2, 3)	(3, 3)	(4, 3)	(5, 3)	(6, 3)
(1, 2)	(2, 2)	(3, 2)	(4, 2)	(5, 2)	(6, 2)
(1, 1)	(2, 1)	(3, 1)	(4, 1)	(5, 1)	(6, 1)

(a)   A total of 7 can occur in six ways:

$$(1, 6), (2, 5), (3, 4), (4, 3), (5, 2), (6, 1).$$

So the probability of a total of 7 is $6/36 = 1/6$.

(b)   A total of 11 can occur in only two ways: (5, 6) and (6, 5). So the probability of a total of 11 is $2/36 = 1/18$. $\blacksquare$

> **CAUTION**
> The set
>
> $$\{2, 3, 4, 5, 6, 7, 8, 9, 10, 11, 12\}$$
>
> consists of the possible totals when rolling two dice, but it cannot be used as the sample space in Example 4 because the outcomes in this set are *not* equally likely. For instance, the total 2 only occurs in one way (namely, (1, 1)) and therefore is less likely than a total of 7, which can occur in six ways.

The solution of many probability problems depends on the counting techniques of Section 7.5.

Example 5   A committee of three people is chosen at random* from a group of four Republicans and six Democrats. Find the probability that the committee consists of

(a)   three Democrats;   (b)   one Republican and two Democrats.

Solution   The sample space S consists of all the possible three-person committees. The number of outcomes in S is the number of ways of choosing three people from a group of ten, namely,

$$_{10}C_3 = \frac{10!}{3!7!} = 120.$$

(a)   The number of ways of choosing three Democrats from the six available is

$$_6C_3 = \frac{6!}{3!3!} = 20$$

so the probability that the committee consists entirely of Democrats is $20/120 = 1/6$.

(b)   There are four choices for the Republican member. The number of ways of choosing two of the six Democrats is $_6C_2 = \frac{6!}{2!4!} = 15$. By the Fundamental Counting Principle, the number of ways of choosing the committee is

$$\begin{pmatrix} \text{Number of ways} \\ \text{of choosing one} \\ \text{Republican} \end{pmatrix} \times \begin{pmatrix} \text{Number of ways} \\ \text{of choosing two} \\ \text{Democrats} \end{pmatrix} = 4 \times {_6C_2} = 4 \cdot 15 = 60.$$

The probability of such a committee being chosen is $60/120 = 1/2$.   ■

---

*This means that each person is equally likely to be chosen.

Mutually Exclusive Events

Two events in the same sample space are said to be **mutually exclusive** if they cannot occur simultaneously. For example, when rolling a single die, the events "rolling a 1" and "rolling an even number" cannot both occur on the same roll, so these events are mutually exclusive. In terms of sets, this means that the sets {1} and {2, 4, 6} have no elements in common. On the other hand, the events of "rolling an odd number" and "rolling a number larger than 4" are *not* mutually exclusive because both occur when a 5 is rolled. In other words, the sets {1, 3, 5} and {5, 6} have an element in common.

Example 6 In the experiment of rolling a single die, let E and F be the mutually exclusive events

$$E = \text{Rolling an even number} = \{2, 4, 6\}$$

$$F = \text{Rolling an odd number greater than } 1 = \{3, 5\}.$$

Find $P(E) + P(F)$ and $P(E \text{ or } F)$, where $P(E \text{ or } F)$ denotes the probability that E or F occurs.

Solution Since the sample space contains six possible outcomes, we have:

$$P(E) = \frac{3}{6} = \frac{1}{2} \quad \text{and} \quad P(F) = \frac{2}{6} = \frac{1}{3},$$

so that

$$P(E) + P(F) = \frac{1}{2} + \frac{1}{3} = \frac{5}{6}.$$

On the other hand, the event "E or F" is the set {2, 3, 4, 5, 6} consisting of the possible outcomes when E or F occurs, so that

$$P(E \text{ or } F) = \frac{5}{6}.$$

Combining these last two results, we see that

$$P(E \text{ or } F) = P(E) + P(F). \quad \blacksquare$$

It can be shown that the conclusion of Example 6 is true for any pair of mutually exclusive events:

Mutually Exclusive Events

If E and F are mutually exclusive events in the same sample space, then

$$P(E \text{ or } F) = P(E) + P(F).$$

CAUTION

The formula in the box is *not* valid when the events E and F are not mutually exclusive. In rolling a die, for instance, the events

$$E = \text{rolling an odd number} = \{1, 3, 5\}$$

and

$$F = \text{rolling a number larger than } 4 = \{5, 6\}$$

are not mutually exclusive because both events contain the outcome 5. The set of possible outcomes for the event "E or F" is $\{1, 3, 5, 6\}$. Consequently,

$$P(E \text{ or } F) = \frac{4}{6}, \quad \text{but} \quad P(E) + P(F) = \frac{3}{6} + \frac{2}{6} = \frac{5}{6}.$$

Thus, $P(E \text{ or } F) \neq P(E) + P(F)$.

Example 7   A five-card hand is dealt from a well-shuffled standard 52-card deck.* What is the probability that it will consist of five hearts or five spades?

Solution   The sample space consists of the possible five-card hands; the number of such hands is

$$_{52}C_5 = \frac{52!}{5!47!} = 2{,}598{,}960.$$

The event E of getting five hearts can occur in as many ways as five cards can be chosen from the 13 hearts in the deck, namely,

$$_{13}C_5 = \frac{13!}{5!8!} = 1287.$$

Similarly, the event F of getting five spades can occur in 1287 ways. Since E and F are mutually exclusive,

$$P(E \text{ or } F) = P(E) + P(F) = \frac{1287}{2{,}598{,}960} + \frac{1287}{2{,}598{,}960} = \frac{2574}{2{,}598{,}960}$$

$$\approx .00099$$

Rounded to three decimal places, this probability is .001. So the chances of being dealt five hearts or five spades are about 1 in 1000. ■

Complements

The **complement** of an event consists of all the outcomes in the sample space that are *not* in the event. In rolling a die, for instance, "rolling a 3 or 4" is the set $\{3, 4\}$. The complement of this event is the set $\{1, 2, 5, 6\}$, which is the event "rolling a 1, 2, 5, or 6." Note that either the event $\{3, 4\}$ or its complement *must* occur whenever the die is rolled.

---

*A picture of a standard deck is on page 553.

The complement of an event E is denoted E'. In every case, E or E' must occur (since every possible outcome is in one or the other). Thus $P(E \text{ or } E') = 1$. Since E and E' are mutually exclusive (no outcome is in both E and E'), we have:

$$P(E \text{ or } E') = P(E) + P(E')$$
$$1 = P(E) + P(E')$$

Rearranging this last equation, we have:

Complements

> If E is an event, then the probability of its complement E' occurring is
>
> $$P(E') = 1 - P(E).$$

Example 8 If a committee of three people is chosen at random from a group of four Republicans and six Democrats, what is the probability that it will have at least one Republican member?

Solution The event "at least one Republican member" is the complement of the event E, "all three committee members are Democrats." In Example 5(a) we saw that $P(E) = 1/6$. Therefore

$$P(E') = 1 - P(E) = 1 - \frac{1}{6} = \frac{5}{6}. \quad \blacksquare$$

Example 9 You are 1 of 35 people at a party. A man offers to bet $100 that at least two people in the room have the same birthday (same month and day, not necessarily the same year). Should you take the bet?

Solution* The event "at least two people in the room have the same birthday" is the complement of the event $E = $ "no people in the room have the same birthday." So we first find the probability of E, which is

$$\frac{\text{Number of ways 35 people can have different birthdays}}{\text{Number of ways 35 people can have birthdays}}.$$

To find the denominator, we note that there are 365 possible birthdays for the first person in the room, 365 for the second, 365 for the third, and so on. By the Fundamental Counting Principle, the number of ways for all 35 birthdays to occur is

$$\underbrace{365 \cdot 365 \cdot 365 \cdots 365}_{35 \text{ factors}} = 365^{35}.$$

To find the numerator, note that there are 365 possible birthdays for the first person in the room, 364 for the second (since they must have different

---

*This solution ignores leap years and assumes that every day of the year is equally likely as a birthday.

birthdays), 363 for the third, and so on. By the Fundamental Counting Principle the number of ways that 35 people can all have different birthdays is

$$365 \cdot 364 \cdot 363 \cdots \quad \text{(35 factors working downward from 365),}$$

which is precisely the number $_{365}P_{35}$ (see the box on page 548). Thus, the probability that no people in the room have the same birthday is

$$\frac{_{365}P_{35}}{365^{35}} \approx .1856,$$

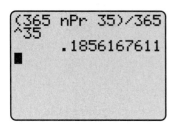

Figure 7–16

as shown in Figure 7–16. Therefore, the probability of the complement (at least two people have the same birthday) is

$$1 - .1856 = .8144.$$

So the man who wants to bet has a better than 80% probability of winning. Don't take the bet! ■

Probability Distributions

In many real-life situations it isn't possible to determine the probability of particular outcomes with the kind of theoretical analysis that we have used with dice, cards, etc. In such cases probabilities are determined empirically, by running experiments, constructing models, or analyzing previous data. To predict the chance of rain tomorrow, for example, a weather forecaster might note that there were 3800 days in the past century with atmospheric conditions similar to those today and that 1600 of these days were followed by rain. It would be reasonable to conclude that the probability of rain tomorrow is approximately $1600/3800 \approx .42$.

In a sample space with n equally likely outcomes, such as those studied earlier, each outcome has a probability of $1/n$ and the sum of these n probabilities is 1. This idea can be generalized to deal with experiments in which the outcomes are not all equally likely. Each outcome in the sample space is assigned a number (its probability) by some reasonable means (using theoretical analysis or an empirical approach), subject to the following conditions:

1. The probability of each outcome is a number between 0 and 1.
2. The sum of the probabilities of all the outcomes is 1.

Such an assignment of probabilities is called a **probability distribution.** It can be shown that the formulas developed above for mutually exclusive events and complements are valid for any probability distribution.

Example 10   Based on survey data, a pollster concludes that Smith has a 42% chance of winning an upcoming election and that Jones has a 38% chance of winning. Assuming these figures are accurate, find the probability that

(a) Smith or Jones will win;

(b) the third candidate, Brown, will win.

Solution   The sample space consists of three outcomes:

$$S = \text{Smith wins}, \qquad J = \text{Jones wins}, \qquad B = \text{Brown wins}.$$

We are given that $P(S) = .42$ and $P(J) = .38$.

(a)   Since Smith and Jones can't both win, S and J are mutually exclusive events; hence

$$P(S \text{ or } J) = P(S) + P(J) = .42 + .38 = .80.$$

Thus there is an 80% chance that Smith or Jones will win.

(b)   The complement of the event "S or J" is B (the only other outcome in the sample space). Consequently,

$$P(B) = 1 - P(S \text{ or } J) = 1 - .80 = .20.$$

Alternatively, $P(B) = .20$ because the sum $P(S) + P(J) + P(B)$ must be 1. In any case, we conclude that Brown has a 20% chance of winning the election.   ■

Exercises 7.6

In Exercises 1–6, list or describe a sample space for the experiment and state the number of outcomes in the sample space.

1. Three coins are flipped.

2. A day in March is chosen at random.

3. A marble is drawn from a jar containing two red and two blue marbles.

4. A coin is flipped and a die is rolled.

5. Two cards are drawn from a stack consisting of four jacks, four queens, four kings, and four aces.

6. Spin both spinners shown here:

In Exercises 7–11, a single die is rolled. Find the probability of rolling

7. An odd number.

8. A number greater than 3.

9. A number other than 5.

10. Any number except 5 or 6.

In Exercises 11–14, three coins are flipped. Find the probability that

11. Exactly one is heads.   12. At least one is heads.

13. At least two are tails.   14. All three are tails.

In Exercises 15–18, a letter is chosen at random from the word

ABRACADABRA.

Find the probability that the chosen letter is

15. A   16. B   17. C or D   18. E

In Exercises 19–22, the 12-sided die shown here is rolled. Each of the numbers 1–6 appears twice (on opposite faces of the die). Find the probability of rolling

19. 6

20. An even number

21. A number greater than 3

22. 3 or 5

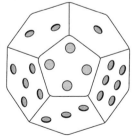

In Exercises 23–28, two dice are rolled, as in Example 4. Find the probability that the sum is

23. Less than 9   24. Greater than 7   25. 2 or 11

26. 8, 9, or 10   27. Even   28. Odd

In Exercises 29–36, a marble is drawn from an urn containing 3 red, 4 white, and 8 blue marbles. Find the probability that this marble is

29. Red   **30.** White   **31.** Blue or white

32. Black   **33.** Not blue   **34.** Not white

35. Red or white   **36.** Not black

In Exercises 37–44, a card is drawn from a well-shuffled standard 52-card deck (see the picture on page 553). Find the probability that this card is

37. An ace   **38.** A heart   **39.** A red queen

40. A face card (king, queen, or jack).

41. A red face card.

42. A red card that is not a face card.

43. Not a face card and greater than 5 (ace is 1).

44. A black card less than 6 (ace is 1).

In Exercises 45–50, a 5-card hand is dealt from a well-shuffled standard 52-card deck (see the picture on page 553). Find the probability that it contains

45. Four aces.   **46.** Four of a kind.

47. Two aces and three kings.   **48.** Five face cards.

49. At least one ace.   **50.** At least one spade.

51. If a student taking a ten-question true-false quiz guesses randomly, find the probability that the student will get
   (a) 10 questions correct.
   (b) 5 questions correct and 5 wrong.
   (c) At least 7 questions correct.

52. A 5-question exam is to be given, with the questions taken from a list of 12 questions. If a student knows the answer to 7 of the 12 questions and does not know the answers to the other 5, what is the probability that the student will correctly answer
   (a) All 5 questions on the exam?
   (b) Exactly 4 questions on the exam?
   (c) No more than 2 questions on the exam?

53. A committee of six people is to be randomly chosen from a group of five men and ten women. Find the probability that the committee consists of
   (a) 3 women and 3 men.
   (b) 4 women and 2 men.
   (c) 6 women.

54. A gift package of 3 items is to be chosen randomly from a collection of 12 books, 6 boxes of candy, and 8 movie tickets. Find the probability that the gift package contains
   (a) All books.
   (b) No movie tickets.

   (c) Two books.
   (d) One book, one box of candy, and one movie ticket.

55. A shipment of 20 stereos contains 5 defective ones. If five stereos in this shipment are chosen at random, what is the probability that exactly two of them will be defective?

56. A package contains 14 fasteners, 2 of which are defective. If three fasteners are chosen at random from the package, what is the probability that all three are not defective?

57. In the Ohio state lottery, the bettor chooses six numbers between 1 and 44. If these six numbers are drawn in the lottery, the player wins a prize of several million dollars. What is the probability that a bettor who buys only one ticket will win the lottery?

58. In another state lottery (see Exercise 57), the bettor chooses six numbers from 1 to 50, but gets two such choices for each ticket bought. What is the probability that a bettor who buys only one ticket will win this lottery?

Exercises 59–62 deal with the birthday problem in Example 9. The same argument used there shows that the probability that no people in a group of n people have the same birthday is $\dfrac{_{365}P_n}{365^n}$.

59. Fill the blanks in the following table. [*Hint:* Use the table feature of a calculator.]

Number of People	Probability That at Least Two Have the Same Birthday
10	
20	
25	
30	
40	
50	

60. What is the smallest number of people for which the probability that at least two of them have the same birthday is at least 1/2?

61. What is the probability that at least two American presidents have had the same birthday? [There have been 42 different presidents, including the one elected in 2000.]

62. **(a)** What is the probability that two members of the U.S. Senate have the same birthday?
(b) What is the probability that two members of the U.S. House of Representatives have the same birthday? [Think!]

63. Based on data going back for several generations, the probability that a baby born to a certain family is a girl is .55. If such a family has four children, what is the probability that they have
(a) Two boys and two girls?
(b) All boys?
(c) At least one boy?

64. A coin is weighted so that when it is flipped it comes up heads 60% of the time. If this coin is flipped three times, what is the probability of getting
(a) Three heads?
(b) At least one head?
(c) At least two tails?

65. If a dart is thrown at the dartboard shown in the figure, what is the probability that it hits
(a) Red?
(b) Blue?
(c) Green?
(d) White?

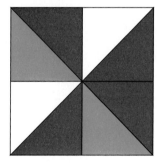

[*Hint:* What is the area of each color on the board? What is the area of the entire board?]

66. In the dartboard shown in the figure, the red inner circle is 3 inches in diameter; the outer edge of the green ring is 10 inches in diameter; the outer edge of the white ring is 18 inches in diameter; and the outer edge of the blue ring is 24 inches in diameter. If a dart is thrown at this board, what is the probability that it hits
(a) Red?
(b) Blue?
(c) Green?
(d) White?

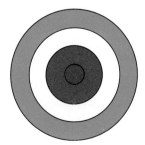

In Exercises 67–70, use the following information from the U.S. Bureau of Labor Statistics on the number of hours worked per week by nonfarm workers in 1997.

Hours per Week	Number of Workers
1–14	5,991,000
15–34	24,207,000
35–40	53,060,000
41 and over	39,001,000

If a nonfarm worker in 1997 is chosen at random, what is the probability that this person

67. Worked at most 14 hours per week?

68. Worked at most 40 hours per week?

69. Worked at least 35 hours per week?

70. Worked less than 15 hours or more than 40 hours per week?

In Exercises 71–74, use the following results of a survey of 800 consumers, which shows the number of people in each income class who watch TV for the stated number of hours each week:

Annual Income	Hours of TV Watched per Week		
	<10	10–20	>20
Less than $20,000	40	120	100
20,000–39,999	70	160	50
40,000–59,999	80	60	20
60,000 or more	40	50	10

If a person from the survey group is chosen at random, what is the probability that this person

71. Has an income of at least $40,000?

72. Watches TV at least 10 hours per week?

73. Has an income under $40,000 and watches TV at least 10 hours per week?

74. Has an income of at least $20,000 and watches TV less than 10 hours per week?

7.7 Mathematical Induction

Mathematical induction is a method of proof that can be used to prove a wide variety of mathematical facts, including the Binomial Theorem and statements such as:

The sum of the first n positive integers is the number $\dfrac{n(n+1)}{2}$.

$2^n > n$ for every positive integer n.

For each positive integer n, 4 is a factor of $7^n - 3^n$.

All of the preceding statements have a common property. For example, a statement such as

The sum of the first n positive integers is the number $\dfrac{n(n+1)}{2}$

or, in symbols,

$$1 + 2 + 3 + \cdots + n = \frac{n(n+1)}{2}$$

is really an infinite sequence of statements, one for each possible value of n:

$$n = 1: \qquad 1 = \frac{1(2)}{2}$$

$$n = 2: \qquad 1 + 2 = \frac{2(3)}{2}$$

$$n = 3 \qquad 1 + 2 + 3 = \frac{3(4)}{2}$$

and so on. Obviously, there isn't time enough to verify every one of the statements on this list, one at a time. But we can find a workable method of proof by examining how each statement on the list is *related* to the *next* statement on the list.

For example, for $n = 50$, the statement is

$$1 + 2 + 3 + \cdots + 50 = \frac{50(51)}{2}.$$

At the moment, we don't know whether or not this statement is true. But just *suppose* that it is true. What could then be said about the next statement, the one for $n = 51$:

$$1 + 2 + 3 + \cdots + 50 + 51 = \frac{51(52)}{2}?$$

Well, *if* it is true that

$$1 + 2 + 3 + \cdots + 50 = \frac{50(51)}{2}$$

then adding 51 to both sides and simplifying the right side would yield these equalities:

$$1 + 2 + 3 + \cdots + 50 + 51 = \frac{50(51)}{2} + 51$$

$$1 + 2 + 3 + \cdots + 50 + 51 = \frac{50(51)}{2} + \frac{2(51)}{2} = \frac{50(51) + 2(51)}{2}$$

$$1 + 2 + 3 + \cdots + 50 + 51 = \frac{(50 + 2)51}{2}$$

$$1 + 2 + 3 + \cdots + 50 + 51 = \frac{51(52)}{2}.$$

Since this last equality is just the original statement for $n = 51$, we conclude that

If the statement is true for $n = 50$, *then* it is also true for $n = 51$.

We have *not* proved that the statement actually *is* true for $n = 50$, but only that *if* it is, then it is also true for $n = 51$.

We claim that this same conditional relationship holds for any two consecutive values of n. In other words, we claim that for any positive integer k,

① *If* **the statement is true for** $n = k$, *then* **it is also true for** $n = k + 1$.

The proof of this claim is the same argument used earlier (with k and $k + 1$ in place of 50 and 51): *If* it is true that

$$1 + 2 + 3 + \cdots + k = \frac{k(k + 1)}{2} \qquad \textit{[Original statement for n = k]}$$

then adding $k + 1$ to both sides and simplifying the right side produces these equalities:

$$1 + 2 + 3 + \cdots + k + (k + 1) = \frac{k(k + 1)}{2} + (k + 1)$$

$$1 + 2 + 3 + \cdots + k + (k + 1) = \frac{k(k + 1)}{2} + \frac{2(k + 1)}{2} = \frac{k(k + 1) + 2(k + 1)}{2}$$

$$1 + 2 + 3 + \cdots + k + (k + 1) = \frac{(k + 2)(k + 1)}{2}$$

$$1 + 2 + 3 + \cdots + k + (k + 1) = \frac{(k + 1)[(k + 1) + 1]}{2}. \quad \textit{[Original statement for n = k + 1]}$$

We have proved that claim ① is valid for each positive integer k. We have *not* proved that the original statement is true for any value of n, but only that *if* it is true for $n = k$, then it is also true for $n = k + 1$. Applying this fact when $k = 1, 2, 3, \ldots$, we see that

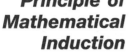

$$
\begin{cases}
\textit{If} \text{ the statement is true for } n = 1, \quad \textit{then} \text{ it is also true for} \\
n = 1 + 1 = 2; \\[4pt]
\textit{If} \text{ the statement is true for } n = 2, \quad \textit{then} \text{ it is also true for} \\
n = 2 + 1 = 3; \\[4pt]
\textit{If} \text{ the statement is true for } n = 3, \quad \textit{then} \text{ it is also true for} \\
n = 3 + 1 = 4; \\[4pt]
\vdots \\[4pt]
\textit{If} \text{ the statement is true for } n = 50, \quad \textit{then} \text{ it is also true for} \\
n = 50 + 1 = 51; \\[4pt]
\textit{If} \text{ the statement is true for } n = 51, \quad \textit{then} \text{ it is also true for} \\
n = 51 + 1 = 52; \\[4pt]
\vdots
\end{cases}
$$

and so on.

We are finally in a position to *prove* the original statement: $1 + 2 + 3 + \cdots + n = n(n + 1)/2$. Obviously, it is *true* for $n = 1$ since $1 = 1(2)/2$. Now apply in turn each of the propositions on list ②. Since the statement *is* true for $n = 1$, it must also be true for $n = 2$, and hence for $n = 3$, and hence for $n = 4$, and so on, for every value of n. Therefore, the original statement is true for *every* positive integer n.

The preceding proof is an illustration of the following principle:

Principle of Mathematical Induction

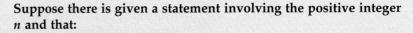

Suppose there is given a statement involving the positive integer n and that:

(i) The statement is true for $n = 1$.

(ii) If the statement is true for $n = k$ (where k is any positive integer), then the statement is also true for $n = k + 1$.

Then the statement is true for every positive integer n.

Property (i) is simply a statement of fact. To verify that it holds, you must prove the given statement is true for $n = 1$. This is usually easy, as in the preceding example.

Property (ii) is a *conditional* property. It does not assert that the given statement *is* true for $n = k$, but only that *if* it is true for $n = k$, then it is also true for $n = k + 1$. So to verify that property (ii) holds, you need only prove this conditional proposition:

If the statement is true for $n = k$, *then* it is also true for $n = k + 1$.

To prove this, or any conditional proposition, you must proceed as in the previous example: Assume the "if" part and use this assumption to prove the "then" part. As we saw earlier, the same argument will usually work for any possible k. Once this conditional proposition has been proved, you can use it *together* with property (i) to conclude that the given statement is necessarily true for every n, just as in the preceding example.

Thus proof by mathematical induction reduces to two steps:

Step 1 Prove that the given statement is true for $n = 1$.

Step 2 Let k be a positive integer. Assume that the given statement is true for $n = k$. Use this assumption to prove that the statement is true for $n = k + 1$.

Step 2 may be performed before step 1 if you wish. Step 2 is sometimes referred to as the **inductive step.** The assumption that the given statement is true for $n = k$ in this inductive step is called the **induction hypothesis.**

Example 1 Prove that $2^n > n$ for every positive integer n.

Solution Here the statement involving n is $2^n > n$.

Step 1 When $n = 1$, we have the statement $2^1 > 1$. This is obviously true.

Step 2 Let k be any positive integer. We assume that the statement is true for $n = k$, that is, we assume that $2^k > k$. We shall use this assumption to prove that the statement is true for $n = k + 1$, that is, that $2^{k+1} > k + 1$.

We begin with the induction hypothesis:* $2^k > k$. Multiplying both sides of this inequality by 2 yields:

$$2 \cdot 2^k > 2k$$

③
$$2^{k+1} > 2k.$$

Since k is a positive integer, we know that $k \geq 1$. Adding k to each side of the inequality $k \geq 1$, we have

$$k + k \geq k + 1$$

$$2k \geq k + 1.$$

Combining this result with inequality ③, we see that

$$2^{k+1} > 2k \geq k + 1.$$

The first and last terms of this inequality show that $2^{k+1} > k + 1$. Therefore, the statement is true for $n = k + 1$. This argument works for any positive integer k. Thus, we have completed the inductive step. By the Principle of Mathematical Induction, we conclude that $2^n > n$ for every positive integer n. ∎

Example 2 Simple arithmetic shows that

$$7^2 - 3^2 = 49 - 9 = 40 = 4 \cdot 10$$

---

*This is the point at which you usually must do some work. Remember that what follows is the "finished proof." It does not include all the thought, scratch work, false starts, and so on that were done before this proof was actually found.

and

$$7^3 - 3^3 = 343 - 27 = 316 = 4 \cdot 79.$$

In each case, 4 is a factor. These examples suggest that

For each positive integer n, 4 is a factor of $7^n - 3^n$.

This conjecture can be proved by induction as follows.

Step 1   When $n = 1$, the statement is "4 is a factor of $7^1 - 3^1$." Since $7^1 - 3^1 = 4 = 4 \cdot 1$, the statement is true for $n = 1$.

Step 2   Let k be a positive integer and assume that the statement is true for $n = k$, that is, that 4 is a factor of $7^k - 3^k$. Let us denote the other factor by D, so that the induction hypothesis is: $7^k - 3^k = 4D$. We must use this assumption to prove that the statement is true for $n = k + 1$, that is, that 4 is a factor of $7^{k+1} - 3^{k+1}$. Here is the proof:

$$
\begin{aligned}
7^{k+1} - 3^{k+1} &= 7^{k+1} - 7 \cdot 3^k + 7 \cdot 3^k - 3^{k+1} && \text{[Since } -7 \cdot 3^k + 7 \cdot 3^k = 0\text{]} \\
&= 7(7^k - 3^k) + (7 - 3)3^k && \text{[Factor]} \\
&= 7(4D) + (7 - 3)3^k && \text{[Induction hypothesis]} \\
&= 7(4D) + 4 \cdot 3^k && [7 - 3 = 4] \\
&= 4(7D + 3^k). && \text{[Factor out 4]}
\end{aligned}
$$

From this last line, we see that 4 is a factor of $7^{k+1} - 3^{k+1}$. Thus, the statement is true for $n = k + 1$, and the inductive step is complete. Therefore, by the Principle of Mathematical Induction the conjecture is actually true for every positive integer n. ∎

Another example of mathematical induction, the proof of the Binomial Theorem, is given at the end of this section.

Sometimes a statement involving the integer n may be false for $n = 1$ and (possibly) other small values of n, but true for all values of n beyond a particular number. For instance, the statement $2^n > n^2$ is false for $n = 1$, 2, 3, 4. But it is true for $n = 5$ and all larger values of n. A variation on the Principle of Mathematical Induction can be used to prove this fact and similar statements. See Exercise 28 for details.

A Common Mistake with Induction

It is sometimes tempting to omit step 2 of an inductive proof when the given statement can easily be verified for small values of n, especially if a clear pattern seems to be developing. As the next example shows, however, *omitting step 2 may lead to error.*

Example 3   An integer (>1) is said to be *prime* if its only positive integer factors are itself and 1. For instance, 11 is prime since its only positive integer factors are 11 and 1. But 15 is not prime because it has factors other than 15 and 1 (namely, 3 and 5). For each positive integer n, consider the number

$$f(n) = n^2 - n + 11.$$

You can readily verify that

$$f(1) = 11, \quad f(2) = 13, \quad f(3) = 17, \quad f(4) = 23, \quad f(5) = 31$$

and that *each of these numbers is prime.* Furthermore, there is a clear pattern: The first two numbers (11 and 13) differ by 2; the next two (13 and 17) differ by 4; the next two (17 and 23) differ by 6; and so on. On the basis of this evidence, we might conjecture:

For each positive integer n, the number $f(n) = n^2 - n + 11$ is prime.

We have seen that this conjecture is true for $n = 1, 2, 3, 4, 5$. Unfortunately, however, it is *false* for some values of n. For instance, when $n = 11$,

$$f(11) = 11^2 - 11 + 11 = 11^2 = 121.$$

But 121 is obviously *not* prime since it has a factor other than 121 and 1, namely, 11. You can verify that the statement is also false for $n = 12$ but true for $n = 13$. ■

In the preceding example, the proposition

If the statement is true for $n = k$, then it is true for $n = k + 1$

is false when $k = 10$ and $k + 1 = 11$. If you were not aware of this and tried to complete step 2 of an inductive proof, you would not have been able to find a valid proof for it. Of course, the fact that you can't find a proof of a proposition doesn't always mean that no proof exists. But when you are unable to complete step 2, you are warned that there is a possibility that the given statement may be false for some values of n. This warning should prevent you from drawing any wrong conclusions.

Proof of the Binomial Theorem

We shall use induction to prove that for every positive integer n,

$$(x + y)^n = x^n + \binom{n}{1}x^{n-1}y$$

$$+ \binom{n}{2}x^{n-2}y^2 + \binom{n}{3}x^{n-3}y^3 + \cdots + \binom{n}{n-1}xy^{n-1} + y^n.$$

This theorem was discussed and its notation explained in Section 7.4.

Step 1 When $n = 1$, there are only two terms on the right side of the preceding equation, and the statement reads $(x + y)^1 = x^1 + y^1$. This is certainly true.

Step 2 Let k be any positive integer and assume that the theorem is true for $n = k$, that is, that

$$(x + y)^k = x^k + \binom{k}{1}x^{k-1}y + \binom{k}{2}x^{k-2}y^2 + \cdots$$

$$+ \binom{k}{r}x^{k-r}y^r + \cdots + \binom{k}{k-1}xy^{k-1} + y^k.$$

[On the right side of this equation, we have included a typical middle term $\binom{k}{r}x^{k-r}y^r$. The sum of the exponents is k, and the bottom part of the binomial coefficient is the same as the y exponent.] We shall use this assumption to prove that the theorem is true for $n = k + 1$, that is, that

$$(x + y)^{k+1} = x^{k+1} + \binom{k+1}{1}x^k y + \binom{k+1}{2}x^{k-1}y^2 + \cdots$$

$$+ \binom{k+1}{r+1}x^{k-r}y^{r+1} + \cdots + \binom{k+1}{k}xy^k + y^{k+1}.$$

We have simplified some of the terms on the right side; for instance, $(k + 1) - 1 = k$ and $(k + 1) - (r + 1) = k - r$. But this is the correct statement for $n = k + 1$: The coefficients of the middle terms are $\binom{k+1}{1}$, $\binom{k+1}{2}$, $\binom{k+1}{3}$, and so on; the sum of the exponents of each middle term is $k + 1$, and the bottom part of each binomial coefficient is the same as the y exponent.

To prove the theorem for $n = k + 1$, we shall need this fact about binomial coefficients: For any integers r and k with $0 \le r < k$,

④
$$\binom{k}{r+1} + \binom{k}{r} = \binom{k+1}{r+1}.$$

A proof of this fact is outlined in Exercise 45 on page 545.

To prove the theorem for $n = k + 1$, we first note that

$$(x + y)^{k+1} = (x + y)(x + y)^k.$$

Applying the induction hypothesis to $(x + y)^k$, we see that

$$(x + y)^{k+1} = (x + y)\left[x^k + \binom{k}{1}x^{k-1}y + \binom{k}{2}x^{k-2}y^2 + \cdots + \binom{k}{r}x^{k-r}y^r \right.$$

$$\left. + \binom{k}{r+1}x^{k-(r+1)}y^{r+1} + \cdots + \binom{k}{k-1}xy^{k-1} + y^k \right]$$

$$= x\left[x^k + \binom{k}{1}x^{k-1}y + \cdots + y^k \right] + y\left[x^k + \binom{k}{1}x^{k-1}y + \cdots + y^k \right].$$

Next we multiply out the right-hand side. Remember that multiplying by x increases the x exponent by 1 and multiplying by y increases the y exponent by 1.

$$(x + y)^{k+1} = \left[x^{k+1} + \binom{k}{1}x^k y + \binom{k}{2}x^{k-1}y^2 + \cdots + \binom{k}{r}x^{k-r+1}y^r \right.$$

$$\left. + \binom{k}{r+1}x^{k-r}y^{r+1} + \cdots + \binom{k}{k-1}x^2 y^{k-1} + xy^k \right]$$

$$+ \left[x^k y + \binom{k}{1}x^{k-1}y^2 + \binom{k}{2}x^{k-2}y^3 + \cdots + \binom{k}{r}x^{k-r}y^{r+1} \right.$$

$$\left. + \binom{k}{r+1}x^{k-(r+1)}y^{r+2} + \cdots + \binom{k}{k-1}xy^k + y^{k+1} \right]$$

$$= x^{k+1} + \left[\binom{k}{1} + 1\right]x^k y + \left[\binom{k}{2} + \binom{k}{1}\right]x^{k-1}y^2 + \cdots$$

$$+ \left[\binom{k}{r+1} + \binom{k}{r}\right]x^{k-r}y^{r+1} + \cdots + \left[1 + \binom{k}{k-1}\right]xy^k + y^{k+1}.$$

Now apply statement ④ to each of the coefficients of the middle terms. For instance, with $r = 1$, statement ④ shows that $\binom{k}{2} + \binom{k}{1} = \binom{k+1}{2}$. Similarly, with $r = 0$, $\binom{k}{1} + 1 = \binom{k}{1} + \binom{k}{0} = \binom{k+1}{1}$, and so on. Then the expression above for $(x + y)^{k+1}$ becomes

$$(x + y)^{k+1} = x^{k+1} + \binom{k+1}{1}x^k y + \binom{k+1}{2}x^{k-1}y^2 + \cdots$$

$$+ \binom{k+1}{r+1}x^{k-r}y^{r+1} + \cdots + \binom{k+1}{k}xy^k + y^{k+1}.$$

Since this last statement says the theorem is true for $n = k + 1$, the inductive step is complete. By the Principle of Mathematical Induction the theorem is true for every positive integer n.

Exercises 7.7

In Exercises 1–18, use mathematical induction to prove that each of the given statements is true for every positive integer n.

1. $1 + 2 + 2^2 + 2^3 + 2^4 + \cdots + 2^{n-1} = 2^n - 1$

2. $1 + 3 + 3^2 + 3^3 + 3^4 + \cdots + 3^{n-1} = \dfrac{3^n - 1}{2}$

3. $1 + 3 + 5 + 7 + \cdots + (2n - 1) = n^2$

4. $2 + 4 + 6 + 8 + \cdots + 2n = n^2 + n$

5. $1^2 + 2^2 + 3^2 + \cdots + n^2 = \dfrac{n(n+1)(2n+1)}{6}$

6. $\dfrac{1}{2} + \dfrac{1}{4} + \dfrac{1}{8} + \cdots + \dfrac{1}{2^n} = 1 - \dfrac{1}{2^n}$

7. $\dfrac{1}{1 \cdot 2} + \dfrac{1}{2 \cdot 3} + \dfrac{1}{3 \cdot 4} + \cdots + \dfrac{1}{n(n+1)} = \dfrac{n}{n+1}$

8. $\left(1 + \dfrac{1}{1}\right)\left(1 + \dfrac{1}{2}\right)\left(1 + \dfrac{1}{3}\right)\cdots\left(1 + \dfrac{1}{n}\right) = n + 1$

9. $n + 2 > n$

10. $2n + 2 > n$

11. $3^n \geq 3n$

12. $3^n \geq 1 + 2n$

13. $3n > n + 1$

14. $\left(\dfrac{3}{2}\right)^n > n$

15. 3 is a factor of $2^{2n+1} + 1$

16. 5 is a factor of $2^{4n-2} + 1$

17. 64 is a factor of $3^{2n+2} - 8n - 9$

18. 64 is a factor of $9^n - 8n - 1$

19. Let c and d be fixed real numbers. Prove that

$$c + (c + d) + (c + 2d) + (c + 3d) + \cdots$$

$$+ [c + (n - 1)d] = \dfrac{n[2c + (n-1)d]}{2}$$

20. Let r be a fixed real number with $r \neq 1$. Prove that

$$1 + r + r^2 + r^3 + \cdots + r^{n-1} = \dfrac{r^n - 1}{r - 1}.$$

   [*Remember:* $1 = r^0$; so when $n = 1$ the left side reduces to $r^0 = 1$.]

21. (a) Write *each* of $x^2 - y^2$, $x^3 - y^3$, and $x^4 - y^4$ as a product of $x - y$ and another factor.
   (b) Make a conjecture as to how $x^n - y^n$ can be written as a product of $x - y$ and another factor. Use induction to prove your conjecture.

22. Let $x_1 = \sqrt{2}$; $x_2 = \sqrt{2 + \sqrt{2}}$;

   $x_3 = \sqrt{2 + \sqrt{2 + \sqrt{2}}}$; and so on. Prove that $x_n < 2$ for every positive integer n.

In Exercises 23–27, if the given statement is true, prove it. If it is false, give a counterexample.

23. Every odd positive integer is prime.

24. The number $n^2 + n + 17$ is prime for every positive integer n.

25. $(n + 1)^2 > n^2 + 1$ for every positive integer n.

26. 3 is a factor of the number $n^3 - n + 3$ for every positive integer n.

27. 4 is a factor of the number $n^4 - n + 4$ for every positive integer n.

28. Let q be a *fixed* integer. Suppose a statement involving the integer n has these two properties:
   (i) The statement is true for $n = q$.
   (ii) *If* the statement is true for $n = k$ (where k is any integer with $k \geq q$), then the statement is also true for $n = k + 1$.
   Then we claim that the statement is true for every integer n greater than or equal to q.
   (a) Give an informal explanation that shows why this claim should be valid. Note that when $q = 1$, this claim is precisely the Principle of Mathematical Induction.
   (b) The claim made before part (a) will be called the *Extended Principle of Mathematical Induction.* State the two steps necessary to use this principle to prove that a given statement is true for all $n \geq q$. (See discussion on pages 570–571.)

In Exercises 29–34, use the Extended Principle of Mathematical Induction (Exercise 28) to prove the given statement.

29. $2n - 4 > n$ for every $n \geq 5$. (Use 5 for q here.)

30. Let r be a fixed real number with $r > 1$. Then $(1 + r)^n > 1 + nr$ for every integer $n \geq 2$. (Use 2 for q here.)

31. $n^2 > n$ for all $n \geq 2$.

32. $2^n > n^2$ for all $n \geq 5$.

33. $3^n > 2^n + 10n$ for all $n \geq 4$.

34. $2n < n!$ for all $n \geq 4$.

Thinkers

35. Let n be a positive integer. Suppose that there are three pegs and on one of them n rings are stacked, with each ring being smaller in diameter than the one below it (see the figure). We want to transfer the stack of rings to another peg according to these rules: (i) Only one ring may be moved at a time; (ii) a ring can be moved to any peg, provided it is never placed on top of a smaller ring; (iii) the final order of the rings on the new peg must be the same as the original order on the first peg.

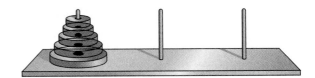

   (a) What is the smallest possible number of moves when $n = 2$? $n = 3$? $n = 4$?
   (b) Make a conjecture as to the smallest possible number of moves required for any n. Prove your conjecture by induction.

36. The basic formula for compound interest $A(t) = P(1 + r)^t$ was discussed on page 358. Prove by induction that the formula is valid whenever t is a positive integer. [*Note:* P and r are assumed to be constant.]

Chapter 7   *Review*

Important Concepts

**Important Facts
and Formulas**

· In an arithmetic sequence $\{a_n\}$ with common difference d:

$$a_n = a_1 + (n - 1)d \qquad \sum_{n=1}^{k} a_n = \frac{k}{2}(a_1 + a_k)$$

$$\sum_{n=1}^{k} a_n = ka_1 + \frac{k(k - 1)}{2}d$$

- In a geometric sequence $\{a_n\}$ with common ratio $r \neq 1$:

$$a_n = r^{n-1}a_1 \qquad \sum_{n=1}^{k} a_n = a_1\left(\frac{1-r^k}{1-r}\right)$$

- $n! = 1 \cdot 2 \cdot 3 \cdot \cdots (n-2)(n-1)n$

- $\dbinom{n}{r} = \dfrac{n!}{r!(n-r)!} = {}_nC_r$

- *The Binomial Theorem:*

$$(x+y)^n = x^n + \binom{n}{1}x^{n-1}y + \binom{n}{2}x^{n-2}y^2 + \binom{n}{3}x^{n-3}y^3 +$$

$$\cdots + \binom{n}{n-1}xy^{n-1} + y^n$$

$$= \sum_{j=0}^{n} \binom{n}{j}x^{n-j}y^j$$

- The number of permutations of n elements is $n!$

- ${}_nP_r = \dfrac{n!}{(n-r)!} = \left(\begin{array}{c}\text{The product of } r \text{ consecutive integers}\\ \text{starting with } n \text{ and working down}\end{array}\right)$

Review Questions

In Questions 1–4, find the first four terms of the sequence $\{a_n\}$.

1. $a_n = 2n - 5$

2. $a_n = 3^n - 27$

3. $a_n = \left(\dfrac{-1}{n}\right)^2$

4. $a_n = (-1)^{n+1}(n-1)$

5. Find the fifth partial sum of the sequence $\{a_n\}$, where $a_1 = -4$ and $a_n = 3a_{n-1} + 2$.

6. Find the fourth partial sum of the sequence $\{a_n\}$, where $a_1 = 1/9$ and $a_n = 3a_{n-1}$.

7. $\displaystyle\sum_{n=0}^{4} 2^n(n+1) = ?$

8. $\displaystyle\sum_{n=2}^{4} (3n^2 - n + 1) = ?$

In Questions 9–12, find a formula for a_n; assume that the sequence is arithmetic.

9. $a_1 = 3$ and the common difference is -6.

10. $a_2 = 4$ and the common difference is 3.

11. $a_1 = -5$ and $a_3 = 7$.

12. $a_3 = 2$ and $a_7 = -1$.

In Questions 13–16, find a formula for a_n; assume that the sequence is geometric.

13. $a_1 = 2$ and the common ratio is 3.

14. $a_1 = 5$ and the common ratio is $-1/2$.

15. $a_2 = 192$ and $a_7 = 6$.

16. $a_3 = 9/2$ and $a_6 = -243/16$.

17. Find the 11th partial sum of the arithmetic sequence with $a_1 = 5$ and common difference -2.

18. Find the 12th partial sum of the arithmetic sequence with $a_1 = -3$ and $a_{12} = 16$.

19. Find the fifth partial sum of the geometric sequence with $a_1 = 1/4$ and common ratio 3.

20. Find the sixth partial sum of the geometric sequence with $a_1 = 5$ and common ratio $1/2$.

21. Find numbers b, c, d such that $4, b, c, d, 23$ are the first five terms of an arithmetic sequence.

22. Find numbers c and d such that $8, c, d, 27$ are the first four terms of a geometric sequence.

23. Is it better to be paid \$5 per day for 100 days or to be paid 5¢ the first day, 10¢ the second day, 20¢ the third day, and have your salary increase in this fashion every day for 100 days?

24. Tuition at Bigstate University is now \$3000 per year and will increase \$150 per year in succeeding years. If a student starts school now, spends four years as an undergraduate, three years in law school, and five years getting a Ph.D., how much tuition will she have paid?

Find the following sums, if they exist.

25. $\displaystyle\sum_{n=1}^{\infty} \frac{1}{2^{n-1}}$ 26. $\displaystyle\sum_{n=1}^{\infty} \left(\frac{-1}{4^n}\right)$

27. Use the Binomial Theorem to show that $(1.02)^{51} > 2.5$.

28. What is the coefficient of u^3v^2 in the expansion of $(u + 5v)^5$?

29. $\dbinom{15}{12} = ?$ 30. $\dbinom{18}{3} = ?$

31. Let n be a positive integer. Simplify $\dbinom{n+1}{n}$.

32. Use the Binomial Theorem to expand $\left(\sqrt{x} + 1\right)^5$. Simplify your answer.

33. $\dfrac{20!5!}{6!17!} = ?$ 34. $\dfrac{7! - 5!}{4!} = ?$

35. Find the coefficient of x^2y^4 in the expansion of $(2y + x^2)^5$.

In Questions 36–40, compute the number.

36. $_7P_3$ 37. $_{24}P_6$

38. $_7C_3$ 39. $_{24}C_{20}$

40. $_kC_t$

41. How many three-digit numbers are there in which the first digit is even and the second is odd?

42. How many possible finishes (first, second, third place) are there in an eight-horse race?

43. A woman has three skirts, four blouses, and two scarves. How many different outfits can she wear (an outfit being a skirt, blouse, and scarf)?

44. A bridge hand consists of 13 cards from a standard 52-card deck. How many different bridge hands are there?

45. How many different committees consisting of two men and two women can be formed from a group of eight women and five men?

46. List a sample space for the experiment of flipping a coin four times in succession.

47. If two dice are rolled, what is the probability that the sum is 8?

48. Which event is more likely: rolling a 3 with a single die or rolling a total of 6 with two dice?

49. Assuming that the probability of a girl being born is the same as that of a boy, find the probability that a family with four children has three girls and a boy.

50. If a card is drawn from a well-shuffled standard 52-card deck, what is the probability that it is a 6 or a king?

51. A committee of four people is to be randomly chosen from a group of 8 mathematicians and 12 chemists. What is the probability that the committee will consist of two mathematicians and two chemists?

52. Two stale candy bars are inadvertently dropped in a box with 18 fresh bars of the same brand. If you take three bars from the box, what is the probability that all of them will be fresh?

53. Prove that for every positive integer n,

$$\frac{1}{3} + \frac{1}{3^2} + \frac{1}{3^3} + \cdots + \frac{1}{3^n} = \frac{3^n - 1}{2 \cdot 3^n}.$$

54. Prove that for every positive integer n,

$$1^3 + 2^3 + 3^3 + \cdots + n^3 = \frac{n^2(n + 1)^2}{4}.$$

55. Prove that for every positive integer n,

$$1 + 5 + 5^2 + 5^3 + \cdots + 5^{n-1} = \frac{5^n - 1}{4}.$$

56. Prove that $2^n \geq 2n$ for every positive integer n.

57. If x is a real number with $|x| < 1$, then prove that $|x^n| < 1$ for all $n \geq 1$.

58. Prove that for any positive integer n,

$$1 + 5 + 9 + \cdots + (4n - 3) = n(2n - 1).$$

59. Prove that for any positive integer n,

$$1 + 4 + 4^2 + 4^3 + \cdots + 4^{n-1} = \frac{1}{3}(4^n - 1).$$

60. Prove that $3n < n!$ for every $n \geq 4$.

61. Prove that for every positive integer n, 8 is a factor of $9^n - 8n - 1$.

Discovery Project 7

Racking, Stacking, and Packing

Arranging objects in orderly space-filling patterns is a common occurrence throughout our experience; we see 50 stars in a rectangular field on the U.S. flag, 15 balls in a poolball rack, and boxes of 88 oranges or 64 crayons. What drives these patterns? Often it is purely aesthetics—we like to have objects arranged in polygons, pyramids, cubes, or other regular objects. First, let's start with a historical pattern change.

In 1959, the number of states in the United States increased from 48 to 50. It's easy to arrange 48 stars in a roughly rectangular pattern; use six rows of eight. Forty-nine stars is also easy; use seven rows of seven, giving a square pattern. Unfortunately, the field of blue on the U.S. flag is not square; it's slightly longer than it is wide. The solution adopted was to offset the lines of stars to make them closer together, as shown below. This general kind of problem is called a packing problem; how do you optimize the organization of regular or irregular objects to fit the most or best into a fixed area?

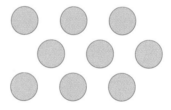

1. How would you arrange fifty objects in offset rows to fit a roughly rectangular pattern? (You can check your solution against any U.S. flag.)

Similar problems occur in gaming. In the game of eight-ball, played on a pool table, balls are prepared for the game by placing 15 numbered balls in an equilateral triangular pattern, called a "rack," with 5 balls on each side.

2. How many balls are there in a game of eight-ball?

3. What is the general formula for the number of pool balls in a rack with n balls on a side?

The exercises become more interesting as you move to three-dimensional forms. For example, spherical objects can be stacked with offset layers in the shape of triangular or square pyramids as long as the bottom layer sits on some type of rack. This is the custom when stacking cannonballs for artillery emplacements. It turns out that triangular pyramids are more stable, so we will consider only those stacks.

4. In a triangular stack, each nested layer is an equilateral triangle. How many cannonballs are there in a triangular stack that is five layers high?

5. One approach to finding a general formula for the number of spheres in a triangular stack with n layers is to calculate $\sum_{j=1}^{n} \left(\sum_{k=1}^{j} k \right)$. Why does this formula work?

6. Reduce the sum in Exercise 5 to a polynomial that tells the number of spheres in a triangular stack. Does this formula produce the same result for a five-layer stack as you got in Exercise 4?

8

Analytic Geometry

Calling all ships!

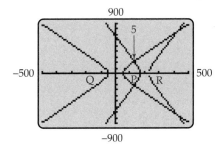

The planets travel in elliptical orbits around the sun and satellites in elliptical orbits around the earth. Parabolic reflectors are used in spotlights, radar antennas, and satellite dishes. The long-range navigation system (LORAN) uses hyperbolas to enable a ship to determine its exact location. See Exercise 48 on page 607.

Interdependence of Sections

The sections of this chapter are essentially independent of one another, but a few exercises in Sections 8.2 and 8.3 assume knowledge of preceding section(s). Sections 8.1–8.3 are prerequisites for Excursion 8.3.A.

The discussion of analytic geometry that was begun in Section 1.1 is continued here, with an examination of conic sections (which have played a significant role in mathematics since ancient times).

When a right circular cone is cut by a plane, the intersection is a curve called a **conic section,** as shown in Figure 8–1.* Conic sections were studied by the ancient Greeks and are still of interest. For instance, planets travel in elliptical orbits, parabolic mirrors are used in telescopes, and certain atomic particles follow hyperbolic paths.

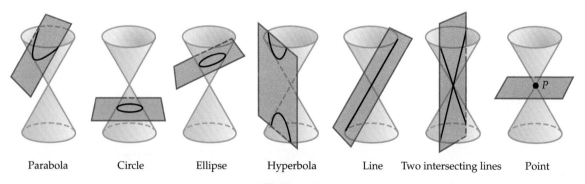

| Parabola | Circle | Ellipse | Hyperbola | Line | Two intersecting lines | Point |

Figure 8–1

Although the Greeks studied conic sections from a purely geometric point of view, the modern approach is to describe them in terms of the coordinate plane and distance, or as the graphs of certain types of equations. This was done for circles in Section 1.1 and will be done here for ellipses, hyperbolas, and parabolas.

In each case the conic is defined in terms of points and distances and its equation determined. The standard form of the equation of a conic includes the key information necessary for a rough sketch of its graph, just as the standard form of the equation of a circle tells you its center and radius.

*A point, a line, or two intersecting lines are sometimes called **degenerate** conic sections.

8.1 Circles and Ellipses

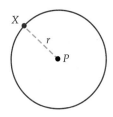

Figure 8–2

Let P be a point in the plane and r a positive number. As we saw in Section 1.1, the circle with center P and radius r consists of all points X in the plane such that

$$\text{Distance from } X \text{ to } P = r$$

as shown in Figure 8–2. When the coordinates of P are known, we can use the distance formula to obtain the following description of a circle (see page 66).

Equation of a Circle

The circle with center (c, d) and radius r is the graph of

$$(x - c)^2 + (y - d)^2 = r^2.$$

For example, the graph of

$$(x - 4)^2 + (y + 3)^2 = 36,$$

which can be written as

$$(x - 4)^2 + [y - (-3)]^2 = 6^2,$$

is the circle with center $(4, -3)$ and radius 6. If the equation of this circle is multiplied out it becomes

$$x^2 - 8x + 16 + y^2 + 6y + 9 = 36$$
$$x^2 + y^2 - 8x + 6y - 11 = 0$$

In many cases, the equation of a circle is given in this latter form and we must determine its center and radius.

The technique for finding the center and radius from the equation is based on this simple algebraic fact:

$$\left(x \pm \frac{b}{2}\right)^2 = x^2 \pm bx + \left(\frac{b}{2}\right)^2,$$

which you can easily verify by multiplying out the left side. Note how the last term on the right is related to the expression $x^2 \pm bx$: it is the square of half the coefficient of x.

Example 1 Show that the graph of $3x^2 + 3y^2 - 12x - 30y + 45 = 0$ is a circle and find its center and radius.

Solution We begin by dividing both sides of the equation by 3 and regrouping the terms:

$$x^2 + y^2 - 4x - 10y + 15 = 0$$

$$(x^2 - 4x) + (y^2 - 10y) = -15$$

We want to add a constant to the expression $x^2 - 4x$ so that the result will be a perfect square. Using the fact discussed before the example, we take half the coefficient of x, namely -2, and square it, obtaining 4. Adding 4 to both sides of the equation, we obtain

$$(x^2 - 4x + 4) + (y^2 - 10y) = -15 + 4,$$

which factors as

$$(x - 2)^2 + (y^2 - 10y) = -11.$$

Similarly, in the expression $y^2 - 10y$, we take half the coefficient of y, namely -5, and square it, obtaining 25. Now add 25 to both sides of the equation and factor:

$$(x - 2)^2 + (y^2 - 10y + 25) = -11 + 25$$

$$(x - 2)^2 + (y - 5)^2 = 14.$$

Since $14 = (\sqrt{14})^2$, the graph of this equation is the circle with center $(2, 5)$ and radius $\sqrt{14}$. ■

The technique of adding the square of half the coefficient of x to $x^2 - 4x$ to obtain the perfect square $(x - 2)^2$ in Example 1 is called **completing the square**. It will be used frequently in this chapter.

Ellipses

Let P and Q be points in the plane and r a number greater than the distance from P to Q. The **ellipse** with **foci*** P and Q is the set of all points X such that

(Distance from X to P) + (Distance from X to Q) = r.

To draw this ellipse, take a piece of string of length r and pin its ends on P and Q. Put your pencil point against the string and move it, keeping the string taut. You will trace out the ellipse, as shown in Figure 8–3.

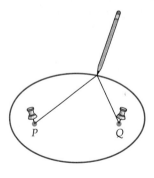

Figure 8–3

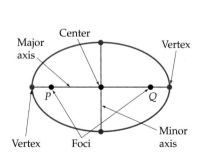

Figure 8–4

*"Foci" is the plural of "focus."

The midpoint of the line segment from P to Q is the **center** of the ellipse. The points where the straight line through the foci intersects the ellipse are its **vertices.** The **major axis** of the ellipse is the line segment joining the vertices; its **minor axis** is the line segment through the center, perpendicular to the major axis, as shown in Figure 8–4.

Next we determine the equation of an ellipse that is centered at the origin. Suppose the foci P and Q are on the x-axis, with coordinates

$$P = (-c, 0) \quad \text{and} \quad Q = (c, 0) \quad \text{for some } c > 0.$$

Let $a = r/2$, so that $2a = r$. Then the point (x, y) is on the ellipse exactly when

$$[\text{Distance from } (x, y) \text{ to } P] + [\text{Distance from } (x, y) \text{ to } Q] = r$$

$$\sqrt{(x + c)^2 + (y - 0)^2} + \sqrt{(x - c)^2 + (y - 0)^2} = 2a$$

$$\sqrt{(x + c)^2 + y^2} = 2a - \sqrt{(x - c)^2 + y^2}$$

Squaring both sides and simplifying (Exercise 60) we obtain

$$a\sqrt{(x - c)^2 + y^2} = a^2 - cx.$$

Again squaring both sides and simplifying, we have

$$(a^2 - c^2)x^2 + a^2y^2 = a^2(a^2 - c^2).$$

To simplify the form of this equation, let $b = \sqrt{a^2 - c^2}$* so that $b^2 = a^2 - c^2$ and the equation becomes

$$b^2x^2 + a^2y^2 = a^2b^2.$$

Dividing both sides by a^2b^2 shows that the coordinates of every point on the ellipse satisfy the equation

$$\frac{x^2}{a^2} + \frac{y^2}{b^2} = 1.$$

Conversely, it can be shown that every point whose coordinates satisfy this equation is on the ellipse. When the equation is in this form the x- and y-intercepts of the graph are easily found. For instance, to find the x-intercepts, we set $y = 0$ and solve:

$$\frac{x^2}{a^2} + \frac{0^2}{b^2} = 1$$

$$x^2 = a^2$$

$$x = \pm a.$$

Similarly, the y-intercepts are $\pm b$.

---

*The distance between the foci is $2c$. Since $r = 2a$ and $r > 2c$ by definition, we have $2a > 2c$ and hence $a > c$. Therefore, $a^2 - c^2$ is a positive number and has a real square root.

A similar argument applies when the foci are on the y-axis and leads to this conclusion:

Standard Equations of Ellipses Centered at the Origin

Let a and b be real numbers with $a > b > 0$. Then the graph of each of the following equations is an ellipse centered at the origin:

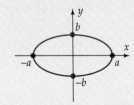

$$\frac{x^2}{a^2} + \frac{y^2}{b^2} = 1 \begin{cases} \text{x-intercepts: $\pm a$} \quad \text{y-intercepts: $\pm b$} \\ \\ \text{major axis on the x-axis, with vertices $(a, 0)$ and $(-a, 0)$} \\ \\ \text{foci: $(c, 0)$ and $(-c, 0)$, where $c = \sqrt{a^2 - b^2}$.} \end{cases}$$

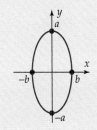

$$\frac{x^2}{b^2} + \frac{y^2}{a^2} = 1 \begin{cases} \text{x-intercepts: $\pm b$} \quad \text{y-intercepts: $\pm a$} \\ \\ \text{major axis on the y-axis, with vertices $(0, a)$ and $(0, -a)$} \\ \\ \text{foci: $(0, c)$ and $(0, -c)$, where $c = \sqrt{a^2 - b^2}$.} \end{cases}$$

In the preceding box $a > b$, but don't let all the letters confuse you: When the equation is in standard form, the denominator of the x term tells you the x-intercepts, the denominator of the y term tells you the y-intercepts, and the major axis is the longer one, as illustrated in the following examples.

Example 2 Identify and sketch the graph of the equation $4x^2 + 9y^2 = 36$.

Solution To identify the graph, we put the equation in standard form:

$$4x^2 + 9y^2 = 36$$

Divide both sides by 36: $\qquad \dfrac{4x^2}{36} + \dfrac{9y^2}{36} = \dfrac{36}{36}$

Simplify: $\qquad \dfrac{x^2}{9} + \dfrac{y^2}{4} = 1$

$$\frac{x^2}{3^2} + \frac{y^2}{2^2} = 1.$$

The graph is now in the form of the first equation in the preceding box, with $a = 3$ and $b = 2$. So its graph is an ellipse with x-intercepts ± 3 and y-intercepts ± 2. Its major axis and foci lie on the x-axis, as do its vertices

(3, 0) and (−3, 0). A hand-sketched graph is shown in Figure 8–5. To graph this ellipse on a calculator, we first solve its equation for y:

$$4x^2 + 9y^2 = 36$$

Subtract $4x^2$ from both sides: $$9y^2 = 36 - 4x^2$$

Divide both sides by 9: $$y^2 = \frac{36 - 4x^2}{9}$$

Taking square roots on both sides, we see that

$$y = \sqrt{\frac{36 - 4x^2}{9}} \quad \text{or} \quad y = -\sqrt{\frac{36 - 4x^2}{9}}.$$

Graphing both of these equations on the same screen, we obtain Figure 8–6. ■

Technology Tip

On Sharp 9600 and TI calculators, you can graph

$$y = \sqrt{\frac{36 - 4x^2}{9}}$$

and

$$y = -\sqrt{\frac{36 - 4x^2}{9}}$$

at the same time by keying in

$$y = \{1, -1\}\sqrt{\frac{36 - 4x^2}{9}}.$$

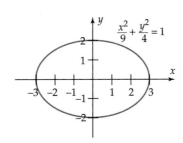

Figure 8–5

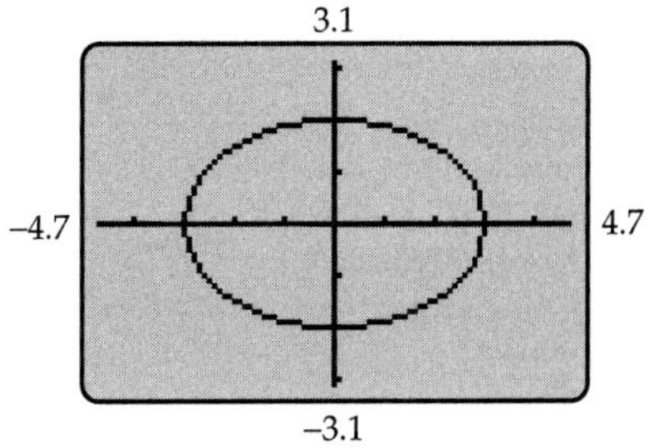

Figure 8–6

Example 3 Find the equation of the ellipse with vertices (0, ±6) and foci $\left(0, \pm2\sqrt{6}\right)$ and sketch its graph.

Solution Since the foci are $\left(0, 2\sqrt{6}\right)$ and $\left(0, -2\sqrt{6}\right)$, the center of the ellipse is (0, 0) and its major axis lies on the y-axis. Hence, its equation is of the form

$$\frac{x^2}{b^2} + \frac{y^2}{a^2} = 1.$$

From the box on page 588 we see that $a = 6$ and $c = 2\sqrt{6}$. Since $c = \sqrt{a^2 - b^2}$, we have $c^2 = a^2 - b^2$, so that

$$b^2 = a^2 - c^2 = 6^2 - \left(2\sqrt{6}\right)^2 = 36 - 4 \cdot 6 = 12.$$

Hence, $b = \sqrt{12}$ and the equation of the ellipse is

$$\frac{x^2}{\left(\sqrt{12}\right)^2} + \frac{y^2}{6^2} = 1, \quad \text{or equivalently,} \quad \frac{x^2}{12} + \frac{y^2}{36} = 1.$$

The graph has x-intercepts $\pm\sqrt{12} \approx 3.46$ and y-intercepts ± 6, as sketched in Figure 8–7. ■

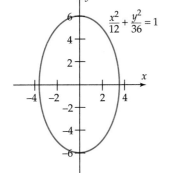

Figure 8–7

Vertical and Horizontal Shifts

We now examine ellipses that have foci on a line parallel to one of the coordinate axes. Recall that in Section 3.4 we saw that replacing the variable x by $x - 5$ in the rule of the function $y = f(x)$ shifts the graph horizontally 5 units to the right, whereas replacing x by $x + 5$ [that is, $x - (-5)$] shifts the graph horizontally 5 units to the left (see the box on page 222).

Similarly, if the rule of a function is given by $y = f(x)$, then replacing y by $y - 4$ shifts the graph 4 units vertically upward because

$$y - 4 = f(x) \quad \text{is equivalent to} \quad y = f(x) + 4$$

(see the box on page 220). For arbitrary equations, we have similar results:

Vertical and Horizontal Shifts

> Let h and k be constants. Replacing x by $x - h$ and y by $y - k$ in an equation shifts the graph of the equation
>
> $|h|$ units horizontally (right for positive h, left for negative h) and
>
> $|k|$ units vertically (upward for positive k, downward for negative k).

Example 4 Identify and sketch the graph of

$$\frac{(x - 5)^2}{9} + \frac{(y + 4)^2}{36} = 1.$$

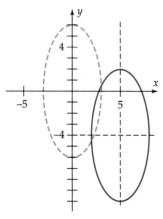

Figure 8–8

Solution This equation can be obtained from the equation $\dfrac{x^2}{9} + \dfrac{y^2}{36} = 1$ (whose graph is known to be an ellipse) as follows:

Replace x by $x - 5$ and replace y by $y - (-4) = y + 4$.

This is the situation described in the previous box with $h = 5$ and $k = -4$. Therefore, the graph is the ellipse $\dfrac{x^2}{9} + \dfrac{y^2}{36} = 1$ shifted horizontally 5 units to the right and vertically 4 units downward, as shown in Figure 8–8. The center of the ellipse is at $(5, -4)$. Its major axis (the longer one) lies on the vertical line $x = 5$, as do its foci. The minor axis is on the horizontal line $y = -4$. ∎

Example 5 Identify and sketch the graph of

$$4x^2 + 9y^2 - 32x - 90y + 253 = 0.$$

Solution We first rewrite the equation:

$$(4x^2 - 32x) + (9y^2 - 90y) = -253$$

$$4(x^2 - 8x) + 9(y^2 - 10y) = -253.$$

CAUTION

Completing the square only works when the coefficient of x^2 is 1. In an expression such as $4x^2 - 32x$, you must first factor out the 4:

$$4(x^2 - 8x)$$

and then complete the square on the expression in parentheses.

The factoring in the last equation was done in preparation for completing the square (see the Caution in the margin). To complete the square on $x^2 - 8x$, we add 16 (the square of half the coefficient of x) and to complete the square on $y^2 - 10y$ we add 25 (the square of half the coefficient of y):

$$4(x^2 - 8x + 16) + 9(y^2 - 10y + 25) = -253 + ? + ?.$$

Be careful here: On the left side we haven't just added 16 and 25. When the left side is multiplied out, we have actually added $4 \cdot 16 = 64$ and $9 \cdot 25 = 225$. To leave the equation unchanged, we must add these numbers on the right:

$$4(x^2 - 8x + 16) + 9(y^2 - 10y + 25) = -253 + 64 + 225$$

Factor and simplify:

$$4(x - 4)^2 + 9(y - 5)^2 = 36$$

Divide both sides by 36:

$$\frac{4(x - 4)^2}{36} + \frac{9(y - 5)^2}{36} = \frac{36}{36}$$

Simplify:

$$\frac{(x - 4)^2}{9} + \frac{(y - 5)^2}{4} = 1.$$

The graph of this equation is the ellipse $\dfrac{x^2}{9} + \dfrac{y^2}{4} = 1$ shifted 4 units to the right and 5 units upward. Its center is at $(4, 5)$. Its major axis lies on the horizontal line $y = 5$ and its minor axis on the vertical line $x = 4$, as shown in Figure 8–9. ■

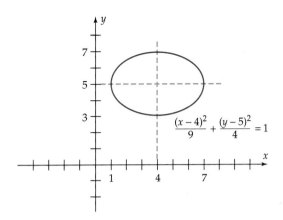

Figure 8–9

By translating the information about ellipses centered at the origin in the box on page 588, we obtain

Standard Equations of Ellipses with Center at (h, k)

Let (h, k) be any point in the plane. If a and b are real numbers with $a > b > 0$, then the graph of each of the following equations is an ellipse with center (h, k).

$$\frac{(x - h)^2}{a^2} + \frac{(y - k)^2}{b^2} = 1 \begin{cases} \text{major axis on the horizontal line } y = k \\ \text{minor axis on the vertical line } x = h \\ \text{foci: } (h - c, k) \text{ and } (h + c, k), \text{ where} \\ \quad c = \sqrt{a^2 - b^2} \end{cases}$$

$$\frac{(x - h)^2}{b^2} + \frac{(y - k)^2}{a^2} = 1 \begin{cases} \text{major axis on the vertical line } x = h \\ \text{minor axis on the horizontal line } y = k \\ \text{foci: } (h, k - c) \text{ and } (h, k + c), \text{ where} \\ \quad c = \sqrt{a^2 - b^2} \end{cases}$$

Technology Tip

The Casio 9850 has a conic section grapher (on the main menu) that produces the graphs of equations in standard form when the various coefficients are entered.

Example 6 Find an appropriate viewing window and graph the ellipse

$$\frac{(x - 3)^2}{40} + \frac{(y + 2)^2}{120} = 1.$$

Solution To graph the ellipse, we first solve its equation for y:

Subtract $\frac{(x - 3)^2}{40}$ from both sides: $\quad \dfrac{(y + 2)^2}{120} = 1 - \dfrac{(x - 3)^2}{40}$

Multiply both sides by 120: $\quad (y + 2)^2 = 120\left[1 - \dfrac{(x - 3)^2}{40}\right]$

Multiply out right side: $\quad (y + 2)^2 = 120 - 3(x - 3)^2$

Take square roots on both sides: $\quad y + 2 = \pm\sqrt{120 - 3(x - 3)^2}$

$$y = \sqrt{120 - 3(x - 3)^2} - 2 \qquad \text{or} \qquad y = -\sqrt{120 - 3(x - 3)^2} - 2$$

Technology Tip

TI-83/86 and Sharp 9600 users can save keystrokes by entering the first equation in Example 6 as y_1 and then using the RCL key to copy the text of y_1 to y_2. On TI-89, use COPY and PASTE in place of RCL. Then only one sign needs to be changed to make y_2 into the second equation of Example 6.

So we should graph both of these last two equations on the same screen.

To determine an appropriate window, look at the original form of the equation:

$$\frac{(x - 3)^2}{40} + \frac{(y + 2)^2}{120} = 1.$$

The center of the ellipse is at $(3, -2)$. Its graph extends a distance of $\sqrt{40}$ to the left and right of the center and a distance of $\sqrt{120}$ above and below the center. Since $\sqrt{40}$ is a bit less than 7 and $\sqrt{120}$ a bit less than 11 (why?), our window should include x-values from $3 - 7$ to $3 + 7$ (that is, $-4 \le x \le 10$) and y-values from $-2 - 11$ to $-2 + 11$ (that is, $-13 \le y \le 9$) So we first try the window in Figure 8–10. In this window, the ellipse looks longer horizontally than vertically. We know from its equation, however, that the major (longer) axis is vertical because the larger constant 120 is the denominator of the y-term. So we change to a square window and obtain the more accurate graph in Figure 8–11.* Even this graph has gaps that shouldn't be there (an ellipse is a connected figure), but is unavoidable because of the limited resolution of the calculator screen.   ■

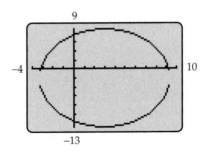

Figure 8–10   **Figure 8–11**

Example 7 Graph the equation $x^2 + 8y^2 + 6x + 9y + 4 = 0$ without first putting it in standard form.

Solution   Rewrite it like this:

$$8y^2 + 9y + (x^2 + 6x + 4) = 0.$$

This is a quadratic equation of the form $ay^2 + by + c = 0$, with

$$a = 8, \qquad b = 9, \qquad c = x^2 + 6x + 4$$

and hence can be solved by using the quadratic formula:

$$y = \frac{-b \pm \sqrt{b^2 - 4ac}}{2a}$$

$$y = \frac{-9 \pm \sqrt{9^2 - 4 \cdot 8 \cdot (x^2 + 6x + 4)}}{2 \cdot 8}$$

$$y = \frac{-9 \pm \sqrt{81 - 32(x^2 + 6x + 4)}}{16}$$

GRAPHING EXPLORATION

Find a complete graph of the original equation by graphing both of the following equations on the same screen. The second Tip on page 592 may be helpful.

$$y = \frac{-9 + \sqrt{81 - 32(x^2 + 6x + 4)}}{16}$$

$$y = \frac{-9 - \sqrt{81 - 32(x^2 + 6x + 4)}}{16}.$$ ■

---

*Figure 8–11 shows a square window for a TI-83. On wide-screen calculators, make the *x*-axis longer to obtain a square window.

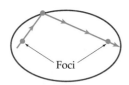

Figure 8–12

Applications

Elliptical surfaces have interesting reflective properties. If a sound or light ray passes through one focus and reflects off an ellipse, the ray will pass through the other focus, as shown in Figure 8–12. Exactly this situation occurs under the elliptical dome of the U.S. Capitol. A person who stands at one focus and whispers can be clearly heard by anyone at the other focus. Before this fact was widely known, when Congress used to sit under the dome, several political secrets were inadvertently revealed by congressmen to members of the other party.

The planets and many comets have elliptical orbits, with the sun as one focus. The moon travels in an elliptical orbit with the earth as one focus. Satellites are usually put into elliptical orbits around the earth.

Example 8   The earth's orbit around the sun is an ellipse that is almost a circle. The sun is one focus and the major and minor axes have lengths 186,000,000 miles and 185,974,062 miles, respectively. What are the minimum and maximum distances from the earth to the sun?

Solution   The orbit is shown in Figure 8–13. If we use a coordinate system with the major axis on the x-axis and the sun having coordinates $(c, 0)$, then we obtain Figure 8–14.

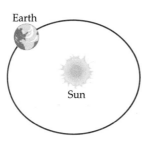

Figure 8–13

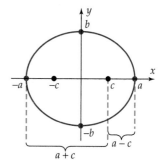

Figure 8–14

The length of the major axis is $2a = 186,000,000$, so that $a = 93,000,000$. Similarly, $2b = 185,974,062$, so that $b = 92,987,031$. As shown earlier, the equation of the orbit is $\dfrac{x^2}{a^2} + \dfrac{y^2}{b^2} = 1$, where

$$c = \sqrt{a^2 - b^2} = \sqrt{(93,000,000)^2 - (92,987,031)^2} \approx 1,553,083.$$

Figure 8–14 suggests a fact that can also be proven algebraically: The minimum and maximum distances from a point on the ellipse to the focus $(c, 0)$ occur at the endpoints of the major axis:

Minimum distance $= a - c \approx 93,000,000 - 1,553,083 = 91,446,917$ miles

Maximum distance $= a + c \approx 93,000,000 + 1,553,083 = 94,553,083$ miles.   ∎

Exercises 8.1

In Exercises 1–6, determine which of the following equations could possibly have the given graph.

$$2x^2 + y^2 = 12, \qquad (x - 4)^2 + (y - 3)^2 = 4,$$
$$x^2 + 6y^2 = 18, \qquad (x + 3)^2 + (y + 4)^2 = 6,$$
$$(x + 3)^2 + y^2 = 9, \qquad x^2 + (y - 3)^2 = 4,$$
$$(x - 3)^2 + (y + 4)^2 = 5, \qquad (x + 2)^2 + (y - 3)^2 = 2.$$

1.

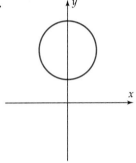

2.

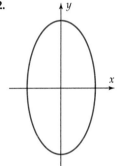

3.

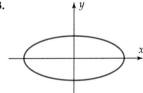

4.

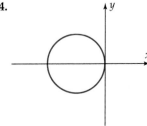

5.

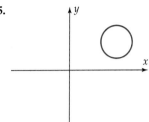

6.

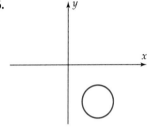

In Exercises 7–12, find the center and radius of the circle whose equation is given, as in Example 1.

7. $x^2 + y^2 + 8x - 6y - 15 = 0$

8. $15x^2 + 15y^2 = 10$

9. $x^2 + y^2 + 6x - 4y - 15 = 0$

10. $x^2 + y^2 + 10x - 75 = 0$

11. $x^2 + y^2 + 25x + 10y = -12$

12. $3x^2 + 3y^2 + 12x + 12 = 18y$

In Exercises 13–20, identify the conic section whose equation is given and find a complete graph of the equation as in Example 2.

13. $\dfrac{x^2}{25} + \dfrac{y^2}{4} = 1$ 　　　　**14.** $\dfrac{x^2}{6} + \dfrac{y^2}{16} = 1$

15. $4x^2 + 3y^2 = 12$ 　　　　**16.** $9x^2 + 4y^2 = 72$

17. $\dfrac{x^2}{10} - 1 = \dfrac{-y^2}{36}$ 　　　　**18.** $\dfrac{y^2}{49} + \dfrac{x^2}{81} = 1$

19. $4x^2 + 4y^2 = 1$ 　　　　**20.** $x^2 + 4y^2 = 1$

In Exercises 21–26, find the equation of the ellipse that satisfies the given conditions.

21. Center $(0, 0)$; foci on x-axis; x-intercepts ± 7; y-intercepts ± 2.

22. Center $(0, 0)$; foci on y-axis; x-intercepts ± 1; y-intercepts ± 8.

23. Center $(0, 0)$; foci on x-axis; major axis of length 12; minor axis of length 8.

24. Center $(0, 0)$; foci on y-axis; major axis of length 20; minor axis of length 18.

25. Center $(0, 0)$; endpoints of major and minor axes: $(0, -7)$, $(0, 7)$, $(-3, 0)$, $(3, 0)$.

26. Center $(0, 0)$; vertices $(8, 0)$ and $(-8, 0)$; minor axis of length 8.

Calculus can be used to show that the area of the ellipse with equation $\dfrac{x^2}{a^2} + \dfrac{y^2}{b^2} = 1$ is πab. Use this fact to find the area of each ellipse in Exercises 27–32.

27. $\dfrac{x^2}{16} + \dfrac{y^2}{4} = 1$ 28. $\dfrac{x^2}{9} + \dfrac{y^2}{5} = 1$

29. $3x^2 + 4y^2 = 12$ 30. $7x^2 + 5y^2 = 35$

31. $6x^2 + 2y^2 = 14$ 32. $5x^2 + y^2 = 5$

In Exercises 33–44, identify the conic section whose equation is given, list its center, and find its graph.

33. $\dfrac{(x - 1)^2}{4} + \dfrac{(y - 5)^2}{9} = 1$

34. $\dfrac{(x - 2)^2}{16} + \dfrac{(y + 3)^2}{12} = 1$

35. $\dfrac{(x + 1)^2}{16} + \dfrac{(y - 4)^2}{8} = 1$

36. $\dfrac{(x + 5)^2}{4} + \dfrac{(y + 2)^2}{12} = 1$

37. $9x^2 + 4y^2 + 54x - 8y + 49 = 0$

38. $4x^2 + 5y^2 - 8x + 30y + 29 = 0$

39. $x^2 + y^2 + 6x - 8y + 5 = 0$

40. $x^2 + y^2 - 4x + 2y - 7 = 0$

41. $4x^2 + y^2 + 24x - 4y + 36 = 0$

42. $9x^2 + y^2 - 36x + 10y + 52 = 0$

43. $25x^2 + 16y^2 + 50x + 96y = 231$

44. $9x^2 + 25y^2 - 18x + 50y = 191$

In Exercises 45–50, find the equation of the ellipse that satisfies the given conditions.

45. Center $(2, 3)$; endpoints of major and minor axes: $(2, -1)$, $(0, 3)$, $(2, 7)$, $(4, 3)$.

46. Center $(-5, 2)$; endpoints of major and minor axes: $(0, 2)$, $(-5, 17)$, $(-10, 2)$, $(-5, -13)$.

47. Center $(7, -4)$; foci on the line $x = 7$; major axis of length 12; minor axis of length 5.

48. Center $(-3, -9)$; foci on the line $y = 9$; major axis of length 15; minor axis of length 7.

49. Center $(3, -2)$; passing through $(3, -6)$ and $(9, -2)$.

50. Center $(2, 5)$; passing through $(2, 4)$ and $(-3, 5)$.

In Exercises 51 and 52, find the equations of two distinct ellipses satisfying the given conditions.

51. Center at $(-5, 3)$; major axis of length 14; minor axis of length 8.

52. Center at $(2, -6)$; major axis of length 15; minor axis of length 6.

In Exercises 53–58, determine which of the following equations could possibly have the given graph.

$$\frac{(x + 3)^2}{4} + \frac{(y + 3)^2}{8} = 1, \qquad \frac{(x - 3)^2}{9} + \frac{(y + 4)^2}{4} = 1,$$

$$2x^2 + 2y^2 - 8 = 0, \qquad 4x^2 + 2y^2 - 8 = 0,$$

$$2x^2 + y^2 - 8x - 6y + 9 = 0,$$

$$x^2 + 3y^2 + 6x - 12y + 17 = 0.$$

53.

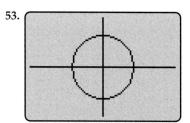

54.

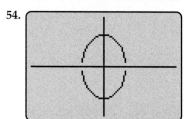

55.

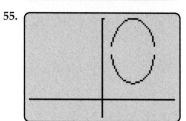

56.

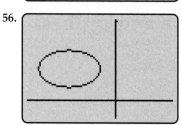

57.

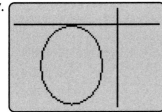

58.

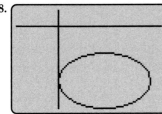

59. Consider the ellipse whose equation is $\dfrac{x^2}{a^2} + \dfrac{y^2}{b^2} = 1$.

Show that if $a = b$, then the graph is actually a circle.

60. Complete the derivation of the equation of the ellipse on page 587 as follows.
   (a) By squaring both sides, show that the equation

$$\sqrt{(x + c)^2 + y^2} = 2a - \sqrt{(x - c)^2 + y^2}$$

   may be simplified as

$$a\sqrt{(x - c)^2 + y^2} = a^2 - cx.$$

   (b) Show that the last equation in part (a) may be further simplified as

$$(a^2 - c^2)x^2 + a^2y^2 = a^2(a^2 - c^2).$$

61. The orbit of the moon around the earth is an ellipse with the earth as one focus. If the length of the major axis of the orbit is 477,736 miles and the length of the minor axis is 477,078 miles, find the minimum and maximum distances from the earth to the moon.

62. Halley's Comet has an elliptical orbit with the sun as one focus and a major axis that is 1,636,484,848 miles long. The closest the comet comes to the sun is 54,004,000 miles. What is the maximum distance from the comet to the sun?

*If $a > b > 0$, then the **eccentricity** of the ellipse*

$$\dfrac{(x - h)^2}{a^2} + \dfrac{(y - k)^2}{b^2} = 1 \quad \text{or} \quad \dfrac{(x - h)^2}{b^2} + \dfrac{(y - k)^2}{a^2} = 1$$

is the number $\dfrac{\sqrt{a^2 - b^2}}{a}$. In Exercises 63–66, find the eccentricity of the ellipse whose equation is given.

63. $\dfrac{x^2}{100} + \dfrac{y^2}{99} = 1$ **64.** $\dfrac{x^2}{18} + \dfrac{y^2}{25} = 1$

65. $\dfrac{(x - 3)^2}{10} + \dfrac{(y - 9)^2}{40} = 1$

66. $\dfrac{(x + 5)^2}{12} + \dfrac{(y - 4)^2}{8} = 1$

67. Based on your answers to Exercises 63–66, how is the eccentricity of an ellipse related to its graph? [*Hint:* What is the shape of the graph when the eccentricity is close to 0? When it is close to 1?]

68. Assuming the same scale was used for the axes, which of these ellipses has the largest eccentricity?

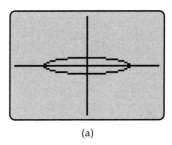

(a)

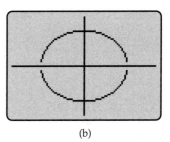

(b)

69. A satellite is to be placed in an elliptical orbit, with the center of the earth as one focus. The satellite's maximum distance from the surface of the earth is to be 22,380 km and its minimum distance 6540 km. Assume that the radius of the earth is 6400 km and find the eccentricity of the satellite's orbit.

70. The first step in landing Apollo 11 on the moon was to place the spacecraft in an elliptical orbit such that the minimum distance from the *surface* of the moon to the spacecraft was 110 km and the maximum distance was 314 km. If the radius of the moon is 1740 km, find the eccentricity of the Apollo 11 orbit.

Thinkers

71. The punch bowl and a table holding the punch cups are placed 50 feet apart at a garden party. A portable fence is then set up so that any guest inside the fence can walk straight to the table, then to the punch bowl, and then return to his or her starting point without traveling more than 150 feet. Describe the longest possible such fence that encloses the largest possible area.

72. An arched footbridge over a 100-foot-wide river is shaped like half an ellipse. The maximum height of the bridge over the river is 20 feet. Find the height of the bridge over a point in the river, exactly 25 feet from the center of the river.

8.2 Hyperbolas

Let P and Q be points in the plane and r a positive number. The set of all points X such that

$$|(\text{Distance from } P \text{ to } X) - (\text{Distance from } Q \text{ to } X)| = r$$

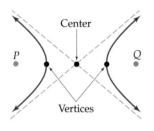

Center

P Q

Vertices

Figure 8–15

is the **hyperbola** with **foci** P and Q; r will be called the **distance difference.** Every hyperbola has the general shape shown by the red curve in Figure 8–15. The dotted straight lines are the **asymptotes** of the hyperbola; it gets closer and closer to the asymptotes, but never touches them. The asymptotes intersect at the midpoint of the line segment from P to Q; this point is called the **center** of the hyperbola. The **vertices** of the hyperbola are the points where it intersects the line segment from P to Q. The line through P and Q is called the **focal axis.**

Another complicated exercise in the use of the distance formula, which will be omitted here, leads to the following algebraic description:

Standard Equations of Hyperbolas Centered at the Origin

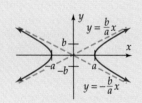

Let a and b be positive real numbers. Then the graph of each of the following equations is a hyperbola centered at the origin:

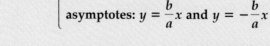

$$\frac{x^2}{a^2} - \frac{y^2}{b^2} = 1$$

x-intercepts: $\pm a$ y-intercepts: none

focal axis on the x-axis, with vertices $(a, 0)$ and $(-a, 0)$

foci: $(c, 0)$ and $(-c, 0)$, where $c = \sqrt{a^2 + b^2}$.

asymptotes: $y = \dfrac{b}{a}x$ and $y = -\dfrac{b}{a}x$

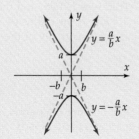

$$\frac{y^2}{a^2} - \frac{x^2}{b^2} = 1$$

x-intercepts: none y-intercepts: $\pm a$

focal axis on the y-axis, with vertices $(0, a)$ and $(0, -a)$

foci: $(0, c)$ and $(0, -c)$, where $c = \sqrt{a^2 + b^2}$.

asymptotes: $y = \dfrac{a}{b}x$ and $y = -\dfrac{a}{b}x$

Once again don't worry about all the letters in the box. When the equation is in standard form with the x term positive and y term negative, the hyperbola intersects the x-axis and opens from side to side. When the x term is negative and the y term positive, the hyperbola intersects the y-axis and opens up and down.

Example 1 Identify and sketch the graph of the equation $9x^2 - 4y^2 = 36$.

Solution We first put the equation in standard form:

$$9x^2 - 4y^2 = 36$$

Divide both sides by 36:
$$\frac{9x^2}{36} - \frac{4y^2}{36} = \frac{36}{36}$$

Simplify:
$$\frac{x^2}{4} - \frac{y^2}{9} = 1$$

$$\frac{x^2}{2^2} - \frac{y^2}{3^2} = 1.$$

Applying the fact in the box with $a = 2$ and $b = 3$ shows that the graph is a hyperbola with vertices $(2, 0)$ and $(-2, 0)$ and asymptotes $y = \frac{3}{2}x$ and $y = -\frac{3}{2}x$. We first plot the vertices and sketch the rectangle determined by the vertical lines $x = \pm 2$ and the horizontal lines $y = \pm 3$. The asymptotes go through the origin and the corners of this rectangle, as shown on the left in Figure 8–16. It is then easy to sketch the hyperbola. ■

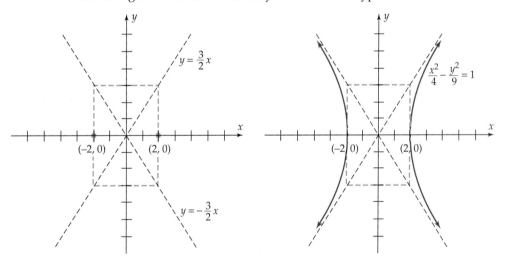

Figure 8–16

Example 2 Find the equation of the hyperbola with vertices $(0, 1)$ and $(0, -1)$ that passes through the point $\left(3, \sqrt{2}\right)$. Then sketch its graph.

Solution The vertices are on the y-axis and the equation is of the form

$$\frac{y^2}{a^2} - \frac{x^2}{b^2} = 1$$

with $a = 1$. Since $\left(3, \sqrt{2}\right)$ is on the graph, we have

$$\frac{\left(\sqrt{2}\right)^2}{1^2} - \frac{3^2}{b^2} = 1$$

Simplify: $\qquad\qquad\qquad\qquad\qquad 2 - \frac{9}{b^2} = 1$

Subtract 2 from both sides: $\qquad\qquad -\frac{9}{b^2} = -1$

Multiply both sides by $-b^2$: $\qquad\qquad 9 = b^2$

Therefore, $b = 3$ and the equation is

$$\frac{y^2}{1^2} - \frac{x^2}{3^2} = 1, \qquad \text{or equivalently,} \qquad y^2 - \frac{x^2}{9} = 1.$$

The asymptotes of the hyperbola are the lines $y = \pm\frac{1}{3}x$. To sketch the graph, we first solve the equation for y:

$$y^2 - \frac{x^2}{9} = 1$$

Add $\dfrac{x^2}{9}$ *to both sides:* $\qquad\qquad\qquad y^2 = 1 + \frac{x^2}{9}$

Take square roots on both sides: $\quad y = \sqrt{1 + \frac{x^2}{9}} \quad \text{or} \quad y = -\sqrt{1 + \frac{x^2}{9}}.$

Graphing these last two equations on the same screen, we obtain Figure 8–17. ∎

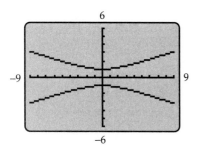

Figure 8–17

Vertical and Horizontal Shifts

Let h and k be constants. As we saw on page 590, replacing x by $x - h$ and y by $y - k$ in an equation shifts the graph of the equation h units horizontally and k units vertically. If the graph of the equation is a hyperbola with center at the origin, then the shifted graph has its center at (h, k). Using this fact and translating the information in the box on page 598, we obtain:

Standard Equations of Hyperbolas with Center at (h, k)

If a and b are positive real numbers, then the graph of each of the following equations is a hyperbola with center (h, k).

$$\frac{(x - h)^2}{a^2} - \frac{(y - k)^2}{b^2} = 1 \begin{cases} \text{focal axis on the horizontal line } y = k \\ \text{foci: } (h - c, k) \text{ and } (h + c, k), \text{ where} \\ \qquad c = \sqrt{a^2 + b^2} \\ \text{vertices: } (h - a, k) \text{ and } (h + a, k) \\ \text{asymptotes: } y = \pm\dfrac{b}{a}(x - h) + k \end{cases}$$

$$\frac{(y - k)^2}{a^2} - \frac{(x - h)^2}{b^2} = 1 \begin{cases} \text{focal axis on the vertical line } x = h \\ \text{foci: } (h, k - c) \text{ and } (h, k + c), \text{ where} \\ \qquad c = \sqrt{a^2 + b^2} \\ \text{vertices: } (h, k - a) \text{ and } (h, k + a) \\ \text{asymptotes: } y = \pm\dfrac{a}{b}(x - h) + k \end{cases}$$

On the left side, we have actually added $6 \cdot 4 = 24$ and $-8 \cdot 9 = -72$, so we must add these numbers on the right to keep the equation unchanged.

$$6(y^2 - 4y + 4) - 8(x^2 + 6x + 9) = 96 + 24 - 72$$

Factor and simplify:
$$6(y - 2)^2 - 8(x + 3)^2 = 48$$

Divide both sides by 48:
$$\frac{(y - 2)^2}{8} - \frac{(x + 3)^2}{6} = 1$$

$$\frac{(y - 2)^2}{8} - \frac{(x - (-3))^2}{6} = 1$$

In this form, we can see that the graph is a hyperbola with center at $(-3, 2)$. ■

Applications

The reflective properties of hyperbolas are used in the design of camera and telescope lenses. If a light ray passes through one focus of a hyperbola and reflects off the hyperbola at a point P, then the reflected ray moves along the straight line determined by P and the other focus, as shown in Figure 8–20.

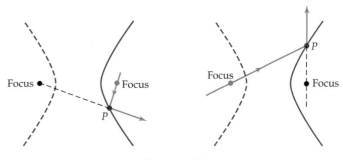

Figure 8–20

Hyperbolas are also the basis of the long-range navigation system (LORAN), which enables a ship to determine its exact location by radio, as illustrated in the next example.

Example 6 Three LORAN transmitters Q, P, and R are located 200 miles apart along a straight line and simultaneously transmit signals at regular intervals. These signals travel at a speed of 980 feet per microsecond. A ship S receives a signal from P and 305 microseconds later a signal from R. It also receives a signal from Q 528 microseconds after the one from P. Determine the ship's location.

Solution Take the line through the LORAN stations as the x-axis, with the origin located midway between Q and P, so that the situation looks like Figure 8–21 on the next page.

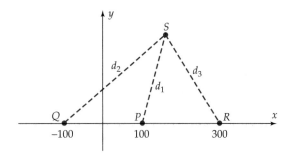

Figure 8–21

If the signal takes t microseconds to go from P to S, then

$$d_1 = 980t \quad \text{and} \quad d_2 = 980(t + 528)$$

so that

$$|d_1 - d_2| = |980t - 980(t + 528)| = 980 \cdot 528 = 517{,}440 \text{ feet.}$$

Since 1 mile is 5280 feet, this means that

$$|d_1 - d_2| = 517{,}440/5{,}280 \text{ miles} = 98 \text{ miles.}$$

In other words,

$$|(\text{Distance from } P \text{ to } S) - (\text{Distance from } Q \text{ to } S)| = |d_1 - d_2| = 98.$$

This is precisely the situation described in the definition of "hyperbola" on page 598: S is on the hyperbola with foci $P = (100, 0)$, $Q = (-100, 0)$, and distance difference $r = 98$. This hyperbola has an equation of the form

$$\frac{x^2}{a^2} - \frac{y^2}{b^2} = 1,$$

where $(\pm a, 0)$ are the vertices, $(\pm c, 0) = (\pm 100, 0)$ are the foci and $c^2 = a^2 + b^2$. Figure 8–22 and the fact that the vertex $(a, 0)$ is on the hyperbola show that

$$|[\text{Distance from } P \text{ to } (a, 0)] - [\text{Distance from } Q \text{ to } (a, 0)]| = r = 98$$

$$|(100 - a) - (100 + a)| = 98$$

$$|-2a| = 98$$

$$|a| = 49.$$

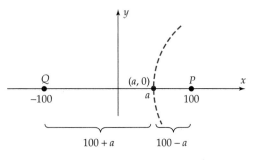

Figure 8–22

Consequently, $a^2 = 49^2 = 2401$ and hence $b^2 = c^2 - a^2 = 100^2 - 49^2 = 7599$. Thus the ship lies on the hyperbola

(∗)
$$\frac{x^2}{2401} - \frac{y^2}{7599} = 1.$$

A similar argument using P and R as foci shows that the ship also lies on the hyperbola with foci $P = (100, 0)$ and $R = (300, 0)$ and center $(200, 0)$, whose distance difference r is

$$|d_1 - d_3| = 980 \cdot 305 = 298{,}900 \text{ feet} \approx 56.61 \text{ miles.}$$

As before you can verify that $a = 56.61/2 = 28.305$ and hence $a^2 = 28.305^2 = 801.17$. This hyperbola has center $(200, 0)$ and its foci are $(200 - c, k) = (100, 0)$ and $(200 + c, k) = (300, 0)$, which implies that $c = 100$. Hence, $b^2 = c^2 - a^2 = 100^2 - 801.17 = 9198.83$ and the ship also lies on the hyperbola

(∗∗)
$$\frac{(x - 200)^2}{801.17} - \frac{y^2}{9198.83} = 1.$$

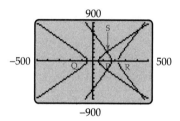

Figure 8–23

Since the ship lies on both hyperbolas, its coordinates are solutions of both the equations (∗) and (∗∗). They can be found algebraically by solving each of the equations for y^2, setting the results equal, and solving for x. They can be found geometrically by graphing both hyperbolas and finding the intersection point. As shown in Figure 8–23, there are actually four points of intersection. However, the two below the x-axis represent points on land in our situation. Furthermore, since the signal from P was received first, the ship is closest to P. So it is located at the point S in Figure 8–23. A graphical intersection finder shows that this point is approximately $(130.48, 215.14)$, where the coordinates are in miles from the origin. ∎

Exercises 8.2

In Exercises 1–6, determine which of the following equations could possibly have the given graph.

$3x^2 + 3y^2 = 12$, $6y^2 - x^2 = 6$,

$x^2 + 4y^2 = 1$, $4x^2 + 4(y + 2)^2 = 12$,

$4(x + 4)^2 + 4y^2 = 12$, $6x^2 + 2y^2 = 18$

$2x^2 - y^2 = 8$, $3x^2 - y = 6$.

1.

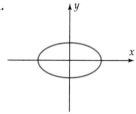

2.

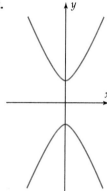

3.

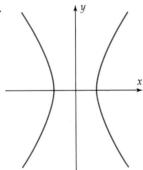

4.

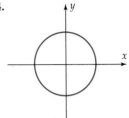

5.

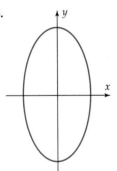

6.

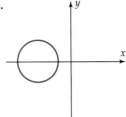

In Exercises 7–14, identify the conic section whose equation is given and find a complete graph of the equation, as in Example 1.

7. $\dfrac{x^2}{6} - \dfrac{y^2}{16} = 1$ **8.** $\dfrac{x^2}{4} - y^2 = 1$

9. $4x^2 - y^2 = 16$ **10.** $3y^2 - 5x^2 = 15$

11. $\dfrac{x^2}{10} - \dfrac{y^2}{36} = 1$ **12.** $\dfrac{y^2}{9} - \dfrac{x^2}{16} = 1$

13. $x^2 - 4y^2 = 1$ **14.** $2x^2 - y^2 = 4$

In Exercises 15–18, find the equation of the hyperbola that satisfies the given conditions.

15. Center $(0, 0)$; x-intercepts ± 3; asymptote $y = 2x$.

16. Center $(0, 0)$; y-intercepts ± 12; asymptote $y = 3x/2$.

17. Center $(0, 0)$; vertex $(2, 0)$; passing through $\left(4, \sqrt{3}\right)$.

18. Center $(0, 0)$; vertex $\left(0, \sqrt{12}\right)$; passing through $(2\sqrt{3}, 6)$.

In Exercises 19–32, identify the conic section whose equation is given, list its center, and find its graph.

19. $\dfrac{(y + 3)^2}{25} - \dfrac{(x + 1)^2}{16} = 1$

20. $\dfrac{(y + 1)^2}{9} - \dfrac{(x - 1)^2}{25} = 1$

21. $\dfrac{(x + 3)^2}{1} - \dfrac{(y - 2)^2}{4} = 1$

22. $\dfrac{(y + 5)^2}{9} - \dfrac{(x - 2)^2}{1} = 1$

23. $(y + 4)^2 - 8(x - 1)^2 = 8$

24. $(x - 3)^2 + 12(y - 2)^2 = 24$

25. $4y^2 - x^2 + 6x - 24y + 11 = 0$

26. $x^2 - 16y^2 = 0$

27. $2x^2 + 2y^2 - 12x - 16y + 26 = 0$

28. $3x^2 + 3y^2 + 12x + 6y = 0$

29. $2x^2 + 3y^2 - 12x - 24y + 54 = 0$

30. $x^2 + 2y^2 + 4x - 4y = 8$

31. $x^2 - 3y^2 + 4x + 12y = 20$

32. $2x^2 + 16x = y^2 - 6y - 55$

In Exercises 33–36, find the equation of the hyperbola that satisfies the given conditions.

33. Center $(-2, 3)$; vertex $(-2, 1)$; passing through $(-2 + 3\sqrt{10}, 11)$.

34. Center $(-5, 1)$; vertex $(-3, 1)$; passing through $\left(-1, 1 - 4\sqrt{3}\right)$.

35. Center $(4, 2)$; vertex $(7, 2)$; asymptote $3y = 4x - 10$.

36. Center $(-3, -5)$; vertex $(-3, 0)$; asymptote $6y = 5x - 15$.

In Exercises 37–42, determine which of the following equations could possibly have the given graph.

$\dfrac{(y - 2)^2}{4} - \dfrac{(x - 3)^2}{9} = 1$, $\dfrac{(x + 3)^2}{3} - \dfrac{(y + 3)^2}{4} = 1$,

$4x^2 - 2y^2 = 8$, $9(y - 2)^2 = 36 + 4(x + 3)^2$,

$3(y + 3)^2 = 4(x - 3)^2 - 12$, $y^2 - 2x^2 = 6$.

37.

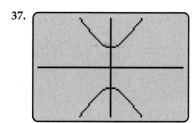

38.

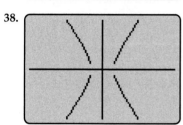

39.

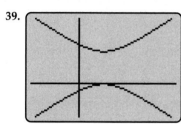

40.

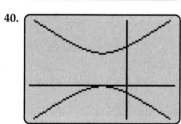

41.

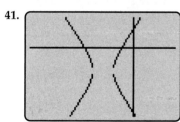

42.

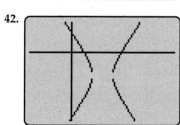

43. Sketch the graph of $\dfrac{y^2}{4} - \dfrac{x^2}{b^2} = 1$ for $b = 2$, $b = 4$, $b = 8$, $b = 12$, and $b = 20$. What happens to the

hyperbola as b takes larger and larger values? Could the graph ever degenerate into a pair of horizontal lines?

44. Find a number k such that $(-2, 1)$, is on the graph of $3x^2 + ky^2 = 4$. Then graph the equation.

45. Show that the asymptotes of the hyperbola $\dfrac{x^2}{a^2} - \dfrac{y^2}{a^2} = 1$ are perpendicular to each other.

46. Find the approximate coordinates of the points where these hyperbolas intersect:
$$\frac{(x-1)^2}{4} - \frac{(y+1)^2}{8} = 1 \quad \text{and} \quad 4y^2 - x^2 = 1.$$

47. Two listening stations 1 mile apart record an explosion. One microphone receives the sound 2 seconds after the other does. Use the line through the microphones as the x-axis, with the origin midway between the microphones, and the fact that sound travels at 1100 feet/second to find the equation of a hyperbola on which the explosion is located. Can you determine the exact location of the explosion?

48. Two transmission stations P and Q are located 200 miles apart on a straight shoreline. A ship 50 miles from shore is moving parallel to the shoreline. A signal from Q reaches the ship 400 microseconds after a signal from P. If the signals travel at 980 feet/microsecond, find the location of the ship (in terms of miles) in the coordinate system with x-axis through P and Q, and origin midway between them.

*If $a > 0$ and $b > 0$, then the **eccentricity** of the hyperbola*
$$\frac{(x-h)^2}{a^2} - \frac{(y-k)^2}{b^2} = 1 \quad \text{or} \quad \frac{(y-k)^2}{a^2} - \frac{(x-h)^2}{b^2} = 1$$

is the number $\dfrac{\sqrt{a^2 + b^2}}{a}$. In Exercises 49–53, find the eccentricity of the hyperbola whose equation is given.

49. $\dfrac{(x-6)^2}{10} - \dfrac{y^2}{40} = 1$ **50.** $\dfrac{y^2}{18} - \dfrac{x^2}{25} = 1$

51. $6(y-2)^2 = 18 + 3(x+2)^2$

52. $16x^2 - 9y^2 - 32x + 36y + 124 = 0$

53. $4x^2 - 5y^2 - 16x - 50y + 71 = 0$

54. (a) Graph these hyperbolas (on the same screen if possible):
$$\frac{y^2}{4} - \frac{x^2}{1} = 1 \qquad \frac{y^2}{4} - \frac{x^2}{12} = 1 \qquad \frac{y^2}{4} - \frac{x^2}{96} = 1$$

(b) Compute the eccentricity of each hyperbola in part (a).

(c) Based on parts (a) and (b), how is the shape of a hyperbola related to its eccentricity?

8.3 Parabolas

Parabolas appeared in Section 4.1 as the graphs of quadratic functions. Parabolas of this kind are a special case of the following more general definition. Let L be a line in the plane and P a point not on L. If X is any point not on L, the distance from X to L is defined to be the length of the perpendicular line segment from X to L. The **parabola** with focus P and **directrix** L is the set of all points X such that

Distance from X to P = Distance from X to L

as shown in Figure 8–24.

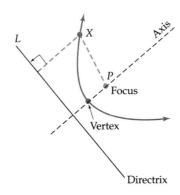

Figure 8–24

The line through P perpendicular to L is called the **axis**. The intersection of the axis with the parabola (the midpoint of the segment of the axis from P to L) is the **vertex** of the parabola, as illustrated in Figure 8–24. The parabola is symmetric with respect to its axis.

Suppose that the focus is on the y-axis at the point $(0, p)$, where p is a nonzero constant and that the directrix is the horizontal line $y = -p$. If (x, y) is any point on the parabola, then the distance from (x, y) to the horizontal line $y = -p$ is the length of the vertical segment from (x, y) to $(x, -p)$, as shown in Figure 8–25.

By the definition of the parabola,

Distance from (x, y) to $(0, p)$ = Distance from (x, y) to $y = -p$

Distance from (x, y) to $(0, p)$ = Distance from (x, y) to $(x, -p)$

$$\sqrt{(x - 0)^2 + (y - p)^2} = \sqrt{(x - x)^2 + [y - (-p)]^2}.$$

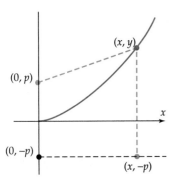

Figure 8–25

Squaring both sides and simplifying, we have

$$(x - 0)^2 + (y - p)^2 = (x - x)^2 + (y + p)^2$$
$$x^2 + y^2 - 2py + p^2 = 0^2 + y^2 + 2py + p^2$$
$$x^2 = 4py.$$

Conversely, it can be shown that every point whose coordinates satisfy this equation is on the parabola.

A similar argument works for the parabola with focus $(p, 0)$ on the x-axis and directrix the vertical line $x = -p$, and leads to this conclusion:

Standard Equations of Parabolas with Vertex at the Origin

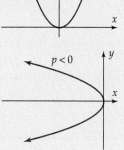

Let p be a nonzero real number. Then the graph of each of the following equations is a parabola with vertex at the origin.

$$x^2 = 4py \begin{cases} \text{focus: } (0, p) \\ \text{directrix: } y = -p \\ \text{axis: } y\text{-axis} \\ \text{opens upward if } p > 0, \text{ downward if } p < 0 \end{cases}$$

$$y^2 = 4px \begin{cases} \text{focus: } (p, 0) \\ \text{directrix: } x = -p \\ \text{axis: } x\text{-axis} \\ \text{opens to right if } p > 0, \text{ to left if } p < 0 \end{cases}$$

Example 1 Show that the graph of the equation is a parabola and find its focus and directrix; then sketch the graph:

(a) $y = -x^2/8$ (b) $x = 3y^2$

Solution

(a) We rewrite the equation so that it matches one of the forms in the preceding box:

$$y = -\frac{x^2}{8}$$

Multiply both sides by -8: $-8y = x^2$

The equation $x^2 = -8y$ is of the form $x^2 = 4px$, with $4p = -8$, so that $p = -2$. Hence, the graph is a downward-opening parabola with focus $(0, p) = (0, -2)$ and directrix $y = -p = -(-2) = 2$, as shown in Figure 8–26.

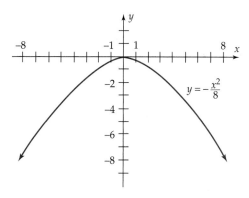

Figure 8–26

(b)   Divide both sides of $x = 3y^2$ by 3, so that the equation becomes

$$y^2 = \frac{x}{3}$$

$$y^2 = \frac{1}{3}x$$

This equation is of the form $y^2 = 4px$, with $4p = 1/3$, so that $p = 1/12$. Therefore, the graph is a parabola with focus $(1/12, 0)$ and directrix $x = -1/12$ that opens to the right. To sketch its graph, we solve the equation $y^2 = x/3$ for y:

$$y = \sqrt{\frac{x}{3}} \quad \text{or} \quad y = -\sqrt{\frac{x}{3}}.$$

Graphing both of these equations on the same screen produces Figure 8–27. ■

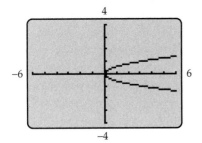

Figure 8–27

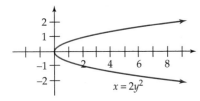

Figure 8–28

Example 2   Find the focus, directrix, and equation of the parabola that passes through the point (8, 2), has vertex (0, 0), and focus on the x-axis.

Solution   The equation is of the form $y^2 = 4px$. Since (8, 2) is on the graph, we have $2^2 = 4p \cdot 8$, so that $p = \frac{1}{8}$. Therefore, the focus is $\left(\frac{1}{8}, 0\right)$ and the directrix is the vertical line $x = -\frac{1}{8}$. The equation is $y^2 = 4\left(\frac{1}{8}\right)x = \frac{1}{2}x$, or equivalently, $x = 2y^2$. Its graph is sketched in Figure 8–28. ■

Vertical and Horizontal Shifts

Let h and k be constants. If we replace x by $x - h$ and y by $y - k$ in the equation of a parabola with vertex at the origin, then the graph of the new equation can be obtained by shifting the parabola vertically and horizontally so that its vertex is at (h, k), as explained on page 590. Using this fact and translating the information in the box on page 609, we obtain:

***Standard
Equations of
Parabolas
with Center
at (h, k)***

If p is a nonzero real number, then the graph of each of the following equations is a parabola with vertex (h, k).

$(x - h)^2 = 4p(y - k)$ $\begin{cases} \text{focus: } (h, k + p) \\ \text{directrix: the horizontal line } y = k - p \\ \text{axis: the vertical line } x = h \\ \text{opens upward if } p > 0, \text{downward if } p < 0 \end{cases}$

$(y - k)^2 = 4p(x - h)$ $\begin{cases} \text{focus: } (h + p, k) \\ \text{directrix: the vertical line } x = h - p \\ \text{axis: the horizontal line } y = k \\ \text{opens to right if } p > 0, \text{to left if } p < 0 \end{cases}$

Example 3 Identify and sketch the graph of $y = 2(x - 3)^2 + 1$.

Solution We first rewrite the equation:

$$y = 2(x - 3)^2 + 1$$

Subtract 1 from both sides $\qquad y - 1 = 2(x - 3)^2$

Divide both sides by 2: $\qquad \dfrac{1}{2}(y - 1) = (x - 3)^2$

$$(x - 3)^2 = \dfrac{1}{2}(y - 1)$$

This is the first form in the preceding box, with $h = 3, k = 1,$ and $4p = 1/2$. Hence, $p = 1/8$ and the graph is an upward-opening parabola with vertex $(3, 1)$, focus $(3, 1 + 1/8) = (3, 9/8)$, and directrix the horizontal line $y = 1 - 1/8 = 7/8$. The graph of

$$(x - 3)^2 = \dfrac{1}{2}(y - 1) \qquad \text{or equivalently,} \qquad y - 1 = 2(x - 3)^2$$

is the graph of $y = 2x^2$ shifted 3 units to the right and 1 unit upward, as shown in Figure 8–29. ∎

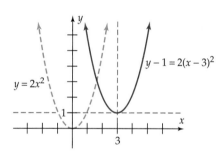

Figure 8–29

Example 4 Graph the equation $x = 5y^2 + 30y + 41$ without putting it in standard form.

Solution There are two methods of graphing this equation on a calculator.

Method 1. Rewrite the equation as

$$5y^2 + 30y + 41 - x = 0.$$

This is a quadratic equation of the form $ay^2 + by + c = 0$, with $a = 5$, $b = 30$, and $c = 41 - x$. It can be solved by using the quadratic formula:

$$y = \frac{-30 \pm \sqrt{30^2 - 4 \cdot 5(41 - x)}}{2 \cdot 5} = \frac{-30 \pm \sqrt{900 - 20(41 - x)}}{10}.$$

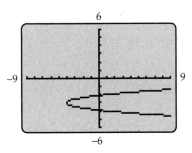

Figure 8–30

Now graph both

$$y = \frac{-30 + \sqrt{900 - 20(41 - x)}}{10} \quad \text{and} \quad y = \frac{-30 - \sqrt{900 - 20(41 - x)}}{10}$$

on the same screen to obtain the parabola in Figure 8–30.

Method 2. Since the equation $x = 5y^2 + 30y + 41$ defines x as a function of y, we can use parametric graphing, as explained in Excursion 3.3.A on page 217. The parametric equations are

$$x = 5t^2 + 30t + 41$$

$$y = t.$$

Graphing these equations in parametric mode also produces Figure 8–30. ■

Example 5   Find the vertex, focus, and directrix of the parabola

$$x = 5y^2 + 30y + 41$$

that was graphed in Example 4.

Solution   We first rewrite the equation:

$$5y^2 + 30y + 41 = x$$

Subtract 41 from both sides: $\qquad 5y^2 + 30y = x - 41$

Factor out 5 on left side: $\qquad 5(y^2 + 6y) = x - 41$

Complete the square* on the expression $y^2 + 6y$ by adding 9 (the square of half the coefficient of y):

$$5(y^2 + 6y + 9) = x - 41 + ?.$$

On the left side we have actually added $5 \cdot 9 = 45$, so we must add the same amount to the right side:

$$5(y^2 + 6y + 9) = x - 41 + 45$$

Factor left side: $\qquad 5(y + 3)^2 = x + 4$

Divide both sides by 5: $\qquad (y + 3)^2 = \frac{1}{5}(x + 4)$

$$[y - (-3)]^2 = \frac{1}{5}[x - (-4)]$$

Thus, the graph is a parabola with vertex $(-4, -3)$. In this case, $4p = 1/5$, so that $p = 1/20 = .05$. Hence the focus is $(-4 + .05, -3) = (-3.95, -3)$ and the directrix is $x = -4 - .05 = -4.05$. ■

Applications

Certain laws of physics show that sound waves or light rays from a source at the focus of a parabola will reflect off the parabola in rays parallel to

---

*Completing the square is explained in Example 1 on page 585.

the axis of the parabola, as shown in Figure 8–31. This is the reason that parabolic reflectors are used in automobile headlights and searchlights.

Conversely, a light ray coming toward a parabola will be reflected into the focus, as shown in Figure 8–32. This fact is used in the design of radar antennas, satellite dishes, and field microphones used at outdoor sporting events to pick up conversation on the field.

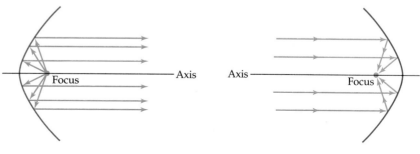

Figure 8–31   **Figure 8–32**

Projectiles follow a parabolic curve, a fact that is used in the design of water slides in which the rider slides down a sharp incline, then up and over a hill, before plunging downward into a pool. At the peak of the hill, the rider shoots up along a parabolic arc several inches above the slide, experiencing a sensation of weightlessness.

Example 6   The radio telescope in Figure 8–33 has the shape of a parabolic dish (a cross section through the center of the dish is a parabola). It is 30 feet deep at the center and has a diameter of 200 feet. How far from the vertex of the parabolic dish should the receiver be placed in order to "catch" all the light rays that hit the dish?

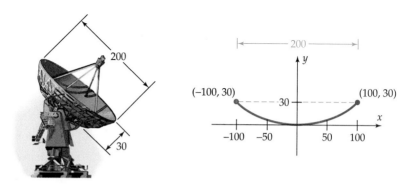

Figure 8–33   **Figure 8–34**

Solution   Light rays hitting the dish are reflected into the focus, as explained above. So the radio receiver must be located at the focus. To find the focus, draw a cross section of the dish, with vertex at the origin, as in Figure 8–34. The equation of this parabola is of the form $x^2 = 4py$. Since the point (100, 30) is on the parabola, we have

$$x^2 = 4py$$

Substitute: $$100^2 = 4p(30)$$

Simplify: $$120p = 100^2$$

Divide both sides by 120: $$p = \frac{100^2}{120}$$

Simplify: $$p = \frac{250}{3}.$$

As we saw in the box on page 609, the focus is the point $(0, p)$, which is p units from the vertex $(0, 0)$. Therefore, the receiver should be placed $250/3 \approx 83.33$ feet from the vertex. ■

Exercises 8.3

In Exercises 1–6, determine which of the following equations could possibly have the given graph.

$y = x^2/4,$ $x^2 = -8y,$ $6x = y^2,$ $y^2 = -4x,$
$2x^2 + y^2 = 12,$ $x^2 + 6y^2 = 18,$
$6y^2 - x^2 = 6,$ $2x^2 - y^2 = 8$

1.
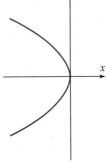

2.
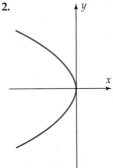

3.
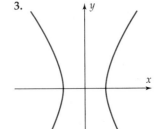

4.
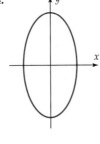

5.
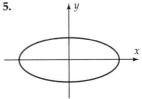

6.
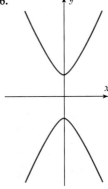

In Exercises 7–10, find the focus and directrix of the parabola.

7. $y = 3x^2$ **8.** $x = .5y^2$
9. $y = .25x^2$ **10.** $x = -6y^2$

In Exercises 11–22, determine the vertex, focus, and directrix of the parabola without graphing *and state whether it opens upward, downward, left, or right.*

11. $x - y^2 = 2$ **12.** $y - 3 = x^2$
13. $x + (y + 1)^2 = 2$ **14.** $y + (x + 2)^2 = 3$
15. $3x - 2 = (y + 3)^2$
16. $2x - 1 = -6(y + 1)^2$
17. $x = y^2 - 9y$
18. $x = y^2 + y + 1$

19. $y = 3x^2 + x - 4$

20. $y = -3x^2 + 4x - 1$

21. $y = -3x^2 + 4x + 5$

22. $y = 2x^2 - x - 1$

In Exercises 23–30, sketch the graph of the equation and label the vertex.

23. $y = 4(x - 1)^2 + 2$ **24.** $y = 3(x - 2)^2 - 3$

25. $x = 2(y - 2)^2$

26. $x = -3(y - 1)^2 - 2$

27. $y = x^2 - 4x - 1$ **28.** $y = x^2 + 8x + 6$

29. $y = x^2 + 2x$ **30.** $x = y^2 - 3y$

In Exercises 31–42, find the equation of the parabola satisfying the given conditions.

31. Vertex $(0, 0)$; axis $x = 0$; $(2, 12)$ on graph.

32. Vertex $(0, 1)$; axis $x = 0$; $(2, -7)$ on graph.

33. Vertex $(1, 0)$; axis $x = 1$; $(2, 13)$ on graph.

34. Vertex $(-3, 0)$; axis $y = 0$; $(-1, 1)$ on graph.

35. Vertex $(2, 1)$; axis $y = 1$; $(5, 0)$ on graph.

36. Vertex $(1, -3)$; axis $y = -3$; $(-1, -4)$ on graph.

37. Vertex $(-3, -2)$; focus $(-47/16, -2)$.

38. Vertex $(-5, -5)$; focus $(-5, -99/20)$.

39. Vertex $(1, 1)$; focus $(1, 9/8)$.

40. Vertex $(-4, -3)$; $(-6, -2)$ and $(-6, -4)$ on graph.

41. Vertex $(-1, 3)$; $(8, 0)$ and $(0, 4)$ on graph.

42. Vertex $(1, -3)$; $(0, -1)$ and $(-1, 5)$ on graph.

In Exercises 43–50, identify the conic section whose equation is given, list its vertex, and find its graph.

43. $x^2 = 6x - y - 5$

44. $y^2 = x - 2y - 2$

45. $3y^2 = x - 1 + 2y$

46. $2y^2 = x - 4y - 5$

47. $3x^2 + 3y^2 - 6x - 12y - 6 = 0$

48. $2x^2 + 3y^2 + 12x - 6y + 9 = 0$

49. $2x^2 - y^2 + 16x + 4y + 24 = 0$

50. $4x^2 - 40x - 2y + 105 = 0$

In Exercises 51–56, determine which of the following equations could possibly have the given graph.

$$y = (x + 5)^2 - 3, \qquad x = (y + 3)^2 + 2,$$
$$y = (x - 4)^2 + 2, \qquad x = -(y - 3)^2 - 2,$$
$$y^2 = 4y + x - 1, \qquad y = -x^2 - 8x - 18.$$

51.

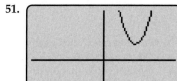

52.

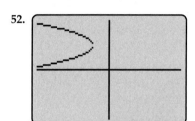

53.

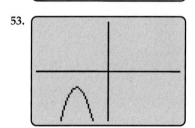

54.

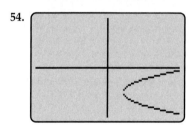

55.

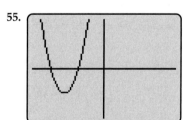

56.

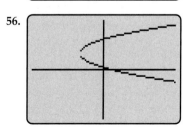

57. Find the number b such that the vertex of the parabola $y = x^2 + bx + c$ lies on the y-axis.

58. Find the number d such that the parabola $(y + 1)^2 = dx + 4$ passed through $(-6, 3)$.

59. Find the points of intersection of the parabola $4y^2 + 4y = 5x - 12$ and the line $x = 9$.

60. Find the points of intersection of the parabola $4x^2 - 8x = 2y + 5$ and the line $y = 15$.

61. A parabolic satellite dish is 4 feet in diameter and 1.5 feet deep. How far from the vertex should the receiver be placed in order to catch all the signals that hit the dish?

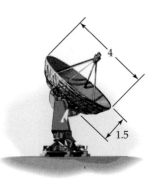

62. A radio telescope has a parabolic dish with a diameter of 300 feet. Its receiver (focus) is located 130 feet from the vertex. How deep is the dish at its center? [*Hint:* Position the dish as in Figure 8–34 and find the equation of the parabola.]

63. A flashlight has a parabolic reflector that is 3 inches in diameter and 1.5 inches deep. For the light from the bulb to reflect in beams that are parallel to the center axis of the flashlight, how far

from the vertex of the reflector should the bulb be located? [*Hint:* See Figure 8–31 and the preceding discussion.]

64. A large spotlight has a parabolic reflector that is 3 feet deep at its center. The light source is located $1\frac{1}{3}$ feet from the vertex. What is the diameter of the reflector?

65. The cables of a suspension bridge are shaped like parabolas. The cables are attached to the towers 100 feet from the bridge surface and the towers are 420 feet apart. The cables touch the bridge surface at the center (midway between the towers). At a point on the bridge, 100 feet from one of the towers, how far is the cable from the bridge surface?

66. At a point 120 feet from the center of a suspension bridge, the cables are 24 feet above the bridge surface. Assume the cables are shaped like parabolas and touch the bridge surface at the center (which is midway between the towers). If the towers are 600 feet apart, how far above the surface of the bridge are the cables attached to the towers?

EXCURSION Rotations and Second-Degree Equations

A **second-degree equation** in x and y is one that can be written in the form

$$Ax^2 + Bxy + Cy^2 + Dx + Ey + F = 0$$

for some constants A, B, C, D, E, F with at least one of A, B, C nonzero.

Example 1 Show that each of the following conic sections is the graph of a second-degree equation:

(a) Ellipse: $\dfrac{x^2}{6} + \dfrac{y^2}{5} = 1$ (b) Hyperbola: $\dfrac{(x+1)^2}{4} - \dfrac{(y-3)^2}{6} = 1$

Solution We need only show that each of these equations is in fact a second-degree equation. In each case, eliminate denominators, multiply out all terms, and gather them on one side of the equal sign.

(a)

$$\frac{x^2}{6} + \frac{y^2}{5} = 1$$

Multiply both sides by 30: $\qquad 5x^2 + 6y^2 = 30$

Rearrange terms: $\qquad 5x^2 + 6y^2 - 30 = 0$

This equation is a second-degree equation because it has the form
$$Ax^2 + Bxy + Cy^2 + Dx + Ey + F = 0$$
with $A = 5$, $B = 0$, $C = 6$, $D = 0$, $E = 0$, and $F = -30$.

(b)

$$\frac{(x + 1)^2}{4} - \frac{(y - 3)^2}{6} = 1$$

Multiply both sides by 12: $\qquad 3(x + 1)^2 + 2(y - 3)^2 = 12$

Multiply out left side: $\qquad 3(x^2 + 2x + 1) + 2(y^2 - 6y + 9) = 12$

$$3x^2 + 6x + 3 + 2y^2 - 12y + 18 = 12$$

Rearrange terms: $\qquad 3x^2 + 2y^2 + 6x - 12y + 9 = 0$

This is a second-degree equation with $A = 3$, $B = 0$, $C = 2$, $D = 6$, $E = -12$, and $F = 9$. ∎

Calculations like those in Example 1 can be used on the equation of any conic section to show that it is the graph of a second-degree equation. Conversely, it can be shown that

> **The graph of every second-degree equation is a conic section**

(possibly degenerate). When the second-degree equation has no xy-term (that is, $B = 0$), as was the case in Example 1, the graph is a conic section in standard position (axis or axes parallel to the coordinate axes). When $B \neq 0$, however, the conic is rotated from standard position, so that its axis or axes are not parallel to the coordinate axes.

Example 2 Graph the equation
$$3x^2 + 6xy + y^2 + x - 2y + 7 = 0.$$

Solution We first rewrite it as:
$$y^2 + 6xy - 2y + 3x^2 + x + 7 = 0$$
$$y^2 + (6x - 2)y + (3x^2 + x + 7) = 0.$$

This equation has the form $ay^2 + by + c = 0$, with $a = 1$, $b = 6x - 2$, and $c = 3x^2 + x + 7$. It can be solved with the quadratic formula:

$$y = \frac{-b \pm \sqrt{b^2 - 4ac}}{2a} = \frac{-(6x - 2) \pm \sqrt{(6x - 2)^2 - 4 \cdot 1 \cdot (3x^2 + x + 7)}}{2 \cdot 1}.$$

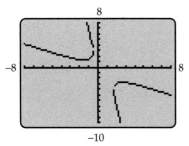

Figure 8–35

The top half of the graph is obtained by graphing

$$y = \frac{-6x + 2 + \sqrt{(6x - 2)^2 - 4(3x^2 + x + 7)}}{2}$$

and the bottom half by graphing

$$y = \frac{-6x + 2 - \sqrt{(6x - 2)^2 - 4(3x^2 + x + 7)}}{2}.$$

The graph is a hyperbola whose focal axis tilts upward to the left, as shown in Figure 8–35. ■

The Discriminant

The following fact, whose proof requires trigonometry, makes it easy to identify the graphs of second-degree equations without actually graphing them.

Graphs of Second-Degree Equations

> If the equation
>
> $$Ax^2 + Bxy + Cy^2 + Dx + Ey + F = 0 \quad (A, B, C \text{ not all } 0)$$
>
> has a graph, then that graph is
>
> A circle or an ellipse (or a point), if $B^2 - 4AC < 0$;
>
> A parabola (or a line or two parallel lines), if $B^2 - 4AC = 0$;
>
> A hyperbola (or two intersecting lines), if $B^2 - 4AC > 0$.

The expression $B^2 - 4AC$ is called the **discriminant** of the equation.

Example 3 Identify the graph of

$$2x^2 - 4xy + 3y^2 + 5x + 6y - 8 = 0.$$

Solution We compute the discriminant with $A = 2$, $B = -4$, and $C = 3$:

$$B^2 - 4AC = (-4)^2 - 4 \cdot 2 \cdot 3 = 16 - 24 = -8.$$

Hence, the graph is an ellipse (possibly a circle or a single point). To find the graph, we rewrite the equation as:

$$3y^2 - 4xy + 6y + 2x^2 + 5x - 8 = 0$$
$$3y^2 + (-4x + 6)y + (2x^2 + 5x - 8) = 0.$$

The equation has the form $ay^2 + by + c = 0$ and can be solved by the quadratic formula:

$$y = \frac{-b \pm \sqrt{b^2 - 4ac}}{2a}$$

$$= \frac{-(-4x + 6) \pm \sqrt{(-4x + 6)^2 - 4 \cdot 3 \cdot (2x^2 + 5x - 8)}}{2 \cdot 3}.$$

The graph can now be found by graphing the last two equations on the same screen.

GRAPHING EXPLORATION

Find a viewing window that shows a complete graph of the equation. In what direction does the major axis run? ■

Example 4 The discriminant of

$$3x^2 + 6xy + 3y^2 + 13x + 9y + 53 = 0$$

is $B^2 - 4AC = 6^2 - 4 \cdot 3 \cdot 3 = 0$. Hence, the graph is a parabola (or a line or parallel lines in the degenerate case).

GRAPHING EXPLORATION

Find a viewing window that shows a complete graph of the equation. ■

Exercises 8.3.A

In Exercises 1–6, assume the graph of the equation is a non-degenerate conic section. Without graphing, determine whether the graph is a circle, ellipse, hyperbola, or parabola.

1. $x^2 - 2xy + 3y^2 - 1 = 0$ **2.** $xy - 1 = 0$

3. $x^2 + 2xy + y^2 + 2\sqrt{2}x - 2\sqrt{2}y = 0$

4. $2x^2 - 4xy + 5y^2 - 6 = 0$

5. $17x^2 - 48xy + 31y^2 + 50 = 0$

6. $2x^2 - 4xy - 2y^2 + 3x + 5y - 10 = 0$

In Exercises 7–24, use the discriminant to identify the conic section whose equation is given and find a viewing window that shows a complete graph.

7. $9x^2 + 4y^2 + 54x - 8y + 49 = 0$

8. $4x^2 + 5y^2 - 8x + 30y + 29 = 0$

9. $4y^2 - x^2 + 6x - 24y + 11 = 0$

10. $x^2 - 16y^2 = 0$

11. $3y^2 - x - 2y + 1 = 0$

12. $x^2 - 6x + y + 5 = 0$

13. $41x^2 - 24xy + 34y^2 - 25 = 0$

14. $x^2 + 2\sqrt{3}xy + 3y^2 + 8\sqrt{3}x - 8y + 32 = 0$

15. $17x^2 - 48xy + 31y^2 + 49 = 0$

16. $52x^2 - 72xy + 73y^2 = 200$

17. $9x^2 + 24xy + 16y^2 + 90x - 130y = 0$

18. $x^2 + 10xy + y^2 + 1 = 0$

19. $23x^2 + 26\sqrt{3}xy - 3y^2 - 16x + 16\sqrt{3}y + 128 = 0$

20. $x^2 + 2xy + y^2 + 12\sqrt{2}x - 12\sqrt{2}y = 0$

21. $17x^2 - 12xy + 8y^2 - 80 = 0$

22. $11x^2 - 24xy + 4y^2 + 30x + 40y - 45 = 0$

23. $3x^2 + 2\sqrt{3}xy + y^2 + 4x - 4\sqrt{3}y - 16 = 0$

24. $3x^2 + 2\sqrt{2}xy + 2y^2 - 12 = 0$

Chapter 8 *Review*

Important Concepts

Important Facts and Formulas

- Equation of ellipse with center (h, k) and axes on the lines $x = h$, $y = k$:

$$\frac{(x - h)^2}{a^2} + \frac{(y - k)^2}{b^2} = 1$$

- Equation of hyperbola with center (h, k) and vertices on the line $y = k$:

$$\frac{(x - h)^2}{a^2} - \frac{(y - k)^2}{b^2} = 1$$

- Equation of hyperbola with center (h, k) and vertices on the line $x = h$:

$$\frac{(y - k)^2}{a^2} - \frac{(x - h)^2}{b^2} = 1$$

- Equation of a parabola with vertex (h, k) and axis $x = h$:

$$(x - h)^2 = 4p(y - k)$$

- Equation of a parabola with vertex (h, k) and axis $y = k$:

$$(y - k)^2 = 4p(x - h)$$

- The discriminant of the equation $Ax^2 + Bxy + Cy^2 + Dx + Ey + F = 0$ (where A, B, C are not all zero) is $B^2 - 4AC$.

  If $B^2 - 4AC < 0$, the graph is a circle or ellipse (or a point).

  If $B^2 - 4AC = 0$, the graph is a parabola (or a line or two parallel lines).

  If $B^2 - 4AC > 0$, the graph is a hyperbola (or two intersecting lines).

Review Questions

In Questions 1–4, find the foci and vertices of the conic and state whether it is an ellipse or a hyperbola.

1. $\dfrac{x^2}{16} + \dfrac{y^2}{20} = 1$

2. $\dfrac{x^2}{9} - \dfrac{y^2}{16} = 1$

3. $\dfrac{(x-1)^2}{7} + \dfrac{(y-3)^2}{16} = 1$

4. $3x^2 = 1 + 2y^2$

5. Find the focus and directrix of the parabola $10y = 7x^2$.

6. Find the focus and directrix of the parabola

$$3y^2 - x - 4y + 4 = 0.$$

In Questions 7–20, sketch the graph of the equation. If there are asymptotes, give their equations.

7. $\dfrac{x^2}{4} + \dfrac{y^2}{25} = 1$

8. $25x^2 + 4y^2 = 100$

9. $\dfrac{(x-3)^2}{9} + \dfrac{(y+5)^2}{4} = 1$

10. $\dfrac{x^2}{9} - \dfrac{y^2}{16} = 1$

11. $\dfrac{(y+4)^2}{25} - \dfrac{(x-1)^2}{4} = 1$

12. $4x^2 - 9y^2 = 144$

13. $x^2 + 4y^2 - 10x + 9 = 0$

14. $9x^2 - 4y^2 - 36x + 24y - 36 = 0$

15. $2y = 4(x-3)^2 + 6$

16. $3y = 6(x+1)^2 - 9$

17. $x = y^2 + 2y + 2$

18. $y = x^2 - 2x + 3$

19. $x^2 + y^2 - 6x + 5 = 0$

20. $x^2 + y^2 - 4x + 6y + 4 = 0$

21. Find the center and radius of the circle whose equation is

$$x^2 + y^2 + 8x + 10y + 33 = 0.$$

22. Find the equation of the circle with center $(-2, 3)$ that passes through the point $(1, 7)$.

23. What is the center of the ellipse $4x^2 + 3y^2 - 32x + 36y + 124 = 0$?

24. Find the equation of the ellipse with center at the origin, one vertex at $(0, 4)$, passing through $\left(\sqrt{3}, 2\sqrt{3}\right)$.

25. Find the equation of the ellipse with center at $(3, 1)$, one vertex at $(1, 1)$, passing through $\left(2, 1 + \sqrt{3/2}\right)$.

26. Find the equation of the hyperbola with center at the origin, one vertex at $(0, 5)$, passing through $\left(1, 3\sqrt{5}\right)$.

27. Find the equation of the hyperbola with center at $(3, 0)$, one vertex at $(3, 2)$, passing through $\left(1, \sqrt{5}\right)$.

28. Find the equation of the parabola with vertex $(2, 5)$, axis $x = 2$, and passing through $(3, 12)$.

29. Find the equation of the parabola with vertex $(3/2, -1/2)$, axis $y = -1/2$, and passing through $(-3, 1)$.

30. Find the equation of the parabola with vertex $(5, 2)$ that passes through the points $(7, 3)$ and $(9, 6)$.

31. The arch shown in the figure has the shape of half of an ellipse. How wide is the arch at a point 8 feet above the ground? [*Hint:* Think of the arch as sitting on the *x*-axis, with its center at the origin, and find its equation.]

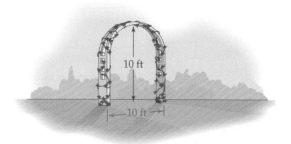

32. A satellite dish with a parabolic cross section is 6 feet in diameter. The receiver is located on the center axis, 1 foot from the base of the dish. How deep is the dish at its center?

In Questions 33–36, assume that the graph of the equation is a nondegenerate conic. Use the discriminant to identify the graph.

33. $3x^2 + 2\sqrt{2}xy + 2y^2 - 12 = 0$

34. $x^2 + y^2 - xy - 4y = 0$ **35.** $4xy - 3x^2 - 20 = 0$

36. $4x^2 - 4xy + y^2 - \sqrt{5}x - 2\sqrt{5}y = 0$

In Questions 37–42, find a viewing window that shows a complete graph of the equation.

37. $x^2 - xy + y^2 - 6 = 0$ **38.** $x^2 + xy + y^2 - 3y - 6 = 0$

39. $x^2 + xy - 2 = 0$ **40.** $x^2 - 4xy + y^2 + 5 = 0$

41. $x^2 + 3xy + y^2 - 2\sqrt{2}x + 2\sqrt{2}y = 0$

42. $x^2 + 2xy + y^2 - 4\sqrt{2}y = 0$

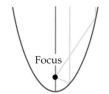

Focus

Designing Light Fixtures

Lighting fixtures are often designed using a parabolic reflector that focuses the light into a relatively parallel beam. This kind of light is referred to as a spotlight, since the beam of light can create a bright circular pool of light on a wall or floor. The effect is created when the light source is placed at the focus deep inside the parabola, sending most of the rays of light in the forward direction.

When the focus is placed closer to the top of the reflector, lots of light leaks out the side, creating a light source with two intensities. This is a popular style with many design advantages; you can accent art objects or create pools of light with a bright center for reading and soft light to surround a chair. We'll explore the factors in designing these lighting fixtures.

To make the job simpler, orient the parabolic reflector so that the axis of symmetry is the y-axis and the vertex sits at the origin. Using the equation $x^2 = 4py$, we also know that the focus of the parabola is at the point $(0, p)$.

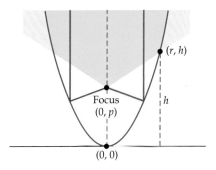

1. Explain why the equation for a nonvertical line through the focus of the parabola is of the form $y = mx + p$.

2. Find two expressions for the value of h, based on the equations of the two graphs that intersect at (r, h).

3. Suppose that you wanted a 4-inch-diameter light where the angle of the diffused light was 45°. This corresponds to a slope of $m = 1$ on the boundary line. How tall is the reflector? How far from the base should the light source (focal point) be?

4. A more focused light would have an angle of diffused light of 30°, corresponding to a slope of $m = 1.73$ on the boundary line. How tall is the reflector? How far from the base should the light source be?

A p p e n d i x

1

Geometry Review

An **angle** consists of two half-lines that begin at the same point P, as in Figure A–1. The point P is called the **vertex** of the angle and the half-lines the **sides** of the angle.

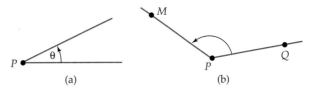

(a)

(b)

Figure A–1

An angle may be labeled by a Greek letter, such as angle θ in Figure A–1(a), or by listing three points (a point on one side, the vertex, a point on the other side), such as angle QPM in Figure A–1(b).

To measure the size of an angle, we must assign a number to each angle. Here is the classical method for doing this:

1. Construct a circle whose center is the vertex of the angle.
2. Divide the circumference of the circle into 360 equal parts (called **degrees**) by marking 360 points on the circumference, beginning with the point where one side of the angle intersects the circle. Label these points $0°$, $1°$, $2°$, $3°$, and so on.
3. The label of the point where the second side of the angle intersects the circle is the degree measure of the angle.

For example, Figure A–2 on the next page shows an angle θ of measure 25 degrees (in symbols, $25°$) and an angle β of measure $135°$.

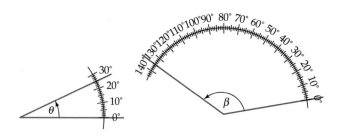

Figure A–2

An **acute angle** is an angle whose measure is strictly between 0° and 90°, such as angle θ in Figure A–2. A **right angle** is an angle that measures 90°. An **obtuse angle** is an angle whose measure is strictly between 90° and 180°, such as angle β in Figure A–2.

A **triangle** has three sides (straight line segments) and three angles, formed at the points where the various sides meet. When angles are measured in degrees,

> **The sum of the measures of all three angles of a triangle is *always* 180°.**

For instance, see Figure A–3.

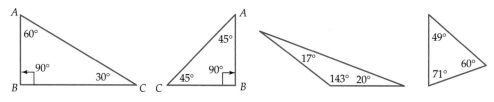

Figure A–3

A **right triangle** is a triangle, one of whose angles is a right angle, such as the first two triangles shown in Figure A–3. The side of a right triangle that lies opposite the right angle is called the **hypotenuse.** In each of the right triangles in Figure A–3, side AC is the hypotenuse.

Pythagorean Theorem

If the sides of a right triangle have lengths a and b and the hypotenuse has length c, then

$$c^2 = a^2 + b^2.$$

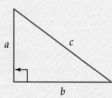

Example 1 Consider the right triangle with sides of lengths 5 and 12, as shown in Figure A–4.

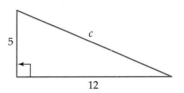

Figure A–4

According to the Pythagorean Theorem the length c of the hypotenuse satisfies the equation: $c^2 = 5^2 + 12^2 = 25 + 144 = 169$. Since $169 = 13^2$, we see that c must be 13. ■

Theorem I

If two angles of a triangle are equal, then the two sides opposite these angles have the same length.

Example 2 Suppose the hypotenuse of the right triangle shown in Figure A–5 has length 1 and that angles B and C measure 45° each.

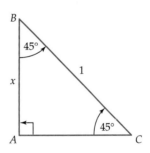

Figure A–5

Then by Theorem I, sides AB and AC have the same length. If x is the length of side AB, then by the Pythagorean Theorem:

$$x^2 + x^2 = 1^2$$

$$2x^2 = 1$$

$$x^2 = \frac{1}{2}$$

$$x = \sqrt{\frac{1}{2}} = \frac{1}{\sqrt{2}} = \frac{\sqrt{2}}{2}.$$

(We ignore the other solution of this equation, namely, $x = -\sqrt{1/2}$, since x represents a length here and thus must be nonnegative.) Therefore, the sides of a $90° - 45° - 45°$ triangle with hypotenuse 1 are each of length $\sqrt{2}/2$. ■

Theorem II

In a right triangle that has an angle of 30°, the length of the side opposite the 30° angle is one-half the length of the hypotenuse.

Example 3 Suppose that in the right triangle shown in Figure A–6 angle B is 30° and the length of hypotenuse BC is 2.

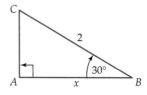

Figure A–6

By Theorem II the side opposite the 30° angle, namely, side AC, has length 1. If x denotes the length of side AB, then by the Pythagorean Theorem:

$$1^2 + x^2 = 2^2$$

$$x^2 = 3$$

$$x = \sqrt{3}. \quad \blacksquare$$

Example 4 The right triangle shown in Figure A–7 has a 30° angle at C, and side AC has length $\sqrt{3}/2$.

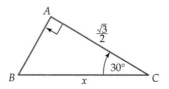

Figure A–7

Let x denote the length of the hypotenuse BC. By Theorem II, side AB has length $\frac{1}{2}x$. By the Pythagorean Theorem:

$$\left(\frac{1}{2}x\right)^2 + \left(\frac{\sqrt{3}}{2}\right)^2 = x^2$$

$$\frac{x^2}{4} + \frac{3}{4} = x^2$$

$$\frac{3}{4} = \frac{3}{4}x^2$$

$$x^2 = 1$$

$$x = 1.$$

Therefore, the triangle has hypotenuse of length 1 and sides of lengths $1/2$ and $\sqrt{3}/2$. ∎

Two triangles, as in Figure A–8, are said to be **similar** if their corresponding angles are equal (that is, $\angle A = \angle D$; $\angle B = \angle E$; and $\angle C = \angle F$). Thus, similar triangles have the same *shape* but not necessarily the same *size*.

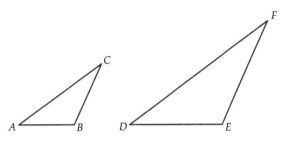

Figure A–8

Theorem III

Suppose triangle *ABC* with sides *a*, *b*, *c* is similar to triangle *DEF* with sides *d*, *e*, *f* (that is, $\angle A = \angle D$; $\angle B = \angle E$; $\angle C = \angle F$).

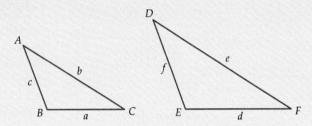

then

$$\frac{a}{d} = \frac{b}{e} = \frac{c}{f}.$$

These equalities are equivalent to:

$$\frac{a}{b} = \frac{d}{e}, \qquad \frac{b}{c} = \frac{e}{f}, \qquad \frac{a}{c} = \frac{d}{f}.$$

The equivalence of the equalities in the conclusion of the theorem is easily verified. For example, since

$$\frac{a}{d} = \frac{b}{e}$$

we have

$$ae = db.$$

Dividing both sides of this equation by be yields:

$$\frac{ae}{be} = \frac{db}{be}$$

$$\frac{a}{b} = \frac{d}{e}.$$

The other equivalences are proved similarly.

Example 5   Suppose the triangles in Figure A–9 are similar and that the sides have the lengths indicated.

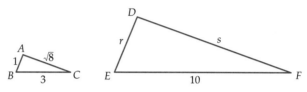

Figure A–9

Then by Theorem III,

$$\frac{\text{Length } AC}{\text{Length } DF} = \frac{\text{Length } BC}{\text{Length } EF}.$$

In other words,

$$\frac{\sqrt{8}}{s} = \frac{3}{10}$$

so that

$$3s = 10\sqrt{8}$$

$$s = \left(\frac{10}{3}\right)\sqrt{8}.$$

Similarly, by Theorem III,

$$\frac{\text{Length } AB}{\text{Length } DE} = \frac{\text{Length } BC}{\text{Length } EF}$$

so that

$$\frac{1}{r} = \frac{3}{10}$$

$$3r = 10$$

$$r = \frac{10}{3}.$$

Therefore the sides of triangle DEF are of lengths 10, $\frac{10}{3}$, and $\frac{10}{3}\sqrt{8}$.   ∎

2

Programs

The programs listed here are of two types: programs to give older calculators some of the features that are built-in on newer ones (such as a table maker) and programs to do specific tasks discussed in this book (such as synthetic division). Each program is preceded by a *Description*, which describes, in general terms, how the program operates and what it does. Some programs require that certain things be done before the program is run (such as entering a function in the function memory); these requirements are listed as *Preliminaries*. Occasionally, italic remarks appear in brackets after a program step; they are *not* part of the program, but are intended to provide assistance when you are entering the program into your calculator. A remark such as "[*MATH NUM menu*]" means that the symbols or commands needed for that step of the program are in the NUM submenu of the MATH menu.

Fraction Conversion (built-in on TI-82/83/85/86 and HP-38)

Description: Enter a repeating decimal; the program converts it into a fraction. The denominator is displayed on the last line and the numerator on the line above.

Sharp 9600

Input N   If fpart (round (N × D, 7)) ≠ 0 Goto 1   [*MATH NUM menu*]

$0 \to D$ int(N × D + .5) → N

Label 1   Print N

$D + 1 \to D$ Print D

Casio 9850

Fix 7   Rnd   [*MATH NUM menu*]

"N = "? → N   (Frac Ans) ≠ 0 ⇒ Goto 1   [*MATH NUM menu*]

$0 \to D$ (Ans + .5) → N

Lbl 1   Norm   [*DISP menu*]

$D + 1 \to D$ (Int N)◢

N × D   D

631

TI-82/83 Quadratic Formula (built-in on other calculators)

Description: Enter the coefficients of the quadratic equation $ax^2 + bx + c = 0$; the program finds all real solutions.

:ClrHome   [*Optional*]

:Disp "AX^2 + BX + C = 0"   [*Optional*]

:Prompt A

:Prompt B

:Prompt C

:(B^2 − 4AC) → S

:If S < 0

:Goto 1

:Disp $(-B + \sqrt{S})/(2A)$

:Disp $(-B - \sqrt{S})/(2A)$

:Stop

:Lbl 1

:Disp "NO REAL ROOTS"

TI-85 Table Maker (built-in on other calculators)

Preliminaries: Enter the function to be evaluated in the function memory as Y_1.

Description: Select a starting point and an increment (the amount by which adjacent x entries differ); the program displays a table of function values. To scroll through the table a page at a time, press "down" or "up." Press "quit" to end the program. [*Note:* An error message will occur if the calculator attempts to evaluate the function at a number for which it is not defined. In this case, change the starting point or increment to avoid the undefined point.]

:Lbl SETUP

:ClLCD   [*Optional*]

Disp " "TABLE SETUP"

:Input "TblMin = ", tblmin

:Input "ΔTbl = ", dtbl   [*CHAR GREEK menu*]

:tblmin → x   [*Use the x-var key for x.*]

:Lbl CONTD

:ClLCD

:Output(1,1,"x")   [*I/O menu*]

:Output(1,10,"y$_1$")

:For (cnt,2,7,1)

:Output(cnt,1,x)

Output(cnt,8," ")

:Output(cnt,9,y$_1$)

:x + dtbl → x

:End

:Menu(1, "Down",CONTD, 2,"UP",CONTU, 4, "Setup," SETUP, 5, "quit", TQUIT)

:Lbl CONTU

:x − 12∗dtbl → x

:Goto CONTD

:Lbl TQUIT

:ClLCD

:Stop

Synthetic Division (built-in on TI-89/92)

Preliminaries: Enter the coefficients of the dividend $F(x)$ (in order of decreasing powers of x, putting in zeros for missing coefficients) as list L_1 (or List 1). If the coefficients are 1, 2, 3, for example, key in {1, 2, 3,} and store it in L_1. The symbols { } are on the keyboard or in the LIST menu. The list name L_1 is on the keyboard of TI-82/83 and Sharp 9600. On TI-85/86 and HP-38, type in L1. On Casio 9850, use "List" in the LIST submenu of the OPTN menu to type in List 1.

Description: Write the divisor in the form $x - a$ and enter a. The program displays the degree of the quotient $Q(x)$, the coefficients of $Q(x)$ (in order of decreasing powers of x), and the remainder. If the program pauses before it has displayed all these items, you can use the arrow keys to scroll through the display; then press ENTER (or OK) to continue.

TI-82/83/85/86

```
:ClrHome [ClLCD on TI-85/86]
:Disp "DIVISOR IS X − A"
:Prompt A
:L₁ → L₂     [See Preliminaries for how to enter list names]
:dim L₁ → N [dimL L₁ on TI-85/86]    [LIST OPS menu]
For (K, 2, N)
:(L₁(K) + A × L₂(K − 1)) → L₂(K)
:End
```

```
:round(L₂(N),9) → R      [MATH NUM menu]
:(N − 1) → dim L₂
:Disp "DEGREE OF QUOTIENT"
:Disp dim L₂ − 1
:Disp "COEFFICIENTS"
:Pause L₂
:Disp "REMAINDER"
:Disp R
```

Sharp 9600

```
Clr T
Print "DIVISOR IS X − A"
Input A
L₁ → L₂     [See Preliminaries for how to enter list names]
dim (L₁) → N     [LIST OPE menu]
2 → K
Label 1
(L₁(K) + A × L₂(K − 1)) → L₂(K)
K + 1 → K
If K ≤ N Goto 1
```

```
round(L₂(N),9) → R      [MATH NUM menu]
(N − 1) → dim L₂
Clr T
Print "DEGREE OF QUOTIENT"
Print dim L₂ − 1
Print "COEFFICIENTS"
Print L₂
Print "REMAINDER"
Print R
```

Casio 9850

"DIVISOR IS X − A"

"$A = $"? $\to A$

List 1 $\to$ List 2   [*See Preliminaries for how to enter list names*]

dim List 1 $\to$ N   [*OPTN LIST menu*]

2 $\to$ K

Lbl 1

(List 1[K] + A × List 2 [K − 1]) $\to$ List 2 [K]

K + 1 $\to$ K

K ≤ N $\Rightarrow$ Goto 1

Fix 9   [*SETUP DISP menu*]

List 2 [N]

Rnd   [*OPTN MATH NUM menu*]

Ans $\to$ R

Norm   [*SETUP DISP menu*]

Seq(List 2 [X], X, 1, N − 1, 1) $\to$ List 2   [*OPTN LIST menu*]

"DEGREE OF QUOTIENT"

dim List 2 − 1◢

"COEFFICIENTS"

List 2◢

"REMAINDER"

R

HP-38

Input A; "SYNDIV", "X − A"; "ENTER A"; 0:

$L_1 \to L_2$:   [*See Preliminaries for how to enter list names*]

Size(L_2) $\to$ N: [*MATH LIST menu*]

For C = 1 to (N − 1)   Step 1; L_2(C) × A + L_2(C + 1) $\to L_2$(C + 1) End:

Makelist (L_2(J), J, 1, N − 1, 1) $\to L_3$:   [*MATH LIST menu*]

Msgbox "DEGREE OF QUOTIENT" Size(L_3) − 1:

Msgbox "COEFFICIENTS" L_3:

Msgbox "REMAINDER" L_2(N):

RREF Program for TI-82 and Casio 9850 (built-in on other calculators)

Preliminaries: Enter the matrix to be reduced as matrix A.

Description: The program transforms matrix A into reduced row echelon form and stores the result as matrix B.

TI-82

:Clr Home
:$[A] \to [B]$
dim $[B] \to L_1$
:$L_1(1) \to M$
:$L_1(2) \to N$
:$1 \to R$
:$1 \to J$
:Lbl 6
:abs $(seq([B](I,J), I, R, M, 1)) \to L_2$
:max$(L_2) \to T$
:If $T < .0000001$
:Goto 3
:$R \to I$
:Lbl 1
:If abs $[B](I,J) = T$
:Goto2
:$I + 1 \to I$
:Goto 1
:Lbl 2
:*row$([B](I,J)^{-1}, [B], I) \to [B]$
:For $(K, 1, I - 1, 1)$
:*row$+(-[B](K,J), [B], I, K) \to [B]$

:End
:For $(K, I + 1, M, 1)$
:*row$+(-[B](K,J), [B], I, K) \to [B]$
:End
:rowSwap $([B], I, R) \to [B]$
:round $([B], 9) \to [B]$
:If $R = min(M, N)$
:Then
:Lbl 7
:Pause round $([B], 6)$ ▶Frac
:Stop
:End
:$R + 1 \to R$
:$J + 1 \to J$
:If $J = N + 1$
:Goto 7
:Goto 6
:Lbl 3
:$J + 1 \to J$
:If $J = N + 1$
:Goto 7
:Goto 6

Casio 9850

Mat A → Mat B

"NUMBER OF ROWS"? → M

"NUMBER OF COLUMNS"? → N

(M, N) → List 3

1 → R

1 → J

Lbl 6

Mat → List(Mat B, J) → List 1

Seq(List 1[X], X, R, M, 1) → List 1

Abs (List 1) → List 1

Max (List 1) → T

T = 0 ⇒ Goto 3

R → I

Lbl 1

Abs (Mat B[I,J]) → S

S = T ⇒ Goto 2

I + 1 → I

Goto 1

Lbl 2

*Row Mat B[I,J]$^{-1}$, B, I

For 1 → K to I − 1

*Row+ −Mat B[K,J], B, I, K

Next

For I + 1 → K to M

*Row+ −Mat B[K,J], B, I, K

Next

Swap B, I, R

If R = Min (list 3)

Then Lbl 7

Lbl 7

Mat B◢

Stop

IfEnd

R + 1 → R

J + 1 → J

J + 1 = N ⇒ 7

Goto 6

Lbl 3

J + 1 → J

J = N + 1 ⇒ 7

Goto 6

HP-38 Rectangular/Polar Conversion Program

Description: Enter the rectangular coordinates of a point in the plane; the program displays the polar coordinates of the point.

Input X; "RECTANGULAR TO POLAR"; "X = "; "ENTER X"; 0:

Input Y: "RECTANGULAR TO POLAR"; "Y = "; "ENTER Y"; 0:

If X > 0

Then (ATAN(Y/X)) → C

Else (ATAN(X,Y) + π) → C:

End:

MSGBOX "R = " $\sqrt{X^2 + Y^2}$ "θ = " C:

HP-38 Polar/Rectangular Conversion Program

Description: Enter the polar coordinates of a point in the plane; the program displays the rectangular coordinates of the point.

Input R; "POLAR TO RECTANGULAR"; "R = "; "ENTER R"; 0:

Input θ; "POLAR TO RECTANGULAR"; "θ = "; "ENTER θ"; 0:   [*CHARS menu*]

MSGBOX "X = " R(cos θ) "Y = " R(sin θ):

Answers to Odd-Numbered Exercises

Chapter 0

Section 0.1, page 11

1.

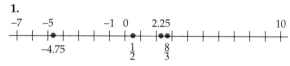

3. $-4 > -8$ **5.** $\pi < 100$ **7.** $y \le 7.5$ **9.** $t > 0$
11. $c \le 3$ **13.** $<$ **15.** $=$ **17.** $>$
19. $\dfrac{2040}{523} < \dfrac{189}{37} < \sqrt{27} < \dfrac{4587}{691} < 6.735 < \sqrt{47}$
21. $b + c = a$
23. a lies to the right of b.
25. $a < b$ **27.** 0 **29.** 10 **31.** 169
33. $\pi - \sqrt{2}$ **35.** $<$ **37.** $>$ **39.** $<$
41. 7 **43.** 14.5 **45.** $\pi - 3$
47. $|p - 25.75| \le 4$ **49.** $|316.3 - 408.6| = 92.3$
51. t^2 **53.** $y^2 + 6y + 9$ **55.** $b - 3$
57. $|x - 5| < 4$ **59.** $|x + 4| \le 17$ **61.** $|c| < |b|$
63. The distance from x to 3 is less than 2 units.
65. The distance from x to -7 is at most 3 units.
67. $x = 1$ or -1 **69.** $x = 1$ or 3
71. $x = -\pi + 4$ or $-\pi - 4$
73. Since $|a| \ge 0$, $|b| \ge 0$, and $|c| \ge 0$, the sum $|a| + |b| + |c|$ is positive only when one or more of $|a|, |b|, |c|$ is positive. But $|a|$ is positive only when $a \ne 0$; similarly for b, c.

Section 0.1.A, page 15

1. $.7777 \cdots$ **3.** $1.6428571428571 \cdots$
5. $.052631578947368421052 \cdots$

7. No; $\dfrac{2}{3} = .6666 \cdots$ **9.** Yes; $\dfrac{1}{64} = .015625$
11. No   **13.** Yes; $\dfrac{1}{.625} = 1.6$
15. $\dfrac{37}{99}$ **17.** $\dfrac{758,679}{9900} = \dfrac{252,893}{3300}$
19. $\dfrac{5}{37}$ **21.** $\dfrac{517,896}{9900} = \dfrac{14,386}{275}$
23. If $d = .74999 \cdots$, then
$1000d - 100d = (749.999 \cdots) - (74.999 \cdots) = 675.$
Hence $900d = 675$ so that $d = \dfrac{675}{900} = \dfrac{3}{4}$. Also
$.75000 \cdots - .75 = \dfrac{75}{100} = \dfrac{3}{4}.$
25. $\dfrac{6}{17} = .35294117647058823529 \cdots$
27. $\dfrac{1}{29} = .03448275862068965517241379310344 \cdots$
29. $\dfrac{283}{47} =$
$6.0212765957446808510638297872340425531914893617 0212 \cdots$
31. (a) One of many possible ways is to use the nonrepeating decimal expansion of π. For instance, with .75, associate $.7531415926 \cdots$; with 6.593 associate $6.59331415926 \cdots$, etc. Thus different terminating decimals correspond to different nonrepeating ones.
(b) As suggested in the *Hint,* associate with $.134134134 \cdots$ the irrational number $.134013400134000134 \cdots$. With $6.17398419841 \cdots$ associate $6.173984109841009841000 9841 \cdots$, etc.

Thus, different repeating decimals lead to different nonrepeating ones.

Section 0.2, page 23

1. 36   **3.** 73   **5.** -5 **7.** -112

9. $\dfrac{81}{16}$ **11.** $\dfrac{129}{8}$ **13.** x^{10} **15.** $.03y^9$

17. $24x^7$ **19.** $9x^4y^2$ **21.** $384w^6$ **23.** ab^3

25. $8x^{-1}y^3$ **27.** $3xy$ **29.** 2^{12} **31.** 2^{-12}

33. x^7 **35.** ce^9 **37.** $a^{12}b^8$ **39.** $\dfrac{1}{c^{10}d^6}$

41. $\dfrac{a^7c}{b^6}$ **43.** c^3d^6

45. Negative   **47.** Negative   **49.** Negative

51. 3^s **53.** $a^{6t}b^{4t}$ **55.** 2.70312×10^8

57. 1.09×10^8 **59.** 6.529×10^6 **61.** 2×10^{-9}

63. 150,000,000,000 m

65. .0000000000000000001602 kg

67. 1×10^5

69. $(1.6384 \times 10^8)(\pi) \approx 5.147 \times 10^8$

71. Her friend will hear it first.

73. $1187.69, $1276.28, $1435.63, $1610.51

75. $2^2 2^3 = 2^5 = 32 \neq 2^6 = 64$

77. $2^{-1} = \frac{1}{2} \neq -2^1 = -2$

Section 0.3, page 30

1. $4\sqrt{5}$ **3.** $7\sqrt{3}$ **5.** $6\sqrt{2}$

7. $\dfrac{-2 + \sqrt{11}}{5}$ **9.** $-\sqrt{2}$ **11.** $15\sqrt{5}$

13. $\dfrac{4a^4}{|b|}$ **15.** $\dfrac{|d^5|}{2\sqrt{c}}$ **17.** $\dfrac{3a^3}{b}$

19. $>$ **21.** $=$ **23.** $(a^2 + b^2)^{1/3}$ **25.** $a^{3/16}$

27. $4t^{27/10}$ **29.** $1000k^{13/4}$ **31.** $x^{9/2}$

33. $\dfrac{1}{3y^{2/3}}$ **35.** $\dfrac{a^{1/2}}{49b^{5/2}}$ **37.** $x^{7/6} - x^{11/6}$

39. $x - y$ **41.** $x + y - (x + y)^{3/2}$

43. $2684.84   **45.** $3166.75

47. 36 million miles   **49.** 886.8 million miles

51. 5.313 million   **53.** 20.004 million

55. (a) 7.65, 15.31, 22.96, 30.61, 45.92
   (b) Doubling capital and labor doubles output. Tripling capital and labor triples output.

57. (a) 13.61, 32.37, 53.73, 76.98, 127.79
   (b) Doubling capital and labor increases output by a factor of $2^{5/4}$. Tripling capital and labor increases output by a factor of $3^{5/4}$.

59. (a) The square (or any even power) of a real number is never negative.

(b) $\sqrt[3]{-8} = -2$, whereas $\sqrt[6]{(-8)^2} = 2$

61. (a) iii is best.
   (b) iii is best, $|c|$

63. $\sqrt{4 + 9} = \sqrt{13} \neq 2 + 3 = 5$,   where $c = 2, d = 3$

65. $\sqrt{16}(\sqrt[3]{16}) = 4(2\sqrt[3]{2}) = 8\sqrt[3]{2} \neq \sqrt[4]{16} = 2$,   where $a = 16$

Section 0.4, page 38

1. $8x$ **3.** $-2a^2b$

5. $-x^3 + 4x^2 + 2x - 3$ **7.** $5u^3 + u - 4$

9. $4z - 12z^2w + 6z^3w^2 - zw^3 + 8$

11. $-3x^3 + 15x + 8$ **13.** $-5xy - x$ **15.** $15y^3 - 5y$

17. $12a^2x^2 - 6a^3xy + 6a^2xy$

19. $12z^4 + 30z^3$ **21.** $12a^2b - 18ab^2 + 6a^3b^2$

23. $x^2 - x - 2$

25. $2x^2 + 2x - 12$ **27.** $y^2 + 7y + 12$

29. $-6x^2 + x + 35$ **31.** $3y^3 - 9y^2 + 4y - 12$

33. $x^2 - 16$ **35.** $16a^2 - 25b^2$

37. $y^2 - 22y + 121$ **39.** $25x^2 - 10bx + b^2$

41. $16x^6 - 8x^3y^4 + y^8$ **43.** $9x^4 - 12x^2y^4 + 4y^8$

45. $2y^3 + 9y^2 + 7y - 3$

47. $-15w^3 + 2w^2 + 9w - 18$

49. $24x^3 - 4x^2 - 4x$ **51.** $x - 25$

53. $9 + 6\sqrt{y} + y$

55. $\sqrt{3}x^2 + 4x + \sqrt{3}$

57. $3ax^2 + (3b + 2a)x + 2b$

59. $abx^2 + (a^2 + b^2)x + ab$

61. $x^3 - (a + b + c)x^2 + (ab + ac + bc)x - abc$

63. $12x^2 - 7x - 10$

65. $8x^3 + 125$

67. (a) Length $= 30 - 2x$, width $= 22 - 2x$
   (b) $4x^3 - 104x^2 + 660x$

69. (a) $16x - 2x^2$
   (b) 14, 24, 30, 32, 30, 24, 14, 0
   (c) 4

71. $4x^2 + 32x$

73. $(x + 2)(x - 2)$ **75.** $(3y + 5)(3y - 5)$

77. $(9x + 2)^2$ **79.** $(\sqrt{5} + x)(\sqrt{5} - x)$

81. $(7 + 2z)^2$ **83.** $(x^2 + y^2)(x + y)(x - y)$

85. $(x + 3)(x - 2)$ **87.** $(z + 3)(z + 1)$

89. $(y + 9)(y - 4)$ **91.** $(x - 3)^2$

93. $(x + 5)(x + 2)$ **95.** $(x + 9)(x + 2)$

97. $(3x + 1)(x + 1)$ **99.** $(2z + 3)(z + 4)$

101. $9x(x - 8)$ **103.** $2(x - 1)(5x + 1)$

105. $(4u - 3)(2u + 3)$ **107.** $(2x + 5y)^2$

109. $(x - 5)(x^2 + 5x + 25)$

111. $(x + 2)^3$

113. $(2 + x)(4 - 2x + x^2)$

115. $(x + 1)(x^2 - x + 1)$

117. $(2x - y)(4x^2 + 2xy + y^2)$

119. $(x^3 + 2^3)(x^3 - 2^3) =$
$(x + 2)(x^2 - 2x + 4)(x - 2)(x^2 + 2x + 4)$

121. $(y^2 + 5)(y^2 + 2)$

123. $(x + z)(x - y)$

125. $(a + 2b)(a^2 - b)$

127. $(x^2 - 8)(x + 4) = (x + \sqrt{8})(x - \sqrt{8})(x + 4)$

129. Example: if $y = 4$, then $3(4 + 2) \neq (3 \cdot 4) + 2$; correct statement: $3(y + 2) = 3y + 6$

131. Example: if $x = 2$, $y = 3$, then $(2 + 3)^2 \neq 2 + 3^2$; correct statement: $(x + y)^2 = x^2 + 2xy + y^2$

133. Example: if $y = 2$, then $2 + 2 + 2 \neq 2^3$; correct statement: $y + y + y = 3y$

135. Example: if $x = 4$, then $(4 - 3)(4 - 2) \neq 4^2 - 5 \cdot 4 - 6$; correct statement: $(x - 3)(x - 2) = x^2 - 5x + 6$

137. If x is the chosen number, then adding 1 and squaring the result gives $(x + 1)^2$. Subtracting 1 from the original number x and squaring the result gives $(x - 1)^2$. Subtracting the second of these squares from the first yields:
$$(x + 1)^2 - (x - 1)^2 =$$
$$(x^2 + 2x + 1) - (x^2 - 2x + 1) = 4x.$$
Dividing by the original number x now gives $\dfrac{4x}{x} = 4$. So the answer is always 4, no matter what number x is chosen.

139. Many correct answers

Section 0.5, page 48

1. $\dfrac{9}{7}$

3. $\dfrac{195}{8}$

5. $\dfrac{x - 2}{x + 1}$

7. $\dfrac{a + b}{a^2 + ab + b^2}$

9. $\dfrac{1}{x}$

11. $\dfrac{29}{35}$

13. $\dfrac{121}{42}$

15. $\dfrac{ce + 3cd}{de}$

17. $\dfrac{b^2 - c^2}{bc}$

19. $\dfrac{-1}{x(x + 1)}$

21. $\dfrac{x + 3}{(x + 4)^2}$

23. $\dfrac{2x - 4}{x(3x - 4)}$

25. $\dfrac{x^2 - xy + y^2 + x + y}{x^3 + y^3}$

27. $\dfrac{-6x^5 - 38x^4 - 84x^3 - 71x^2 - 14x + 1}{4x(x + 1)^3(x + 2)^3}$

29. 2

31. $2/(3c)$

33. $3y/x^2$

35. $\dfrac{12x}{x - 3}$

37. $\dfrac{5y^2}{3(y + 5)}$

39. $\dfrac{u + 1}{u}$

41. $\dfrac{35}{24}$

43. $\dfrac{u^2}{vw}$

45. $\dfrac{x + 3}{2x}$

47. $\dfrac{cd(c + d)}{c - d}$

49. $\dfrac{-3y + 3}{y}$

51. $\dfrac{-2x - h}{x^2(x + h)^2}$

53. $\dfrac{3\sqrt{2}}{4}$

55. $\dfrac{3\sqrt{3} - 3}{4}$

57. $\dfrac{2\sqrt{x} - 4}{x - 4}$

59. $\dfrac{1}{\sqrt{x + h + 1} + \sqrt{x + 1}}$

61. $\dfrac{2x + h}{\sqrt{(x + h)^2 + 1} + \sqrt{x^2 + 1}}$

63. (a) $298.62
(b) $2333.78

65. (a) $503.43
(b) $109,236.40

67. $\dfrac{\pi}{4} \approx .785$, or 78.5% chance

69. $\dfrac{1}{25} = .04$, or 4% chance

71. (a) 25%
(b) 25, 33, 32, 29, 25, 22
(c) At $t = 1$
(d) It declines.

73. $\dfrac{1}{2} + \dfrac{1}{3} \neq \dfrac{1}{5}$ $\qquad \dfrac{1}{a} + \dfrac{1}{b} = \dfrac{a + b}{ab}$

75. $\left(\dfrac{1}{\sqrt{4} + \sqrt{9}}\right)^2 = \left(\dfrac{1}{5}\right)^2 = \dfrac{1}{25} \neq \dfrac{1}{13}$
$\left(\dfrac{1}{\sqrt{a} + \sqrt{b}}\right)^2 = \dfrac{1}{a + b + 2\sqrt{ab}}$

77. $\dfrac{1}{2 + 1} = \dfrac{1}{3} \neq \dfrac{1}{2}$
$\dfrac{x}{2 + x}$ is already simplified.

79. $(\sqrt{4} + \sqrt{9})\left(\dfrac{1}{\sqrt{4} + \sqrt{9}}\right) = 1 \neq 13$
$\sqrt{x} + \sqrt{y}\left(\dfrac{1}{\sqrt{x} + \sqrt{y}}\right) = 1$

Chapter 0 Review, page 51

1. (a) $>$ (b) $<$ (c) $<$ (d) $>$ (e) $=$

3. (a) $-10 < y < 0$
(b) $0 \leq x \leq 10$

5. (a) $|x + 7| < 3$
(b) $|y| > |x - 3|$

7. $x = 2$ or 8

9. $x = -1/2$ or $-11/2$

11. (a) $7 - \pi$
(b) $\sqrt{23} - \sqrt{3}$

13. c **15.** 28/99   **17.** $\dfrac{c^{15}}{d^{15}e^5}$ **19.** $\dfrac{2}{u}$

21. (a) 1.232×10^{16}
(b) 7.89×10^{-11}

23. c^2 **25.** $a^{10/3}b^{42/5}$ **27.** $u^{1/2} - v^{1/2}$

29. $\dfrac{c^2 d^4}{2}$ **31.** d

33. $-2p^3 - 4p^2 + 3p + 9$

35. $12z^3 + 14z^2 - 7z + 5$

37. $12k^2 - 20k + 3$

39. $12x^3 - 24x^2 + 12x = 12x(x^2 - 2x + 1) =$
$12x(x - 1)^2$

41. $p(x) = -.0027x^4 + .25x^3 - 6.2x^2 + 70x - 520$

43. $(x + 1)(x + 4)$ **45.** $(x - 4)(x + 2)$

47. $(2x + 3)(2x - 3)$ **49.** $(2x + 3)(2x - 5)$

51. $x(3x - 1)(x + 2)$ **53.** $(x^2 + 1)(x + 1)(x - 1)$

55. e **57.** $1 + 2a$ **59.** $\dfrac{6x - 11}{24x}$ **61.** $\dfrac{-y}{x - y}$

63. $\dfrac{x - 1}{x - 3}$ **65.** $\dfrac{4}{9}$ **67.** $\dfrac{2}{r + 2}$ **69.** $\dfrac{x + 1}{x - 1}$

71. $\dfrac{2}{\sqrt{2x + 2h + 1} + \sqrt{2x + 1}}$

Chapter 1

Section 1.1, page 67

1. $A(-3, 3)$; $B(-1.5, 3)$; $C(-2.3, 0)$; $D(-1.5, -3)$;
$E(0, 2)$; $F(0, 0)$; $G(2, 0)$; $H(3, 1)$; $I(3, -1)$

3. $P(-6, 3)$ **5.** $P(4, 2)$

7.

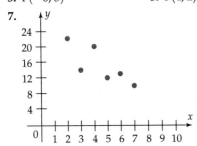

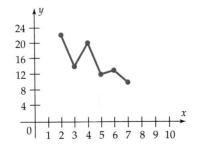

9. (a) About $0.94 in 1987 and $1.19 in 1995
(b) About 26.6%
(c) In the first third of 1985 and from 1989 onward

11. (a) Quadrant IV
(b) Quadrants III or IV

13. (a)

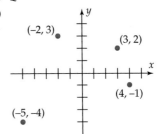

(b)

(c) They are mirror images of each other, with the
x-axis being the mirror. In other words, they lie
on the same vertical line, on opposite sides of
the x-axis, the same distance from the axis.

15. $13; \left(-\dfrac{1}{2}, -1\right)$ **17.** $\sqrt{17}; \left(\dfrac{3}{2}, -3\right)$

19. $\sqrt{6 - 2\sqrt{6}} \approx 1.05; \left(\dfrac{\sqrt{2} + \sqrt{3}}{2}, \dfrac{3}{2}\right)$

21. $\sqrt{2}|a - b|; \left(\dfrac{a + b}{2}, \dfrac{a + b}{2}\right)$

23. (a)

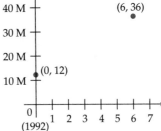

(b)

(c) In 1995, about 24 million personal computers were sold. We must assume that sales increased steadily.

25. (a) 45.03 yards, (20, 10), $(48\frac{1}{3}, 45)$
  (b) $(34\frac{1}{6}, 27\frac{1}{2})$

27. Yes   **29.** Yes   **31.** No

33. $(x + 3)^2 + (y - 4)^2 = 4$

35. $x^2 + y^2 = 2$

37.

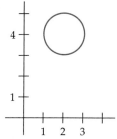

39.

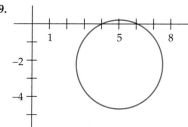

41. Hypotenuse from (1, 1) to (2, −2) has length $\sqrt{10}$; other sides have lengths $\sqrt{2}$ and $\sqrt{8}$. Since $(\sqrt{2})^2 + (\sqrt{8})^2 = (\sqrt{10})^2$, this is a right triangle.

43. Hypotenuse from (−2, 3) to (3, −2) has length $\sqrt{50}$; other sides have lengths $\sqrt{5}$ and $\sqrt{45}$. Since $(\sqrt{5})^2 + (\sqrt{45})^2 = (\sqrt{50})^2$, this is a right triangle.

45. $(x - 2)^2 + (y - 2)^2 = 8$

47. $(x - 1)^2 + (y - 2)^2 = 8$

49. $(x + 5)^2 + (y - 4)^2 = 16$

51. $(x - 2)^2 + (y - 1)^2 = 5$

53. $(-3, -4)$ and $(2, 1)$

55. $x = 6$ **57.** B   **59.** C

61. (a) 66 years
  (b) 75 years
  (c) 1920–1940

63. M has coordinates $(s/2, r/2)$ by the midpoint formula. Hence the distance from M to $(0, 0)$ is
$$\sqrt{\left(\frac{s}{2} - 0\right)^2 + \left(\frac{r}{2} - 0\right)^2} = \sqrt{\frac{s^2}{4} + \frac{r^2}{4}},$$
and the distance from M to $(0, r)$ is the same:

$$\sqrt{\left(\frac{s}{2} - 0\right)^2 + \left(\frac{r}{2} - r\right)^2} = \sqrt{\left(\frac{s}{2}\right)^2 + \left(-\frac{r}{2}\right)^2}$$
$$= \sqrt{\frac{s^2}{4} + \frac{r^2}{4}}$$

as is the distance from M to $(s, 0)$:

$$\sqrt{\left(\frac{s}{2} - s\right)^2 + \left(\frac{r}{2} - 0\right)^2} = \sqrt{\left(-\frac{s}{2}\right)^2 + \left(\frac{r}{2}\right)^2}$$
$$= \sqrt{\frac{s^2}{4} + \frac{r^2}{4}}.$$

65. The circle $(x - k)^2 + y^2 = k^2$ has center $(k, 0)$ and radius $|k|$ (the distance from $(k, 0)$ to $(0, 0)$). So the family consists of every circle that is tangent to the y-axis *and* has center on the x-axis.

67. The points are on opposite sides of the origin because one first coordinate is positive and one is negative. They are equidistant from the origin because the midpoint of the line segment joining them is
$$\left(\frac{c + (-c)}{2}, \frac{d + (-d)}{2}\right) = (0, 0).$$

Section 1.2, page 82

1. $P = (3, 5); Q = (-6, 2)$

3. $P = (-12, 8); Q = (-3, -8.5)$

5–10. Answers vary.

11.

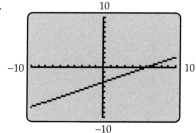

13.

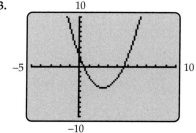

15.

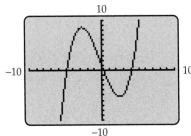

17. c   **19.** d   **21.** e   **23.** (c)

25. $0 \le x \le 42$ and $0 \le y \le 4500$, with x-scl $= 5$, y-scl $= 500$. Both x (time) and y (concentration) must be nonnegative and this window shows the part of the graph where they are.

27. $0 \le x \le 4000$ and $0 \le y \le 100$, with x-scl $= 500$, y-scl $= 10$. Both x and y (numbers of barrels) must be nonnegative and this window shows the part of the graph where they are.

29. Maximum at about $(.7922, 4.48490)$
Minimum at about $(4.2078, -3.4849)$

31. 1993   **33.** 1961

35. Not the same

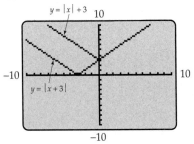

37. Same

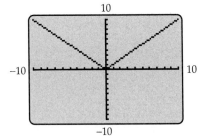

39. Not the same

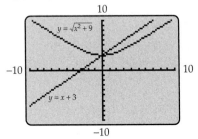

41. (a) Graphs appear identical.

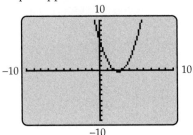

(b) Graphs are different.

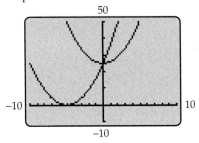

43. Possibly true

In Answers 45–49, the graphs shown were made on a TI-83. For wide-screen calculators such as TI-86/89, Sharp 9600, Casio 9850, and HP-38, the x-axis should be longer than this one to have a square window with the same y-axis as here.

45.

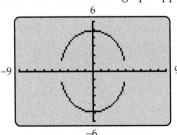

47. The two halves of the graph should be connected, but the ends of the two pieces are almost vertical so the calculator could not plot enough points near them to make the graph appear connected.

49.

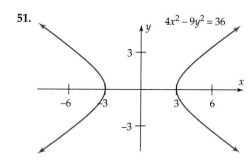

51.

$4x^2 - 9y^2 = 36$

53.

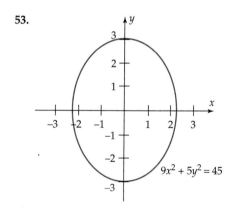

$9x^2 + 5y^2 = 45$

55. $-5 \leq x \leq 5$ and $-100 \leq y \leq 100$

57. $-10 \leq x \leq 10$ and $-2 \leq y \leq 20$ [Where is the right half of the graph?]

59. $-6 \leq x \leq 12$ and $-100 \leq y \leq 250$

61. (a)

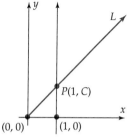

(c) Many correct answers, including $0 \leq x \leq 1.5$ and $-3 \leq y \leq -1$

63. (a)

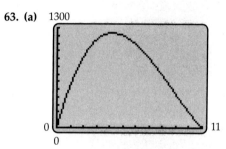

(b) The x-coordinate is the length of the side of the square to be cut from the corners. The y-coordinate is the volume of the resulting box.

(c) $x \approx 4.182$

Section 1.3, page 90

1. 3   **3.** 3   **5.** 2

7. $x = -2.42645$ **9.** $x = 1.1640$ **11.** $x = 1.23725$

13. $x = 1.1921$ **15.** $x = -1.3794$ **17.** c

19. $x = 1.8608059$ **21.** $x = 2.1017$

23. $x = -1.7521$ **25.** $x = .9505$

27. $x = 0$ or 2.2074 **29.** $x = 2.3901454$

31. $x = -.6513878188$ or 1.151387819

33. $x = 7.033393042$ **35.** $x = 2/3$

37. $x = 1/12$ **39.** $x = \sqrt{3}$

41. $x = 1.4528$ or -112 **43.** $x = .0499$ or 1.9097

45. 2001   **47.** 1991

Section 1.4, page 99

1. (a) C   **(b)** B   **(c)** B   **(d)** D

3. Slope 2; y-intercept 5

5. Slope $-3/7$; y-intercept $-11/7$

7. Slope $5/2$

9. Slope 4

11.

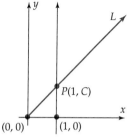

Slope of $L = \dfrac{C - 0}{1 - 0} = C$

13. $y = x + 2$ **15.** $y = -x + 8$

17. $y = -x - 5$ **19.** $y = -\frac{7}{3}x + \frac{34}{9}$

21. Perpendicular

23. Parallel

25. Parallel

27. Slope from $(-5, -2)$ to $(-3, 1)$ is 1.5; slope from $(3, 0)$ to $(5, 3)$ is 1.5; slope from $(-5, -2)$ to $(3, 0)$ is .25; slope from $(-3, 1)$ to $(5, 3)$ is .25.

29. Yes

31. $y = 3x + 7$ **33.** $y = \frac{3}{2}x$

35. $y = x - 5$ **37.** $y = -x + 2$

39. $y = -\frac{3}{4}x + \frac{25}{4}$ **41.** $y = -\frac{x}{2} + 6$

43. Both have slope $-A/B$.

45. **(a)** $y = 20{,}000x + 120{,}000$
  (b) $x = 5; y = \$220{,}000$

47. **(a)** $y = 310x + 2125$
  (b) $x = 5; y = 3675$ billion (1991)
    $x = 14; y = 6465$ billion (2000)

49. $\$375{,}000; \$60{,}000$

51. **(a)** $y = 58.2x + 698$
  (b) $x = 3; y = 872.6$ billion (1993)
    $x = 9; y = 1221.8$ billion (1999)

53. **(a)** $c(x) = 25x + 180{,}000$
  (b) $r(x) = 40x$
  (c) $p(x) = 15x - 180{,}000$
  (d) $x = 12{,}000$

55. **(a)** $y = 8.50x + 50{,}000$
  (b) $\$11, \$9.50, \$9$ per hat

57. **(a)** $\dfrac{145x + 120{,}000}{x}$
  (b) 4000

59. **(a)** $x = 10$
  (b) $x = 30$

61. **(a)**

Women's Target Weight	
x inches over 5 ft	Weight (lb)
0	100
1	105
2	110
3	115
6	130
12	160
15	175

Men's Target Weight	
x inches over 5 ft	Weight (lb)
0	106
1	112
2	118
3	124
6	142
12	178
15	196

  (b) $y = 5x + 100$ **(c)** 145 lb
  (d) $y = 6x + 106;$ 160 lb

63. Let $y = mx + b$ and $y = mx + c$ be equations of lines with same slope m, and $b \neq c$. Suppose (x_1, y_1) is an arbitrary point lying on both lines. Then $y_1 = mx_1 + b$ and $y_1 = mx_1 + c$. So,

$$mx_1 + b = mx_1 + c \text{ and,}$$

$$b = c, \text{ a contradiction.}$$

Thus, the lines share no point in common so must be parallel.

65.

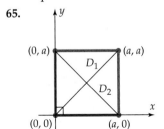

Equation of D_1: $y = x$, with slope 1
Equation of D_2: $y = -x + a$, with slope -1
(slope D_1)(slope D_2) = $(1)(-1) = -1$

Section 1.5, page 113

1. **(a)** $y = \frac{3}{4}x + \frac{5}{4}$
  (b) Sum of squares = $2\frac{3}{8}$
  Model C still has least error.

3. **(a)** Slope = 1.05640540541
  (b) $y = 1.05640540541x + 21.0778918918$
  (c) Line described in (b) predicts a higher number of workers.

5. Positive   **7.** Very little

9. **(a)** Linear   **(b)** Positive

11. **(a)** Nonlinear

13. **(a)** Linear   **(b)** Negative

15. (a)

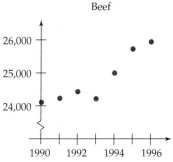

Beef

(b) y_1, poultry; y_2, beef

17. (a) $y = -3.866929638x + 542.7452878$
(b) 2090

19. (a) $y = 4x + 18$
(b) Slope indicates a positive correlation between time and percentage; y-intercept shows that regression line passes through actual plotted values.
(c) 100%

21. (a)

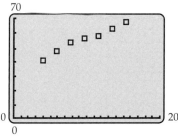

(b) $y = 1.714285714x + 37.42857143$
(c) About 60, about 72

23. (a)

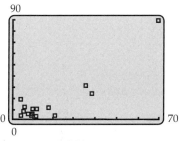

(b) $y = 1.089256088x - 1.413259148$

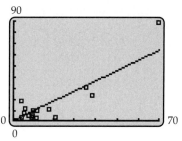

(c) United States, Russia, China
(d) The country uses more energy than it produces.
(e) The country produces more energy than it uses.
(f) United States, Russia, China, Saudi Arabia, Canada, Japan

Chapter 1 Review, page 119

1. $\sqrt{58}$ **3.** $\sqrt{c^2 + d^2}$ **5.** $\left(d, \dfrac{c + 2d}{2} \right)$

7. (a) $\sqrt{17}$ **(b)** $(x - 2)^2 + (y + 3)^2 = 17$

9.

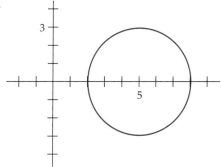

11. (b) and **(d)**

13.

15. (a) a, d
(b) b, c do not show peaks and valleys near the origin; e crowds the graph onto the y-axis.
(c) a or $-4 \le x \le 6$ and $-10 \le y \le 10$

17. (a) None of them
(b) a, b, d do not show any peaks or valleys; c does not show the valleys; e shows only one

point on the graph (which can't be distinguished because it's on the y-axis).
(c) $-7 \le x \le 11$ and $-1000 \le y \le 500$

19. (a) b, c
(b) a shows no peaks or valleys; d is too crowded horizontally; e shows only one point.
(c) $-10 \le x \le 10$ and $-150 \le y \le 150$

21. 1979   **23.** False

25.

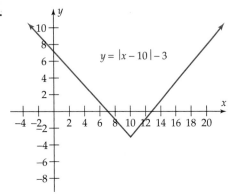

27.

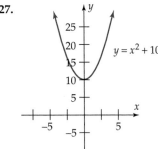

29. $x = 3.2678$ **31.** $x = -3.2843$ **33.** $x = 1.6511$

35. 1994

37. (a) 1
(b) 4/5

39. $y = 3x - 7$ **41.** $x - 5y = -29$

43. 25,000 ft   **45.** False   **47.** False

49. False   **51.** False

53. (d)

55. (e)

57. (a) $y = .25x + 62.9$
(b) $x = 40, y = 72.9$ years

59. C; slope 75   **61.** D; slope 20

63. (a)

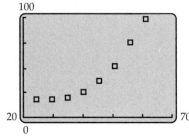

(b) Nonlinear

65. (a)

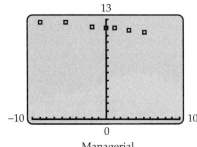

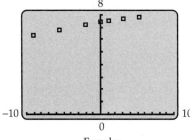

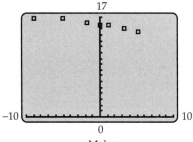

(b) Managerial, y_2 (negative slope, y-intercept at 11.74); Female, y_1 (positive slope, y-intercept at 7.34); Male, y_3 (negative slope, y-intercept at 15.48)

67. (a) $y = 0.3055833333x + 6.912$
(b) \$10.27; \$13.33; \$0.05

69. (a) $y = 6.274090909x + 47.46409091$
(b) 141.58; 185.49

Chapter 2

Section 2.1, page 138

1. $x = 3/2$
3. $x = -5$
5. $x = 8$
7. $x = -5/6$
9. $y = -32$
11. $z = -1/13$
13. $x = -3$
15. $x = 5/6$
17. No solution
19. $x = 1$
21. No solution
23. $x = -5/2$
25. $x \approx -1.239697$
27. $x = 3.765$
29. $x \approx .24361$
31. $Y = 800$
33. $Y = 1300$
35. $Y = 1500$
37. $Y = 3040$
39. $y = \dfrac{x + 5}{3}$

41. $y = \dfrac{3x}{4} - 18$

43. $b = \dfrac{2A - hc}{h}$, with $h \neq 0$

45. $h = \dfrac{4V}{\pi d^2}$, with $d \neq 0$

47. $c = -3$

49. x; $229 + x$; $\dfrac{229 + x}{4}$

51. L; W; $2L + 2W$; $L = 2W$
53. $1.08x = 1593$
55. $9 + x = 3x - 1$
57. 17.88
59. 14.40
61. $366.67 at 12% and $733.33 at 6%
63. 14%
65. 96
67. 30 lb
69. $2\frac{2}{3}$ quarts
71. 65 mph
73. 4:40 P.M.
75. 60 mph
77. 45 hours
79. 30 minutes
81. 90 minutes
83. 60 by 20 meters
85. $1437.50
87. 84 years

Section 2.2, page 151

1. $x = 3$ or 5
3. $x = -2$ or 7
5. $y = 1/2$ or -3
7. $t = -2$ or $-1/4$
9. $u = 1$ or $-4/3$
11. $x = 1/4$ or $-4/3$
13. $x = 2 \pm \sqrt{3}$
15. $x = -3 \pm \sqrt{2}$
17. No real number solutions

19. $x = \dfrac{1}{2} \pm \sqrt{2}$
21. $x = 1 \pm \dfrac{\sqrt{3}}{2}$

23. $u = \dfrac{-4 \pm \sqrt{6}}{5}$

25. 2
27. 2
29. 1
31. $x = -5$ or 8
33. $x = \dfrac{-5 \pm \sqrt{57}}{8}$

35. $x = -3$ or -6
37. $x = \dfrac{-1 \pm \sqrt{2}}{2}$

39. $x = 5$ or $-3/2$
41. No real number solutions
43. No real number solutions
45. $x \approx 1.824$ or $.47$
47. $x \approx 13.79$
49. $k = 10$ or -10
51. $k = 16$
53. Solution is rational.

55. $c = \pm\sqrt{\dfrac{E}{m}}$; $m \neq 0$, $E/m \geq 0$

57. $r = \dfrac{-\pi h \pm \sqrt{\pi^2 h^2 + 4A\pi}}{2\pi}$; $\pi h^2 + 4A \geq 0$

59. $k = 4$
61. L; W; $2L + 2W$; $2L + 2W = 45$; LW; $LW = 112.5$; W = 7.5 cm, L = 15 cm
63. L; W; LW; $LW = 5.7$; $L = W + 2.3$; W = 1.5 ft, L = 3.8 ft
65. 3 cm, 4 cm, 5 cm
67. 4 cm
69. Approximately 1.753 ft
71. 2 meters
73. 2.5 yards
75. 2 in. by 2 in.
77. 1.48 inches
79. 9 mph; 16 mph
81. Red Riding Hood, 54 mph; wolf, 48 mph
83. Approximately 132.7 feet
85. 12 hours
87. (a) Approximately 6.3 seconds
    (b) Approximately 4.9 seconds
89. (a) Approximately 4.4 seconds
    (b) After 50 seconds
91. 2000 meters

Section 2.3, page 161

1. $y = \pm 1$ or $\pm\sqrt{6}$
3. $x = \pm\sqrt{7}$
5. $y = \pm 2$ or $\pm\sqrt{2}/2$
7. $x = \pm\sqrt{5}/5$
9. $x = 7$
11. $x = 4$
13. $x = -2$
15. $x = \pm 3$
17. $x = -1$ or 2
19. $x = 9$
21. $x \approx 1/2$
23. $x = 1/2$ or -4
25. $x = 6$
27. $x = 3$ or 7

29. $b = \sqrt{\dfrac{a^2}{A^2 - 1}}$
31. $u = \sqrt{\dfrac{x^2}{1 - K^2}}$

33. $x = 3$ or -6
35. $x = 3/2$

37. $x = 2$ **39.** $x = 3/2$

41. $x = -5$ or -3 or -1 or 1

43. 50 by 120 cm

45. $r = 4.658$ **47.** $r \approx 8.019$

49. 11.47 ft or 29.91 ft

51. Approximately 6.205 miles

53. 2.2 by 4.4 by 4 ft high

Section 2.4, page 170

1. $x \le 3/2$ **3.** $x > -2$

5. $x \le -8/5$ **7.** $x > 1$

9. $2 < x < 4$ **11.** $-3 \le x < 5/2$

13. $x < 4/7$ **15.** $x \ge -7/17$

17. $-1 \le x < 1/8$ **19.** $x \ge 5$

21. $x < \dfrac{b + c}{a}$ **23.** $c < x < a + c$

25. $-300 \le t \le 840$

27. (a) $T = 350 + 7.5x$
    (b) $3150 \le 350 + 7.5x \le 4575$
    (c) $373 < x \le 563$

29. $x > -1.43$ **31.** B   **33.** D

35. $-4/3 \le x \le 0$ **37.** $7/6 < x < 11/6$

39. $x < -2$ or $x > -1$

41. $x \le -11/20$ or $x \ge -1/4$

43. $x < -53/40$ or $x > -43/40$

45. $\left| p - 25\frac{3}{4} \right| \le 4$

47. Approximately 8.608 cents per kwh

49. More than \$12,500

51. Between \$4000 and \$5400

Section 2.5, page 178

1. $1 \le x \le 3$ **3.** $x \le -7$ or $x \ge -2$

5. $x \le -3$ or $x \ge 3$

7. $x \le -2$ or $x \ge 3$

9. $x \le -1/2$ or $x \ge 2/3$

11. $x < \dfrac{-\sqrt{97} - 11}{2}$ or $x > \dfrac{\sqrt{97} - 11}{2}$

13. $\dfrac{5 - \sqrt{45}}{2} < x < \dfrac{5 + \sqrt{45}}{2}$

15. All real x

17. $-1 \le x \le 0$ or $x \ge 1$

19. $x < -1$ or $0 < x < 3$

21. $-2 < x < -1$ or $1 < x < 2$

23. $x \le -7$ or $x \ge -3$

25. $x \le -1$ or $x = 3$

27. $-3 - \sqrt{3} < x < -3 + \sqrt{3}$ or $x > 0$

29. $\dfrac{3 - \sqrt{29}}{2} \le x \le \dfrac{3 + \sqrt{29}}{2}$

31. $-2.26 \le x \le 0.76$ or $x \ge 3.51$

33. $.5 < x < .84$

35. $-3.79 \le x \le -.60$ or $.44 \le x \le 8.95$

37. $x < -3.87$ or $-1.56 < x < -1.18$ or $.06 < x < 2.29$

39. $x < -1/3$ or $x > 2$

41. $-2 < x < -1$ or $1 < x < 3$

43. $x > 1$

45. $x \le -4.5$ or $x > -3$

47. $-3 < x < 1$ or $x \ge 5$

49. $-\sqrt{7} < x < \sqrt{7}$ or $x > 5.34$

51. $1 < x < 19$ and $y = 20 - x$

53. $10 < x < 35$ **55.** $1 \le t \le 4$

57. $2 < t < 2.25$

59. $x < -1.32$ or $0.03 < x < 4.05$ or $x > 47.24$

61. $x < -3$ or $.5 < x < 5$

Chapter 2 Review, page 180

1. $x = 44/7$ **3.** $x = 5$

5. $r = \dfrac{b - a}{2Q}$ **7.** $x = \dfrac{3 + 2y}{1 - y}$

9. 3/11 ounces of gold; 8/11 ounces of silver

11. $2\frac{2}{9}$ hours   **13.** 9.6 feet

15. No real solutions

17. $y = -1$ or $5/3$

19. None

21. $x = 3$ or -3 or $\sqrt{2}$ or $-\sqrt{2}$

23. $x = -1$ or $5/3$

25. No. If $x + y = 2$ and $xy = 2$, then $y = 2 - x$ and $2 = xy = x(2 - x)$. Verify that the equation $2 = x(2 - x)$ has no real solutions.

27. 4 feet   **29.** $x = \pm\sqrt{5}$

31. $x = 2$ or -1 **33.** $x = \dfrac{5 - \sqrt{5}}{2}$

35. No real solutions   **37.** $s = \dfrac{gt^2}{2}$

39. $-9/2 < x < 2$

41. $x \le -4/3$ or $x \ge 0$

43. $x < -2$ and $x > -1/3$

45. $0 \le I < 360$

47. $x \le -1$ or $0 \le x \le 1$

49. (e)

51. $x \leq 0$ or $x \geq 4/5$

53. $x \leq -7$ or $x > -4$

55. $x < -2\sqrt{3}$ or $-3 < x < 2\sqrt{3}$

57. $x < \dfrac{1 - \sqrt{13}}{6}$ or $x > \dfrac{1 + \sqrt{13}}{6}$

Chapter 3

Section 3.1, page 191

1. Yes. Each input produces only one output.

3. No. The value -5 produces two outputs.

5. 6   **7.** -2 **9.** -17

11. y is a function of x.   **13.** x is a function of y.

15. y is a function of x and x is a function of y.

17. Neither x nor y is a function of the other.

19.

21.

23. (500, 0); (1509, 0); (3754, 35.08); (6783, 119.15); (12500, 405); (55342, 2547.10)

25. Each input (income) yields only one output (tax).

27. Postage is a function of weight since each weight determines one and only one postage amount. But weight is *not* a function of postage since a given postage amount may apply to several different weights. For instance, *all* letters under 1 oz use just one first-class stamp.

29. (a)

Average Man	
Drinks in 1 hour	**Blood Alcohol Level**
2	.03
3	.05
4	.07
5	.10

Average Woman	
Drinks in 1 hour	**Blood Alcohol Level**
2	.05
3	.08

(b) Yes. For average man chart, the domain consists of all x such that $2 \leq x \leq 5$ and the range of all y such that $.03 \leq y \leq .10$. For the average woman the domain consists of all x such that $2 \leq x \leq 3$ and the range of all y such that $.05 \leq y \leq .08$.

31. (a) No, since $p(40)$ gives two different values.
(b) Yes

33. (a) $A = \pi r^2$ (b) $A = \frac{1}{4}\pi d^2$

35. $V = 4x^3$

37. $D = 400 - 16t^2$; range consists of all D such that $0 \leq D \leq 400$.

39. (a) $y = 18{,}932.1429 + 2{,}833.0357x$
(b) 33,100; 61,400; 89,800

41. (a) All positive numbers that can be entered in your calculator
(b) All numbers between -1 and 1 (inclusive) that can be displayed on your calculator

43. (a) The integer part of a positive number c is the integer that is closest to the number and less than or equal to the number, that is, $[c]$.
(b) All negative integers: $-1, -2, -3, \dots$
(c) All negative numbers that are not integers

Section 3.2, page 201

1. (a) $-4/5$
(b) $-3/4$
(c) $-2/5$
(d) $-1/8$
(e) 0

3. $\sqrt{3} + 1$ **5.** $\sqrt{\sqrt{2} + 3} - \sqrt{2} + 1$

7. 4   **9.** $\dfrac{34}{3}$ **11.** $\dfrac{59}{12}$

13. $(a + k)^2 + \dfrac{1}{a + k} + 2$

15. $(2 - x)^2 + \dfrac{1}{2 - x} + 2 = 6 - 4x + x^2 + \dfrac{1}{2 - x}$

17. 8   **19.** -1 **21.** $(s + 1)^2 - 1 = s^2 + 2s$

23. $t^2 - 1$ **25.** $c = -2$ **27.** $d = 2$

29. $x = -1.099, x = 9.099$

31. $x = -3.5, x = 3.0$

33. 1   **35.** 3   **37.** $-2x - h + 1$

39. $\dfrac{1}{\sqrt{x + h} + \sqrt{x}}$

41. (a) False   **(b)** True

43. (a) All x such that $x \le 20$
  (b) 3
  (c) -1
  (d) 1
  (e) 2

45. (iii) or (v)

47. All real numbers

49. All real numbers

51. All nonnegative real numbers

53. All nonzero real numbers

55. All real numbers

57. All real numbers except -2 and 3

59. All x such that $6 \le x \le 12$

61. Many possible answers, including $f(x) = x^2$ and $g(x) = |x|$

63. $f(x) = \sqrt{2x - 5}$

65. $f(x) = \dfrac{x^3 + 6}{5}$

67. $d(t) = \begin{cases} 55t & \text{if } 0 \le t \le 2 \\ 110 + 45(t - 2) & \text{if } t > 2, \end{cases}$
  where t = time (in hours) and $d(t)$ = distance (in miles)

69. (a)

	(i) $22.2x + 509$	(ii) $-1.03x^2 + 60.8x + 302$	(iii) $-.03x^3 + .74x^2 + 36.7x + 484$
0	509	302	484
25	1064	1178.25	1364
36	1308.2	1155.92	1271.25

Clearly (ii) gives the best approximation.
  (b) Approximately 1,186,700

71. (a) $f(x) = 5x + 200; g(x) = 10x$
  (b) 300 and 200; 375 and 350; 450 and 500
  (c) 45

73. $C(x) = 5.75x + \dfrac{45,000}{x}$

75. (a) $y = 677.4571429x + 8884.942857$
  (b) 10,917; 12,272; 13,627
  All are within \$60 of the actual figure.
  (c) 14,982

Section 3.3, page 212

1.

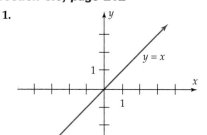

3.

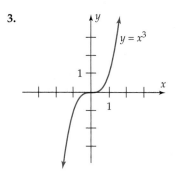

5.

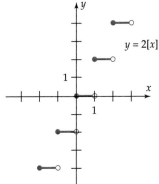

$y = 2[x]$

7.

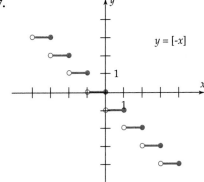

$y = [-x]$

9.

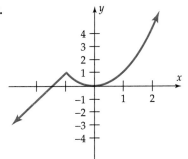

11.

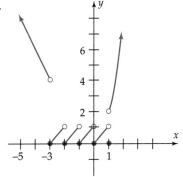

13. (a) Several possibilities, including

$$p(x) = \begin{cases} [x] & \text{if } x \text{ is an integer} \\ [x] + 1 & \text{if } x \text{ is not an integer} \end{cases}$$

or $p(x) = -[-x]$, with $x > 0$ in all cases,
where x is the weight in ounces

(b) Graph for $0 < x \le 4$:

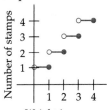

Weight in ounces

(c)

15. (a) $f(x) = \begin{cases} x + 2 & x \ge 0 \\ -x + 2 & x < 0 \end{cases}$

(b)

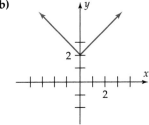

17. (a) $h(x) = \begin{cases} \dfrac{x}{2} - 2 & x \ge 0 \\[2mm] -\dfrac{x}{2} - 2 & x < 0 \end{cases}$

(b)

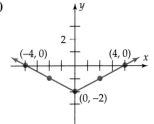

19. Minimum at $x = .57735$; maximum at $x = -.57735$

21. Minimum at $x = -1$; maximum at $x = 1$

23. Minimum at $x = 0.7633$; maximum at $x = 0.4367$

25. Increasing when $-2.5 < x < 0$ and $1.7 < x < 4$; decreasing when $-6 < x < -2.5$ and $0 < x < 1.7$

27. Constant when $x \le -1$ and $x \ge 1$; decreasing when $-1 < x < 1$

29. Increasing when $-5.7936 \le x \le .46028$; decreasing when $x < -5.7936$ and $x > .46028$

31. Increasing when $0 < x < .867$ and $x > 2.883$; decreasing when $x < 0$ and $.867 < x < 2.883$

33. Many correct answers, including

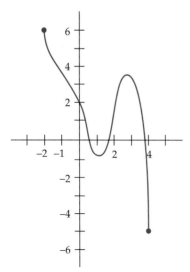

35. **(a)** $2x + 2z = 100$
   (b) $A(x) = 50x - x^2$
   (c) $x = 25, z = 25$

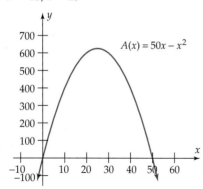

37. **(a)** $S = 2x^2 + 4xh$
   (b) $867 = hx^2$
   (c) $S(x) = 2x^2 + \dfrac{3468}{x}$
   (d) $x = h = 9.5354$

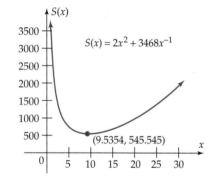

39. **(a)** iv   **(c)** v   **(e)** ii
   (b) i   **(d)** iii

41. Many correct answers, including:

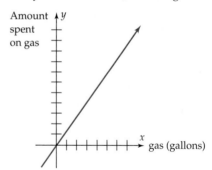

Approximate domain: $0 \le x \le 150$. Approximate range: $0 \le y \le 300$.

43. Many correct answers, including

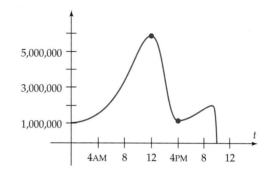

45.

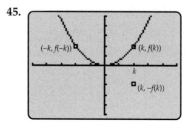

47. **(a)** 7%
   (b) 15%
   (c) 7%
   (d) 1993; 1990

49. All x such that $-3 \le x \le 5$

51. 4   **53.** 3.5   **55.** 4.5   **57.** 1, 5

59. All x such that $-8 \le x \le 9$

61. $-7, -3, 0, 3, 7$ **63.** $-7, -3, 0, 3, 7$

65. Domain f: All x such that $-6 \le x \le 7$; domain g: All x such that $-8 \le x \le 9$

67. Approximately -1.5 and $-.2$

69. $-2 \le x \le -1$ and $3 \le x \le 7$

71. Approximately $-\$13,000$ (that is, a loss of $\$13,000$)

73. Approximately 12,300

75. (a)

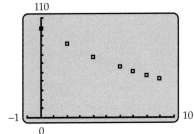

(b) $y = 94.749 - 6.054x$

(c) $f(1) = 88.70, f(5) = 64.48$; off by 60 cents and $3.00, respectively

(d) After 9.86 years, or during 1997

77. At 15 minutes she stopped for 5 minutes, then continued running steadily for 10 minutes. At 30 minutes she turned back and jogged home without stopping, arriving home 55 minutes after her start.

Excursion 3.3.A, page 219

1. $-15 \le x \le 30$ and $-10 \le y \le 10$ $(-10 \le t \le 10)$

3. $-10 \le x \le 6$ and $-7 \le y \le 7$ $(-7 \le t \le 7)$

5. $-15 \le x \le 0$ and $0 \le y \le 4$ $(0 \le t \le 4)$

7. $-5 \le x \le 45$ and $-65 \le y \le 65$

9. Entire graph: $-2 \le x \le 32$ and $-10 \le y \le 75$; near the origin: $-2 \le x \le 5$ and $-10 \le y \le 10$

11. Entire graph: $-16 \le x \le 2$ and $-62 \le y \le 60$; near the origin: $-2 \le x \le 2$ and $-4 \le y \le 2$

13. 7 times

Section 3.4, page 226

1. H   **3.** F   **5.** K   **7.** C

9.

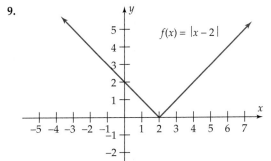

11.

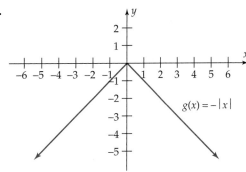

13. Viewing window: $-10 \le x \le 10$ and $-36 \le y \le 42$

15. Viewing window: $-13 \le x \le 12$ and $-2 \le y \le 14$

17.

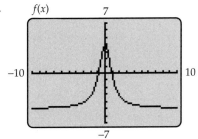

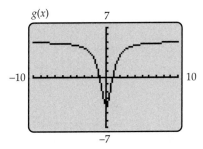

19. Shift the graph of f horizontally 3 units to the right; then shift it 2 units vertically upward.

21. Reflect the graph of f in the x-axis; shrink it toward the x-axis by a factor of $1/2$; shift it vertically 6 units downward.

23. $g(x) = f(x + 5) + 4 = (x + 5)^2 + 2 + 4 = (x + 5)^2 + 6$

25. $g(x) = 2f(x - 6) - 3 = 2\sqrt{x - 6} - 3$

27.

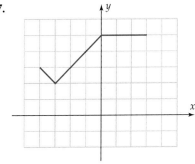

29.

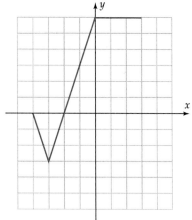

31.

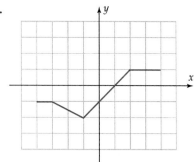

33.

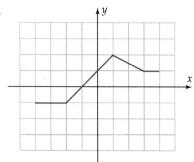

35.

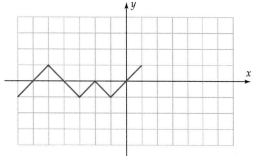

37.

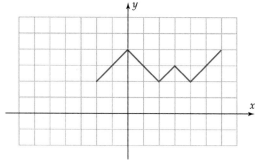

39.

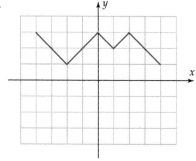

41.

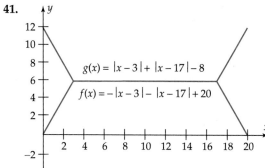

$$g(x) = |x - 3| + |x - 17| - 8$$

$$f(x) = -|x - 3| - |x - 17| + 20$$

43.

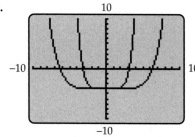

45.

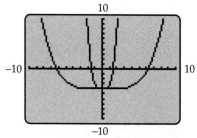

47.

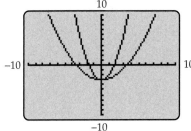

49.

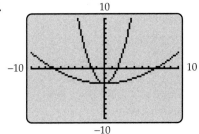

Excursion 3.4.A, page 233

1. Symmetric with respect to the y-axis

3. No axis or origin symmetry

5. Odd   **7.** Even   **9.** Even   **11.** Even

13. Neither   **15.** Yes   **17.** No   **19.** Origin

21. Origin   **23.** y-axis

25.

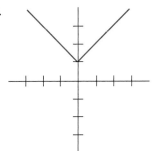

27.

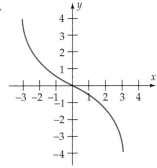

29.

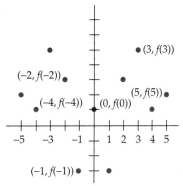

31. Many correct graphs, including the one shown here:

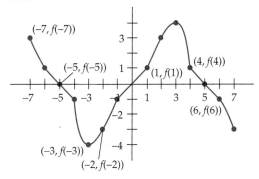

33. Suppose the graph is symmetric to the x-axis and the y-axis. If (x, y) is on the graph, then $(x, -y)$ is on the graph by x-axis symmetry. Hence, $(-x, -y)$ is on the graph by y-axis symmetry. Therefore, (x, y) on the graph implies that $(-x, -y)$ is on the graph, so the graph is symmetric with respect to the origin. Next suppose that the graph is symmetric to the y-axis and the origin. If (x, y) is on the graph, then $(-x, y)$ is on the graph by y-axis symmetry. Hence, $(-(-x), -y) = (x, -y)$ is on the graph by origin symmetry. Therefore, (x, y) on the graph implies that $(x, -y)$ is on the graph, so the graph is symmetric with respect to the x-axis. The proof of the third case is similar to that of the second case.

Section 3.5, page 239

1. $(f + g)(x) = x^3 - 3x + 2;$
$(f - g)(x) = -x^3 - 3x + 2;$
$(g - f)(x) = x^3 + 3x - 2$

3. $(f + g)(x) = \dfrac{1}{x} + x^2 + 2x - 5;$

$(f - g)(x) = \dfrac{1}{x} - x^2 - 2x + 5;$

$(g - f)(x) = x^2 + 2x - 5 - \dfrac{1}{x}$

5. $(fg)(x) = -3x^4 + 2x^3$;

$$\left(\frac{f}{g}\right)(x) = \frac{-3x + 2}{x^3};$$

$$\left(\frac{g}{f}\right)(x) = \frac{x^3}{-3x + 2}$$

7. $(fg)(x) = x^2\sqrt{x - 3} - 3\sqrt{x - 3}$

$$\left(\frac{f}{g}\right)(x) = \frac{x^2 - 3}{\sqrt{x - 3}}$$

$$\left(\frac{g}{f}\right)(x) = \frac{\sqrt{x - 3}}{x^2 - 3}$$

9. Domain of fg: all real numbers except 0; domain of f/g: all real numbers except 0

11. Domain of fg: All real numbers x such that $-4/3 \le x \le 2$; domain of f/g: All real numbers x such that $-4/3 < x \le 2$

13. 0   **15.** 30   **17.** 49; 1; -8 **19.** -3; -3; 0

21. $(f \circ g)(x) = (x + 3)^2$; $(g \circ f)(x) = x^2 + 3$; domain of $f \circ g$ and $g \circ f$ is all real numbers.

23. $(f \circ g)(x) = 1/\sqrt{x}$; $(g \circ f)(x) = 1/\sqrt{x}$; domain of $f \circ g$ and $g \circ f$ consists of all positive real numbers.

25. $(ff)(x) = x^6$; $(f \circ f)(x) = x^9$

27. $(ff)(x) = \dfrac{1}{x^2}$; $(f \circ f)(x) = x$

29. $(f \circ g)(x) = f\left(\dfrac{x - 2}{9}\right) = 9\left(\dfrac{x - 2}{9}\right) + 2 = x$ and $(g \circ f)(x) = g(9x + 2) = \dfrac{(9x + 2) - 2}{9} = x$

31. $(f \circ g)(x) = f(x - 2)^3 = \sqrt[3]{(x - 2)^3} + 2 = x$ and $(g \circ f)(x) = g(\sqrt[3]{x} + 2) = (\sqrt[3]{x} + 2 - 2)^3 = x$

33.

x	-4	-3	-2	-1	0
$f(x)$	-3	-1	0	1/2	1
$g(x) = f(f(x))$	-1	1/2	1	1.2	1.3

x	1	2	3	4
$f(x)$	1.3	1	-2	-2
$g(x) = f(f(x))$	1.35	1.3	0	0

35.

x	1	2	3	4	5
$(g \circ f)(x)$	4	2	5	4	4

37.

x	1	2	3	4	5
$(f \circ f)(x)$	1	3	3	5	1

In Answers 39–45, the given function is $B \circ A$, where A and B are the functions listed here. In some cases other correct answers are possible.

39. $A(x) = x^2 + 2, B(x) = \sqrt[3]{x}$

41. $A(x) = 7x^3 - 10x + 17, B(x) = x^7$

43. $A(x) = 3x^2 + 5x - 7, B(x) = \dfrac{1}{x}$

45. $A(x) = x + 5, B(x) = x^2 + \dfrac{1}{x}$

47. Not the same

49. (a) One day, .00012246 square inches; one week, .0000025 square inches; one 31-day month, .00000013 square inches

(b) No; no. The model is probably reasonable until the puddle is about the size of a period, with a radius of about .01. This occurs after approximately 15 hours.

51. (a) $A = \pi\left(\dfrac{d}{2}\right)^2 = \pi \cdot \dfrac{d^2}{4} = \dfrac{\pi}{4}\left(6 - \dfrac{50}{t^2 + 10}\right)^2$

(b) $\pi/4 \approx .7854$ square inches; 22.2648 square inches

(c) In approximately 11.39 weeks

53. $V = 256\pi t^3/3$; 17, 157.28 cm^3

Section 3.6, page 247

1. (a) 14 ft/sec
(b) 54 ft/sec
(c) 112 ft/sec
(d) $93\frac{1}{3}$ ft/sec

3. (a) \$692.5 million/year
(b) \$3400 million/year
(c) \$7881.1 million/year
(d) \$3868.7 million/year
(e) Between 1980 and 1997

5. (a) 14 cents/year
(b) -10.5 cents/year
(c) No change
(d) -9 cents/year

7. (a) 250 ties/month
(b) 438 ties/month
(c) 500 ties/month
(d) 563 ties/month
(e) -188 ties/month
(f) -750 ties/month
(g) -1500 ties/month
(h) -375 ties/month

9. (a) .709 gallons/inch
(b) 2.036 gallons/inch

11. (a) 10 feet/second
(b) 30 feet/second
(c) 80 feet/second

13. -72 **15.** -2 **17.** -1 **19.** 1.5858

21. (a) C, 62.5 ft/sec; D, 75 ft/sec
   (b) Approximately $t = 4$ to $t = 9.8$ sec
   (c) The average speed of car D from $t = 4$ to
   $t = 10$ sec is the slope of the secant line joining
   the (approximate) points (4, 100) and (10, 600),
   namely, $\dfrac{600 - 100}{10 - 4} \approx 83.33$ ft/sec. The average
   speed of car C is the slope of the secant line
   joining the (approximate) points (4, 475) and
   (10, 800), namely, $\dfrac{800 - 475}{10 - 4} \approx 54.17$ ft/sec.

23. (a) From day 0 until any day up to day 94, the
   average growth rate is positive.
   (b) From day 0 to day 95
   (c) -28, meaning that the population is decreasing
   at a rate of 28 chipmunks per day
   (d) 20, -20, and 0 chipmunks per day

25. (a)

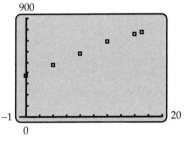

   (b) $y = 388.443956 + 23.690109x$
   (c) \$23.69/year
   (d) The data gives an average of \$23.71/year
   change over the period 1980 to 1997. However,
   the change in the period 1992 to 1997 is only
   \$16.40/year.

Excursion 3.6.A, page 254

1. 1   **3.** $2x + h$

5. $2t + h - 8000$ **7.** $2\pi r + \pi h$

9. (a) Average rate of change is -7979.9, which
   means that water is leaving the tank at a rate
   of 7979.9 gallons/minute.
   (b) -7979.99
   (c) -7980

11. (a) 6.5π
   (b) 6.2π
   (c) 6.1π
   (d) 6π
   (e) It's the same.

Chapter 3 Review, page 256

1. (a) -3 **(b)** 1755   **(c)** 2   **(d)** -14

3.

x	0	1	2	-4	t	k
$f(x)$	7	5	3	15	$7 - 2t$	$7 - 2k$

x	$b - 1$	$1 - b$	$6 - 2u$
$f(x)$	$9 - 2b$	$5 + 2b$	$-5 + 4u$

5. Many possible answers, including:
   (a) $f(x) = x^2$, $a = 2$, $b = 3$;
   $f(a + b) = f(2 + 3) = 5^2 = 25$, but
   $f(a) + f(b) = f(2) + f(3) = 2^2 + 3^2 = 13$, so the
   statement is false.
   (b) $f(x) = x + 1$, $a = 0$, $b = 1$; $f(ab) = f(0) = 1$, but
   $f(a)f(b) = f(0)f(1) = 1 \cdot 2 = 2$, so the statement
   is false.

7. All $r \geq 4$

9. $(t + 2)^2 - 3(t + 2) = t^2 + t - 2$

11. $2\left(\dfrac{x}{2}\right)^3 + \left(\dfrac{x}{2}\right) + 1 = \dfrac{x^3}{4} + \dfrac{x}{2} + 1$

13. (a) $f(t) = 50\sqrt{t}$
   (b) $g(t) = 2500\pi t$
   (c) Radius: 150 meters; area: 70,685.83 square
   meters
   (d) 12.73 h

15.

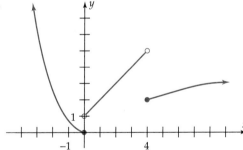

17. (b)

19. No local maxima; minimum at $x = -.5$; increasing
   when $x > -.5$; decreasing when $x < -.5$.

21. Maximum at $x \approx -5.0704$; minimum at
   $x \approx -.2629$. Increasing when $x > -5.0704$ and
   $x < -.2629$; decreasing when
   $-5.0704 < x < -.2629$

23.

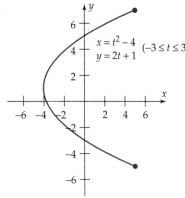

$x = t^2 - 4$
$y = 2t + 1$ $(-3 \le t \le 3)$

25. Many correct answers, including

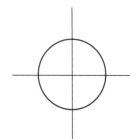

27. x-axis, y-axis, origin **29.** Even

31. Odd **33.** y-axis

35.

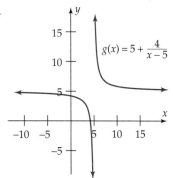

$g(x) = 5 + \dfrac{4}{x - 5}$

37. Approximately $-3 \le x \le 3.8$

39. Many correct answers, including $x = -2$; all x such that $2.5 < x < 3.8$; all x such that $5 \le x \le 6$

41. 1 **43.** -3 **45.** True

47. $x = 4$ **49.** $x \le 3$ **51.** $x < 3$

53. (a) King Richard's car
(b) King Richard's
(c) Fireball Bob's

55. Shrink the graph of g toward the x-axis by a factor of .25, then shift the graph vertically 2 units upward.

57. Shift the graph of g horizontally 7 units to the right; then stretch it away from the x-axis by a

factor of 3; then reflect it in the x-axis; finally, shift the graph vertically 2 units upward.

59. (e)

61. (a) $1/3$ (b) $(x - 1)\sqrt{x^2 + 5}$ $(x \ne 1)$

(c) $\dfrac{\sqrt{c^2 + 2c + 6}}{c}$

63.

x	-4	-3	-2	-1	0
$g(x)$	1	4	3	1	-1
$h(x)$	-3	-3	-4	-3	1

x	1	2	3	4
$g(x)$	-3	-2	-4	-3
$h(x)$	4	3	1	4

65. $\dfrac{82}{27}$ **67.** $\dfrac{1}{x^3} + 3$ **69.** $\dfrac{1}{4}$

71. $(f \circ g)(x) = f(x^2 - 1) = \dfrac{1}{x^2 - 1}$;

$(g \circ f)(x) = g\left(\dfrac{1}{x}\right) = \dfrac{1}{x^2} - 1$

73. All nonnegative numbers except 1

75. (a) $-1/3$ (b) $5/8$

77. 6 **79.** 3 **81.** $2x + h$

83. (a) \$290/ton (b) \$230/ton (c) \$212/ton

85. (a) Approximately day 45 to day 50
(b) Approximately any 10-day interval between day 20 and day 35
(c) Approximately day 30 to day 40

Chapter 4

Section 4.1, page 274

1. (5, 2), upward

3. (1, 2), downward

5. (3, -6), upward

7. ($-3/2$, 15/4), upward

9. I **11.** K **13.** J **15.** F

17. $f(x) = 3x^2$

19. $f(x) = 3(x - 2)^2 + 5$

21. $f(x) = 2(x + 3)^2 + 4$

23. $f(x) = \frac{1}{2}(x - 4)^2 + 3$

25. (-3, -21); $x = -6.24, .24$

27. (4, 14); $x = .26, 7.74$

29. (2/3, 19/3); $x = -.79, 2.12$

31. $(1/2, 1/4)$; $x = 0, 1$

33. $g(x) = 2x^2 - 5$; vertex $(0, -5)$

35. $h(x) = 2(x - 3)^2 + 4$; vertex $(3, 4)$

37. **(a)** 20,000; 100,000
     (b) 200; 550

39. $t = 2.5$ sec; $h = 196$ feet

41. 22 feet

43. **(a)** $t = 125/8$ seconds; $h = 3906.25$ feet
     (b) $t = 31.25$ seconds

45. Height = 841.92 feet; distance = 13,986.52 feet

47. **(a)** At 30 km/h, 30 meters; at 100 km/h, 170 meters
     (b) 50 km/h

49. \$3.50

51. \$3.67 (but if tickets must be priced in multiples of .20, then \$3.60 is best)

53. Minimum product is -4; numbers are 2 and -2.

55. 125/2 by 125/2

57. Area = 704.17 square feet

59. Two 50-ft sides and one 100-ft side

61. **(a)** $f(x) = \dfrac{1}{1152}x^2 - 2$
     (b) Approximately 33.94 inches from the center

Section 4.2, page 285

1. Quotient $3x^3 - 3x^2 + 5x - 11$; remainder 12

3. Quotient $x^2 + 2x - 6$; remainder $-7x + 7$

5. Quotient $5x^2 + 5x + 5$; remainder 0

7. No   **9.** Yes   **11.** 0, 2

13. $2\sqrt{2}, -1$ **15.** 2   **17.** 6

19. -30 **21.** 170, 802

23. No   **25.** Yes

27. $(x + 4)(2x - 7)(3x - 5)$

29. $(x - 3)(x + 3)(2x + 1)^2$

31. $f(x) = (x + 2)(x + 1)(x - 1)(x - 2)(x - 3)$
     $= x^5 - 3x^4 - 5x^3 + 15x^2 + 4x - 12$

33. $f(x) = x(x + 1)(x - 1)(x - 2)(x - 3)$
     $= x^5 - 5x^4 + 5x^3 + 5x^2 - 6x$

35. Many correct answers, including
     $(x - 1)(x - 7)(x + 4)$

37. Many correct answers, including
     $(x - 1)(x - 2)^2(x - \pi)^3$

39. $f(x) = \dfrac{17}{100}(x - 5)(x - 8)x$

41. $x = \pm 1$ or -3 **43.** $x = \pm 1$ or -5

45. $x = -4, 0, 1/2,$ or 1

47. $x = -3$ or 2   **49.** $x = 2$

51. $x = -5, 2,$ or 3   **53.** $(x - 2)(2x^2 + 1)$

55. $x^3(x^2 + 3)(x + 2)$ **57.** $(x - 2)(x - 1)^2(x^2 + 3)$

59. **(a)** The only possible rational roots of
     $f(x) = x^2 - 2$ are ± 1 or ± 2 (why?). But $\sqrt{2}$ is a root of $f(x)$ and $\sqrt{2} \neq \pm 1$ or ± 2. Hence, $\sqrt{2}$ is irrational.
     (b) $\sqrt{3}$ is a root of $x^2 - 3$ whose only possible rational roots are ± 1 or ± 3 (why?). But $\sqrt{3} \neq \pm 1$ or ± 3.

61. **(a)** 8.6378 people per 100,000
     (b) 1995

63. **(a)** 6 degrees/day at the beginning; 6.6435 degrees/day at the end
     (b) Day 2.0330 and day 10.7069
     (c) Day 5.0768 and day 9.6126

Excursion 4.2.A, page 290

1.
$$
\begin{array}{r|rrrrr}
2 & 3 & -8 & 0 & 9 & 5 \\
  &   & 6 & -4 & -8 & 2 \\
\hline
  & 3 & -2 & -4 & 1 & \boxed{7}
\end{array}
$$
quotient $3x^3 - 2x^2 - 4x + 1$;
remainder 7

3.
$$
\begin{array}{r|rrrrr}
-3 & 2 & 5 & 0 & -2 & -8 \\
   &   & -6 & 3 & -9 & 33 \\
\hline
   & 2 & -1 & 3 & -11 & \boxed{25}
\end{array}
$$
quotient $2x^3 - x^2 + 3x - 11$;
remainder 25

5.
$$
\begin{array}{r|rrrrr}
7 & 5 & 0 & -3 & -4 & 6 \\
  &   & 35 & 245 & 1694 & 11,830 \\
\hline
  & 5 & 35 & 242 & 1690 & \boxed{11,836}
\end{array}
$$
quotient $5x^3 + 35x^2 + 242x + 1690$;
remainder 11,836

7.
$$
\begin{array}{r|rrrrr}
2 & 1 & -6 & 4 & 2 & -7 \\
  &   & 2 & -8 & -8 & -12 \\
\hline
  & 1 & -4 & -4 & -6 & \boxed{-19}
\end{array}
$$
quotient $x^3 - 4x^2 - 4x - 6$;
remainder -19

9. Quotient $3x^3 + \dfrac{3}{4}x^2 - \dfrac{29}{16}x - \dfrac{29}{64}$; remainder $\dfrac{483}{256}$

11. Quotient $2x^3 - 6x^2 + 2x + 2$; remainder 1

13. $g(x) = (x + 4)(3x^2 - 3x + 1)$

15. $g(x) = \left(x - \dfrac{1}{2}\right)(2x^4 - 6x^3 + 12x^2 - 10)$

17. Quotient $x^2 - 2.15x + 4$; remainder 2.25

19. $c = -4$

Section 4.3, page 296

1. Yes   **3.** Yes   **5.** No

7. Degree 3, yes; degree 4, no; degree 5, yes

9. No

11. Degree 3, no; degree 4, no; degree 5, yes

13. The graphs have the same *shape* in the window with $-40 \le x \le 40$ and $-1000 \le y \le 5000$ but don't look identical.

15. -2 is a root of odd multiplicity, as are 1 and 3.

17. -2 and -1 are roots of odd multiplicity; 2 is a root of even multiplicity.

19. (e)   **21.** (f)   **23.** (c)

25. The graph in the standard viewing window does not rise at the far right as does the graph of the highest degree term x^3, so it is not complete.

27. The graph in the standard viewing window does not rise at the far left and far right as does the graph of the highest degree term $.005x^4$, so it is not complete.

29. $-9 \le x \le 3$ and $-20 \le y \le 40$

31. $-6 \le x \le 6$ and $-60 \le y \le 320$

33. Left half: $-33 \le x \le -2$ and $-50{,}000 \le y \le 260{,}000$; right half: $-2 \le x \le 3$ and $-20 \le y \le 30$

35. (a) The graph of a cubic polynomial (degree 3) has at most $3 - 1 = 2$ local extrema. When $|x|$ is large, the graph resembles the graph of ax^3, that is, one end shoots upward and the other end downward. If the graph had only one local extremum, both ends of the graph would go in the same direction (both up or down). Thus, the graph of a cubic polynomial has either two local extrema or none.

(b) These are the only possible shapes for a graph that has 0 or 2 local extrema, 1 point of inflection, and resembles the graph of ax^3 when $|x|$ is large.

37. (a) Odd   **(b)** Positive   **(c)** $-2, 0, 4$, and 6
(d) 5

39. (d)

41.

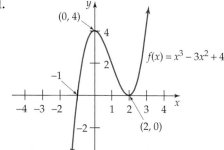

43.

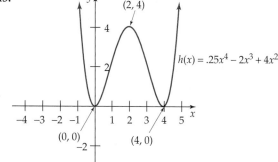

45.

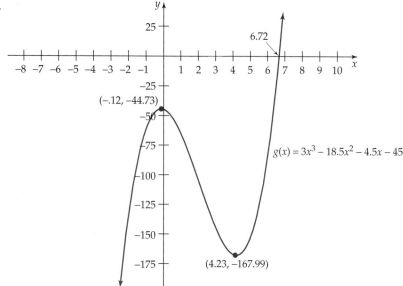

47. 1973

49. (a) The solutions are roots of

$$g(x) - 4 = .01x^3 - .06x^2 + .12x - .08.$$

This polynomial has degree 3 and hence has at most 3 roots.

(b) $1 \le x \le 3$ and $3.99 \le y \le 4.01$

(c) Suppose $f(x)$ has degree n. If the graph of $f(x)$ had a horizontal segment lying on the line $y = k$ for some constant k, then the equation $f(x) = k$ would have infinitely many solutions (why?). But the polynomial $f(x) - k$ has degree n (why?) and thus has at most n roots. Hence the equation $f(x) = k$ has at most n solutions, which means the graph cannot have a horizontal segment.

51. (a)

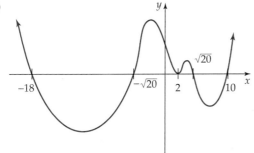

(b)–(d) Depends on the calculator

Excursion 4.3.A, page 302

1. (a) $R(x) = 29x$

(b) $P(x) = 29x - (.001x^3 + .06x^2 - 1.5x)$
$= -.001x^3 - .06x^2 + 30.5x$

(c) 82,794; $1546.39

3. (a) 600

(b) 958

5. (a) 4.4267 by 4.4267 inches

(b) 4 by 4 inches (because no side can exceed 12 inches)

7. 39.1487 by 39.1487 by 19.5743

9. (a) Approximately 206

(b) Approximately 269; approximately $577

Section 4.4, page 307

1. Cubic   **3.** Quadratic

5. (a) $y = -2.134090909x^3 + 52.00963203x^2 - 359.2162338x + 5512.618182$

(b) 1987: 4814.58 per 100,000; 1995: 4623.99

(c) 3951.94

(d) Answers vary; perhaps through the year 2000.

7. (a)

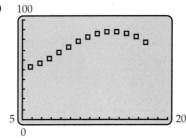

(b) $y = -.5179820180x^2 + 14.65684316x - 20.88711289$

(c) Noon: 80°; 9 A.M.: 69°; 2 P.M.: 83°

9. (a)

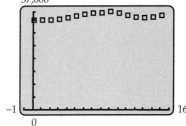

(b) Quartic

(c) $y = 2.397660405x^4 - 73.93709223x^3 + 691.3121614x^2 - 1762.930366x + 33,079.5044$

(d) $66,475; no; this is double that of 1994.

(e) 1998

11. (a)

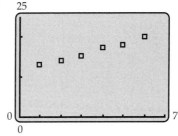

(b) $y = 1.32x + 12.9133$; a linear model, chosen from the scatter plot, although a quadratic or cubic model fits the data points reasonably well. A quartic model fits these data points, but indicates a very large increase after 1996, which seems unlikely.

13. (a) Cubic: $y = -.01897x^3 + .56273x^2 - 5.13900x + 21.94073$;
Quartic: $y = -.00504x^4 + .17273x^3 - 2.01756x^2 + 9.29074x - 6.03968$

(b) No

Section 4.5, page 319

1. All real numbers except $-5/2$

3. All real numbers except $3 + \sqrt{5}$ and $3 - \sqrt{5}$

5. All real numbers except $-\sqrt{2}$, 1, and $\sqrt{2}$

7. Vertical asymptotes: $x = -1$ and $x = 6$

9. Vertical asymptotes: $x = 0$ and $x = -1$

11. Vertical asymptotes: $x = -2$ and $x = 2$

13. $y = 3$; any window with $-115 \le x \le 110$

15. $y = -1$; any window with $-31 \le x \le 35$

17. $y = 5/2$; any window with $-40 \le x \le 42$

19.

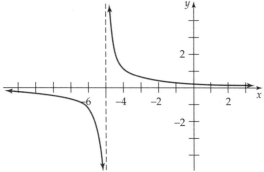

vertical asymptote $x = -5$
horizontal asymptote $y = 0$

21.

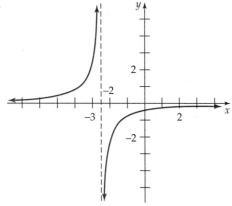

vertical asymptote $x = -2.5$
horizontal asymptote $y = 0$

23.

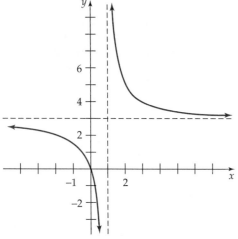

vertical asymptote $x = 1$
horizontal asymptote $y = 3$

25.

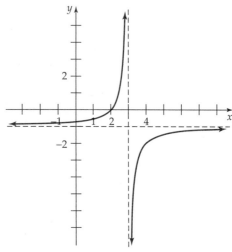

vertical asymptote $x = 3$
horizontal asymptote $y = -1$

27.

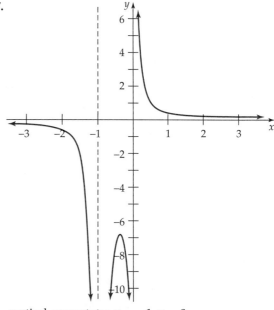

vertical asymptotes $x = -1, x = 0$
horizontal asymptotes $y = 0$

29.

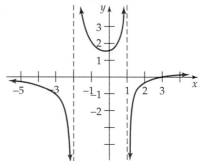

vertical asymptotes $x = -2, x = 1$
horizontal asymptotes $y = 0$

31.

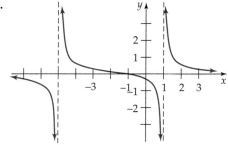

vertical asymptotes $x = -5, x = 1$
horizontal asymptote $y = 0$

33.

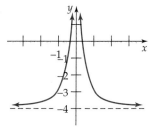

vertical asymptote $x = 0$
horizontal asymptote $y = -4$

35.

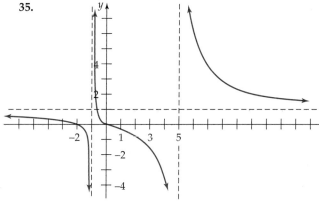

vertical asymptotes $x = -1, x = 5$
horizontal asymptote $y = 1$

37. Overall: $-5 \le x \le 4.4$ and $-8 \le y \le 4$; hidden area near origin: $-2 \le x \le 2$ and $-.5 \le y \le .5$; hidden area near $x = -5$: $-15 \le x \le -3$ and $-.07 \le y \le .02$

39. For vertical asymptotes and x-intercepts: $-4.7 \le x \le 4.7$ and $-8 \le y \le 8$; to see graph get close to the horizontal asymptote: $-40 \le x \le 35$ and $-2 \le y \le 3$

41. Overall: $-4.7 \le x \le 4.7$ and $-3 \le y \le 3$; hidden area near $x = 4$: $3 \le x \le 15$ and $-.02 \le y \le .01$

43. For vertical asymptote:
$$0.5 \le x \le 1 \text{ and } -2 \le y \le 2$$

45. 8.4343 in. $\times$ 8.4343 in. $\times$ 14.057 in.

47. (a) $c(x) = \dfrac{2800 + 3.5x^2}{x}$ **(b)** $13.91 \le x \le 57.52$
(c) 28.28 mph

49. (a) $p(x) = \dfrac{500 + x^2}{x}$ **(b)** $10 \le x \le 50$
(c) $x = 22.36$; 22.36 m by 11.18 m

51. (a) $h_1 = h - 2$
(b) $h_1 = \dfrac{150}{\pi r^2} - 2$ (because $\pi r^2 h = 150$)
(c) $V = \pi(r - 1)^2\left(\dfrac{150}{\pi r^2} - 2\right)$
(d) The walls are 1 ft thick.
(e) $r \approx 2.88$ ft; $h \approx 5.76$ ft

53. (a) $g(0) = 9.801$ m/sec^2
(b)

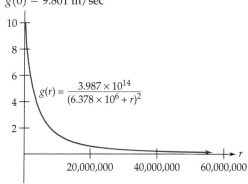

$$g(r) = \dfrac{3.987 \times 10^{14}}{(6.378 \times 10^6 + r)^2}$$

(c) There are no r-intercepts because the numerator is never zero. So you can never completely escape the pull of gravity.

Excursion 4.5.A, page 325

1. Asymptote: $y = x$; window: $-40 \le x \le 40$ and $-40 \le y \le 40$

3. Asymptote: $y = x^2 - x$; window: $-15 \le x \le 6$ and $-40 \le y \le 240$

5.

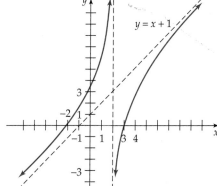

$y = x + 1$

vertical asymptote $x = 2$
oblique asymptote $y = x + 1$

7.

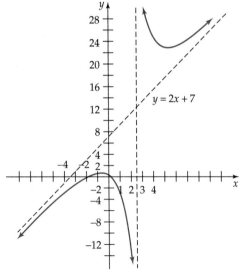

vertical asymptote $x = 5/2$
oblique asymptote $y = 2x + 7$

9.

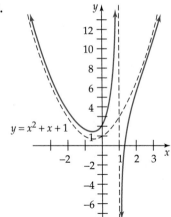

vertical asymptote $x = 1$
parabolic asymptote $y = x^2 + x + 1$

11.

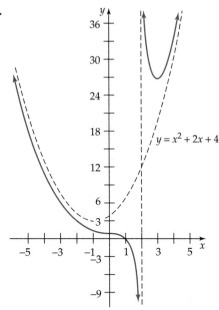

vertical asymptote $x = 2$
parabolic asymptote $y = x^2 + 2x + 4$

13. $-15.5 \leq x \leq 8.5$ and $-16 \leq y \leq 8$

15. $-4.7 \leq x \leq 4.7$ and $-12 \leq y \leq 8$

17. Overall: $-13 \leq x \leq 7$ and $-20 \leq y \leq 20$; hidden
area near the origin: $-2.5 \leq x \leq 1$ and
$-.02 \leq y \leq .02$

19. (b) Approximately $.06 \leq x \leq 2.78$

Section 4.6, page 330

1. $8 + 2i$ **3.** $-2 - 10i$

5. $-\dfrac{1}{2} - 2i$ **7.** $\left(\dfrac{\sqrt{2} - \sqrt{3}}{2}\right) + 2i$

9. $1 + 13i$ **11.** $-10 + 11i$

13. $-21 - 20i$ **15.** 4 **17.** $-i$

19. i **21.** i

23. $\dfrac{5}{29} + \dfrac{2}{29}i$ **25.** $-\dfrac{1}{3}i$ **27.** $\dfrac{12}{41} - \dfrac{15}{41}i$

29. $\dfrac{-5}{41} - \dfrac{4}{41}i$ **31.** $\dfrac{10}{17} - \dfrac{11}{17}i$ **33.** $\dfrac{7}{10} + \dfrac{11}{10}i$

35. $-\dfrac{113}{170} + \dfrac{41}{170}i$ **37.** $6i$ **39.** $\sqrt{14}i$

41. $-4i$ **43.** $11i$ **45.** $(\sqrt{15} - 3\sqrt{2})i$

47. $2/3$ **49.** $-41 - i$

51. $(2 + 5\sqrt{2}) + (\sqrt{5} - 2\sqrt{10})i$

53. $\dfrac{1}{3} - \dfrac{\sqrt{2}}{3}i$ **55.** $x = 2, y = -2$

57. $x = -3/4, y = 3/2$ **59.** $x = \dfrac{1}{3} \pm \dfrac{\sqrt{14}}{3}i$

61. $x = -\dfrac{1}{2} \pm \dfrac{\sqrt{7}}{2}i$ **63.** $x = \dfrac{1}{4} \pm \dfrac{\sqrt{31}}{4}i$

65. $x = \dfrac{3 \pm \sqrt{3}}{2}$

67. $x = 2, -1 + \sqrt{3}i, -1 - \sqrt{3}i$

69. $x = 1, -1, i, -i$ **71.** -1

73. $\bar{z} = a - bi$, so that $\bar{\bar{z}} = a - (-bi) = a + bi = z$

75. $\dfrac{1}{z} = \left(\dfrac{a}{a^2 + b^2}\right) + \left(\dfrac{-b}{a^2 + b^2}\right)i$

Section 4.7, page 336

1. 2 **3.** 6 **5.** -30

7. $x = 0$ (multiplicity 54); $x = -4/5$ (multiplicity 1)

9 $x = 0$ (multiplicity 15); $x = \pi$ (multiplicity 14); $x = \pi + 1$ (multiplicity 13)

11. $x = 1 + 2i$ or $1 - 2i$; $f(x) = (x - 1 - 2i)(x - 1 + 2i)$

13. $x = -\dfrac{1}{3} + \dfrac{2\sqrt{5}}{3}i$ or $-\dfrac{1}{3} - \dfrac{2\sqrt{5}}{3}i$; $f(x) = \left(x + \dfrac{1}{3} - \dfrac{2\sqrt{5}}{3}i\right)\left(x + \dfrac{1}{3} + \dfrac{2\sqrt{5}}{3}i\right)$

15. $x = 3$ or $-\dfrac{3}{2} + \dfrac{3\sqrt{3}}{2}i$ or $-\dfrac{3}{2} - \dfrac{3\sqrt{3}}{2}i$; $f(x) = (x - 3)\left(x + \dfrac{3}{2} - \dfrac{3\sqrt{3}}{2}i\right)\left(x + \dfrac{3}{2} + \dfrac{3\sqrt{3}}{2}i\right)$

17. $x = -2$ or $1 + \sqrt{3}i$ or $1 - \sqrt{3}i$; $f(x) = (x + 2)(x - 1 - \sqrt{3}i)(x - 1 + \sqrt{3}i)$

19. $x = 1$ or i or -1 or $-i$; $f(x) = (x - 1)(x - i)(x + 1)(x + i)$

21. $x = \sqrt{5}$ or $-\sqrt{5}$ or $\sqrt{2}i$ or $-\sqrt{2}i$; $f(x) = (x - \sqrt{5})(x + \sqrt{5})(x - \sqrt{2}i)(x + \sqrt{2}i)$

23. Many correct answers, including
$$(x - 1)(x - 7)(x + 4)$$

25. Many correct answers, including
$$(x - 1)(x - 2)^2(x - \pi)^3$$

27. $f(x) = 2x(x - 4)(x + 3)$

In Exercises 29–40, there are many correct answers, including the following.

29. $x^2 - 4x + 5$ **31.** $(x - 2)(x^2 - 4x + 5)$

33. $(x + 3)(x^2 - 2x + 2)(x^2 - 2x + 5)$

35. $x^2 - 2x + 5$ **37.** $(x - 4)^2(x^2 - 6x + 10)$

39. $(x^4 - 3x^3)(x^2 - 2x + 2)$

41. $3x^2 - 6x + 6$ **43.** $-2x^3 + 2x^2 - 2x + 2$

45. Many correct answers, including
$$x^2 - (1 - i)x + (2 + i)$$

47. Many correct answers, including
$$x^3 - 5x^2 + (7 + 2i)x - (3 + 6i)$$

49. $3, -\dfrac{1}{2} + \dfrac{\sqrt{3}}{2}i, -\dfrac{1}{2} - \dfrac{\sqrt{3}}{2}i$

51. $i, -i, -1, -2$ **53.** $1, 2i, -2i$

55. $i, -i, 2 + i, 2 - i$

57. **(a)** Since $z + w = (a + c) + (b + d)i$, $\overline{z + w} = (a + c) - (b + d)i$. Since $\bar{z} = a - bi$ and $\bar{w} = c - di$, $\bar{z} + \bar{w} = (a - bi) + (c - di) = (a + c) - (b + d)i$. Hence $\overline{z + w} = \bar{z} + \bar{w}$.

 (b) Since $zw = (ac - bd) + (ad + bc)i$, $\overline{zw} = (ac - bd) - (ad + bc)i$. Since $\bar{z} = a - bi$ and $\bar{w} = c - di$, $\bar{z} \cdot \bar{w} = (a - bi)(c - di) = (ac - bd) - (ad + bc)i$. Hence $\overline{zw} = \bar{z} \cdot \bar{w}$.

59. **(a)** $\overline{f(z)} = \overline{az^3 + bz^2 + cz + d}$ [Definition of $f(z)$]
$= \overline{az^3} + \overline{bz^2} + \overline{cz} + \bar{d}$ Exercise 57(a)
$= \bar{a}\overline{z^3} + \bar{b}\overline{z^2} + \bar{c}\bar{z} + \bar{d}$ Exercise 57(b)
$= a\overline{z^3} + b\overline{z^2} + c\bar{z} + d$ $\bar{r} = r$ for r real
$= a\bar{z}^3 + b\bar{z}^2 + c\bar{z} + d$ Exercise 57(b)
$= f(\bar{z})$ Definition of f

 (b) Since $f(z) = 0$, we have $0 = \bar{0} = \overline{f(z)} = f(\bar{z})$. Hence $\bar{z}$ is a root of $f(x)$.

61. If $f(z)$ is a polynomial with real coefficients, then $f(z)$ can be factored as $g_1(z)g_2(z)g_3(z)\cdots g_k(z)$, where each $g_i(z)$ is a polynomial with real coefficients and degree 1 or 2. The rules of polynomial multiplication show that the degree of $f(z)$ is the sum: degree $g_1(z)$ + degree $g_2(z)$ + degree $g_3(z)$ + $\cdots$ + degree $g_k(z)$. If all of the $g_i(z)$ have degree 2, then this last sum is an even number. But $f(z)$ has odd degree, so this can't occur. Therefore, at least one of the $g_i(z)$ is a first-degree polynomial and hence must have a real root. This root is also a root of $f(z)$.

Chapter 4 Review, page 338

1. $(2, 3)$ **3.** $(4, -4)$ **5.** $(1.5, -5.75)$

7. **(a)** $y = 260 - x$
 (b) $A = -x^2 + 260x - 3500$
 (c) $x = 130$ ft; $y = 130$ ft

9. $x = 30$ ft, $y = 60$ ft

11. (a), (c), (e), (f) **13.** 0

15.
$$\begin{array}{r|rrrrrrr} 2 & 1 & -5 & 8 & 1 & -17 & 16 & -4 \\ & & 2 & -6 & 4 & 10 & -14 & 4 \\ \hline & 1 & -3 & 2 & 5 & -7 & 2 & \boxed{0} \end{array}$$
other factor: $x^5 - 3x^4 + 2x^3 + 5x^2 - 7x + 2$

17. Many correct answers, including
$$f(x) = 5(x - 1)^2(x + 1) = 5x^3 - 5x^2 - 5x + 5$$

19. -1 and $5/3$

21. $x = -2$, or $\dfrac{4 \pm \sqrt{3}}{3}$

23. (a) $1, -1, 3, -3, 1/2, -1/2, 3/2, -3/2$

(b) 3

(c) $3, \dfrac{1 + \sqrt{3}}{2}, \dfrac{1 - \sqrt{3}}{2}$

25. d

27. $2x + h + 1$

29. Many correct answers

31. (c)

33. $-3 \le x \le 9$ and $-35 \le y \le 15$

35. $-2 \le x \le 18$ and $-500 \le y \le 1100$

37. (a)

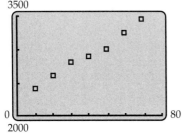

(b) $c(x) \approx 0.008515x^3 - 1.1094x^2 + 56.2583x + 2017.2576$

(c) $\$26.80$

(d) Average cost of 35: $\$85.49$; average cost of 75: $\$47.84$

39.

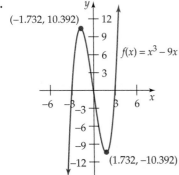

41.

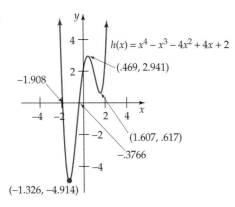

43.

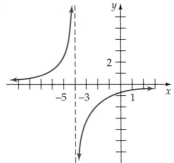

45.

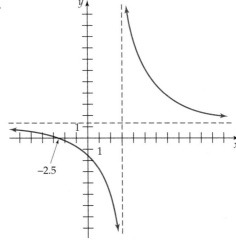

47. Overall: $-4.7 \le x \le 4.7$ and $-5 \le y \le 5$ (Adjust x range to avoid erroneous vertical lines on widescreen calculators.) Near $x = 3$: $1 \le x \le 25$ and $-.5 \le y \le .1$

49. $-19 \le x \le 19$ and $-8 \le y \le 8$

51. More than 400 bags (priced at $\$1.75$ each) and less than 2500 bags (priced at $\$0.70$ each)

53. (a) $T = \dfrac{40}{x} + \dfrac{110}{x + 25}$, where x is the speed of the car and $0 < x < 60$ (speed limit)

(b) At least 44.08 mph

55. $\dfrac{1}{(x + h + 1)(x + 1)}$

57. $x = \dfrac{-3 \pm \sqrt{31}i}{2}$

59. $x = \dfrac{3 \pm \sqrt{31}i}{10}$

61. $x = \sqrt{2/3}$ or $-\sqrt{2/3}$ or i or $-i$

63. $x = -2$ or $1 + \sqrt{3}i$ or $1 - \sqrt{3}i$

65. $i, -i, 2, -1$

67. Many correct answers, including $x^4 - 2x^3 + 2x^2$

Chapter 5

Section 5.1, page 355

1.

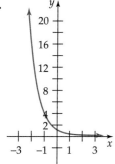

3.

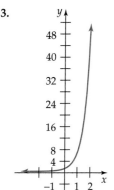

5.

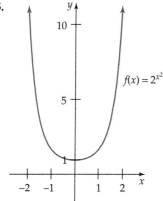

$f(x) = 2^{x^2}$

7. Shift the graph of h vertically 5 units downward.

9. Stretch the graph of h away from the x-axis by a factor of 3.

11. Shift the graph of h horizontally 2 units to the left, then vertically 5 units downward.

13. $f(x)$: C; $g(x)$: A; $h(x)$: B

15. Neither   **17.** Even

19. Even   **21.** 11   **23.** .8

25. $\dfrac{10^{x+h} - 10^x}{h}$ **27.** $\dfrac{(2^{x+h} + 2^{-x-h}) - 2^x - 2^{-x}}{h}$

29. $-3 \le x \le 3$ and $0 \le y \le 12$

31. $-4 \le x \le 4$ and $0 \le y \le 10$

33. $-10 \le x \le 10$ and $0 \le y \le 20$

35. $-10 \le x \le 10$ and $0 \le y \le 6$

37. The x-axis is a horizontal asymptote for the left side of the graph; local minimum at $(-1.44, -0.53)$.

39. No asymptotes; local minimum at $(0, 1)$

41. The x-axis is a horizontal asymptote; local maximum at $(0, 1)$.

43. **(a)** $t = 0.75, C = 8; t = 1, C = 16$
     (b) $C(t) = 2^{4t}$

45. $f(x) = -315(e^{0.00418x} + e^{-0.00418x}) + 1260$

47. **(a)** About 520 in 15 days; about 1559 in 25 days
     (b) In 29.3 days

49. **(a)** 1980: 74.06; 2000: 76.34
     (b) 1930

51. **(a)**

Folds	0	1	2	3	4
Thickness	.002	.004	.008	.016	.032

     (b) $f(x) = .002(2^x)$ **(c)** 2097.15 in. = 174.76 ft
     (d) 43

53. **(a)** 100,000 now; 83,527 in 2 months; 58,275 in 6 months
     (b) No. The graph continues to decrease toward zero.

55. **(a)** The current population is 10, and in 5 years it will be about 149.
     (b) After about 9.55 years

57. Many correct answers: $f(x) = a^x$ for any nonnegative constant a

59. **(a)** The graph of f is the mirror image of the graph of g.
     (b) $k(x) = f(x)$; see (a).

Section 5.2, page 365

1. Annually: $1469.33; quarterly: $1485.95; monthly: $1489.85; weekly: $1491.37

3. $585.83   **5.** $610.40   **7.** $639.76

9. $563.75   **11.** $582.02   **13.** $3325.29

15. $3359.59   **17.** $6351.16   **19.** $568.59

21. Fund C   **23.** $385.18

25. $1,162,003.14

27. Take $4000 in four years

29. About 5.00%   **31.** About 5.92%

33. (a) 9 years
    (b) 9 years
    (c) 9 years
    (d) Investment and doubling time are independent.

35. 9.9 years

37. (a) 12.55%
    (b) 12.55%; 12.68%; 12.75%

39. (a) $f(x) = 6(3^x)$ or $f(x) = 18(3^{x-1})$
    (b) 3
    (c) No; yes

41. (a) $g(x) = 67.4(1.026)^x$
    (b) 112.62 million

43. (a) $f(x) = 5550(1.0368^x)$
    (b) \$7966.05
    (c) 1996–1997

45. About 256; about 654

47. (a) $f(x) = 100(.75^x)$
    (b) About 8 feet

49. (a) $f(x) = 100(.5^{x/1620})$
    (b) 71.0 mg; 50.4 mg; 25.4 mg

51. 7171.3 years old

51.

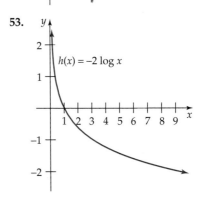

53.

55. $0 \le x \le 9.4$ and $-6 \le y \le 6$ (vertical asymptote at $x = 1$)

57. $-10 \le x \le 10$ and $-3 \le y \le 3$

59. $0 \le x \le 20$ and $-6 \le y \le 3$

61. .5493   **63.** $-.2386$

65. (a) $\dfrac{\ln(3 + h) - \ln 3}{h}$ **(b)** $h \approx 2.2$

67. (a) Both are acid.   **(b)** .01   **(c)** Yes

69. $A = -9; B = 10$

71. About 4392 meters

73. (a) 77
    (b) 66; 59
    (c) About 14 weeks

75. (a) 9.9 days
    (b) About 6986

77. $n = 30$ gives an approximation with a maximum error of 0.00001 when $-.7 \le x \le .7$

79. (a) No advertising: about 120 bikes; \$1000: about 299 bikes; \$10,000: about 513 bikes
    (b) \$1000, yes; \$10,000; yes
    (c) Yes; yes (but not as worthwhile as spending \$1000)

Section 5.3, page 376

1. 4   **3.** -2.5 **5.** $10^3 = 1000$

7. $10^{2.88} = 750$ **9.** $e^{1.0986} = 3$ **11.** $e^{-4.6052} = .01$

13. $x^2 + 2y = e^{z+w}$ **15.** $\log .01 = -2$

17. $\log 3 = .4771$ **19.** $\ln 25.79 = 3.25$

21. $\ln 5.5527 = 12/7$ **23.** $\ln w = 2/r$ **25.** $\sqrt{43}$

27. 15   **29.** 1/2   **31.** 931   **33.** $x + y$

35. x^2 **37.** $x > -1$ **39.** $x < 0$

41. (a) For all $x > 0$
    (b) According to the fourth property of natural logarithms on page 373, $e^{\ln x} = x$ for every $x > 0$.

43. They are the same for $x > 0$. If $x < 0$, $g(x)$ does not exist, while $f(x)$ is symmetric with respect to the y-axis.

45. Stretch the graph of g away from the x-axis by a factor of 2.

47. Shift the graph of g horizontally 4 units to the right.

49. Shift the graph of g horizontally 3 units to the left, then shift it vertically 4 units downward.

Section 5.4, page 384

1. $\ln(x^2 y^3)$ **3.** $\log(x - 3)$ **5.** $\ln(x^{-7})$

7. $3\ln(e - 1)$ **9.** $\log(20xy)$ **11.** $2u + 5v$

13. $\dfrac{1}{2}u + 2v$

15. $\dfrac{2}{3}u + \dfrac{1}{6}v$

17. False; the right side is not defined when $x < 0$, but the left side is.

19. True by the Power Law

21. False; the graph of the left side differs from the graph of the right side.

23. Many possible answers. For example $a = 2, b = 3$.

25. e

27. 2

29. Approximately 2.54

31. 20 decibels

33. Approximately 66 decibels

35. Twice as loud

37. (a) 1.255; 1.255
(b) 2.659; 2.659
(c) 3.952; 3.952
(d) $\log x = \dfrac{\ln x}{\ln 10}$ for all $x > 0$

39. (a) .9422
(b) 1.9422
(c) 2.9422
(d) 3.9422
(e) 4.9422
(f) If we have x_n, then $x_{n+1} = 10x_n$; if we have $\log x_n$, then $\log x_{n+1} = \log 10x_n = 1 + \log x_n$; each increase of the logarithmic scale by 1 corresponds to multiplying by 10. Also, an increase by y in the logarithmic scale corresponds to multiplying by 10^y.

Excursion 5.4.A, page 390

1.

x	0	1	2	4
$f(x) = \log_4 x$	Not defined	0	.5	1

3.

x	1/36	1/6	1	216
$h(x) = \log_6 x$	-2	-1	0	3

5.

x	0	1/7	$\sqrt{7}$	49
$f(x) = 2\log_7 x$	Not defined	-2	1	4

7.

x	-2.75	-1	1	29
$h(x) = 3\log_2(x+3)$	-6	3	6	15

9. $\log .01 = -2$

11. $\log \sqrt[3]{10} = 1/3$

13. $\log r = 7k$

15. $\log_7 5{,}764{,}801 = 8$

17. $\log_3(1/9) = -2$

19. $10^4 = 10{,}000$

21. $10^{2.88} \approx 750$

23. $5^3 = 125$

25. $2^{-2} = \dfrac{1}{4}$

27. $10^{z+w} = x^2 + 2y$

29. $\sqrt{97}$

31. $x^2 + y^2$

33. 1/2

35. 6

37. $b = 3$

39. $b = 20$

41. 5

43. 3

45. 4

47. $\log \dfrac{x^2 y^3}{z^6}$

49. $\log(x^2 - 3x)$

51. $\log_2(5c)$

53. $\log_4\!\left(\dfrac{1}{49c^2}\right)$

55. $\ln\!\left[\dfrac{(x+1)^2}{x+2}\right]$

57. $\log_2(x)$

59. $\ln(e^2 - 2e + 1) = \ln(e-1)^2 = 2\ln(e-1)$

61. 3.3219

63. .8271

65. 1.1115

67. 1.6199

69. True

71. True

73. False

75. 397^{398}

77. $\log_b u = \dfrac{\log_a u}{\log_a b}$

79. $\log_{10} u = 2\log_{100} u$

81. $\log_b x = \dfrac{1}{2}\log_b v + 3 = \log_b \sqrt{v} + \log_b b^3 = \log_b(b^3\sqrt{v})$; hence $x = b^3\sqrt{v}$.

83. $f(x) = g(x)$ only when $x \approx .123$, so the statement is false.

85.

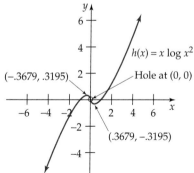

$h(x) = x \log x^2$

Hole at $(0, 0)$

$(-.3679, .3195)$

$(.3679, -.3195)$

Section 5.5, page 397

1. $x = 4$

3. $x = 1/9$

5. $x = 1/2$ or -3

7. $x = -2$ or $-1/2$

9. $x = (\ln 5)/(\ln 3) \approx 1.465$

11. $x = (\ln 3)/(\ln 1.5) \approx 2.7095$

13. $x = \dfrac{\ln 3 - 5\ln 5}{\ln 5 + 2\ln 3} \approx -1.825$

15. $x = \dfrac{\ln 2 - \ln 3}{3\ln 2 + \ln 3} \approx -.1276$

17. $x = (\ln 5)/2 \approx .805$

19. $x = (-\ln 3.5)/1.4 \approx -.895$

21. $x = 2\ln(5/2.1)/\ln 3 \approx 1.579$

23. If $\ln u = \ln v$, then $e^{\ln u} = e^{\ln v}$, so $u = v$

25. $x = 9$ **27.** $x = 5$

29. $x = 6$ **31.** $x = 3$

33. $x = \dfrac{-5 + \sqrt{37}}{2}$ **35.** $x = \dfrac{9}{e - 1}$

37. $x = 5$ **39.** $x = \pm\sqrt{10{,}001}$

41. $x = \sqrt{\dfrac{e + 1}{e - 1}}$

43. .444 billion years   **45.** 14.95 days

47. 9081.8 years old   **49.** 6.8 years

51. Yes, because $1.073^{10} > 2$

53. 79.36 years

55. (a) About 2.1548%
   (b) In the year 2013

57. (a) $k \approx 21.459$
   (b) $t \approx .182$

59. (a) There are 20 bacteria at the beginning and 2500 three hours later.
   (b) $\dfrac{\ln 2}{\ln 5} \approx .43$ hr ≈ 25.8 min

61. (a) $k \approx .229, c \approx 83.3$
   (b) 12.43 weeks

Section 5.6, page 405

1. Cubic, logistic

3. Exponential, quadratic, cubic

5. Exponential, logarithmic, quadratic, cubic

7. Quadratic, cubic   **9.** Quadratic, cubic

11. Ratios range from approximately 5.050 to 5.076; exponential is appropriate.

13. (a) For large values of x the term $54e^{-.0228x}$ is close to zero so the quantity $(1 + 54e^{-.0228x})$ is slightly larger than 1, which means $\dfrac{384.57}{1 + 54e^{-.0228x}}$ is always less than (but very close to) 384.57.

   (b)

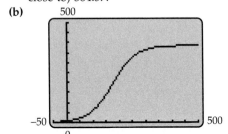

15. (a)

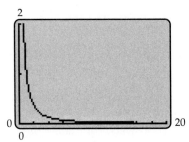

   (b)

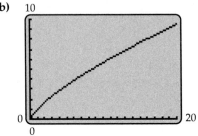

   (c)

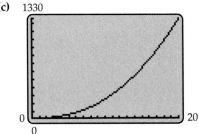

17. Power model

19. Power or logarithmic model

21. (a)

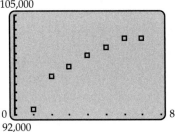

   (b)

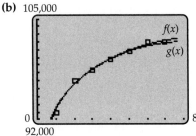

(c)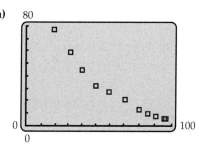

23. (a)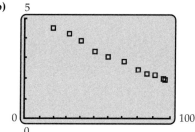

(b)

(c) Exponential.
$y = 154.55(.97^x)$

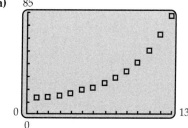

25. (a)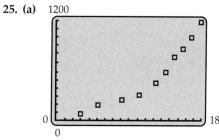

(b) $y = 5.57x^2 - 43.12x + 205.61$

(c) $y = \dfrac{4539.4}{1 + 85.61e^{-.1944x}}$

(d) 1,571,210; 1,648,400

27. (a)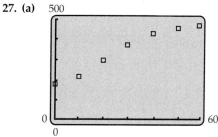

(b) $y = -.0025x^3 + .167x^2 + 3.784x + 170.411$
This model is not reasonable because it predicts smaller and smaller farm sizes after 1997, decreasing to 0 in about 30 years.

(c) $y = \dfrac{519.79}{1 + 2.236e^{-.056x}}$

(d) 482.8 acres

(e) Never; the model levels off at approximately 519.8 acres.

29. (a)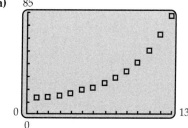

(b) $y = 10.48(1.16^x)$

(c–d)

Year	Worldwide Shipments (thousands)	Predicted Number Shipments (thousands)	Worldwide Shipments Ratio (current to previous)
1985	14.7	12.2	
			1.03
1986	15.1	14.1	
			1.11
1987	16.7	16.4	
			1.08
1988	18.1	19	
			1.18
1989	21.3	22	
			1.11
1990	23.7	25.5	
			1.14
1991	27	29.6	
			1.2
1992	32.4	34.4	
			1.2
1993	38.9	39.9	
			1.23
1994	47.9	46.2	
			1.26
1995	60.2	53.6	
			1.18
1996	70.9	62.2	
			1.19
1997	84.3	72.2	

An exponential model may not be appropriate.

Section 5.7, page 418

1. $g(x) = -x$

3. $g(x) = \dfrac{x + 4}{5}$

5. $g(x) = \sqrt[3]{\dfrac{5 - x}{2}}$

7. $g(x) = \dfrac{x^2 + 7}{4}, \quad (x \geq 0)$

9. $g(x) = \dfrac{1}{x}$

11. $g(x) = \dfrac{1}{2x} - \dfrac{1}{2}$

13. $g(x) = \sqrt[3]{\dfrac{5x + 1}{1 - x}}$

15. $(f \circ g)(x) = f(g(x)) = f(x - 1) = (x - 1) + 1 = x$
$(g \circ f)(x) = g(f(x)) = g(x + 1) = (x + 1) - 1 = x$

17. $(f \circ g)(x) = f\left(\dfrac{1 - x}{x}\right) = \dfrac{1}{\left(\dfrac{1 - x}{x}\right) + 1} =$

$$\dfrac{1}{\dfrac{(1 - x) + x}{x}} = x$$

and

$$(g \circ f)(x) = g\left(\dfrac{1}{x + 1}\right) = \dfrac{1 - \dfrac{1}{x + 1}}{\dfrac{1}{x + 1}} =$$

$$\dfrac{\dfrac{(x + 1) - 1}{x + 1}}{\dfrac{1}{x + 1}} = x$$

19. $(f \circ g)(x) = f(\sqrt[5]{x}) = (\sqrt[5]{x})^5 = x$ and
$(g \circ f)(x) = g(x^5) = \sqrt[5]{x^5} = x$

21. No

23. Yes

25. Yes

27. No

29. $(f \circ f)(x) = f(f(x)) = \dfrac{2f(x) + 1}{3f(x) - 2}$

$= \dfrac{2\left[\dfrac{2x + 1}{3x - 2}\right] + 1}{3\left[\dfrac{2x + 1}{3x - 2}\right] - 2}$

$= \dfrac{\dfrac{2(2x + 1) + (3x - 2)}{3x - 2}}{\dfrac{3(2x + 1) - 2(3x - 2)}{3x - 2}} = \dfrac{\dfrac{7x}{3x - 2}}{\dfrac{7}{3x - 2}} = x$

31.

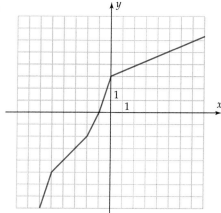

33.

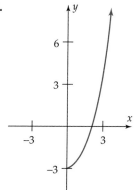

35.

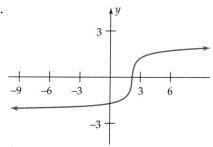

37.

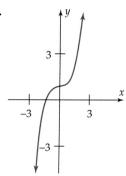

39. One restricted function is $h(x) = |x|$ with $x \geq 0$ [so that $h(x) = x$]; inverse function $g(y) = y$ with $y \geq 0$.

41. One restricted function is $h(x) = -x^2$ with $x \leq 0$; inverse function $g(y) = -\sqrt{-y}$ with $y \leq 0$. Another restricted function is $h(x) = -x^2$ with $x \geq 0$; inverse function $g(y) = \sqrt{-y}$ with $y \leq 0$.

43. One restricted function is $h(x) = \dfrac{x^2 + 6}{2}$ with $x \geq 0$; inverse function $g(y) = \sqrt{2y - 6}$ with $y \geq 3$.

45. One restricted function is $f(x) = \dfrac{1}{x^2 + 1}$ with $x \leq 0$; inverse function $g(y) = -\sqrt{\dfrac{1}{y} - 1} = -\sqrt{\dfrac{1 - y}{y}}$ with $0 < y \leq 1$.

47. (a) $f^{-1}(x) = \dfrac{x - 2}{3}$ **(b)** $f^{-1}(1) = -1/3$ and $1/f(1) = 1/5$

49. Let $y = f(x) = mx + b$. Since $m \neq 0$, we can solve for x and obtain $x = \dfrac{y - b}{m}$. Hence, the rule of the inverse function g is $g(y) = \dfrac{y - b}{m}$, and we have:

$(f \circ g)(y) = f(g(y)) = f\left(\dfrac{y - b}{m}\right)$

$= m\left(\dfrac{y - b}{m}\right) + b = y$ and

$(g \circ f)(x) = g(f(x))$

$= g(mx + b) = \dfrac{(mx + b) - b}{m} = x.$

51. (a) Slope $= \dfrac{a - b}{b - a} = \dfrac{-(b - a)}{b - a} = -1.$

(b) The line $y = x$ has slope 1 (why?) and by (a), line PQ has slope -1. Since the product of their slopes is -1, the lines are perpendicular.

(c) Length $PR = \sqrt{(a-c)^2 + (b-c)^2}$
$= \sqrt{a^2 - 2ac + c^2 + b^2 - 2bc + c^2}$
$= \sqrt{a^2 + b^2 + 2c^2 - 2ac - 2bc};$
Length $RQ = \sqrt{(c-b)^2 + (c-a)^2}$
$= \sqrt{c^2 - 2bc + b^2 + c^2 - 2ac + a^2}$
$= \sqrt{a^2 + b^2 + 2c^2 - 2ac - 2bc}.$
Since the two lengths are the same, $y = x$ is the perpendicular bisector of segment PQ.

53. (a) Suppose $a \neq b$. If $f(a) = f(b)$, then $g(f(a)) = g(f(b))$. But $a = g(f(a))$ by (1) and $b = g(f(b))$. Hence, $a = b$, contrary to our hypothesis. Therefore, it cannot happen that $f(a) = f(b)$, that is, $f(a) \neq f(b)$. Hence, f is one to one.
(b) If $g(y) = x$, then $f(g(y)) = f(x)$. But $f(g(y)) = y$ by (2). Hence, $y = f(g(y)) = f(x)$.
(c) If $f(x) = y$, then $g(f(x)) = g(y)$. But $g(f(x)) = x$ by (1). Hence, $x = g(f(x)) = g(y)$.

Chapter 5 Review, page 421

1. $-4 \leq x \leq 4$ and $-1 \leq y \leq 9$

3. $-4 \leq x \leq 4$ and $-3 \leq y \leq 9$

5. $-3 \leq x \leq 3$ and $0 \leq y \leq 2$

7. $-5 \leq x \leq 20$ and $-4 \leq y \leq 4$

9. (a) About 2.03
(b) About 31.97
(c) Approximately 6 to 10 months
(d) Never; however, at the end of 18 months about 99.6% of the program will be mastered.

11. (a) $1341.68
(b) $541.68

13. (a) $2357.90
(b) After 32.65 years

15. $f(x) = 56,000(1 + .065)^x$

17. 3.75 grams

19. $\ln 756 = 6.628$ **21.** $\ln(u + v) = r^2 - 1$

23. $\log 756 = 2.8785$ **25.** $e^{7.118} = 1234$

27. $e^t = rs$ **29.** $5^u = cd - k$ **31.** 3   **33.** 3/4

35. $2 \ln x$ **37.** $\ln(9y/x^2)$ **39.** $\ln(1/x^{10})$

41. 2   **43.** (c)

45. (c)   **47.** $x = \pm\sqrt{2}$ **49.** $x = \dfrac{3 \pm \sqrt{57}}{4}$

51. $x = -1/2$ **53.** $x = e^{(u-c)/d}$ **55.** $x = 2$

57. $x = 101$ **59.** About 1.64 mg

61. Approximately 12 years

63. $452.89   **65.** 7.6

67. (a) 12°F
(b)

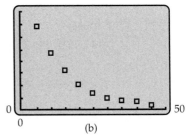

(c) The points $(x, \ln(y))$ are approximately linear.

69. $g(x) = 5 - (x - 7)^2 = -x^2 + 14x - 44; x \geq 7$

71.

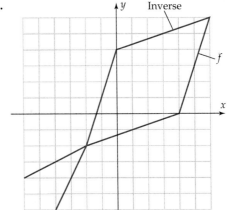

73. The graph of f passes the horizontal line test and hence has an inverse function. It is easy to verify either geometrically [by reflecting the graph of f in the line $y = x$] or algebraically [by calculating $f(f(x))$] that f is its own inverse function.

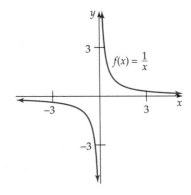

75. There is no inverse function because the graph of f fails the horizontal line test (use the viewing window with $-10 \leq x \leq 20$ and $-200 \leq y \leq 100$).

Chapter 6

Section 6.1, page 438

1. Yes   3. Yes   5. No

7. $x = 11/5, y = -7/5$

9. $x = 2/7, y = -11/7$

11. $r = 5/2, s = -5/2$

13. $x = \dfrac{3c}{2}, y = \dfrac{-c + 2d}{2}$

15. $x = 28, y = 22$ 17. $x = 2, y = -1$

19. Inconsistent

21. $x = b, y = \dfrac{3b - 4}{2}$, where b is any real number

23. $x = b, y = \dfrac{3b - 2}{4}$, where b is any real number

25. Inconsistent

27. $x = -6, y = 2$ 29. $x = 66/5, y = 18/5$

31. $x = \dfrac{rd - sb}{ad - bc}, y = \dfrac{as - cr}{ad - bc}$

33. $p = 6, x = 8$ 35. $p = 180, x = 4$

37. $x = \dfrac{3c + 8}{15}, y = \dfrac{6c - 4}{15}$ is the only solution.

39. $c = -3, d = 1/2$ 41. $x = 120$

43. (a) $R(x) = 60x$
     (b) 400

45. No   47. Yes

49. 140 adults, 60 children

51. Boat speed 18 mph; current speed 2 mph

53. (a) Electric: $y = 960x + 2000$;
     Solar: $y = 114x + 14,000$
     (b) Electric: \$6800; Solar: \$14,570
     (c) Costs same in fifteenth year; electric; solar

55. (a) $y = 7.50x + 5000$
     (b) $y = 8.20x$
     (c) 130,000

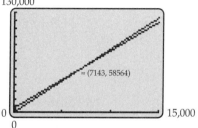

= (7143, 58564)

Costs equal at approximately 7143 cases.
     (d) The company should buy from the supplier any number of cases less than 7143 and produce their own beyond that quantity.

57. \$14,450 at 9% and \$6450 at 11%

59. 3/4 lb cashews and 9/4 lb peanuts

61. 12 liters of the 18%; 18 liters of the 8%

63. 24 g of 50% alloy; 16 g of 75% alloy

65. (a) $x = -\dfrac{1}{2} z_1 - \dfrac{2}{3} z_2 + 62,000$

     (b) $y = -\dfrac{1}{2} z_1 - z_2 + 76,000$

Excursion 6.1.A, page 444

1. $x = 3, y = 9$ or $x = -1, y = 1$

3. $x = \dfrac{-1 + \sqrt{73}}{6}, y = \dfrac{37 - \sqrt{73}}{18}$ or

     $x = \dfrac{-1 - \sqrt{73}}{6}, y = \dfrac{37 + \sqrt{73}}{18}$

5. $x = 7, y = 3$ or $x = 3, y = 7$

7. $x = 0, y = -2$ or $x = 6, y = 1$

9. $x = 2, y = 0$ or $x = 4, y = 2$

11. $x = 4, y = 3$ or $x = -4, y = 3$ or
     $x = \sqrt{21}, y = -2$ or $x = -\sqrt{21}, y = -2$

13. $x = -1.6237, y = -8.1891$ or
     $x = 1.3163, y = 1.0826$ or
     $x = 2.8073, y = 2.4814$

15. $x = -1.9493, y = .4412$ or
     $x = .3634, y = .9578$ or
     $x = 1.4184, y = .5986$

17. No solutions

19. $x = \dfrac{13 - \sqrt{105}}{8}, y = \dfrac{-3 - \sqrt{105}}{8}$ or

     $x = \dfrac{13 + \sqrt{105}}{8}, y = \dfrac{-3 + \sqrt{105}}{8}$

21. There is no solution when the wire is 70 ft long because the graphs of $y = 70 - 2x$ and $y^2 = 4\pi (100 - x^2/8)$ do not intersect.

23. Two possible boxes: 2 by 2 by 4 meters and approximately 3.123 by 3.123 by 1.640 meters

25. -4 and -12 27. 1.6 and 2.6

29. 15 and -12 31. 12 ft by 17 ft

33. 8×15 inches

Section 6.2, page 451

1. $x = -68, y = 13, z = -6, w = 3$

3. $x = 20, y = -9, z = 7.5, w = 3$

5. $\begin{aligned} x \qquad\quad -3z &= 2 \\ 2x - 4y + 5z &= 1 \\ 5x - 8y + 7z &= 6 \\ 3x - 4y + 2z &= 3 \end{aligned}$

7. $\begin{aligned} 3x \quad\;\; + z + 2w + 18v &= 0 \\ -4x + y \quad\;\; - w - 24v &= 0 \\ 7x - y + z + 3w + 42v &= 0 \\ 4x \quad\;\; + z + 2w + 24v &= 0 \end{aligned}$

9. $\begin{aligned} x + y + 2z + 3w &= \;\; 1 \\ - y - z - 2w &= -1 \\ 3x + y + 4z + 5w &= \;\; 2 \end{aligned}$

11. $\begin{aligned} x + 12y - 3z + 4w &= 10 \\ 2y + 3z + w &= \;\; 4 \\ - z &= -7 \\ 6y - 2z - 3w &= \;\; 0 \end{aligned}$

13. $x = 3/2, y = 3/2, z = -3/2$

15. $x = -14, y = -6, z = 2$

17. $x = 100, y = 50, z = 50$

19. $x = 1, y = 2$

21. $x = 12/5, y = 9/5$

23. No point is on all three lines; hence, no real solutions.

25. \$10,000 in AAA bonds, \$20,000 in A bonds, and \$5,000 in B bonds.

27. 40 lb of corn chips, 46.67 lb of nuts, and 13.33 lb of pretzels.

29. 400 box seats, 5400 grandstand seats, and 1200 bleacher seats.

Section 6.3, page 461

1. $\begin{pmatrix} 2 & -3 & 4 & 1 \\ 1 & 2 & -6 & 0 \\ 3 & -7 & 4 & -3 \end{pmatrix}$

3. $\begin{pmatrix} 1 & -1/2 & 7/4 & 0 \\ 2 & -3/2 & 5 & 0 \\ 0 & -2 & 1/3 & 0 \end{pmatrix}$

5. $\begin{aligned} 2x - 3y &= 1 \\ 4x + 7y &= 2 \end{aligned}$

7. $\begin{aligned} x \quad\;\; + z \quad\;\; &= 1 \\ x - y + 4z - 2w &= 3 \\ 4x + 2y + 5z \quad\;\; &= 2 \end{aligned}$

9. $x = 3/2, y = 5, z = -2, w = 0$

11. $x = 2 - t, y = -3 - 2t, z = 4, w = t$, where t is any real number

13. $z = t, y = -1 + \dfrac{1}{3}t, x = 2 - \dfrac{4}{3}t$, where t is any real number

15. $x = 4/3, y = 11/3, z = -2/3$

17. No solutions

19. $x = 1/3, y = 29/3, z = 10/3$

21. $z = t, y = \dfrac{1}{2} - 2t, x = t$, where t is any real number

23. No solutions

25. $z = t, y = t - 1, x = -t + 2$, where t is any real number

27. $x = 5, y = 6, z = 0, w = -1$

29. No solutions

31. $x = -1, y = 1, z = -3, w = -2$

33. $z = t, w = 0, y = 2 - 3t, x = -3 + 5t$

35. $x = 3/5, y = 1/5$

37. $x = -1, y = 2$ **39.** $A = -1, B = 2$

41. $A = -3/25, B = 3/25, C = 7/5$

43. $A = 2, B = 3, C = -1$

45. $x = 1/2, y = 1/3, z = -1/4$

47. 10 quarters; 28 dimes; 14 nickels

49. \$3000 from her friend; \$6000 from the bank; \$1000 from the insurance company

51. \$15,000 in the mutual fund; \$30,000 in bonds; \$25,000 in food franchise

53. Three possible solutions:
18 bedroom, 13 living room, 0 whole house
16 bedroom, 8 living room, 2 whole house
14 bedroom, 3 living room, 4 whole house

55. 8 h for Tom; 24 h for Dick; 12 h for Harry

57. 2000 chairs; 1600 chests; 2500 tables

Section 6.4, page 474

1. AB defined, 2×4; BA not defined

3. AB defined, 3×3; BA defined, 2×2

5. AB defined, 3×2; BA not defined

7. $\begin{pmatrix} 3 & 0 & 11 \\ 2 & 8 & 10 \end{pmatrix}$ **9.** $\begin{pmatrix} 1 & -3 \\ 2 & -1 \\ 5 & 6 \end{pmatrix}$

11. $\begin{pmatrix} 1 & -1 & 1 & 2 \\ 4 & 3 & 3 & 2 \\ -1 & -1 & -3 & 2 \\ 5 & 3 & 2 & 5 \end{pmatrix}$

13. $AB = \begin{pmatrix} 17 & -3 \\ 33 & -19 \end{pmatrix}$; $BA = \begin{pmatrix} -4 & 9 \\ 24 & 2 \end{pmatrix}$

15. $AB = \begin{pmatrix} 8 & 24 & -8 \\ 2 & -2 & 6 \\ -3 & -21 & 15 \end{pmatrix}$; $BA = \begin{pmatrix} 19 & 9 & 8 \\ -10 & 2 & 0 \\ 0 & 0 & 0 \end{pmatrix}$

17. $\begin{pmatrix} -2 & 1 \\ 3/2 & -1/2 \end{pmatrix}$ **19.** No inverse

21. No inverse   **23.** $\begin{pmatrix} -3 & 2 & -4 \\ -1 & 1 & -1 \\ 8 & -5 & 10 \end{pmatrix}$

25. $x = -1, y = 0, z = -3$

27. $x = -8, y = 16, z = 5$

29. $x = -.5, y = -2.1, z = 6.7, w = 2.8$

31. $x = 10.5, y = 5, z = -13, v = 32, w = 2.5$

33. $x = -1149/161, y = 426/161, z = -1124/161,$
$w = 579/161$

35. $x = 0, y = 0, z = 0, v = 0, w = 0$

37. Inconsistent system; no solutions

39. $x = -2 - 3w, y = 5/4 - (1/4)z$, where z and w are any real numbers.

41. $a = 35/8, b = 13/4, c = -117/8$

43. $a = 2, b = -3, c = 1$

45. $a = 3, b = -1, c = 2$

47. $a = 1, b = -4, c = 1$

49. Jeans are \$34.50; jackets are \$72; sweaters are \$44; shirts are \$21.75

51. 15,000 of A; 18,000 of B; 54,000 of C

Excursion 6.4.A, page 484

1. $A + B = \begin{pmatrix} 10 & -3 \\ 3 & 7 \end{pmatrix}$; $AB = \begin{pmatrix} 17 & -3 \\ 33 & -19 \end{pmatrix}$;

$BA = \begin{pmatrix} -4 & 9 \\ 24 & 2 \end{pmatrix}$; $2A - 3B = \begin{pmatrix} -15 & 19 \\ 16 & -16 \end{pmatrix}$

3. $A + B = \begin{pmatrix} 2 & 1/2 \\ 13/2 & 9/2 \end{pmatrix}$; $AB = \begin{pmatrix} 23/4 & -1/4 \\ 43/4 & -5/2 \end{pmatrix}$;

$BA = \begin{pmatrix} -21/4 & -17/4 \\ 31/4 & 17/2 \end{pmatrix}$; $2A - 3B = \begin{pmatrix} 3/2 & 17/2 \\ 1/2 & 4 \end{pmatrix}$

5. $A + B = \begin{pmatrix} 3 & 6 \\ 8 & 9 \end{pmatrix}$; $AB = \begin{pmatrix} 7 & 9 \\ 3 & 5 \end{pmatrix}$; $BA = \begin{pmatrix} 5 & 3 \\ 9 & 7 \end{pmatrix}$;

$2A - 3B = \begin{pmatrix} -9 & -13 \\ -19 & -27 \end{pmatrix}$

7. $A + B = \begin{pmatrix} 4 & 4 \\ 6 & -5 \end{pmatrix}$; $AB = \begin{pmatrix} 1 & 2 \\ -6 & 10 \end{pmatrix}$;

$BA = \begin{pmatrix} 1 & -4 \\ 3 & 10 \end{pmatrix}$; $2A - 3B = \begin{pmatrix} 13 & -2 \\ -3 & 10 \end{pmatrix}$

9. $\begin{pmatrix} -2 & 0 \\ 1 & 1 \end{pmatrix}$

11. $\begin{pmatrix} 7 & 0 \\ -3 & -2 \end{pmatrix}$

13. $\begin{pmatrix} 0 & -8 & -2 \\ 4 & 16 & 18 \end{pmatrix}$

15. $\begin{pmatrix} 8 & -8 \\ 5 & 2 \\ -3 & 25 \end{pmatrix}$

17. $\begin{pmatrix} -6 & 11 \\ 10 & 3 \end{pmatrix}$

19. $\begin{pmatrix} 4 & -7/2 \\ 4 & 21/2 \end{pmatrix}$

21. $\begin{pmatrix} 5/2 & 1 \\ -6 & 11/2 \end{pmatrix}$

23. $A + B = \begin{pmatrix} 1 & 2 \\ 3 & 0 \end{pmatrix} + \begin{pmatrix} -1 & 2 \\ 3 & 4 \end{pmatrix} = \begin{pmatrix} 0 & 4 \\ 6 & 4 \end{pmatrix}$ and

$B + A = \begin{pmatrix} -1 & 2 \\ 3 & 4 \end{pmatrix} + \begin{pmatrix} 1 & 2 \\ 3 & 0 \end{pmatrix} = \begin{pmatrix} 0 & 4 \\ 6 & 4 \end{pmatrix}$

25. By Exercise 23, $c(A + B) = 2\begin{pmatrix} 0 & 4 \\ 6 & 4 \end{pmatrix} = \begin{pmatrix} 0 & 8 \\ 12 & 8 \end{pmatrix}$

and $cA + cB = 2\begin{pmatrix} 1 & 2 \\ 3 & 0 \end{pmatrix} + 2\begin{pmatrix} -1 & 2 \\ 3 & 4 \end{pmatrix} = \begin{pmatrix} 2 & 4 \\ 6 & 0 \end{pmatrix} +$

$\begin{pmatrix} -2 & 4 \\ 6 & 8 \end{pmatrix} = \begin{pmatrix} 0 & 8 \\ 12 & 8 \end{pmatrix}$

27. $cdA = 2 \cdot 3 \begin{pmatrix} 1 & 2 \\ 3 & 0 \end{pmatrix} = 6 \begin{pmatrix} 1 & 2 \\ 3 & 0 \end{pmatrix} = \begin{pmatrix} 6 & 12 \\ 18 & 0 \end{pmatrix}$ and

$c[dA] = 2\left[3\begin{pmatrix} 1 & 2 \\ 3 & 0 \end{pmatrix}\right] = 2\begin{pmatrix} 3 & 6 \\ 9 & 0 \end{pmatrix} = \begin{pmatrix} 6 & 12 \\ 18 & 0 \end{pmatrix}$

29. $A(B + C) = \begin{pmatrix} 1 & 2 \\ 3 & 0 \end{pmatrix}\left[\begin{pmatrix} -1 & 2 \\ 3 & 4 \end{pmatrix} + \begin{pmatrix} 2 & 3 \\ 1 & 2 \end{pmatrix}\right] =$

$\begin{pmatrix} 1 & 2 \\ 3 & 0 \end{pmatrix}\begin{pmatrix} 1 & 5 \\ 4 & 6 \end{pmatrix} = \begin{pmatrix} 9 & 17 \\ 3 & 15 \end{pmatrix}$ and

$AB + AC = \begin{pmatrix} 1 & 2 \\ 3 & 0 \end{pmatrix}\begin{pmatrix} -1 & 2 \\ 3 & 4 \end{pmatrix} + \begin{pmatrix} 1 & 2 \\ 3 & 0 \end{pmatrix}\begin{pmatrix} 2 & 3 \\ 1 & 2 \end{pmatrix} =$

$\begin{pmatrix} 5 & 10 \\ -3 & 6 \end{pmatrix} + \begin{pmatrix} 4 & 7 \\ 6 & 9 \end{pmatrix} = \begin{pmatrix} 9 & 17 \\ 3 & 15 \end{pmatrix}$

31. $k = r$ and $n = t$

33. $AB = \begin{pmatrix} 0 & 0 \\ 0 & 0 \end{pmatrix} = AC$, but $B \neq C$

35. $(A + B)(A - B) = \begin{pmatrix} 5 & 0 \\ 7 & -1 \end{pmatrix}\begin{pmatrix} 1 & 2 \\ -3 & -7 \end{pmatrix} =$

$\begin{pmatrix} 5 & 10 \\ 10 & 21 \end{pmatrix}$, but

$A^2 - B^2 = \begin{pmatrix} 11 & -1 \\ -2 & 18 \end{pmatrix} - \begin{pmatrix} -1 & -5 \\ 25 & 4 \end{pmatrix} = \begin{pmatrix} 12 & 4 \\ -27 & 14 \end{pmatrix}$

37. $A^t = \begin{pmatrix} a & c \\ b & d \end{pmatrix}$ and $B^t = \begin{pmatrix} r & u \\ s & v \end{pmatrix}$

39. $A = \begin{pmatrix} a & b \\ c & d \end{pmatrix}$, $A^t = \begin{pmatrix} a & c \\ b & d \end{pmatrix}$, $(A^t)^t = \begin{pmatrix} a & b \\ c & d \end{pmatrix}$

41. $\begin{pmatrix} 220 & 400 & 280 \\ 360 & 300 & 225 \\ 180 & 380 & 300 \end{pmatrix}\begin{pmatrix} 2.75 \\ 3.25 \\ 2.20 \end{pmatrix} = \begin{pmatrix} \$2521.00 \\ \$2460.00 \\ \$2390.00 \end{pmatrix}$

43. $\begin{pmatrix} 340 & 60 \\ 108 & 292 \end{pmatrix}$

45. 158, −125, 103, 151, 174, −166, 115, 116, 145, −159, 89, 18, 131, −86, 77, 115, 130, −89, 58, 11, 138, −51, 32, −51

47. Tom has a new job.

49. 222, 104, 41, 128, 120, 279, 172, 227, 102, 47, 199, 10, 15, −10, 218, 160, 60, −23, 80, 150, 374, 174, 165, 114, 36

51. (a) 81, 189, 166, 113
   (b) Our matrix in (a) has no inverse. It must have one if we hope to decode our message.

Section 6.5, page 493

1.
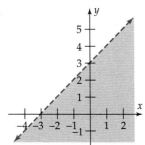

3.
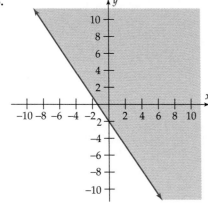

5.
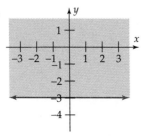

7.
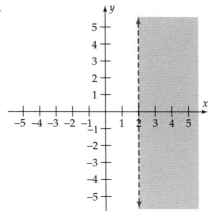

9.
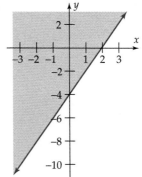

11.
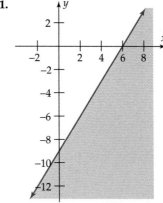

13.
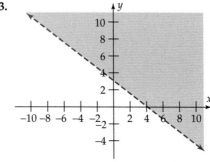

15. F   **17.** A   **19.** E

21.

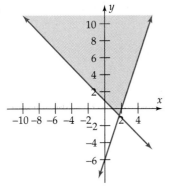

23.

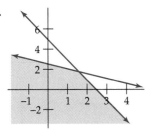

25.

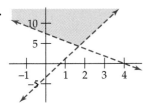

27.

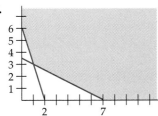

29.

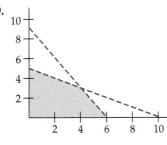

31.

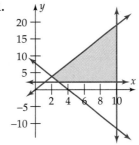

33.

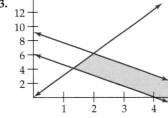

35.

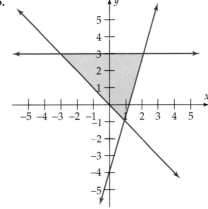

37.

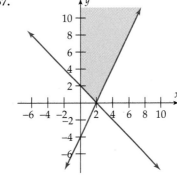

39.

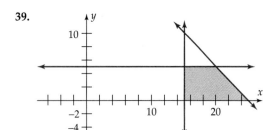

41.

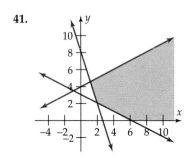

43. $y \leq -\dfrac{4}{3}x + 8$

$y \leq 4$

$x \geq 0, y \geq 0$

45. $x \leq 5;$ $y \leq (3/4)x + 2;$ $y \geq 2 - x;$ $x \geq 0, y \geq 0$

47. $x > 2;$ $x < 7;$ $y > -1;$ $y < 3$

49. $x < 2;$ $-2x + 3y < 8;$ $6y + x > -4$

51. $2r + 3p \leq 12;$ $3r + 3.5p \leq 16$

Regular

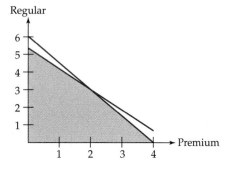

53. $4t + 2s \leq 20;$ $3t + 4s \leq 20$

Silverware

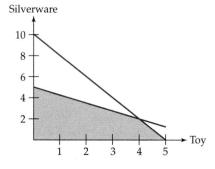

55. $2x + 3y \leq 12;$ $2x + y \geq 2;$ $y \geq 1$

Meat

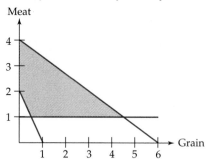

Section 6.6, page 501

1. Mininum is $F = 19$ at $(3, 2)$; maximum is $F = 49$ at $(5, 12)$.

3. Minimum is $F = 11$ at $(3, 2)$; maximum is $F = 27$ at $(5, 12)$.

5. Minimum is $F = 5$ at $(1, 1)$; maximum is $F = 19$ at $(5, 2)$.

7. Minimum is $F = 7$ at $(1, 1)$; maximum is $F = 16$ at $(0, 4)$.

9. Q has coordinates $(8/3, 4/3)$; minimum is $F = 0$ at $(0, 0)$; maximum is $F = 128/3$ at $(8/3, 4/3)$.

11. Q has coordinates $(8, 3)$; minimum is $F = 0$ at $(0, 0)$; maximum is $F = 42$ at $(12, 0)$.

13. Minimum is $F = 0$ at $(0, 0)$; maximum is $F = 15$ at $(0, 5)$.

15. Minimum is $F = 23$ at $(5, 4)$; no maximum.

17. Minimum is $F = 15$ at $(3, 6)$; maximum is $F = 28$ at $(8, 4)$.

19. Minimum is $F = 5$ at $(5, 0)$; maximum is $F = 27$ at $(3, 8)$.

21. Minimum is $F = 0$ at $(0, 0)$; maximum is $F = 22$ at $(2, 4)$.

23. No solutions.

25. Six of the four-person rafts and 18 of the two-person rafts should be made each week for a maximum profit of $780.

27. Thirty boxes of Minto-Mix (and none of Choco-Mix) should be made, which will generate the maximum revenue of $480.

29. Thirty Supervites and 120 Vitahealth pills will meet the requirements at a minimum cost of $11.40.

31. Eight pounds of feed A and 3 pounds of feed B will meet the requirements at a minimum cost of $2.80.

33. Using 28 tons of oats (and no bran) will minimize the cost at $6300.

35. If she takes 5 humanities courses and 12 science courses, she will earn the maximum of 235 quality points.

37. Because the inequalities have no common solution (that is, there are no feasible solutions): Adding $2x + y \leq 10$ and $x + 2y \leq 8$ gives $3x + 3y \leq 18$, or $x + y \leq 6$, which contradicts $x + y \geq 8$.

39. Taking three humanities and four science courses gives an expected grade point average of 2.91. Corresponding figures for the other vertices are: (3, 12): 2.96; (5, 12): 2.94; (15, 4): 2.81. Thus (3, 12) earns the highest grade point average, whereas (5, 12) earns the greatest number of quality points. There is no contradiction; the grade point average and quality point functions are different, though related, so there is no reason why they should have their maxima at the same vertex. Since the grade point average function is not linear, we cannot even be sure that the vertex (3, 12) gives the maximum grade point average!

Chapter 6 Review, page 506

1. $x = -5, y = -7$

3. $x = 0, y = -2$

5. $x = -35, y = 140, z = 22$

7. $x = 2, y = 4, z = 6$

9. $x = -t + 1, y = -2t, z = t$ for any real number t

11. 37 and -19

13. (c)

15. 100

17. $x = 3, y = 9$ or $x = -1, y = 1$

19. $x = 1 - \sqrt{7}, y = 1 + \sqrt{7}$ or $x = 1 + \sqrt{7}, y = 1 - \sqrt{7}$

21. $x = -1.692, y = 3.136$ or $x = 1.812, y = 2.717$

23. $\begin{pmatrix} 1 & -2 & 3 & 4 \\ 2 & 1 & -4 & 3 \\ -3 & 4 & -1 & -2 \end{pmatrix}$

25. $\begin{pmatrix} 2 & -1 & -2 & 2 \\ 1 & 3 & -2 & 1 \\ -1 & 4 & 2 & -3 \end{pmatrix}$

27. $\dfrac{3}{x + 2} + \dfrac{1}{x - 3}$

29. $\begin{pmatrix} -2 & 3 \\ -4 & -1 \end{pmatrix}$

31. Not defined

33. $\begin{pmatrix} -9 & 7 \\ -4 & 3 \end{pmatrix}$

35. $\begin{pmatrix} 1 & 2 & -2 \\ -1 & 3 & 0 \\ 0 & -2 & 1 \end{pmatrix}$

37. $x = -1/85, y = -14/85, z = -21/34, w = 46/85$

39. $y = 5x^2 - 2x + 1$

41. 30 lb corn, 15 lb soybeans, 40 lb by-products

43. $\begin{pmatrix} 4 & 5 \\ 24 & 1 \\ 18 & -7 \end{pmatrix}$

45. $\begin{pmatrix} -5 & 1 \\ -6 & -5 \end{pmatrix}$

47. Not defined

49. $\begin{pmatrix} 227 & 124 & 49 \\ 266 & 40 & 94 \end{pmatrix}$

51.

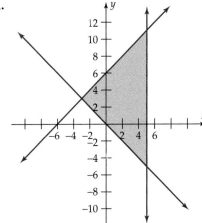

53.

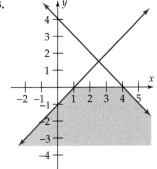

55. Minimum value of F is .72 at $x = 4, y = 8$; there is no maximum value of F.

57. 15 containers, 35 barrels; no unused space on the truck

Chapter 7

Section 7.1, page 522

1. 8, 10, 12, 14, 16

3. $1, \dfrac{1}{8}, \dfrac{1}{27}, \dfrac{1}{64}, \dfrac{1}{125}$

5. $-\sqrt{3}, 2, -\sqrt{5}, \sqrt{6}, -\sqrt{7}$

7. 3.9, 4.01, 3.999, 4.0001, 3.99999

9. 2, 7, 8, 13, 14

11. $\displaystyle\sum_{i=1}^{11} i$

13. $\displaystyle\sum_{i=7}^{13} \frac{1}{2^i}$ (Other answers possible)

15. 45   **17.** 224   **19.** 15,015

21. $a_n = (-1)^n$ **23.** $a_n = \dfrac{n}{n+1}$ **25.** $a_n = 5n - 3$

27. 4, 11, 25, 53, 109   **29.** 1, −2, 3, 2, 3

31. $2, 3, 3, \dfrac{9}{2}, \dfrac{27}{4}$ **33.** 3, 1, 4, 1, 5

35. Third − 10; sixth − 2

37. Third 5; sixth 0

39. $\displaystyle\sum_{n=1}^{6} \frac{1}{2n+1}$ (Other answers possible)

41. $\displaystyle\sum_{n=8}^{12} (-1)^n\left(\frac{n-7}{n}\right)$ (Other answers possible)

43. 2.613035   **45.** $\dfrac{1,051,314}{31,415} \approx 33.465$

47. 1.5759958

49. (a) 1993: $0.34; 1995: $0.45; 1998: $0.615
(b) $3.38

51. (a) 1993: 17 billion; 2000: 21.2 billion
(b) 185.4 billion

53. (b) 59, 61, 67, 71

55. 4, 9, 25, 49, 121   **57.** 3, 7, 13, 19, 23

59. (a) 1, 1, 2, 3, 5, 8, 13, 21, 34, 55
(b) 1, 2, 4, 7, 12, 20, 33, 54, 88, 143
(c) nth partial sum $= a_{n+2} - 1$

61. $n = 1$: $5(1)^2 + 4(-1)^1 = 1 = 1^2$;
$n = 2$: $5(1)^2 + 4(-1)^2 = 9 = 3^2$;
$n = 3$: $5(2)^2 + 4(-1)^3 = 16 = 4^2$;
$n = 4$: $5(3)^2 + 4(-1)^4 = 49 = 7^2$; etc.

63. We have $a_1 = a_3 - a_2; a_2 = a_4 - a_3; a_3 = a_5 - a_4; \ldots$
$a_{k-1} = a_{k+1} - a_k; a_k = a_{k+2} - a_{k+1}$. If these
equations are listed vertically, the sum of the left-
side terms is $\displaystyle\sum_{n=1}^{k} a_k$. On the right side, one term in
each line is the same as a term in the next line,
except for sign. So the sum of the right-side terms
is $a_{k+2} - a_2$. Since $a_2 = 1$, conclude that

$$\sum_{n=1}^{k} a_n = a_{k+2} - 1.$$

Section 7.2, page 527

1. $13; a_n = 2n + 3$ **3.** $5; a_n = n/4 + 15/4$

5. $8; a_n = -n/2 + 21/2$ **7.** 87

9. −21/4   **11.** 30

13. $a_n - a_{n-1} = (3 - 2n) - [3 - 2(n-1)] = -2$;
arithmetic with $d = -2$

15. $a_n - a_{n-1} = \dfrac{5+3n}{2} - \dfrac{5+3(n-1)}{2} = \dfrac{3}{2}$; arithmetic
with $d = \dfrac{3}{2}$

17. $a_n - a_{n-1} = (c + 2n) - [c + 2(n-1)] = 2$;
arithmetic with $d = 2$

19. $a_5 = 14; a_n = 2n + 4$ **21.** $a_5 = 25; a_n = 7n - 10$

23. $a_5 = 0; a_n = -15/2 + 3n/2$

25. 710   **27.** $156\frac{2}{3}$ **29.** 2550   **31.** 20,100

33. $77,500 in tenth year; $437,500 over ten years

35. 428   **37.** 23.25, 22.5, 21.75, 21, 20.25, 19.5, 18.75

Section 7.3, page 532

1. Arithmetic   **3.** Geometric

5. Arithmetic   **7.** Geometric

9. $a_6 = 160; a_n = 2^{n-1} \cdot 5$

11. $a_6 = \dfrac{1}{256}; a_n = \dfrac{1}{4^{n-2}}$

13. $a_6 = -\dfrac{5}{16}; a_n = \dfrac{(-1)^{n-1} \cdot 5}{2^{n-2}}$

15. 315/32   **17.** 381

19. $\dfrac{a_n}{a_{n-1}} = \dfrac{\left(-\dfrac{1}{2}\right)^n}{\left(-\dfrac{1}{2}\right)^{n-1}} = -\dfrac{1}{2}$; geometric with $r = -\dfrac{1}{2}$

21. $\dfrac{a_n}{a_{n-1}} = \dfrac{5^{n+2}}{5^{(n-1)+2}} = 5$; geometric with $r = 5$

23. $a_5 = 1; a_n = \dfrac{(-1)^{n-1} 64}{4^{n-2}} = \dfrac{(-1)^{n-1}}{4^{n-5}}$

25. $a_5 = \dfrac{1}{16}; a_n = \dfrac{1}{4^{n-3}}$

27. $a_5 = -\dfrac{8}{25}; a_n = -\dfrac{2^{n-2}}{5^{n-3}}$

29. 254   **31.** $-\dfrac{4921}{19,683}$ **33.** $\dfrac{665}{8}$

35. (a) Since for all n, the ratio r is
$$\dfrac{a_{n+1}}{a_n} = \dfrac{1.71(1.191^{n+1})}{1.71(1.191^n)} = \dfrac{1.71(1.191^n)(1.191)}{1.71(1.191^n)}$$
$$= 1.191, \text{ the sequence is geometric.}$$
(b) $217.47

37. 23.75 ft

39. $\displaystyle\sum_{n=1}^{31} 2^{n-1} = \dfrac{1-2^{31}}{1-2} = (2^{31} - 1)$ cents =
$21,474,836.47

41. $1898.44

43. $\log a_n - \log a_{n-1} = \log \dfrac{a_n}{a_{n-1}} = \log r$

45. The sequence is $\{2^{n-1}\}$ and $r = 2$. So for any k, the kth term is 2^{k-1}, and the sum of the preceding terms is the $(k-1)$th partial sum of the sequence,

$$\sum_{n=1}^{k-1} 2^{n-1} = \frac{1 - 2^{k-1}}{1 - 2} = 2^{k-1} - 1.$$

47. 37 payments

Excursion 7.3.A, page 538

1. 1   **3.** $.06/.94 = 3/47$ **5.** $\dfrac{500}{.6} = 833\dfrac{1}{3}$

7. $4 + 2\sqrt{2}$ **9.** $2/9$ **11.** $597/110$

13. $10{,}702/4995$ **15.** $174{,}067/99{,}900$

17. (a) $a_1 = 2(1.5)^1 = 3$. For each $n \geq 2$, $a_{n-1} = 2(1.5)^{n-1}$ and $a_n = 2(1.5)^n$, so that the ratio $\dfrac{a_n}{a_{n-1}}$ is $\dfrac{2(1.5)^n}{2(1.5)^{n-1}} = 1.5$. Therefore, this is a geometric series.
   (b) $S_k = -6 + 6(1.5)^k$ and $f(x) = -6 + 6(1.5)^x$
   (c) The graph shows that as x gets large, $f(x)$ gets huge, so there is no horizontal asymptote. Hence, the series does not converge.

Section 7.4, page 544

1. 720   **3.** 220   **5.** 0   **7.** 64   **9.** 3,921,225

11. $x^5 + 5x^4y + 10x^3y^2 + 10x^2y^3 + 5xy^4 + y^5$

13. $a^5 - 5a^4b + 10a^3b^2 - 10a^2b^3 + 5ab^4 - b^5$

15. $32x^5 + 80x^4y^2 + 80x^3y^4 + 40x^2y^6 + 10xy^8 + y^{10}$

17. $x^3 + 6x^2\sqrt{x} + 15x^2 + 20x\sqrt{x} + 15x + 6\sqrt{x} + 1$

19. $1 - 10c + 45c^2 - 120c^3 + 210c^4 - 252c^5 + 210c^6 - 120c^7 + 45c^8 - 10c^9 + c^{10}$

21. $x^{-12} + 4x^{-8} + 6x^{-4} + 4 + x^4$

23. 56   **25.** $-8i$ **27.** $10x^3y^2$ **29.** $35c^3d^4$

31. $\dfrac{35}{8}u^{-5}$ **33.** 4032   **35.** 160

37. (a) $\dbinom{9}{1} = \dfrac{9!}{1!8!} = 9; \dbinom{9}{8} = \dfrac{9!}{8!1!} = 9$

   (b) $\dbinom{n}{1} = \dbinom{n}{n-1} = \dfrac{n!}{1!(n-1)!} = \dfrac{n(n-1)!}{(n-1)!} = n$

39. $2^n = (1+1)^n = 1^n + \dbinom{n}{1}1^{n-1} \cdot 1 + \dbinom{n}{2}1^{n-2} \cdot 1^2 +$
   $\dbinom{n}{3}1^{n-3} \cdot 1^3 + \cdots + \dbinom{n}{n-1}1^1 \cdot 1^{n-1} + 1^n =$
   $\dbinom{n}{0} + \dbinom{n}{1} + \dbinom{n}{2} + \dbinom{n}{3} + \cdots + \dbinom{n}{n-1} + \dbinom{n}{n}$

41. (a) $f(x+h) - f(x) = (x+h)^5 - x^5$
   $= \left[x^5 + \dbinom{5}{1}x^4h + \dbinom{5}{2}x^3h^2 + \dbinom{5}{3}x^2h^3 + \dbinom{5}{4}xh^4 + h^5\right]$
   $- x^5 = \dbinom{5}{1}x^4h + \dbinom{5}{2}x^3h^2 + \dbinom{5}{3}x^2h^3 + \dbinom{5}{4}xh^4 + h^5$

   (b) $\dfrac{f(x+h) - f(x)}{h} = \dbinom{5}{1}x^4 + \dbinom{5}{2}x^3h +$
   $\dbinom{5}{3}x^2h^2 + \dbinom{5}{4}xh^3 + h^4$

   (c) When h is *very* close to 0, so are the last four terms in part (b), so $\dfrac{f(x+h) - f(x)}{h} \approx$
   $\dbinom{5}{1}x^4 = 5x^4.$

43. $\dfrac{f(x+h) - f(x)}{h} = \dfrac{(x+h)^{12} - x^{12}}{h} = \dbinom{12}{1}x^{11} +$
   $\dbinom{12}{2}x^{10}h + \dbinom{12}{3}x^9h^2 + \dbinom{12}{4}x^8h^3 + \cdots +$
   $\dbinom{12}{10}x^2h^9 + \dbinom{12}{11}xh^{10} + h^{11} \approx \dbinom{12}{1}x^{11}$
   $= 12x^{11}$, when h is very close to 0.

45. (a) $(n-r)! =$
   $(n-r)(n-r-1)(n-r-2)(n-r-3)\cdots 2\cdot 1$
   $= (n-r)[n-(r+1)][n-(r+1)-1]\cdots 2\cdot 1$
   $= (n-r)[n-(r+1)]!$
   (b) Since $(n+1) - (r+1) = n - r$,
   $[(n+1)-(r+1)]! = (n-r)!$
   (c) $\dbinom{n}{r+1} + \dbinom{n}{r} = \dfrac{n!}{(r+1)![n-(r+1)]!} +$
   $\dfrac{n!}{r!(n-r)!} = \dfrac{n!(n-r) + n!(r+1)}{(r+1)!(n-r)!}$
   $= \dfrac{n!(n+1)}{(r+1)!(n-r)!}$
   $= \dfrac{(n+1)!}{(r+1)![(n+1)-(r+1)]!} = \dbinom{n+1}{r+1}$
   (d) For example, rows 2 and 3 of Pascal's triangle are

   1   2   1

   1   ③   3   1

   that is,

   $\dbinom{2}{0}$ $\dbinom{2}{1}$ $\dbinom{2}{2}$

   $\dbinom{3}{0}$ $\dbinom{3}{1}$ $\dbinom{3}{2}$ $\dbinom{3}{3}$

   The circled 3 is the sum of the two closest entries in the row above: $1 + 2$. But this just says that $\dbinom{3}{1} = \dbinom{2}{0} + \dbinom{2}{1}$, which is part (c) with $n = 2$ and $r = 0$. Similarly, in the general case, verify that the two closest entries in the row above $\binom{n+1}{r+1}$ are $\binom{n}{r}$ and $\binom{n}{r+1}$ and use part (c).

Section 7.5, page 551

1. 24 **3.** 40,320 **5.** 132

7. 91 **9.** 3,764,376 **11.** $n(n-1) = n^2 - n$

13. n **15.** $\dfrac{(n^2 - n)}{2}$ **17.** $n = 11$

19. $n = 9$ **21.** $n = 7$ **23.** 1,000,000,000

25. 35,152 **27.** 864 **29.** 1,860,480

31. 256 **33.** 40,320 **35.** 24

37. 3024 **39.** 56 **41.** 20,358,520

43. 26,334 **45.** 43,545,600 **47.** 64

49. 576

51. (a) 59,280
(b) Is 4-12-30 the same "combination" as 30-12-4?

53. 13,983,816 **55.** 24

57. 56 **59.** 11,550

61. (a) 630 (b) 840

63. (a) 2 (b) 6 (c) 24 (d) $(n-1)!$

65. 479,001,600 **67.** 76,204,800

Excursion 7.5.A, page 556

1. 12 **3.** 50,400 **5.** 1,396,755,360

Section 7.6, page 565

1. {HHH, HHT, HTH, HTT, THH, THT, TTH, TTT}

3. $\{r, b\}$

5. {JJ, QQ, KK, AA, JQ, JK, JA, QK, QA, QJ, KJ, KQ, KA, AJ, AQ, AK}

7. 1/2 **9.** 5/6 **11.** 3/8

13. 1/2 **15.** 5/11 **17.** 2/11

19. 1/6 **21.** 1/2 **23.** 13/18

25. 1/12 **27.** 1/2 **29.** 1/5

31. 4/5 **33.** 7/15 **35.** 7/15

37. 1/13 **39.** 1/26 **41.** 3/26

43. 5/13 **45.** $\dfrac{1}{54,145}$ **47.** $\dfrac{1}{108,290}$

49. The probability of a hand with no aces is $\dfrac{35,673}{54,145} \approx .66$; so the probability of a hand with at least one ace (the complement of an aceless hand) is approximately $1 - .66 = .34$.

51. (a) $\dfrac{1}{1024}$ (b) $\dfrac{252}{1024} \approx .246$ (c) $\dfrac{176}{1024} \approx .172$

53. (a) $\dfrac{1200}{5005} \approx .24$ (b) $\dfrac{2100}{5005} \approx .42$ (c) $\dfrac{210}{5005} \approx .04$

55. Approximately .29

57. $\dfrac{1}{7,059,052}$

59. .12, .41, .57, .71, .89, .97

61. .91

63. (a) .37 (b) .04 (c) .91

65. (a) 3/8 (b) 2/8 (c) 1/8 (d) 2/8

67. .05 **69.** .75 **71.** .325 **73.** .5375

Section 7.7, page 575

1. *Step 1:* For $n = 1$ the statement is $1 = 2^1 - 1$, which is true. *Step 2:* Assume that the statement is true for $n = k$: that is, $1 + 2 + 2^2 + 2^3 + \cdots + 2^{k-1} = 2^k - 1$. Add 2^k to both sides, and rearrange terms:
$$1 + 2 + 2^2 + 2^3 + \cdots + 2^{k-1} + 2^k = 2^k - 1 + 2^k$$
$$1 + 2 + 2^2 + 2^3 + \cdots + 2^{k-1} + 2^{(k+1)-1} = 2(2^k) - 1$$
$$1 + 2 + 2^2 + 2^3 + \cdots + 2^{k-1} + 2^{(k+1)-1} = 2^{k+1} - 1$$
But this last line says that the statement is true for $n = k + 1$. Therefore, by the Principle of Mathematical Induction the statement is true for every positive integer n.

Note: **Hereafter, in these answers, step 1 will be omitted if it is trivial (as in Exercise 1), and only the essential parts of step 2 will be given.**

3. Assume that the statement is true for $n = k$: $1 + 3 + 5 + \cdots + (2k - 1) = k^2$. Add $2(k + 1) - 1$ to both sides:
$1 + 3 + 5 + \cdots + (2k - 1) + [2(k + 1) - 1] = k^2 + 2(k + 1) - 1 = k^2 + 2k + 1 = (k + 1)^2$.
The first and last parts of this equation say that the statement is true for $n = k + 1$.

5. Assume that the statement is true for $n = k$:
$$1^2 + 2^2 + 3^2 + \cdots + k^2 = \frac{k(k+1)(2k+1)}{6}$$
Add $(k + 1)^2$ to both sides:
$$1^2 + 2^2 + 3^2 + \cdots + k^2 + (k+1)^2$$
$$= \frac{k(k+1)(2k+1)}{6} + (k+1)^2$$
$$= \frac{k(k+1)(2k+1) + 6(k+1)^2}{6}$$
$$= \frac{(k+1)[k(2k+1) + 6(k+1)]}{6}$$
$$= \frac{(k+1)(2k^2 + 7k + 6)}{6}$$

$$= \frac{(k+1)(k+2)(2k+3)}{6}$$

$$= \frac{(k+1)[(k+1)+1][2(k+1)+1]}{6}$$

The first and last parts of this equation say that the statement is true for $n = k + 1$.

7. Assume that the statement is true for $n = k$:

$$\frac{1}{1 \cdot 2} + \frac{1}{2 \cdot 3} + \cdots + \frac{1}{k(k+1)} = \frac{k}{k+1}.$$

Adding $\dfrac{1}{(k+1)[(k+1)+1]} = \dfrac{1}{(k+1)(k+2)}$ to both sides yields:

$$\frac{1}{1 \cdot 2} + \frac{1}{2 \cdot 3} + \cdots + \frac{1}{k(k+1)} + \frac{1}{(k+1)(k+2)}$$

$$= \frac{k}{k+1} + \frac{1}{(k+1)(k+2)}$$

$$= \frac{k(k+2)+1}{(k+1)(k+2)} = \frac{k^2+2k+1}{(k+1)(k+2)}$$

$$= \frac{(k+1)^2}{(k+1)(k+2)} = \frac{k+1}{k+2} = \frac{k+1}{(k+1)+1}$$

The first and last parts of this equation show that the statement is true for $n = k + 1$.

9. Assume the statement is true for $n = k$: $k + 2 > k$. Adding 1 to both sides, we have: $k + 2 + 1 > k + 1$, or equivalently, $(k + 1) + 2 > (k + 1)$. Therefore, the statement is true for $n = k + 1$.

11. Assume the statement is true for $n = k$: $3^k \geq 3k$. Multiplying both sides by 3 yields: $3 \cdot 3^k \geq 3 \cdot 3k$, or equivalently, $3^{k+1} \geq 3 \cdot 3k$. Now since $k \geq 1$, we know that $3k \geq 3$ and hence that $2 \cdot 3k \geq 3$. Therefore, $2 \cdot 3k + 3k \geq 3 + 3k$, or equivalently, $3 \cdot 3k \geq 3k + 3$. Combining this last inequality with the fact that $3^{k+1} \geq 3 \cdot 3k$, we see that $3^{k+1} \geq 3k + 3$, or equivalently, $3^{k+1} \geq 3(k + 1)$. Therefore, the statement is true for $n = k + 1$.

13. Assume the statement is true for $n = k$: $3k > k + 1$. Adding 3 to both sides yields: $3k + 3 > k + 1 + 3$, or equivalently, $3(k + 1) > (k + 1) + 3$. Since $(k + 1) + 3$ is certainly greater than $(k + 1) + 1$, we conclude that $3(k + 1) > (k + 1) + 1$. Therefore, the statement is true for $n = k + 1$.

15. Assume the statement is true for $n = k$; then 3 is a factor of $2^{2k+1} + 1$; that is, $2^{2k+1} + 1 = 3M$ for some integer M. Thus, $2^{2k+1} = 3M - 1$. Now $2^{2(k+1)+1} = 2^{2k+2+1} = 2^{2+2k+1} = 2^2 \cdot 2^{2k+1} = 4(3M - 1) = 12M - 4 = 3(4M) - 3 - 1 = 3(4M - 1) - 1$. From the first and last terms of this equation we see that $2^{2(k+1)+1} + 1 =$

$3(4M - 1)$. Hence, 3 is a factor of $2^{2(k+1)+1} + 1$. Therefore, the statement is true for $n = k + 1$.

17. Assume the statement is true for $n = k$: 64 is a factor of $3^{2k+2} - 8k - 9$. Then $3^{2k+2} - 8k - 9 = 64N$ for some integer N so that $3^{2k+2} = 8k + 9 + 64N$. Now $3^{2(k+1)+2} = 3^{2k+2+2} = 3^{2+(2k+2)} = 3^2 \cdot 3^{2k+2} = 9(8k + 9 + 64N)$. Consequently,

$$3^{2(k+1)+2} - 8(k+1) - 9 = 3^{2(k+1)+2} - 8k - 8 - 9$$

$$= 3^{2(k+1)+2} - 8k - 17$$

$$= [9(8k + 9 + 64N)] - 8k - 17$$

$$= 72k + 81 + 9 \cdot 64N - 8k - 17$$

$$= 64k + 64 + 9 \cdot 64N = 64(k + 1 + 9N).$$

From the first and last parts of this equation we see that 64 is a factor of $3^{2(k+1)+2} - 8(k+1) - 9$. Therefore, the statement is true for $n = k + 1$.

19. Assume the statement is true for $n = k$: $c + (c + d) + (c + 2d) + \cdots + [c + (k-1)d] = \dfrac{k[2c + (k-1)d]}{2}$. Adding $c + kd$ to both sides, we have

$$c + (c + d) + (c + 2d) + \cdots + [c + (k-1)d] + (c + kd)$$

$$= \frac{k[2c + (k-1)d]}{2} + c + kd$$

$$= \frac{k[2c + (k-1)d] + 2(c + kd)}{2}$$

$$= \frac{2ck + k(k-1)d + 2c + 2kd}{2}$$

$$= \frac{2ck + 2c + kd(k-1) + 2kd}{2}$$

$$= \frac{(k+1)2c + kd(k-1+2)}{2}$$

$$= \frac{(k+1)2c + kd(k+1)}{2} = \frac{(k+1)(2c+kd)}{2}$$

$$= \frac{(k+1)(2c + [(k+1)-1]d)}{2}$$

Therefore, the statement is true for $n = k + 1$.

21. **(a)** $x^2 - y^2 = (x - y)(x + y)$;
    $x^3 - y^3 = (x - y)(x^2 + xy + y^2)$;
    $x^4 - y^4 = (x - y)(x^3 + x^2y + xy^2 + y^3)$

   (b) *Conjecture:* $x^n - y^n = (x - y)(x^{n-1} + x^{n-2}y + x^{n-3}y^2 + \cdots + x^2y^{n-3} + xy^{n-2} + y^{n-1})$.
   Proof: The statement is true for $n = 2, 3, 4$, by part (a). Assume that the statement is true for $n = k$:

$$x^k - y^k =$$

$$(x - y)(x^{k-1} + x^{k-2}y + \cdots + xy^{k-2} + y^{k-1}).$$

Now use the fact that $-yx^k + yx^k = 0$ to write $x^{k+1} - y^{k+1}$ as follows:

$$\begin{aligned}
x^{k+1} - y^{k+1} &= x^{k+1} - yx^k + yx^k - y^{k+1} \\
&= (x^{k+1} - yx^k) + (yx^k - y^{k+1}) \\
&= (x - y)x^k + y(x^k - y^k) \\
&= (x - y)x^k + y(x - y)(x^{k-1} + x^{k-2}y \\
&\quad + x^{k-3}y^2 + \cdots + xy^{k-2} + y^{k-1}) \\
&= (x - y)x^k + (x - y)(x^{k-1}y + \\
&\quad x^{k-2}y^2 + x^{k-3}y^3 + \cdots + xy^{k-1} + y^k) \\
&= (x - y)[x^k + x^{k-1}y + x^{k-2}y^2 + \\
&\quad x^{k-3}y^3 + \cdots + xy^{k-1} + y^k]
\end{aligned}$$

The first and last parts of this equation show that the conjecture is true for $n = k + 1$. Therefore, by mathematical induction, the conjecture is true for every integer $n \geq 2$.

23. False; counterexample: $n = 9$

25. True: *Proof:* Since $(1 + 1)^2 > 1^2 + 1$, the statement is true for $n = 1$. Assume the statement is true for $n = k$: $(k + 1)^2 > k^2 + 1$. Then $[(k + 1) + 1]^2 = (k + 1)^2 + 2(k + 1) + 1 > k^2 + 1 + 2(k + 1) + 1 = k^2 + 2k + 2 + 2 > k^2 + 2k + 2 = k^2 + 2k + 1 + 1 = (k + 1)^2 + 1$. The first and last terms of this inequality say that the statement is true for $n = k + 1$. Therefore, by induction the statement is true for every positive integer n.

27. False; counterexample: $n = 2$

29. Since $2 \cdot 5 - 4 > 5$, the statement is true for $n = 5$. Assume the statement is true for $n = k$ (with $k \geq 5$): $2k - 4 > k$. Adding 2 to both sides shows that $2k - 4 + 2 > k + 2$, or equivalently, $2(k + 1) - 4 > k + 2$. Since $k + 2 > k + 1$, we see that $2(k + 1) - 4 > k + 1$. So the statement is true for $n = k + 1$. Therefore, by the Extended Principle of Mathematical Induction, the statement is true for all $n \geq 5$.

31. Since $2^2 > 2$, the statement is true for $n = 2$. Assume that $k \geq 2$ and that the statement is true for $n = k$: $k^2 > k$. Then $(k + 1)^2 = k^2 + 2k + 1 > k^2 + 1 > k + 1$. The first and last terms of this inequality show that the statement is true for $n = k + 1$. Therefore, by induction, the statement is true for all $n \geq 2$.

33. Since $3^4 = 81$ and $2^4 + 10 \cdot 4 = 16 + 40 = 56$, we see that $3^4 > 2^4 + 10 \cdot 4$. So the statement is true for $n = 4$. Assume that $k \geq 4$ and that the statement is true for $n = k$: $3^k > 2^k + 10k$. Multiplying both sides by 3 yields: $3 \cdot 3^k > 3(2^k + 10k)$, or equivalently, $3^{k+1} > 3 \cdot 2^k + 30k$. But

$$3 \cdot 2^k + 30k > 2 \cdot 2^k + 30k = 2^{k+1} + 30k.$$

Therefore, $3^{k+1} > 2^{k+1} + 30k$. Now we shall show that $30k > 10(k + 1)$. Since $k \geq 4$, we have $20k \geq 20 \cdot 4$, so that $20k > 80 > 10$. Adding $10k$ to both sides of $20k > 10$ yields: $30k > 10k + 10$, or equivalently, $30k > 10(k + 1)$. Consequently,

$$3^{k+1} > 2^{k+1} + 30k > 2^{k+1} + 10(k + 1).$$

The first and last terms of this inequality show that the statement is true for $n = k + 1$. Therefore, the statement is true for all $n \geq 4$ by induction.

35. (a) 3 (that is, $2^2 - 1$) for $n = 2$; 7 (that is, $2^3 - 1$) for $n = 3$; 15 (that is, $2^4 - 1$) for $n = 4$.

(b) *Conjecture:* The smallest possible number of moves for n rings is $2^n - 1$. *Proof:* This conjecture is easily seen to be true for $n = 1$ or $n = 2$. Assume it is true for $n = k$ and that we have $k + 1$ rings to move. To move the *bottom* ring from the first peg to another peg (say, the second one), it is first necessary to move the top k rings off the first peg *and* leave the second peg vacant at the end (the second peg will have to be used *during* this moving process). If this is to be done according to the rules, we will end up with the top k rings on the third peg in the *same* order they were on the first peg. According to the induction assumption, the least possible number of moves needed to do this is $2^k - 1$. It now takes one move to transfer the bottom ring [the $(k + 1)$st] from the first to the second peg. Finally, the top k rings now on the third peg must be moved to the second peg. Once again by the induction hypothesis, the least number of moves for doing this is $2^k - 1$. Therefore, the smallest total number of moves needed to transfer all $k + 1$ rings from the first to the second peg is $(2^k - 1) + 1 + (2^k - 1) = (2^k + 2^k) - 1 = 2 \cdot 2^k - 1 = 2^{k+1} - 1$. Hence, the conjecture is true for $n = k + 1$. Therefore, by induction it is true for all positive integers n.

Chapter 7 Review, page 578

1. $-3, -1, 1, 3$ **3.** $1, \dfrac{1}{4}, \dfrac{1}{9}, \dfrac{1}{16}$ **5.** -368

7. 129   **9.** $a_n = 9 - 6n$

11. $a_n = 6n - 11$ **13.** $a_n = 2 \cdot 3^{n-1}$

15. $a_n = \dfrac{3}{2^{n-8}}$ **17.** -55

19. $\dfrac{121}{4}$ **21.** 8.75, 13.5, 18.25

23. Second method is better.   **25.** 2

27. $(1.02)^{51} = (1 + .02)^{51} = 1^{51} + \binom{51}{1} 1^{50} (.02) +$

$\binom{51}{2} 1^{49}(.02)^2 +$ other positive terms $= 2.53 +$

other positive terms > 2.5

29. 455   **31.** $n + 1$ **33.** 1140   **35.** 80

37. 96,909,120   **39.** 10,626   **41.** 200   **43.** 24

45. 280   **47.** 5/36   **49.** 1/4

51. Approximately .38

53. True for $n = 1$. If the statement is true for $n = k$,

then $\dfrac{1}{3} + \dfrac{1}{3^2} + \dfrac{1}{3^3} + \cdots + \dfrac{1}{3^k} = \dfrac{3^k - 1}{2 \cdot 3^k}$. Hence,

$\dfrac{1}{3} + \dfrac{1}{3^2} + \dfrac{1}{3^3} + \cdots + \dfrac{1}{3^k} + \dfrac{1}{3^{k+1}} = \dfrac{3^k - 1}{2 \cdot 3^k} + \dfrac{1}{3^{k+1}} =$

$\dfrac{3^k - 1}{2 \cdot 3^k} + \dfrac{1}{3 \cdot 3^k} = \dfrac{3(3^k - 1)}{2 \cdot 3^{k+1}} + \dfrac{2}{2 \cdot 3^{k+1}} =$

$\dfrac{3^{k+1} - 3 + 2}{2 \cdot 3^{k+1}} = \dfrac{3^{k+1} - 1}{2 \cdot 3^{k+1}}$.

Hence, the statement is true for $n = k + 1$ and therefore true for all n by induction.

55. True for $n = 1$. If the statement is true for $n = k$,

then $1 + 5 + \cdots + 5^{k-1} = \dfrac{5^k - 1}{4}$ so that

$1 + 5 + \cdots + 5^{k-1} + 5^k = \dfrac{5^k - 1}{4} + 5^k =$

$\dfrac{5^k - 1 + 4 \cdot 5^k}{4} = \dfrac{5 \cdot 5^k - 1}{4} = \dfrac{5^{k+1} - 1}{4}$.

Hence, the statement is true for $n = k + 1$ and therefore true for all n by induction.

57. Since the statement is obviously true for $x = 0$, assume $x \neq 0$. Then the statement is true for $n = 1$. If the statement is true for $n = k$, then $|x^k| < 1$. Then $|x^k| \cdot |x| < |x|$. Thus, $|x^{k+1}| = |x^k| \cdot |x| < |x| < 1$. Hence, the statement is true for $n = k + 1$ and therefore true for all n by induction.

59. True for $n = 1$. If the statement is true for $n = k$,

then $1 + 4 + \cdots + 4^{k-1} = \dfrac{1}{3}(4^k - 1)$. Hence,

$1 + 4 + \cdots + 4^{k-1} + 4^k = \dfrac{1}{3}(4^k - 1) + 4^k =$

$\dfrac{1}{3}(4^k - 1) + \dfrac{3 \cdot 4^k}{3} = \dfrac{4^k - 1 + 3 \cdot 4^k}{3} = \dfrac{4 \cdot 4^k - 1}{3} =$

$\dfrac{1}{3}(4^{k+1} - 1)$.

Hence, the statement is true for $n = k + 1$ and therefore for all n by induction.

61. If $n = 1$, then $9^n - 8n - 1 = 0$. Since $0 = 0 \cdot 8$, the statement is true for $n = 1$. If the statement is true

for $n = k$, then $9^k - 8k - 1 = 8D$, so that $9^k - 1 = 8k + 8D = 8(k + D)$. Consequently, $9^{k+1} - 8(k + 1) - 1 = 9^{k+1} - 8k - 8 - 1 = 9^{k+1} - 9 - 8k = 9(9^k - 1) - 8k = 9[8(k + D)] - 8k = 8[9(k + D) - k]$. Thus, 8 is a factor of $9^{k+1} - 8(k + 1) - 1$ and the statement is true for $n = k + 1$. Therefore it is true for all n by induction.

Chapter 8

Section 8.1, page 595

1. $x^2 + (y - 3)^2 = 4$ **3.** $x^2 + 6y^2 = 18$

5. $(x - 4)^2 + (y - 3)^2 = 4$

7. Center $(-4, 3)$, radius $2\sqrt{10}$

9. Center $(-3, 2)$, radius $2\sqrt{7}$

11. Center $(-12.5, -5)$, radius $\sqrt{169.25}$

13. Ellipse

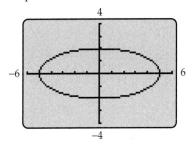

15. Ellipse

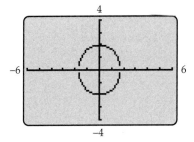

17. Ellipse

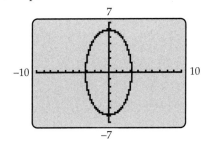

19. Circle

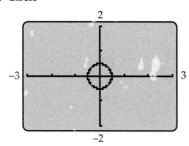

21. $\dfrac{x^2}{49} + \dfrac{y^2}{4} = 1$ **23.** $\dfrac{x^2}{36} + \dfrac{y^2}{16} = 1$

25. $\dfrac{x^2}{9} + \dfrac{y^2}{49} = 1$

27. 8π **29.** $2\sqrt{3}\pi$ **31.** $\dfrac{7\pi}{\sqrt{3}}$

33. Ellipse; center $(1, 5)$

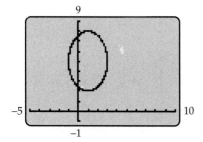

35. Ellipse; center $(-1, 4)$

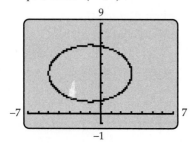

37. Ellipse; center $(-3, 1)$

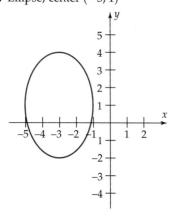

39. Circle; center $(-3, 4)$

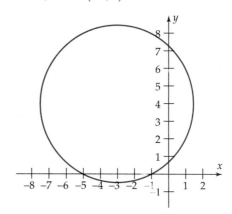

41. Ellipse; center $(-3, 2)$

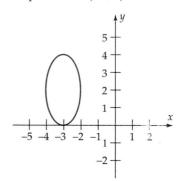

43. Ellipse; center $(-1, -3)$

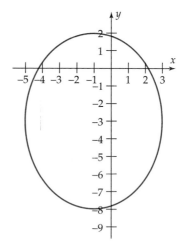

45. $\dfrac{(x-2)^2}{4} + \dfrac{(y-3)^2}{16} = 1$

47. $\dfrac{4(x-7)^2}{25} + \dfrac{(y+4)^2}{36} = 1$

49. $\dfrac{(x-3)^2}{36} + \dfrac{(y+2)^2}{16} = 1$

51. $\dfrac{(x + 5)^2}{49} + \dfrac{(y - 3)^2}{16} = 1$ or

$\dfrac{(x + 5)^2}{16} + \dfrac{(y - 3)^2}{49} = 1$

53. $2x^2 + 2y^2 - 8 = 0$

55. $2x^2 + y^2 - 8x - 6y + 9 = 0$

57. $\dfrac{(x + 3)^2}{4} + \dfrac{(y + 3)^2}{8} = 1$

59. If $a = b$, then $\dfrac{x^2}{a^2} + \dfrac{y^2}{a^2} = 1$. Multiplying both sides by a^2 gives $x^2 + y^2 = a^2$, the equation of a circle of radius a with center at the origin.

61. Approximately 226,335 miles and 251,401 miles

63. Eccentricity $= .1$

65. Eccentricity $= \dfrac{\sqrt{3}}{2} \approx .87$

67. If eccentricity is close to 0, the ellipse is almost circular. If the eccentricity is close to 1, the ellipse is elongated.

69. Eccentricity $\approx .38$

71. Let P denote the punch bowl and Q the table. In the longest possible trip starting at point X, the sum of the distance from X to Q and the distance from X to P must be 100 (since the distance from Q to P is 50). Thus, the fence should be an ellipse with foci P and Q and $r = 100$, as described on pages 586–587 (with $c = 25$). Verify that the length of its major axis is 100 ft and the length of its minor axis is approximately 86.6 ft.

Section 8.2, page 605

1. $x^2 + 4y^2 = 1$ **3.** $2x^2 - y^2 = 8$

5. $6x^2 + 2y^2 = 18$

7. Hyperbola

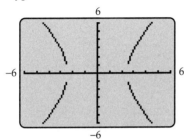

Because of limited resolution, this calculator-generated graph does not show that the top and bottom halves of the graph are connected.

9. Hyperbola

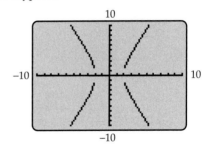

Because of limited resolution, this calculator-generated graph does not show that the top and bottom halves of the graph are connected.

11. Hyperbola

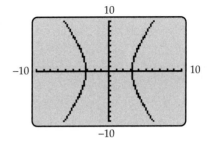

13. Hyperbola

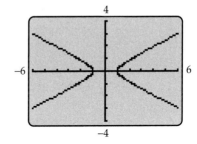

15. $\dfrac{x^2}{9} - \dfrac{y^2}{36} = 1$

17. $\dfrac{x^2}{4} - y^2 = 1$

19. Hyperbola; center $(-1, -3)$

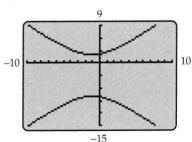

21. Hyperbola; center $(-3, 2)$

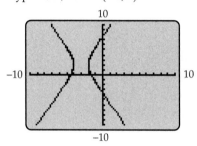

23. Hyperbola; center $(1, -4)$

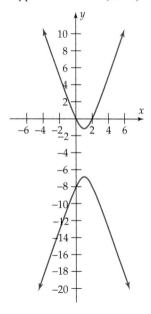

25. Hyperbola; center $(3, 3)$

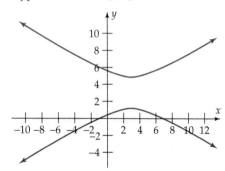

27. Circle; center $(3, 4)$

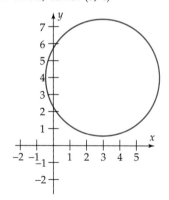

29. Ellipse; center $(3, 4)$

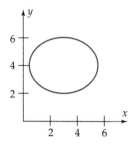

31. Hyperbola; center $(-2, 2)$

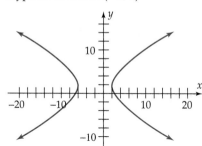

33. $\dfrac{(y-3)^2}{4} - \dfrac{(x+2)^2}{6} = 1$

35. $\dfrac{(x-4)^2}{9} - \dfrac{(y-2)^2}{16} = 1$

37. $y^2 - 2x^2 = 6$

39. $\dfrac{(y-2)^2}{4} - \dfrac{(x-3)^2}{9} = 1$

41. $\dfrac{(x+3)^2}{6} - \dfrac{(y+3)^2}{8} = 1$

43. The two branches of the hyperbola are very "flat" when b is large. With very large b and a small viewing window, the hyperbola may look like two horizontal lines, but its asymptotes $y = \pm\dfrac{2}{b}x$ are not horizontal (their slopes, $\pm 2/b$, are close to, but not equal to, 0 when b is large).

45. The asymptotes of $\dfrac{x^2}{a^2} - \dfrac{y^2}{a^2} = 1$ are $y = \pm\dfrac{a}{a}x$ or $y = \pm x$, with slopes 1 and -1. Since $(1)(-1) = -1$, these lines are perpendicular.

47. $\dfrac{x^2}{1,210,000} - \dfrac{y^2}{5,759,600} = 1$ (measurement in feet). The exact location cannot be determined from the given information.

49. $\sqrt{5} \approx 2.24$ **51.** 1.73   **53.** 1.50

Section 8.3, page 614

1. $6x = y^2$ **3.** $2x^2 - y^2 = 8$

5. $x^2 + 6y^2 = 18$

7. Focus: $(0, 1/12)$, directrix: $y = -1/12$.

9. Focus: $(0, 1)$, directrix: $y = -1$.

11. Vertex: $(2, 0)$, focus: $(9/4, 0)$, directrix: $x = 7/4$, opens right

13. Vertex: $(2, -1)$, focus: $(7/4, -1)$, directrix: $x = 9/4$, opens left

15. Vertex: $(2/3, -3)$, focus: $(17/12, -3)$, directrix: $x = -1/12$, opens right

17. Vertex: $(-81/4, 9/2)$, focus: $(-20, 9/2)$, directrix: $x = -41/2$, opens right

19. Vertex: $(-1/6, -49/12)$, focus: $(-1/6, -4)$, directrix: $y = -25/6$, opens up

21. Vertex: $(2/3, 19/3)$, focus: $(2/3, 25/4)$, directrix: $y = 77/12$, opens down

23.

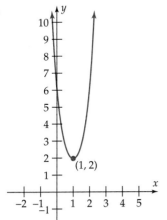

25.

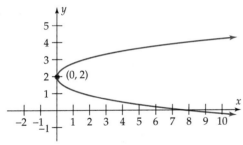

27.

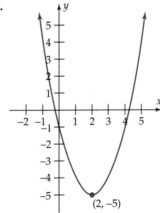

29.

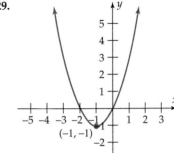

31. $y = 3x^2$ **33.** $y = 13(x - 1)^2$

35. $x - 2 = 3(y - 1)^2$ **37.** $x + 3 = 4(y + 2)^2$

39. $2(x - 1)^2 = y - 1$ **41.** $(y - 3)^2 = x + 1$

43. Parabola

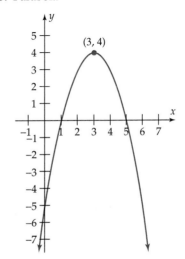

45. Parabola

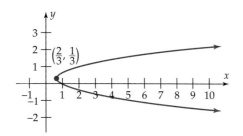

47. Circle

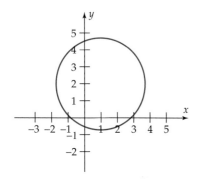

49. Hyperbola; vertex $(-4, 2)$

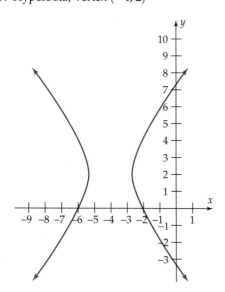

51. $y = (x - 4)^2 + 2$ **53.** $y = -x^2 - 8x - 18$

55. $y = (x + 5)^2 - 3$ **57.** $b = 0$

59. $\left(9, -\dfrac{1}{2} + \dfrac{1}{2}\sqrt{34}\right), \left(9, -\dfrac{1}{2} - \dfrac{1}{2}\sqrt{34}\right)$

61. 2/3 of a foot from the vertex

63. .375 inches from the vertex

65. 27.44 feet above the bridge surface

Excursion 8.3.A, page 619

1. Ellipse   **3.** Parabola   **5.** Hyperbola

7. Ellipse; $-6 \le x \le 3$ and $-2 \le y \le 4$

9. Hyperbola; $-7 \le x \le 13$ and $-3 \le y \le 9$

11. Parabola; $-1 \le x \le 8$ and $-3 \le y \le 3$

13. Ellipse; $-1.5 \le x \le 1.5$ and $-1 \le y \le 1$

15. Hyperbola; $-15 \le x \le 15$ and $-10 \le y \le 10$

17. Parabola; $-19 \le x \le 2$ and $-1 \le y \le 13$

19. Hyperbola; $-15 \le x \le 15$ and $-15 \le y \le 15$

21. Ellipse; $-6 \le x \le 6$ and $-4 \le y \le 4$

23. Parabola; $-9 \le x \le 4$ and $-2 \le y \le 10$

Chapter 8 Review, page 621

1. Ellipse, foci: $(0, 2)$, $(0, -2)$, vertices: $(0, 2\sqrt{5})$, $(0, -2\sqrt{5})$

3. Ellipse, foci: $(1, 6)$, $(1, 0)$, vertices: $(1, 7)$, $(1, -1)$

5. Focus: $(0, 5/14)$, directrix: $y = -5/14$

7.

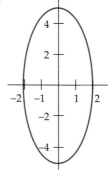

9.

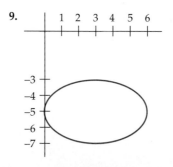

11. Asymptrotes: $y + 4 = \pm\frac{5}{2}(x - 1)$

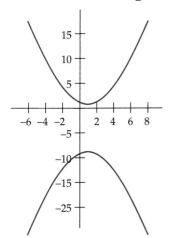

13.

15.

17.

19.

21. Center: $(-4, -5)$; radius: $\sqrt{8}$

23. Center: $(4, -6)$

25. $\dfrac{(x - 3)^2}{4} + \dfrac{(y - 1)^2}{2} = 1$

27. $\dfrac{y^2}{4} - \dfrac{(x - 3)^2}{16} = 1$

29. $\left(y + \dfrac{1}{2}\right)^2 = -\dfrac{1}{2}\left(x - \dfrac{3}{2}\right)$

31. 6 feet

33. Ellipse

35. Hyperbola

37. $-6 \le x \le 6$ and $-4 \le y \le 4$

39. $-9 \le x \le 9$ and $-6 \le y \le 6$

41. $-15 \le x \le 10$ and $-10 \le y \le 20$

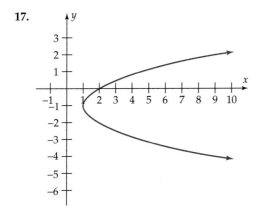

My sincere thanks go to Michael Robinson, who examined (and corrected where necessary) this answer section. Additional thanks go to Mathew Foss for communicating any errors he found in the answer section while completing the Instructor's Resource Manual and Student's Resource Manual. Their work has greatly improved this book.

Index of Applications

Index

extraneous solution, 131
extrema, local, 294

F
factor(s)
  common, 35
  and remainders, 279
  theorem, 281
factorial, *n*, 538
factoring
  patterns, 35, 36
  solution method, 141
factorization
  over complex numbers, 332
  over real numbers, 335
feasible
  region, 497
  solution, 497
Fibonacci
  numbers, 518, 523
  sequence, 518, 523
first-degree equation, 130
fixed costs, 95
focal axis, 598
foci, *see* focus
focus
  of ellipse, 586
  of hyperbola, 598
  of parabola, 608
FOIL, 35
formula, quadratic, 144
fraction(s)
  arithmetic with, 41
  partial, 460
fractional expression, 41
function(s), 187
  absolute value, 207
  average rate of change of, 243,
    246, 252
  composition of, 237
  constant on an interval, 208
  continuous, 292
  decreasing, 208, 415
  defined by equations, 188
  difference of, 235
  domain of, 187
  evaluation of, 190, 191
  even, 232
  exponential, 346–349
  graph of, 204
  greatest integer, 188, 204
  increasing, 208, 415
  inverse, 410
  linear rational, 312
  logarithmic, 373, 385

function(s) (*continued*)
  natural exponential, 352
  notation, 194, 196
  objective, 497
  odd, 233
  one-to-one, 414
  operations on, 235
  piecewise-defined, 199, 205
  polynomial, 277, 291
  product of, 235
  quadratic, 266, 269
  quotient of, 235
  range of, 187
  rational, 309, 315, 321
  rule of, 187
  step, 204
  sum of, 235
functional notation, 194, 196
  mistakes with, 198
Fundamental
  Counting Principle, 546
  Theorem of Algebra, 332

G
Gauss-Jordan method, 456
Gaussian elimination, 446, 449
geometric
  progression, 529
  sequence, 529
  series, 535
geometry, *see* Geometry Review
    Appendix
graph(s), 65
  complete, 79
  of conic sections, 584
  contraction of, 223
  of equation, 65
  expansion of, 223
  of function, 204
  horizontal shifts of, 222
  of inverse functions, 416
  line, 62
  of linear rational functions, 312
  parametric, 217
  polynomial, 291
  reading, 210
  reflections of, 224, 225
  of second-degree equations, 618
  symmetry of, 229, 232
  transformations of, 219
  vertical shifts of, 220
graphical
  intersection finder, 87
  maximum/minimum finder, 76
  root finder, 86

graphing exploration, 73
greater than, 3
greatest integer function, 188, 204
growth, exponential, 351, 362

H
half-life, 364
homogeneous system, 458
horizontal
  asymptote, 311, 315
  line test, 414
  shift, 222, 590
hyperbola, 598
  eccentricity of, 607
hypotenuse, 626
hypothesis, inductive, 571

I
i (complex number), 326
identity matrix, 468
imaginary number, 326
inconsistent system, 434, 456
increasing
  function, 415
  on an interval, 208
index, summation, 519
induction
  hypothesis, 571
  mathematical, 568, 570, 576
inductive
  hypothesis, 571
  step, 571
inductively defined sequence, 517
inequalities
  absolute value, 168
  basic principles for solving, 164
  equivalent, 163
  linear, 163, 487
  polynomial, 171, 175
  rational, 175
  system of, 430, 489
inequality, triangle, 8
infinite series, 534
input, 187
instantaneous rate of change, 253
integers, 2
  prime, 523, 572
integral exponents, 16
intercepts, 84, 92, 293, 313
  and solutions, 85
interest
  compound, 358
  continuously compounded, 361
  effective rate of, 366
  simple, 135